Lecture Notes in Electrical Engineering

Volume 685

The book series *Lecture Notes in Electrical Engineering* (LNEE) publishes the latest developments in Electrical Engineering - quickly, informally and in high quality. While original research reported in proceedings and monographs has traditionally formed the core of LNEE, we also encourage authors to submit books devoted to supporting student education and professional training in the various fields and applications areas of electrical engineering. The series cover classical and emerging topics concerning:

- Communication Engineering, Information Theory and Networks
- Electronics Engineering and Microelectronics
- Signal, Image and Speech Processing
- Wireless and Mobile Communication
- Circuits and Systems
- Energy Systems, Power Electronics and Electrical Machines
- Electro-optical Engineering
- Instrumentation Engineering
- Avionics Engineering
- Control Systems
- Internet-of-Things and Cybersecurity
- Biomedical Devices, MEMS and NEMS

For general information about this book series, comments or suggestions, please contact leontina.dicecco@springer.com.

To submit a proposal or request further information, please contact the Publishing Editor in your country:

China

Jasmine Dou, Associate Editor (jasmine.dou@springer.com)

India, Japan, Rest of Asia

Swati Meherishi, Executive Editor (Swati.Meherishi@springer.com)

Southeast Asia, Australia, New Zealand

Ramesh Nath Premnath, Editor (ramesh.premnath@springernature.com)

USA, Canada:

Michael Luby, Senior Editor (michael.luby@springer.com)

All other Countries:

Leontina Di Cecco, Senior Editor (leontina.dicecco@springer.com)

**** Indexing: The books of this series are submitted to ISI Proceedings, EI-Compendex, SCOPUS, MetaPress, Web of Science and Springerlink ****

More information about this series at http://www.springer.com/series/7818

Dario Fernando Cortes Tobar ·
Vo Hoang Duy · Tran Trong Dao
Editors

AETA 2019 - Recent Advances in Electrical Engineering and Related Sciences: Theory and Application

Springer

Editors
Dario Fernando Cortes Tobar
Electronic Engineering program,
Corporación Unificada de Educación
Superior CUN
Faculty of Engineering
Bogotá, Colombia

Vo Hoang Duy
Department of Automatic Control
Ton Duc Thang University
Ho Chi Minh, Vietnam

Tran Trong Dao
International Cooperation,
Research and Training Institute
Ton Duc Thang University
Ho Chi Minh, Vietnam

ISSN 1876-1100 ISSN 1876-1119 (electronic)
Lecture Notes in Electrical Engineering
ISBN 978-3-030-53020-4 ISBN 978-3-030-53021-1 (eBook)
https://doi.org/10.1007/978-3-030-53021-1

This Springer imprint is published by the registered company Springer Nature Switzerland AG
The registered company address is: Gewerbestrasse 11, 6330 Cham, Switzerland

Preface

Modern world is based on vitally important technologies that merge, e.g. electronics, cybernetics, computer science, telecommunication and physics together. Since the beginning of our technologies, we have been confronted with numerous technological challenges such as finding the optimal solution of various problems including controlling technologies, power sources construction, robotics, etc. Technology development of those and related areas has had and continues to have a profound impact on our civilization and our future lifestyle.

Therefore, this proceedings book containing articles of international conference AETA 2019, held in Bogota, Colombia, between 6 and 8 November 2019 and edited by Dario Cortés Tobar (Colombia), Vo Hoang Duy (Vietnam) and Tran Trong Dao (Vietnam), is a timely volume to be welcomed by the community focused on telecommunication, power control and optimization as well as computational science community and beyond.

This proceedings book consists of 12 topic areas of selected papers like telecommunication, power systems, digital signal processing, robotics, control system, renewable energy, power electronics, soft-computing and more. Readers can find interesting papers of those areas about optoelectronic oscillator at S-band and C-band for 5G telecommunications, neural networks identification of eleven types of faults in high-voltage transmission lines, cyber-attack mitigation on smart low-voltage distribution grids, optimum load of a piezoelectric-based energy harvester and others. All selected papers represent interesting ideas and state-of-the-art overview.

Participations were carefully selected and reviewed; hence, this proceedings book certainly is one of the few discussing the benefit from the intersection of those modern and fruitful scientific fields of research. We hope that the proceedings book will be an instructional material for senior undergraduate and entry-level graduate students working in the area of electronic, power technologies, energy distribution, control and robotics, etc. The proceedings book will also be resource and material for practitioners who want to apply discussed topics to solve real-life problems in their challenging applications. The important part of proceedings book is participation of keynote speakers from Spain, Brazil and Colombia.

The decision to organize AETA conference and to create this proceedings book was based on facts that technologies mentioned above, their use and impact on life are an interesting area, which is under intensive research from many other branches of science today. This proceedings book is written to contain simplified versions of experiments with the aim to show how, in principle, problems about power systems can be solved.

It is obvious that this proceedings book does not encompass all aspects of discussed topics due to limited space and time of the conference. Only the main ideas and results of selected papers are reported here. The authors and editors hope that the readers will be inspired to do their own experiments and simulations, based on information reported in this proceedings book, thereby moving beyond the scope of it.

This proceedings book is devoted to the studies of common and related subjects in intensive research fields of modern electric, electronic and related technologies. For these reasons, we believe that this proceedings book will be useful for scientists and engineers working in the above-mentioned fields of research and applications.

At the end, we **would like to thank** Corporación Unificada de Educación Superior CUN (Bogotá, Colombia) Ton Duc Thang University (Ho Chi Minh City, Vietnam), VŠB-Technical University (Ostrava, Czech Republic) and Pukyong National University (Busan, Korea) for interest and strong support in AETA conference organization. Also **many thanks** belong to Springer publishing company for its highly professional, precise and quick production process. Without all of this, it would be impossible to organize successful conference joining American, European and Asian participants.

November 2019 Dario Fernando Cortes Tobar

This conference was supported by the Corporación Unificada de Educación Superior CUN (Bogotá, Colombia), Ton Duc Thang University (Ho Chi Minh City, Vietnam), VŠB—Technical University (Ostrava, Czech Republic) and Pukyong National University (Busan, Korea) and co-sponsored by Universidad Central (Bogotá, Colombia).

Contents

Identification of Crack Tip Location Using Edge Detection Algorithms and Artificially Generated Displacement Fields

Jorge G. Díaz[1(✉)] and Deisy C. Paez[2]

[1] Mechanical Engineering Department, Universidad Industrial de Santander, Cra 27 calle 9, Ciudad Universitaria, Bucaramanga, Colombia
jorge.diaz5@correo.uis.edu.co

[2] Mechatronics Engineering Department, Universidad Santo Tomás, Cra 18 #9-27, Bucaramanga, Colombia

Abstract. The present paper shows a way to identify crack tip location (CTL) using edge detection algorithms. Arbitrary crack tip coordinates were fed to Linear Elastic Facture mechanics (LEFM) analytical models to simulate displacement fields. Then, first degree edge detection algorithms were applied to the obtained displacement fields to identify crack edges. Results show excellent correlation between random and identified crack tip coordinates. The paper proves the method and that CTL can be identified not only using photographs but also the associated displacement fields in a sample loaded in different axis.

Keywords: Edge detection · LEFM · Crack growth · Mixed mode loading

1 Introduction

Linear elastic fracture mechanics (LEFM) deals with flawed bodies under a loading condition. There are a handful of analytical solutions, being the solution for an infinite plate under remote loading probably the most widely used. Irwin's solution [1] is based on a complex function, whereas Williams' [2] solution is based on a trigonometric series. Although they use different mathematics, both arrive to the same solution. Both of them, however, place a coordinate system at the crack tip to describe the stress field based on the stress intensity factor (SIF). Therefore, one must know with low uncertainty where the crack tip location (CTL) is to obtain an accurate SIF, so a subsequent comparison to material toughness can be made. Failing to do so, renders an experimental error.

Different methods have been used to pinpoint CTL, especially under fatigue loading that produces crack growth, changing constantly the crack tip coordinates. One of the most used is the direct current (D.C.) potential drop technique (as required by ASTM E1820), which estimates crack advance by measuring the voltage drop in a cracked coupon test subjected to a known electrical potential. This method requires a tedious calibration, it is very noise sensitive, and it has been cited for giving erroneous readings when in presence of crack blunting [7, 8] which is common under overloads or

D. F. Cortes Tobar et al. (Eds.): AETA 2019, LNEE 685, pp. 1–8, 2021.
https://doi.org/10.1007/978-3-030-53021-1_1

multiaxial loads [9]. Another commonly used method measures compliance with a back strain gage [10] if a solution (analytical, numerical or empirical) is known to look for the crack length that balances the equation. Another method follows the crack front advance with a traveling microscope mounted on a graduated scale [10–12]. This, however, is limited to flat specimens or when crack path is known in advance, such in pure mode I loading. Lopez-Crespo et al. [13] used the Sobel algorithm to identify CTL but results were very noisy for low loads, probably because the crack was not fully open or experienced blunting [11] to comply with theoretical models. Zhang and He [14] introduced rigid body and rotation components into Williams' model by fitting DIC displacements fields from a deformed Aluminum C(T) sample to find coordinate locations. Harilal [15] approached CTL finding as an optimization problem with the objective function to be minimized the sum of square error difference between the measured displacements and curve fitted displacements using multi-parameter displacement field equation. Gonzales et al. [11] fitted 2D displacement fields to William´s equation into the possible locations measured by DIC. Diaz et. at. [17] used the results presented in [9] to calculate SIF using different approaches and comparing numerical and experimental results stressing the importance of finding an appropriate CTL. All in all, it can be said that more recent methods [9, 13–16] take advantage of measured stress, strain or displacement fields to perform curve fitting to analytical solutions.

An early survey of edge detection algorithms can be found in [3]. Nowadays, they are used in a wide number of applications. Bernal-Romero [4] used the Euclidean distance to reconstruct a skull 3D tomography, and Salazar et al. [5] used maximum and minimum algorithm to detect color bands. A through reference about image treatment techniques can be found in [6].

This paper shows how displacement fields can be fed to edge detection algorithms to identify CTL accurately.

2 Theoretical Background

This section contains the relevant information to explain the experimental technique and to perform the mentioned calculations.

2.1 LEFM

Williams' solution for a cracked body is expressed as an infinite series of n terms [1], as shown in Eq. (1) for displacement in three independent axis.

$$\begin{aligned}
u &= \sum_{n=1}^{\infty} \frac{r^{n/2}}{2G} \begin{pmatrix} a_n\left\{\left[k + \frac{n}{2} + (-1)^n\right]Cos\frac{n\theta}{2} - \frac{n}{2}Cos\left(\frac{n-4}{2}\right)\theta\right\} \\ -b_n\left\{\left[k + \frac{n}{2} - (-1)^n\right]Sin\frac{n\theta}{2} - \frac{n}{2}Sin\left(\frac{n-4}{2}\right)\theta\right\} \end{pmatrix} \\
\mathrm{v} &= \sum_{n=1}^{\infty} \frac{r^{n/2}}{2G} \begin{pmatrix} a_n\left\{\left[k - \frac{n}{2} + (-1)^n\right]Sin\frac{n\theta}{2} + \frac{n}{2}Sin\left(\frac{n-4}{2}\right)\theta\right\} \\ +b_n\left\{\left[k - \frac{n}{2} + (-1)^n\right]Cos\frac{n\theta}{2} + \frac{n}{2}Cos\left(\frac{n-4}{2}\right)\theta\right\} \end{pmatrix} \\
w &= \sum_{n=1}^{\infty} \left\{ \frac{2r^{n-\frac{1}{2}}}{G} c_n Sin\left(n - \tfrac{1}{2}\right)\theta \right\}
\end{aligned} \tag{1}$$

where $a_I = K_I/\sqrt{2\pi}$, $b_I = K_{II}/\sqrt{2\pi}$, $c_I = K_{III}/\sqrt{2\pi}$, $a_2 = \sigma_{ox}/4$, G is shear modulus, ν is Poisson´s modulus, (r,θ) are polar coordinates based on CTL, K_I, K_{II}, and K_{III} are the stress intensity factors in opening mode I, in-plane sliding mode II, out-of-plane sliding mode III respectively, and k is the Kolosov constant given by Eq. (2).

$$k = \begin{cases} 3 - 4\nu; \ \varepsilon_{pl} \\ \frac{3-\nu}{1+\nu}; \ \sigma_{pl} \end{cases} \tag{2}$$

From Eq. (1), one can observe the importance of identifying well CTL to establish an appropriate SIF.

2.2 Image and Edge Detection

The goal of image detection algorithms is to line drawing of a photograph. Digitally acquired images are maps of bytes *f(x, y)* that represents the level of gray in any pixel. Therefore, when one has an image, one already has a matrix of gray levels in case of black and white photographs. So, by using discrete derivatives one can observe abrupt changes in grey. Then, a border is represented as a point, or a line, of rapid change in the intensity function of the image. Equation (3) shows the approximation of a first degree derivative using pixel unity as increment.

$$\begin{aligned} \frac{\partial f}{\partial x} &= f_{(i,j+1)} - f_{(i,j)} \\ \frac{\partial f}{\partial y} &= f_{(i,j)} - f_{(i+1,j)} \end{aligned} \tag{3}$$

The gradient *I*, on the other hand, provides information about the intensity of the edge and its direction of orthogonal to the direction of the edge. It can be expressed as Eq. (4).

$$\nabla f = \vec{I} = \begin{Bmatrix} \frac{\partial f}{\partial x} \\ \frac{\partial f}{\partial y} \end{Bmatrix} = \begin{Bmatrix} Mx \\ My \end{Bmatrix} \tag{4}$$

The gradient *I* has a linear norm and an angle θ at which it occurs shown in Eq. (5).

$$\begin{aligned} \left\| \vec{I} \right\| &= I = \sqrt{Mx^2 + My^2} \\ \theta &= \tan^{-1} \frac{My}{Mx} \end{aligned} \tag{5}$$

So, one can search for the convolute between the original image and the chosen mask at each pixel as shown in Eq. (6). These masks are small matrixes used for edge detection, and they are also known as kernels or filters.

$$F(x, y) = f(x, y) * M(x, y) \tag{6}$$

The Roberts operator finds edges at those points where the gradient of *I* is maximum presented as a mask in Eq. (7).

$$Mx = \begin{bmatrix} 1 & 0 \\ 0 & -1 \end{bmatrix}; \; My = \begin{bmatrix} 0 & -1 \\ 1 & 0 \end{bmatrix} \tag{7}$$

One can extend the analysis beyond the next neighbor pixels, so a mask will look like Eq. (8).

$$Mx = c\begin{bmatrix} -1 & 0 & 1 \\ -1 & 0 & 1 \\ -1 & 0 & 1 \end{bmatrix}; \; My = c\begin{bmatrix} -1 & -1 & -1 \\ 0 & 0 & 0 \\ 1 & 1 & 1 \end{bmatrix} \tag{8}$$

The Prewitt algorithm uses c = 1, whereas the Sobel uses c = 2 returning edges at those points where the gradient of I is maximum.

3 Materials and Methods

The displacement fields used to apply the edge detection algorithms were obtained from Eq. (1). The perpendicular-to-crack (v), parallel-to-crack (u), and out-of-plane (w) displacement fields are shown in Fig. 1, where the sudden drop in all fields represents the crack edges. The fields were obtained with a stress intensity factor $\mathrm{K}i = 1\ \mathrm{MPa}\sqrt{\mathrm{m}}$ and $\mathrm{K}j = 0\ \mathrm{MPa}\sqrt{\mathrm{m}}$.

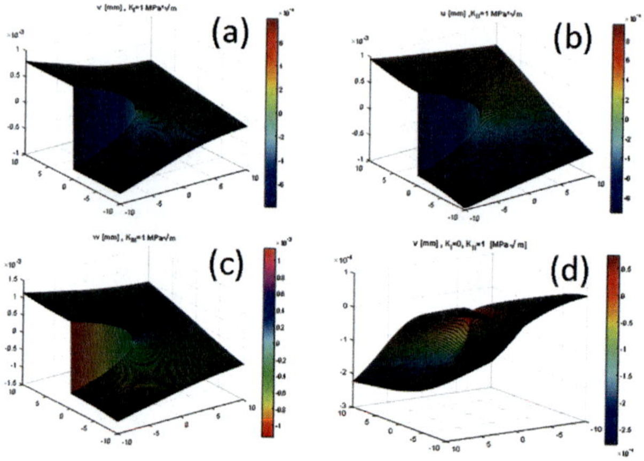

Fig. 1. Distribution of displacement fields from Eq. (1), a) v, perpendicular-to-crack $\mathrm{K_I} = 1$, b) u, parallel-to-crack $\mathrm{K_{II}} = 1$, c) w, out-of-plane $\mathrm{K_{III}} = 1$, d) v, perpendicular-to-crack $\mathrm{K_{II}} = 1$; $\mathrm{K_i}$ in $\mathrm{MPa}\sqrt{\mathrm{m}}$

Although the cases here are simulated, there is enough experimental evidence of each mode occurring separately [2, 7, 9, 16, 17] or combined. The effect and simulation of each one has been studied in the past [13, 18]. Moreover, because the fields in this study are simulated, there is no need for them to be filtered. In case the algorithms are applied to measured fields, such as the ones presented in [9, 11–16], some type of filtering should be applied [6, 19].

4 Results

Figure 2 shows the result of applying Prewitt, Sobel, and Roberts algorithms to the perpendicular-to-crack (v) displacement field with $K_I = 1$ and $K_{II} = 0$ MPa$\sqrt{m}$. It can be seen all of them identify the CTL at the convergence of crack edges but also, they draw lines ahead of the crack front that do exist.

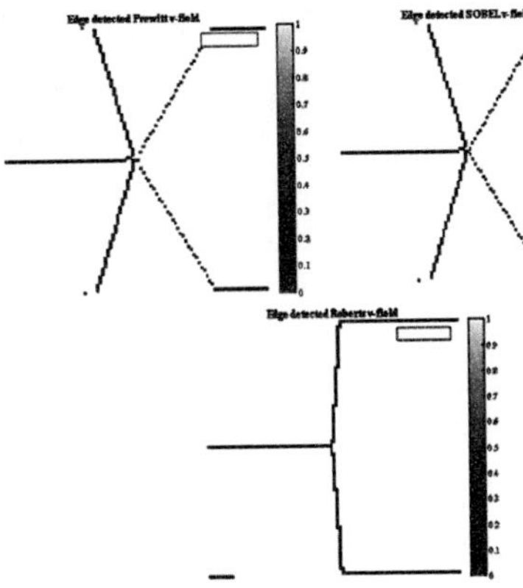

Fig. 2. Detected edges by Prewitt, Sobel, and Roberts algorithms in pure opening mode I, v field.

Figure 3 shows the result of applying Prewitt, Sobel, and Roberts algorithms to the parallel-to-crack (u) displacement field with $K_I = 0$ and $K_{II} = 1$ MPa$\sqrt{m}$. It can be seen the Prewitt and Sobel do identify the CTL correctly and they draw a line right ahead of the crack front that do exist. Roberts identifies the crack edges well, but also it draws a line in front of the CTL but the edges seem to be mirrored around a vertical axis.

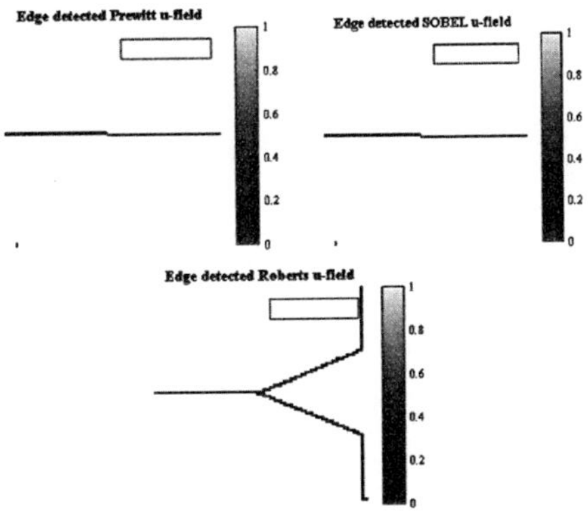

Fig. 3. Detected edges by Prewitt, Sobel, and Roberts algorithms in pure opening mode II, u field

Figure 4 shows the result of applying Prewitt, Sobel, and Roberts algorithms to the out-of-plane (w) displacement field with K_I, $K_{II} = 0$ and $K_{III} = 1$ MPa$\sqrt{m}$. As seen in

Eq. (1) the u and v fields do not depend on K_{III}, therefore they present zero displacement so they were not analyzed. However, the Prewitt and Sobel algorithms seem to interpret the coordinate system´s axis as edges. Roberts identifies the crack edges well (although mirrored around a vertical axis), but also it draws a line in front of the CTL.

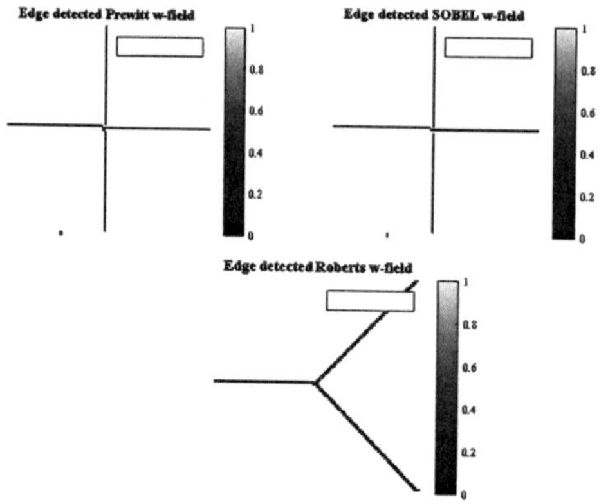

Fig. 4. Detected edges by Prewitt, Sobel, and Roberts algorithms in pure opening mode III, w field

Figure 5 shows the result of applying Prewitt, Sobel, and Roberts algorithms to the parallel-to-crack (u) displacement field with $K_I = 1$ and K_{II}, $K_{III} = 0$ MPa$\sqrt{m}$. As seen in Eq. (1) the u and v fields not depend both on K_I and K_{II}, therefore when one is present both displacement fields u, v are non-zero. However, the Prewitt and Sobel algorithms seem to interpret the crack´s edges as being apart from each other and mirrored around a vertical axis. Roberts finds the edges and CTL well.

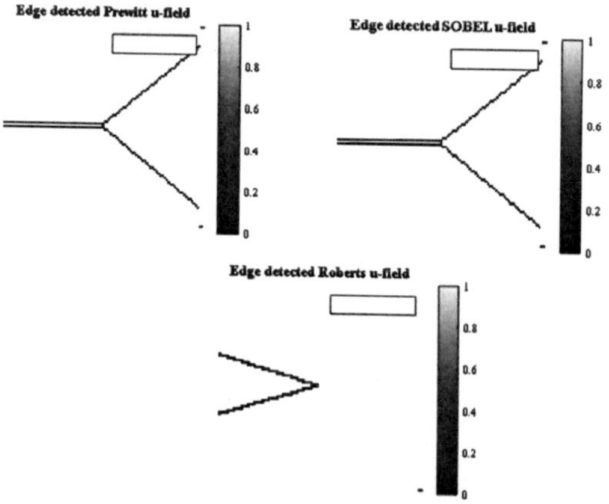

Fig. 5. Detected edges by Prewitt, Sobel, and Roberts algorithms in pure opening mode I for u field

A summary of crack tip coordinates for the above cases is presented in Table 1. Although the CTL is closely identified in most cases, it is seen that the Sobel and Prewitt algorithms line draw a square crack tip for the u-field, Fig. 5. Extensive experimental evidence shows the crack tip is sharp [1, 2, 6, 10, 17, 18] or in some cases [9, 11, 16] blunted, but it is never square. This is needed to take into account when analyzing the results presented in Table 1, for opening mode I in the u field for the Sobel and Prewitt algorithms which show there are 2 coordinates for the y position.

Table 1. Coordinates (x, y) found for different loading conditions.

Opening mode, field	Prewitt	Sobel	Roberts
I, v	50, 48	50, 48	50, 50
II, u	-, 52	-, 52	49, 51
III, w	50, 53	50, 52	50, 50
I, u	50, 48/52	50, 48/52	50, 50

5 Conclusion

A method to line draw crack edges using not photographs but displacement fields was tested and compared to known solution. Results show good correlation for Roberts although it the simplest, and less computationally expensive, of the analyzed algorithms.

The surveyed edge detection algorithms do not seem to find a difference when identifying crack edges between mode I (opening) and mode II (sliding), at least in the simulated fields. However, when measured displacement fields there is a documented influence of crack roughness, and other phenomena such as cyclic plasticity and non-proportional hardening that hinders sliding in mode II and mode III. Moreover, for mode III, the Prewitt and Sobel algorithms did not seem to find the edges correctly as they interpreted an edge ahead of the crack tip.

References

1. Irwin, G.R.: Analysis of stresses and strains near the end of a crack traversing a plate. Appl. Mech. **24**, 361–364 (1957)
2. Williams, M.L.: On the stress state at the base of a stationary crack. J. Appl. Mech. **24**, 109–114 (1957). https://doi.org/10.1115/1.3640470
3. Peli, T., Malah, D.: A study of edge detection algorithms. Comput. Graph. Image Process. **20**, 1–21 (1982). https://doi.org/10.1016/0146-664X(82)90070-3
4. Bernal Romero, Ó., Molina Prado, M.L., Arias Hernandez, N.A.: Simulación de entorno 3D y cálculo a punto en radioterapia por procesamiento de imágenes diagnósticas. ITECKNE **11**, 129–139 (2014). https://doi.org/10.15332/iteckne.v11i2.722
5. Salazar-Centeno, C., Niño, C., Díaz-Suárez, R.: Color bands detection on a gel electrophoresis image in one dimension applying a location algorithm based on maximums and minimums. ITECKNE **14**, 122–130 (2017). https://doi.org/10.15332/iteckne.v14i2.1766

6. Gonzalez & Woods: Digital Image Processing, 3rd edn, p. 2008. Prentice Hall, Upper Saddle River (2008)
7. Tong, J.: Notes on direct current potential drop calibration for crack growth in compact tension specimens. J. Test. Eval. **29**, 402–406 (2001). https://doi.org/10.1520/JTE12269J
8. Chen, F., Nanstad, R.K., Sokolov, M.A.: Application of direct current potential drop for the j-integral vs. crack growth resistance curve characterization. In: Evaluation of Existing and New Sensor Technologies for Fatigue, Fracture and Mechanical Testing (2015). https://doi.org/10.1520/STP158420140057
9. Vormwald, M., Hos, Y., Freire, J.L., et al.: Crack tip displacement fields measured by digital image correlation for evaluating variable mode-mixity during fatigue crack growth. Int. J. Fatigue **115**, 53–66 (2018). https://doi.org/10.1016/j.ijfatigue.2018.04.030
10. Castro, J.T., Meggiolaro, M.A., Ortiz Gonzalez, J.A.: Can Delta Keff be assumed as the driving force for fatigue crack growth? Frat. ed Integrità Strutt. **33**, 97–104 (2015). https://doi.org/10.3221/IGF-ESIS.33.13
11. Gonzáles, G.L.G., Díaz, J.G., González, J.A.O., et al.: Determining SIF using DIC considering Crack closure and blunting. In: Zhu, Y., Zehnder, A.T. (eds.) Experimental and Applied Mechanics, vol. 4, pp. 25–36. Springer, Orlando (2016). https://doi.org/10.1007/978-3-319-42028-8_4
12. Díaz, J.G., Gonzáles, G.L., González, J.A.O., de Freire, J.L.: Analysis of mixed-mode stress intensity factors using digital image correlation displacement fields. In: 24th COBEM (2017). https://doi.org/10.26678/abcm.cobem2017.cob17-0684
13. Lopez-Crespo, P., Shterenlikht, A., Patterson, E.A., et al.: The stress intensity of mixed mode cracks determined by digital image correlation. J. Strain Anal. Eng. Des. **43**, 769–780 (2008). https://doi.org/10.1243/03093247JSA419
14. Zhang, R., He, L.: Measurement of mixed-mode stress intensity factors using digital image correlation method. Opt. Lasers Eng. **50**, 1001–1007 (2012). https://doi.org/10.1016/j.optlaseng.2012.01.009
15. Harilal, R., Vyasarayani, C.P., Ramji, M.: A linear least squares approach for evaluation of crack tip stress field parameters using DIC. Opt. Lasers Eng. **75**, 95–102 (2015). https://doi.org/10.1016/j.optlaseng.2015.07.004
16. Vormwald, M., Hos, Y., Freire, J.L.F., et al.: Variable mode-mixity during fatigue cycles – crack tip parameters determined from displacement fields measured by digital image correlation. Frat. ed Integrita Strutt. **41**, 320–328 (2017). https://doi.org/10.3221/IGF-ESIS.41.42
17. Diaz, J.G., Nazare, L.F., Guzman, R.: Mixed-mode stress intensity factors for tubes under pure torsion loading. Key Eng. Mater. **74**, 373–378 (2017). https://doi.org/10.4028/www.scientific.net/KEM.774.373
18. Kibey, S., Sehitoglu, H., Pecknold, D.A.: Modeling of fatigue crack closure in inclined and deflected cracks. Int. J. Fatigue **129**, 279–308 (2004). https://doi.org/10.1023/B:FRAC.0000047787.94663.c8
19. Basto-Pineda, J.C., Plata-Gómez, A.: Comparación y evaluación de métodos de supresión de ruido en imágenes de origen astronómico utilizando wavelets. UIS Ingenierías **9**(2), 227–235 (2010)

Sequential Convex Optimization for the OPF in Isolated DC-grids

Alejandro Garcés(✉)

Universidad Tecnológica de Pereira, AA: 97, 660003 Pereira, Colombia
alejandro.garces@utp.edu.co

Abstract. This paper presents a sequential convex-optimization method for the optimal power flow in dc-grids under island operation. The proposed method is general for both high and low power applications. The former refers to multi-terminal high voltage direct current transmission and the latter include dc-distribution and dc-microgrids. Our analysis is based on the characteristics of the power flow equations, represented as an n-dimensional manifold embedded in $\mathbb{R}^n$. This formalism allows a geometric representation for each step in the optimization process as well as a geometric interpretation of the results. The problem is divided into several convex optimization problems defined on the tangent space. Then, results are projected to the manifold defining a new tangent space. Simulations results in Matlab complement the theoretical analysis for the CIGRE-multiterminal HVDC and the a microgrid test systems.

Keywords: Non-linear optimization · Optimal power flow · Multi-terminal high voltage direct current transmission · DC microgrids · Island operation

1 Introduction

Modern power systems include components in dc, for both high-power and low-power applications. This is driven by the increasing penetration of renewable energy resources as well as energy storage devices. It is therefore a natural step, to create grids that operate entirely in dc [1]. These grids can be operated in island as well as connected to a main grid. However, island operation is more challenging since the grid requires to achieve an optimal equilibrium trusting entirely on its internal resources [2].

In order to achieve this equilibrium, modern dc-grids rely on a hierarchical strategy of primary/secondary/tertiary control, similar to the automatic generation control used for the frequency stabilization of ac grids [3]. Primary control seeks stabilization while secondary control search for a suitable operation point

A. Garcés—This work was partially supported by the project 111077657914 funded by the Colombian Ministry of Science, and the program of Electric Power Engineering at Universidad Tecnológica de Pereira.

D. F. Cortes Tobar et al. (Eds.): AETA 2019, LNEE 685, pp. 9–19, 2021.
https://doi.org/10.1007/978-3-030-53021-1_2

and tertiary control tries to maintain an optimal voltage profile with reduced losses [4]. The last control in the hierarchy is closely related to the optimal power flow problem (OPF) which is a classic optimization problem for power systems operation. However, the OPF for tertiary control requires to include the droop constant from the primary control and the constrains given by the primary resource. The resulting optimization problem is challenging due to the non-linearity of the equations which are defined on a non-affine space. Although the optimization problem has some convenient features, such as convex objective function and a set of box constraints, the space in which it is defined, is non-affine meaning that the paradigm of convex optimization cannot be applied.

The power flow equations define an n-dimentional manifold embedded in $\mathbb{R}^n$. This type of space has some geometric features such as continuity and compactness that can be used in order to analyze the problem and propose solution algorithms. This paper analyzes the power flow equations of dc-grids under this paradigm. In addition, we present a sequential convex optimization algorithm for solving the problem. The main idea is to define a convex optimization problem on the tangent space of the power-flow-manifold. After that, the resulting vector is projected to the main manifold in which we define a new tangent space and repeat the process. Our approach is general for high and low power applications. High power applications include multiterminal high voltage direct current transmission which leads to the development of supergrids [5]. Low power applications include dc-distribution and microgrids [6], two emergent technologies in modern systems. Convex optimization has been an active research area for the optimal power flow in ac and dc grids. Semidefinite and second order cone approximations have been proposed for this problem [7]. Linear approximations have been also proposed [8]. These methods have advantages and disadvantages. Among the advantages, a convex approximation grantees optimality and convergence but only on the approximation. As disadvantage, the final solution could be distant from the real solution. We require therefore, a method that takes the advantages of the convex approach but solving the complete non-linear problem and not only an approximation.

On the other hand, manifold optimization is a generalization of non-linear programming which considers the geometry of the model allowing a better understanding of the problem. These type of algorithms seek for the optimal solution of the problem in a non-affine space. Therefore, we can not grantee global optimality as in the case of convex optimization, although the method analyzes the entire problem considering its complexities. This type of methods have been studied for computer science and machine learning applications [9], however, there are few works on this formalism for microgrid applications. The rest of the paper is organized as follows: Sect. 2 presents the general model for a dc grid and the formalization of this model. After that, Sect. 3 presents the proposed sequential convex optimization algorithm and some features of each sub-problem. Finally, simulations results and conclusions.

2 Problem Definition

Let us consider a dc-grid under island operation where step nodes are eliminated by a Kron's reduction, meaning that each node has a constant power with a droop control as depicted in Fig. 1, hence the model of the grid is given by (1):

$$p_k - h_k(1 - v_k) = \sum_{m=1}^{n} g_{km} v_k v_m \tag{1}$$

where p_k are generated or demanded power, h_k is the droop constant, v_k, v_m are the nodal voltages and g_{km} comes from the nodal admittance matrix. Notice this is a non-affine function that is challenging for optimization problems.

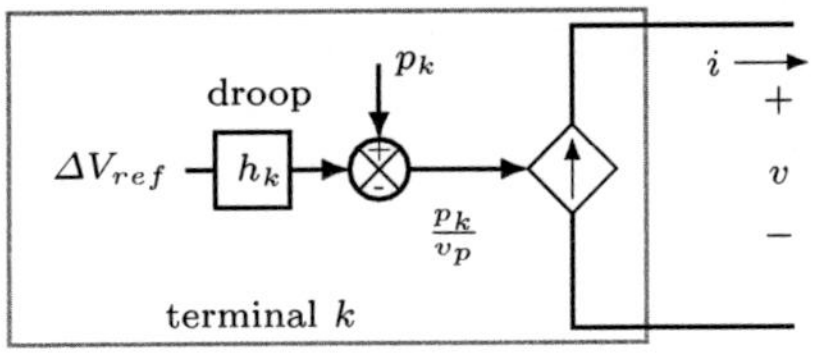

Fig. 1. Representation of a generic terminal for island operation of dc grids

The droop constant h_k must be designed in such a way that the system achieves stability in a feasible equilibrium point. However, this equilibrium could be inefficient from the point of view of the power loss or the power sharing between components. Therefore, we require an additional control in a upper layer of the hierarchy that minimizes the power loss of the grid. This control can be formulated as the following optimization model.

Model 1 (General OPF for dc grids)

$$\min P_L(v, p) \tag{2}$$

$$p_k - h_k(1 - v_k) = \sum_{m=1}^{n} g_{km} v_k v_m \tag{3}$$

$$p_{k(min)} \leq p_k - h_k(1 - v_k) \leq p_{k(max)} \tag{4}$$

$$-f_{km(max)} \leq \frac{v_k - v_m}{r_{km}} \leq f_{km(max)} \tag{5}$$

$$v_{min} \leq v_k \leq v_{max} \tag{6}$$

where $P_L(v, p)$ are the power loss of the grid, $p_{k(max)}, p_{k(min)}$ are the limits of each terminal, r_{km} is the resistance of each line segment and $f_{km(max)}$ is its maximum capability. This model is evidently non-convex since it is defined in a non-affine space, represented by (1). However, the objective function is convex as well as the inequality constraints. In order to solve this model, we require a better understanding of the space in which the problem is defined.

3 Manifold Formalism

In order to analyze the problem, we use the theoretical framework of differentiable manifolds. Intuitively, a manifold is a non-linear continuous subset of $\mathbb{R}^n$ which behaves locally as the Euclidean space. It can be also regarded as a generalization of a surface in a high dimensional space. In our case, we are interested in a particular type of manifolds, namely, manifolds representable by a set of equations:

Definition 1 (Manifold [10]**).** *Let $\mathcal{V} \subset \mathbb{R}^n$ be an open subset, and $\phi : \mathcal{V} \to \mathcal{P} \subset \mathbb{R}^n$ a differentiable mapping with Lipschitz derivative $D\phi(v)$. Let $\mathcal{M} \subset \mathcal{P} \times \mathcal{V}$ be the set of solutions to the equation $\phi(v) = p$. If $D\phi(v)$ is onto, then $\mathcal{M}$ is an $n-$dimentional manifold embedded in $\mathbb{R}^n$.*

Let us apply this definition to our problem:

Lemma 1. *Equation (1) defines a manifold embedded in $\mathbb{R}^n$, that is $\mathcal{M} = \{v \in \mathbb{R}^n : \phi(v) - p = 0\}$ with*

$$\phi_k(v) = h_k(1 - v_k) + \sum_{m=1}^{n} g_{km} v_k v_m = p_k \tag{7}$$

we call this the power-flow-manifold.

Proof. Just calculate the Jacobian as presented in [11].

Geometric interpretation of the power-flow-manifold is given in Fig. 2. There are two spaces embedded in $\mathbb{R}^n$ namely $\mathcal{P}$ and $\mathcal{V}$. The function ϕ is a bijection between these two spaces since it takes values from $\mathcal{V}$ and gives values on $\mathcal{P}$ but it is also possible to take values from $\mathcal{P}$ and obtain a unique vector on $\mathcal{V}$, by using ϕ^{-1} which is nothing but the procedure of the power flow. It is clear that ϕ^-1 gives a unique solution in $\mathcal{V}$ as demonstrated in [12]. So, for each point p there is a unique value of $v \in \mathcal{M}$ and any optimization algorithm that moves in $\mathcal{P}$ will have an equivalent movement on $\mathcal{M}$. The space $\mathcal{M}$ is evidently non-affine and hence, any optimization model will be non-convex. However, it is always possible to define an affine space around each point of $\mathcal{M}$ such that the convex optimization paradigm is applicable.

Definition 2 (Tangent space and tangent bundle [13]**).** *Let $\mathcal{M}$ be an n-manifold as in Definition 1, $v \in \mathcal{M}$ and $p \in \mathcal{P}$ a point such that $\phi(v) = p$. The set of all vectors $\Delta v \in \mathbb{R}^n$ of the form*

$$\mathcal{T}_v\mathcal{M} = \{\Delta v \in \mathbb{R}^n : [D\phi(v)][\Delta v] = 0\} \tag{8}$$

is called the tangent space to $\mathcal{M}$ at v and is denoted as $\mathcal{T}_v\mathcal{M}$. The set of all tangent spaces of $\mathcal{M}$ is called the tangent bundle of $\mathcal{M}$

In our case, $\mathcal{T}_v\mathcal{M}$ is a linear space spanned by $D\phi(v)$ where the zero of the space is the point v (see Fig. 2).

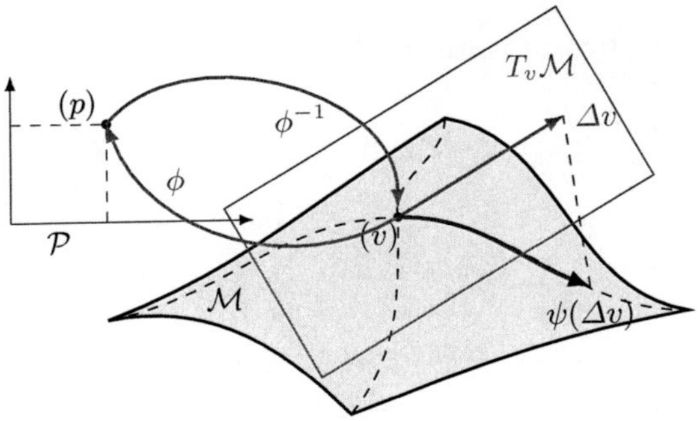

Fig. 2. Schematic representation of the power-flow-manifold. In this case $\mathcal{P} \subset \mathbb{R}^n$ is the space of the nodal powers and $\mathcal{V} \subset \mathbb{R}^n$ is the set of voltages.

Lemma 2. *The tangent space of the power-flow-manifold denoted by $\mathcal{T}_v\mathcal{M}$ is given by the kernel of $D\phi(v)$ that generates the following linear space*

$$-\left[diag(h) - diag(v)g - diag(gv)\right][\Delta v] = \Delta p \tag{9}$$

Proof. This lemma comes from the evaluation of the Jacobian of (1).

We can define a convex optimization model on each tangent space associated to the tangent bundle of $\mathcal{M}$. The tangent space can be understood as a linearization of the space around a particular point. Writting this in coordinates, we have

$$h_k \Delta v_k - \sum_{m=1}^{n} g_{km} v_k \Delta v_m - \sum_{m=1}^{n} g_{km} v_m \Delta v_k = 0 \tag{10}$$

This space is affine and hence we can use the paradigm of convex optimization. However, the result of the optimization method in this space is just an approximation of the original problem. We require a way to return to the manifold in order to solve the original problem as explained in the next section.

4 Convex Formulation of the OPF in $\mathcal{T}_v\mathcal{M}$

We replace v_k in Model 1 by $v_k + \Delta v_k$ resulting in the following convex optimization problem

Model 2 (Convex formulation of the OPF in $\mathcal{T}_v\mathcal{M}$)

$$\min \sum_{m=1}^{n}\sum_{k=1}^{n} g_{km}(v_k + \Delta v_k)(v_m + \Delta v_m) \tag{11}$$

$$h_k \Delta v_k - \sum_{m=1}^{n} g_{km} v_k \Delta v_m - \sum_{m=1}^{n} g_{km} v_m \Delta v_k = -\Delta p \tag{12}$$

$$p_{k(min)} \leq p_k + \Delta p_k - h_k(1 - v_k) + h_k \Delta v_k \leq p_{k(max)} \tag{13}$$

$$- f_{km(max)} \leq f_{km} + \frac{\Delta v_k - \Delta v_m}{r_{km}} \leq f_{km(max)} \tag{14}$$

$$v_{min} \leq v_k + \Delta v_k \leq v_{max} \tag{15}$$

Remark 1. The decision variables in this optimization problem are Δv and Δp_k being v_k, p_k constants.

Lemma 3. *Model 2 is convex and hence, the optimum is global.*

Proof. Demonstration of this lemma is straightforward taking into account that the conductance matrix $G = (g_{km})$ is positive semidefinite and the feasible set are box constraints in an affine space.

At this point we have a solution of the optimization problem in the tangent space. This solution, given by $v+\Delta v, p+\Delta p$, is an approximation of the original problem since it is defined in the tangent space and not the manifold itself. However, due to Lemma 3 we can guarantee there is a unique solution for each point in the tangent space. Therefore, we can generate sequential optimization problems in each tangent space in order to solve the problem. In addition, we have the following observation that complement the idea of the algorithm.

Lemma 4. *In the case in which $\Delta v = 0$ and $\Delta p = 0$ are the optimal solutions of Model 2, then the values of v, p are local solutions of Model 1.*

Proof. Just notice that $f(\tilde{v}+\Delta v) \geq f(\tilde{v})$ which is the definition of local optimum.

This lemma gives a stopping criteria for the algorithm, that is, a solution $\tilde{v}, \tilde{p}$ is optimal for Model 1 if the corresponding solutions for Model 2 is zero.

5 Retraction to $\mathcal{M}$

We require an additional operator that allows to return to the original manifold. This operator is defined as follows

Definition 3 (Retraction [9]**).** *A retraction on an embedded n-manifold $\mathcal{M}$ is a smooth mapping $\psi : T\mathcal{M} \to \mathcal{M}$ given by*

$$\phi(\Delta v, \Delta p) = \pi(\phi(v + \Delta v, p + \Delta p)) \tag{16}$$

where $\phi : \mathcal{V} \times \mathcal{M} \to \mathcal{M} : (v, p) \mapsto (v)$ is the projection onto the first component.

A retraction is basically a function that takes a vector in the tangent space and returns a point in the manifold (see Fig. 2). It considers therefore, the non-linear nature of the feasible space. In the particular case of the power-flow-manifold we can define the following retraction

Lemma 5. *Let v, p a point in $\mathcal{M}$, then, a retraction for the power-flow-manifold can be defined by two new points u, q given by*

$$u_k = v_k + \Delta v_k \tag{17}$$

$$q_k = h_k(1 - v_k - \Delta v_k) + \sum_{m=1}^{n} g_{km}(v_k + \Delta v_k)(v_m + \Delta v_m) \tag{18}$$

This lemma whose proof is strait forward from the previous results, gives a simple method to return to the manifold from a solution in the tangent space. Notice that the power flow algorithm is not explicitly required in order to obtain a new point in the manifold since the retraction is defined in function of Eq. (1). This is an advantage since many optimal power methodologies require an iterative algorithm of power flow. In this case, we only require a fast implementation of Model 2 (a convex model) and simple calculation of the retraction which is just an equation.

6 Sequential Convex Algorithm

With the aforementioned background, we can define a sequential convex optimization algorithm for the optimal power flow of dc-grids. The main idea is quite simple: We solve iteratively Model 2 and retract the solution to the main manifold. Since each sub-problem is convex, we can guarantee to find efficiently a solution which is global in the tangent space (Lemma 3). In addition, we have a well defined stopping criteria for the algorithm (Lemma 4) and a method to find a unique value in the manifold (Lemma 5). The pseudo-for the algorithm is presented as Algorithm 1. We do not require to solve a power flow in any iteration. We only require to solve a convex optimization model which is quite simple since it is a quadratic model defined in affine space with box constraints.

Algorithm 1. Sequential convex optimizationion for the OPF

Require: $G, p_{max}, p_{min}, v_{min}$
1: $v \leftarrow 1pu$
2: $p \leftarrow \phi(v)$
3: $\epsilon \leftarrow \infty$
4: **while** $\epsilon >$ *tolerance* **do**
5: $\quad \Delta v \leftarrow$ Solve Model 1
6: $\quad v, p \leftarrow \psi(\Delta v)$
7: $\quad \epsilon \leftarrow \|\Delta v, \Delta p\|$
8: Calculate P_{loss}

7 Results

Two sets of simulations were performed in order to demonstrate the usage of the proposed method in dc-microgrids and in multiterminal HVDC transmision. The test systems are depicted in Fig. 3. The dc-grid consists on nine nodes connected to a main grid via an ac/dc converter. However, we assume this converter is disconnected and hence the grid is operated in island. Distances of the cable are depicted in the figure and nominal values are 380 V/1 kW with a constant resistance of 1.5 mΩ/m. Numerical calculations were performed in Matlab and each sub-optimization model (Model 2) was solved using cvx which is a package for specifying and solving convex programs [14].

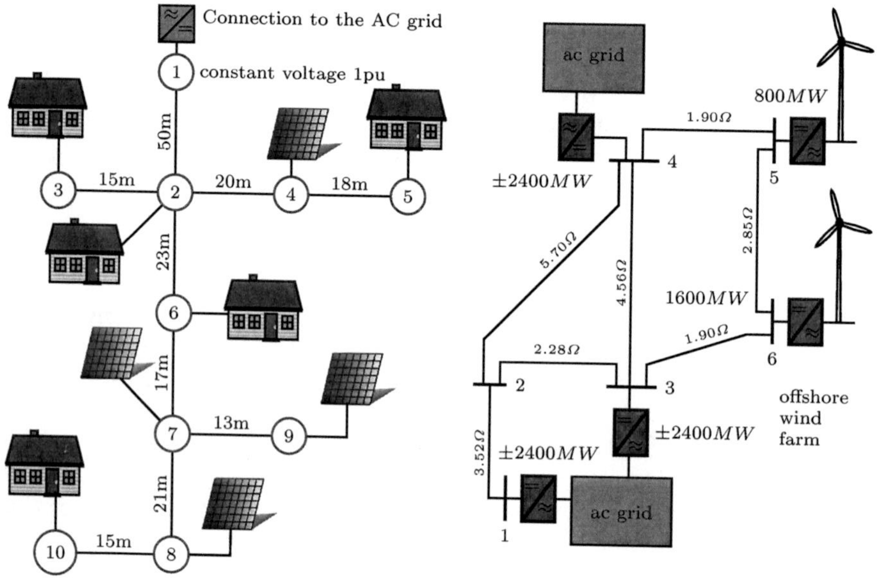

Fig. 3. Test systems: a) microgrid, b) multi-terminal HVDC transmision

The algorithm achieved convergence after 4 iterations. Nodal voltages are shown in Fig. 4. Due to the high penetration of solar energy, the voltages increased close to the maximum admissible which was set in 1.1 pu for this case.

A second set of numerical calculations were performed in the five terminals HVDC grid proposed by the CIGRE [15]. Two offshore wind farms were considered at nodes 5 and 6 injecting 800 MW and 1600 MW respectively. This power requires to be delivered to the ac grid through converters placed in Nodes 1, 3 and 4. Each of these converters has a capacity of ±2400 M and a droop constant of 5%. The optimization model must determine how to distributed the generated power in order to minimize power loss. Other objective functions can be considered. Nodal voltages are presented in Fig. 4. Just as in the previous case, the optimization model seeks for a high voltage profile in order to reduce total losses.

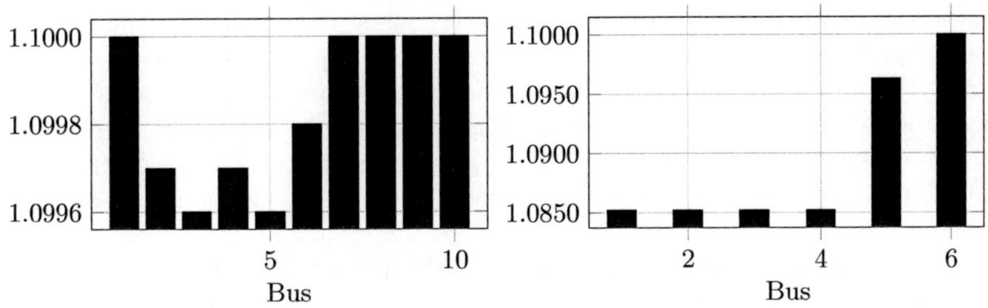

Fig. 4. Nodal voltages after optimization for a) DC microgrid, b) MT-HVDC grid

In both cases, voltages were initialized in 1 p.u. but the algorithm achieves convergence even from different initializations. The convergence of the algorithm is shown in Fig. 5. In both case, the sequential algorithm achieves an optimal solution in less than 4 iterations. In fact, the approximation given by the first iteration is close enough to the optimal value and can be considered a solution for practical applications.

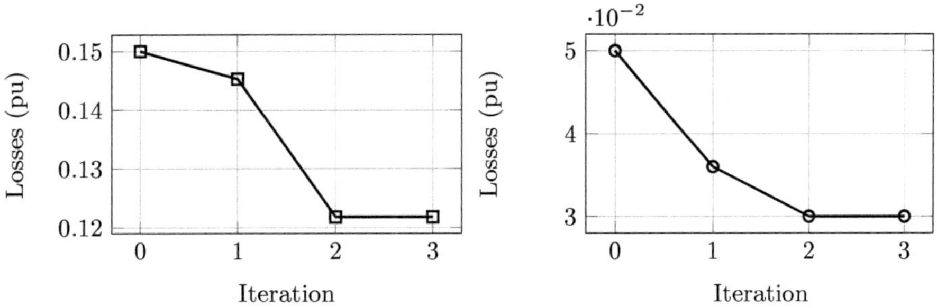

Fig. 5. Convergence of the sequential convex algorithm for microgrids and MT-HVDC

An additional advantage of the proposed method is its implementation in cvx. The model has a transparent implementation in cvx due to the geometric intuition behind the algorithm. It is worth to notice that cvx solves only convex problems but here we are solving a non-convex problem by a sequential application of convex models.

8 Conclusions

A numerical method for the optimal power flow in island operated dc-grids was presented. The proposed method is based on the key observation that the feasible space defines an n-manifold embedded in $\mathbb{R}^n$ while the objective function and the inequality constraints are convex. Therefore, a sequential solution of convex optimization models defined in the tangent space of the manifold are proposed. A retraction is also defined from the tangent bundle to the manifold allowing to return to the non-linear space. Some properties related to convexity of the solutions and global optimality of the models defined in each step complement the analysis.

The proposed method is applicable to microgrids, dc-distribution and multi-terminal high voltage direct current transmission. In all cases, a general droop control is considered. Therefore, the optimal power method can be regarded as a tertiary control on the hierarchical structure for the operation of these type of grids. The sequential convex optimization algorithm allows to find a local solution from the original non-linear non-convex problem.

References

1. Wang, Y., Qingshan, X., Liu, M., Zheng, J.: A novel system operation mode with flexible bus type selection method in DC power systems. Int. J. Electr. Power Energy Syst. **103**, 1–11 (2018)
2. Krishan, R., Verma, A., Mishra, S.: Loadability analysis of DC distribution systems. Int. J. Electr. Power Energy Syst. **103**, 176–184 (2018)
3. Bidram, A., Davoudi, A.: Hierarchical structure of microgrids control system. IEEE Trans. Smart Grid **3**(4), 1963–1976 (2012)
4. Abdali, A., Noroozian, R., Mazlumi, K.: Simultaneous control and protection schemes for dc multi microgrids systems. Int. J. Electr. Power Energy Syst. **104**, 230–245 (2019)
5. Xydis, G.: Comparison study between a renewable energy supply system and a supergrid for achieving 100% from renewable energy sources in islands. Int. J. Electr. Power Energy Syst. **46**, 198–210 (2013)
6. Elsayed, A.T., Mohamed, A.A., Mohammed, O.A.: DC microgrids and distribution systems: an overview. Electr. Power Syst. Res. **119**, 407–417 (2015)
7. Gan, L., Low, S.H.: Optimal power flow in direct current networks. IEEE Trans. Power Syst. **29**(6), 2892–2904 (2014)
8. Montoya, O.D., Gil-Gonzalez, W., Garces, A.: Optimal power flow on DC microgrids: a quadratic convex approximation. IEEE Trans. Circ. Syst. II Exp. **66**, 1018–1022 (2018)
9. Absil, P.-A., Mahony, R., Sepulchre, R.: Optimization Algorithms on Matrix Manifolds, 1st edn. Princeton University Press, Princeton (2008)
10. John Hamal Hubbard and Barbara Burke Hubbard: Vector Calculus, Linear Algebra, and Differential Forms a Unified Approach. Prentice Hall, Upper Saddle River (1999)
11. Garcés, A.: On the convergence of Newton's method in power flow studies for DC microgrids. IEEE Trans. Power Syst. **33**(5), 5770–5777 (2018)

12. Garces, A.: Uniqueness of the power flow solutions in low voltage direct current grids. Electr. Power Syst. Res. **151**, 149–153 (2017)
13. Munkres, J.: Analisis on Manifolds, 1st edn. Addison-Wesley publishing, Redwood city (1991)
14. Inc. CVX Research. Cvx: Matlab software for disciplined convex programming, version 2.0, August 2012. http://cvxr.com/cvx
15. Vrana, T.-K., Yang, Y., Jovcic, D., Dennetiere, S., Jardini, J., Saad, H.: The cigre b4 dc grid test system. Electra **270**, 10–19 (2013)

Reference Model Supported in Academic Analytic for the Collection and Analysis of Data from Students and Teachers

Fredys A. Simanca H.[1](✉), Fabian Blanco Garrido[1], Pablo E. Carreño H.[1], Alexandra Abuchar[2], and Pedro Rivera[3]

[1] Universidad Libre, Bogotá, Colombia
fredysa.simancah@unilibre.edu.co
[2] Universidad Distrital Francisco José de Caldas, Bogotá, Colombia
[3] Universidad Minuto de Dios, Bogotá, Colombia

Abstract. Recently, the interest in the analysis of learning and the processes that surround it has been increasing, especially the analysis of academic processes (Academic Analytic), this methodology deals with combining techniques and predictive models in the extraction of academic-administrative data, for decision making within educational institutions. Academic Analytic is also a field in which several areas converge, from academic analysis to data mining. This document describes a reference model supported in Academic Analytic for the collection and subsequent analysis of the data of students and teachers of Universidad Libre. As part of the model, we created a software that allows the collection and subsequent analysis of data from students and teachers that are already in the information system of the educational institution; after that, different analyzes were applied to this information. The model yielded important information that served as a basis for decision making by the members of the academic community in this institution.

Keywords: Academic Analytic · Big data and education · Data analysis

1 Introduction

Over time, information and communication technologies have been moving forward in the process of structuring different spaces, they are implemented in the resolution of problems and decision making; data mining also known as Knowledge Discovery in Databases allows to reveal new and very useful information to whoever requests it and in large amounts. Its purpose is to have the necessary characteristics in decision making, turning in this way the data into a very useful tool in the diagnosis and identification of variables [1].

In doing so, it is necessary to implement software developments for decision-making, in this case at Universities where different patterns of student behavior must be evaluated [2], such as subjects with higher loss rate, time slots in which better scores are obtained, areas in which they have better performance according to student, enrollment and student desertion.

D. F. Cortes Tobar et al. (Eds.): AETA 2019, LNEE 685, pp. 20–29, 2021.
https://doi.org/10.1007/978-3-030-53021-1_3

It is necessary to develop software tools in this area in order to have bases in the decision making, in this case in the Universities in which different patterns of student behavior must be evaluated, such as: subjects with higher loss rates, time-slots in which students obtain better scores, areas in which they have better performance depending on the student, enrollment and student desertion [3], etc.

Based on the above, the answer to these questions could lead to the improvement of the structure, schedules, school support, and decrease in student desertion, thus leading to the need to use data mining for the benefit of the Universidad Libre de Colombia, which is a source of information that will help managers, teachers, and students as it is a technology that tries to help with the understanding of the contents that are in the database, allowing filtering and obtaining the desired information [4]. This model can be replicated later to other educational institutions or to other faculties and/or programs.

According to [5] the true value of data mining is: (a) the ability to extract useful information, decision making or exploration, and (b) the understanding of the governing phenomenon in the data source. This first one is important, because thanks to it, positive results can be generated in front of a data, yielding answers of the advancement in a determined period.

On the other hand [6], describes how the analysis of mass data generated by different causes such as the restructuring and reordering of educational institutions and the growth of school and teaching population contributes to decision making in education; being mining data a tool that is used to visualize the results of progress, and growth or decrease of a business.

For [7], decision making is a process that begins by recognizing the existence of a problem to be treated (by necessity or imposition) and culminates in a conscious and rational choice of one possibility among various alternatives, in order to solve the problem. According to [8], business intelligence is the corporate ability to make decisions, and it is achieved through the use of methodologies, applications, and technologies, allowing to gather, debug, transform data, and apply analytical techniques of knowledge extraction. They must be executed correctly and used at the right time in order to take full advantage of the updated information, allowing companies to make changes in the strategies that they have implemented, to analyze their results, or to carry out performance tests, ensuring that the beneficiary is comfortable with the service that has been provided, thus obtaining an extension of coverage and scope.

Bearing in mind that higher education has a number of challenges in the management of the educational processes of its students, such as lowering the dropout rate, the academic quality, among others, different technological tools that could contribute to the decision making have been sought, with a view to have a quality education together with the different institutions of competence in the same field, reasons why Academic Analytic is created, and within its characteristics, it combines large sets of data, techniques, and predictive models in the extraction of institutional data, in order to make administrative decisions from the particular to the general of a company.

According to [9], it is important because it makes decisions based on data, where the information is used to support and justify decisions at all levels, reason why they must have an estimated amount of data, so that better strategies can be proposed, with which safe diagnoses of what is necessary for a future, in this case in higher education, can be made, analyzing the activity of students and other members in order to make adjustments that contribute a quality of teaching-learning that is collected from data sources for each one of the levels.

The process of analysis involves the collection and organization of the information from which five steps are presented for the analytics, these are: to capture, report, predict, act, and refine, each of them of vital importance in the academic analysis using data from within and outside the institution, which are taken by modeling the same data, or the so-called data mining [10]; after this step, the data are used to make reports, showing what was obtained in a graphic manner. Here, the comparison with the goals or objectives is made so that it can be analyzed more easily, then the data is analyzed using statistics with several data points and statistical algorithms which generate predictions. This is done with a group charged with certain characteristics that are related to what is being done, what is expected to be achieved, and knowledge about the subject.

This is how the research question is posed. How to develop a Software that allows processing the data based on Academic Analytic, making it easier for the educational community to acquire statistics that lead to the improvement of the academic process? To answer this question, we raised the possibility of designing a reference model supported in Academic Analytic that would allow its analysis by means of the data that was obtained in Excel files (XSLX), Comma separated values (CSV), through the Information System of Universidad Libre de Colombia (SIUL by its abbreviation in Spanish), and in this way capture and process the information.

This article shows how the reference model that allows processing the data was developed, based on Academic Analytics for the collection and analysis of data from students and teachers of Universidad Libre de Colombia, organized as following: in Sect. 2, materials and methods, the study of the problem and the process of analysis, design, and development of the reference model is explained; in Sect. 3, results and discussion, the results of the analysis of the reference model are presented with the information that was obtained by means of the SIUL information system, and finally, the conclusions that were reached in the process of implementing the reference model are established.

1.1 State of the Art

There is a variety of investigations that have worked with the management of the information that is stored in databases for decision making in higher education using ICT. In the research conducted by [11], the work is shown at the institutional level (Academic Analytic) in which models are developed by means of Big Data optimizing the dropout in the university environment, performing follow-ups to the students where the stored information is used in order to generate and implement models allowing to fulfill the proposed objectives; this in order to increase the graduation rates.

The academic analysis presented by [12] analyzes and processes the collected data proposing improvements and solutions in the learning process in which each of the parties involved in an institution has results based on the handling of information such as: teaching practices, teacher behavior, and follow-up to teacher coordination, in which the tools being used are responsible for processing the information that has been collected and applying the corrections to the difficulties that arise in university education.

Likewise, a visual analysis model was created by [13], developing the Vela system (Visual eLearning Analytics) being a Java visualization framework that includes data that are stored in an LMS through Moodle, based on the general opinion, the representation of the school, university, or course depends on the data that is restored or obtained, providing a semantic Tagcloud analysis of the contents that are published by students and teachers.

MADlib is an open code free library source of database analytical methods, made by [14] whose objective is to eventually play a role for scalable database systems that is similar to the CRAN library for R: a community repository of statistical methods.

The uses of technologies have allowed to obtain tangible and intangible benefits by giving the users an individual experience with which they can improve by generating higher production in higher education, since it is based on the information of the students' data by doing different investigations and providing an excellent service, generated from the experiences of all those who make up the institution [15].

2 Materials and Methods

This section describes the study materials and methods that were used for the development of the web application for the collection and analysis of data supported in Academic Analytic, detailing the methodology and design that were used to carry out this tool.

During the development of this application, we used the SCRUM agile project development methodology, which is based on the book [16]; also, during the analysis of the information requirements, we made a systems proposal to synthesize the findings with the help of use cases based on [17].

2.1 Design of the Model

In this phase, we collected the information that was necessary in order to carry out the study of the problematic situation, in which Universidad Libre wishes to have a system that is capable of making statistics on the historical data that was obtained during the academic process. This application will not have direct access to the databases where this information is stored, but it will have the capacity to process the reports that are created by the applications that generate them, such as the Information System of Universidad Libre (SIUL by its abbreviation in Spanish).

The data that were taken into account for analysis from this system were: a) Subjects by study plan, b) List of students enrolled by program and level, c) Historical consolidated semi-annual evaluation report, d) List of classes, and e) Historical Results report by course.

For the ideal functioning of the tool to be developed, we propose five subprojects, each one in charge of a specific task at the level of programmatic operations. The creation of different projects aims at encapsulating codes that may undergo changes, as proposed by the SRR principle from the group of principles S.O.L.I.D.

Subproject LUA.WebClient. It will contain the Front End of the Web application developed in ASP.NET technology. In this project, solutions will be implemented in terms of micro-services and service-oriented architecture (SOA) [18], in order to comply with the non-functional requirement "Modern Interface", through a CSS Framework. Materialize CSS includes JavaScript libraries, which simulate the design regulations that are focused on the visualization of the Android operating system, web implemented by Google design language called Material Design.

Subproject LUA.Business. It will contain the logic of the business and will be the intermediary layer between the Front End and the Data layer (LUA.Data).

Subproject LUA.Util. It will encapsulate all the utilitarian methods for the good operation of the application, such as obtaining mails from names, the conversion of types of data, the insertion of exceptions in the Log of errors or in the use of encryption for passwords. This project will be referenced in all other projects, but it will not reference any of them itself.

Subproject LUA.Document.Factory. It implements the same design patterns as LUA.Data.Factory, this allows, by means of the reflection (Technology used by .Net to instantiate objects without making reference to them), the possibility to create classes to process different types of format without the need to recompile the code.

Subproject LUA.Data. It will represent the data layer of the application. This project will make use of the DBEngineAdapter Object (it should be able to provide methods that allow the interaction with any data base) provided by LUA.Data.Factory.

Subproject LUA.Data.Factory. It implements the "Abstract Factory" and "Factory" design patterns, defined below:

Abstract Factory provides an interface to create families of related or independent objects without specifying their specific class.

Factory defines an interface to create an Object, but lets the subclasses define which classes to instantiate [19].

Distribution Graphics. For the distribution graphics (Fig. 1) we used the strategy described by [20] called exploratory data analysis from graphics in order to capture the distribution of a categorical variable in a faster manner.

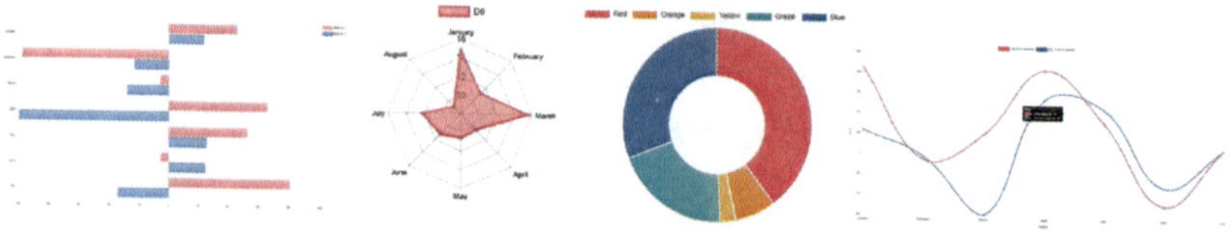

Fig. 1. Types of distribution graphics

Figure 2 graphically structures the design of the reference model for the analysis of the information of students and teachers of Universidad Libre, supported in Academic Analytic techniques.

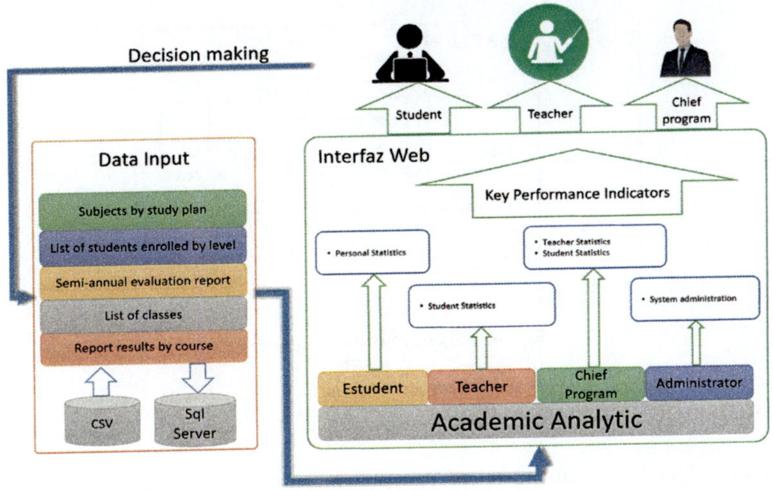

Fig. 2. Schematization of the reference model supported in Academic Analytic for the collection and analysis of data from students and teachers of Universidad Libre

Database Design. The design of the database was based on the Entity-Relationship model shown in Fig. 3, the database engine selected for its development was the SQL Server.

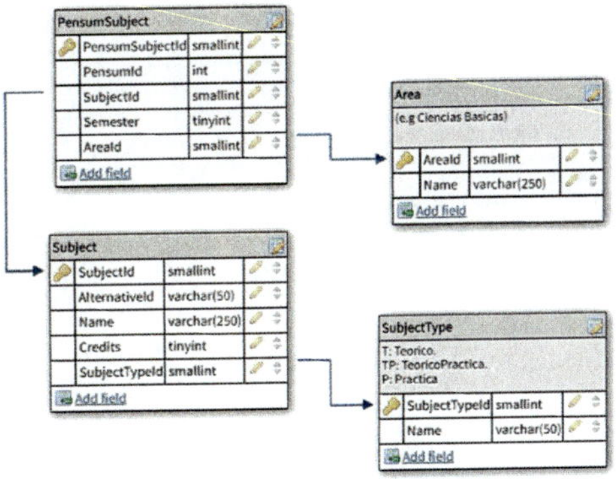

Fig. 3. Definition of subjects, ER model

Programmatic Design. In this phase, a diagram of components is created (see Fig. 4) where the relations and cardinality that exist between the different proposed projects will be seen.

User Profiles. The Software has four user profiles: Student, Teacher, Program Director, and Administrator.

Each user will have a defined role with permissions in accordance with their position.

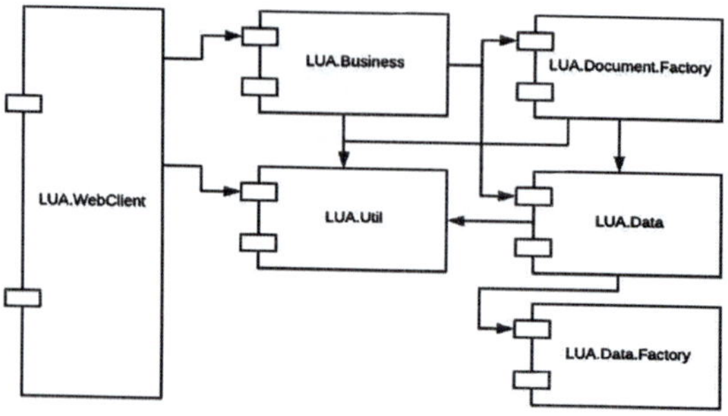

Fig. 4. Diagram of components.

3 Results and Discussion

Each module that has been presented can provide the required information with the different statistical samples that have been presented.

3.1 Individual Statistics by Student

In this module, the student can see basic information and the statistics of the results (Fig. 5), in addition to the detailed information of each academic period, its average, general progress by subject, and teacher.

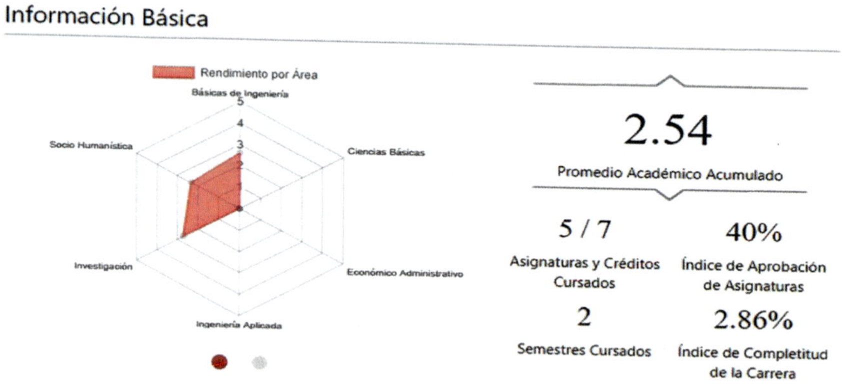

Fig. 5. Basic information, individual statistics by student

3.2 Individual Statistics by Teacher

In this module, the teachers will be able to see basic information according to their profile and the statistics of the results that were obtained, as in the student profile. In addition, they will be able to see the performance of the teacher and student assessments in detail and the statistics resulting from the information of students in the desired period (See Fig. 6) according to their need.

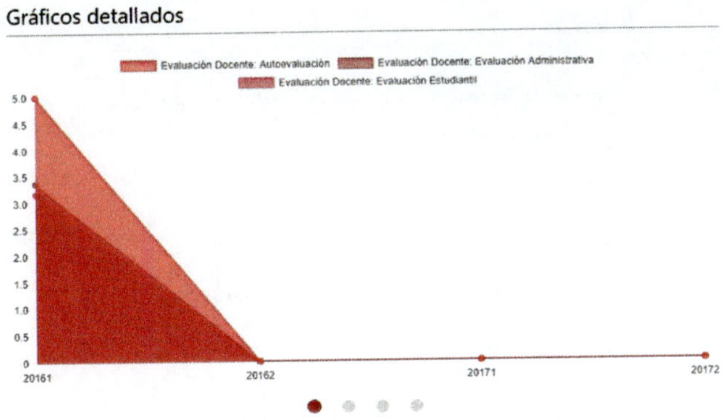

Fig. 6. Detailed graphics, individual statistics by teacher

3.3 Global Statistics of Students

In this module, the global statistics of all the students with the academic information and the results obtained during the period (Fig. 7) by area, by subject, and averages.

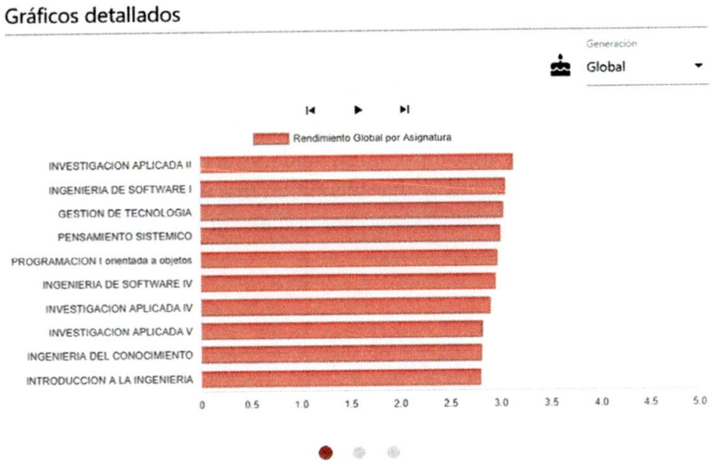

Fig. 7. Detailed graphics, global statistics from students.

3.4 Global Statistics from Teachers

This module shows the overall results of all teachers, with the information that is relevant to their modality as well as the statistics of the results (see Fig. 8) according to the course, performance, evaluation, active teachers, and category.

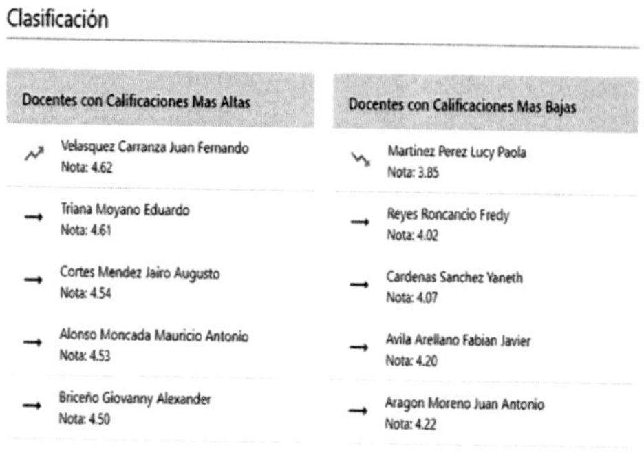

Fig. 8. Classification, global statistics by teachers

4 Conclusions

The design of the reference model supported in Academic Analytic for the collection and analysis of the data from students and teachers of Universidad Libre was successfully achieved. This model was translated into a web environment development in the .NET platform (ASPX) with the SQL Server database engine.

Likewise, it was possible to validate its use in the Systems Engineering program, with its students, teachers, and program chief, evidencing that these technological resources are important for decision-making at the academic administrative management level; but it also good for teachers and students.

This is a short advance in this subject, as evidenced in the survey of the state of the art, as the developments in this area are very few, and universities and educational centers have not dimensioned the great utility that there is in this type of tools and analysis of the data from the students and teachers who are in their information systems.

References

1. Timarán Pereira, S.R., Hernández Arteaga, I., Caicedo Zambrano, S.J., Hidalgo Troya, A., Alvarado Perez, J.C.: El proceso de descubrimiento de conocimiento en bases de datos. de Descubrimiento de patrones de desempeño académico con árboles de decisión en las competencias genéricas, Bogota, Ediciones Universidad Cooperativa de Colombia, pp. 63–86 (2016)
2. Simanca, H.F., González Crespo, R., Burgos, D., Rodriguez-Baena, L.: Identifying students at risk of failing a subject by using learning analytics for subsequent customized tutoring. Appl. Sci. **9**(33), 1–17 (2019)
3. Simanca, H.F.A., Gonzalez Crespo, R., Burgos, D., Rodriguez-Baena, L.: Automation of the tutoring process in online environments through the analytics of learning. In: 13th Iberian Conference on Information Systems and Technologies (CISTI), Cáceres (2018)
4. Gutierrez, J.A., Molina, B.: Identificación de la minería de datos para apoyar la toma de decisiones en la solución de problemas empresariales. Ontare **3**(2), 33–51 (2016)
5. Riquelme, J.C., Ruiz, R., Gilbert, K.: Minería de Datos: Conceptos y Tendencias. Revista Iberoamericana de Inteligencia Artificial **10**(29), 11–18 (2006)

6. Tejada Escobar, F., Murrieta Marcillo, R., Villao Santos, F., Garzón Balcázar, J.: Big Data en la Educación: Beneficios e Impacto de la Analítica de Datos. Revista Científica y Tecnológica UPSE **5**(2), 80–88 (2018)
7. Rozenfarb: Impacto de la Business Intelligence enel Proceso de Toma de Decisiones. Conexión abierta (2009)
8. Vásquez Castrillón, J.B., Sucerquia Osorio, A.: La Inteligencia de Negocios: Etapas del proceso (2011)
9. Ferreira, S.A., Andrade, A.: Analítica académicos: Mapping the Genoma de la Universidad. Revista Iberoamericana de Tecnologías del Aprendizaje, vol. 9, no. 3 (2014)
10. Campbell, J.P., Oblinge, D.G.: Analítica académicos. Educause (2007)
11. Bollatti, R.: Big Data en la educación. XV Workshop de Investigadores en Ciencias de la Computación. Entre rios, pp. 1196–1198 (2013)
12. Cantabella Sabater, M.: Modelos y herramientas para la representación y análisis de datos en LMS para enseñanzas universitarias. Universidad Católica de Murcia, Murcia (2018)
13. Conde, M.A., García Peñalvo, F.J., Gómez Aguilar, D.A., Therón, R.: Exploring software engineering subjects by using visual learning analytics techniques. IEEE Revista Iberoamericana de Tecnologías del Aprendizaje (IEEE RITA) **10**(4), 242–252 (2015)
14. Hellerstein, J.M., Ré, C., Schoppmann, F., Zhe Wang, D., Fratkin, E., Gorajek, A., Siong Ng, K., Welton, C., Feng, X., Li, K., Kumar, V.: The MADlib analytics library: or MAD skills, the SQL. ACM Digit. Libr. **5**(12), 1700–1711 (2012)
15. Koon Ong, V.: Business intelligence and big data analytics for higher education: cases from UK higher education institutions. Inf. Eng. Exp. Int. Inst. Appl. Inform. **2**(1), 65–75 (2016)
16. Stellman, A., Greene, J.: Learning Agile: Understanding Scrum, XP, Lean, and Kanban. O Reilly; Media, USA (2014)
17. Kendall, K.: Análisis y diseño de sistemas. México, Pearson (2005)
18. Pattankar, M., Hurbuns, M.: Mastering ASP.NET Web API. Packt Publishing, Birmingham (2017)
19. Gamma, E., Helm, R., Johnson, R., Vlissides, J.: Design Patterns: Elements of Reusable Object-Oriented Software. Addison-Wesley Professional, USA (1994)
20. Moore, D.: Estadística aplicada básica. Editorial Antoni Bosch, España (2000)

Electronic System for Protection of People Victims of Domestic Violence in Areas of Interior and Exterior

Sergio Daniel Díaz[1(✉)], César Orlando Díaz[2], and Darío Fernando Cortés T.[1]

[1] Corporación Unificada de Educación Superior – CUN, Bogotá D.C., Colombia
{sergio_diaz,dario_cortes}@cun.edu.co
[2] Universidad Jorge Tadeo Lozano, Bogotá, Colombia
cesaro.diazb@utadeo.edu.co

Abstract. This paper presents the results of a research project in which a prototype monitoring and control architecture for victims of gender violence, using mobile technology to determine the location of their potential aggressors previously identified by the penal system develops. Hardware, software and configuration needed to implement a viable technological solution is presented, taking into account the legal constraints, to allow victims of abuse or domestic violence to determine the location of his assailant ensuring that restraining orders meet in public places not covered by the prison system.

Keywords: Electronic control · Location · Bluetooth · Gender violence

1 Introduction

The denominated "domestic and gender violence" it has been converted in one of the main problems that faces our society. Do not understand frontiers, social classes, cultures, ethnicities or religions. The United Nations agency denominated World Health Organization [1], discloses the principal expression of violence at global scale on its "violence and health" report.

The methods of electronic surveillance are alternatives to prison and feature advantages to judicial and penitentiary system because they result less expensive, allows lighten the jails' occupation, besides the guarantee of the human rights fulfillment of those are deprived of freedom and allows the subject remain in its social-work field.

J. Sergio Daniel Díaz—Telecommunication and electronic engineer of the Universidad Católica de Colombia. Researcher of Corporación Unificada Nacional de Educación Superior – CUN.
C. O. Díaz—Doctor of Computer Science, University of Luxembourg. Researcher of Universidad Jorge Tadeo Lozano.
T. Darío Fernando Cortés—Ing. Electronics and telecommunications. Member of the researcher team group IDECUN. Researcher of Corporación Unificada de Educación Superior – CUN.

D. F. Cortes Tobar et al. (Eds.): AETA 2019, LNEE 685, pp. 30–41, 2021.
https://doi.org/10.1007/978-3-030-53021-1_4

In Colombia the system came into operation in February of 2009, year which had provided the installation of 4.962 surveillance devices, but it's not reached the goal due to the conditions of penal legislation have not allowed. The backwardness in the process is related to specific exigencies provided by the law for beneficiaries. However, on 6th February of 2009 was implemented the first electronic surveillance device to domiciliary prisoners.

GENSA has been the entity commissioned to total implementation of the system in Colombia, which include the construction of electronic monitoring centers, with related systems for optimal operation, the supply of monitoring equipment, bracelets and all required technologies, among other reasons, by the experience it managed accredit in the construction, for entities of the state, of monitoring centers and telemetry systems [2].

The final report of the operations evaluation of the Electronic Surveillance System projects (SVE, for its initials in Spanish) [3], published the three of February of 2012, it recommended, in last, reengineering of the project and also exposed the problems it has been presented, the electronic control system in the legal and technological field.

2 Methodology

The structure of the platform can be divided into 2 parts, the first part is focused on the indoor environments such as shopping malls, airports or any different to the home's victim, where will have the central device (hereafter called BEACON), the mobile device of the victim and the aggressor's bracelet; in this environment it will make use of the Bluetooth/WiFi technology to the (common to both sides) mobile application to perform calculation of the positioning in relation to the position of the victim and the aggressor. The second part, focused on outdoor environments, count again with the mobile device of the victim and the aggressor handle electronics, but the mobile application will use GPS technology to make the necessary calculations.

Discrimination or differentiation of use of both technologies (Bluetooth/WiFi and GPS) is mainly due to the GPS technology is not very reliable in indoor environments, and this is where comes into play the Bluetooth/WiFi communication, becoming the key differentiator of the platform, and thus ensuring greater reliability and precision.

The Fig. 1 explain the internal operation of the system, is taken example in which it is assumed that the aggressor attacks a victim. The perpetrator is prosecuted and found guilty, so proceeds according to the Colombian penal law, and implementing our solution to the problem:

1. The aggressor is assigned a device that cannot be removed from the aggressor and a cell phone with the system software to the victim.
2. The data of the two persons entering the system through a browser by which is accessed the Web page location system. To create individuals and relations must be the system administrator user ID and password.
3. After creating the two individuals, aggressor-victim relationship is established, also the maximum legal distance and is assigned a case manager.
4. At any time, it has a record in the database of the last location in any of the two individuals.

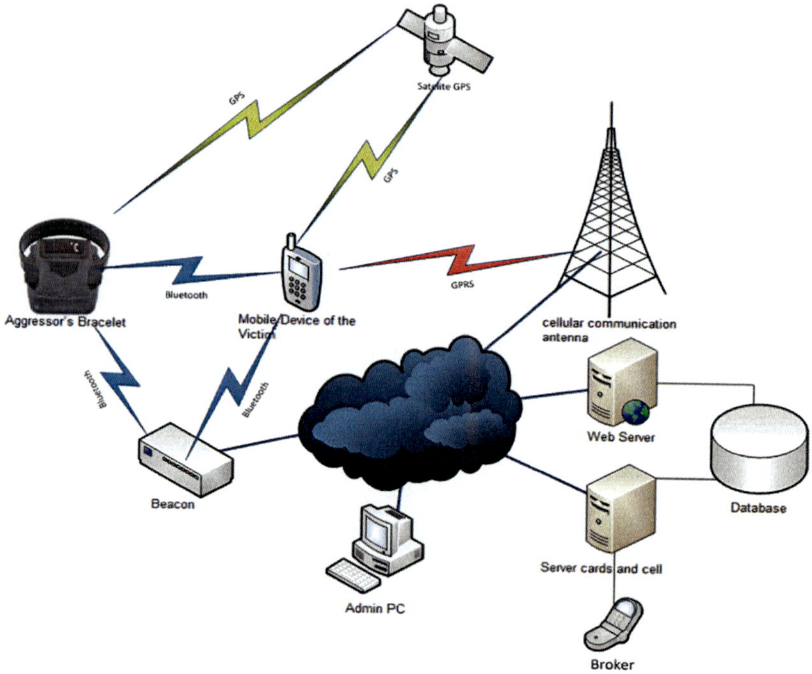

Fig. 1. General scheme system

5. In case of outdoors environments, aggressor's device sends a message every minute giving his position, which obtained from GPS module of his device. This message reaches the mobile server that is responsible for updating its position in the database.
6. In case of indoor environments, there are 2 possibilities:

   - The first is that instead get into the place, a mall for example, but a beacon that scans nearby Bluetooth devices every 20 s, because it is the minimum time interval in which scanning is performed unnecessarily have to remove the cache data and send its location using a TCP message to the cell server that also updates its position.
   - If the place where the person enter does not have the beacon, the watch alert mechanism would be the Bluetooth that scans devices around and sends through GPRS to the cell server which processes the information.

Based on this scheme we can make the definition of the system requirements:

1. The system will continuously search for mobile devices of victims and aggressors.
2. The system must ensure real-time location of both the aggressor and the victim.
3. The system must immediately inform the victim and the management system when victim and aggressor are within a secure space.
4. The beacon will have a copy of the database management system where all devices of victims and aggressors registered in the system relate.

5. The beacon and the device of the victim have redundant GPRS and WiFi communication with the management system to guarantee that there is no loss of communication.
6. The application software of the device of the victim should prevail in its operation over other smartphone applications and cannot be disabled by the user.
7. The system will keep a register of the last location of aggressor's bracelet.

After analyzing the different existing technologies in the market, potential solutions to implement the three devices are (Tables 1, 2 and 3):

Table 1. Hardware beacon

Beacon	
Device	Function
Raspberry Pi	Embedded system technology that will host the communications and control system
WiFi module	Allows Wireless communications standard IEEE 802.11 g
Bluetooth module	Allows Wireless communications standard IEEE 802.15
RF module	Allows Wireless communications in the 434 MHz range
GSM/GPRS module	Allows data transmission on bands of 900 MHz, 1800 MHz, 1900 MHz

Table 2. Hardware aggressor device

Aggressor's Device	
Device	Function
Microcontroller	Performs the tasks of monitoring devices and resource management
Lithium battery	Assure the battery of the device at least 3 days
Bluetooth module	Allows Wireless communication IEEE 802.15 standard
Rf module	Allows Wireless communications in the 434 MHz range
GSM/GPRS module	Allows data transmission on bands of 900 MHz, 1800 MHz, 1900 MHz

Table 3. Hardware victim device

Victim's device	
Device	Function
Smartphone	Phone with GSM/GPRS, GPS and Bluetooth connectivity that will host the application location The mobile must be high-end to avoid wrong GPS locations

The description of the proposed architecture is the following:

In the place that will want to implement the surveillance system a beacon will be installed, embedded system based on microcontroller which will have connectivity GPS, GPRS, WiFi, RF and Bluetooth and whose function is to make a permanent monitoring of the place in search of devices of victims and aggressors assigned bracelets. This beacon has wireless communication standards needed to ensure that it can locate these devices when the GPS cease to be reliable for being in a closed environment.

The victim of aggression has a smartphone that has GPS connectivity for outdoor location and Bluetooth standard. The beacon constantly searching for Bluetooth devices and when it locates the smartphone sends notification of the situation to the victim and the central management system, while searching for aggressor's bracelet related with that victim to verify whether it is in the same place or not.

The aggressor is assigned a monitoring bracelet anchored in his forearm which also features GPS, GPRS, Bluetooth and RF standards that can be identified by the beacon as soon as the aggressor entering the place. Once identified, the beacon searches its database device related with the victim and tries to locate it in the place to notify the victim and control system.

3 Experimental Setup

3.1 Communication

For the implementation of the beacon and communication with mobile locating devices it is necessary that the Raspberry Pi works with NMEA GPS, quadband GPRS, Bluetooth Class 1 and 802.11 g Wi-Fi modules. The Raspberry Pi have the programming to initialize the search for indoor devices using Bluetooth, confirming the position with GPS, sending alerts to the victim through Bluetooth and GPRS, and transmission of information to the control center via GPRS and Wi-Fi.

In the diagram below (See Fig. 2) the implementation that is done for our own bracelet location is explained. As a central part is the microcontroller, who is in charge of making all instructions and apply math algorithms for data either supplied by the GPS or GSM module.

3.2 Sustained Development Platform on Raspberry PI

Raspberry Pi is an embedded system, therefore, is a motherboard controlled by a microprocessor which already has USB ports, HDMI, LAN and GPIO communication, making virtually all communication modules that are used are plug and play and only need scripts to develop software to automate its operation.

The interaction of the system components shown in Fig. 3.

Bluetooth and GPS testing devices based on the Raspberry and serves as a locator beacon module is carried out.

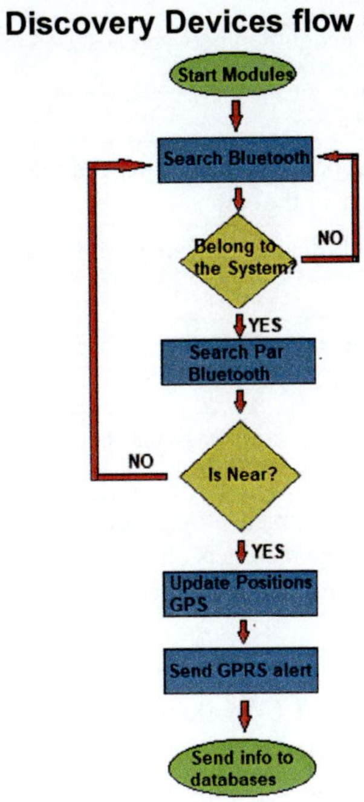

Fig. 2. Discovery devices

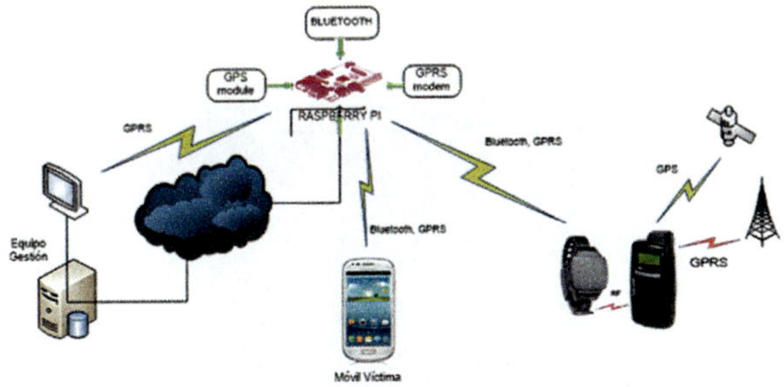

Fig. 3. Interaction between system elements

Identification of Bluetooth Devices. Here can be viewed the search for Bluetooth devices from the Raspberry Pi.

In a first series of tests a scenario in which the unit is active and the victim's cell phone is off but within the protected area is contemplated. Once the phone is turned on, we expect to recognize the beacon and if we do, also how long it takes (Table 4).

Table 4. Test 1: recognition between beacon – cellphone

Recognition 1 between beacon - cell	
Description	Detection Test between the beacon and the victim's cell phone when the phone goes from off to on
Elements	1 Cell Android, Beacon implemented in the Raspberry Pi
Number of attempts	10
Expected results	Recognition in less than 1 min
Results obtained	Average time of 55 s
Observations	The mobile of the victim was identified in all attempts, always with time in the same range
Conclusions	Although the recommendation is always keep turned on mobile, given the case where the mobile is off and turn it on in the surveillance area, the system ensures that it will be found and associate in less than 1 min

This test showed that for scenario 1 (Mobile initially turned off), the system detects the mobile 100% of the time and in all cases takes less than 1 min to do so.

A second detection test shows the expected normal operation scenario, it working the beacon and the victim's cell phone is turned on to the protected area.

For these tests the beacon was located at a central point in the "El Claustro" (See Fig. 4) headquarter of the Catholic University of Colombia, as in the 2nd floor of Block L on the IEEE room.

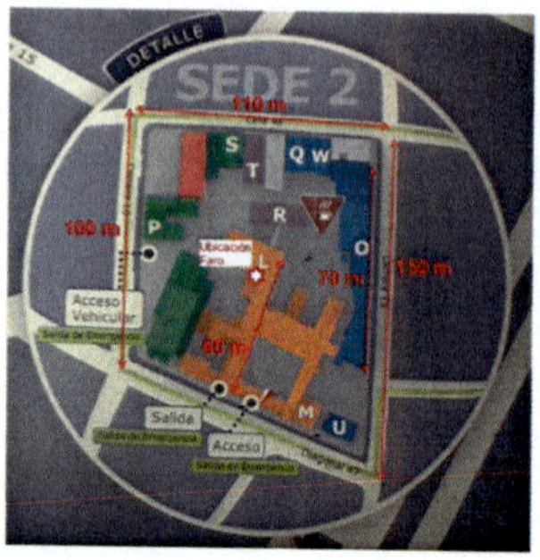

Fig. 4. 'El Claustro' place plane

Given that the "El Claustro" headquarter has a complex architecture, labyrinth type L and M buildings, tests were divided into two sub-scenarios, one entering the parking lot of the career 16 and another entering the main access diagonal 47. Both scenarios show very different conditions because entering through the parking we have line of sight between the mobile and beacon, while when we walked through the diagonal access 47 we have a network of walls and hallways that represent an obstacle to direct communication between devices (Table 5).

Table 5. Test 2: recognition between beacon – cellphone

Recognition 2 between beacon - cell			
Description	Detection Test between the beacon and the victim's cell phone when the mobile moves toward the beacon		
Elements	1 Cell Android, Beacon implemented in the Raspberry Pi, metro		
Number of attempts	6		
Expected results	Detection at 40 m		
Results obtained	Higher detection distance of 40 m		
Observations	**Distance**	**Obstacles**	**Results**
	40 m	No	It is detected
	50 m	No	It is detected
	60 m	No	It is detected
	30 m	*Yes	It is detected
	40 m	*Yes	It is detected
	45 m	*Yes	It is not detected
Conclusions	Despite the complex architecture of the "El Claustro" headquarter, we guarantee that under the worst scenario will have a maximum distance of 40 m recognition		

Greeting Delivery. Once the Bluetooth device has been identified and recognized as belonging the system the beacon sends a flat file containing a welcome message.

In Fig. 5 can be seen an example of the message received by the mobile of the victim.

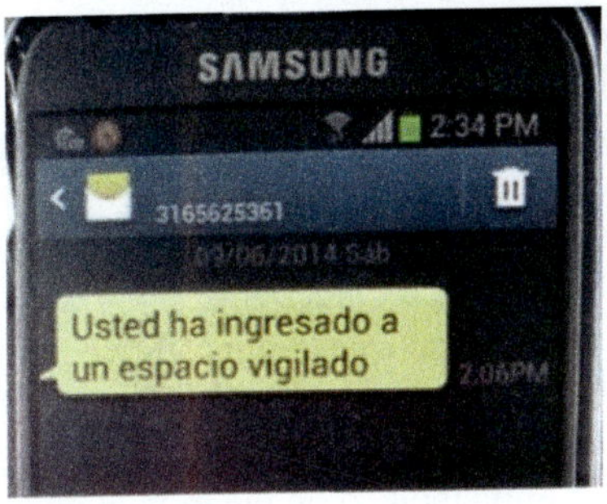

Fig. 5. Welcome message to the system

3.3 Implementation of Wireless Module in Control Bracelet

Initially the components are deployed in a breadboard to make a series of preliminary tests and validate or not the design (Fig. 6).

Similar to what happens with Raspberry way, Microchip has a community of developers who put their code to the service of those who want it. In the Microchip website we find the needed to put to work the NMEA GPS module: [11].

Fig. 6. Implementation of modules for the bracelet

Here can be viewed (Fig. 7) the results of GPS location with the system running on the breadboard. It can be seen that the location given by the GPS module bracelet is the same as that provided by the GPS beacon in point 6.

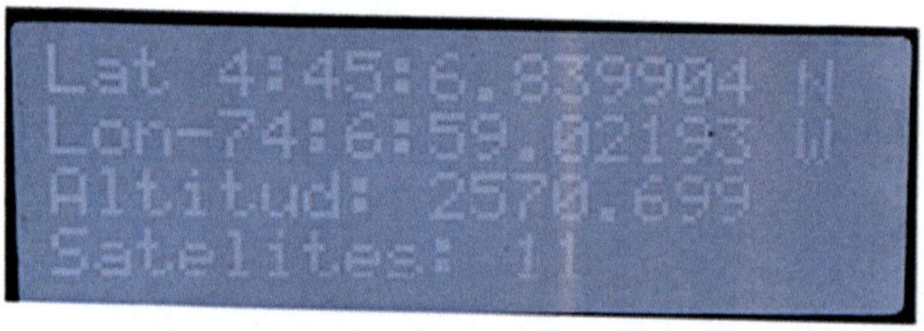

Fig. 7. GPS location results on the bracelet

Detection Bracelet Tests. In this section tests are done to determine if the implemented system (Beacon) is able to detect the RF band currently used by the authorities.

The module implemented in the beacon is the Microchip MRF49, operating in the bands 433 and 915 MHz, compatible with the bracelet ElmoTech used in government the Electronic Surveillance System (Table 6).

A second test considers our own bracelet and determine the maximum distance at which the moving of the victim is able to recognize.

The beacon on Raspberry designed system has proved effective both to locate and identify the victim's cell phone, the bracelet Elmotech currently used by the authorities and the new bracelet designed in this project (Table 7).

The software application implemented on the mobile device of the victim is able to associate with the beacon, identify the proximity of the bracelet and receive SMS both welcome and warning messages.

The bracelet designed for this project has shown better range than the bracelet used by the authorities, with the added advantage that we have another extra location system such as Bluetooth is, not just RF bracelets as currently used by the judicial authorities (Fig. 8).

Table 6. Maximum recognition distance between beacon – bracelet

Maximum recognition distance beacon - bracelet			
Description	Test to determine the maximum distance that the beacon identifies the bracelet ElmoTech		
Elements	Metro, Raspberry Pi with RF module, bracelet Elmotech		
Number of attempts	7		
Expected results	Scope at least 15 m		
Results obtained	Recognition distance is higher than 50 m		
Observations	**Distance**	**Obstacles**	**Results**
	5 m	No	It is detected
	10 m	No	It is detected
	15 m	No	It is detected
	20 m	*Yes	It is detected
	25 m	*Yes	It is detected
	30 m	*Yes	It is not detected
	35 m	*Yes	It is not detected
Conclusions	With state devices, the Beacon recognizes the bracelet 25 m away.		

Table 7. Maximum recognition distance between cellphone – bracelet

Maximum recognition distance cell - bracelet			
Description	Test to determine the maximum distance that the victim's phone system identifies the bracelet		
Elements	Metro, Cell Android, Bracelet of our system		
Number of attempts	7		
Expected results	Scope at least 15 m		
Results obtained	Recognition distance is higher than 50 m		
Observations	**Distance**	**Obstacles**	**Results**
	10 m	No	It is detected
	10 m	*Yes	It is detected
	15 m	*Yes	It is detected
	20 m	*Yes	It is detected
	30 m	*Yes	It is detected
	40 m	*Yes	It is detected
Conclusions	The mobile of the victim running with our application is able to detect the bracelet designed by us at least 40 m away		

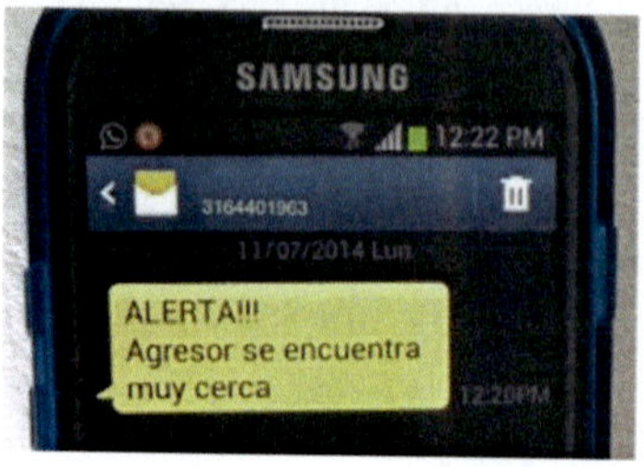

Fig. 8. Alert by proximity

3.4 Conclusions

In the last decade the world has experienced unprecedented technological explosion. Technology has revolutionized the way of life of society and individual customs of each person, there are technological devices developed for different purposes, from the workplace to entertainment, through activities that were unthinkable a few decades ago using technology. The rise shown by mobile phone networks and the Internet make it important to use this technology to provide services that can help people improve their quality of life, facilitating their daily activities and provide security through TICs.

The goals set at the beginning of the project were successfully completed in its entirety since it has demonstrated the viability of producing a system of indoor positioning by Bluetooth communication that complements the GPS systems used in the domestic penal system, it has verified the potential as far as communications are concerned the proposed system and its applicability in other fields and groups.

The technological solution presented is simple and easy to implement both in the location and the receiver device that carries the victim. But as to the device that must wear the batterer, should make a complete study of economic and legislative viability to determine which option is the most feasible, whether to develop a completely new bracelet or adapt acquired system by the Colombian government.

Bibliography

1. World Health Organization. The World report on violence injury prevention. http://www.who.int/violence_injury_prevention
2. Alliance G Consultants & Center for Law Studies, Justice and Society. Final Assessment Project operations of electronic surveillance systems, February 2012 report
3. Cortez, D.F.: Electronic platform for assistance to victims of gender violence indoor environments: Development of software tool based on a Linux operating system. Master's Thesis in Electronics, signal processing and communications. University of Seville (2010). http://www.dinel.us.es/grupos/aceti/docs/Documento3.pdf
4. Ortiz, C.M.: Detection and Tracking Device Inspection and Maintenance (DIM) using the Global Positioning System (GPS) PEMEX Networking Product. University of the State of Hidalgo, Pachuca (2007)
5. Toloza, J.M.: Algorithms and Techniques for Real-Time Increased Relative Positional Accuracy Using Standard GPS Receivers. National University of La Plata, La Plata (2012)
6. Implementation of a Bluetooth wireless network, Universidad del Valle, 2003 (2003)

7. Zhou, S., Pollard, J.K.: Position measurement using bluetooth. IEEE Trans. Consum. Electron. **52**(2), 555–558 (2006)
8. www.raspberrypi.org/
9. Meier, J.D., Farre, C., Bansode, P., Barber, S., Rea, D.: Performance Testing Guidance for Web Applications. Microsoft Corporation, September 2007
10. Toledo, F., Reina, M., de Uvarow, S., López, H.: Performance Testing Methodology. Institute of Computer Science - Faculty of Engineering University of the Republic Montevideo, Uruguay (2008)
11. GPSController.py. GiTHub.martinohanlon/pelmetcam. https://github.com/martinohanlon/pelmetcam/blob/master/GPSController.py
12. Arregui, M.: Tutorial de UML. Dept. of Languages and Systems, University Jaume I. Castellón (2004)
13. Peña, L.: Oriented Analysis and Design Using UML Object. Autonomous University of Colombia. Cali, Faculty of Engineering (2004)

Designing a Controller for Autonomous Underwater Vehicle Using Decoupled Model and Fuzzy Logic

Long Le Ngoc Bao[1], Pham Viet Anh[2], and Duy Anh Nguyen[1](✉)

[1] National Key Lab of Digital Control and System Engineering, VNU-HCM, Ho Chi Minh University of Technology, Ho Chi Minh City, Vietnam
duyanhnguyen@hcmut.edu.vn
[2] University of Transport, Ho Chi Minh City, Vietnam

Abstract. The increasing demands of exploring and researching underwater are still developing worldwide. Since people can hardly reach such environments, Autonomous Underwater Vehicles (AUVs) are invented to take place instead. There are many types of controller and models that often used in AUV control, but in this paper, we are going to discuss about decoupled model, which can be expressed at a combination of three separate linearized sub-systems that control three main movements of an AUV, called surge control, steering control and depth control. The control torques are determine through three separate Fuzzy Logic Controllers (FLCs) to finally achieve the state with acceptable errors. In the simulation part, we will show how the FLCs control the sub-systems in decouple model separately, as well as the whole system and evaluate their performances.

Keywords: REMUS · AUV dynamics · Linearization model · Decoupled model · Fuzzy logic controller

1 Introduction

In this paper, we are going to simulate performance of REMUS AUV using decoupled dynamic model. REMUS AUV is a model developed and associated at the Woods Hole Oceanographic Institute [1]. The non-zero coefficients used in the decoupled model (mentioned later in Sect. 2) were found from experiments and are listed in Table 1 [1].

From the decoupled model, the whole system would be divided into three independent sub-systems that controls three main motions of an AUV: Surge speed control (surge motion), steering control (sway/yaw motion) and depth control (heave motion). For each sub-system, we are going to build a FLC (mentioned in Sect. 3) to meet our accuracy requirements, and in Sect. 4 we will simulate their performances, as well as their combination in a whole system and make comparisons between individual controls and whole control.

D. F. Cortes Tobar et al. (Eds.): AETA 2019, LNEE 685, pp. 42–51, 2021.
https://doi.org/10.1007/978-3-030-53021-1_5

Table 1. Coefficients of REMUS AUV used in decoupled dynamic model

Coefficient	Value	Coefficient	Value
Added mass $X_{\dot{u}}$	-9.30×10^{-1} [kg]	Added mass $Z_{\dot{q}}$	-1.93×10^{0} [kg.m/rad]
Axial drag $X_{u\|u\|}$	-1.62×10^{0} [kg/m]	Added mass $M_{\dot{w}}$	-1.93×10^{0} [kg.m/rad]
Added mass $Y_{\dot{v}}$	-3.55×10^{1} [kg]	Added mass $M_{\dot{q}}$	-4.88×10^{0} [kg.m^2/rad]
Added mass $Y_{\dot{r}}$	1.93×10^{0} [kg.m/rad]	Combined term Z_w	-6.66×10^{1} [kg/s]
Added mass $N_{\dot{v}}$	1.93×10^{0} [kg.m]	Combined term Z_q	-9.67×10^{0} [kg.m/s]
Added mass $N_{\dot{r}}$	-4.88×10^{0} [kg.m^2/rad]	Combined term M_w	3.07×10^{1} [kg.m/s]
Added mass $Z_{\dot{w}}$	-3.55×10^{1} [kg]	Combined term M_q	-6.87×10^{0} [kg.m^2/s]
Fin Lift $Y_{uu\delta r}$	9.64×10^{0} [kg/(m.rad)]	Fin Lift $N_{uu\delta r}$	-6.15×10^{0} [kg/rad]
Fin Lift $Z_{uu\delta s}$	-9.64×10^{0} [kg/(m.rad)]	Fin Lift $M_{uu\delta s}$	-6.15×10^{0} [kg/rad]

2 Decoupled Dynamic Model of an AUV

In this section, we are going to define what decoupled dynamic model is and how to separate it into three sub-systems as mentioned in Sect. 1.

2.1 Linearized Dynamic Equations

We will only discuss briefly how to achieve linearized dynamic equations of a six DOF (Degree of freedom) since it has been explained clearly in [2]. The general and complete form of the set of dynamic equations are expressed in (1) and (2):

$$\dot{x} = Ax + Bu \tag{1}$$

$$\begin{bmatrix} \dot{x}_1 \\ \dot{x}_2 \end{bmatrix} = \begin{bmatrix} -M^{-1}(C+D) & -M^{-1}g(t) \\ J_{B/E}(t) & 0 \end{bmatrix} \begin{bmatrix} x_1 \\ x_2 \end{bmatrix} + \begin{bmatrix} M^{-1} \\ 0 \end{bmatrix} \tau_c \tag{2}$$

where $x_1(t) = \Delta\nu(t) = \nu(t) - \nu_{ref}(t)$ and $x_2(t) = \Delta\eta(t) = \eta(t) - \eta_{ref}(t)$ are the perturbations from the referenced state; M, C, D, g, $J_{B/E}$ and τ_C are mass matrix, Coriolis force matrix, hydrodynamic damping matrix, gravitational force matrix, transformation matrix from Body-fixed frame {B} to Earth-fixed frame {E}, and control torques matrix, respectively [2].

2.2 Decoupled Model

Harley and Marco [3] supposed that a full DOF dynamic model could be divided into three independent sub-systems, called surge speed control, steering control and depth control [2–5]. Their equations are all derived from the main equation (2), where all unrelated terms in each sub-system are neglected.

Surge Speed Control. This system would mainly control the surge speed u, with the differential equation shown in (3) [2]:

$$(m - X_{\dot{u}})\dot{u} = X_{u|u|}u|u| + X_{thruster} \tag{3}$$

where $X_{thruster}$ is the force generated from the single thruster. The thrusting force $X_{thruster}$ can be calculated from the control velocity u_C as shown in (4) [1]:

$$X_{thruster} = -X_{u|u|}u_C|u_C| \tag{4}$$

Steering Control. This system would mainly control the sway and yaw motion with related quantities v, r and ψ. The differential equation are shown in (5) [2]:

$$\begin{bmatrix} m - Y_{\dot{v}} & mx_G - Y_{\dot{r}} & 0 \\ mx_G - N_{\dot{v}} & I_{zz} - N_{\dot{r}} & 0 \\ 0 & 0 & 1 \end{bmatrix} \begin{bmatrix} \dot{v} \\ \dot{r} \\ \dot{\psi} \end{bmatrix} + \begin{bmatrix} -Y_v & mu_{ref} - Y_r & 0 \\ -N_v & mx_G - N_r & 0 \\ 0 & -1 & 0 \end{bmatrix} \begin{bmatrix} u \\ v \\ \psi \end{bmatrix} = \begin{bmatrix} Y_{uu\delta r}u_{ref}^2 \\ N_{uu\delta r}u_{ref}^2 \\ 0 \end{bmatrix} \delta_r \tag{5}$$

where u_{ref} is the desired velocity and δ_r is the rudder angle from the hull of AUV.

Depth Control. This system would mainly control the heave motion with related quantities w, q, θ and z. The differential equation are shown in (6):

$$\begin{bmatrix} m - Z_{\dot{w}} & mx_G - Z_{\dot{q}} & 0 & 0 \\ mx_G - M_{\dot{w}} & I_{zz} - M_{\dot{q}} & 0 & 0 \\ 0 & 0 & 1 & 0 \\ 0 & 0 & 0 & 1 \end{bmatrix} \begin{bmatrix} \dot{w} \\ \dot{q} \\ \dot{\theta} \\ \dot{z} \end{bmatrix} + \begin{bmatrix} -Z_w & mu_{ref} - Z_q & 0 & 0 \\ -M_w & mx_G u_{ref} - M_q & z_G W & 0 \\ 0 & -1 & 1 & 0 \\ -1 & 0 & u_{ref} & 0 \end{bmatrix} \begin{bmatrix} w \\ q \\ \theta \\ z \end{bmatrix} = \begin{bmatrix} Z_{uu\delta s}u_{ref}^2 \\ M_{uu\delta s}u_{ref}^2 \\ 0 \\ 0 \end{bmatrix} \delta_s \tag{6}$$

where δ_s is the stern angle from the hull of AUV.

3 Designing Fuzzy Logic Controllers for the Sub-systems

In this section, we are going to design three non-interacting FLCs for three sub-systems mentioned in Sect. 2. The inputs and outputs of each FLC are dependent on the representing differential equation as demonstrated in (3), (5) and (6). To reduce the computational cost and the complexity of the system, we only manage to design FLCs in zero-order Sugeno model, with maximum of five membership functions for each input and output. The defuzzification method for all FLCs is chosen as weighted average, which can be formulated as below:

$$z = \frac{\sum w_i z_i}{\sum w_i} \tag{7}$$

where z_i and $w_i \in [0, 1]$ is the crisp output and output level for each fuzzy rule i, respectively.

3.1 Surge Control Sub-system

From (3), we notice that the surge velocity u only depends on the thruster force $X_{thruster}$, so we are going to design a FLC with only one input and one output. The error $e_u = u_{ref} - u$ would be imported to the FLC in order to export a control velocity u_C, expressed as an amount of change k_u due to current value. The control velocity u_C will be then converted into $X_{thruster}$ through (4), and then we solve for u based on Euler's numerical integration method [1, 6]. However, since we have a limited maximum surge velocity $u_{MAX} = 1.5$ m/s, the change has to ensure that u_C would never exceed u_{MAX}.

As we desire to control the surge velocity with accuracy of ±0.01 m/s, we suggest an FLC with the properties of input and output illustrated in Fig. 1 and Table 2. From the crisp value k_u defuzzificated, we calculate the control velocity u_C and the control thruster force $X_{thruster}$ using Eq. (8). From (8), we guarantee that u would be always limited in range $[0, u_{MAX}]$. Combining (3), (4) and (8) and using Euler's method [1, 6] to solve numerically, we can compute the next values of u to continue the loop, until it reaches u_C.

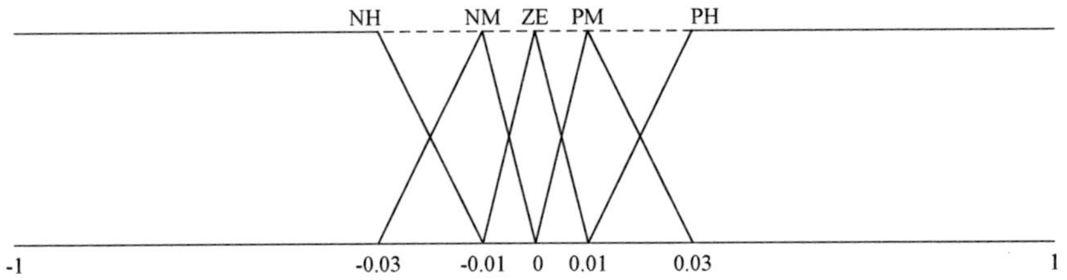

Fig. 1. Membership functions for input value e_u, in [m/s]

Table 2. Singleton output values for amount of change k_u

NH	NM	ZE	PM	PH
−1	−0.6	0	0.6	1

$$u_C = \begin{cases} u + k_u(u_{MAX} - u), & k_u > 0 \\ u, & k_u = 0 \\ u + k_u u, & k_u < 0 \end{cases} \tag{8}$$

3.2 Steering Control Sub-system

The linearized equation in (5) indicates that the steering control sub-system would use the rudder angle δ_r to control the steering state $[v, r, \psi]^T$, but in general we usually choose yaw angle for the output. With a desired checkpoint, we can calculate the necessary yaw angle to steer the AUV. However, unlike the surge control, we decided to use the yaw error e_ψ and rate of yaw error Δe_ψ as inputs of the FLC. Moreover, we also limit δ_r in range [−20°, 20°] due to mechanical constraint. From those points, we suggest a FLC with properties illustrated in Fig. 2 and Table 3, to achieve an accuracy ±0.5° of yaw angle. Note that although all angle values in the computation are in radian, we manage to express e_ψ and δ_r in degree as they are easier to visualize and display later on.

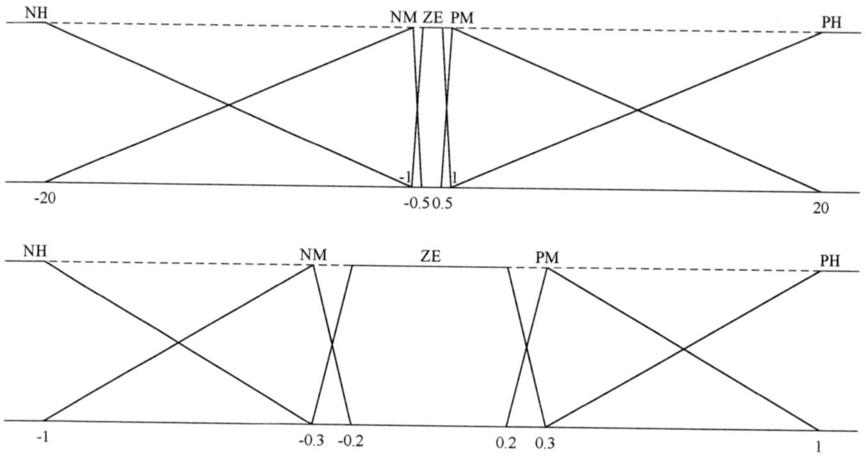

Fig. 2. a. Membership functions for input value e_ψ, in [deg] b. Membership functions for input value $\Delta e_\psi \times 10^3$, in [rad]

Table 3. Singleton output values for rudder angle δ_r, in [deg]

NH	NM	ZE	PM	PH
−20	−10	0	10	20

3.3 Depth Control Sub-system

The linearized equation in (6) indicates that the depth control sub-system would use the stern angle δ_s to control the sinking state $[w, q, \theta, z]^T$. In general this sub-system only guarantee the depth sinking, leaving the pitch angle uncontrolled. To make sure that the pitch angle is small enough and close to equilibrium point, we limit δ_s tighter than δ_r, in range [−12°, 12°].

Similarly to steering control sub-system, we also use the depth error e_z and rate of depth error Δe_z as inputs of the FLC. From those points, we suggest a FLC with properties illustrated in Fig. 3 and Table 4 below, to achieve an accuracy ±0.01 m of depth:

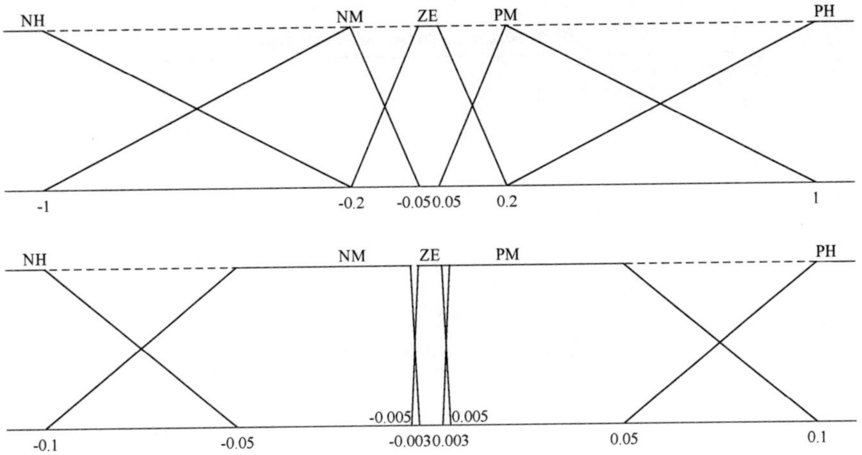

Fig. 3. a. Membership functions for input value e_Z, in [metre] b. Membership functions for input value Δe_Z, in [metre]

Table 4. Singleton output values for stern angle δ_S, in [deg]

NH	NM	ZE	PM	PH
−12	−6	0	6	12

4 Simulation Results

In this section, we are going to simulate the performance of the sub-systems using the FLCs that we designed in Sect. 3. First of all, we would test each FLC separately, and then we integrate them in a complete system to evaluate the impacts and the differences between individual controls and overall control. For all simulation progresses, we assume that the initial position and orientation of the AUV are fixed at $O[x_0, y_0, z_0, \varphi_0, \theta_0, \psi_0]^T = [0, 0, 1, 0, 0, 0]^T$. We also create a reference point A with position $[x_A, y_A, z_A]^T = [40, 20, 5]^T$. This point would be used to evaluate each individual control as well as the overall control when integrating three sub-systems mentioned above. All simulations use the same sample time $t_s = 0.01$ s within duration of 40 s. Distance variables are measured in metre, while velocities are measured in metre per second and radian per second.

4.1 Simulating Individual Controllers

Surge Speed Controller. In surge speed control, we are going to simulate the accelerating process from 0.5 to 1 m/s, and the decelerating process from 1 to 0.5 m/s. The results of surge velocity and thruster force in each case are plotted in Fig. 4. We can see that the FLC designed has made the goal predetermined. However, since our thruster could only generate positive force to boost up, the only way to decelerate is to turn off the thurster motor. That is the reason why accelerating process happens much faster than decelerating one.

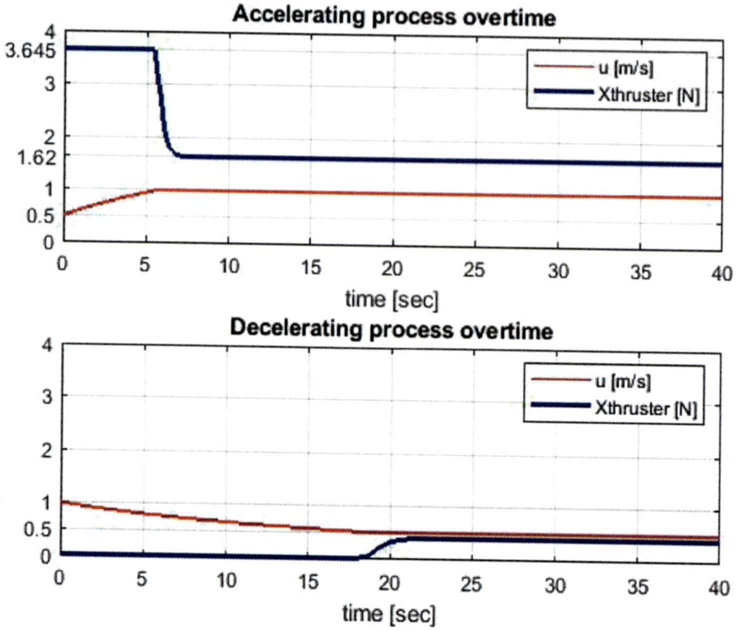

Fig. 4. Performance of surge control in accelerating and decelerating cases

Steering Controller. In steering control, we are going to simulate the steering ability with constant speed $u_{ref} = 1$ m/s, by manipulating the yaw angle to steer from O to reference point A. Since steering control only outputs the yaw angle, we could only test how it performs to achieve a reference yaw angle $\psi_A = 26.5651°$ derived from vector $\overrightarrow{OA}$. The yaw angle in degree is plotted in Fig. 5. In result, we obtain an overshoot of 12.0932° and a settling error of 0.2106° in about 10 s, which satisfied our accuracy requirement of ±0.5°. Note that this controller depends significantly on u_{ref} and the maximum amplitude of rudder. In fact, most AUVs need to slow down when performing steering and sinking actions to increase the precision, so we suppose that this overshoot could reduce when we steer with lower surge speed.

Depth Controller. In depth control, we are going to simulate the sinking process from original depth $z_0 = 1$ m to reference depth $z_A = 5$ m, with constant surge speed $u_{ref} = 1$ m/s. The results of depth and pitch angle are plotted in Fig. 6. In result, we obtain almost zero overshoot and a settling error of 0.0145 m in about 40 s, which satisfied our accuracy requirements of ±0.05 m. The maximum pitch angle when sinking is −17.2697°, which is acceptable in a short duration. The depth controller is also dependent strongly in reference speed u_{ref} and the maximum amplitude of stern. As the depth controller could not control the pitch angle, the maximum amplitude of stern angle has to be limited more strictly than of rudder angle, which results in a longer sinking time, but better overshoot and precision.

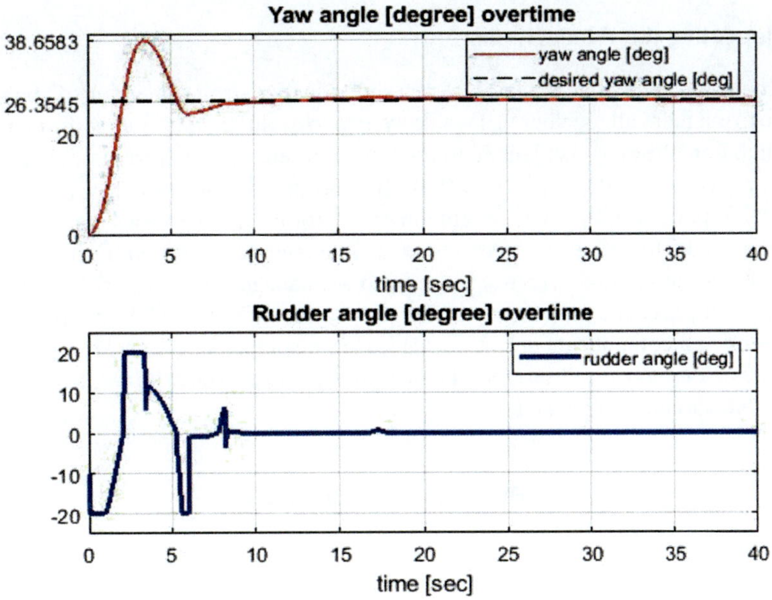

Fig. 5. Performance of steering controller

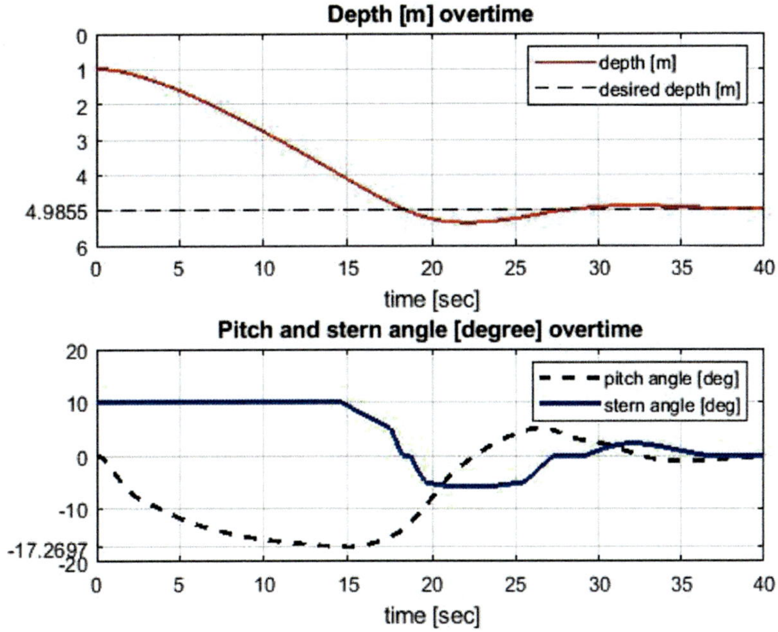

Fig. 6. Performance of depth controller

4.2 Simulating the Whole System

After confirming the effectiveness of each individual controller mentioned above, we manage to put them all together in a whole system to evaluate an overall performance. In this simulation, the AUV is planned to track the reference point A from initial position $[x_0, y_0, z_0, \varphi_0, \theta_0, \psi_0]^T = [0, 0, 1, 0, 0, 0]^T$ and initial velocity $[u_0, v_0, w_0, p_0, q_0, r_0]^T = [0.5, 0, 0, 0, 0, 0]^T$. The maximum errors for each dimensions are expected as $[\pm 0.05$ m, ± 0.05 m, ± 0.05 m$]^T$. As we can see from the steering control's performance, the overshoot was too high with $u_{ref} = 1$ m/s, so we manage to reduce it to $u_{ref} = 0.7$ m/s, in order to improve the performance. The real trajectory of the AUV are displayed in Fig. 7, while all components in state variables are displayed in Fig. 8. Since three sub-systems cannot control roll motion, the variables p and φ would be recalculated from computable variables q, r, θ and ψ.

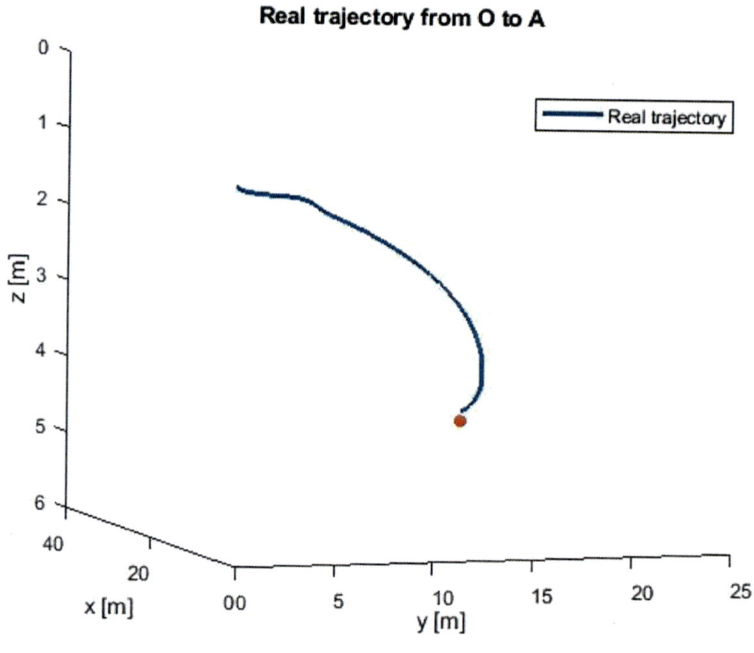

Fig. 7. Trajectory of the AUV model

From the simulation, we achieved the final position $[39.9569, 20.0002, 4.9088]^T$ in 66.39 s, meaning the final errors only guarantee accuracy in xy-plane. It can be explained that, when combining three sub-systems, the computable variables have some impacts to each other as well as the uncontrollable ones. The heave velocity w is too small compared to surge velocity u and sway velocity v, resulting in a small value of roll angle φ and roll rate p. In fact, the influences of uncontrolled variables to controlled ones are usually treated as noises in decoupled model, so the result can be improve by noise-filtering approaches. However, it might lead to some unexpected errors as the decoupled model

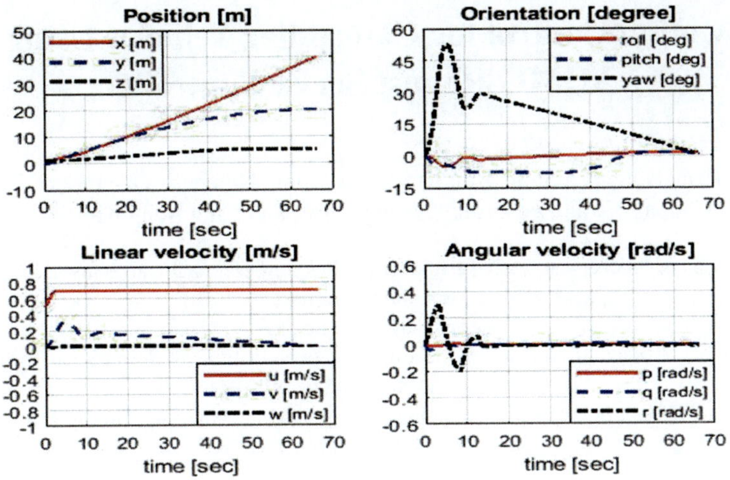

Fig. 8. State of AUV overtime

itself already has linearization error in comparison to original dynamic model, which are usually nonlinear.

5 Conclusion

In this paper, we successfully three FLCs with guaranteed errors for each sub-system control in decoupled and linearized model. However, the combination of them still could not satisfy the predefined accuracy in 3-D area. It might be caused by errors in linearization process and error of numerical integration method.

In future, we would manage to improve the performance of the controllers by trying to solve the differential equations by better methods and better linearization, or maybe we would manage to handle the dynamic model in nonlinear aspect to increase the generality and reality.

References

1. Prestero, T.: Verification of a Six-Degree of Freedom Simulation Model for the REMUS Autonomous Underwater Vehicle. University of California at Davis (1994)
2. Shahaji, L.: Some studies on Control of Autonomous Underwater Vehicles. Swami Ramanand Teerth Marathwada University (2017)
3. Londhe, P., Santhakumar, M., Patre, B., Waghmare, L.: Task space control of an autonomous underwater vehicle manipulator system by robust single-input fuzzy logic control scheme. IEEE J. Oceanic Eng. **42**(1), 13–28 (2017)
4. Fossen: Handbook of Marine Craft Hydrodynamics and Motion Control (2001)
5. Regardt, B.: Modelling and Simulation of an Autonomous Underwater Vehicle. University of Stellenbosch, South Africa (2009)
6. Numerical Methods for Solving Differential Equations, Lecture from San Joaquin Delta College. http://calculuslab.deltacollege.edu

Motion Control for Caterpillar Vehicles Using a MIMO Robust Servo Controller

Van Lanh Nguyen[1], Sung Won Kim[1], Huy Hung Nguyen[3], Dae Hwan Kim[1], Choong Hwan Lee[2], Hak Kyeong Kim[1], and Sang Bong Kim[1(✉)]

[1] Department of Mechanical Design Engineering, Pukyong National University, Busan 48513, Republic of Korea
kimsb@pknu.ac.kr

[2] Department of Machine System Engineering, Dongwon Institute of Science and Technology, Yangsan 50598, Republic of Korea

[3] Faculty of Electronics and Telecommunication, Sai Gon University, Ho Chi Minh City, Vietnam

Abstract. This paper proposes a MIMO robust servo controller design for Mobile Robots as Caterpillar Vehicles with an external disturbance to track desired linear displacement and orientation references using a linear shift invariant differential (LSID) operator. To do this tasks, the followings are done. Firstly, MIMO modeling of the Caterpillar Vehicle is presented. Secondly, by operating the LSID operator to the state space model and the output error vector, a new extended system and a new control law are obtained. Thirdly, a proposed MIMO robust servo controller for the given the Caterpillar Vehicle is designed by using the pole assignment approach. Fourthly, by applying the inverse LSID operator, a servo compensator and a control law for the MIMO system are obtained. Finally, in order to verify the effectiveness of the proposed MIMO robust servo controller, the simulation results are shown. The simulation results show the good tracking performance of the proposed MIMO robust servo controller under as a step type of external disturbance and step types of linear displacement and angular reference signals. This simulation results of the proposed MIMO robust servo controller are compared with the adaptive backstepping control method proposed by P.S. Pratama in 2013. The proposed MIMO robust servo controller shows the faster and better tracking performance than the that adaptive backstepping control method.

Keywords: LSID · MIMO · Robust servo controller · External disturbance · Mobile robot · Caterpillar Vehicle

1 Introduction

Recently, autonomous robots may action instead of human beings. The mobile robots are able to accomplish many tasks in dangerous places where humans cannot enter. These tasks may take place in unsafe environment, for example nuclear waste facilities, such sites where harmful gases or high temperature are present a hard environment for humans. There are several types of methods that used for control the mobile robot.

D. F. Cortes Tobar et al. (Eds.): AETA 2019, LNEE 685, pp. 52–63, 2021.
https://doi.org/10.1007/978-3-030-53021-1_6

M. V. Gomes et al. [1] proposed a PID controller for trajectory tracking of the mobile robot. S. C. Yuan et al. [2] proposed another control method such as fuzzy control. Razvan Solea et al. [3] proposed the sliding mode controller design for trajectory tracking of a wheeled mobile robot in presence of uncertainties. P. S. Pratama et al. [4, 5] proposed a adaptive backtepping control method to track the trajectory. However, most of previous methods had slow response and were not robust due to the external disturbance as a friction and slip force. To solve the robust servo system design problem, S. B. Kim et al. [6] introduced a new design concept for the robust servo control system by using the polynomial differential operator (PDO). D. H. Kim et al. [7] suggested a servo system design for speed control of AC induction motors using PDO. S.B. Kim et al. [8] developed perfectly this PDO design concept using a linear shift invariant differential (LSID) operator. However, these papers were not applied to mobile robots as Caterpillar Vehicles.

Therefore, this paper proposes a MIMO robust servo controller design for the mobile robot as a Caterpillar Vehicle using the LSID operator. The main advantages of LSID are fast response and strong robustness against external disturbances. In order to verify the effectiveness of the proposed MIMO robust servo controller, the simulation results are shown. These results of the proposed MIMO robust servo controller are compared with the adaptive backstepping control method proposed by P.S. Pratama in [4].

2 System Modeling

A Mobile robot as a Caterpillar Vehicle used in this paper is shown in Fig. 1. This Caterpillar Vehicles uses a differential drive wheeled configuration.

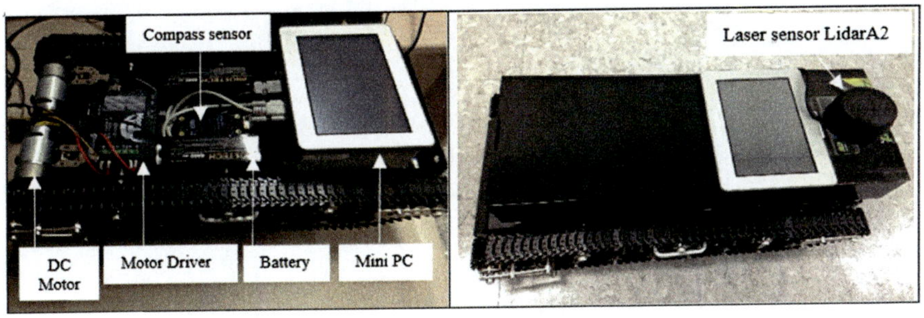

Fig. 1. Struture of the Caterpillar Vehicle platform

In this paper, the Caterpillar vehicle is considered to be two wheeled vehicle. The schematic modeling of a Caterpillar Vehicle is shown in Fig. 2. *OXY* is a global coordinate frame. *Cxy* is the moving coordinate frame attached on the Caterpillar Vehicle platform. This Caterpillar Vehicles has two driving standard wheels *W1* and *W2* with radius *r*, and *W3* and *W4* are passive wheels. (X_A, Y_A) is the Caterpillar Vehicle position vector in global coordinate *OXY* with orientation angle θ_A. The Caterpillar Vehicle moves with linear velocity V_A and angular velocity ω_A. v_R and v_L are the right and left wheel linear velocities, respectively, $\dot{\phi}_R$ and $\dot{\phi}_L$ are the right and left wheel angular velocities, respectively. b is a distance between wheels and Cx axis of the Caterpillar Vehicle. D is a mass center of the Caterpillar Vehicle. d is the distance between C and D.

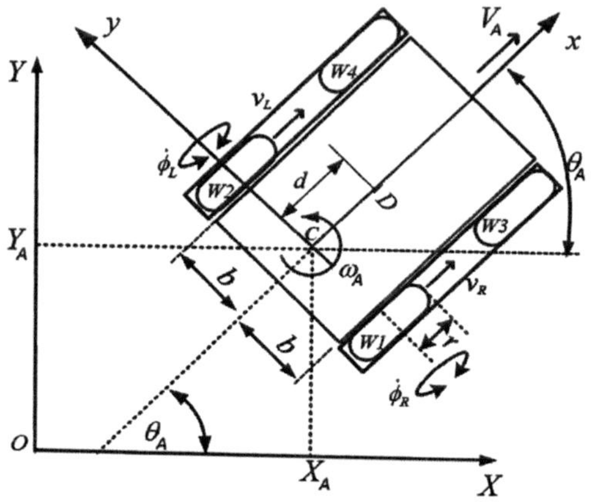

Fig. 2. Schematic diagram of the Caterpillar Vehicle

The kinematic equation of the Caterpillar Vehicle as shown in Fig. 2 can be expressed as follows:

$$\dot{\boldsymbol{q}} = \boldsymbol{H}(\boldsymbol{q})\boldsymbol{z} \tag{1}$$

$$\boldsymbol{q} = \begin{bmatrix} X_A \\ Y_A \\ \theta_A \end{bmatrix}, \boldsymbol{H}(\boldsymbol{q}) = \begin{bmatrix} \cos\theta_A & 0 \\ \sin\theta_A & 0 \\ 0 & 1 \end{bmatrix}, z = \begin{bmatrix} V_A \\ \omega_A \end{bmatrix} \tag{2}$$

where $\boldsymbol{q}$ is the posture vector of the Caterpillar Vehicle in global coordinate OXY, $\boldsymbol{H(q)}$ is full rank matrix, $\boldsymbol{z}$ is the velocity vector of the Caterpillar Vehicle.

Relation of the linear velocity V_A and angular velocity ω_A with the right and left wheel angular velocities $(\dot{\phi}_R, \dot{\phi}_L)$ can be expressed as:

$$z = \boldsymbol{T}\dot{\boldsymbol{\Phi}} \text{ for } \boldsymbol{T} = \begin{bmatrix} \frac{r}{2} & \frac{r}{2} \\ \frac{r}{2b} & \frac{-r}{2b} \end{bmatrix}, \dot{\boldsymbol{\Phi}} = \begin{bmatrix} \dot{\phi}_R \\ \dot{\phi}_L \end{bmatrix} \tag{3}$$

The dynamic equations of the Caterpillar Vehicle can be described by Euler-Lagrange formulation as follows [9]:

$$\overline{\boldsymbol{M}}\,\dot{z} + \overline{\boldsymbol{V}}z = \boldsymbol{u} \tag{4}$$

where

$$\bar{\boldsymbol{M}} = \begin{bmatrix} \frac{r^2}{4b^2}(m_t b^2 + I) + I_w & \frac{r^2}{4b^2}(m_t b^2 - I) \\ \frac{r^2}{4b^2}(m_t b^2 - I) & \frac{r^2}{4b^2}(m_t b^2 + I) + I_w \end{bmatrix}, \bar{\boldsymbol{V}} = \begin{bmatrix} 0 & \frac{r^2}{2b} m_c d\dot{\theta}_A \\ -\frac{r^2}{2b} m_c d\dot{\theta}_A & 0 \end{bmatrix},$$

$$\boldsymbol{I} = m_c d^2 + 2m_w b^2 + I_c + 2I_m.$$

$z = \left[V_A \; \omega_A \right]^T$ is the velocity vector of the Caterpillar Vehicle, V_A is the linear velocity and ω_A is the angular velocity of the Caterpillar Vehicle, $\boldsymbol{u} = \left[\tau_R \; \tau_L \right] \in R^m (m = 2)$ is the torque input vector applied to the two wheels of the Caterpillar Vehicle, $m_t = m_c + 2m_w$ is the total mass of the Caterpillar Vehicle, m_c is the mass of the body without driving wheels, m_w is the mass of each wheel, I_m is the moment of inertia of each motor, I_w is the moment of inertia of each wheel, I_c is the moment of inertia of the body.

The Caterpillar Vehicle model can be described by the following MIMO linear model with a disturbance as:

$$\begin{cases} \dot{\boldsymbol{x}} = \boldsymbol{A}\boldsymbol{x} + \boldsymbol{B}\boldsymbol{u} + \boldsymbol{\varepsilon} \\ \boldsymbol{y} = \boldsymbol{C}\boldsymbol{x} \end{cases} \tag{5}$$

$$\boldsymbol{A} = \begin{bmatrix} \boldsymbol{L} & \boldsymbol{0} \\ \boldsymbol{I_2} & \boldsymbol{0} \end{bmatrix}, \; \boldsymbol{B} = \begin{bmatrix} \boldsymbol{\Lambda} \\ \boldsymbol{0} \end{bmatrix}, \; \boldsymbol{I_2} = \begin{bmatrix} 1 & 0 \\ 0 & 1 \end{bmatrix}, \; \boldsymbol{L} = (\bar{\boldsymbol{M}}\boldsymbol{T}^{-1})^{-1}\bar{\boldsymbol{V}}\boldsymbol{T}^{-1}, \; \boldsymbol{\Lambda} = (\bar{\boldsymbol{M}}\boldsymbol{T}^{-1})^{-1},$$

$$\boldsymbol{C} = \begin{bmatrix} 0\,0\,1\,0 \\ 0\,0\,0\,1 \end{bmatrix} = \left[\boldsymbol{c}_1^T \; \boldsymbol{c}_2^T \; \boldsymbol{c}_3^T \; \boldsymbol{c}_4^T \right]^T, \; \boldsymbol{x} = \left[V_A \; \omega_A \; l_A \; \theta_A \right]^T \in R^n (n = 4),$$

where $\boldsymbol{A}$, $\boldsymbol{B}$ and $\boldsymbol{C}$ are parameter matrices for the Caterpillar Vehicle system, $\boldsymbol{x}$ is the system state vector, l_A is linear displacement and θ_A is the orientation of the Caterpillar Vehicle, $\boldsymbol{y} = \left[l_A \; \theta_A \right]^T \in R^p (p = 2)$ is the output vector of the Caterpillar Vehicle, and $\boldsymbol{\varepsilon} = \left[\varepsilon_1 \; \varepsilon_2 \; \varepsilon_3 \; \varepsilon_4 \right]^T \in R^n$ is the disturbance vector.

3 MIMO Robust Servo Controller Design

An output error vector is defined as the difference between the output vector $\boldsymbol{y}$ and reference input vector $\boldsymbol{y_r}$:

$$\boldsymbol{e} = \left[e_1 \; e_2 \right]^T = \boldsymbol{y} - \boldsymbol{y_r} \in R^p \tag{6}$$

where $\boldsymbol{y_r} = \left[y_{r1} \; y_{r2} \right]^T \in R^p (p = 2)$ is a reference input vector.

It is supposed that the following differential equation forms:

$$L_q(D)\boldsymbol{y}_r(t) = 0 \text{ and } L_q(D)\boldsymbol{\varepsilon}(t) = 0 \tag{7}$$

where $L_q(D)$ is a LSID operator for reference and disturbance signals.

A MIMO robust servo controller design method is implemented as follows [8]:

Firstly, to eliminate the effect of the disturbance in Eq. (5), operating a LSID operator $L_q(D)$ to both sides of Eq. (5) can be written as:

$$\frac{d}{dt}\{L_q(D)\boldsymbol{x}\} = \boldsymbol{A}L_q(D)\boldsymbol{x} + \boldsymbol{B}L_q(D)\boldsymbol{u} \tag{8}$$

$$L_q(D) = \prod_{i=1}^{q} (D - \beta_i I_d) = \prod_{i=1}^{q} P_{\alpha i}(D) \tag{9}$$

where $P_{\alpha i}(D)$ is the first order LSID operator of the ith factorized term and $(D - \beta_i I_d)$ is the ith factor of $L_q(D)$ with $dim\{L_q(D)\} = q$, $\beta_i \in C, D = d/dt$ and I_d are derivative and identity operators, respectively.

The ith output error can be obtained as follows:

$$e_i(t) = \boldsymbol{\gamma} L_q^{-1}(D) z_i(t) \tag{10}$$

$$\boldsymbol{\gamma} = \left[\gamma_0\ \gamma_1 \cdots \gamma_{q-1}\right]^T, \gamma_0 = \prod_{i=1}^{q} \beta_i, \gamma_1 = \prod_{j \neq i, i=1}^{q} \beta_{ij}, \cdots, \gamma_q = \sum_{i=1}^{q} \beta_i,$$

$$z_i = \left[e_i\ e_i^{(1)} \ldots e_i^{(q-1)}\right]^T \in R^q,$$

where $L_q^{-1}(D)$ is the inverse LSID operator.

Secondly, by operating $L_q(D)$ to Eq. (6), and combining with Eqs. (7)–(9), an extended system can be obtained as follows:

$$\dot{\boldsymbol{x}}_e = \boldsymbol{A}_e \boldsymbol{x}_e + \boldsymbol{B}_e \boldsymbol{v} \tag{11}$$

$$\boldsymbol{A}_e = \begin{bmatrix} \boldsymbol{A} & \boldsymbol{0} \\ \boldsymbol{M} & diag(\boldsymbol{N})_p \end{bmatrix} \in R^{(n+pq)\times(n+pq)}, \boldsymbol{B}_e = \begin{bmatrix} \boldsymbol{B} \\ \boldsymbol{0} \end{bmatrix} \in R^{(n+pq)\times m}, \boldsymbol{x}_e = \begin{bmatrix} L_q(D)\boldsymbol{x} \\ \boldsymbol{z} \end{bmatrix} \in R^{n+pq},$$

$$\boldsymbol{M} = \left[\boldsymbol{M}_1^T\ \boldsymbol{M}_2^T \cdots \boldsymbol{M}_p^T\right]^T, diag(\boldsymbol{N})_p \in R^{pq\times pq}, \boldsymbol{z} = \left[\boldsymbol{z}_1^T\ \boldsymbol{z}_2^T \cdots \boldsymbol{z}_p^T\right]^T \in R^{pq},$$

$$\boldsymbol{N} = \begin{bmatrix} \boldsymbol{0} & & \boldsymbol{I}_{q-1} \\ \cdots & \cdots & \cdots \\ & -\boldsymbol{\gamma}^T & \end{bmatrix} \in R^{q\times q}, \boldsymbol{M}_i = \left[\boldsymbol{0} \cdots \mathbf{c}_i^T\right]^T \in R^{q\times n}$$

where $\boldsymbol{x}_e \in R^{n+pq}$ is the extended system state variable vector, $\boldsymbol{v} \in R^m$ is a new control law for the extended system, $\boldsymbol{z} \in R^{pq}$ is an error variable vector for the extended system, and $diag(\boldsymbol{N})_p$ has p elements of matrix N on the diagonal parts.

A new control law for the extended system is defined by the following form:

$$\boldsymbol{v} = L_q(D)\boldsymbol{u} = -\boldsymbol{F}\boldsymbol{x}_e \in R^m \text{ for } \boldsymbol{F} = \left[\boldsymbol{F}_x\ \boldsymbol{F}_z\right] \in R^{m\times(n+pq)} \tag{12}$$

where $\boldsymbol{F}_x \in R^{m\times n}$ and $\boldsymbol{F}_z \in R^{m\times pq}$ are feedback control gain matrices for $L(D)\boldsymbol{x}$ and $\boldsymbol{z}$, respectively.

A new error variable vector for the extended system can be defined as:

$$\boldsymbol{\zeta} = L^{-1}(D)\boldsymbol{z} \text{ for } \boldsymbol{\zeta} = \left[\zeta_1^T \; \zeta_2^T \cdots \zeta_p^T\right]^T \in R^{pq} \tag{13}$$

Using Eq. (12), the control law of Eq. (5) can be obtained as follows:

$$\boldsymbol{u} = L_q^{-1}(D)\boldsymbol{v} = -\boldsymbol{F}L_q^{-1}(D)\boldsymbol{x}_e = -\boldsymbol{F}\boldsymbol{x}_\zeta \tag{14}$$

where $\boldsymbol{x}_\zeta = L_q^{-1}(D)\boldsymbol{x}_e = \left[\boldsymbol{x}^T \; \boldsymbol{\zeta}^T\right]^T \in R^{n+pq}$ is a new extended system variable vector.

From Eq. (11) and Eq. (12), a closed loop system is obtained as:

$$\dot{\boldsymbol{x}}_e = (\boldsymbol{A}_e - \boldsymbol{B}_e\boldsymbol{F})\boldsymbol{x}_e \tag{15}$$

Thirdly, by operating the inverse LSID operator $L_q^{-1}(D)$ for Eq. (11) and using the new input vector $\boldsymbol{v}$ of Eq. (12), the following equations can be obtained:

$$\frac{d}{dt}\boldsymbol{x} = [\boldsymbol{A} - \boldsymbol{B}\boldsymbol{F}_x]\boldsymbol{x} - \boldsymbol{B}\boldsymbol{F}_z\boldsymbol{\zeta} + \boldsymbol{\varepsilon} \tag{16}$$

$$\frac{d\boldsymbol{\zeta}}{dt} = \boldsymbol{N}_z\boldsymbol{\zeta} + \boldsymbol{I}_\zeta\boldsymbol{e} \; for \; L_q(D)\boldsymbol{\varepsilon} = \boldsymbol{0} \tag{17}$$

where

$$\boldsymbol{N}_z = \begin{bmatrix} \boldsymbol{N} & \boldsymbol{0} & \cdots & \boldsymbol{0} \\ \boldsymbol{0} & \boldsymbol{N} & \cdots & \boldsymbol{0} \\ \vdots & \vdots & \ddots & \vdots \\ \boldsymbol{0} & \boldsymbol{0} & \cdots & \boldsymbol{N} \end{bmatrix} \in R^{pq\times pq}, \boldsymbol{I}_\zeta = \begin{bmatrix} \boldsymbol{\lambda} & \boldsymbol{0} & \cdots & \boldsymbol{0} \\ \boldsymbol{0} & \boldsymbol{\lambda} & \cdots & \boldsymbol{0} \\ \vdots & \vdots & \ddots & \vdots \\ \boldsymbol{0} & \boldsymbol{0} & \cdots & \boldsymbol{\lambda} \end{bmatrix} \in R^{pq\times n},$$

$$and \;\; \boldsymbol{\lambda} = \left[0 \cdots 0 \; 1\right]^T \in R^q.$$

The servo compensator of Eq. (17) includes the model of reference and disturbance signals since the matrix $\boldsymbol{N}_z$ is composed of the least common multiple model of two signals.

The configuration of the proposed MIMO robust servo controller under the disturbance can be described as shown in Fig. 3.

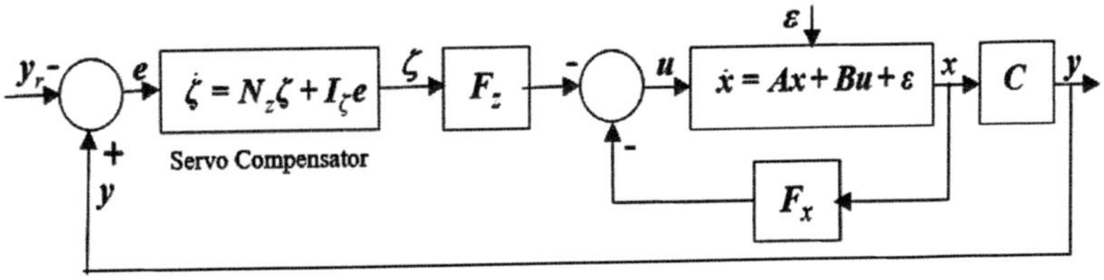

Fig. 3. Configuration of the proposed MIMO robust servo controller

4 Simulation Results

Simulations are done to verify the effectiveness of the proposed MIMO robust servo controller for a given Caterpillar Vehicle. This results of the proposed MIMO robust servo controller are compared with the adaptive backstepping control method proposed by P.S. Pratama in [4].

4.1 Proposed MIMO Robust Servo Controller

The given MIMO system with $\boldsymbol{u} = \left[u_1\ u_2\right]^T$ is a torque input vector, and an output vector $\boldsymbol{y} = \left[y_1\ y_2\right]^T$ is the posture output vector of the Caterpillar Vehicle. The reference input vector consisting of linear displacement and orientation reference are chosen as follows: $\boldsymbol{y}_r = \left[2.24\,\text{m}\ 0.46\,\text{rad}\right]^T$. It is assumed that the step type of disturbance vector is given as the following $\boldsymbol{\varepsilon} = \left[0.05\,\text{m/s}\ 0.01\,\text{rad/s}\ 0.1\,\text{m}\ 0.05\,\text{rad}\right]^T$.

Tables 1, 2, 3 and 4 show parameters of the Caterpillar Vehicle, the matrices' values of MIMO system, the initial values and parameter values for the proposed controller.

Table 1. Parameter values of Caterpillar Vehicle

Parameters	Values	Parameters	Values
b	0.12 m	d	0 m
r	0.04 m	I_w	0.08 kgm^2
m_w	1 kg	I_m	0.48 kgm^2
m_c	20 kg	I_c	2.4 kgm^2

Table 2. Matrices' values of given MIMO system

Matrices	$\boldsymbol{A}$	$\boldsymbol{B}$	$\boldsymbol{C}$
Values	$\boldsymbol{A} = \begin{bmatrix} 0&0&0&0 \\ 0&0&0&0 \\ 1&0&0&0 \\ 0&1&0&0 \end{bmatrix}$	$\boldsymbol{B} = \begin{bmatrix} 183.882 & 101.770 \\ 164.503 & 60.047 \\ 0 & 0 \\ 0 & 0 \end{bmatrix}$	$\boldsymbol{C} = \begin{bmatrix} 0&0&1&0 \\ 0&0&0&1 \end{bmatrix} = \left[\boldsymbol{c}_1^T\ \boldsymbol{c}_2^T\ \boldsymbol{c}_3^T\ \boldsymbol{c}_4^T\right]^T$

Table 3. Initial values of the proposed MIMO robust servo controller

Initial values	$\boldsymbol{x}(0)$	$\boldsymbol{\zeta}(0)$
Values	$\left[0\,0\,0\,0\right]^T$	$\left[0\,0\right]^T$

Table 4. Matrices' values of the proposed MIMO robust servo controller

Matrices	Values
F_x, F_z	$F_x = \begin{bmatrix} -0.4469 & 0.6747 & -6.2026 & 8.4644 \\ 1.1796 & -1.2300 & 15.8669 & -15.5667 \end{bmatrix}, F_z = \begin{bmatrix} -28.1496 & 35.2550 \\ 70.1422 & -65.3878 \end{bmatrix}$
N_z, I_ζ	$N_z = \begin{bmatrix} 0 & 0 \\ 0 & 0 \end{bmatrix}, I_\zeta = \begin{bmatrix} 1 & 0 \\ 0 & 1 \end{bmatrix}$

4.2 Adaptive Backstepping Controller

The adaptive backstepping controller design for the Caterpillar Vehicle was described in [4]. This adaptive backstepping controller was designed based on Lyapunov stability theory and an adaptive backstepping control theory to track a desired trajectory. This backstepping controller is applied for the Caterpillar Vehicle with the parameters and gain values are shown in Table 5.

Table 5. Parameter values of adaptive backstepping controller

Parameters	Values	Parameters	Values
b	0.12 m	$X_A(0)$	0 m
r	0.04 m	$Y_A(0)$	0 m
k_1	2	$\theta_A(0)$	0 rad
k_2	5	$V_A(0)$	0 m/s
k_3	21.5	$\omega_A(0)$	0 rad/s

where k_1, k_2, and k_3 are positive constant values.

4.3 Simulation Results

Simulation results using the step reference input vector $\boldsymbol{y}_r$ for the both controllers are shown in Figs. 4, 5, 6, 7, 8, 9, 10, 11, 12, 13, 14, 15, 16, 17, 18 and 19. Figures 4 and 5 show the control law vector using both controllers. The control laws u_1 and u_2 of the proposed MIMO robust servo controller converge to zero after 0.75 s, while the control laws u_1 and u_2 of the adaptive backstepping controller converge to zero after 3 s and 0.5 s, respectively. Figures 6 and 7 show that the output y_1 and y_2 of both controllers. Figure 6 shows the output y_1 of both controllers track the step reference input y_{r1} well after about 0.75 s and 3 s, respectively. Figure 7 shows the output y_2 of both controllers track the step reference input y_{r2} well after about 0.75 s and 0.5 s, respectively. Figures 8 and 9 show that the output X_A on the X axis and Y_A on the Y axis of both controllers. Figure 8 shows the output X_A of both controllers converge to $X = 2$ m after about 0.75 s and 3 s, respectively. Figure 9 shows the output Y_A of both controllers converge to $Y = 1$ m after about 0.75 s and 3 s, respectively.

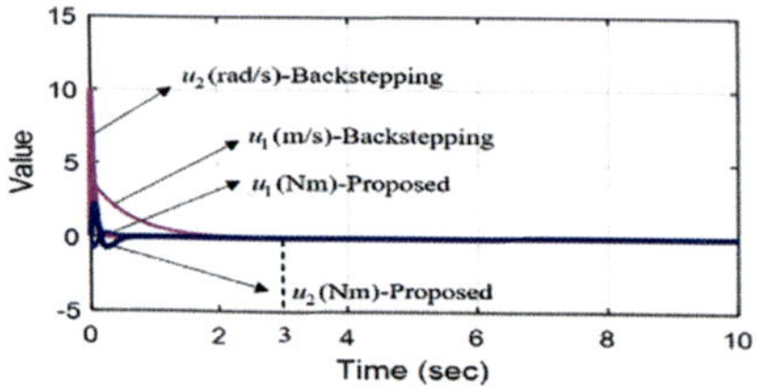

Fig. 4. Control law vector $\boldsymbol{u}$

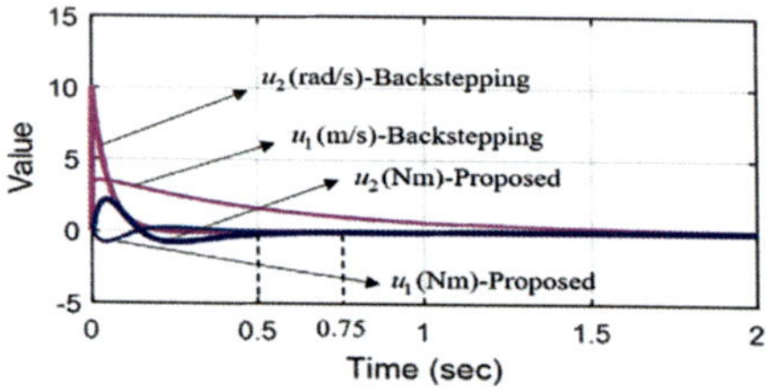

Fig. 5. Control law vector $\boldsymbol{u}$ at beginning

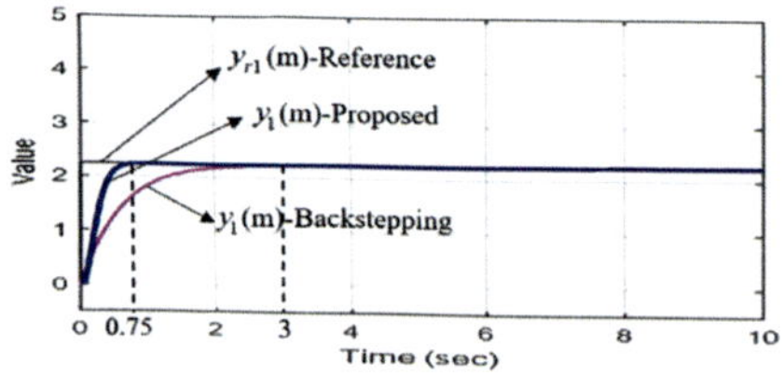

Fig. 6. Output y_1 for the step reference input

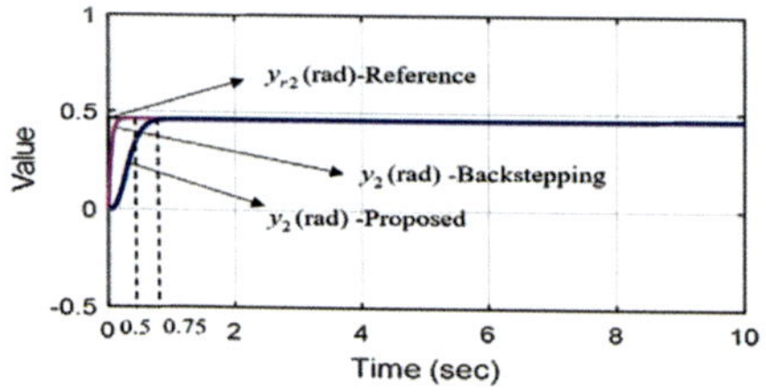

Fig. 7. Output y_2 for the step reference input

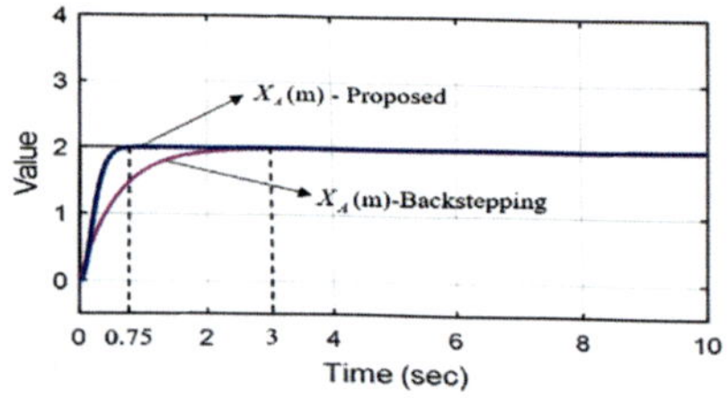

Fig. 8. Output X_A on the X axis of the Caterpillar Vehicle

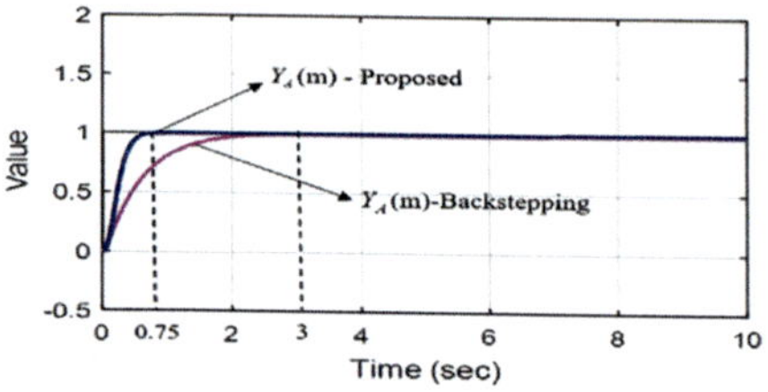

Fig. 9. Output Y_A on the Y axis of the Caterpillar Vehicle

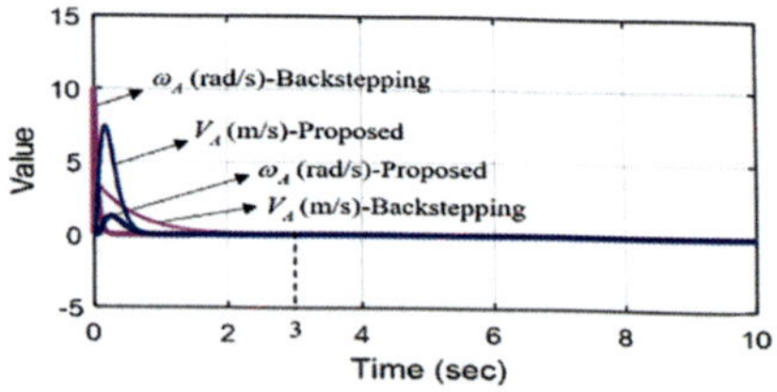

Fig. 10. Velocity state vector for the step reference input

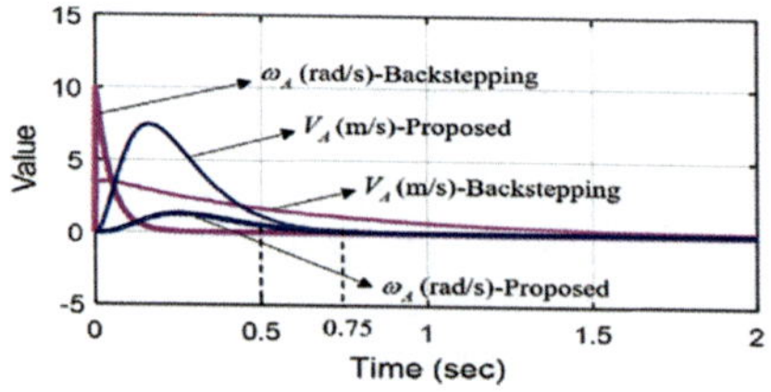

Fig. 11. Velocity state vector at beginning

Figures 10 and 11 show that the velocity state vector of both controllers. The linear velocity and the angular velocity of the Caterpillar Vehicle in the velocity state vector of the proposed MIMO robust servo controller converge to zero after about 0.75 s, while the linear velocity and the angular velocity of the Caterpillar Vehicle of the adaptive backstepping controller converge to zero after about 3 s and 0.5 s, respectively.

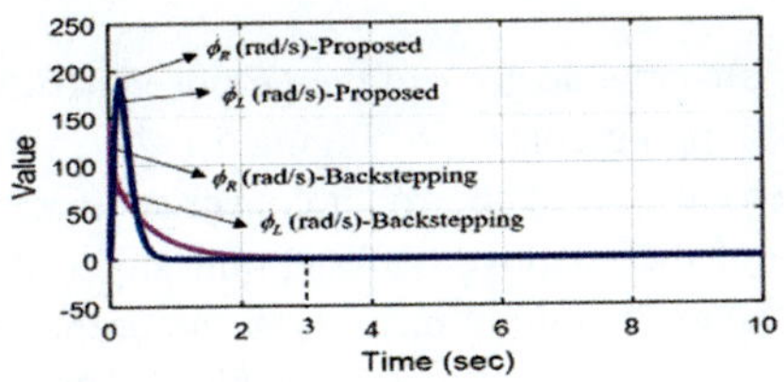

Fig. 12. Angular velocity vector of the right and left wheels

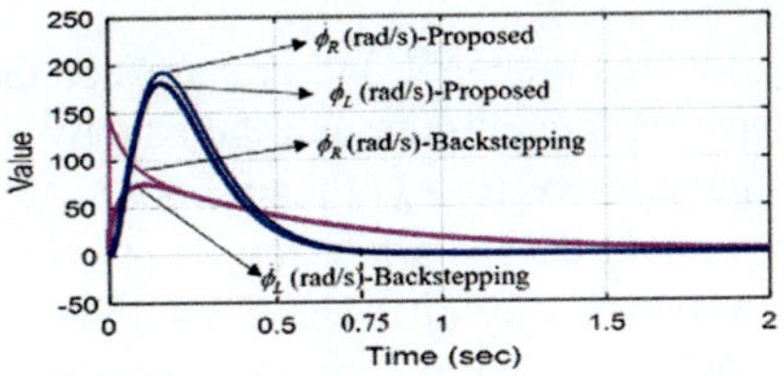

Fig. 13. Angular velocity vector of the right and left wheels at beginning

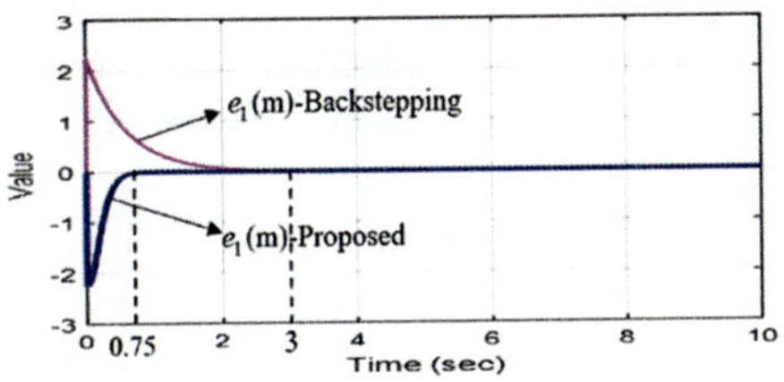

Fig. 14. Output error e_1 for the step reference input

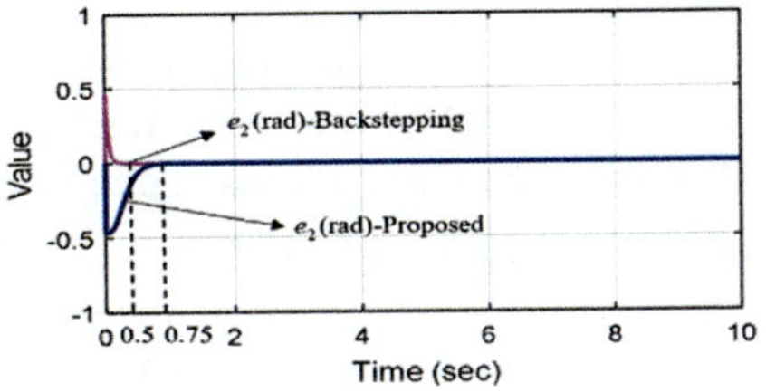

Fig. 15. Output error e_2 for the step reference input

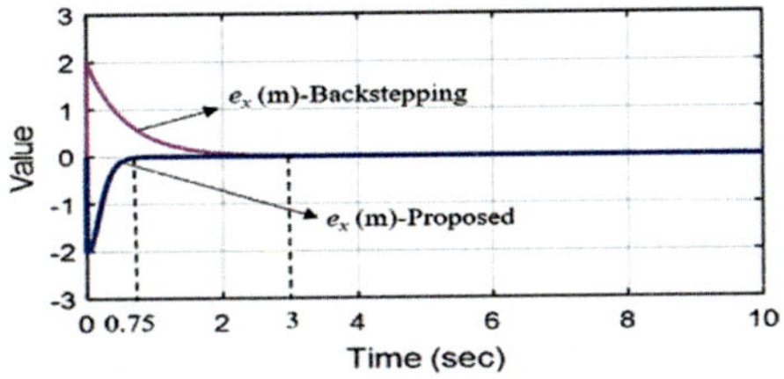

Fig. 16. Output error e_X on the X axis

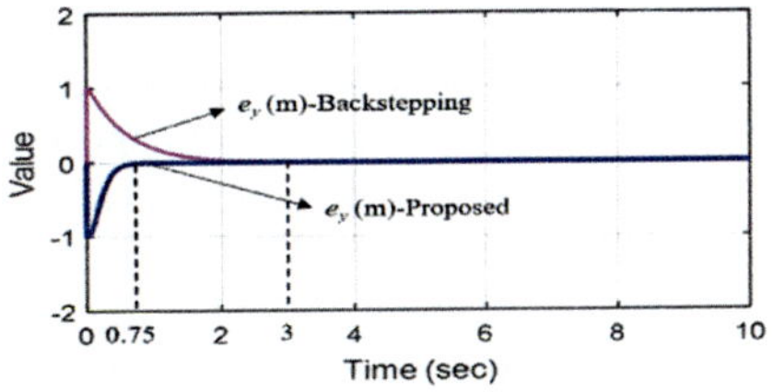

Fig. 17. Output error e_Y on the Y axis

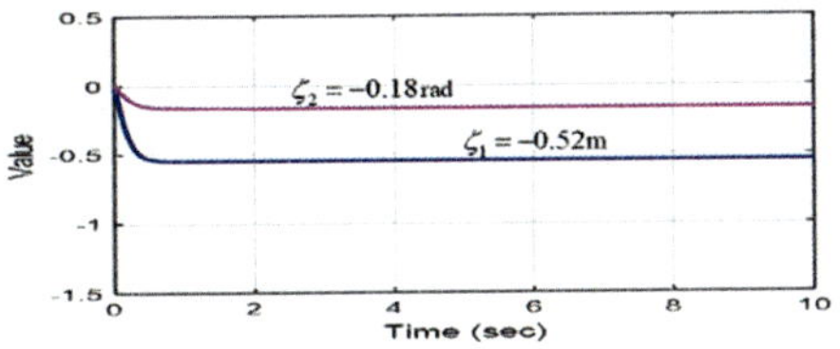

Fig. 18. Servo compensator for the MIMO robust servo controller

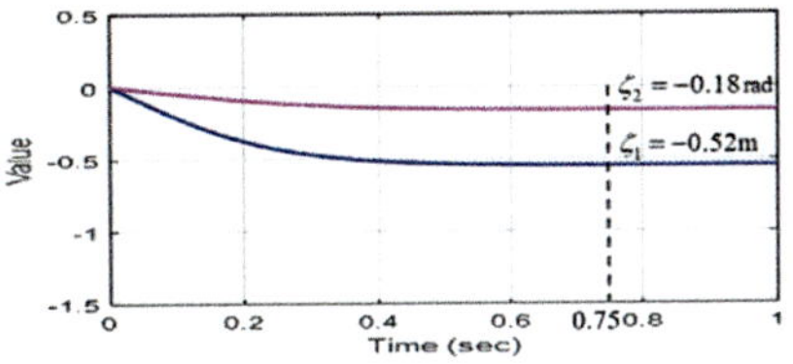

Fig. 19. Servo compensator of the MIMO robust servo controller at beginning

Figures 12 and 13 shows that the angular velocity vector of right and left wheels of both controllers. The angular velocity vector of right and left wheels of the proposed MIMO robust servo controller converges to zero after about 0.75 s, while the angular velocity vector of right and left wheels of the adaptive backstepping controller converges to zero after about 3 s, respectively.

Figures 14, 15, 16 and 17 show that the output error vector $\boldsymbol{e}$ of the proposed MIMO robust servo controller and the adaptive backstepping controller. Figure 14 shows the output error vector e_1 of the proposed MIMO robust servo controller converges to zero after about 0.75 s, while the output error e_1 of the adaptive backstepping controller converges to zero after about 3 s. Figure 15 shows the output error e_2 of the proposed MIMO robust servo controller converges to zero after about 0.75 s, while the output error vector e_2 of the adaptive backstepping controller converges to zero after about 0.5 s. Figures 16 and 17 show the output error e_x on the X axis and the output error e_y on the Y axis of the proposed MIMO robust servo controller converges to zero after about 0.75 s, while the output error e_x on the X axis and the output error e_y on the Y axis of the adaptive backstepping controller converges to zero after about 3 s. Therefore, the proposed MIMO robust servo controller tracks the step reference input well and better than the adaptive backstepping controller [4]. Figures 18 and 19 show the servo compensator output vector $\boldsymbol{\zeta} = \left[\zeta_1\ \zeta_2\right]^T$ for the step reference input vector of the proposed MIMO robust servo controller. The servo compensator output ζ_1 converges to -0.52 m after about 0.75 s, while the servo compensator output ζ_2 converges to $-$ 0.18 rad after about 0.75 s.

5 Conclusions

A MIMO robust servo controller design for a Caterpillar Vehicle with the friction and slip force disturbance to track desired linear displacement and orientation references using a linear shift invariant differential (LSID) operator was proposed. The simulation results showed that the proposed MIMO robust servo controller had good tracking performance for a step type of reference input vector and a step type of disturbance vector. The output vector of the MIMO robust servo controller tracked the step type of the reference input vector well after about 0.75 s, while the output vector of the adaptive backstepping controller [4] tracked the step type of reference input vector after about 3 s. Therefore, the proposed MIMO robust servo controller showed the faster tracking performance than the adaptive backstepping controller [4].

Future Works

The practical effectiveness and applicability of the proposed MIMO robust servo controller for the Caterpillar Vehicle of this paper will be verified by experiments.

References

1. Gomes, M.V., Bassora, L.A., Morandin Jr., O., Vivaldini, K.C.T.: PID control applied on a line-follower AGV using a RGB camera. In: IEEE 19th International Conference on Intelligent Transportation Systems (ITSC), pp. 194–198 (2016)
2. Solea, R., Filipescu, A., Nunes, U.: Sliding-mode control for trajectory-tracking of a wheeled mobile robot in presence of uncertainties. In: 7th Asian Control Conference. IEEE (2009)
3. Yuan, S.C., Yao, L.: Robust type-2 fuzzy control of an automatic guided vehicle for wall-following. In: Proceedings of International Conference of Soft Computing and Pattern Recognition, pp. 172–177 (2009)

4. Pratama, P.S., Luan, B.T., Tran, T.T., Kim, H.K., Kim, S.B.: Trajectory tracking algorithm for automatic guided vehicle based on adaptive backstepping control method. In: Lecture Notes in Electrical Engineering, vol. 282, pp. 535–544. Springer, Heidelberg (2013)
5. Pratama, P.S., Nguyen, T.H., Kim, H.K., Kim, D.H., Kim, S.B.: Positioning and obstacle avoidance of automatic guided vehicle in partially known environment. Int. J. Control Autom. Syst. **14**(6), 1572–1581 (2016)
6. Kim, S.B., Kim, D.H., Pratama, P.S., Kim, J.W., Kim, H.K., Oh, S.J., Jung, Y.S.: MIMO robust servo controller design based on internal model principle using polynomial differential operator. In: Lecture Notes in Electric Engineering, vol. 371, pp. 469–484. Springer, Cham (2015)
7. Kim, D.H., Nguyen, T.H., Pratama, P.S., Gulakari, A.V., Kim, H.K., Kim, S.B.: Servo system design for speed control of AC induction motors using polynomial differential operator. Int. J. Control Autom. Syst. **15**(3), 1207–1216 (2017)
8. Kim, S.B., Nguyen, H.H., Kim, D.H., Kim, H.K.: Robust servo controller design for MIMO systems based on linear shift invariant differential operator. J. Inst. Control Robot. Syst. **24**(6), 501–511 (2018)
9. Hung, N., Im, J.S., Jeong, S.K., Kim, H.K., Kim, S.B.: Design of a sliding mode controller for an automatic guided vehicle and its implementation. Int. J. Control Autom. Syst. **8**(1), 81–90 (2010)

Estimation of Stator Voltage of Inverter-Supplied Induction Motor Using Kalman Filter

Pavel Karlovsky(✉), Ondrej Lipcak, and Jiri Lettl

Department of Electric Drives and Traction, Czech Technical University in Prague, Technicka 2, Prague, Czech Republic
{karlopav,lipcaond,lettl}@fel.cvut.cz

Abstract. Most of the control algorithms of variable speed drives with induction motor require the knowledge of the stator voltage vector applied to the motor terminals. This vector is usually reconstructed from the known microcontroller's PWM signals or the commanded voltage for the inverter is used within the control algorithm. However, these solutions require a DC-link voltage sensor and compensation of the nonlinear inverter behavior. In this paper, stator voltage estimator based on the Extended Kalman filter is proposed. This approach requires neither the knowledge of the DC-link voltage nor the nonlinear model of the inverter. Only the knowledge of the stator currents and the rotational speed is needed. The proposed estimator is verified within the simulation of predictive-torque control of induction motor drive in Matlab Simulink where the comparison of the applied and the estimated voltage vector is presented along with their harmonic analysis. The accuracy of the estimated voltage vector shows its suitability for further inverter nonlinearities investigation.

Keywords: Induction motor drive · Voltage estimation · Kalman filter

1 Introduction

Development in modern power electronics devices along with today's high computational power of digital signal processors (DSP) has led to a massive expansion of induction motors (IM) in variable speed drives both in the industry and every day's applications [1]. Such drives are in the vast majority supplied by a two-level voltage-source inverter (VSI) [1]. In some of the control algorithms, it is necessary to know the voltage applied to the stator terminals [2–4]. However, direct evaluation of the stator voltage becomes almost an impossible task due to the nature of VSI [5–8]. Power electronics devices do not permit operation in the linear region [1]. Therefore, the fundamental wave of the voltage is formed by a defined switching of the semiconductor elements [1–4].

The non-ideal behavior of the VSI leads to a deviation between the commanded voltage and the actual voltage applied to the motor terminals [5–8]. The first aspect that distorts the output voltage is the necessary protective time, i.e. dead-time inserted by the microcontroller or the VSI driver in order to prevent the shoot through of the VSI

D. F. Cortes Tobar et al. (Eds.): AETA 2019, LNEE 685, pp. 64–73, 2021.
https://doi.org/10.1007/978-3-030-53021-1_7

DC-link. Then there is a propagation delay between the signal from the microcontroller and the actual output from the VSI driver. If the delay times for the switching on and off are significantly different, then they can negatively contribute to the voltage distortion. The behavior of the VSI is then also aggravated by the nonlinear semiconductor devices, i.e. delayed and finite current and voltage transitions and the voltage drop in the on state [5–8].

One approach of the evaluation of the stator voltage lies within its reconstruction from the switching signals and the DC-link voltage [2]. Another one is to work with the reference voltage vector [5–8]. The commonly used compensation strategies fall into the category of so-called instantaneous average voltage methods [5–8]. These approaches can be quite simple and effective, but they require many VSI parameters that are device dependent [7, 8]. The compensation parameters are usually obtained from the direct measurement on the VSI, either manual or automatic [6–8].

In this paper, an estimation of the VSI output voltage based on the Kalman filter (KF) [9] is presented. The use of KF in the electric drive is not new. However, it is typically used for the estimation of the rotor flux or rotor mechanical speed [10], change of the motor parameters [11] or characteristics of the superior mechanical system such as wheel slip [12]. The possibility of the VSI voltage vector reconstruction by KF was already presented and verified in the simulations on a PMSM motor drive [13, 14]. In this case, the inverter nonlinearities were estimated, and the applied voltage vector reconstructed out of the knowledge of the DC-link voltage and the PWM signals. On the contrary, this paper does not assume any prior knowledge of the DC-link voltage or the switching combinations. The voltage vector is estimated from the measured stator currents and the rotational speed of the rotor only.

The proposed method is tested in a simulation environment Matlab Simulink on the mathematical model of an induction motor. The predictive torque control (PTC) [15] is utilized to control the speed of the motor. However, the presented voltage estimation method is not dependent on the employed control method because the KF is used as an independent observer. The waveform of the estimated VSI voltage vector is presented and compared with the applied voltage. The precision of the voltage vector estimation is then shown on the harmonic analysis. The accuracy of the estimated voltage vector shows its suitability for further inverter nonlinearities investigation.

2 Induction Motor Model

Generally, there are many ways of modeling the induction motor [4]. The application's requirements always determine the model to be used. The paper uses the so-called T-equivalent circuit [4], [16] where the stator voltage is the input, and the stator current is the output. The state variables are then the rotor flux and stator current. The following vector differential equations then describe the system

$$\frac{\mathrm{d}\bar{i}_\mathrm{s}}{\mathrm{d}t} = \bar{i}_\mathrm{s}\frac{R_\mathrm{s}L_\mathrm{r}^2 + L_\mathrm{m}^2 R_\mathrm{r}}{L_\mathrm{r}\left(L_\mathrm{m}^2 - L_\mathrm{s}L_\mathrm{r}\right)} + \mathrm{j}\bar{\psi}_\mathrm{r}\omega\frac{L_\mathrm{m}}{L_\mathrm{m}^2 - L_\mathrm{s}L_\mathrm{r}} - \bar{\psi}_\mathrm{r}\frac{L_\mathrm{m}R_\mathrm{r}}{L_\mathrm{r}(L_\mathrm{m}^2 - L_\mathrm{s}L_\mathrm{r})} - \bar{v}_\mathrm{s}\frac{L_\mathrm{r}}{L_\mathrm{m}^2 - L_\mathrm{s}L_\mathrm{r}}, \quad (1)$$

$$\frac{\mathrm{d}\bar{\psi}_\mathrm{r}}{\mathrm{d}t} = \mathrm{j}\omega\bar{\psi}_\mathrm{r} - \frac{R_\mathrm{r}}{L_\mathrm{r}}\bar{\psi}_\mathrm{r} + \frac{R_\mathrm{r}L_\mathrm{m}}{L_\mathrm{r}}\bar{i}_\mathrm{s}, \quad (2)$$

where $\bar{\psi}_{\mathrm{r}}$ is the rotor magnetic flux space vector, $\bar{i}_{\mathrm{s}}$ is the stator current space vector, $\bar{v}_{\mathrm{s}}$ is the stator voltage space vector, ω is the electrical speed of the rotor, j denotes the imaginary unit, L_{r} is the rotor leakage inductance, L_{m} is the magnetizing inductance, R_{s} is the stator resistance, L_{s} is the stator leakage inductance, R_{r} is the rotor resistance and σ is the leakage factor calculated as

$$\sigma = 1 - \frac{L_{\mathrm{m}}^2}{L_{\mathrm{s}} L_{\mathrm{r}}}.$$

The convenient form of the system description is the state-space representation

$$\dot{\mathbf{x}} = \mathbf{A}\mathbf{x} + \mathbf{B}\mathbf{u} \tag{3}$$

$$\mathbf{y} = \mathbf{C}\mathbf{x} + \mathbf{D}\mathbf{u} \tag{4}$$

where **x** is the state vector, **u** is the input vector, **y** is the output vector, **A** is the system matrix, **B** is the input matrix, and **C** is the output matrix. For the chosen IM representation, the matrices are defined as

$$\mathbf{A} = \begin{bmatrix} \frac{L_{\mathrm{m}}^2 R_{\mathrm{r}} + R_{\mathrm{s}} L_{\mathrm{r}}^2}{-L_{\mathrm{r}}^2 L_{\mathrm{s}} \sigma} & 0 & \frac{L_{\mathrm{m}}}{L_{\mathrm{r}}^2 L_{\mathrm{s}} \sigma} & \frac{L_{\mathrm{m}} n_{\mathrm{p}} \omega_{\mathrm{m}}}{L_{\mathrm{r}} L_{\mathrm{s}} \sigma} \\ 0 & \frac{L_{\mathrm{m}}^2 R_{\mathrm{r}} + R_{\mathrm{s}} L_{\mathrm{r}}^2}{-L_{\mathrm{r}}^2 L_{\mathrm{s}} \sigma} & -\frac{L_{\mathrm{m}} n_{\mathrm{p}} \omega_{\mathrm{m}}}{L_{\mathrm{r}} L_{\mathrm{s}} \sigma} & \frac{L_{\mathrm{m}}}{L_{\mathrm{r}}^2 L_{\mathrm{s}} \sigma} \\ \frac{L_{\mathrm{m}} R_{\mathrm{r}}}{L_{\mathrm{r}}} & 0 & -\frac{R_{\mathrm{r}}}{L_{\mathrm{r}}} & -n_{\mathrm{p}} \omega_{\mathrm{m}} \\ 0 & \frac{L_{\mathrm{m}} R_{\mathrm{r}}}{L_{\mathrm{r}}} & n_{\mathrm{p}} \omega_{\mathrm{m}} & -\frac{R_{\mathrm{r}}}{L_{\mathrm{r}}} \end{bmatrix},$$

$$\mathbf{B} = \begin{bmatrix} \frac{1}{L_{\mathrm{s}} \sigma} & 0 \\ 0 & \frac{1}{L_{\mathrm{s}} \sigma} \\ 0 & 0 \\ 0 & 0 \end{bmatrix}, \mathbf{C} = \begin{bmatrix} 1\,0\,0\,0 \\ 0\,1\,0\,0 \end{bmatrix}, \mathbf{D} = \begin{bmatrix} 0\,0 \\ 0\,0 \end{bmatrix}, \tag{5}$$

where n_{p} is the number of pole pairs and ω_{m} is the mechanical rotor shaft speed. The system input **u**, state **x** and output **y** are defined as (6)

$$\mathbf{x} = \begin{bmatrix} i_{\mathrm{s}\alpha} \\ i_{\mathrm{s}\beta} \\ \psi_{\mathrm{r}\alpha} \\ \psi_{\mathrm{r}\beta} \end{bmatrix}, \mathbf{u} = \begin{bmatrix} v_{\mathrm{s}\alpha} \; v_{\mathrm{s}\beta} \end{bmatrix}, \mathbf{y} = \begin{bmatrix} i_{\mathrm{s}\alpha} \\ i_{\mathrm{s}\beta} \end{bmatrix}, \tag{6}$$

where the subscripts α and β refer to the real and imaginary axes of the stator-fixed coordinate system. Due to the variable speed, the matrix **A** is variant.

3 Extended Model for the Voltage Estimation

The algorithm aims to estimate the input voltage vector, and therefore, in the particular application, the system is treated to be without inputs. Instead, the voltage is considered as an additional state of the system. The new system description has current, rotor flux and voltage space vector components as the state variables, current components as the output and zero vector as the input

$$\mathbf{x} = \begin{bmatrix} i_{s\alpha} \\ i_{s\beta} \\ \psi_{r\alpha} \\ \psi_{r\beta} \\ v_{s\alpha} \\ v_{s\beta} \end{bmatrix}, \mathbf{u} = \mathbf{O}, \mathbf{y} = \begin{bmatrix} i_{s\alpha} \\ i_{s\beta} \end{bmatrix}, \tag{7}$$

where $\mathbf{O}$ denotes the zero vector. The state space matrices change too as

$$\mathbf{A} = \begin{bmatrix} \frac{L_m^2 R_r + R_s L_r^2}{-L_r^2 L_s \sigma} & 0 & \frac{L_m}{L_r^2 L_s \sigma} & -\frac{L_m n_p \omega_m}{L_r L_s \sigma} & \frac{1}{L_s \sigma} & 0 \\ 0 & \frac{L_m^2 R_r + R_s L_r^2}{-L_r^2 L_s \sigma} & -\frac{L_m n_p \omega_m}{L_r L_s \sigma} & \frac{L_m}{L_r^2 L_s \sigma} & 0 & \frac{1}{L_s \sigma} \\ \frac{L_m R_r}{L_r} & 0 & -\frac{R_r}{L_r} & -n_p \omega_m & 0 & 0 \\ 0 & \frac{L_m R_r}{L_r} & n_p \omega_m & -\frac{R_r}{L_r} & 0 & 0 \\ 0 & 0 & 0 & 0 & 0 & 0 \\ 0 & 0 & 0 & 0 & 0 & 0 \end{bmatrix}, \mathbf{B} = \mathbf{O}, \tag{8}$$

Because neither the value nor the behavior of the voltage is known, the last two rows in the state matrix $\mathbf{A}$ are filled with zeroes. The voltage vector (the previous $\mathbf{B}$ matrix) now appears as additional column to the right side of the matrix $\mathbf{A}$.

For the implementation, the state-space representation discrete form is required. The following discretization of the state matrix $\mathbf{A}_d$ and the output matrix $\mathbf{C}_d$ were used

$$\mathbf{A}_d = e^{AT_s} \approx \mathbf{I} + \mathbf{A}T_s, \tag{9}$$

$$\mathbf{C}_d = \mathbf{C}, \tag{10}$$

where T_s is the sampling period. The matrices change to

$$\mathbf{A}_d = \begin{bmatrix} 1 + T_s \frac{L_m^2 R_r + L_r^2 R_s}{L_r(L_m^2 - L_r L_s)} & 0 & T_s \frac{L_m R_r}{L_r(L_r L_s - L_m^2)} & T_s \frac{n_p L_m}{L_r L_s - L_m^2} \omega & T_s \frac{L_r}{L_r L_s - L_m^2} & 0 \\ 0 & 1 + T_s \frac{L_m^2 R_r + L_r^2 R_s}{L_r(L_m^2 - L_r L_s)} & -T_s \frac{n_p L_m}{L_r L_s - L_m^2} \omega & T_s \frac{n_p L_m}{L_r L_s - L_m^2} \omega & 0 & T_s \frac{L_r}{L_r L_s - L_m^2} \\ T_s \frac{L_m R_r}{L_r} & 0 & 1 - T_s \frac{R_r}{L_r} & -T_s n_p \omega & 0 & 0 \\ 0 & T_s \frac{L_m R_r}{L_r} & T_s n_p \omega & 1 - T_s \frac{R_r}{L_r} & 0 & 0 \\ 0 & 0 & 0 & 0 & 1 & 0 \\ 0 & 0 & 0 & 0 & 0 & 1 \end{bmatrix}, \tag{11}$$

$$\mathbf{C}_d = \begin{bmatrix} 1\,0\,0\,0\,0\,0 \\ 0\,1\,0\,0\,0\,0 \end{bmatrix}, \tag{12}$$

4 Kalman Filter Design

The KF algorithm consists of two stages – prediction and correction. The calculation of the first estimate utilizes the knowledge of the system description given by the matrix $\mathbf{A}_d$ and $\mathbf{B}_d$ that is, in this case, zero matrix

$$\mathbf{x}(k|k-1) = \mathbf{A}_d(\mathbf{x}(k-1|k-1))\mathbf{x}(k-1|k-1), \tag{13}$$

where (k|k − 1) refer to the variables calculated at the time instant k based on the values at the time instant k − 1. The first estimate (a priori) of the covariance matrix $\mathbf{P}$ is

$$\mathbf{P}(k|k-1) = \mathbf{A}_d(k-1)\mathbf{P}(k-1|k-1)\mathbf{A}_d(k-1)' + \mathbf{Q}, \tag{14}$$

where $\mathbf{Q}$ is the state noise matrix. After the first step, the a priori estimate is corrected by the Kalman gain. The optimal Kalman gain is defined by the matrix $\mathbf{K}$ as

$$\mathbf{K}(k) = \mathbf{P}(k|k-1)\mathbf{C}_d'(\mathbf{C}_d\mathbf{P}(k|k-1)\mathbf{C}_d + \mathbf{R})^{-1}, \tag{15}$$

where $\mathbf{R}$ is the output noise matrix. The final (a posteriori) estimate is calculated from the first estimate, Kalman gain and the error between the real and the a priori estimate

$$\mathbf{x}(k|k) = \mathbf{x}(k|k-1) + \mathbf{K}(k)(\mathbf{y}(k) - \mathbf{C}_d\mathbf{x}(k|k-1)), \tag{16}$$

Within the final step, the state covariance matrix $\mathbf{P}$ is updated

$$\mathbf{P}(k|k) = (\mathbf{I}_6 - \mathbf{K}(k)\mathbf{C}_d)\mathbf{P}(k|k-1), \tag{17}$$

where $\mathbf{I}_6$ is the identity matrix with rank six.

The matrix $\mathbf{Q}$ is a diagonal matrix. It factorizes the noise in the system (how much the noise influences the states). In other words, how much the states are dependent on the system description given by the matrix $\mathbf{A}$. Since the voltage is unknown and the system description does not correctly track the voltage, the coefficients in $\mathbf{Q}$ connected with the voltage must be sufficiently high. On the other hand, the coefficients connected with the current and flux components shall be close to zero since the system description determines them adequately. The choice of the matrix values is crucial for the proper operation of the algorithm. The $\mathbf{Q}$ matrix is expressed as

$$\mathbf{Q} = \begin{bmatrix} q_1 & 0 & 0 & 0 & 0 & 0 \\ 0 & q_2 & 0 & 0 & 0 & 0 \\ 0 & 0 & q_3 & 0 & 0 & 0 \\ 0 & 0 & 0 & q_4 & 0 & 0 \\ 0 & 0 & 0 & 0 & q_5 & 0 \\ 0 & 0 & 0 & 0 & 0 & q_6 \end{bmatrix}, \tag{18}$$

where the constraints for the matrix coefficients are chosen as q1 = q2, q3 = q4 and q5 = q6. The coefficients were tuned experimentally.

The matrix $\mathbf{R}$ is also a diagonal one. It reflects the measurement noise. In the proposed algorithm, multiple values were examined. The reason is, that, in our case, the subject of interest is the fundamental wave of the stator voltage rather than its instantaneous value. Increasing the coefficients of the matrix **R** leads to greater filtration of the estimated voltage and, therefore, the estimated waveforms are smoother. The matrix **R** is expressed as

$$\mathbf{R} = \begin{bmatrix} r & 0 \\ 0 & r \end{bmatrix}, \tag{19}$$

The covariance matrix **P** is a diagonal matrix and is modified within every iteration. In the paper, the identity matrix has been chosen as the initial matrix **P**.

5 Simulation Results

The proposed voltage estimator based on KF was implemented in Matlab Simulink on mathematical model of four-pole 5.5 kW induction motor drive. The IM supply converter was modeled as the conventional two-level VSI. The IM was controlled by the PTC algorithm. Kubota observer [17] was used to estimate the fluxes required by the PTC algorithm. The simulation topology is depicted in Fig. 1.

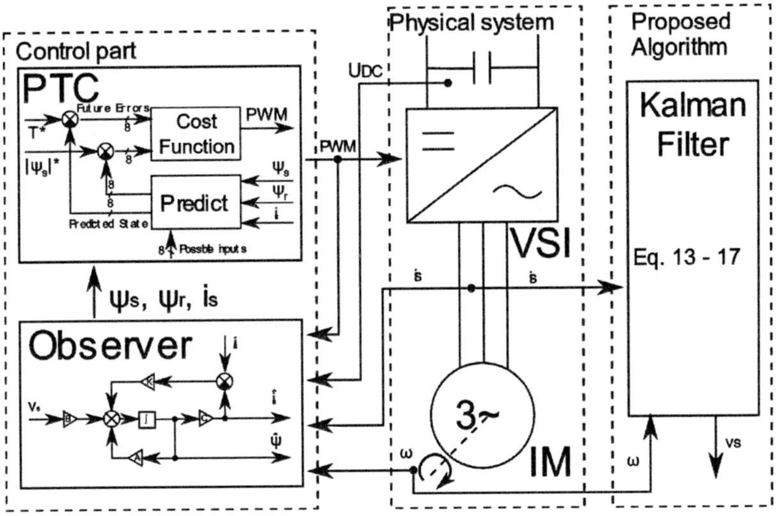

Fig. 1. Simulation topology

The discrete-time control part (observer, PTC, KF) was simulated with the step time 100 μs, which corresponds to the computational power of modern DSPs. The part representing the physical system (VSI, IM) was simulated with the 1 μs step. The motor ran at constant speed 1000 rpm in order to preserve the frequency of the fundamental wave of the voltage. The motor was loaded with the nominal torque 37 Nm.

Figure 2 shows, from the left top, the IM speed and developed torque and the α components of the actual and estimated stator current, rotor flux and voltage space vectors. The simulations were run for three cases of the output noise matrix R (r = 0.01, r = 1 and r = 100). In all the cases, the current and flux were estimated correctly since their behavior is well described by the system matrices. The figure also shows a small phase shift between the real and the estimated rotor flux. This phase shift is not observed in the current waveforms because the currents are used as the filter feedback. The behavior of the voltage is not described by the system matrices, so the shape of the estimated voltage waveforms differs for the three examined R matrixes.

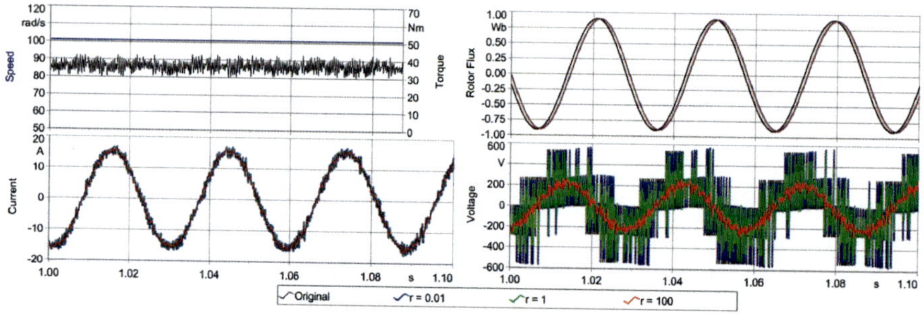

Fig. 2. Simulation run overview

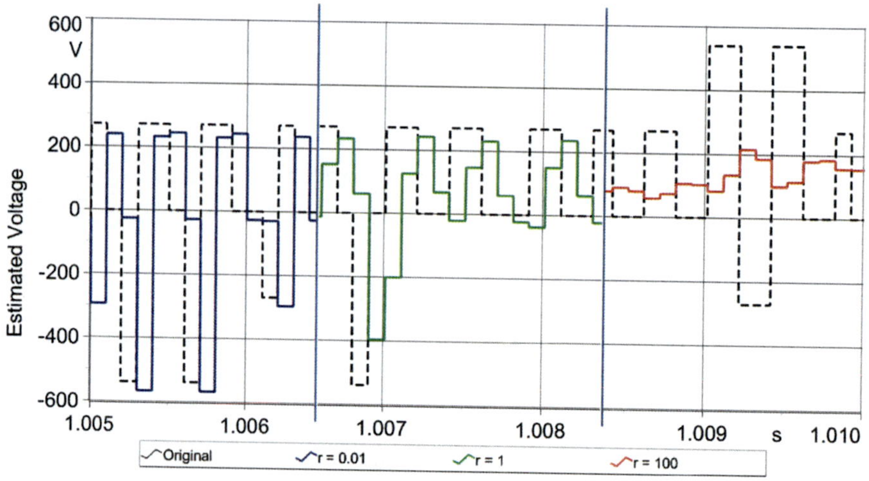

Fig. 3. The comparison of the applied and estimated voltages for different coefficients.

Figure 3 shows in detail the α component of the actual and estimated stator voltage. The curve of the estimated voltage is shifted and depends on the **R** matrix coefficients selection. The figures show the comparison for the same three values as in Fig. 2.

The voltage waveforms FFT analyses were performed. Figure 4 shows the harmonic composition of the applied and of the estimated voltages for the three different noise matrixes R (r = 0.01, r = 1 and r = 100). The analyses were run in order to compare

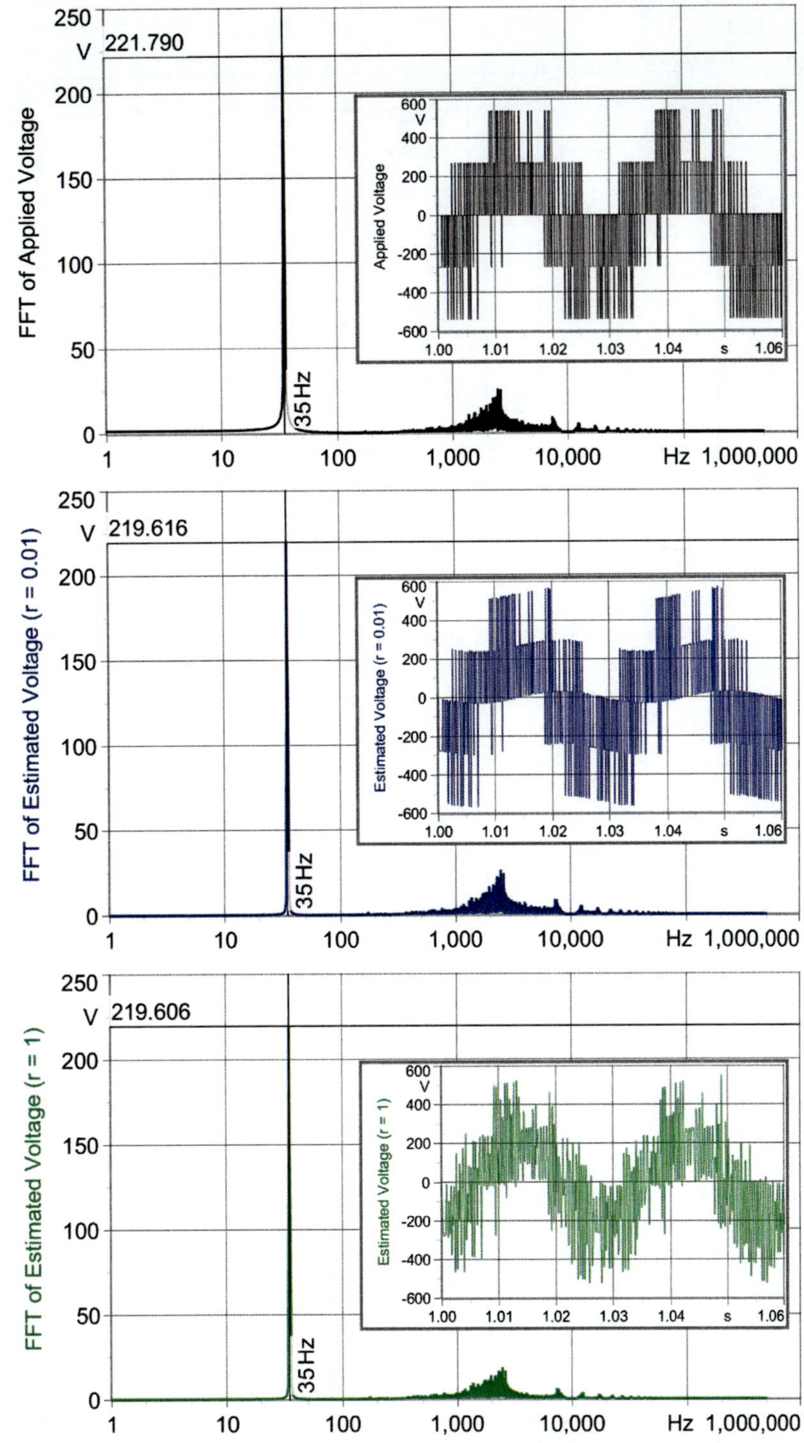

Fig. 4. Harmonic analysis of the applied and estimated voltage with different coefficients.

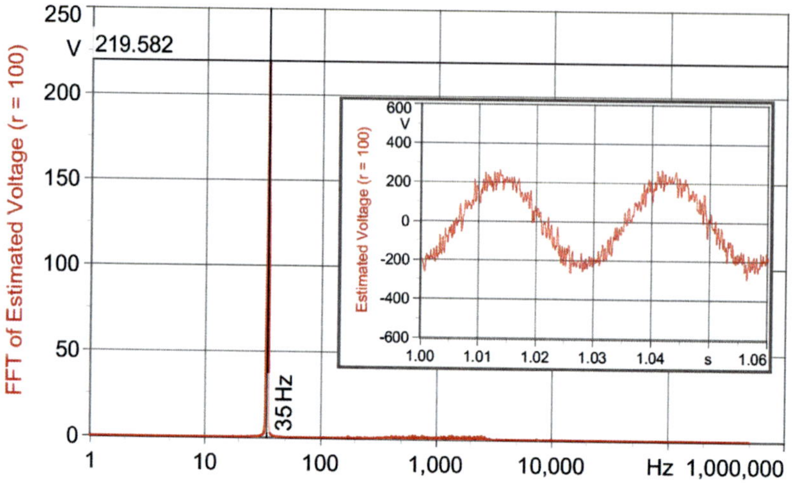

Fig. 4. (*continued*)

the actual and estimated voltage. The fundamental wave frequency (35 Hz) is estimated correctly in all the cases. However, the amplitudes of the estimated voltage fundamental waves are in all the cases a bit smaller than the one of the actual voltage. The FFT is shown only for the α component. Based on the results, it is concluded that the voltage vector or at least its fundamental frequency can be reconstructed by the proposed KF algorithm based on the knowledge of the stator currents and the rotor speed.

6 Conclusion

In this paper, online algorithm based on KF for determination of the two-level VSI output voltage has been presented. The proposed method was verified within the PTC of IM supplied by the VSI in Matlab Simulink. The KF algorithm requires neither the knowledge of the DC-link voltage nor the knowledge of the switching combinations. It only utilizes measured stator currents and rotor speed.

The simulations were carried out for three different output noise matrices. The frequencies of the fundamental wave of the estimated voltage have been determined accurately in all the cases. The amplitudes were estimated only with a small deterioration of few volts. It has been shown that with small filtration coefficients, the voltage is estimated correctly with only one step delay. On the other hand, the higher filtration coefficients lead to smoother estimated waveforms.

The advantage of the proposed solution lies within the elimination of the DC-link voltage sensor. It also presents the possibility of identification of the nonlinear VSI behavior. The method disadvantages are the necessity of the speed sensor and the phase shift of the estimated quantities.

Acknowledgement. This research was supported by the Technology Agency of the Czech Republic under the grant for Competence Centers Program, project No. TE02000103, and on the work supported by the Student Grant Competition of the Czech Technical University in Prague under grant No. SGS18/132/OHK3/2T/13.

References

1. Bose, B.K.: Modern Power Electronics and AC Drives. PHI Learning Private Limited, New Jersey (2013)
2. Quang, N.P., Dittrich, J.-A.: Vector Control of Three-Phase AC Machines, 1st edn. Springer, Heidelberg (2015)
3. Vas, P.: Sensorless Vector and Direct Torque Control. Oxford University Press, Oxford (1998)
4. Popescu, M.: Induction Motor Modelling for Vector Control Purposes. Report, 144 p. Helsinki University of Technology, Laboratory of Electromechanics, Espoo (2000)
5. Anuchin, A., Gulyaeva, M., Briz, F., Gulyaev, I.: Modeling of AC voltage source inverter with dead-time and voltage drop compensation for DPWM with switching losses minimization. In: International Conference on Modern Power Systems (MPS), Romania (2017)
6. Wang, Y., Xie, W., Wang, X., Gerling, D.: A precise voltage distortion compensation strategy for voltage source inverters. IEEE Trans. Ind. Electron. **65**(1), 59–66 (2018)
7. Salt, D.E., Drury, D., Holliday, D., Griffo, A., Sangha, P., Dinu, A.: Compensation of inverter nonlinear distortion effects for signal-injection-based sensorless control. In: IEEE Trans. Ind. Appl. **47**(5), 2084–2092 (2011)
8. Shen, G., Yao, W., Chen, B., Wang, K., Lee, K., Lu, Z.: Automeasurement of the inverter output voltage delay curve to compensate for inverter nonlinearity in sensorless motor drives. IEEE Trans. Power Electron. **29**(10), 5542–5553 (2014)
9. Grewal, M.S., Andrews, A.P.: Kalman Filtering Theory and Practice Using MATLAB. Wiley-IEEE Press (2008)
10. Barut, M., Bogosyan, S., Gokasan, M.: Speed-sensorless estimation for induction motors using extended Kalman filters. IEEE Trans. Ind. Electron. **54**(1), 272–280 (2007)
11. Aksoy, S., Mühürcü, A., Kizmaz, H.: State and parameter estimation in induction motor using the extended Kalman filtering algorithm. In: 2010 Modern Electric Power Systems, Wroclaw, pp. 1–5 (2010)
12. Pichlík, P., Zděnek, J.: Extended Kalman filter utilization for a railway traction vehicle slip control. In: 2017 International Conference on Optimization of Electrical and Electronic Equipment (OPTIM) & 2017 International Aegean Conference on Electrical Machines and Power Electronics (ACEMP), Brasov, pp. 869–874 (2017)
13. Buchta, L., Otava, L.: Adaptive compensation of inverter non-linearities based on the Kalman filter. In: IECON 2016 - 42nd Annual Conference of the IEEE Industrial Electronics Society, Florence, pp. 4301–4306 (2016)
14. Buchta, L., Otava, L.: Compensation of dead-time effects based on Kalman filter for PMSM drives. IFAC-PapersOnLine **51**(6), 18–23 (2018)
15. Rodriguez, J., Cortes, P.: Predictive Control of Power Converters and Electrical Drives. Wiley-IEEE Press (2012)
16. Fligl, S., Bauer, J., Vlcek, M., Lettl, J.: Analysis of induction machine T and Γ circuit coequality for use in electric drive controllers. In: 2012 13th International Conference on Optimization of Electrical and Electronic Equipment (OPTIM), Brasov, pp. 659–664 (2012)
17. Kubota, H., Matsuse, K., Nakano, T.: DSP-based speed adaptive flux observer of induction motor. IEEE Trans. Ind. Appl. **29**(2), 344–348 (1993)

EEG Vowel Silent Speech Signal Discrimination Based on APIT-EMD and SVD

S. I. Villamizar[1(✉)], L. C. Sarmiento[2(✉)], O. López[2], J. Caballero[1], and J. Bacca[1]

[1] Facultad de Ingeniería, Universidad Nacional de Colombia, Bogotá, Colombia
sivillamizard@unal.edu.co

[2] Departamento de Tecnología, Universidad Pedagógica Nacional, Bogotá, Colombia
lsarmiento@pedagogica.edu.co

Abstract. A Brain-Computer Interface (BCI) System captures the neural activity of the Central Nervous System (CNS) and delivers an output which replaces the natural output of the CNS [1]. That helps who have lost their ability to speak, spelling words in a monitor or helps to recover the movements for people who have suffered some amputation of their limbs or a motor paralysis of their body. The main objective of the project is to control an upper limb using a Myohand twin Ottobock prosthesis ref 8E38 = 7 [2] using EEG silent speech signals. On this research, we will focus on a novel methodology that attempts to classify imagined speech based on vowels, which uses as the primary technique for artifact removal the Adaptive-Projection Intrinsically Transformed Multivariate Empirical Mode Decomposition (APIT-MEMD) [3], and for the feature generation stage the Singular Value Decomposition (SVD) [4]. For the classification stage, two classifiers where tested, the Extremely Randomized Trees classifier (ET) [5], and the Adaboost (ADB) [6]. The overall accuracy achieved per subject and per vowels' pairwise classification, was 91.54% using ET. For a multiclass classifier, the overall accuracy over the eighteen subjects of the database was 79.06%.

Keywords: EEG silent speech · Singular Value Decomposition · Vowels · APIT-MEMD

1 Introduction and State of the Art

Among several upper prosthetic limbs available on the market, the electromyography' (EMG) prosthesis are the most popular [7], using the EMG signals from a specific part of the body to control it. However, the training process would be demanding and increase the degrees of freedom is a tough task. On the other hand, there is the neuro-prosthesis [8] which takes the cortex brain activity using implanted sensors through surgery, but there is a risk involved in the sensor positioning procedure. The above makes the EEG signals as the right candidate for a BCI system because they are noninvasive and their high time resolution.

On the EEG field there are mainly four kinds, the steady-state visual evoked potentials (SSVEP) [9], the P300 [10], the motor imagery (MI) [11], and the imagined speech signals (ISS) [12–15]. On the ISS the subject thinks in speaking a vowel but do not move

D. F. Cortes Tobar et al. (Eds.): AETA 2019, LNEE 685, pp. 74–83, 2021.
https://doi.org/10.1007/978-3-030-53021-1_8

any muscle. The ISS is noninvasive, do not require any external stimulus, because using the alphabet do not require training procedure and can be straightforward to increase the degrees of freedom. The EEG-ISS are produced by the synchronization of groups of neuron in certain parts of the brain [16] which makes that the captured signals are non-stationary signals with high levels of noise, non-linearities and low levels of power because of scalp, skull, and brain. Also, the EEG-ISS has power disbalances between channels, and intra and inter subjects. Due to the before, the processing of these type of signals requires robust and adaptive techniques and that the methodology would be easily scalable to an online application extracting the features trial to trial.

On the state of the art of pattern recognition of the EEG-SSI researchers have been to discriminate silent vowels or states as /a/ vs /u/ vs control [17], and five vowels /a/, /e/, /i/, /o/, /u/ [12–15]. Several techniques was used as a features extractor stage as Power Spectral Density (PSD) [18], mean, variance, standard deviation, and entropy [12], sparse regression model [13], Autorregressive Coefficients (AR) [15], Common Spatial Pattern (CSP) and Empirical Model Decomposition (EMD) [14, 17].

On this research, we proposed a novel methodology for capturing and processing the EEG-SSI. For the signal acquisition stage, the location for the electrodes was settled to maximize the capturing of the brain activity of the language zones (Wernicke's area, Broca's Area, and Motor's area) in contrast with the 10–20 system. For the preprocessing stage, an adaptive robust algorithm APIT-MEMD clean from artifacts the signal. The SVD extract the discriminative features trial to trial, and finally, the ET or ADB assign a label to each observation.

2 Methodologies

Figure 1 shows the proposed methodology, which consists of three main stages, the artifact removal, the feature generator, and the translator. On the first one stage, the input trial is standardized subtracting the trial mean and dividing by its standard deviation. After, the APITP-MEMD decomposes the signal on their oscillating modes and choose the ones which central frequency is between the frequency range of [18–49] Hz. On the feature generator stage, the multivariate signal is mapped to the feature space using the SVD. The features vector is conforming using staking each right singular vector multiply by their explanation of data. The final stage is to assign a label to each trial. After classifier have been trained, cross-validation is performed to ensure the statically validity of results.

2.1 Database

The testing database was one captured for one of the authors on their Ph.D. thesis named "Neurophysiology database - (NDB)" [18]. The experimentation was in the Clinical Electrophysiology Laboratory from the National University of Colombia using a NicoletOneTM V32 amplifier of 21 channels plus reference and ground. On the NDB database, the participants were comfortable seated one meter away from a LED light. The subjects had to think in the vowel previously said once the light was on for three seconds. After they have a rest time of three seconds for a total time of six seconds per trial. The before

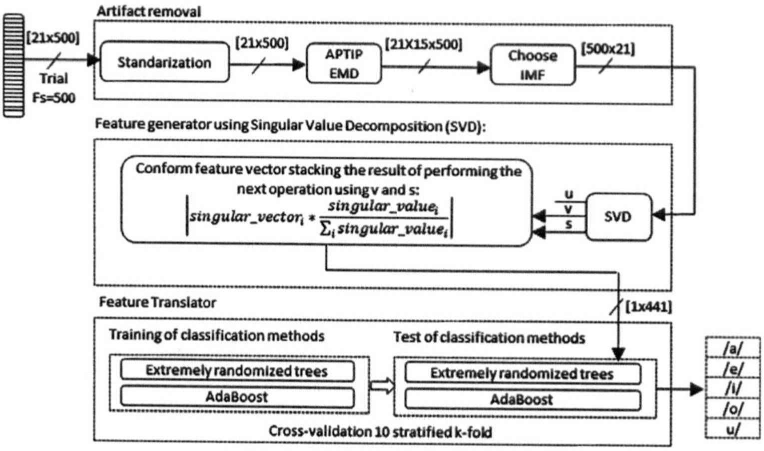

Fig. 1. Schematic of the proposed method

is repeated ten times per vowel and per subject which precede a five-minute rest for the next thinking task of another vowel. Finally, split the active data (where LED was on) into trials of 1 s, with 500 ms of overlapping to get a total of 50 trials per subject and vowel.

The database consist of five Spanish vowels /a/, /e/, /i/, /o/, and /u/, which according to the degree of opening of the oral cavity, /a/ is open, /e/ is medium, /i/ is closed, /o/ is medium and /u/ is closed for the vowels [18]. The sampling frequency was 500 Hz, and, the impedance was under five kΩ to reduce the electrode artifacts. For the test, all subjects were right-handed and healthy and were three female eight and thirteen because of d seventeen male. The final database was of eighteen subjects, excluding the participants 8 and 13 because of ground or acquisition problems. To maximizing the silent speech brain activity, the ubication of the electrodes was placed on the Broca and Wernicke zones [18] as you can see in Fig. 2.

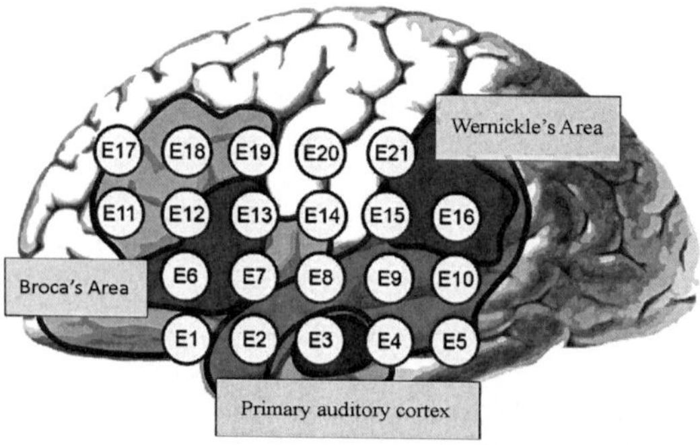

Fig. 2. Electrode' position of NDB database, taken from [18].

2.2 Techniques

Due to the scalp, skull, and brain, the EEG signals captured contains several noises, and distortions on their frequency bands like blinking artifacts [lows freqs], muscular artifacts [50 Hz–150 Hz], power-line noise [50 Hz–60 Hz], cardiac noise [1 Hz–2 Hz], and bias noise [<1 Hz].

Because of the above, the signals have nonstationary and distortion behavior with high variability. Our first step is to standardized signals, which are highly recommended for when data contains variance offset and power disbalances [19]. Because of the non-stationary nature of the EEG data on this case, each trial is standardizing through the Z-normalization [20] as $x_{zt} = \frac{x_{trial} - mean(x_{trial})}{std(x_{trial})}$ where x_{trial} means one trial of the database. Based on the above, the artifact removal is performing making a decomposition and choosing the right frequency band of the signals. A method that can handle signals from the EEG nature is the promising empirical mode decomposition (EMD) technique since this tool is intuitive, direct application, and adaptive to input data. For this research, the input is a multivariate signal with dimensions of [21 × 500], then a multivariate EMD (MEMD) [3] is necessary. The EMD algorithm base its decomposition on the fact that the data are composed of several simple intrinsic modes functions (IMF) of oscillation. The linear or nonlinear IMFs are quasi-stationary (same number of ends and zero crossings and zero mean) and are constructed through a sieving process. The summation of all IMFs reconstructs the original data [3].

The EEG-ISS signals contain power disbalances and high correlation between channels. To make an optimal decomposition and capturing the multivariate data dynamics, it is required that the decomposition technique be adaptive to input data and be able to reveal "intrinsic and correlated" inter-channel information. The Adaptive-projection intrinsically transformed multivariate empirical mode decomposition (APIT-MEMD) [3] address several problems that appear when working with real data as a mix of modes, a lot of IMFs for unbalanced data, and high computational cost for dispersing inputs. To be adaptive with input data APIT-MEMD propose to project data into pre-generated steering vectors around the first principal component of the covariance matrix and project into others steering vectors around a vector that points in the diametrically opposite direction of the first eigenvector [3]. These direction vector serves each sieving process to estimate the local mean to finally calculate the residue as $d(t) = x(t) - m(t)$ and repeat the sieving process until the residue meets the conditions of an IMF for multivariable signals. The complete explanation of the APIT-MEMD algorithm can be found in [3]. The choosing of the IMFs that remains in the frequency band range between [18–40] Hz concerning three main aspects: achieves the major discriminability, the blinking artifacts remains on the low frequencies IMFs (different of the range chosen), and that frequency boundary eliminates most of the other artifacts.

Now, On the feature's generation stage, the SVD extract discriminative features trial by trial. The SVD is an eigendecomposition made for no-squared matrices. The SVD of a $m \times n$ matrix E is the factorization of E on three matrices as $E = U \Lambda V^T$ as it can see in (1), where the columns of U $m \times m$ and V $n \times n$ are orthonormal called singular vectors u_i, v_i so Λ is a diagonal $m \times n$ matrix with real and positive values called singular values which corresponds to the root of the positive eigenvalues of EE^T or E^TE. U and V are invertible for any input matrix, and they are orthonormal [4].

$$\overset{U}{\begin{pmatrix} u_{11} & \cdots & u_{m1} \\ \vdots & \ddots & \vdots \\ u_{1m} & \cdots & u_{mm} \end{pmatrix}} \overset{\Lambda}{\begin{pmatrix} \lambda_1 & & \\ & \lambda_r & \\ & & 0 \end{pmatrix}} \overset{V}{\begin{pmatrix} v_{11} & \cdots & v_{n1} \\ \vdots & \ddots & \vdots \\ v_{1n} & \cdots & v_{nn} \end{pmatrix}} \tag{1}$$

Then, taking the right singular vector matrix V and stacking the factor of each singular vector v_i (Columns of V) with the explanation of data $\lambda_i \Big/ \sum_j \lambda_j$ form the final features vector of dimensions $[1 \times 441]$ per trial.

On the classification stage, two classifiers were tested to assign a label to a testing set (like the Extremely Randomized Trees classifier (ET) or the Adaboost (ADB)). These two belongs to the ensemble' type classifiers, which group different techniques like Decision Tree (DT) [5] to predict the final mapping rules reducing bias, variance, and increasing the success rate. The training of the ensemble techniques could be in parallel like ET or sequentially like ADB. On the first, the training of each estimator is made regardless of others. On the contrary, on a sequential form, the base classifiers are training concerning the classification accuracy of the before step. The general idea of ADB is to make an accurate prediction of infrequent trials based on weighted training samples and base classifiers per iteration. The complete explanation of the training process is in [6]. The ET group several DT and perform the training in parallel to then make a final prediction considering the vote of each tree to assign a label. The training procedure is in [5].

3 Results

Due to the programming language used was the Python open-source programming language, it was mandatory to migrate the APIT-MEMD to Python and optimize several subroutines. The final work consists of two parts programming in Python. The one for training the classifiers and the one for testing. As follows t will be showed several results from the testing part. All processing of data was offline.

3.1 Artifact Removal

The objective of this block is to clean the signal from undesired artifacts. The defined input for this stage is a matrix which contains a multivariate signal where rows are the sensors and columns are time samples. The output is a standardize signal with their majority power between the frequency bands of 18 Hz to 49 Hz. On these blocks, there are only two parameters to set up, the alpha value, and the number of projection vectors of the APIT-MEMD. If the alpha is zero, the APIT-MEMD behaves like the traditional MEMD, and if alpha is one, the projections vectors are relocated close to the first eigenvector and its opposite. In this research, the alpha value was 0.7, and the number of decompositions was in 16 on the test.

3.2 Features Generators

The propose of this stage is to obtain from the cleaned signals discriminative features that serve as input for the classification stage. On the methodology proposed the right singular vectors, and the singular values of a trial input multivariate signal were used to form the features vector per vowel. In Fig. 3, an illustration of all feature's vectors from one subject is shown for two different vowels, only 150 features are showed. There are several differences between then in some features that make more accessible the classification among them.

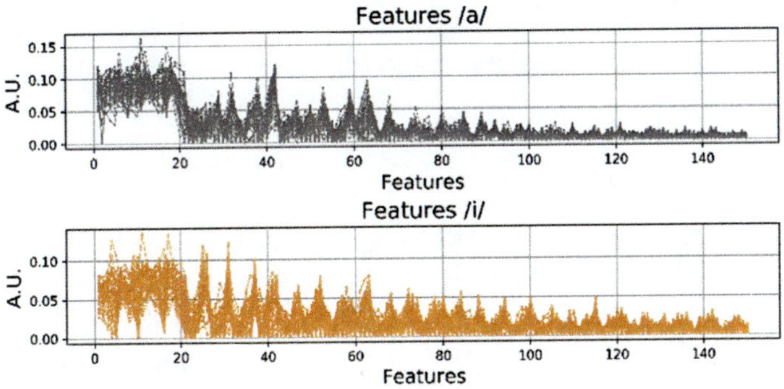

Fig. 3. All trial for subject 12 - features vectors from one subject of the vowels /a/, /i/.

3.3 Classification Stage

The implementation of the ET of the scikit-learn from Python was used as classifier [21]. From default values in the library, only the number of estimators was modified setting up in 2000. Even though the training took a long time, the classification test time is short. The testing of the database was done using the cross-validation (CV) technique to obtain a statically validity score, non-overfitting, and non-biased. Using the CV, it will give us an insight into how the methodology works facing unknown datasets due to the testing subset are not involve in the training stage. On the CV process, the k fold selected was 10, then, nine subsets were for training and the left to test. Later include the one used for test to the training subsets and doing the test with one group which still has not yet been tested. Continue until all sets have been verified and calculate an average of the score (AVG), and a standard deviation (STD).

Pairwise Classification

For the pairwise classification, the CV was per subject and binary classifier. For the database, it has ten pairwise classifiers. Figure 4 shows the averaging of the scores and STD over eighteen subjects of the database, and in Fig. 5 the results per subject are displayed over 10-fold cross-validation. Figure 4. Overall 18 subjects Pairwise accuracy over 10-fold cross-validation

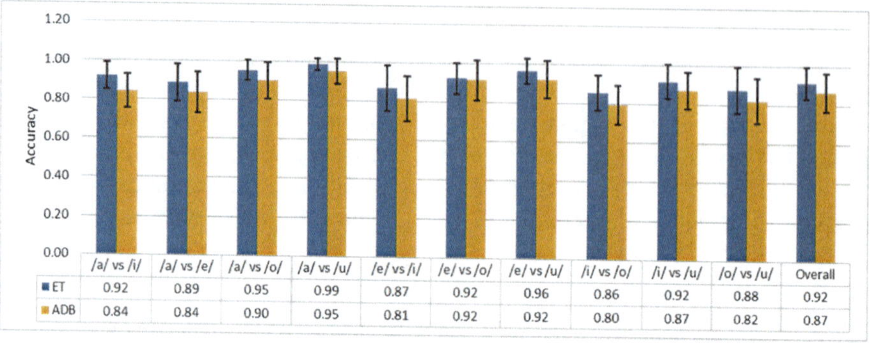

Fig. 4. Overall 18 subjects Pairwise accuracy over 10-fold cross-validation

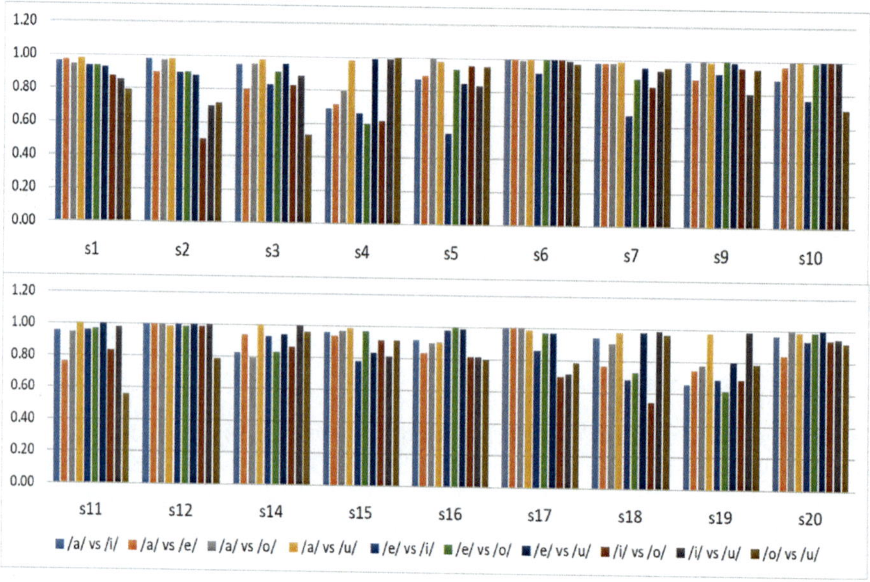

Fig. 5. ET Classifier eighteen subjects pairwise accuracy over 10-fold cross-validation

Multiclass Classification

Two tests were made to test the reliability and robustness of the multi-class classifier. The first perform a multiclass classifier (/a/ vs /e/ vs /i/ vs /o/ vs /u/) per subject using the 10-fold CV to validate the methodology ability to predict the label of data. On Fig. 6, the results are shown. The second test split the database in 70 percent for training and 30 percent for test and plot the confusion matrix obtained as it can see in Fig. 7, Only matrices were showed from subject 4 and 12. The implementation of the classifiers was done using the scikit-learn library from Python. The number of estimators of the classifier was set up in 2000 for ADB and ET.

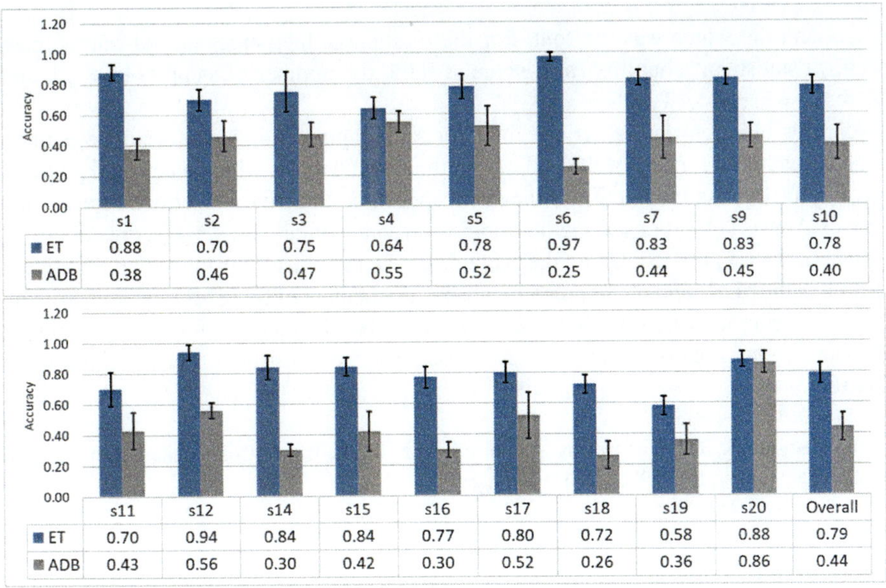

Fig. 6. Multiclass Classifier /a/ vs /e/ vs /i/ vs /o/ vs /u/ per-subject over 10-fold cross-validation

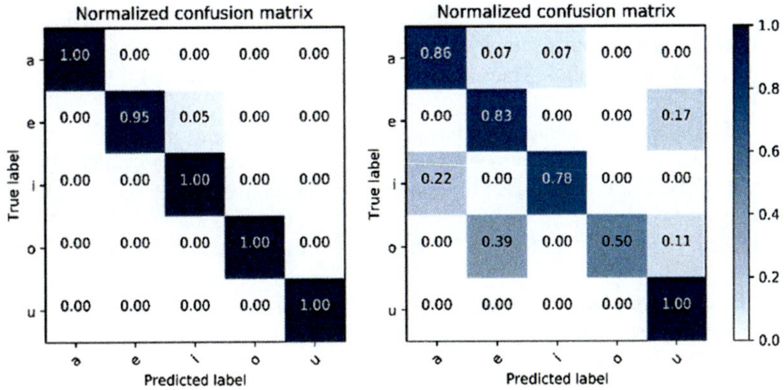

Fig. 7. Normalized Confusion matrix using stratified data - 70% for training, and 30% for testing. Left subject 4, and right subject 12.

4 Discussion and Conclusion

As it an see in Fig. 4, the best pairwise classifier was */a/ vs /u/* with an accuracy score of 0.99 ± 0.03. The worst classifier was */i/ vs /o/* from the ADB with an accuracy of 0.80 ± 0.10. Observing Fig. 5, the best subject is the six, and the worst subject is the four. For the database, the percentage of randomness to affirm statistical reliability (at p-value < 0.05, the number of samples $n = 50$ (per subject), and classes $c = 2$) is 58%. Now considering the results for the five classes classifier, the best classifier through all tests was the ET with an overall accuracy of 0.79 ± 0.07, and again the best subject was the six and the worse was the four. For the multiclass, the percentage of randomness to affirm statistical reliability (at p-value < 0.05, the number of samples $n = 50$, and classes $c = 5$) is 24.4%.

On this research multiclass classifier for a real application was developed using a database captured for one of the authors on their Ph.D. thesis which electrode' position was carefully chosen to maximize the silent speech brain activity (Wernicke's area, and Broca's Area). The proposed methodology overcomes in all cases, the scores obtained in state of the art [12–15]. Using the APIT-MEMD, the noises introduced by traditional filters like the ripple and phase noise are absent. Also, the APIT-MEMD allow us to clean the signal from undesired noises like the bias, power-line, and even blinking artifacts. The binary classificatory tests serve as a basis for the construction of the multi-class classifier.

Acknowledgment. This work was supported with funds from (Colciencias) according to financing agreement RC 838 code: 110177758402.

References

1. Saeid, S., Chambers, J.A.: EEG Signal Processing. Cardiff University, Wiley (2007)
2. Ottobock: Myo terminal device digital twin system electric hand, Austria (2018)
3. Hemakom, A., Goverdovsky, V., Looney, D., Mandic, D.P.: Adaptive-projection intrinsically transformed multivariate empirical mode decomposition in cooperative brain-computer interface applications. Philos. Trans. R. Soc. A Math. Phys. Eng. Sci. **374**(2065), 20150199 (2016)
4. Harner, R.N.: Singular value decomposition—a general linear model for analysis of multivariate structure in the electroencephalogram. Brain Topogr. **3**(1), 43–47 (1990)
5. Geurts, P., Ernst, D., Wehenkel, L.: Extremely randomized trees. Mach. Learn. **63**(1), 3–42 (2006)
6. Zhu, J., Zou, H., Rosset, S., Hastie, T.: Multi-class AdaBoost. Stat. Interface **2**, 349–360 (2009)
7. Farina, D., et al.: The extraction of neural information from the surface EMG for the control of upper-limb prostheses: emerging avenues and challenges. IEEE Trans. Neural Syst. Rehabil. Eng. **22**(4), 797–809 (2014)
8. Muzumdar, A. (ed.): Powered Upper Limb Prostheses. Springer, Heidelberg (2004)
9. Zhao, B., Chen, X., Gao, X., Xu, S., Wang, Y.: Control of a 7-DOF robotic arm system with an SSVEP-based BCI. Int. J. Neural Syst. **28**(08), 1850018 (2018)
10. Kaongoen, N., Jo, S.: A novel hybrid auditory BCI paradigm combining ASSR and P300. J. Neurosci. Methods **279**, 44–51 (2017)
11. D'Croz-Baron, D., Ramirez, J.M., Baker, M., Alarcon-Aquino, V., Carrera, O.: A BCI motor imagery experiment based on parametric feature extraction and fisher criterion. In: CONIELECOMP 2012, 22nd International Conference on Electrical Communications and Computers, pp. 257–261 (2012)
12. Aguila, M.J., Basilio, H.D.V., Suarez, P.V.C., Dueñas, J.P.E., Prado, S.V.: Comparative study of linear and nonlinear features used in imagined vowels classification using a backpropagation neural network classifier. In: Proceedings of the 7th International Conference on Bioscience, Biochemistry and Bioinformatics – ICBBB, pp. 7–11 (2017)
13. Min, B., Kim, J., Park, H., Lee, B.: Vowel imagery decoding toward silent speech BCI using extreme learning machine with electroencephalogram. Biomed. Res. Int. **2016**, 1–11 (2016)

14. Kim, J., Lee, S.K., Lee, B.: EEG classification in a single-trial basis for vowel speech perception using multivariate empirical mode decomposition. J. Neural Eng. **11**(3), 036010 (2014)
15. Riaz, A., Akhtar, S., Iftikhar, S., Khan, A.A., Salman, A.: Inter comparison of classification techniques for vowel speech imagery using EEG sensors. In: 2014 2nd International Conference System Informatics, ICSAI, pp. 712–717 (2015)
16. Stam, C.J.: Nonlinear dynamical analysis of EEG and MEG review of an emerging field. Clin. Neurophysiol. **116**(10), 2266–2301 (2005)
17. DaSalla, C.S., Kambara, H., Sato, M., Koike, Y.: Single-trial classification of vowel speech imagery using common spatial patterns. Neural Netw. **22**(9), 1334–1339 (2009)
18. Sarmiento, L.: Interfaces cerebro-computador para el reconocimiento automático del habla silenciosa. Universidad Pedagógica Nacional, Bogota (2019)
19. Ding, H., Trajcevski, G., Scheuermann, P., Wang, X., Keogh, E.: Querying and mining of time series data: experimental comparison of representations and distance measures. Proc. VLDB Endow. **1**(2), 1542–1552 (2008)
20. Gogolou, A., Tsandilas, T., Palpanas, T., Bezerianos, A.: Comparing similarity perception in time series visualizations. IEEE Trans. Vis. Comput. Graph. **25**(1), 523–533 (2018)
21. Klikauer, T.: Scikit-learn: machine learning in python. TripleC **14**(1), 260–264 (2016)

Modelling and Control System Design for UAV Tri-Rotor

Hernando González[1](✉), Carlos Arizmendi[1], César Valencia[2], Diego Valle[1], Brayam Bernal[1], and Alhim Vera[1]

[1] Universidad Autónoma de Bucaramanga, Bucaramanga, Colombia
{hgonzalez7,carizmendi,dva-lle,bbernal45,avera488}@unab.edu.co
[2] Universidad Santo Tomás, Bucaramanga, Colombia
chvn00@gmail.com

Abstract. The drones are devices of high technology that work thanks to the rotors that allow them to remain and to be propelled in the air. Their movements are controlled by adjusting the propulsion of each propeller and this is done with a device called a flight controller. The paper presents two control system for a UAV-Tricopter. The control techniques proposed are: proportional derivative and integral controller (PID) and Linear Quadratic Gaussian Controller (LQG). These techniques force the tricopter to follow with precision the given trajectory. The simulink toolbox was used to make the designs of each controller and then see the behavior of each control strategy.

Keywords: UAV-Tricopter · PID controller · LQG controller

1 Introduction

In recent years, technological developments in the sensor, actuator and microcon- troller industries have facilitated the path from automatic pilots to unmanned aerial ve- hicles (UAVs) where the pilot is no longer needed [1], there are numerous types such as: bicopter, tricopter, quadrotor and the conventional helicopter. Of these, the tricopter has 3 engines, and is typically configured in a *Y* shape, where the arms are generally separated by 120°. But it can also be found in a *T* shape. Two propellers on the front arms rotate in the opposite direction to counteract each other. The rear engine can be tilted to the left and right by a servo to allow movement on the Z- axis, so it does not lose thrust on yaw [2]. They are mainly used in military applications, where human life is protected in very risky and dangerous actions, capable of maintaining a controlled and sustained flight level. In addition, UAVs can fly at very low speeds if necessary, allowing images to be obtained. They also stand out in uses of civil protection, ecology and public works monitoring.

Due to the non-linearity and complexity of the dynamic model, the problem of drone (Tricopter) control has been widely considered. In addition, the under-performance defined as having more degrees of freedom than actuators - in the dynamic model makes

D. F. Cortes Tobar et al. (Eds.): AETA 2019, LNEE 685, pp. 84–93, 2021.
https://doi.org/10.1007/978-3-030-53021-1_9

the control problem more challenging [3]. Currently, several non-linear methods are proposed for UAV control such as the backlash controller, adaptive slide mode controllers [4], direct adaptive controller, adaptive PID controller and neural network based controllers [5]. In particular, different adaptive controllers' approaches are pro- posed for attitude control of unmanned aircraft with uncertain parameters [6]. In this document, a mathematical model expressed in the space of states with their respective linearization is proposed. Later, PID and LQG controllers are designed so that they can be simulated by means of MATLAB software and the use of the Simulink tool.

2 Model of the UAV-Tricopter

Mathematical models are used to analyze the relationship between two or more variables. They can be used to understand natural phenomena, depending on the objective sought and the design of the same model can serve to predict the value of variables in the future, make hypotheses, evaluate the effects of an activity, among others. The dynamic behavior and the movement of the motors can be evidenced in the Fig. 1.

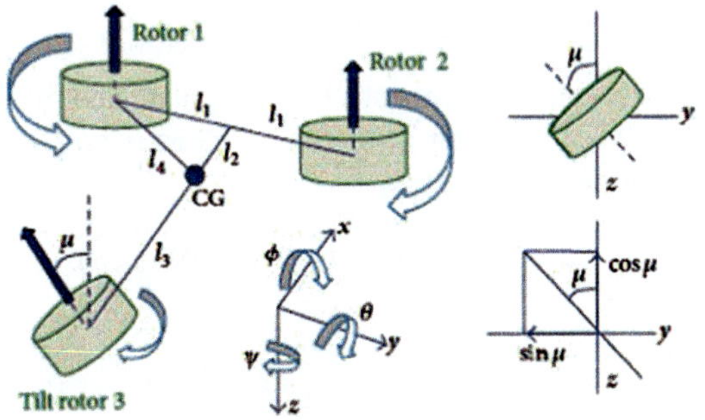

Fig. 1. Movement of the motors of a tricoper

This type of drones is composed of three motors, three regulators, a servomotor and four propellers. The three motors are located at the end of the three arms, in the same place, we can find the sensorics component of the tricopter, the handling of this UAV is done by means of the propellers that direct the regulators, which are in the trunk of the object. The Newton-Euler formulation is used to develop the model of the tri-rotor. Table 1 shows the variables of dynamic model.

$$F_x = m(\dot{u} - rv - qw) + mgsin(\theta) \tag{1}$$

$$F_y = m(\dot{v} - ru - pw) - mgsin(\phi)coscos(\theta) \tag{2}$$

$$F_z = m(\dot{w} + pv - qr) - mgcos(\phi)coscos(\theta) \tag{3}$$

where *Fx*, *Fy and Fz* are the sums of the forces and *g* is the gravity.

$$L = I_{xx}\dot{p} - \left(I_{yy} - I_{zz}\right)qr \tag{4}$$

Table 1. Variables of dynamic model

Symbol	Description
m	Total mass
$[\phi, \theta, \psi]^T$	Euler angles
$[p, q, r]^T$	Angular velocities of the body
$[\dot{p}, \dot{q}, \dot{r}]^T$	Angular accelerations of the body
$[u, v, w]^T$	Linear velocities in space
$[\dot{u}, \dot{v}, \dot{w}]^T$	Linear accelerations in space

$$M = I_{yy}\dot{q} - (I_{zz} - I_{xx})pr \tag{5}$$

$$N = I_{zz}\dot{r} - (I_{xx} - I_{yy})pq \tag{6}$$

where L, M and N are the external moments, they are calculated by Newton's third law and $I = diag([I_{xx}I_{yy}I_{zz}])$ is the matrix of inertia of the body with respect to the centre of gravity. The kinematics equations are:

$$\dot{\phi} = p + qsen(\phi)tan(\theta) + rcos(\phi)tan(\theta) \tag{7}$$

$$\dot{\theta} = qcos(\phi) - rsin(\phi) \tag{8}$$

$$\dot{\psi} = \frac{qsen(\phi) + rcos(\phi)}{cos(\theta)} \tag{9}$$

$$\ddot{\phi} = \dot{p} = qr\left(\frac{I_{yy} - I_{zz}}{I_{xx}}\right) + \frac{LU_1}{I_{xx}} \tag{10}$$

$$\ddot{\theta} = \dot{q} = pr\left(\frac{I_{zz} - I_{xx}}{I_{yy}}\right) + \frac{LU_2}{I_{yy}} \tag{11}$$

$$\ddot{\psi} = \dot{r} = pq\left(\frac{I_{xx} - I_{yy}}{I_{zz}}\right) + \frac{LU_3}{I_{zz}} \tag{12}$$

The entries U_1 U_2 U_3 are assumed to be zero, since in this way the adjustment point of the UAV-tricopter engines can be found. In [5], the moments generated by the angular velocity of three rotors and the tilt angle of rotor 3 are represented by the drive thrust as shown below.

$$U_1 = b_1\left(\Omega_1^2 + \Omega_2^2\right) + b_2\Omega_3^2\cos(\mu) \tag{13}$$

$$U_2 = l_1 b_1\left(\Omega_1^2 - \Omega_2^2\right) \tag{14}$$

$$U_3 = l_2 b_1\left(\Omega_1^2 + \Omega_2^2\right) - l_3 b_2 \Omega_3^2 \cos(\mu) \tag{15}$$

$$U_t = -l_4 d_1 \Omega_1^2 + l_4 d_2 \Omega_2^2 + l_3 d_3 \Omega_3^2 + l_3 b_3 \Omega_3^2 \sin(\mu) \tag{16}$$

U_t is the total force of the UAV in the z direction, U_x is the moment of the roll angle in the x direction (L), U_y is the moment of the pitch angle in the y direction (M), U_z is the moment of the yaw angle in the z direction (N); $\Omega_1, \Omega_2, \Omega_3$ are the velocities of the three motors respectively, μ is the tilt angle of the tail, b is the thrust coefficient and d is the lift coefficient; l_1, l_2, l_3, l_4 are lengths with respect to the center of mass as shown in Fig. 1. Table 2 shows the parameters value used in this paper.

Table 2. Parameters value of dynamic model

Parameter	Valor
I_{xx}	0.0057211 [kgm^2]
I_{yy}	0.0073933 [kgm^2]
I_{zz}	0.012545 [kgm^2]
$l1$	0.19 [m]
$l2$	0.122 [m]
$l3$	0.23 [m]
$l4$	0.225 [m]
B	1.3678e−4 $\left[Ns^2\right]$
D	2.4323e−5 $\left[Nms^2\right]$

3 Controller Design

3.1 State Space Representation

The mathematical model is described by a set of inputs, outputs and state variables related by differential equations that are combined in a first-order matrix differential equation. To visualize inputs, outputs and states, variables are expressed as vectors and algebraic equations are written in matrix form. States are derived from the mathematical model; thus, states and control inputs are described respectively by:

$$x = \left[\phi\dot{\phi}\theta\dot{\theta}\psi\dot{\psi}\right]^T \tag{17}$$

$$u = \left[U_x U_y U_z\right]^T \tag{18}$$

Thus, the reasons for change are described below.

$$\dot{x}_1 = x_2 \tag{19}$$

$$\dot{x}_2 = x_4 x_6 \left(\frac{I_{yy} - I_{zz}}{I_{xx}} \right) + \frac{U_1}{I_{xx}} \tag{20}$$

$$\dot{x}_3 = x_4 \tag{21}$$

$$\dot{x}_4 = x_2 x_6 \left(\frac{I_{zz} - I_{xx}}{I_{yy}} \right) + \frac{U_2}{I_{yy}} \tag{22}$$

$$\dot{x}_5 = x_6 \tag{23}$$

$$\dot{x}_6 = x_2 x_4 \left(\frac{I_{xx} - I_{yy}}{I_{zz}} \right) + \frac{U_3}{I_{zz}} \tag{24}$$

The points of equilibrium x_{10}, x_{30} and x_{50} can take any value, since these tell us the angles of the UAV-Tricopter that respectively are roll, pitch and yaw; the others points are zero. A linear system is a relationship between input and output defined by a linear equation.

$$A = \begin{bmatrix} 0 & 1 & 0 & 0 & 0 & 0 \\ 0 & 0 & 0 & \frac{x_6(I_{yy}-I_{zz})}{I_{xx}} & 0 & \frac{x_4(I_{yy}-I_{zz})}{I_{xx}} \\ 0 & 0 & 0 & 1 & 0 & 0 \\ 0 & -\frac{x_6(I_{xx}-I_{zz})}{I_{yy}} & 0 & 0 & 0 & -\frac{x_2(I_{xx}-I_{zz})}{I_{yy}} \\ 0 & 0 & 0 & 0 & 0 & 1 \\ 0 & \frac{(x_4(I_{xx}-I_{yy}))}{I_{zz}} & 0 & \frac{x_2(I_{xx}-I_{yy})}{I_{zz}} & 0 & 0 \end{bmatrix} \tag{25}$$

$$B = \begin{bmatrix} 0 & 0 & 0 & 0 \\ \frac{1}{I_{xx}} & 0 & 0 & 0 \\ 0 & 0 & 0 & 0 \\ 0 & \frac{1}{I_{yy}} & 0 & 0 \\ 0 & 0 & 0 & 0 \\ 0 & 0 & 0 & \frac{1}{I_{zz}} \end{bmatrix}$$

$$C = \begin{bmatrix} 1\,0\,0\,0\,0\,0 \\ 0\,0\,1\,0\,0\,0 \\ 0\,0\,0\,0\,1\,0 \end{bmatrix}$$

3.2 Proportional, Integral and Derivative Regulator (PID)

PID control has been used successfully in many industrial control systems for over half a century. The basic principle of the PID control scheme is the sum of three control actions: proportional control action (where the control action is proportional to the acting error signal, which is the difference between the input and the feedback signal); integral control action (where the control action is proportional to the integral of the acting error signal); and derivative control action (where the control action is proportional to the derivative of the acting error signal). The next equation is the continuous expression of the PID controller (26), where *e(t)* is the input to the controller (acting error signal), *m(t)* is the controller output (signal manipulated), *K* is the proportional gain, T_i is the integral time (or readjustment time) and T_d is the derivative time (or advance time).

$$m(t) = K\left[e(t) + \frac{1}{T_i}\int_0^t e(t)dt + T_d\frac{de(t)}{dt}\right] \tag{26}$$

To obtain the pulse transfer function from the digital PID controller [7], the equation is discretized (26), so it would be established in the form:

$$M(z) = \left[K_p + \frac{K_i}{1 - z^{-1}} + K_d\left(1 - z^{-1}\right)\right]E(z) \tag{27}$$

where proportional gain, integral gain and derivative gain can be expressed from the following equations:

$$K_p = K - \frac{KT}{2T_i} \tag{28}$$

$$K_i = \frac{KT}{T_i} \tag{29}$$

$$K_d = \frac{KT_d}{T} \tag{30}$$

Each of the transfer functions associated with the degrees of freedom is discretized with a sampling period of $T_m = 6.25$ [ms] and an ideal PD control is designed, considering the proportional and derivative constant, $K_P = 0.61$ and $K_D = 60.39$, which is a regulation system that tries to take advantage of each of the basic action controllers.

It has the advantage of offering a very fast response and immediate compensation for disturbances. Equations 31, 32 and 33 present the structures of each controller obtaining a setting time of 0.518 [s], 0.567 [s] and 0.568 [s] for the roll, pitch and yaw, respectively; and with an overshoot of less than 4.5% in both controllers.

$$C_{roll} = \frac{30(z - 0.99)}{z} \tag{31}$$

$$K_{pitch} = \frac{36(z - 0.99)}{z} \tag{32}$$

$$K_{yaw} = \frac{61(z - 0.99)}{z} \tag{33}$$

3.3 Linear Quadratic Gaussian Control (LQG)

The LQG controller is the combination of a state feedback gain and a Kalman-type state estimator [9]. The first step in designing the LQG controller is to determine a gain matrix

$\hat{K}$ that minimizes the cost function J, (34). The matrices Q and R are positive defined matrices, generally diagonal, which determine the balance between the tracking of a path and the energy of the control signal required by the system to achieve the control objectives.

$$J = \frac{1}{2}\sum_{k=0}^{\infty}\left(\xi^T Q\xi + u_e R u_e\right) \tag{34}$$

To estimate the states a Kalman filter is used, which assumes the presence of Gaussian noise in the control signal and the variable to be measured in the process. Known the variances of the noise of these two signals, Qn y Rn, a gain vector L is determined that minimizes the error in the estimation, $x - \hat{x}$, using the Riccatti equation. Mathematically, states are calculated from the following expression:

$$\hat{x}(k+1) = G_L\hat{x}(k) + H_L u(k) + L\left(y(k) - C_L\hat{x}(k)\right) \tag{35}$$

where $u(k)$ is the control signal and $y(k)$ the variable to be controlled in the process. For the design of the LQG control system, the representation of state space is discretized with a sampling period of $T_m = 6.25$ [ms]. The following diagonal matrices are defined for controller design Q and R.

$$Q = diag\left[\left(0.5, 0.5, 0.5, 0.5, 0.5, 0.5, 1e^4, 1e^4, 1e^4\right)\right] \tag{36}$$

$$R = diag[(0.1, 0.1, 0.04)]$$

By solving the Riccatti equation, you get the matrix $\hat{K}$

$$K = \begin{bmatrix} 5.198 & 0.351 & 0 & 0 & 0 & 0 & -0.252 & 0 & 0 \\ 0 & 0 & 5.585 & 0.389 & 0 & 0 & 0 & 0.262 & 0 \\ 0 & 0 & 0 & 0 & 9.003 & 0.633 & 0 & 0 & -0.418 \end{bmatrix} \tag{37}$$

The following variances are assumed for the observer system:

$$Qn = diag[(0.03, 0.05, 0.01)] \tag{38}$$

$$Rn = diag[(0.01, 0.005, 0.03)] \tag{39}$$

From these values the matrix L is obtained. The closed-loop system has an establishment time of 4.56 [s], 4.79 [s] and 4.79 [s] for the roll, pitch and yaw, respectively, and with an overshoot of less than 2.5% in both controllers.

$$L = \begin{bmatrix} 59.1 & 0 & 0 \\ 174.83 & 0 & 0 \\ 0 & 52 & 0 \\ 0 & 135.14 & 0 \\ 0 & 0 & 40 \\ 0 & 0 & 800 \end{bmatrix} \tag{40}$$

4 Results

When designing a control system for a physical process, one or more parameters must be selected to choose the best performance. For this purpose, a measure known as error rates is used. The three criteria according to Graham and Lathrop are: reliability, applicability, and selectivity [10]. The error rates taken into account were: absolute error integral (IAE), quadratic error integral (ISE), time integral by absolute error (ITAE), and time integral by quadratic error (ITSE), defined by Eqs. 41 to 44, where *e(k)* is the vector of the error and *t(k)* the time vector.

$$ISE = \sum_{k=0}^{\infty} e(k)^2 \tag{41}$$

$$ITSE = \sum_{k=0}^{\infty} t(k)e(k)^2 \tag{42}$$

$$ESE = \sum_{k=0}^{\infty} |e(k)| \tag{43}$$

$$ISE = \sum_{k=0}^{\infty} t(k)|e(k)| \tag{44}$$

The transient response of the two controllers and the non-linear model were simulated against a step input and the four error rates are estimated. Table 3 shows the results of the four indices, IAE, ISE, ITAE and ITSE, and highlights that for roll, pitch, yaw the LQG control has the lowest values.

Table 3. Comparison of different errors, ITAE, ITSE, IAE, ISE as performance index

Variable Roll	*IAE*	*ITAE*	*ISE*	*ITSE*
Control PID	0.3685	0.1485	0.1529	0.04776
Control LQG	0.1496	0.01668	0.1007	0.006511
Variable Pitch	*IAE*	*ITAE*	*ISE*	*ITSE*
Control PID	0.03392	0.002593	0.01558	0.00212
Control LQG	0.1356	0.01226	0.0977	0.005876
Variable Yaw	*IAE*	*ITAE*	*ISE*	*ITSE*
Control PID	0.2309	0.08963	0.06626	0.01742
Control LQG	0.1501	0.01611	0.1036	0.006726

Figure 2 shows the transient response of roll angle, pitch angle and yaw angle for the linear model and no-linear model (measured in degree) for a maximum variation of $-\pi$ to π, despite being evaluated for more abrupt changes the system is still fast, the three axes of rotation were evaluated with asynchronous inputs, and the system responded optimally with minimal disturbances and quickly stabilized. Figure 3 shows the control signal of the LQG controller, the signals presents fast action, this is natural considering that it is a system that requires answers of this type.

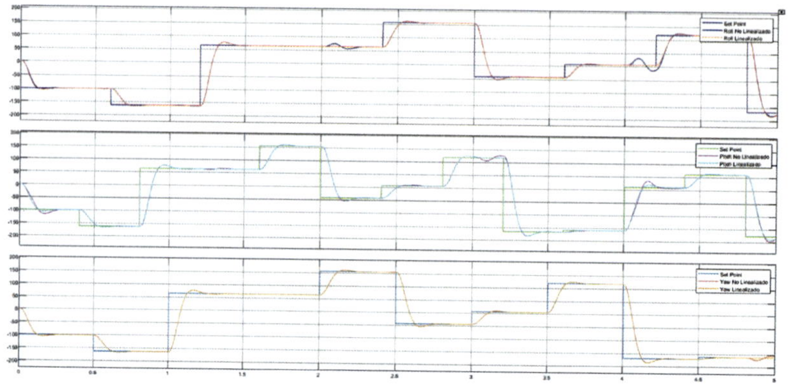

Fig. 2. LQG Controller for linear and no linear model. a) Roll angle, b) Pitch angle, c) Yaw angle

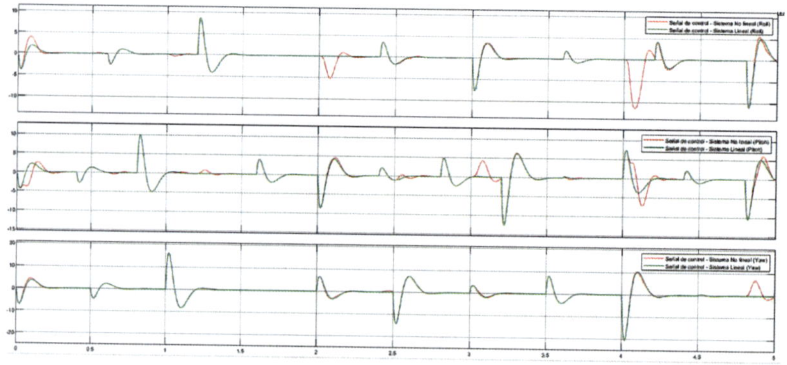

Fig. 3. Control signal for linear and no linear model

5 Conclusions

The paper describes the dynamic model and design of two control strategies, PID and LQG, for a tricopter. Simulations show that the UAV is capable of tracking any trajectory in 3D space with the least error possible. To evaluate the two control strategies two tests were performed, the first test was to evaluate the response of the system to a step input and calculates the error rates IAE, ISE, ITAE and ITSE, the second test was to simulate a trajectory and check the overshoot, raised time and setting time, using SIMULINK's tools, obtaining that the LQG controller turned out to be the best control strategy, since it presents a shorter settling time and overshoot, compared to the PID regulator. On the other hand, it minimizes the effect of variations in the reference signal of one link with respect to the other, being a fundamental factor in this type of systems.

References

1. Sarabakha, A., Kayacan, E.: Y6 tricopter autonomous evacuation in an indoor environment us-ing Q-learning algorithm. In: IEEE 55th Conference on Decision and Control (CDC), pp. 5992–5997 (2016)
2. Sarabakha, A., Changhong, F., Kayacan, E., Kumbasar, T.: Type-2 fuzzy logic controllers made even simpler: from design to deployment for UAVs. IEEE Trans. Ind. Electronics. **65**(6), 5069–5077 (2018)
3. Sabatino, F.: Quadrotor control, modeling, nonlinear control design, and simulation. Master's Degree Project. Stockholm, Sweden, June 2015
4. Zhi, L., Xin, M., Zhigang, X., Yafang, W.: Chattering free sliding adaptive attitude control for quadrotor. In: IEEE Chinese Guidance, Navigation and Control Conference, pp. 707– 712 (2016)
5. Antonelli, G., Cataldi, E., Arrichiello, F., Giordano, P.: Adaptive trajectory tracking for quadrotor MAVs in presence of parameter uncertainties and external disturbances. IEEE Trans. Control Syst. Technol. **99**, 1–7 (2017)
6. Tomashevich, S., Fradkow, A., Andrievsky, B., Belyavskyi, K.: Simple adaptive control of quadrotor attitude. Algorithms and experimental results. In: 25th Mediterranean Conference on Control and Automation, pp. 933–938 (2017)
7. Ogata, K.: Sistema de control en tiempos discretos. Prentice Hall Hispanoamericas, S.A, México (2000)
8. Clark, R.: Control System Dynamic. Cambridge University Press, Cambridge (1996)
9. Burns, R.: Advanced Control Engineering. Butterworth Heinemann, Roland (2001)
10. Basquel, P., Burke, R., Curran, P.: Optimal closed-loop transfer functions for non-standard performance indices. In: 28th Irish Signals and Systems Conference (ISSC), Killarney (2017)
11. Graham, D., Lathrop, R.: The synthesis of "Optimum" transient response: criteria and standard forms. Trans. Am. Inst. Electr. Eng. **72**(5), 273–288 (1953)

Thermal Study of the Variation and Distribution of Temperature in the Cabin of a Car Exposed to Direct Solar Radiation

Pedro Pablo Diaz[1](✉), Diego Ricardo Páez[1], Maria Camila Isaza[1], and Handel Andrés Matínez[2]

[1] Faculty of Mechatronics Engineering, University Santo Tomás, Cra 18 #9-27, Bucaramanga, Colombia
{pedro.diaz,diego.paez,camila.isaza}@ustabuca.edu.co
[2] Faculty of Mechanical Engineering, University of Campinas – UNICAMP, Sao Paulo, Brazil
handelsarache@fem.unicamp.br

Abstract. The accumulation of thermal energy inside the car is the result of heat transfer generated by solar radiation that permeates through the windshield, walls and roof of the vehicle. It has been determined that the internal temperature of the car cabin outdoors, reaches up to 80 °C on a sunny day. To design a heat extraction system, applicable to enclosures that are under the direct effect of the sun's rays, it is necessary to know the thermal mass, the temperature distribution and the cooling load; the objective of this work was to study the variation and distribution of temperature in the cabin of a car exposed to direct solar radiation. To design a heat extraction system, applicable to enclosures that are under the direct effect of the sun's rays, it is necessary to know the thermal mass, the temperature distribution and the cooling load; the objective of this work was to study the variation and distribution of temperature in the cabin of a car exposed to direct solar radiation. To achieve the objective, a simulated computational model was created in ANSYS and a prototype was developed to experimentally measure the internal and external environmental conditions of a car cabin. The variation of the temperature inside the cabin of a vehicle exposed to the sun was studied throughout the day, it was evaluated during 15 days with temperatures that oscillated between 22 and 34 °C, with the windows of the vehicle closed. The results obtained, both for the computational model and for the experimental measurements, show that a variation of the internal temperature of the vehicle is represented at a heating rate that depends directly on the intensity of the radiation from the external environment and is distributed with different intensities according to the physical and environmental conditions of the model.

Keywords: Solar radiation · Thermal profile · Car cabin cooling system · Temperature distribution

1 Introduction

The use of the private vehicle as the main means of transport for the development of daily tasks increases every year. The main function of the car is to transport the user

D. F. Cortes Tobar et al. (Eds.): AETA 2019, LNEE 685, pp. 94–106, 2021.
https://doi.org/10.1007/978-3-030-53021-1_10

quickly, comfortably and safely between a point a and b, once that goal is completed, the vehicle is usually parked and exposed to solar radiation for much of the day, causing the temperature of the car's cabin is increased.

The increase in temperature inside a car (Sunawar and Garniwa 2017) under the effect of sunlight, is mainly due to heat transfer due to solar radiation, through the windshield, of the walls and roof of the vehicle body. The carriage glasses allow solar radiation to enter freely, but block the infrared radiation emitted by the interior surfaces. This causes an increase in temperature inside as a result of the accumulation of thermal energy (Molina and Benitez 2012).

Previous studies based on thermal models have determined that the internal temperature of an outdoor parked car cabin can reach up to 80 °C on a sunny day (Al-Kayiem et al. 2010), (Saidur et al. 2009). The increase of the temperature has multiple implications in the life of the users that go beyond the comfort, in (Ahmad and Turi 2018) the following disadvantages are described: Children and pets left in a car parked for 30 min suffer stress by the heat and the possible death; the windows of cars act like a greenhouse, trapping sunlight and heat, and approximately twenty five children die each year as a result of being left or trapped in hot vehicles; Heat stroke in an adult person can occur when the temperature reaches more than 50 °C. Additionally (Fedoruk and Kerger 2003) has determined that the high temperatures in the cabin of the automobiles produce volatile organic compounds that can affect the human health. Increasing the temperature in a closed enclosure reduces the amount of water vapor causing a reduction in relative humidity (Dianne Dotson 2018). Studies conducted by (Lowen and Steel 2014; Zeng et al. 2017) conclude that exposure to changes from humid to dry environments favors the transmission of viruses such as influenza, which spread more easily in dry environments and also increases the risk of death from cardiovascular system failures.

The research conducted by McLaren et al. (2005) evaluated the rate of temperature increase inside a vehicle cabin when there were days with high and moderate temperatures and evaluated the effect of having car windows open in similar conditions. The temperature was recorded inside a dark sedan continuously for 16 days, two windows were left open 3.8 cm. Each day the temperature was measured in 60-min periods with ambient temperatures ranging from 22 to 35 °C.

The results of the study showed an average increase in cabin temperature of 2° every 5 min, with 80% of the increase during the first 30 min. The rate of temperature increase inside the car remained independent of ambient temperature and variations in window opening. The authors concluded that even at relatively cold ambient temperatures, the temperature increase in vehicles is significant on clear, sunny days.

Basically, the cooling load of the air conditioning system of a site, consists of the energy that the equipment must remove from the interior of the site, so that the pre-established comfort conditions are maintained. In this way, the cooling load considers the thermal contribution that each heat source generates towards the premises, these sources are classified as external and internal.

The external sources are mainly composed of, incidental solar radiation and thermal convection between the external air and the surface that covers the premises (walls, doors and windows). On the other hand, the internal sources of heat are due to electrical elements, lighting and people. In the present work, these charges were not considered, since the case study consists of a stationary car, so it is assumed that there will be no occupants or electrical elements operating during the period of evaluation of the cooling load.

To design a heat extraction system, applicable to enclosures that are under the direct effect of sunlight, you must know the thermal mass, the temperature distribution and the cooling load. As an initial phase, the purpose of this work is to create a simulated model of the enclosure and develop an electronic prototype to experimentally measure the temperature, solar radiation and relative humidity in a car cabin. The car was parked in a large space with dirt floor, without the presence of large shadows like nearby buildings or trees. A thermal model of the variation and distribution of the temperature is established; Defining the heating rate, radiation intensity and its behavior, validating by computer simulation and field measurements of the experimental electronic prototype.

2 Model of the Car Cabin

2.1 Characteristics of the Cabin

In order to develop a simulated model of the thermal behavior in different environments of the cabin to be studied, the conceptual design of the cabin of a real vehicle is established, which represents the geometric model of study.

The vehicle on which the measurements and experimental calculations were made is a Nissan New XTrail model 2018, which is equipped with a top window or sunroof that allows light and ventilation to enter through the roof. For the purposes of the calculations made, the sunroof was considered closed and covered with its insulating layer, so that it was not considered as a window. The CAD design of the vehicle's cab model is presented in Fig. 1.

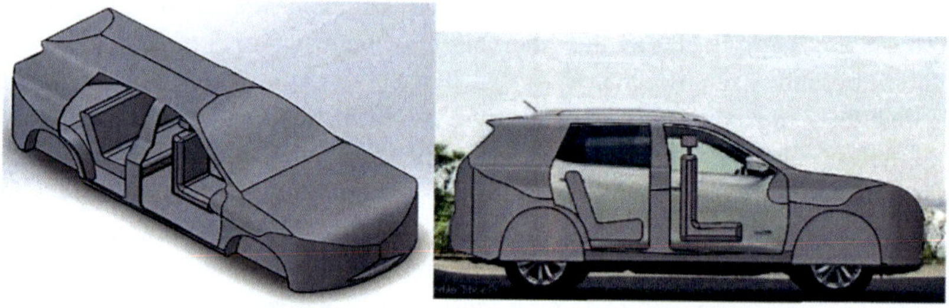

Fig. 1. CAD representation of the vehicle's cab model.

Figure 2 presents a simplified diagram of a car's cabin, showing the various heat flows through the walls surrounding the cabin. In this figure, the regions in the red control volume represent the surfaces over which the heat flow is considered (this includes side elements such as doors and windows); while the green lines refer to assumed adiabatic regions (regions with zero heat transfer) as a consequence of the high resistance to heat transmission due to empty internal spaces.

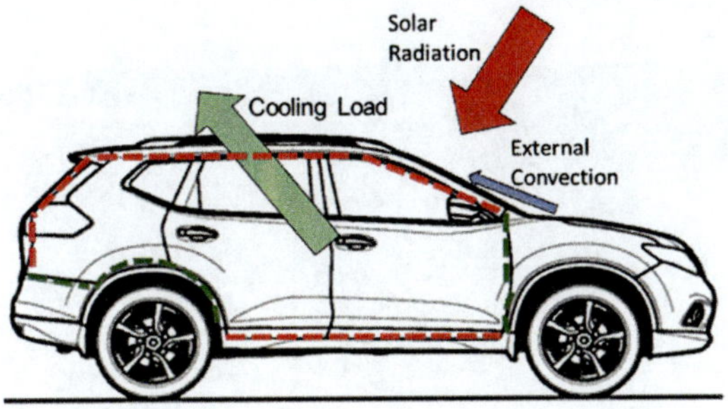

Fig. 2. Main energy flows over the cab of the car.

3 Environmental Monitoring System

3.1 Configuration of the Experiment

To carry out the experimentation of this project, a Nissan Xtrail car was used, the vehicle was parked in the uncovered parking lot of the Santo Tomas University section Bucaramanga (Colombia) for 15 days between 9 am and 5 pm from Monday to Friday, with the windows closed and located with the front part of the car directed to the west, that is, during the morning the sun radiated on the trunk of the car. The experiment consisted of recording the temperature and humidity inside the vehicle cabin, in addition to recording the ambient temperature under solar radiation.

Table 1. Location of temperature sensors inside the vehicle.

Temperature sensor	Location within the vehicle
ST1	Right rear window
ST2	Right front window
ST3	Center of the passenger chairs
ST4	Gear lever
ST5	Center of the trunk
ST6	Temperature Ambient to the sun
ST7	Front left window
ST8	Rear left window
ST9	Panoramic glass

The internal temperature was measured at different points in the vehicle cabin using 8 sensors (see Table 1) distributed as shown in Fig. 3. The location of the sensors was defined in such a way as to record the increase in temperature caused by solar radiation though the day. A sensor was placed on the outside of the vehicle to measure the external temperature.

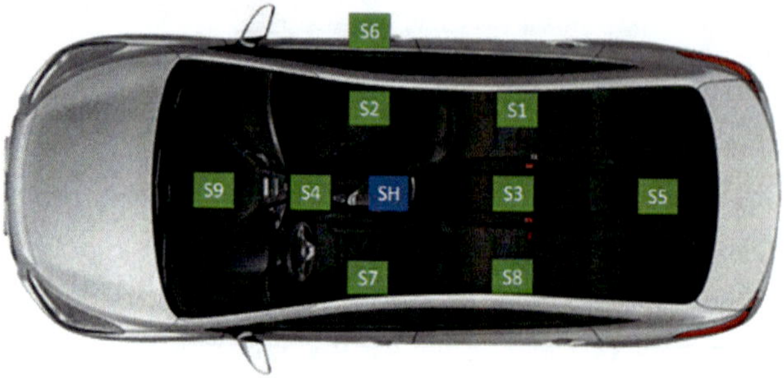

Fig. 3. Distribution of sensors in the vehicle cabin.

3.2 Hardware

The developed hardware consists of a data logging module and a sensor module (see Fig. 4). The sensor module consists of temperature and humidity sensors, the temperature sensor DS18B20 (Maxim Integrated 2015) has a user-configurable resolution ranging from 9 to 12 bits that generate increments of 0.5 °C to 0.0625 °C respectively and has an operating range of −55 °C to 125 °C. The sensor has a unique id and communicates with the processing board through a single cable using the 1-Wire protocol, facilitating the connection of multiple temperature sensors without using Analog Digital Converters (ADC).

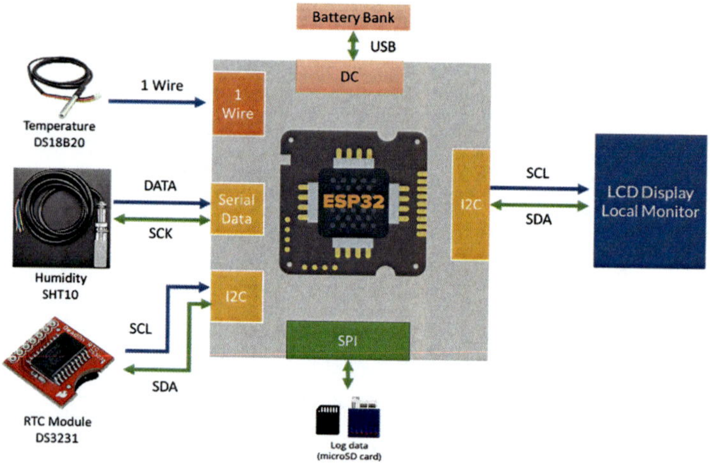

Fig. 4. Description of the environmental monitoring system.

The relative humidity sensor SHT10 (Summary 2011) is linked to the processing card via a two-wire serial protocol. The data acquired by the sensor is converted to digital values by means of an internal ADC and generates a calibrated digital signal, the calibration coefficients are stored in the chip's internal memory.

The data logging module is integrated by a microSD memory adapter and a Real Time Clock (RTC) module. They are used together to store the environmental variables measured during the operation of the system, the RTC keeps the time variables of the microcontroller updated and the microSD adapter gives access to a flash memory for the storage of the variables of interest.

3.3 Firmware

The firmware designed for the ESP32 board (Espressif, n.d.) manages the recording of data in plain text files. For the implementation of the firmware, the libraries associated with the temperature and humidity sensors, the RTC module and the data read/write manager in the microSD card are imported. Afterwards, the necessary variables are defined to work with the sensors, the clock and the memory. With the defined variables, it is necessary to initialize and configure both the text file in which the data will be stored and the alarm that is activated every 5 min to record the data. The main process is to evaluate if the alarm linked to the data record is activated and indicate to the user with a led if the storage of data in the microSD memory was successful. The firmware is described in the flow diagram presented in Fig. 5.

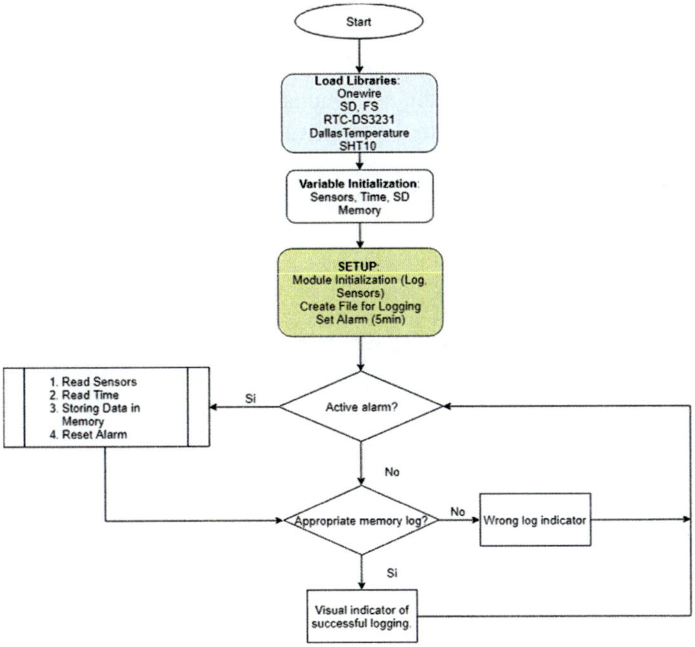

Fig. 5. Flow diagram of environmental monitoring system firmware.

To access the data, three options were enabled, the first one is to connect the microSD memory to a computer through a USB adapter. The second is based on the transmission of data via Bluetooth to a mobile device through an application. Finally, the third option is a local monitor in which, by means of an LCD display, the current environmental variables and an average of the day's history are presented to the system operator (see Fig. 6).

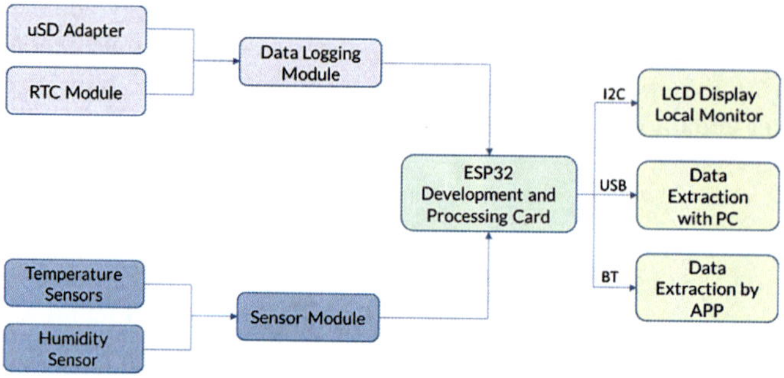

Fig. 6. Architecture of the environmental monitoring system

4 Computational Model Simulation

4.1 Transient Analysis in Ansys

With the model defined for the simulation of thermal behavior, using the Ansys Worckbench 2019 R1 software, a transient analysis of the heat transfer by radiation from the sun through the external air, by conduction through the walls of the vehicle cabin and by convection in the interior air of the cabin of the vehicle under study was conducted. For the meshing the Ansys computational simulation tool was used, for the automatic generation of unstructured meshes in the implementation of the finite element method and calculation of heat transfer in transitory regime. Figure 7 shows the cabin modeled in ANSYS and the mesh defined for the study.

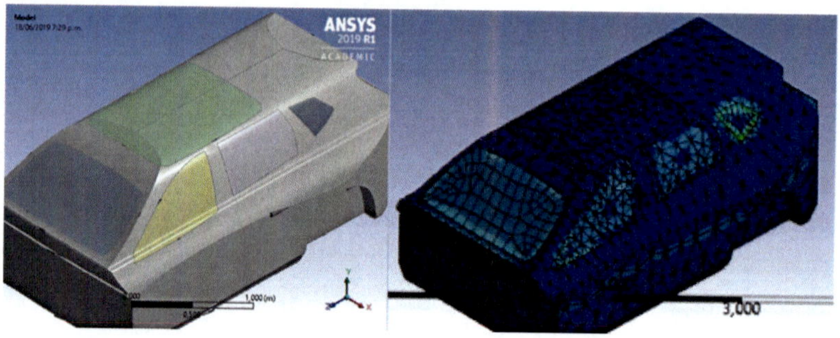

Fig. 7. Model for transient analysis of the thermal behavior of the cabin.

Initially in the interface Thermal Transient is created to perform the analysis in transient state. The Engineering Data option is selected with which the materials of the pieces to work are added; 2 materials were added: that of the structure and that of the windows. For the structure the density is 1100 kg/m^3, thermal conductivity 0.13 w/m * K and specific heat 580 J/Kg * K. For the windows the values of density 2500 kg/m^3, thermal conductivity 0.9 w/m * K and specific heat 800 J/Kg * K were entered.

The design with scdoc extension is added, the Model with Multiple System - Mechanical is entered to enter the required values within the study of the variables for the study.

The meshing of the assembly (car cabin) is generated with the option Generate Mesh, to later see the behavior of the temperature applied. In the Transient Thermal option, the temperature values of the environment (exterior of the cabin) and its variation as a function of time are added, simulating a sunny day from 9:00 AM to 5:00 PM, in Analysis Setting the analysis time is considered in seconds, the values of ambient temperature and solar radiation are added. The solar radiation values are added by area of the model (see Table 2) and it's considered a critical situation for maximum ambient temperature 34 °C (condition of September de 2018, warmest month of the year, specifically September 21).

Table 2. Physical and thermal properties of the walls and windows of the car cabin.

Description	A [m^2]	L [mm]	ρ* [kg/m^3]	k* [W/m.K]	Cp* [J/kgK]	α*	ε*	Heat critical radiation [W]
Right side windows	0,265	4	2500	0,9	800	0,39	0,80	38,5
Rear side windows	0,285	4	2500	0,9	800	0,39	0,80	41,4
Windshield	1,179	5	2400	0,6	830	0,43	0,80	890,8
Sunroof	0,924	5	2400	0,6	830	0,43	0,80	134,6
Rear window**	0,90	4	2500	0,9	800	0,39	0,80	182,8
The ceiling	1,376	35	270	0,054	560	0,78	0,85	95,9
Front side doors	0,991	20	1100	0,13	580	0,78	0,85	27,6
Rear side doors	0,824	20	1100	0,13	580	0,78	0,85	22,9
Front side pillars	0,075	20	1100	0,13	580	0,78	0,85	7,2
Rear side pillars	0,189	20	1100	0,13	580	0,78	0,85	5,3
Floor	1,715	30	480	0,041	800	0,78	0,85	105,4
Back door	1,10	20	1100	0,13	580	0,78	0,85	75,8

* Taken from (6) ** Not considered in this study

The data was taken and calculated with parameters offered by ASHRAE; for the case of the city of Bucaramanga, which is positioned at 7.1362° N and 73.1282° W, at a height of 959 m above sea level and the time zone of the city is Central Standard Time (−90° E out of phase with respect to the Greenwhich meridian).

4.2 Properties of Vehicle Surfaces and Parameters Used

Table 2 presents the identification, surface, thickness, physical and thermal properties of each of the walls that cover the car cabin. It is worth mentioning the great difficulty for the lifting of the physical and thermal properties of each one of the walls, both opaque, and windows. In the case of opaque surfaces such as doors and roof, the thermal and optical properties were taken from the work done by Pokorny et al. (Pokorny 2014), where the authors estimated, for a Skoda Felicia Combi car, the properties of each wall considering the composition of the same (number of layers, empty spaces and materials).

The Material specifications for the cart in SolidWorks for windows: Cutting module: 318.9 N/mm^2, Mass density: 2500 kg/m^3, Tensile limit: 30 N/mm^2, Thermal Conductivity: 0.9 W/(m * K), Specific heat: 800 J/(Kg * K); for Bodywork: Sustainability: Ductile iron (SN) in SolidWorks materials, Mass density: 1100 kg/m^3, Thermal Conductivity: 0.13 W/(m * K), Specific heat: 580 J/(Kg * K).

5 Results

5.1 Temperature Distribution with the Experimental Method

Results of the recording of internal and external variables of the vehicle cabin are presented; with field measurements taken by the electronic device designed and constructed for the present study. The measurements were taken between the months of June and July in the northern part of the city of Bucaramanga, department of Santander, Colombia. The vehicle was located from east to west, with the front facing west.

Figure 8 shows the variation of temperature and humidity inside the cab of the vehicle on a day of registration, taken from the highest of the field tests carried out with the prototype of the designed device. It is observed that the temperature of the trunk suffers an increase of 10 °C with respect to the other sensors located in the cabin from 8:30 to 10:48, reaching a peak of 68.8 °C.

The sensor that records the temperature variation in the panoramic has the highest temperature in the same time zone described previously, after 10:48 h it is maintained at a temperature that varies ±4 °C around 59 °C.

The ambient temperature presents a difference of approximately 20 °C with the temperature of the vehicle cabin between 12:00 and 13:12 h, evidencing the greenhouse effect inside the cabin and its increase.

In Fig. 8 it is observed that the humidity inside the cabin falls during the whole day and it is observed that at the end of the afternoon (15:50 h) the percentage of humidity inside the vehicle grows back again.

Figure 9 shows the result of the measurement with a reference instrument, where the internal temperature of the vehicle (T1) and the ambient temperature under sun (T2) are observed. When comparing the data with the experimental reference device in the same time slot, an error of 10% is observed. Which allows to conclude that it is necessary to perform a greater number of calibration tests with a certified device.

Figure 10 shows the web page on which the data is updated every time a new measurement is made.

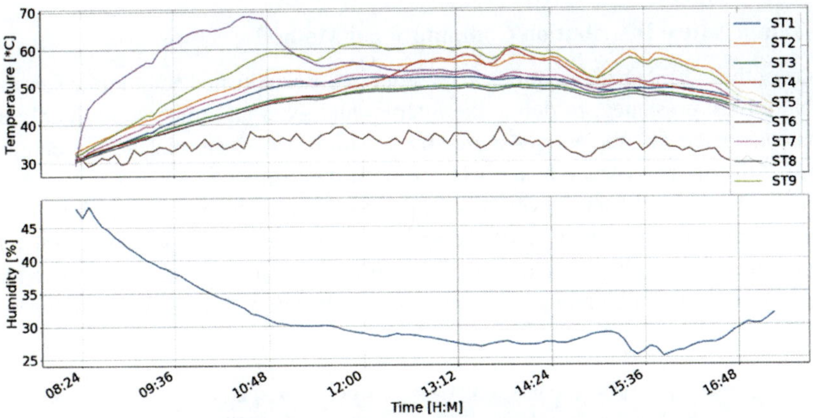

Fig. 8. Record of temperature and humidity taken with the device in the vehicle cabin.

Fig. 9. Reference reading temperature in the vehicle cabin and room temperature (3:30 p.m.).

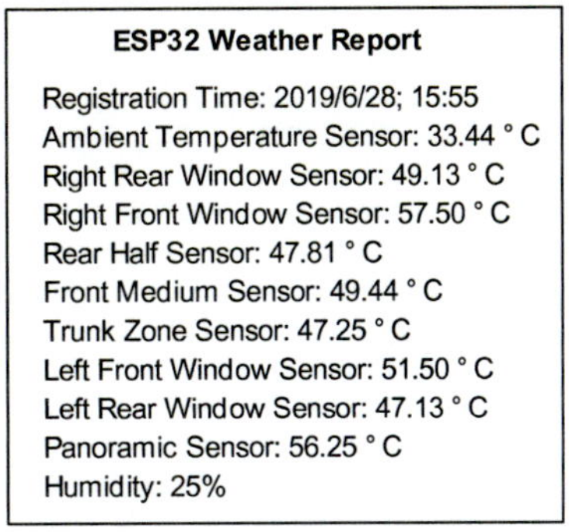
ESP32 Weather Report

Registration Time: 2019/6/28; 15:55
Ambient Temperature Sensor: 33.44 ° C
Right Rear Window Sensor: 49.13 ° C
Right Front Window Sensor: 57.50 ° C
Rear Half Sensor: 47.81 ° C
Front Medium Sensor: 49.44 ° C
Trunk Zone Sensor: 47.25 ° C
Left Front Window Sensor: 51.50 ° C
Left Rear Window Sensor: 47.13 ° C
Panoramic Sensor: 56.25 ° C
Humidity: 25%

Fig. 10. Web page for monitoring environmental variables record from the device.

5.2 Temperature Distribution Computational Method

Results of the thermal behavior of the car model of the established automobile simulated in ANSYS are presented, to define the internal and external temperature distribution. It was considered that the energy source by radiation, which simulates solar energy, is located at the top and perpendicular to the roof of the vehicle.

Figure 11 shows the thermal effect of the simulation on the model with the mesh, by means of transient analysis in ANSYS, where the increase of temperature in the cabin is appreciated by the effect of the heat transfer product of solar radiation simulated. The average maximum temperature in the car cabin produced by the simulation in the range of 55 °C.

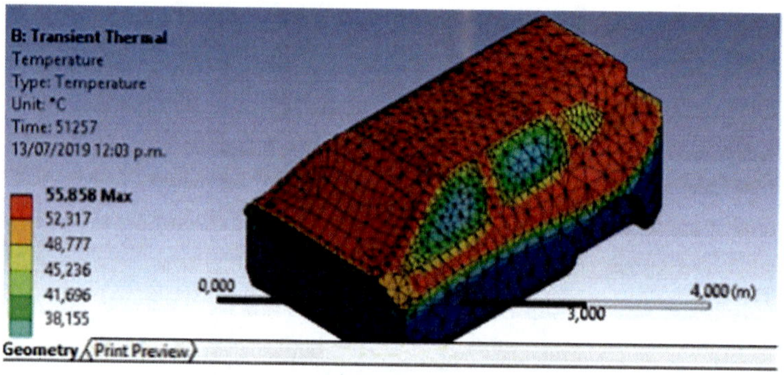

Fig. 11. Simulation of temperature distribution in the car cabin.

Taking as a reference the part of the front panoramic glass of the model and plotting the results of the simulation for this area throughout the day, the temperature distribution curve presented in Fig. 12 is obtained. It is observed that the maximum temperature value is presents between 13:00 and 14:00 h, with a value of 55,9 °C.

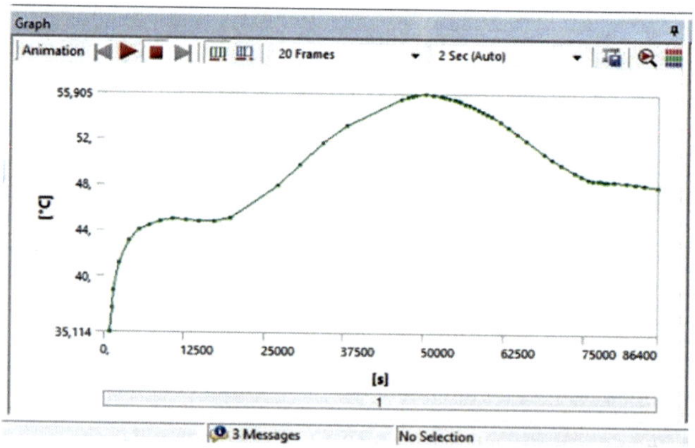

Fig. 12. Graph for the simulation of the thermal variation in the front glass of the car cabin throughout the day.

6 Conclusions

The results show approximate values to those obtained in other investigations (Al-Kayiem et al. 2010; Molina and Benitez 2012; Sunawar and Garniwa 2017), where they enunciate the considerable and important increase inside the cabin of the automobile, although some articles mention that temperatures near 80 °C, are reached; it is important to emphasize that in the majority of the measurements of field, due to the time of the year that they have presented little sunny days and in some occasions with rains, therefore, it is considered that it has not been possible to record the highest temperature the vehicle can reach. However, the results obtained from experimentation and simulation are very approximate.

It can be concluded that the hottest time of day and the larger glass of the vehicle favours the increase in temperature within the cabin, which indirectly causes the stored water vapor to be reduced, i.e. the relative humidity is reduced as the temperature increases. The results confirm the above mentioned by (Dianne Dotson 2018) in which the relative humidity decreases with increasing temperature and increases with decreasing temperature, this can be observed in the measurements presented in the results.

It should be noted that simulating atmospheric effects and climatic variations towards the interior of the vehicle because of solar radiation has its differences with the actual measurements and with the error percentages that these may present.

The time in which the highest temperature inside the cabin of the automobile is presented is the one between the 1:00 and 3:00 of the afternoon, being the frontal part the one of greater affectation by the area of the panoramic one and the greater direct exposure to the rays of the sun. Although it is worth mentioning that these results depend on the location of the vehicle, with respect to the cardinal points and the time of year.

In the simulation, the behavior of the temperature distribution in the hottest hour of the day can be observed, as recorded by the measuring device, concluding that the curve reflects the areas of greatest heat transfer in the vicinity of the front of the cabin.

References

Ahmad, M., Turi, J.A.: Study of air temperature propagation of parked car. Int. J. Sci. Eng. Res. **9**(3), 1503–1513 (2018)

Al-Kayiem, H.H., Sidik, M.F.B.M., Munusammy, Y.R.A.: Study on the thermal accumulation and distribution inside a parked car cabin. Am. J. Appl. Sci. **7**(6), 784–789 (2010). https://doi.org/10.3844/ajassp.2010.784.789

Espressif. ESP32 DevKit (n.d.). https://docs.espressif.com/projects/esp-idf/en/latest/get-started/get-started-devkitc.html. Accessed 26 May 2019

Fedoruk, M.J., Kerger, B.D.: Measurement of volatile organic compounds inside automobiles. J. Expo. Anal. Environ. Epidemiol. **13**(1), 31–41 (2003). https://doi.org/10.1038/sj.jea.7500250

Dianne Dotson, J.: How Temperature & Humidity are Related (2018). https://sciencing.com/weather-affect-us-23423.html. Accessed 3 Aug 2019

Lowen, A.C., Steel, J.: Roles of humidity and temperature in shaping influenza seasonality. J. Virol. **88**(14), 7692–7695 (2014). https://doi.org/10.1128/jvi.03544-13

Maxim Integrated. Datasheet DS18B20 digital thermometer. Maxim Integr. **92** 20 (2015). www.maximintegrated.com

McLaren, C.: Heat stress from enclosed vehicles: moderate ambient temperatures cause significant temperature rise in enclosed vehicles. Pediatrics **116**(1), e109–e112 (2005). https://doi.org/10.1542/peds.2004-2368

Molina, J., Benitez, V.: Sistema de refrigeración con energía alterna para acondicionar la temperatura interior de una cabina de automóvil (2012). http://www.irsitio.com/refbase/documentos/49_MolinaGonzalez+BenitezBaltazar2012.pdf

Pokorny, J., Fiser, J., Jicha, M.: A parametric study of influence of material properties on car cabin environment. In: EPJ Web of Conferences, vol. 67 (2014)

Saidur, R., Masjuki, H.H., Hasanuzzaman, M.: Performance of an improved solar car ventilator. Int. J. Mech. Mater. Eng. **4**(1), 24–34 (2009). http://ejum.fsktm.um.edu.my/article/755.pdf

Summary, P.: Datasheet SHT1x (SHT10, SHT11, SHT15) 5. www.sensirion.com. Accessed 1–12 Dec 2011

Sunawar, A., Garniwa, I.: Maximum potential of the car cabin temperature in the outdoor parking conditions as a source of energy in thermoelectric generator. In: IOP Conference Series: Materials Science and Engineering, vol. 180, p. 012104. IOP Publishing (2017). https://doi.org/10.1088/1757-899X/180/1/012104

Zeng, J., Zhang, X., Yang, J., Bao, J., Xiang, H., Dear, K., Liu, Q., Lin, S., Lawrence, W.R., Lin, A., Huang, C.: Humidity may modify the relationship between temperature and cardiovascular mortality in Zhejiang province, China. Int. J. Environ. Res. Pub. Health **14**(11) (2017). https://doi.org/10.3390/ijerph14111383

Cyber-Attack Mitigation on Low Voltage Distribution Grids by Using a Novel Distribution System State Estimation Approach

Efrain Bernal Alzate[1(✉)], Diana Lancheros-Cuesta[1], and Zihao Huang[2]

[1] Department of Automation Engineering, Universidad de La Salle, Bogota 111711, Colombia
ebernalal@unisalle.edu.co

[2] Research Institute, Wuhan Branch, Hang Lung Group, Wuhan 430017, China

Abstract. Distribution electrical power systems are routinely designed and built to resist a variety of natural disruptions and continue to operate. However, if terrorist attacks targeted massive components such as smart meters, attackers can bypass traditional anomaly detection tests and influence negatively the stability of the system. In this paper, a novel distribution state estimation method (DSSE) to mitigate the impact of cyber attacks is presented. The new method for distribution state estimation uses few readings from smart meters, doing it robust, even for ill conditioned systems. To validate the method and show its robustness a test was made for a 30 smart home low-voltage (LV) distribution system. Results show how the proposed method is worthwhile in case of communication failure or attacks.

Keywords: Cyber security · False data injection attack · Distribution state estimation · Smart grid

1 Introduction

Power distribution systems are complex systems connecting electric power generators to consumers through power transmission and distribution networks across a large geographical area. System monitoring is necessary to ensure the reliable operation of power grids. For the longest period of time, cyber security was not an issue with the electric grid. However, the integration of renewable resources into the electrical grid implies a transformation process of the traditional electrical grid to improve power energy chain that runs from a major power generator plant all the way inside the customer domain. This is referred as the smart grid concept. Moreover, while the law of physics is not changing, everything else is changing; with the smart grid, several changes have been happening [1]: newer systems such as Distribution Management System (DMS) are being implemented. These and some of the pre-existing systems identified above are now becoming more connected and becoming less isolated from the Internet. Moreover, sensors and controls (including smart meters) that are being installed

D. F. Cortes Tobar et al. (Eds.): AETA 2019, LNEE 685, pp. 107–116, 2021.
https://doi.org/10.1007/978-3-030-53021-1_11

along the transmission and distribution systems are being moved from proprietary mechanisms to IP-based mechanisms. These changes have brought incredible flexibility to the electric utility industry, but, they also have brought in the same vulnerabilities that have plagued the rest of the computer based systems. Normally, voltage-control studies do not include the effect of the communication network involved with the central control station. Furthermore, throughput variations, packages loss, or false data injection can affect overall grid performance. This can have important economic impacts and technical implications for distribution and transmission systems, such as increasing losses and line overloading, unwanted reactive power fluctuations, and consequently, a systems blackout. The paper is organized as follows: Section 2 introduces the implemented distribution state estimation method (DSSE) by using the available information from smart meters. Section 3 presents the attack considerations and the methodology for the attack detection and identification. Section 4 shows a comparative analysis and the performance under outages. Finally, Sect. 5 summarizes the results and provides some conclusion.

2 Distribution State Estimation Method (DSSE)

The power system state estimation (SE) as a part of the smart grid concept, as pointed out by Massoud Amin -considered the father of the smart grid- in his classical paper [2], is not new. State estimation is used in system monitoring to best estimate the power grid state through analysis of meter measurements and power system models. The recent inclusion of information and communication technology (ICT) components into the low-voltage (LV) distribution grid was a turning point regarding the possibility of including a state estimator in the distribution system too. However, the methods and assumptions used in transmission-level SE are often not valid when considering distribution systems, and many of well-established techniques used in typical SE cannot be applied directly [3–6]. However, recent widespread deployment of high quality residential smart meter provides the control centers with the necessary information for the monitoring and control in real-time of smart LV distribution. In fact, smart meters are demonstrated to be able to serve as synchro-sensors for smart distribution grid control applications [7], for the customer classification, load profiling and demand response applications [8,9], as well as to support optimal operation of distribution systems [10–12]. Dynamic state estimator methods are complex and require several simplifications for modeling and implementation. Therefore, static state estimator methods are most popular and currently used in commercial SE software. The goal of static state estimation method is to find the values of a set of state variables in the vector x that fix to the basic equation:

$$z = h(x) + r \tag{1}$$

where z is a vector for the given m number of measurement values with $z \in \Re^m$, x is the n number of state variables vector to estimate with $x \in \Re^n$, $h(\cdot)$ is the nonlinear functions vector that relates the m measurements to the n state variables with $h(\cdot) \in \Re^m$, and r is the (measurement) error vector on the data ordinates.

2.1 WLS Estimator

Most of the DSE algorithms are based on weighted least square (WLS) approaches and differ mainly in the choice of state variables and measurements used for the estimation. Bus voltage magnitude and phase angle are widely used as state variables in the WLS estimator. As in the case of non-linear least squares problem, the objective function is a quadratic function as

$$f(z - h(x)) = (z - h(x))^T W (z - h(x)) \tag{2}$$

where W is the weighting matrix, which is chosen to be the inverse of the covariance matrix of the measurement error vector, positive and defined, and T index represents the transpose of a matrix o vector. As $h(x)$ is a nonlinear function, x cannot be calculated directly, the Newton Raphson method is the most widely method used to solve the problem iteratively. The mathematical formulation of this problem is to find the zeros of the derivate of $J(x)$ as described in 3 and 4:

$$J(x) = \frac{1}{2} \sum_{i=1}^{m} \left[(z_i - h_i(x))^2 W_{ii} \right] \tag{3}$$

$$\frac{\partial J(x)}{\partial x} = - \sum_{i=1}^{m} \left[(z_i - h_i(x)) W_{ii} \frac{\partial h_i(x)}{\partial x} \right] = 0 \tag{4}$$

Then, the iterative and correction process is conducted by 5 and 6 until the objective function reaches a minimum, as follows:

$$x^{k+1} = x^k - \Delta x^k \tag{5}$$

$$\Delta x^k = \left[G(x^k) \right]^{-1} H(x^k)^T W \Delta Z(x^k) \tag{6}$$

where k denotes the iteration index, $H(x)$ is the Jacobian matrix of the measurement function $h(x)$, and $G(x)$ is the gain matrix obtained by squaring the Jacobian matrix $H(x)$ as

$$G(x^k) = H(x^k)^T W H(x^k) \tag{7}$$

ΔZ is the residual vector, used to update the state variable values as follow:

$$\Delta Z(x^k) = z - h(x^k) \tag{8}$$

2.2 Implemented DDSE Approach

WLS is usual in transmission grid, which is meshed, while the LV distribution systems are radially with an unbalanced load operation. On the other hand, as large amount of injection measurements are needed for a good estimation in WLS, the transmission grid is usually observed by an adequate number of measurement points; however, in the case of LV distribution systems these are limited and the system cannot be observable. Fortunately, in the case of smart LV distribution systems; the high penetration of smart meters offers the possibility to control the system, [13], having a measurement point at each bus, that means that the number of unknown variables m equals the number of known variables n, but the gain matrix is nearly singular, as an ill-conditioned case. To avoid the weakness of the singular matrix, the Levenberg-Marquardt method [14], [15] is used as part of the distribution state estimation method. Rao [16] proposed the incorporation of the Levenberg-Marquardt algorithm in the WLS formulation adding a positive definite diagonal matrix to the gain matrix $G(x)$ for the case of power transmission systems. An extension of this method to SE in real-time of smart LV distribution systems was done, as follows. The main idea is to add a positive definite diagonal matrix to the gain matrix 7 to avoid the problems presented above as

$$G_{LM}(x^k) = H(x^k)^T W H(x^k) + \mu_k I \tag{9}$$

where I is the identity matrix, and μ_k is the so-called damping parameter with $\mu_k \in \Re \mid \mu_k > 0$. Consequently, the iterative and correction process is given by

$$\Delta x^k = \left[G_{LM}(x^k)\right]^{-1} H(x^k)^T W \Delta Z(x^k) \tag{10}$$

Since G_{LM} is positive definite for observable systems, there is no need to worry about pivoting to preserve numerical stability. If μ is large enough, then iterative and correction process can be written as

$$\Delta x^k \approx \left[\mu_k I\right]^{-1} H(x^k)^T W \Delta Z(x^k). \tag{11}$$

which is an incremental step along the direction steepest descent of the weighted sum of squares.

3 Communication Failure and False Data Injection Attack

The communication failure means that the control center does not get data from some nodes. To solve the unknown variables, the equation system needs at least $2 * (N-1)$ equations. If no data is supplied by a smart meter, there are only $2*(N-2)$ equations with $2*(N-1)$ variables, in which case there certainly will not be a unique solution. One possibility is to use the smart meter installed on the node of transformer station, where the active and reactive power, as well as the node voltage are known. Thus the equations of the Slack node are usable and again $2*(N-1)$ equations with $2*(N-1)$ variables. The extra data, of voltage or current amounts from the rest of the smart meters provides redundancy to make the solution safe in case of communication failure or false data injection attack.

3.1 Communication Failure

The failure can easily be determined by additional data. Because the voltage magnitude of one node is not equal to zero even though the load is in operation or not. If the amount of stress from one node is abnormal, there is necessarily a failure and this node is marked as unknown. In practice, a failure detection is performed as follows: Consider the distribution system with n buses described by the voltage-current relationships of the system in terms of the bus (node) voltages $u = [U_1, U_2, \ldots, U_n] \in \mathbb{C}^n$, the currents $i = [I_1, I_2, \ldots, I_n] \in \mathbb{C}^n$, the active powers $p = [P_1, P_2, \ldots, P_n] \in \mathbb{R}^n$, the reactive powers $q = [Q_1, Q_2, \ldots, Q_n] \in \mathbb{R}^n$, and the admittance matrix Y. The sample time in the control station Ts, for the simulation case is $Ts = 30\,\mathrm{s}$. Here it is important to note, as described before, that the slack node compensates for the power balance. But in the present case, the unknown node is not the slack node, but a load node whose states must be safely triggered. Therefore the power flow calculation is not always suitable here. The DSSE has to be used in this case.

Input: u, Y, p, q; Obtained data from smart meters at Ts
Output: $ufail, unew$; voltage status and corrected voltage vectors
`FailureDetection`(u, N)
$ufail \longleftarrow 0$
foreach U_i **do**
  if $U_i = 0Vac$ **then**
    $Ufail_i \longleftarrow 1$;
  end
end
return $ufail$
if $ufail > 0$ **then**
  `DSSE(`u, $ufail$, Y, p, q`)`; Use DSSE to supply missing data
  return $unew$; Return voltage vector including missing data
end

Algorithm 1: Algorithm for Detection of Communication Failure

3.2 Detection of a False Data Injection Attack

In the low voltage network, the attacker's goal is to manipulate the power usage data from the smart meter so that it can destroy the power grid or steal the power. Some attacks against the normal control system are presented in [17,18], in which they present a method for detection and identification. One of the attacks is the wrong Date Injection Attack, which is especially damaging to the state calculation. The attack model used in this paper is based on that presented in [19], as follows:

The false data injection attack assumes that the state estimation uses a model of the power grid. Let z_a represent the vector of observed measurements that may contain malicious data. z_a can be represented as

$$z_a = H * x + a \tag{12}$$

Here, H represents the Jacobs matrix, x the readings from smart meters, and a denote the attack vector. The estimated solution is:

$$X = (H^T WH)^{-1} H^T WZ \tag{13}$$

Here, x represents the estimated states. The traditional method for bad data detection is evaluation of the value of the objective function $J(x)$. $J(x)$ is a distributed function according to the premise (the errors have Gaussian distribution) for x^2. Therefore, there is no bad data or attack, when $J(x) < x^2_{(m-n),p}$. Here m is the number of measured values, n represents the number of variables or states and p the confidence intervals. p is a given value $(0 < p < 1)$ regularly equal to 0.95. If $J(x) > x^2_{(m-n),p}$ then there is attack. The index of attack is going through

$$R_{residue} = ||z - H * x|| \tag{14}$$

If there is an element of $R_{residue} > \tau$, where τ is the detection threshold, an attack has occurred in the site. Therefore, the measurement values of the corresponding node are removed and the state estimation repeats to fix the bad data. The model with wrong data injection attack is defined as $a = Hc$. AS, the vector a represents the attack, if there is a real vector c, the estimated states variables with attack x_{bad} can now be described as $x + c$. The deviation between measured data and estimated values is now:

$$||z_a - H * x_{bad}|| = ||z + aH * (x + c)|| = ||zHx + (aHc)|| = ||zHx|| \tag{15}$$

That is, in false data injection attacks, the attack vector a should satisfy the condition $(a - Hc) = 0||$, while the generalized false data injection attacks relax this condition so that any vector a that satisfies $(aHc) < ||$ can be used as the attack vector. The attack is usually defined as Bad Data. That is, because the target of the low-voltage network attack is usually to steal electricity by falsifying the readings, or looking for destabilize voltage-reactive controllers; in this work the attack is defined as $p_{attack} = p + a$. In order to detect and identify the attack, the following considerations are consider to run an algorithm for false data injection attack detection: The basic equations for the power flow calculation in distribution systems have to be reached. The estimated state values are safe if there is no attack. The attacker can only hack low number of smart meters. Attacks detection is more complex with LV distribution systems as they are considered Ill-conditioned problems. The structure of the system is ill-conditioned when there are too few measurements, doing the Jacobean matrix singular, and naturally lead typical DSSE methods to slow convergence or failure to converge at all. The more singular a matrix is, the more ill-conditioned its associated system will be. That is the advantage of the proposed DSSE method used in this paper which was introduced and validated in [12]. Taking that in consideration, the final algorithm is presented as follows:

Table 1. Distribution of samples households by urban, intermediate and rural areas

Attack	$J(x)$	Δx
Non attack	2.99e−05	1.44e−06
One bad data	2e−4	4e−5

Input: u, Y, p, q, N; Obtained data from smart meters at Ts
Output: $ufail, unew$; voltage status and corrected voltage vectors
`AttackDetection(`u, N`)`

$$J(x) \longleftarrow \sum_{i=1}^{N} ||\Delta U_i||$$

if $J(x) >$ **then**
  Find p_{attack}
  $Ufail_{p_{attack}} \longleftarrow 1$
  `DSSE(`$u,\ ufail,\ Y,\ p,\ q$`)`; Use DSSE to supply bad data
  return $unew$; Return voltage vector including baddata ;**end**
  return $ufail, unew$

Algorithm 2: Algorithm for Detection of False Data Injection

4 Results

Although there is a conceptual difference between state calculation by conventional load flow and the state estimator, both solve the basic power flow equations. A 30 bus LV distribution systems is used to compare the proposed DSSE method with respect to Newton-Raphson method, Gauss Seidel method, and the conventional WLS method. The proposed DSSE method and comparison methods were implemented on a personal computer with Core i5 and 6-GB of RAM running Matlab R2014a. To ensure the validity of the results obtained, we compare the results of the estimation in terms of voltage magnitude and voltage angle in each bus, and the active an reactive power. Results obtained for each bus were in agreement with conventional power flow methods and WLS method. We evaluate the methods when there are one, two and three communication failures in the communication system. As presented in Fig. 1, typical WLS method and the proposed DSSE method work when there is one fail, but the error of WLS method is higher. Moreover, in the case of more failures only the DSSE method converges, showing the robustness of the proposed DSSE method. In order to investigate the method of detection and identification of attacks, two attacks scenarios we done: Pattacke = 5% Preal and Pattacke = 10% Preal. Table 1 shows that the deviation from stress is very small and the value of $J(x)$ is only around e−4 if there is no attack. So the *tao* parameteras 0.01 is selected. As shown in Tables 2 and 3, if there is attack and one smart meter with a communication

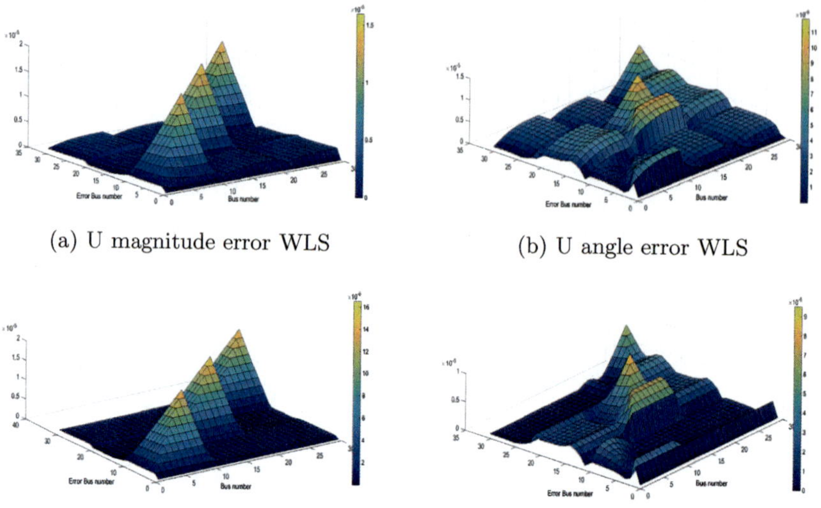

(a) U magnitude error WLS

(b) U angle error WLS

(c) U magnitude error DSSE

(d) U angle error DSSE

Fig. 1. Comparison of errors obtained at all the buses of the evaluated test case.

Table 2. Attack 5% and one smart meter with a communication failure

Attack										
Node attack	2	3	4	5	6	7	8	9	10	11
$J(x)$	0.11	0.03	0.013	0.16	0.87	0.38	0.06	0.02	0.21	1.03
Node attack	12	13	14	15	16	17	18	19	20	21
$J(x)$	0.11	0.48	0.04	0.016	0.18	0.99	0.43	0.07	0.02	0.24
Node attack	22	23	24	25	26	27	28	29	30	31
$J(x)$	0.11	0.47	0.03	0.016	0.18	0.98	0.42	0.06	0.02	0.23

failure, the value $J(x)$ is greater than *tao*, and the node with the attack can be identified correctly by the residuals. Figure 2 shows the voltage deviations from the power grid with an attack Pattacke = 10% Preal. As shown in the figure, if the attack is not corrected, the attack affects the estimated values of the nodes that are in the same subnet. The same methodology was used to evaluate the scenario with attack and two or three smart meters with total communication failure. In the following figure the obtained errors are presented for all the different possible combinations. As can be seen, the DSSE method reach the values very closed to scenario without false data injection attacks and communication failures.

Table 3. Attack 10% and one smart meter with a communication failure

Attack										
Node attack	2	3	4	5	6	7	8	9	10	11
$J(x)$	0.55	0.15	0.07	0.79	4.34	1.90	0.30	0.10	1.05	5.15
Node attack	12	13	14	15	16	17	18	19	20	21
$J(x)$	0.53	2.39	0.19	0.08	0.90	4.91	2.13	0.34	0.11	1.19
Node attack	22	23	24	25	26	27	28	29	30	31
$J(x)$	0.53	2.39	0.20	0.08	0.90	4.91	2.13	0.34	0.12	1.19

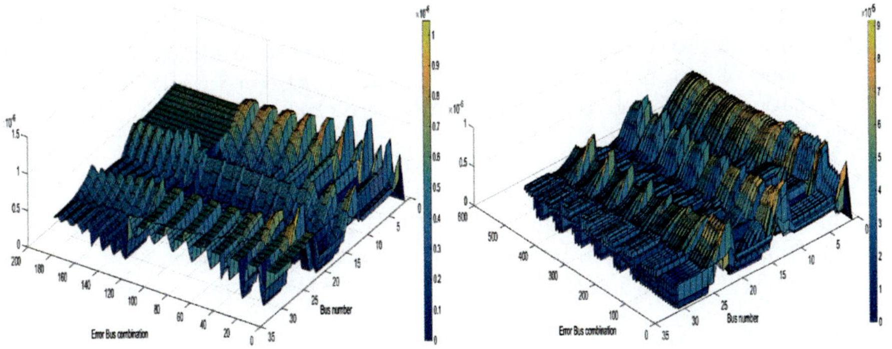

(a) 2 missing smart meters and 10% attack (b) 3 missing smart meters and 10% attack

Fig. 2. (a) 2 missing smart meters and 10% attack (b) 3 missing smart meters and 10% attack

5 Conclusion

A novel distribution state estimation method (DSSE) to mitigate the impact of cyber attacks was presented. The new method for distribution state estimation uses few metrics, and it is also very robust in the case of LV distribution systems. Results show how the proposed method is able to estimate the data from smart meters even with 10% of communication failure or attack.

References

1. Line, M.B., Tøndel, I.A., Jaatun, M.G.: Cyber security challenges in smart grids. In: 2011 2nd IEEE PES International Conference and Exhibition on Innovative Smart Grid Technologies, pp. 1–8. IEEE (2011)
2. Massoud Amin, S., Wollenberg, B.F.: Toward a smart grid: power delivery for the 21st century. IEEE Power Energy Mag. **3**(5), 34–41 (2005)

3. Li, H., Yang, M.: A branch-current-based state estimation for distribution systems non-measurement loads. In: 2004 IEEE Power Engineering Society General Meeting, pp. 440–444. IEEE (2004)
4. Pereira Barbeiro, P.N., Krstulovic, J., Teixeira, H., Pereira, J., Soares, F.J., Iria, J.P.: State estimation in distribution smart grids using autoencoders. In: 2014 IEEE 8th International Power Engineering and Optimization Conference (PEOCO), pp. 358–363. IEEE (2014)
5. Hayes, B., Prodanovic, M.: State estimation techniques for electric power distribution systems. In: 2014 European Modelling Symposium (EMS), pp. 303–308. IEEE (2014)
6. Han, X., You, S., Thordarson, F., Tackie, D.V., Østberg, S.M., Pedersen, O.M., Bindner, H.W., Nordentoft, N.C.: Real-time measurements and their effects on state estimation of distribution power system. In: IEEE PES ISGT Europe 2013, pp. 1–5. IEEE (2013)
7. Vasudevan, K., Atla, C.S.R., Balaraman, K.: Improved state estimation by optimal placement of measurement devices in distribution system with DERs. In: 2015 International Conference on Power and Advanced Control Engineering (ICPACE), pp. 253–257. IEEE (2015)
8. Grigoras, G., Ivanov, O., Gavrilas, M.: Customer classification and load profiling using data from smart meters. In: 2014 12th Symposium on Neural Network Applications in Electrical Engineering (NEUREL), pp. 73–78. IEEE (2014)
9. Degefa, M.Z., Millar, R.J., Koivisto, M., Humayun, M., Lehtonen, M.: Load flow analysis framework for active distribution networks based on smart meter reading system (2013)
10. Grigoraş, G., Scarlatache, F.: Use of data from smart meters in optimal operation of distribution systems. In: 2014 International Conference on Optimization of Electrical and Electronic Equipment (OPTIM), pp. 179–184. IEEE (2014)
11. Alimardani, A., Therrien, F., Atanackovic, D., Jatskevich, J., Vaahedi, E.: Distribution system state estimation based on nonsynchronized smart meters. IEEE Trans. Smart Grid **6**(6), 2919–2928 (2015)
12. Alzate, E.B., Bueno, M., Jian, X., Strunz, K.: Distribution system state estimation to support coordinated voltage-control strategies by using smart meters. IEEE Trans. Power Syst. (2019)
13. Alzate, E.B., Li, Q., Xie, J.: A novel central voltage-control strategy for smart LV distribution networks. In: International Workshop on Data Analytics for Renewable Energy Integration, pp. 16–30. Springer (2015). © 2015. License Number: 4195880235037
14. Levenberg, K.: A method for the solution of certain non-linear problems in least squares. Q. Appl. Math. **2**(2), 164–168 (1944)
15. Marquardt, D.W.: An algorithm for least-squares estimation of nonlinear parameters. J. Soc. Ind. Appl. Math. **11**(2), 431–441 (1963)
16. Rao, N.D., Tripathy, S.C.: Power system static state estimation by the Levenberg-Marquardt algorithm. IEEE Trans. Power Appar. Syst. **2**, 695–702 (1980)
17. Pasqualetti, F., Dörfler, F., Bullo, F.: Attack detection and identification in cyber-physical systems–Part I: Models and fundamental limitations. arXiv preprint arXiv:1202.6144 (2012)
18. Pasqualetti, F., Dörfler, F., Bullo, F.: Attack detection and identification in cyber-physical systems. IEEE Trans. Autom. Control **58**(11), 2715–2729 (2013)
19. Liu, Y., Ning, P., Reiter, M.K.: False data injection attacks against state estimation in electric power grids. ACM Trans. Inf. Syst. Secur. (TISSEC) **14**(1), 13 (2011)

Horswill Algorithm Application to Avoid Obstacles

José León León[1,2(✉)], Beatriz Nathalia Serrato Panqueba[2], and José Manuel Wilches[1,2]

[1] Universidad Distrital Francisco José de Caldas, Bogotá, Colombia
jleonl@correo.udistrital.edu.co
[2] Universidad Católica de Colombia, Bogotá, Colombia
{jleon,bnserrato}@ucatolica.edu.co

Abstract. This article shows Horswill [1] simplified algorithm application in a mobile platform to avoid obstacles using the information given by a monocular vision sensor to drive it remotely. It was necessary to find the direct kinematic model of the mobile platform to know the translation and rotation coordinates required to move it between two points. The proposed system is able to avoid obstacles whose area exceeds 0.064 m^2, whose limit is given by the image processing algorithms and illumination effects.

Keywords: Artificial vision · Horswill algorithm · Mobile platform · Remote manipulation

1 Introduction

Collision avoidance is one of the most common problems addressed when referring to mobile platforms. That is why diverse mathematical models have been developed, which based on a defined or no defined environment allow to find a safe path between a start and ending point. However, one of the main drawbacks of using vision sensors deal with the shapes and sizes of the obstacles because the segmentation effects, the morphological treatments and other images processing algorithms generate variations in the test scenario. In mobile robotics, one of the challenges or difficulties with applications of navigation in unknown terrains is collision avoidance with different obstacles. When the robot collides it can get stuck not being able to navigate the terrain afterwards. The avoidance of obstacles in mobile robotics is a field of study due to the diversity of applications that this type of solutions can generate.

An important component in a navigation system is the identification of obstacles along the path of a mobile platform. There are many alternatives to identify obstacles no matter the environment it is in. In [2] an ultrasonic sensor is used in the platform DaNI 2.0, which detects, maps and avoids collisions using potential fields in the environment. In [3] a bidimensional laser is used on a non-manned helicopter to avoid collisions with an algorithm in real time. In [4, 5] a RGB-D camera is used in obstacles avoidance applications. In [6] the collision avoidance approach on non-manned aquatic is based

D. F. Cortes Tobar et al. (Eds.): AETA 2019, LNEE 685, pp. 117–125, 2021.
https://doi.org/10.1007/978-3-030-53021-1_12

on artificial vision and as an alternative in [7] two different obstacle detection systems are used: one of stereoscopic vision and another with an ultrasonic sensor, which work as backup systems in case one of the sensors fails.

When the obstacle is identified, the platform has to perform an action to avoid the collision so it can continue navigating in the terrain. There are algorithms that make these decisions based on the information provided by the sensors. In [8] Branch and Lift algorithms are used, which are optimization algorithms applied to the obstacle avoidance. In [9] an obstacle avoidance system is proposed based on the reaction of microscopic bacteria-powered robots to electric fields. In [10] an obstacle avoidance strategy is proposed in a dynamic environment using a control algorithm based on the collision cone approach, in [11] the choice of turning is made based on a virtual obstacle which is drawn in a depth map and in [12] a double fuzzy controller, one detects the obstacle and the other moves the robot to the goal.

Each mobile platform has particular features according to its design, one of them is the type of locomotion. In [13] a differential drive steering system is used, in [14] a car parking system for Ackerman steering configuration is used and in [15] a kinematic analysis of the omnidirectional platform of the mobile robot KUKA youBot is carried out.

In mobile robotics the navigation of platforms in unknown environments has applications in diverse fields, such as the search and rescue of victims trapped under debris [16–18], cleaning robots [19, 20], cargo transportation [21], exploration missions [22] and wheelchair robots [23].

2 Materials and Methods

To implement an obstacle avoidance algorithm, we had to define first the mobile platform that we would be using during the testing and the validation processes. We used the platform DaNI 2.0 which is a robotics starter kit with an ultrasonic sensor, a Single-Board RIO, two motors with their encoders and a Pitsco TETRIX erector, which allowed us to integrate other components such as sensors and actuators, which helped to the exploration of the environment and facilitated its locomotion.

DaNI 2.0 can be programmed in graphical programming making it easier to develop algorithms to localize and plan routes, which allows the mobile robot move in an unknown environment avoiding obstacle collisions. We used the LabVIEW software to programming and testing the performance of the algorithm in the mobile platform.

To be able to recognize the environment we used a digital camera of 5Mpx resolution, which has a 30 cm visual field of height and 26 cm of width. The digital processing tests were carried out with yellow light of 150 lm and 102 lm white light, that lighting setting was uniform for the living space. However, it was necessary to define a comparative strategy with zoned illumination in the three space channels, R (red), G (green) and B (blue) to establish which channel provided more information during the segmentation stage offering a cleaner map of the environment.

Afterwards, we implemented the control system of simple reaction, which was developed based on the virtual obstacles taking into account their vortices slopes and according to the characteristics the robot knew which space was free managing to avoid the obstacle.

2.1 Kinematic Model of the Testing Mobile Platform

Considering that the robot used during the tests was of differential locomotion, the movement generalized equations can be represented based on [24] given the Eqs. (1), (2) and (3) that correspond to the rotations and translations respectively.

$$\dot{x} = \frac{(w_d + w_i) * r}{2} \cos\varphi \quad (1)$$

$$\dot{y} = \frac{(w_d + w_i) * r}{2} \sin\varphi \quad (2)$$

$$\dot{\varphi} = \frac{(w_d - w_i) * \mathrm{r}}{2l} \quad (3)$$

Where.

w_i = left wheel angular velocity
w_d = right wheel angular velocity
r = spokes of the car wheels
l = separation between the middle axle of the car and the drive wheel
φ = angle of orientation
$\dot{x}$ = linear velocity component x
$\dot{y}$ = linear velocity component y
$\dot{\varphi}$ = angular velocity

To validate that the standard model corresponds to the studied platform, the equations were applied to define the error values of the ideal system compared to the real performance of DaNI 2.0. This process was carried out in 4 tests in which the robots moves one meter to different angular velocities (see Table 1). It is noteworthy to mention that such velocities were the same one in each of the wheels.

Table 1. Tests of wi and wd to 1, 2, 3 y 4 rad/s. Source: authors

Test A	X(m)	Y(m) W = 1 rad/s	Y(m) W = 2 rad/s	Y(m) W = 3 rad/s	Y(m) W = 4 rad/s
1	1	0,0015	0,0008	0,0004	0,0004
2	1	0,005	0,0004	0,0004	0,0013
3	1	0,0095	0,0012	0,0002	0,0002
4	1	0,0035	0,0002	0,0011	0,0014
5	1	0,003	0,001	0,0014	0,0001
Average	1	0,0045	0,00072	0,0007	0,00068

Having in mind that the simulation of the kinematic model compared to the real performance of the platform, it is evidenced from the error (see Table 2) to greater velocity, the error between the model and the robot narrows. The aim when finding such model is to obtain a function able to predict the trajectory of the robot depending on the velocities among its wheels.

Table 2. Error in different velocities. Source: authors

W (rad/s)	m (error)
1	0,0045
2	0,00072
3	0,0007
4	0,00068

2.2 Image Processing Algorithm

For this stage we worked with the LabVIEW software which allowed us to control the mobile platform wirelessly following the TCP/IP protocol. On the other hand, we used a mobile telephone camera to receive the information about the environment, which sent the information by means of IP and we used LabVIEW to capture the image. To do this we used the IP Camera Adapter software which converts the camera of a mobile phone in a peripheral device as any other.

On the other hand, after getting the images, we proceeded to process the image. This consisted on the conversion of the colors RGB to blacks and whites. This binarized image facilitates the morphological and filtering processing of it. To extract the relevant information that would serve as the obstacle avoidance algorithm analysis scenario, we used the Otsu's thresholding method, which analyzes the global entourage of the image and determines the greatest variance between pixels giving as a result the cut value generated to separate binarized data.

2.3 Implementation of Obstacles Avoidance Algorithm

The algorithm we chose to avoid the objects of the environment was Horswill due to its low computational cost and because it offered the possibility to do an in-depth analysis of two dimensional environments. Its implementation is described step by step in Algorithm 1. In principle, this algorithm was proposed to be initially implemented in MATLAB software. However, to facilitate the interaction with the DaNI 2.0 platform we had to migrate the algorithm to LabVIEW looking for the improvement in wireless communication thanks to the use of peripheral devices connected to the TCP/IP network.

Algorithm I: Obstacle avoidance
1: Image acquisition.
2: Brightness and contrast adjust.
3: Setting thresholds and image binarization.
4: 0 to free space and 255 to obstacles.
5: Remove noise.
6: The screen is split into three sections; left, center and right.
7: Obtain coordinates of obstacle.
8: xi, xc, xr = X coordinate center of the sections
9: yi, yc, yr = Y coordinate location pixel close to lower.
10: Draw a virtual obstacle connect [(xi,yi),(xc,yc)], [(xc,yc),(xr,yr)] and [(xi,yi),(xr,yr)].
11: Find the gradient of the vertices.
12: Turning depend on sign and magnitude of the gradient.

To detect obstacles, Horswill proposes that those elements represented by pixels of the images in the lower part of these, is because the obstacle are close to the robot, likewise, which appear on the upper part, would be farther. To do this, each image taken by the camera which was set on the mobile platform will become a depth map (see Fig. 1).

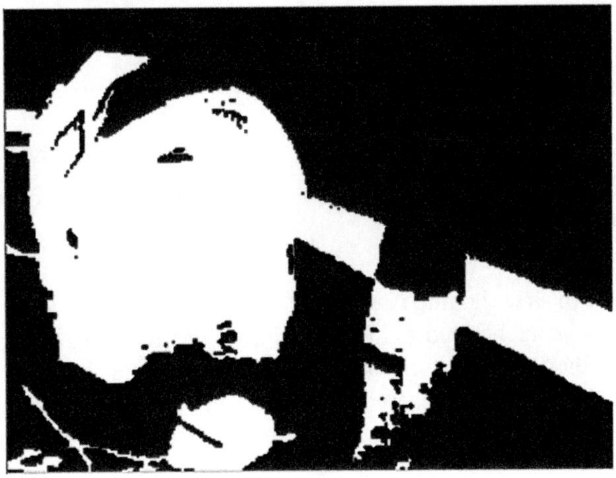

Fig. 1. Depth map. Source: authors.

Afterwards, the algorithm find the empty spaces so the robot can decide where to turn, in [11] proposes to segment the image of the depth map into three screens so as to find the coordinates of the pixels of the nearest obstacle to the robot and in this way build a virtual obstacle that will allow the robot to find the free space to move.

3 Analysis of Results

The tests were carried out in different scenarios with changing conditions. The first one was in a closed space whose ground had black and white spots, which confused the algorithm because the segmentation process did not generate defined zones as obstacles properly. Afterwards, the robotic platform was taken to a testing environment with different lighting conditions and a more color uniformity surface. On that opportunity, it was possible to demonstrate that the algorithm worked accurately from midday until late afternoon (see Fig. 2). The trajectory is observed in Fig. 2b.

a) Testing environment

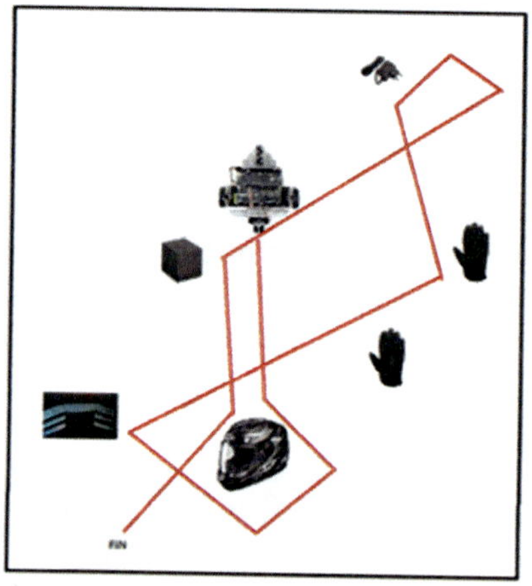

b) DaNI 2.0 path

Fig. 2. Testing environment with changing lighting conditions and uniform surface. Source: authors.

To validate the diverse lighting scenarios, tests were carried out with yellow light of 150 lm; with the aforementioned lighting conditions test the wheels velocity changed when looking for their stability limit point, which was found at 10 rad\s. Additionally, it was verified that the smallest size which was verified that the algorithm manages to detect obstacles with an approximate area of 0.064 m^2 (see Fig. 3). The trajectory is observed in Fig. 3b.

With white light of 102 lm the algorithm did not work properly due to the fact that on some occasions it did not detect the obstacles accurately and it generated a trajectory through them (see Fig. 4). The trajectory is observed in Fig. 4b.

Finally, it is evident that the algorithm is not only affected by the intensity of light but by its color, this is because the wavelength of each one affects the camera sensors in a different way and in turn the behavior of the algorithm. It is recommended to work in environments with lighting over 150 lm.

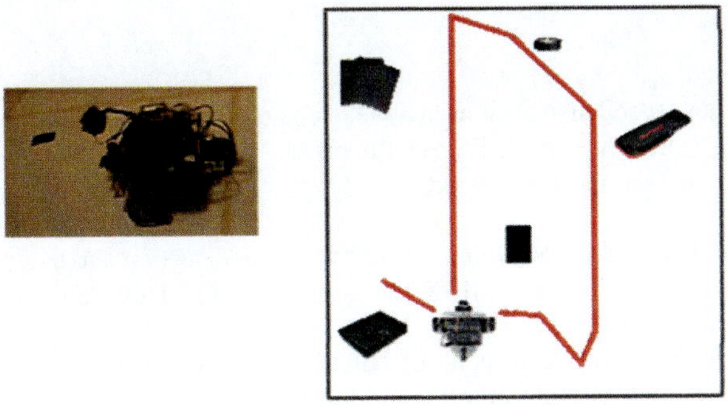

a) Testing environment b) DaNI 2.0 path

Fig. 3. Tests with yellow light. Source: authors.

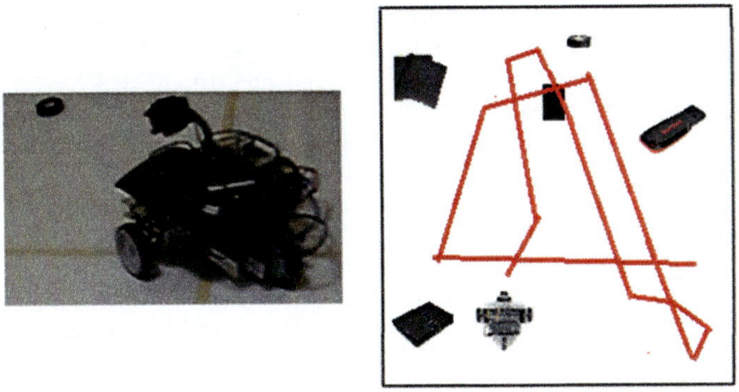

a) Testing environment b) DaNI 2.0 path

Fig. 4. Tests with white light. Source: authors.

4 Conclusions

In the robotics field, there are different possibilities to determine the planning of trajectories to avoid obstacles that might appear in an unknown environment. It is proposed to use the Horswill algorithm due to its easiness of implementation and high reliability in obstacles avoidance. However, among the requirements of the algorithm is to count on a proper testing scenario provided by the segmentation and morphological processing done to extract the information needed.

In the validation model it is evident that to a greater velocity in the wheels, the error is reduced, due to its kinematics in relation with the image processing velocity and the application of the algorithm. It is noteworthy to mention that in this research it is verified that the light intensity and its shades generate very important features when segmenting the scenario, which will be the starting point to identify a near and close obstacle. For future research, it is suggested the implementation of algorithms of adaptive or sectorized segmentation to diminish the negative effects of the lighting on the global environment.

References

1. Horswill, I.: Visual collision avoidance by segmentation. In: Proceedings of the IEEE/RSJ/GI International Conference on Intelligent Robots and Systems 1994, IROS 1994. Advanced Robotic Systems and the Real World, vol. 2, pp. 902–909 (1994). https://doi.org/10.1109/IROS.1994.407486
2. Ojeda, D.L.A., Vera, Y.G., Manzano, M.A.I.: Obstacle detection and avoidance by a mobile robot using probabilistic models. IEEE Lat. Am. Trans. **13**(1), 69–75 (2015). https://doi.org/10.1109/TLA.2015.7040630
3. Zheng, J., Liu, B., Meng, Z., Zhou, Y.: Integrated real time obstacle avoidance algorithm based on fuzzy logic and L1 control algorithm for unmanned helicopter. In: 2018 Chinese Control and Decision Conference (CCDC), pp. 1865–1870 (2018). https://doi.org/10.1109/CCDC.2018.8407430
4. Thapa, V., Capoor, S., Sharma, P., Mondal, A.K.: Obstacle avoidance for mobile robot using RGB-D camera. In: 2017 International Conference on Intelligent Sustainable Systems (ICISS), pp. 1082–1087 (2017). https://doi.org/10.1109/ISS1.2017.8389347
5. Zhang, Y., Wang, G.: An improved RGB-D VFH+; obstacle avoidance algorithm with sensor blindness assumptions. In: 2017 2nd International Conference on Robotics and Automation Engineering (ICRAE), pp. 408–414 (2017). https://doi.org/10.1109/ICRAE.2017.8291420
6. Park, J., Cho, Y., Yoo, B., Kim, J.: Autonomous collision avoidance for unmanned surface ships using onboard monocular vision. In: OCEANS 2015 - MTS/IEEE Washington, pp. 1–6 (2015). https://doi.org/10.23919/OCEANS.2015.7404502
7. Yu, Y., Tingting, W., Long, C., Weiwei, Z.: Stereo vision based obstacle avoidance strategy for quadcopter UAV. In: 2018 Chinese Control and Decision Conference (CCDC), pp. 490–494 (2018). https://doi.org/10.1109/CCDC.2018.8407182
8. Feng, X., Villanueva, M.E., Chachuat, B., Houska, B.: Branch-and-lift algorithm for obstacle avoidance control. In: 2017 IEEE 56th Annual Conference on Decision and Control (CDC), pp. 745–750 (2017). https://doi.org/10.1109/CDC.2017.8263749
9. Kim, H., Kim, M.J.: Electric field control of bacteria-powered microrobots using a static obstacle avoidance algorithm. IEEE Trans. Robot. **32**(1), 125–137 (2016). https://doi.org/10.1109/TRO.2015.2504370
10. Mallik, G.R., Sinha, A.: A novel obstacle avoidance control algorithm in a dynamic environment. In: 2013 IEEE Symposium on Computational Intelligence for Security and Defense Applications (CISDA), pp. 57–63 (2013). https://doi.org/10.1109/CISDA.2013.6595428
11. Vision based collision avoidance by plotting a virtual obstacle on depth map. In: IEEE Conference Publication. https://ieeexplore.ieee.org/document/5512394/. Accessed 12 Jul 2018
12. Cui, S., Su, X., Zhao, L., Bing, Z., Yang, G.: Study on ultrasonic obstacle avoidance of mobile robot based on fuzzy controller. In: 2010 International Conference on Computer Application and System Modeling (ICCASM 2010), vol. 4, pp. V4-233–V4-237 (2010). https://doi.org/10.1109/ICCASM.2010.5620069
13. Tang, J., Li, J., Liu, Y.: Differential drive steering research on multi-axle in-wheel motor driving vehicle. In: 2014 IEEE Conference and Expo Transportation Electrification Asia-Pacific (ITEC Asia-Pacific), pp. 1–6 (2016). https://doi.org/10.1109/ITEC-AP.2014.6941234
14. Gupta, A., Divekar, R., Agrawal, M.: Autonomous parallel parking system for Ackerman steering four wheelers. In: 2010 IEEE International Conference on Computational Intelligence and Computing Research, pp. 1–6 (2010). https://doi.org/10.1109/ICCIC.2010.5705869
15. Adamov, B.I.: Influence of mecanum wheels construction on accuracy of the omnidirectional platform navigation (on exanple of KUKA youBot robot). In: 2018 25th Saint Petersburg International Conference on Integrated Navigation Systems (ICINS), pp. 1–4 (2018). https://doi.org/10.23919/ICINS.2018.8405889

16. Doroodgar, B., Liu, Y., Nejat, G.: A learning-based semi-autonomous controller for robotic exploration of unknown disaster scenes while searching for victims. IEEE Trans. Cybern. **44**(12), 2719–2732 (2014). https://doi.org/10.1109/TCYB.2014.2314294
17. Liu, Y., Nejat, G., Doroodgar, B.: Learning based semi-autonomous control for robots in urban search and rescue. In: 2012 IEEE International Symposium on Safety, Security, and Rescue Robotics (SSRR), pp. 1–6 (2012). https://doi.org/10.1109/SSRR.2012.6523902
18. Lima, P.U.: Search and rescue robots: the civil protection teams of the future. In: 2012 Third International Conference on Emerging Security Technologies, pp. 12–19 (2012). https://doi.org/10.1109/EST.2012.40
19. Kleiner, A., Baravalle, R., Kolling, A., Pilotti, P., Munich, M.: A solution to room-by-room coverage for autonomous cleaning robots. In: 2017 IEEE/RSJ International Conference on Intelligent Robots and Systems (IROS), pp. 5346–5352 (2017). https://doi.org/10.1109/IROS.2017.8206429
20. Huang, H.Y., Hu, Y.W., Lu, M.F., Chen, S.S., Jeng, J.T., Chen, W.P.: An obstacle avoidance of large-scale indoor tricycle drive cleaning robot using laser scanner. In: 2017 Joint 17th World Congress of International Fuzzy Systems Association and 9th International Conference on Soft Computing and Intelligent Systems (IFSA-SCIS), pp. 1–6 (2017). https://doi.org/10.1109/IFSA-SCIS.2017.8023244
21. Cabral, K.M., dos Santos, S.R.B., Givigi, S.N., Nascimento, C.L.: Design of model predictive control via learning automata for a single UAV load transportation. In: 2017 Annual IEEE International Systems Conference (SysCon), pp. 1–7 (2017). https://doi.org/10.1109/SYSCON.2017.7934800
22. Lehner, P., et al.: Mobile manipulation for planetary exploration. In: 2018 IEEE Aerospace Conference, pp. 1–11 (2018). https://doi.org/10.1109/AERO.2018.8396726
23. Wu, B.F., Jen, C.L., Li, W.F., Tsou, T.Y., Tseng, P.Y., Hsiao, K.T.: RGB-D sensor based SLAM and human tracking with Bayesian framework for wheelchair robots. In: 2013 International Conference on Advanced Robotics and Intelligent Systems, pp. 110–115 (2013). https://doi.org/10.1109/ARIS.2013.6573544
24. Camarena, J.G.: Análisis Cinemático, Dinámico y Control en Tiempo Real de un Vehículo Guiado Automáticamente. M.S. thesis, cenidet, México, Cuernavaca (2009)

A Study on the AC Power Control Using PIC Microcontroller

Sang Won Ji[1] and Seung Hun Han[2](✉)

[1] Department of Mechanical System Engineering, Pukyong National University, Busan 48513, Republic of Korea
shhan@gnu.ac.kr
[2] Department of Mechanical System Engineering, Gyeongsang National University, Tongyeong 53064, Republic of Korea

Abstract. The PIC microcontroller is responsible for the control of the object system. It takes input/output of data, selects and controls of peripheral IC, communication control of data and executes the program. It has flash program memory and EEPROM, so it can modify and extend the program. In addition, it has function of A/D conversion that coverts direct analog value into digital value and calculate data. Therefore, the PIC microprocessor chip having such functions is possible to simplify the existing complicated peripheral circuits to design the entire system easily. In this study, we design the AC power controller using PIC microcontroller and traic. Through the change of A/D value input directly to the micropreces-sor, the power controller was verified by experiment to control brightness of the incandescent lamp. As a result, it is confirmed that the AC power controller is well-controlled by phase control employing PIC microcontroller and traic.

Keywords: PIC microcontroller · AC power control system · Incandescent lamp · Phase control

1 Introduction

The PIC microprocessor is a one-chip microcontroller of Microchip Technology Inc. widely used for general purpose in the industries. It is composed of an arithmetic logic unit, a register program counter, a command decoder, and a control circuit, and its control programs are based on C language [1].

In general, the PIC microcontroller plays the role of system controller. It is responsible for data input and output, selection and control of peripheral IC, and data communication control, and executes programs. Furthermore, it has a flash program memory and EEPROM, which enables program modification and expansion. Furthermore, since it has A/D conversion function, it can directly convert analog values to digital values for operation and comparison [2–4]. Therefore, the existing complex peripheral circuits can be simplified by using the PIC microprocessor one-chip, and it enables simple design of the total systems.

In this study, an AC power control system was designed through PIC microcontroller and the phase control of traic. Furthermore, the AC power control system was verified by

D. F. Cortes Tobar et al. (Eds.): AETA 2019, LNEE 685, pp. 126–135, 2021.
https://doi.org/10.1007/978-3-030-53021-1_13

conducting a simulation experiment that controls the brightness of an incandescent lamp of the load side by directly inputting the A/D conversion value in the microprocessor through a variable resistor.

2 System Configuration and Controller Design

Figure 1 shows the AC power control system used in this study. It is a testing device for adjusting the brightness of an incandescent lamp through a traic by inputting an A/D conversion value through a variable resistor. As shown in Fig. 1, a zero-crossing detection circuit was designed. The microprocessor generates an interrupt whenever the AC waveform passes through the zero point in the circuit. When an interrupt occurs, the AC power was controlled by arbitrarily adjusting the gate current signal applied to the traic by using the SFR (special function register) Timer 1 of the microprocessor PIC18F4520.

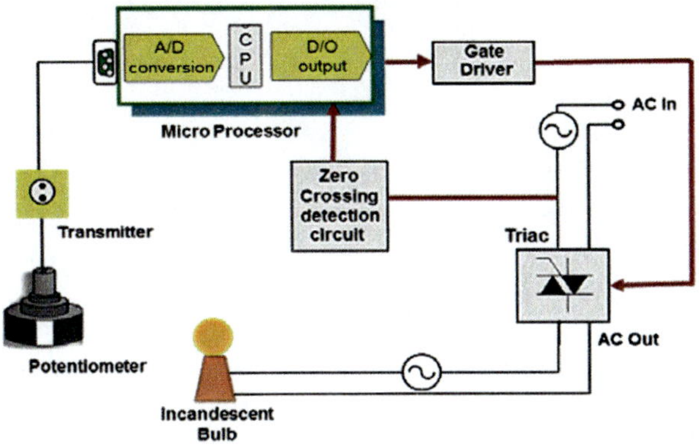

Fig. 1. Schematic diagram of the AC power control system

2.1 Zero-Crossing Detection Circuit

Figure 2 shows the total zero-crossing detection circuit. It plays the role of a reference point for adjusting the gate current signal in the traic. Since the base current of the transistor does not flow near 0 V of AC 220 V, + pulse is output. The waveform of the output was 500 μs. To detect the accurate zero-crossing point, it was designed to output pulses with a width of 5 μs by using the 74LS123 device.

Figure 3 shows the (a), (b), (c), and (d) output waveforms of the zero-crossing detection circuit in Fig. 2 In the zero-crossing circuit, (a) is the AC input waveform and (b) is the waveform of a full wave rectification circuit. When this waveform enters the base of the transistor, the output waveform of (c) is a 500 μs trapezoidal pulse waveform.

The 74LS14 device was used to rectify this waveform, and the 74LS123 device was connected to find the accurate zero point. Furthermore, the pulse width was reduced by adjusting the time constant value and it was designed to output approximately 5 μs waveform as shown in (d). As the gate turn-on time of a general traic is 2 μs, a double time was given to prevent the turn-on failure of traic [5].

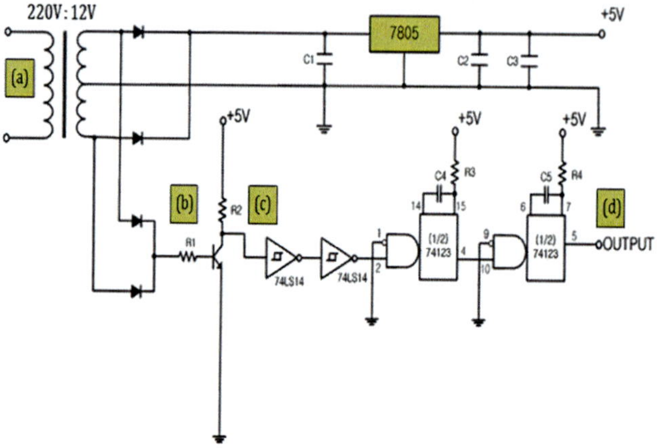

Fig. 2. Zero-crossing detection circuit

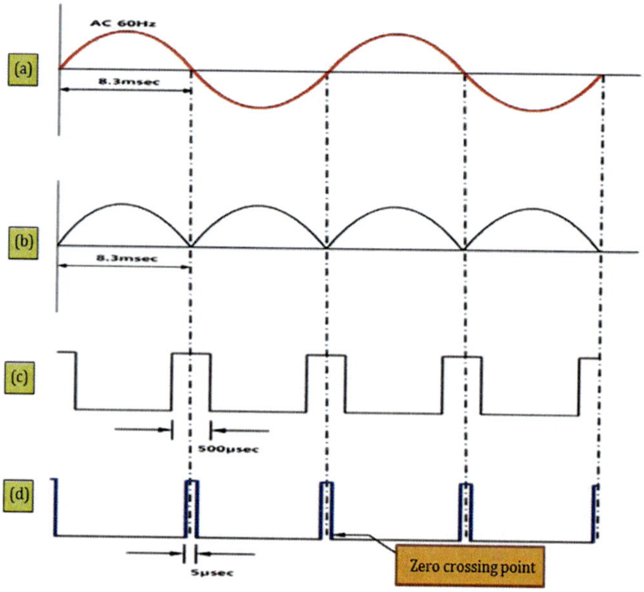

Fig. 3. Waveform of Zero-crossing detection circuit

2.2 Power Control by Traic

Figure 4 shows a power control circuit by traic. For traic, a power traic (BTA41-600B) was used, and a photo coupler (TLP 560J) device was used to isolate the ground on the control and power sides. TLP 560J is composed of a photo diode and optical traic, and the ground is separated by light. The Snubber circuit was designed to protect the traic when a surge occurs in both ends of the traic [6].

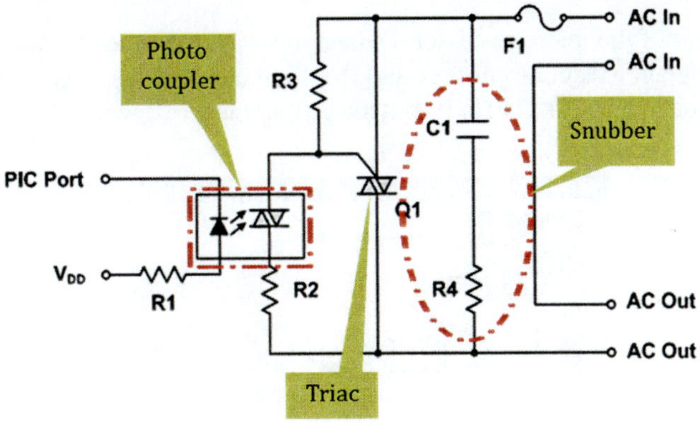

Fig. 4. Circuit of traic switching

The zero point of the AC power was detected for the switching of traic. This signal is input to the external input interrupt of the microprocessor and the switching time of traic is calculated internally. For this, Timer 1 overflow interrupt was used. The internal clock was used as source, and the prescaler was used at 1:3. Since the zero-crossing clock is input every half-period, it becomes 120 Hz [3, 7]. Furthermore, the main clock used in the microprocessor is 20 MHz, and inside the processor, the input clock is synchronized at 1/4. When the number of overflows per AC power half-cycle is calculated based on this, it is expressed as Eq. (1) [5].

$$\frac{20\ \text{MHz}}{120\,\text{Hz} \times 8 \times 4} = 5208 \tag{1}$$

Figure 5 shows the Timer 1 interrupt flowchart.

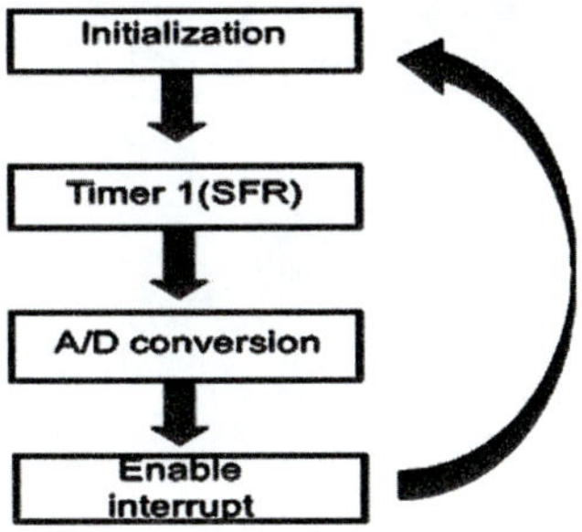

Fig. 5. Flow chart of Timer 1 interrupt

3 Experiment Result and Discussion

In this study, the CCS-C compiler was used, which is a C compiler for PIC produced by CCS (Customs Computer Service) in the U.S. Figure 6 shows the control program of the ICD unit of the microprocessor. Debugging was performed by writing a source file in the program execution process and the computer and the microprocessor were connected through the ICD-U40 unit, and the coding file was downloaded and executed.

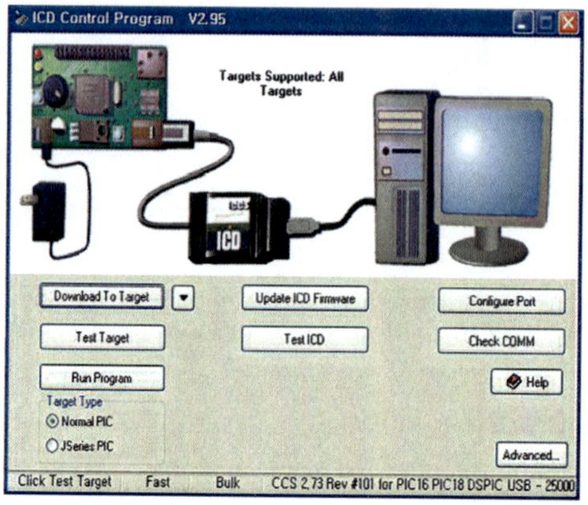

Fig. 6. Image of ICD control program

Figure 7 shows the AC control system fabricated in this study. The switching on signal of the traic calculated through Timer1 of the microprocessor PIC 18F4520 is input to the gate of the optically isolated traic through the input/output port, and the power is controlled according to the conduction time input to the gate.

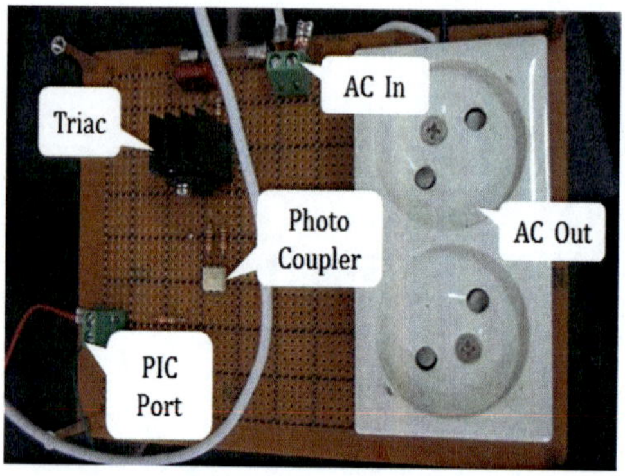

Fig. 7. Photo of AC power controller

In this experiment, to measure voltage waveforms of the positive (+) and negative (−) sides, 100[V] was applied as input instead of 220[V] to the AC power controller.

For the power control data value, 10 bit values from 0 to 1023 were input by adjusting the variable resistor. When the power control data value is 0, the incandescent lamp is turned off; when it is 1023, the power becomes the maximum and the lamp becomes the brightest. In other words, the power is matched to the overflow count from 0 to 5208 per half-cycle of the AC power, and this corresponds to the gate switching time of the traic.

Figures 8, 9, 10, 11 and 12 are graphs showing the output voltage according to the phase control of the traic, representing the measured waveforms at the phase angles of 177.2°, 135°, 90°, 45°, and 2.8°. For the 2.8° phase angle section at the beginning and end, the value of Timer1 was set so that it would not work for the definite turn-on and turn-off of the triac [5]. In Figs. 8, 9, 10, 11 and 12 (a), Ch1 is the voltage waveform of the traic output terminal, and Ch2 is the voltage waveform input to the load side. Time/Div is 4.0 [ms], and Volts/Div is 100 [V]. In Figs. 8, 9, 10, 11 and 12 (b), the waveform of Ch1 is the traic gate signal, and Ch2 is the voltage waveform input to the load side. Time/Div is 4.0 [ms], and Volts/Div is 5 [V] and 100 [V].

Figure 8 (a) shows the waveform obtained when the power control phase angle is 177.2°. It can be seen that the power is very small because the traic was conducted at the end of the AC power. Figures 9, 10, 11 (a) show the waveforms obtained when the power control phase angle was 135°, 90°, and 45°. It can be seen that the power increased as the conduction time of the traic became faster.

Figure 12 (a) shows the waveform obtained when the power control phase angle is 2.8°. It can be seen that as soon as the AC power passed the zero point, the traic was conducted and the power increased to the maximum.

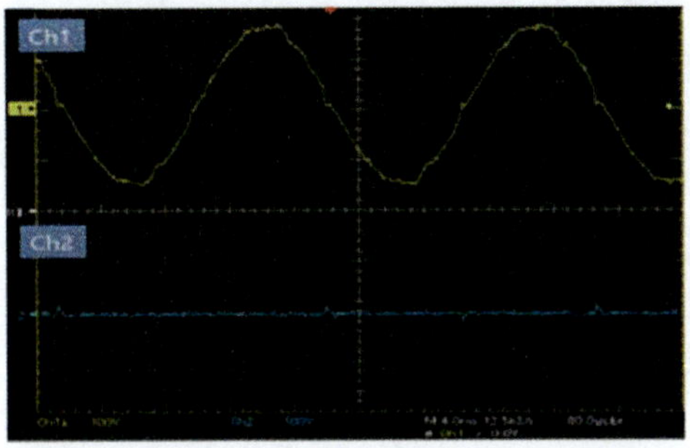

(a) Traic output

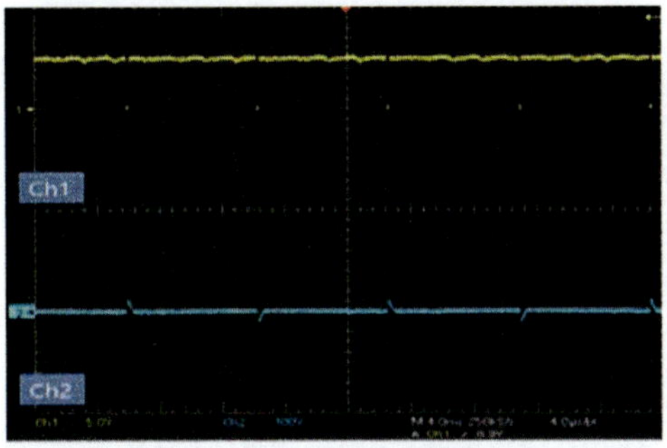

(b) Traic gate

Fig. 8. Voltage waveforms of traic output & gate and load at 177.2°

As can be seen in Fig. 9, 10, 11 and 12 (b), the power control test according to the conduction signal of traic confirmed that the power was controlled well according to the preset power data.

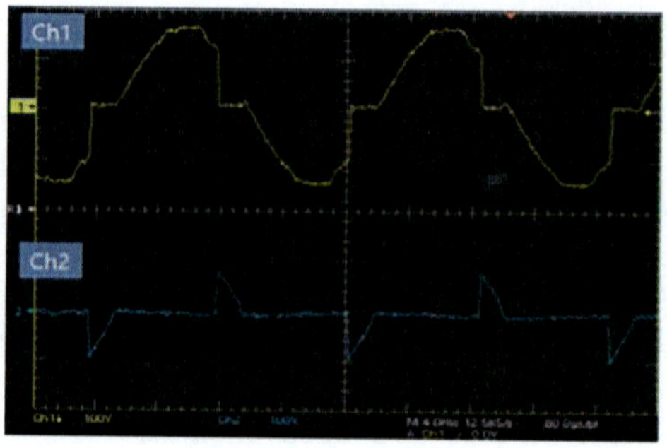

(a) Traic output

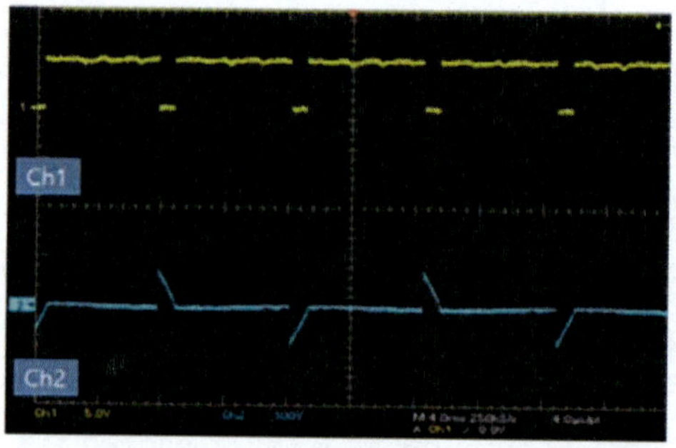

(b) Traic gate

Fig. 9. Voltage waveforms of traic output & gate and load at 135°

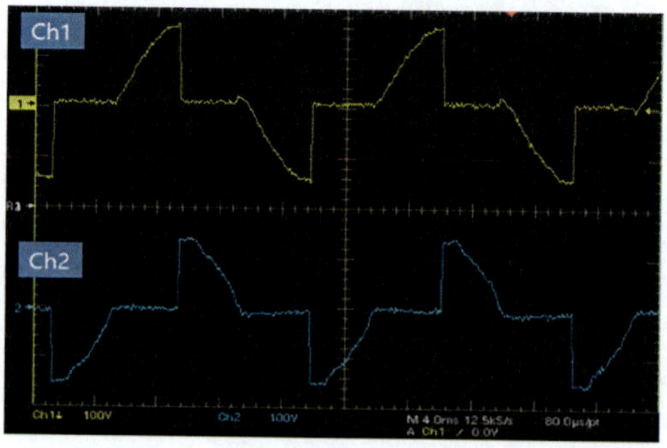

(a) Traic output

Fig. 10. Voltage waveforms of traic output & gate and load at 90°

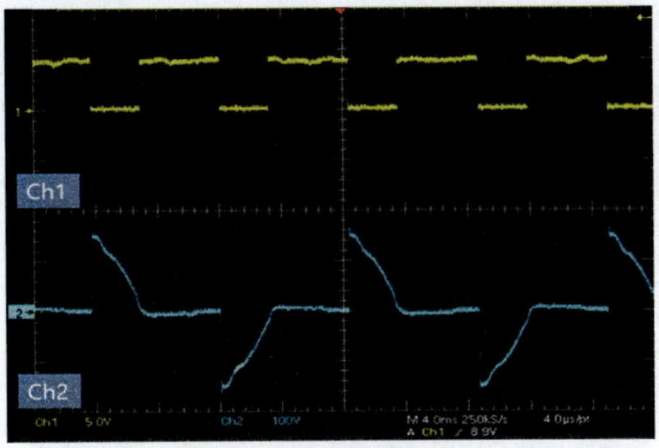

(b) Traic gate

Fig. 10. (*continued*)

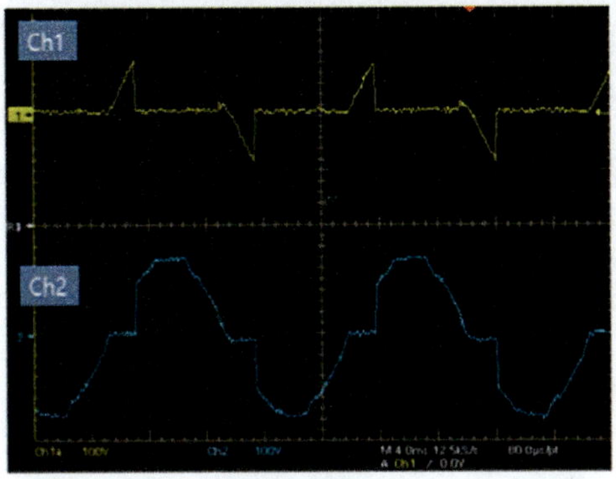

(a) Traic output

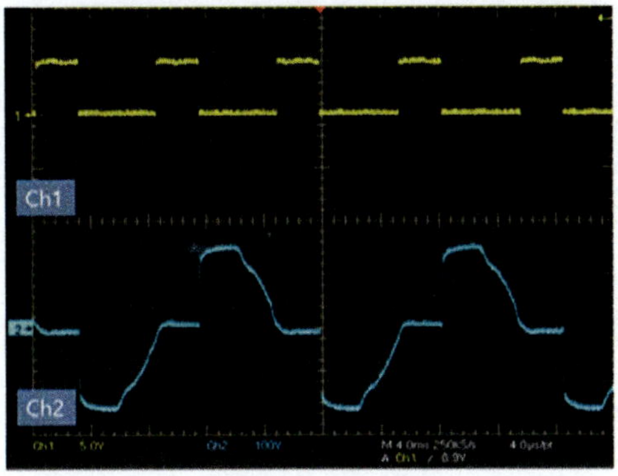

(b) Traic gate

Fig. 11. Voltage waveforms of traic output & gate and load at 45°

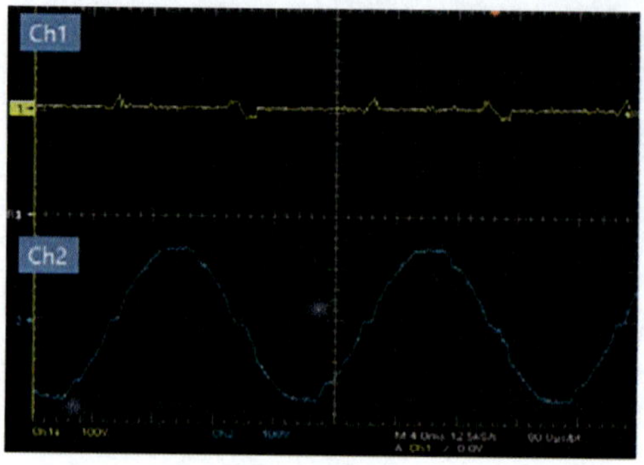

(a) Traic output

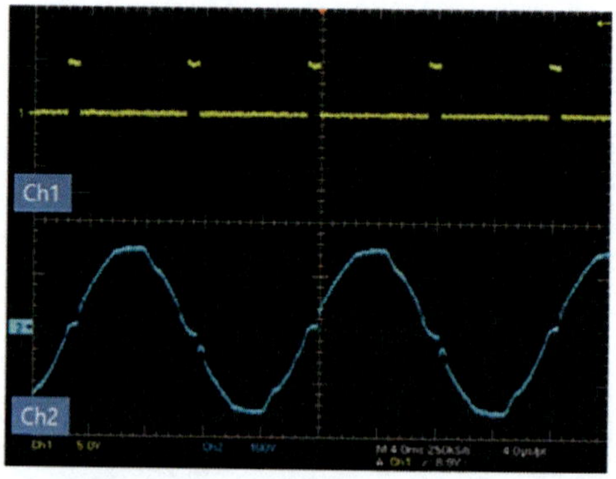

(b) Traic gate

Fig. 12. Voltage waveforms of traic output & gate and load at 2.8°

4 Conclusions

In this study, an AC power control system using PIC microcontroller was designed and the AC power was controlled by adjusting the gate current signals applied to the traic. The achievements of this study are as follows.

1. A zero-crossing detection circuit that plays the role of a reference point for the gate current signal adjustment of the traic was designed. Furthermore, the 74LS123 device was used for accurate turn-on signal of the traic, and it was designed to output pulses with a width of 5 μs of the zero-crossing point.
2. It was verified that the switching on signal of the traic calculated through the Timer1 of the microprocessor PIC 18F4520 was input to the gate of the optically isolated traic through the input/output port, and the power was controlled well according to the gate conduction time.
3. The results of this study are expected to be applicable to the automatic voltage controller and battery charger of the generator for ships through the phase control of the traic.

Acknowledgement. This study was funded by the Gyeongsang National University Development Fund Foundation in 2019.

References

1. Yang, Y.S., Phuc, D.P., Ahn, B.W., Bae, C.O.: Design of traffic light controller using PIC microcontroller. In: Proceeding of the Korean Society of Marine Environment & Safety Conference, pp. 97–99 (2012)
2. Choi, K.H., Lee, Y.C., Kwon, T.K.: Development of keypad test system using PIC contorller. J. Korean Soc. Precis. Eng. **21**(10), 94–101 (2004)
3. Bae, C.O., Park, Y.S.: A study on the luminosity control of bulbs by using PIC. J. Korean Soc. Marine Environ. Saf. **13**(3), 235–240 (2007)
4. Microchip technology inc. PIC18F2420/2520/4420/4520/ Data Sheet, pp. 1–390 (2004)
5. Han, S.H., Bae, C.O., Ahn, B.W.: A study on the temperature control of insulated open-end water vessel. J. Korean Soc. Marine Environ. Saf. **36**(8), 1097–1103 (2012)
6. Jung, J.H., Song, W.H., Nho, E.C., Kim, I.D., Kim, H.G., Chun, T.W., Yoo, D.W.: Overvoltage protection snubber for diode-clamped 3-level IGBT inverter. Proc. Korean Inst. Power Electron. 277–279 (2009)
7. Barnetted, R.C., Cox, S., Ocull, L.: Embedded C programing and the Micro PIC. Thomson Delmar Learning, pp. 11–102 (2003)

Finding Elements with a Continuous Point to Point Spectrum Analysis: A New Technique for Finding Elements in a Vibration Propagation System Using LabView

Herberth Gracia-León(✉) and Leonardo Rodríguez-Urrego

Universidad EAN, Bogotá, Colombia
{hgracia,lrodriguezu}@universidadean.edu.co

Abstract. This paper purpose the design and development of a testing bench to analyze the vibration spectrum of some static elements. In the paper, we are going to describe the application of dynamic vibration analysis into static tests analysis of elements as concrete, glass or wood, the design and methodology of the testing bench and the software development to help the point to point continuous analysis (CP2P) used to evaluate de frequencies in the vibration wave. Finally it is exposed the tests done with those elements, the results and the future research in topics as data size or time to processing. This paper open the vibration analysis research to others application fields different to rotary machinery, in this specific case to a new type of scanner for the products verification.

Keywords: Vibration · Spectrum · Comparison · Point-to-point

1 Introduction

Vibration is considered a physical phenomenon with mechanical characteristics [1], and is composed by the oscillation of an object with respect to a point of equilibrium, this being the position that the object adopts when the forces that intervene in the oscillation return to rest. This phenomenon, for its mathematical analysis, it is built as a sinusoidal wave and constituted by characters such as frequency (f), amplitude (A), period (T) and displacement, all of these under the time domain [2].

The vibratory effect has untapped its industrial potential especially in the canons of predictive maintenance, having active participation in dynamic systems [3] permanently analyzing "the way of vibrating" that systems have, deducting an effective database and analysis of trends with which possible to project, in an optimal way, the preventive maintenance before the damages reach a critical state. The analysis is based on changing the domain of the vibratory wave of time to its amplitude or frequency, this is possible mathematically thanks to the Fourier theory [4]. Which in turn has its pillars in Parseval's Theorem, then, delimiting the wave in The domain of its frequency identifies certain harmonics that determine an incidence within the system [5].

D. F. Cortes Tobar et al. (Eds.): AETA 2019, LNEE 685, pp. 136–144, 2021.
https://doi.org/10.1007/978-3-030-53021-1_14

The analysis of vibrations is study in a great variety of equipment, allowing the monitoring of machines through the integration of CMS and SCADA systems [6]. The applications of monitoring systems can be used in steam, gas, hydraulic and wind turbines, reciprocating compressors, electric motors, gearboxes, fans, pumps, cooling towers, among others; acquiring data such as wear, overloads, imbalances, misalignments, failures, lubrication problems, etc. [7].

Academically it is evident the current interest in the analysis of vibrations in international dynamic systems, finding mechanical failures within bearings from variations in the frequency of vibration [8], to the point of deciphering the affected bearing within the system to then perform the relevant maintenance [9]. Also the response of flexibility in structures controlled with piezoelectric, where thanks to the vibration it is possible to analyze the mechanical differences within the structure for decision making [10]. Nationally, the analysis of vibrations has achieved the mathematical breakdown in the identification of the places where a bearing can be damaged, thanks to the design of a test prototype and the capture of vibrating signals by means of this [11].

It is important to emphasize the developments in programmatic matters regarding the analysis of vibrations, in the structural field. This has also shown progress as shown by models for low impact analysis [12], also analysis in electromagnetic wave interactions [13], and different achievements in acoustics [14]. All the advances in the field were develop from the new technologies and show the scope that the analysis of vibrations can have within different scenarios, ending as products in applications for different sectors of the global industry.

Achieving the interpretation of vibrations comprises challenges currently covered by certain systems and measuring equipment, the acquisition of vibratory signals in one of them, this is possible thanks to the use of transducers recommended for acceleration [5], the vast majority are used piezoelectric sensors for the interpretation of these signals. The handling, visualization and supervision of the signals produced by the accelerometer must be part of a SCADA software [15], in which the signal generate the information for the required analysis. for this purpose, LabView is one of the tools chosen for the management of information, thanks to its adaptability with different interfaces that enable the built of an entire sensory system that accesses the interpretation of the vibratory wave [7, 9, 10].

In this article, we intend to design a mathematical method in LabView that allows the analysis and comparison of vibratory spectra from a point-to-point evaluation, thus defining an index of relationship or similarity that allows identifying oscillations of frequency and similar amplitude. The mathematical method it is going to describe the elements that are components of the test bench, either if they are permanent or if we want to analyze the system with them. Section 2 describes the test bench that will lead to the identification of the spectra, then, in the third, the processing of the signals and its analysis, in the fourth chapter the application of continuous analysis point to point in spectra of vibration and finally the conclusions regarding the work done are given.

The test are going to be with element as concrete, glass and wood and pretend to identify those into the tests with an analysis point to point in a spectrum comparison.

2 Test Bench Design

Before developing the mathematical algorithm in Labview software [16], it is necessary to design a test bench in which the acquisition of spectra is possible. The experiment is not based on rotary elements so the device for vibration acquisition will not have as motor components or bearings, since these elements generate specific vibrations within the spectrum that do not intend to be analyzed within the tests [17].

The design will been based on two resonant boxes, an external one to control the propagation of the vibration and an internal one with the task of absorption, that will be responsible for receiving the impact and generate movement on itself, for this will be anchored to eight axes at the top and bottom of it. That will be responsible for acting as zero coordinate within the vibration.

It is important to emphasize that the caliper of the sheet will be 5 mm, this, with the intention of being able to drill and incorporate the stem of a sensor in charge of collecting the qualitative characterization of the vibration. Below is a 3D view of the basic prototype.

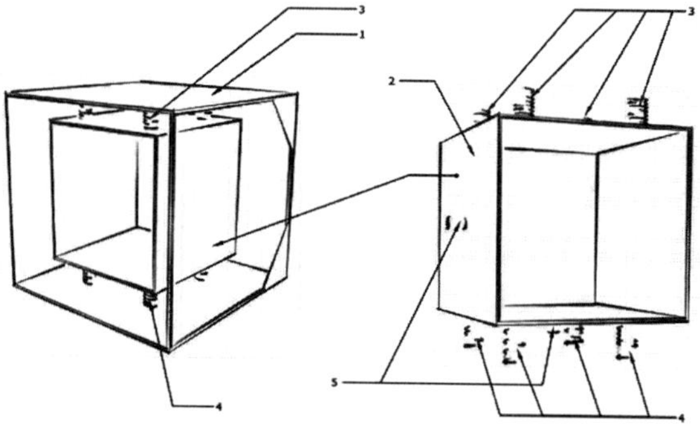

Fig. 1. Test bench

The Fig. 1 shows the test bench with its parts listed below.

1. External Box
2. Internal Box
3. Top Springs
4. Bottom Springs
5. Accelerometers

The methodology of the process begins with a blow on the test sheet that generates a vibration on the material; this blow will be execute with a hammer that includes a piezoelectric that measures the force with which the impact is made. Finally, the wave generated by the vibration of the material will be acquire by means of accelerometers, thus enabling its characterization and treatment [18].

The methodology for the extraction process of a spectrum is carried out according to: locate the product or item to which you want to extract the spectrum within the vibrating box, then start of spectrum saving software and impact the vibrating box with a percussion system, capture the vibration signal in the software using the sensors of vibration or accelerometers. Repeat the process two more times in order to average the results and complete the database of each spectrum, then verify the n signatures and their similarities in the quantitative characterization in frequency to approve the saved for finally save the resulting signature of the product or element.

3 Vibration Processing

The algorithm for the processing of vibration signals has been develop in LabView software, a specialized program for the development of engineering projects, visualization and analysis of information from sensors [19].

To program the wave processing first, were make a simulation [20] and processing the signal with a Fourier transformation [21]. The simulation of signals was done characterizing both, the frequency and the amplitude, the objective was the identification of sinusoidal waves within the program, its graphic and numerical visualization.

Operations were also develop with signals in order to treat them and you learn ways to interact with them within the platform. For the acquisition, calibration and visualization of the signals we started the program in LabView with the DAQ Assistant, this allows the connectivity of the cDAQ (Data Acquisition Device) with the PC. Inside the DAQ Assistant, we configure everything related to the signals of vibration acquired by the device, number of samples, sampling flow, units of the signal and certain pre-filtered. To make the connectivity between the software and our test bench we use National Instruments equipment [22].

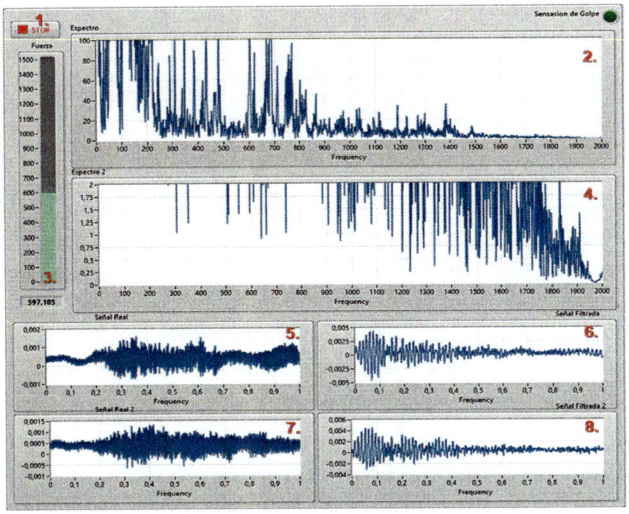

Fig. 2. Visualization of the spectrum

In Fig. 2, the visualization platform is shown and it is the result of the programming of the wave and spectra acquisition. The platform is compose by the next items.

1. Start and Stop button test
2. Axial Spectrum
3. Impact Force
4. Radial Spectrum
5. Axial original wave
6. Axial filtered wave
7. Radial original wave
8. Radial filtered wave

In Fig. 2, show the software take two signals (axial, radial) from the same test divided into axial and radial, these signals are first filter to ensure that the amplitudes of the main frequencies are actually those indicated for the analysis. The diagram made in the Lab View software for the processing of these signals is show below.

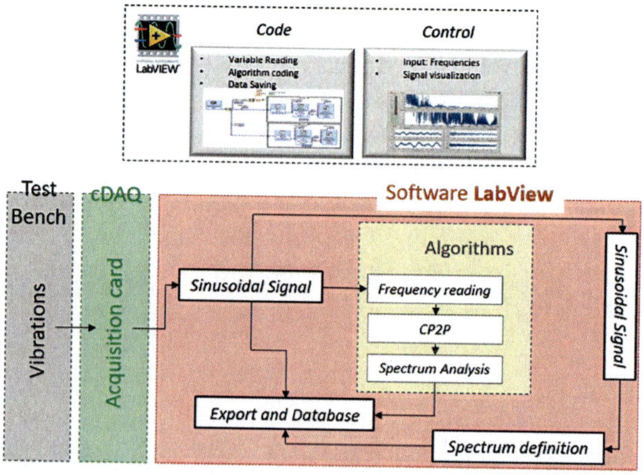

Fig. 3. Information schema

In previous figure, the filters used to clean the signal is shown; in addition, the internal preprocessing is shown to use the Perceval theorem and to command the sinusoidal vibration signals in a spectrum. This will be represent under the frequency domain in order to identify the specific points in which the amplitude is relevant for the identification of common vibration marks [23]. Figure 3 explain the data flow into the comparison and analysis software, it also shown the inputs and outputs of the code sections to help the understanding of the processes and methodology of the test.

4 Continuous Point-to-Point Spectrum Analysis

It's important to say that now at days one of the algorithms used to find an optimal alignment between two given sequences, is Dynamic Time Wrapping (DTW), the problem to work with it in the experiment its that we want to correlate points of frequency to identify that the element of our first test it now in de n test. The DTW work in the domain of time and we were working on de domain of frequency [24].

To dynamic systems the differences of the propose make that the methodology that is used to identify elements into the system don't work in our experiment. For example, in dynamic systems in a rotor unbalance you have to pay attention in the harmonics, to describe a fault or take some conclusion in the system [25]. In our case we have to filter the line and the harmonic frequencies to make a clean zero signal that at the time of excite the system describe in the spectrum the elements that are vibrating in it.

Using the lab view software, we achieved mathematical analysis to make comparisons between spectra [26]. This analysis is called continuous point-to-point analysis, since each basic point of frequency of the spectrum one (Sp1) is evaluated against the basic point of the same number of the spectrum two (Sp2). The comparison is been made by evaluating the At this time the prototype is being in the patent filing process in Colombia and therefore the complete mechanical scheme can not be published amplitude of the two points resulting in the deviation between them to result in a rating with respect to this value.

$$x = f_n(Sp_1) - f_n(Sp_2) \tag{1}$$

$$L_S(R_n) < x < L_i(R_n) \rightarrow x \in R_n \tag{2}$$

Equation 1 shows the difference between a certain frequency (f_n) for the spectra to be evaluated (Sp_n), then in Eq. 2 the resulting value (x) is evaluated within the upper (L_s) and lower (L_i) limits of a defined range (R_n) to identify which of the ranges it belongs to.

In Eq. 3, we see that then, according to the defined range, a qualification (Q) will be given to that comparison between the amplitudes of that frequency (f_n)

$$\boldsymbol{If x \in R_n \rightarrow Q\left(fn_{(Sp_{(1,2)})}\right)} = \omega \tag{3}$$

Finally, within the spectral analysis performed, 12,999 basis points in frequency are measured; therefore, after applying Eqs. 1, 2 and 3, to perform the comparison and evaluation between spectra, we apply a measure of central tendency between the qualifications that has had each deviation within all frequencies to obtain the percentage of similarity between samples. It is worth noting that the ratings given for Eq. 5 regardless of the number of stipulated ranges must be according to Eq. 4.

$$0 \leq \omega \leq 100 \tag{4}$$

$$Correlationindex = \sum\nolimits_{f_n=0}^{Maxf_n} \omega \Big/ Max\ f_n \tag{5}$$

The result is an index of correlation between the points of a spectrum based on a continuous point-to-point analysis; the algorithm was make in Lab View. The problem with the spectra is that they cannot be analyzed as graphs or lines since each frequency point can mean a different element within the vibratory movement therefore the frequency behavior is vertical, in the amplitude of each point, not horizontal to compare de value of the amplitude in a frequency set to be compare. With the described method, we were able to make horizontal, from the unit qualifications, its analysis and application of measures of central tendency to enable its comparison. As mentioned above, it is interesting to apply novel techniques such as DTW, mathematical regressions and speculative analysis within the proposed methodology associated with the test bench of the experiment. This will clearly develop tests on different elements in order to identify within the resulting spectra.

It is important to indicate that our method has a problem regarding the time it takes the test, when using programming within the labview program the algorithm processes are sequential. Therefore, the process becomes slow when it is composed of as set of data of large quantities, a total test with 12799 frequency points takes approximately 25 min, which gives us several opportunities for improvement in future experiments. Interesting techniques can be applied in large quantities of data such as the PCA Principal Component Analysis or techniques as Genetic Algorithms and Fuzzy Logic, to compare the data analysis and the time processing [27].

5 Conclusions

The work done demonstrates the possibility of doing a thorough monitoring of the evaluation and comparison of spectra, by not leaving the analysis in line frequencies, harmonics and exited areas due to elements in the vibratory movement, it is possible to design a System that compares spectra and read them as vibration fingerprints.

The vibration as a dynamic phenomenon can be evaluate in any system if it is possible to quantify the excitation of it and differentiate the participating components within the spectrum. The test shows an opportunity for improvement including other algorithms in the programming, trying to decrease the time in each test to be able to perform with greater continuity. First, we must compare the results of these alternative algorithms in the tests and then combine the techniques with better performance to develop a better analysis and obtain a more specific result in tests with static elements.

The continuous analysis point to point in the analysis of the spectrum of vibrations makes feasible the identification of components in their natural frequency, the study shows that any change in the spectrum for the element in the test, using the correlation of spectra, makes possible the existence of an element in a test.

Finally, the development of this type of analysis presents great opportunities to deepen and to open possibilities in vibrational systems areas where it has not been deepened. In this case, a novel type of scanner from vibrations has been developed and is in patent process after several experiments and studies. This scanner will allow in the future not only to identify shapes and colors like the current scanners, but also to identify the percentage of the materials that are inside the element to be analyzed.

6 Appendices

The method contained in this paper was the base to develop a vibration scanner presented in the patent solicitation NC2018/0009118 in Colombia and the PCT international patent solicitation NC2019/0005957.

References

1. Rao, S.S.: Vibraciones Mecanicas. Pearson, London (2016)
2. White, G.: Introduccion al analisis de vibraciones. Azima DLI (2010)
3. Bogacz, R.: Recent investigations in dynamics of continuous systems subjected to moving load, vol. 23. Virago (2008)
4. Falconi, F., Park, Y., Azaña, J., Bogoni, A., Malacarne, A.: Evolution and performance of high-speed a-scan based on real-time optical spectrum fourier transformation. In: Conference on Lasers and Electro-Optics, p. JW2A.143 (2018)
5. Fernandez, F.J.: Mantenimiento Industrial Avanzado. SP:FC Editorial, Madrid (2010)
6. Lind, P., Vera-Tudela, L., Wächter, M., Kühn, M., Peinke, J.: Normal behaviour models for wind turbine vibrations: comparison of neural networks and a stochastic approach. Energies **10**(12), 1944 (2017)
7. Robles, F.B.: Monitorizaci6n de maquinas por analisis de vibraci6n. Ind. Quim. Equipos y plantas proceso, pp. 48–54
8. Zhang, Z., Shankar, K., Morozov, E.V., Tahtali, M.: Vibration-based delamination detection in composite beams through frequency changes. J. Vib. Control **22**(2), 496–512 (2016)
9. Naynjyoti Boro, H.D.: Vibrational signal analysis for bearing fault detection in mechanical systems. In: Foundations and Frontiers in Computer, Communication and Electrical Engineering, pp. 63–67 (2016)
10. Purohit, P.P.: Vibration control of structures using piexoelectric material. In: Multi-Disciplinary Sustainable Engineering, pp. 111–115 (2015)
11. Jimenez, J.A.: Diseño, calculos y construccion del prototipo para analisis de vibraciones en rodamientos. Perspect. en Tecnol. pp. 21–26 (2006)
12. Bagnold, E., Łodygowski, T.: Numerical modeling of fracture in brittle material under impact loading. Vib. Phys. Syst. **22**, 280 (1988)
13. Bagnold, E., Hrytsyna, O.: Mechanical and electromagnetic wave interaction in linear isotropic dielectrics with local mass displacement and polarization inertia. Vib. Phys. Syst. **24**, 227–232 (2010)
14. Bagnold, E., Cempel, C.: Acoustic attenuation performance of Helmholtz resonator and spiral duct. Vib. Phys. Syst. **23**, 247–252 (2008)
15. Hafaifa, A., Guemana, M., Daoudi, A.: Vibrations supervision in gas turbine based on parity space approach to increasing efficiency. J. Vib. Control **21**(8), 1622–1632 (2015)
16. Collins, L., et al.: Multifrequency spectrum analysis using fully digital G Mode-Kelvin probe force microscopy. Nanotechnology **27**(10), 105706 (2016)
17. Rodriguez-Urrego, L., Moreno, E.G., Anglada, F.M., Salvador, A.C., Cucarella, E.Q.: Hybrid analysis in the latent nestling method applied to fault diagnosis. IEEE Trans. Autom. Sci. Eng. **10**(2), 415–430 (2013)
18. Patel, V.K., Patel, M.N.: Development of Smart Sensing Unit for Vibration Measurement by Embedding Accelerometer with the Arduino Microcontroller | Viral Patel - Academia.edu. Int. J. Instrum. Sci. (2017)
19. Nandhini, K.: Analysis and experimental investigation of mechanical vibration and acoustic noise in four phase switched reluctance motor using PI, PID, fuzzy logic controller. In: 2017 IEEE International Conference on Electrical, Instrumentation and Communication Engineering (ICEICE), pp. 1–7 (2017)
20. Hall, L.: Simulations and analyses of train-induced ground vibrations in finite element models. Soil Dyn. Earthq. Eng. **23**(5), 403–413 (2003)
21. Jothi Saravanan, T., Gopalakrishnan, N., Prasad Rao, N.: Detection of damage through coupled axial–flexural wave interactions in a sagged rod using the spectral finite element method. J. Vib. Control **23**(20), 3345–3364 (2017)

22. Ribeiro, D.R., Kubinhetz, H., Campo, A.B., Sousa, J.S., Alencar, J.P.: Data acquisition system for turbojet engine using LabVIEW programming and PCI extensions for instrumentation. In: 33rd AIAA Aerodynamic Measurement Technology and Ground Testing Conference (2017)
23. Tsypkin, M.: Induction motor condition monitoring: Vibration analysis technique — diagnosis of electromagnetic anomalies. In: 2017 IEEE AUTOTESTCON, pp. 1–7 (2017)
24. Müller, M.: Dynamic time warping. In: Information Retrieval for Music and Motion, pp. 69–84. Springer, Heidelberg (2007)
25. Xu, M., Marangoni, R.D.: Vibration analysis of a motor-flexible coupling-rotor system subject to misalignment and unbalance, Part II: experimental validation. J. Sound Vib. **176**(5), 681–691 (1994)
26. Gani, A., Salami, M.J.E.: A LabVIEW based data acquisition system for vibration monitoring and analysis. In: Student Conference on Research and Development, pp. 62–65
27. Mohammad, K.S., Al-Dmour, A.S.: Vibration control of unknown flexible beam structures. Modal Anal. **8**(1), 69–86 (2002)

Signal Analysis in Power Systems with High Penetration of Non-conventional Energy Sources

J. M. Sanabria-Villamizar[1], M. Bueno-López[1(✉)], and Efrain Bernal Alzate[2]

[1] Department of Electrical Engineering, Universidad de la Salle, Bogotá, Colombia
{jsanabria16,maxbueno}@unisalle.edu.co
[2] Department of Automation, Engineering, Universidad de la Salle, Bogotá, Colombia
ebernalal@unisalle.edu.co

Abstract. This paper describes the development and implementation of a methodology for the signal analysis in non-conventional energy systems. The proposed methodology involves the Hilbert-Huang transform supported by Empirical Mode Decomposition (EMD) to decompose one signal in its intrinsic mode function (IMF). A computational tool designed in MATLAB is used to detect oscillations and different frequencies of a non-linear and non-stationary system.

Keywords: Smart grid · Hilbert-Huang transform · Hilbert spectrum · Power quality · Harmonics · Energy · Non-conventional · Instant frequency · Event detection

1 Introduction

The new technologies, such as non-conventional sources of renewable energies (NCSRE), Non-linear loads and electronic devices, are producing critical variations in the behavior of the transmitted wave form, therefore it is necessary to develop new methodologies for the analysis of signals which have a wide range of characteristics [1]. The results with the traditional methods are not the best because the time-frequency resolution has some problems in the detection of low frequencies and small variations in the oscillations. The quality in power systems is focus in ensure a voltage supply with an excellent waveform and reliability for the correct operation of the electronic equipment, which are handling critical processes of great importance for the customers.

In modern power systems it is necessary to analyze harmonic signals whose behaviour is non-linear and non-stationary, as these are used to the control of multiple variables. As for example System Average Interruption Frequency Index and System Average Interruption Duration Index (SAIFI and SAIDI, respectively) are Used to detect failures in the network, also, indicate the contributions of the distributed generators on the dynamic behavior of the power system, and in a general way in the optimization of the distributed power systems [2].

D. F. Cortes Tobar et al. (Eds.): AETA 2019, LNEE 685, pp. 145–154, 2021.
https://doi.org/10.1007/978-3-030-53021-1_15

The process of data analysis and monitoring becomes complex in systems with high penetration of non-linear loads, mostly in renewable energy systems, which hinders the estimation of power required by the users. One of the most common strategies for signal analysis in power systems has been the Fast Fourier transform (FFT). However, a limitation of FFT is the resolution in the time domain. Therefore, new methods are needed to ensure good resolution in both time and frequency domain, involving the concept of instantaneous frequency for accurate and rapid detection of disturbances. The Hilbert-Huang transform has emerged as an aid to solve this problem [3,4] and [5].

HHT is an adaptive method for analysis in both domains, that allows to be used to work on signals with the behavior previously described [3], based on the principle of instantaneous frequency, seeking an accurate and immediate detection of disturbances. When comparing the HHT with the FFT, it is possible to say that the first can detect behavior patterns more easily in signals with strong oscillations, in less computational time [6].

In the present research an analysis methodology is developed capable of extract the behavior characteristics of a signal transmitted in the time-frequency domain. Based on the HHT, the proposed methodology not only can display the estimation of the instantaneous frequency-amplitude, but also the fundamental frequency of operation of the system. Furthermore, with the advantage of requiring less computational time with greater efficiency compared to the other conventional methods of analysis for this type of signals.

2 Non-linear and Non-stationary Signal Analysis

2.1 Hilbert-Huang Transform (HHT)

The Hilbert-Huang transform of a function $x(t)$ is a concept presented by Norden E. Huang in [7]. The HHT consists of two important parts: empirical modal decomposition (EMD) and Hilbert spectrum (HS). EMD consists of decomposing the original signal in several signals, which allow to detect different oscillations, called IMFs (intrinsic mode function), This decomposition should be developed to find a monotonic function that indicates the inability to extract more IMFs. The IMF, which according to Huang [7], should be a function that fulfils two conditions: 1) The number of ends and the number of crosses by zero should be different from just one, and 2) Its local medial is zero. Once the decomposition is finished, the HT is applied to each of the IMFs that compose the original signal and with this process the HS is obtained.

For a signal $x(t)$, the analysis signal $z(t)$ is defined as:

$$z\left(t\right)=x\left(t\right)+iy\left(t\right)=a\left(t\right)e^{i\theta(t)} \tag{1}$$

where $y(t)$ is the Hilbert transform of $x(t)$, so that:

$$y\left(t\right)=\frac{1}{\pi}P\int_{-\infty}^{\infty}\frac{x\left(\tau\right)}{t-\tau}\,d\tau \tag{2}$$

Where P is the “Cauchy principal value”. Of (1) amplitude and phase are:

$$a\left(t\right)=\sqrt{x^{2}+y^{2}} \tag{3}$$

$$\theta\left(t\right)=\arctan\left(\frac{y\left(t\right)}{x\left(t\right)}\right) \tag{4}$$

From the phase the instantaneous frequency is extracted, as a function of the time, is defined by:

$$f\left(t\right)=\frac{1}{2\pi}\frac{d\theta\left(t\right)}{dt} \tag{5}$$

Masking Signal. In 2005, Deering and Kaiser [8] Define an improvement method for EMD, based on the use of mask signals. The algorithm is defined as follows:

- A mask signal is constructed, $s(t)$, from the original signal information, $x(t)$.
- The EMD is made to the following two signals, obtained from the original and the Mask:

$$x^{+}\left(t\right)=x\left(t\right)+s\left(t\right) \tag{6}$$

$$x^{-}\left(t\right)=x\left(t\right)-s\left(t\right) \tag{7}$$

- Subsequently, IMF^{+} and IMF^{-} are obtained.
- Where the resulting IMF is defined as:

$$IMF\left(t\right)=\frac{IMF^{+}\left(t\right)+IMF^{-}\left(t\right)}{2} \tag{8}$$

- So on, applying each iteration to the resulting residue.

Table 1. Comparison of time-frequency analysis methodologies

	FFT	WVD	WT	HHT
Base	Priority	Priority	Priority	Adaptive
Non-linear	x	✓	x	✓
Non-stationary	x	x	✓	✓
Feature extraction	x	x	x	✓

Table 1 shows the advantages and disadvantages of the methodologies of the analysis in time frequency. The FFT has the problem of aliasing which consists in the inability to break down a wave in a monotonic signal, being a low-efficiency strategy for non-linear and non-stationary wave signals. The WT and WVD are like the FFT, decomposing a signal in a series of basic functions, but with different problems. One of the most critical aspects in the WT is the selection of the type of wavelet (window) that best matches the signal to be analyzed, to conform to the components of low and high frequencies of the signal. The WVD has a better resolution in different applications, but like the FFT does not have the capacity to break down a signal in monotonic signals, therefore, it generates a distortion of the frequency spectrum.

However, HHT is not free of problems, its weakness is in the heuristic way of selecting EMD type, like the WT it depends on a mother function; although the time frequency results have a higher resolution and decomposition capacity for any type of signal, If the appropriate base function is not selected, the result of the IMFs will not be indicated.

3 Case Study

3.1 Synthetic Signal

To clarify the concepts presented above, we present the analysis based on the EMD and the EMD masking, on a synthetic signal. The signals present a phenomenon called mixture of modes, to solve this applies the methodology of the signal mask in [8].

The synthetic signal is described in Eqs. (9) and (10), also shown in Fig. 1, with its corresponding frequency modes.

$$x\left(t\right) = sen\left(2\pi \cdot 6t\right) + sen\left(2\pi \cdot 12t\right) + S_3 \tag{9}$$

$$S_3 = \begin{cases} sen(2\pi \cdot 18t) & si \quad 1 < t < 2 \\ 0 & otherwise \end{cases} \tag{10}$$

3.2 Power System

The analysis based on HHT and EMD, as shown formerly, is also applied to signals from electrical systems. In this case, the study will be carried out to a photo-voltaic system of a residential house [9], with the capacity to supply power to the distribution network, as shown in Fig. 2.

The data correspond to the voltage of the electrical system, taken at a sampling frequency of 16.7 kHz for 186 days. For the purposes of the study, 2 s of the data are analyzed, as shown in Fig. 3.

Unlike the synthetic case, no characteristic of the wave is known (Fig. 3). The voltage signal presents a switch that shows the changes of supply to the network in the first instance the system consumes the generated power, on the other hand, about 1s the system commutates to deliver power to the grid.

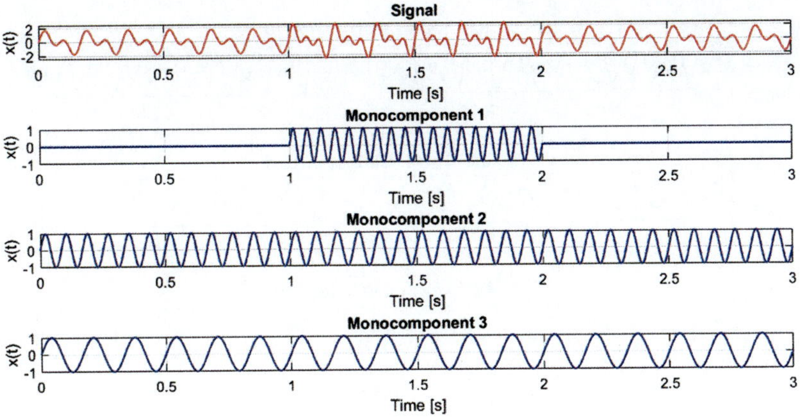

Fig. 1. Synthetic signal with their respective frequencies.

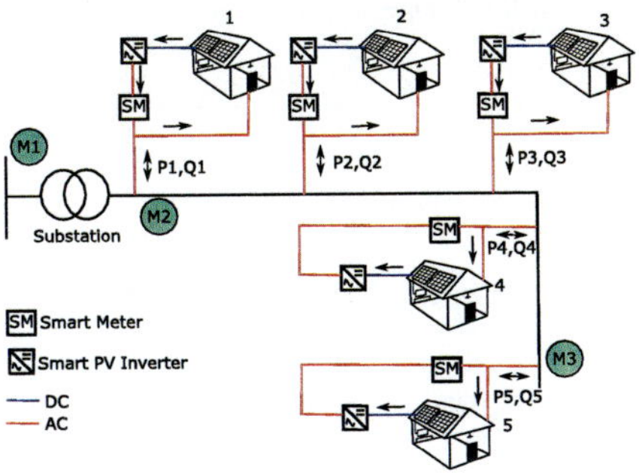

Fig. 2. Low voltage distribution system

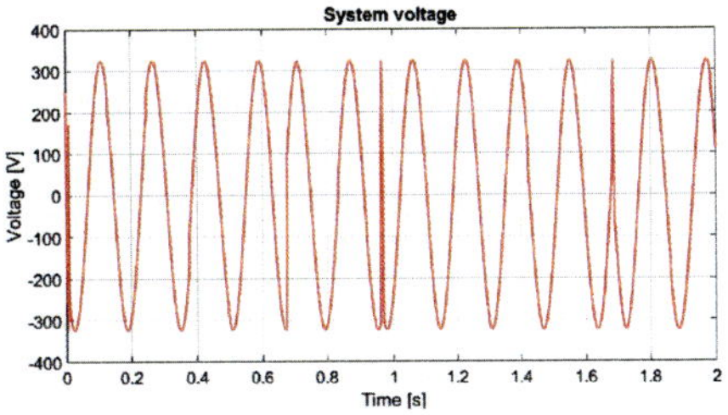

Fig. 3. Voltage wave obtained from the test system in the point M2

4 Case Study Results

4.1 Synthetic Signal Results

The empirical modal decomposition is shown in Fig. 4, where the first IMF shows a mixture of modes between the frequencies of 12 and 18 Hz, the following IMF also presents a mixture of modes between 6 and 12 Hz and finally the third is the 4 Hz frequency component. Similarly, in Fig. 5 the instantaneous frequencies corresponding to each IMF are shown.

It should be noted that, when knowing the signal to be analyzed, it is expected that in the decomposition of the signal the separation of the frequency modes that make up the signal will be visible. For this reason, the mask signal is applied with the form $s(t) = A_0 sen(2\pi ft)$, where a_0 and f is obtained according to Deering, As shown in [8]. In this case the mask signal that was used was $s(t) = 2sen(2\pi 28T)$.

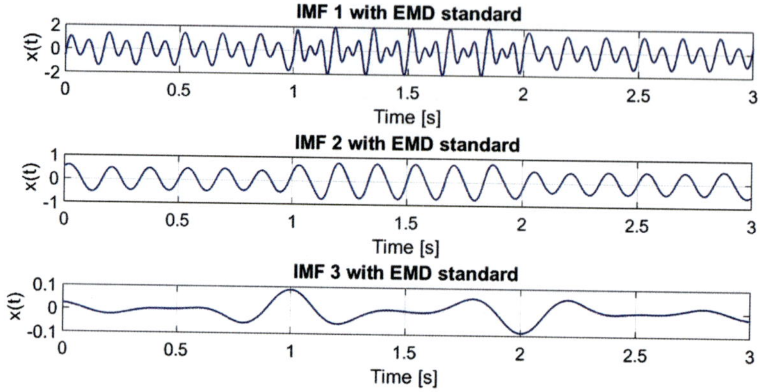

Fig. 4. Decomposition using EMD standard.

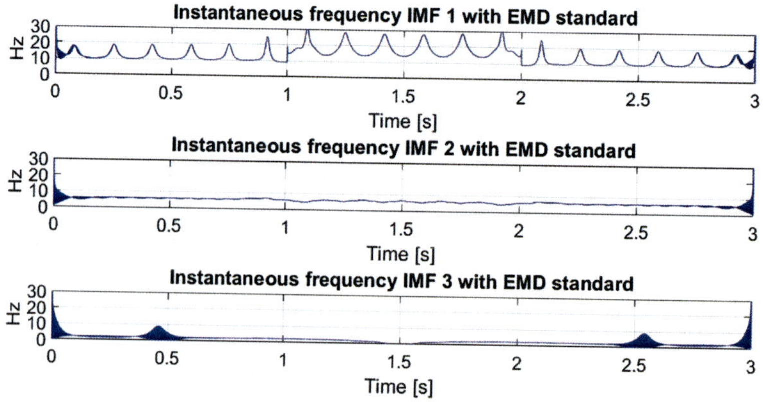

Fig. 5. Instantaneous frequency obtained with EMD standard.

The results of this process can be seen in Figs. 6 and 7, which show the IMFs obtained by performing the empirical decomposition and the respective instantaneous frequencies. In Fig. 6, it may be noted that the frequency components are extracted as expected, the frequencies of 18, 12 and 6 appear in the IMFs 1, 2 and 3 respectively.

4.2 Power System Results

The empirical decomposition in modes is shown in Fig. 8, where in the first IMF there is a noise in the wave, in the same way a mixture of modes in the IMF 1, 2 and 3, finally the IMF 4 is with the frequency component of 1 Hz. Similarly, in Fig. 9 the instantaneous frequencies corresponding to each IMF are shown.

By not knowing the characteristics of the signal, The mask signal is applied with the form $s(t) = A_0 sen(2\pi ft)$ in the same way as the previous case. In this case the mask signal that was used was $s(t) = 2sen(2\pi 70T)$.

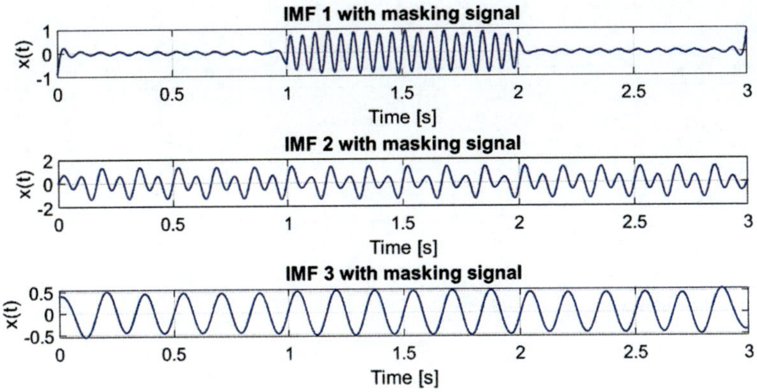

Fig. 6. Decomposition using EMD with Masking signal.

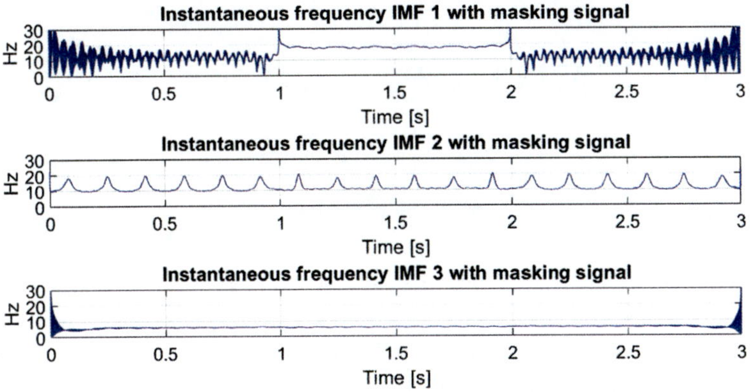

Fig. 7. Instantaneous frequency obtained with EMD - Masking signal

The results of this process can be seen in Figs. 10 and 11, which show the IMFs obtained by performing the empirical decomposition and the respective instantaneous frequencies. Note that the frequency components and signal noise are extracted from Fig. 10. IMFs 1 and 2 shown in Fig. 10, correspond to the noise of the signal, the following IMF describes the behavior of the wave, according to Fig. 11; finally, in the fourth IMF is characterized by being responsible for varying the amplitude of the wave.

Moreover, a problem is detected since the implementation of the standard EMD. In Fig. 4 the Fig. 3 IMFs obtained from the synthetic study signal are observed. In the first and second IMFs, a mixture of modes is observed in $t = 1\,\mathrm{s}$ and $t = 2\,\mathrm{s}$. This phenomenon is visible in Fig. 5 where the corresponding instantaneous frequencies are observed. To give solution to this problem has been created a masking signal according to the methodology proposed in [9]. Figure 6 shows without difficulty how each of the IMFs obtained observes and

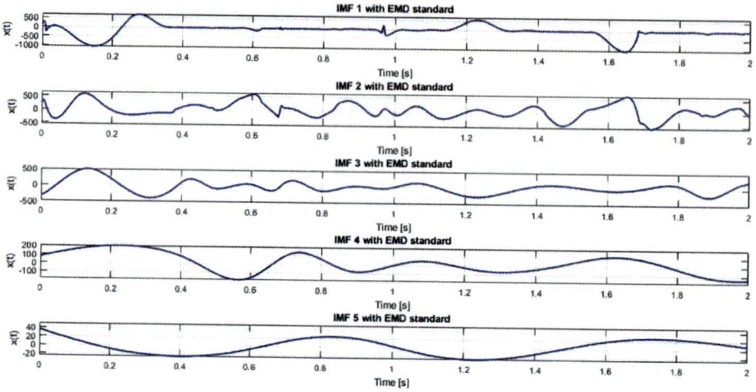

Fig. 8. Voltage wave decomposition using EMD standard.

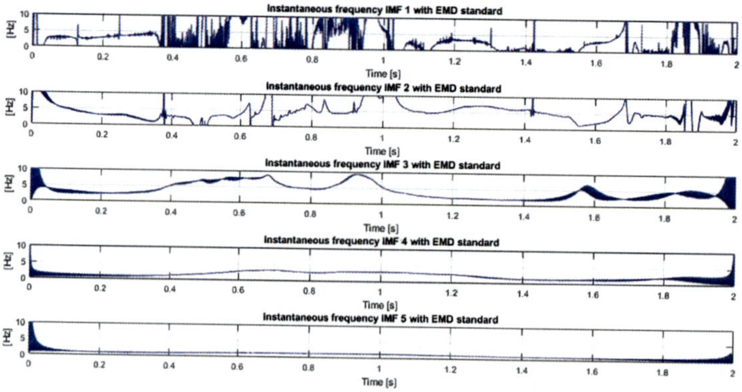

Fig. 9. Instantaneous frequency of the voltage wave obtained using EMD standard.

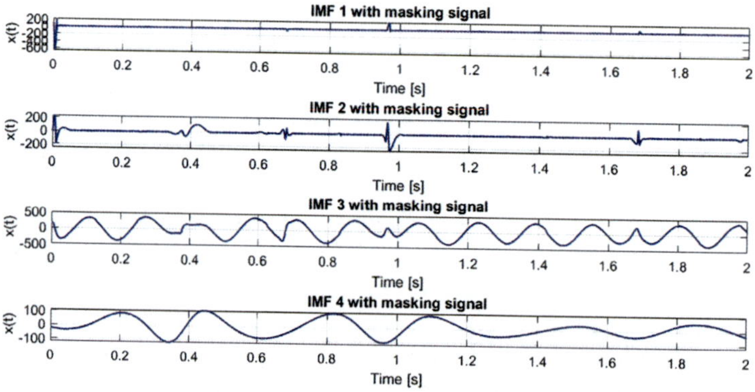

Fig. 10. Voltage wave decomposition using EMD with Masking signal

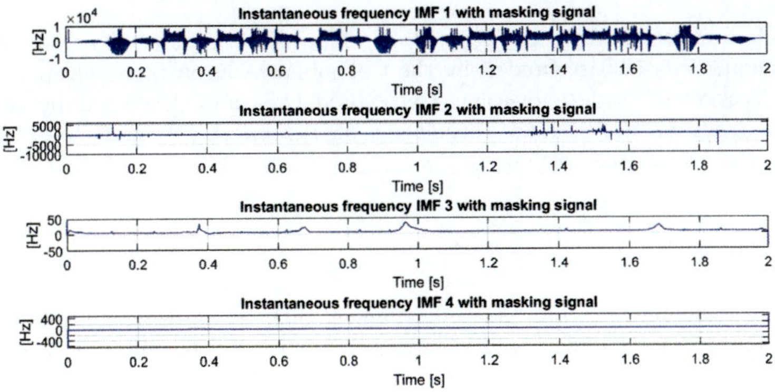

Fig. 11. Instantaneous frequency of the voltage wave obtained by EMD with Masking signal

identifies a single type of oscillation and this is confirmed in Fig. 7 where the instantaneous frequency can be deduced with greater precision than Fig. 5.

5 Conclusions

In this paper we have discussed the importance when using different methods for signal analysis in power systems with high penetration of non-conventional energy sources. It has been possible to decompose the original signal into a set of components that allow analyzing the behavior of a physical phenomenon related to each IMF, for the specific case of the electrical system one of the main applications that will be investigated is the detection of faults in real time. The procedure is effective when you know in advance the frequency components that appear, if on the contrary these are not known you would not have the necessary information to develop the methodology. Therefore, the analysis of the case study was based on a small part of the strategy described in [8] and on the average of the frequencies found from the standard EMD method shown in Fig. 9. On the other hand, this strategy emerges as an alternative for those signals where there is a set of data, knowing its amplitude value, but not its frequency, the results obtained, which are presented in this article, give beginning to a new strategy for the choice of the mask signal. However, the Hilber-Huang transform presents difficulties in mixing modes and edge effect.

The implementation of the standard EMD and EMD with the mask signal showed excellent results. Figures 10 and 11 showed how the number of IMFs is reduced by applying the method with the mask signal, therefore, it is indicated that the frequency components of the signal are grouped with greater precision. Besides, it is possible to observe the oscillations with greater clarity from the third IMF obtained, describing better the behavior of the system, the first two IMFs show the characteristic noise of the signal of a photo-voltaic system.

Acknowledgment. This paper is part of the project number 111077657914 and contract number 031-2018, funded by the Colombian Administrative Department of Science, Technology and Innovation (COLCIENCIAS) and developed by the ICE3 Research Group at Universidad Tecnologica de Pereira (UTP) and CALPOSALLE Group at Universidad de La Salle.

References

1. Kulia, G., Molinas, M., Lundheim, L., Larsen, B.B.: Towards a real-time measurement platform for microgrids in isolated communities. Procedia Eng. **159**, 94–103 (2016)
2. Puliafito, V., Vergura, S., Carpentieri, M.: Fourier, Wavelet, and Hilbert-Huang transforms for studying electrical users in the time and frequency domain. Energies **10**(2), 188 (2017)
3. Huang, N.E.: Introduction to the Hilbert–Huang transform and its related mathematical problems. In: Hilbert–Huang Transform and its Applications, pp. 1–26, World Scientific (2014)
4. Antonino-Daviu, J.A., Riera-Guasp, M., Pineda-Sanchez, M., Perez, R.B.: A critical comparison between DWT and Hilbert-Huang-based methods for the diagnosis of rotor bar failures in induction machines. IEEE Trans. Ind. Appl. **45**(5), 1794–1803 (2009)
5. O'Toole, J.M., Mesbah, M., Boashash, B.: A new discrete analytic signal for reducing aliasing in the discrete Wigner-Ville distribution. IEEE Trans. Signal Process. **56**(11), 5427–5434 (2008)
6. Peng, L., Jing, G., Duo, X., Chang, W., Xavier, Y.: Hilbert-huang transform with adaptive waveform matching extension and its application in power quality disturbance detection for microgrid. J. Modern Power Syst. Clean Energy **4**(1), 19–27 (2016)
7. Huang, N.E., Shen, Z., Long, S.R., Wu, M.C., Shih, H.H., Zheng, Q., Yen, N.-C., Tung, C.C., Liu, H.H.: The empirical mode decomposition and the Hilbert spectrum for nonlinear and non-stationary time series analysis. Proc. R. Soc. London Ser. A Math. Phys. Eng. Sci. **454**(1971), 903–995 (1998)
8. Deering, R., Kaiser, J.F.: The use of a masking signal to improve empirical mode decomposition. In: 2005 Proceedings of IEEE International Conference on Acoustics, Speech, and Signal Processing, (ICASSP 2005), vol. 4, pp. iv–485. IEEE (2005)
9. Alzate, E.B., Li, Q., Xie, J.: A novel central voltage-control strategy for smart LV distribution networks. In: International Workshop on Data Analytics for Renewable Energy Integration, pp. 16–30. Springer (2015)

Software-Defined Device for the Industrial IoT in a Product Assembly Context

Emir Cuk[1,2(✉)], Florian Grabi[1,2], and Purnima Dutt[1,2]

[1] Fraunhofer Institute for Manufacturing Engineering and Automation IPA, Stuttgart, Germany

[2] Fraunhofer Institute for Manufacturing Engineering and Automation IPA, University of Stuttgart, Stuttgart, Germany

{emir.cuk,florian.grabi,purnima.dutt}@ipa.fraunhofer.de

Abstract. To make assembly processes transparent, there is a need to track the present-day events. These processes are not standardized and can vary a lot, therefore they are poorly and/or partly documented [1]. Tracking assembly processes manually is costly and only possible for processes with a specific duration. There are new approaches to collect data of processes via human activity recognition, but it is implied that they violate the worker's privacy. Other available products in the market also show disadvantages in certain requirements. This paper is focused on creating a device ("thing" in the context of the Industrial Internet of Things), which is attached to a tool or component involved in the investigated assembly processes. This device has to meet certain requirements to ensure the non-interruption of value creation. Therefore, we combine consumer hardware with custom firmware and a framework to create a so-called "software-defined device". We compare our results with other competing products regarding the set requirements.

Keywords: IoT · Software-defined device · Minitag · ESP8266 · Inertial Measurement Unit (IMU) · Microcontroller Unit (MCU) · Data processing · Software engineering

1 Introduction

From experience, we know that there is a high optimization potential in manual assembly processes, especially if these processes have a duration of several days up to weeks. Having insights into the duration of the processes and their sub-processes can reveal ways to optimize the assembly line. To make assembly processes transparent, there is a need for data acquisition [1]. Process data acquisition systems can also show optimization potential, but the acquisition of the required data is often the difficult part [2]. The most pragmatic way of recording sub-process data is the manual way with pen and paper or an Excel datasheet maintained by the worker. The recorded data is often incomplete, and there is also a lack of standardized wording for the tracked events. This leads

D. F. Cortes Tobar et al. (Eds.): AETA 2019, LNEE 685, pp. 155–164, 2021.
https://doi.org/10.1007/978-3-030-53021-1_16

to human misinterpretation of the data, which impacts the overall quality of the data itself. Also, the steadily rising amount of variants of a single product affects the amount of possible assembly sub-process combinations in a product assembly process [3]. Thus it becomes more difficult to define all the planned times for product assembly processes. Seamless and efficient sub-process data tracking is not technically and economically feasible with the manual pen and paper approach. An approach to not violate the worker's privacy [4] is to track the environment around the worker. Our application's goal is to track "events", i.e., distinct sub-processes through wireless devices placed on tools or components used in the assembly process. Therefore, we built a device with a focus on the following requirements: low cost, small size, good and easy connectivity, low energy consumption, and detection flexibility. Low cost is important, given that the device deployment is as economical as possible. These devices have to be small and wireless to not interrupt value creation. Also, energy consumption has to be very low to decrease maintenance intervals and increase usability. Detection flexibility is needed so that a wide range of specific applications can be covered without time-consuming redeployment steps. Current solutions do not combine all these features and are often limited to one specific use case. Our goal is to build a device with a low cost per unit and long battery run time to track events in the assembly process with all the requirements mentioned above.

2 State of the Art

Currently, significant changes are happening to control processes in the industrial sector, and innovative technologies for automation are implemented to reduce operational costs. Considering the concept of IoT in Industry 4.0, the internet has caused a massive change and has revolutionized the monitoring and control of systems in the industry. This will transform the manufacturing and production sectors by enabling interaction between the virtual and the real world. In the vision of IoT, numerous devices and things will be connected to the internet, providing data, information, and services [5].

2.1 Industrial Internet of Things

The IoT is the philosophy of devices (things) being connected to each other, providing a higher degree of information available through the collection and analysis of data. These devices can be sensors, actuators or data processing nodes. Some of the possible applications involve sensing, tracking and collecting data [6]. The Industrial Internet of Things (IIoT) differs from IoT in the goals rather than in the applications. It uses IoT technology in an industrial setting to reach the goals of the industry [7]. These ideas heavily depend on the devices connected to the network which are introduced in the next section.

2.2 Hardware

The current IoT market landscape comprises a large number of prototype devices and developer boards, in addition to custom industrial solutions. The basis for most IoT prototypes are the Raspberry Pi Single Board Computer (SBC) and Arduino ecosystems [8], using chip-sets like the Broadcom BCM2837, espressif ESP8266/ESP32, or Nordic Semiconductor nRF52840. Development kits for IIoT hardware applications are available from multiple vendors including Bosch [9] and Texas Instruments [10]. The hobbyist IoT scene has led to a wide variety of environmental and inertial sensors becoming as easily available as the prototype breakout boards [11], while the mobile and gaming market lead to inertial and environmental Integrated Circuit (IC) sensors becoming smaller and more power-efficient.

2.3 Operating System (OS)

As IoT is a wide field, there is a large number of applied operating systems driving devices. This ranges from Linux variants or embedded Windows for more capable devices like the Raspberry Pi (3), Real-Time Operating System (RTOS) for faster MCU like the ESP8266/ESP32, or "bare-metal" binary firmware compiled with the respective embedded tool-chain for lower-tier hardware.

2.4 Data Processing

Data processing refers to the conversion process of raw data to meaningful information. It is a crucial step which is required in almost all the fields, using smart devices with varying levels of complexity [12] or in most cases data post-processing using a dedicated IoT service. Most available IoT data processing services are cloud-based (public or on-premise private cloud), which poses a conflict with the focus on security and safety apparent in the industry [14]. An alternative to this is local data processing on the IoT devices. An example of this is available end-user IoT devices and services that promise to improve human health, wellness and recognize the human activity. These include smartphones, smartwatches, and wearables, such as fitness trackers that monitor heart rates, track the number of steps or estimate the number of calories burnt [13]. These devices employ different ways to locally process data collected from the sensors available in them. However, all of those devices and applications are designed for human-centered information gathering. Our work which focuses on a device that does not violate worker's privacy is explained in the following sections.

3 Research Gap

Our research shows that there is a wide range of existing IoT devices which can be used to track industrial assembly processes. However, all reviewed devices (see Table 1 and Fig. 1) exhibit shortcomings regarding our requirements: The Bosch XDK and Sens'it Discovery are too large to be mounted on most tools, the TI SimpleLink Sensortag is not fully programmable and in its smaller version not available in the authors' country due to legislative issues. Other solutions are not programmable and were not reviewed further. Our approach is to design a device that fits our requirements better, especially regarding application flexibility. By implementing a multipurpose device with an unchanging hardware configuration, but variable application defined by its onboard software, the hardware development costs are kept low.

Table 1. Sample IoT devices comparison

Device	Cost (€)	Weight (g)	Size (mm)	Volume (cm^3)	Battery	Connectivity
Bosch XDK	200	43	60 × 42 × 22	5.544	560	WiFi, BLE
TI Sensortag BLE	30	40	40 × 50 × 15	3	225	BLE, Zigbee, 6LoWPAN
Sens'it Discovery	70	110	110 × 80 × 25	30.8	250	Sub-1GHz
Own Prototype	50*	27	65 × 30 × 15.5	3.0225	120	WiFi

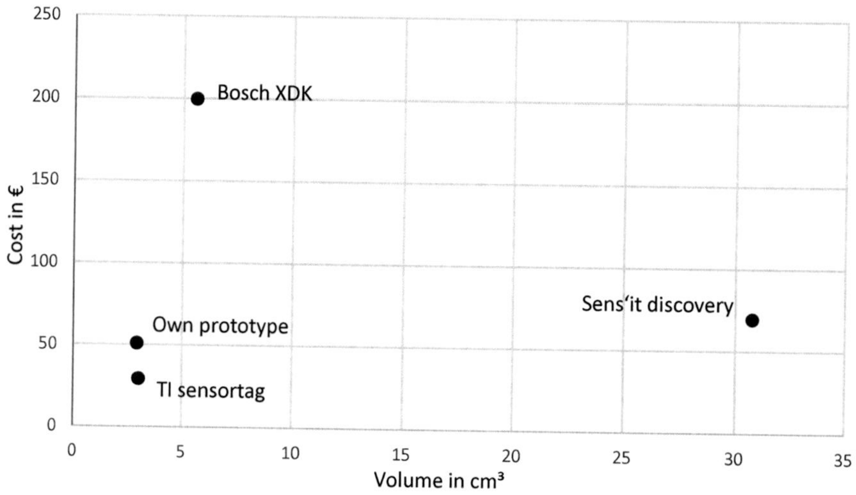

Fig. 1. IoT devices in cost-volume space

4 Implementation

We start with the hardware architecture and then the choice of the operating system. Then, we describe the firmware architecture, including the four system Finite State Machine (FSM) states and the data flow which enables the device's flexibility. Finally, we implement a possible data processing algorithm on the device.

4.1 Hardware Architecture

The prototypical wireless device used for this paper is custom made assembly of Commercial off-the-shelf (COTS) components. Functional components can be divided into sensors/data acquisition, data processing and formatting, wireless communication, and a Power Supply Unit (PSU) with a rechargeable battery. The collected data is bundled after processing and sent to a WiFi-enabled receiver device, where it is stored in a human-readable format.
Sensor Data - To keep the device versatile, only inertial measurements are collected in this first hardware version. For this, the Invensense MPU-9250 IMU is used, enabling the collection of acceleration, rotational speed, and absolute magnetic heading data on all 3 axes.
Processing and Wireless Communication - To keep the device small, a WiFi (2.4 Ghz)-enabled single-core MCU, the espressif ESP8266, is used in a variant with 4 Megabyte (MB) memory and a clock frequency of 80 Mega-hertz (MHz). This device offers fast prototyping abilities and partial code portability, due to its support of the Arduino toolchain and its widespread use in the hobbyist community. Also, a wide variety of pre-certified modules combining this chip with peripherals and an antenna is available, further minimizing development cost.

The ESP8266, as a main current consumer in the device, has a theoretical average current draw of 80 milliamperes (mA) [14] at 3.3 V supply voltage. To maximize battery life the device firmware makes use of the built-in sleep modes with theoretical currents of 20 microamperes (uA).

4.2 Operating System

The ESP8266 runs the FreeRTOS RTOS which is abstracted by the Arduino toolchain to a simple C++ runtime environment without preemptive task management.

4.3 Firmware Architecture

Running on top of the OS is the presented data processing framework, with a fixed number of system states and a modular data processing feature. Regardless of the particular processing algorithm, the device cycles through different states following a FSM approach. The data has to be recorded, processed and sent. To meet low energy requirements, an idle state is added. The idle state is entered when the sensor data does not change over a certain amount of time and activates the deep sleep mode of the device to save power, as shown in Fig. 3.

The software implementation is divided among 4 states with state transitions as depicted in Fig. 3.

1. **Idle** - This state is primarily used as a power-saving mode. As mentioned before, when the sensor data does not change over a certain amount of time, this state becomes the active state and makes most of the device go to deep

sleep, while the IMU enters a low-power movement alert mode with wake-up rules depending on the application. Also, other basic configurations such as enabling debug outputs and initialization of the IMU are done in this state. This is the default state on code startup.
2. **Record** - This state is entered on wake-up and active during data recording. In this state, the device collects and stores raw sensor data of the IMU to be processed later. It is necessary to keep the raw sensor data buffer (as shown in Fig. 2) as empty as possible to prevent buffer wraparound and raw data loss. Therefore, the Process state has to be entered as often as possible when raw data is available.
3. **Process** - This state is used to process the stored raw sensor data. We will employ different algorithms for data processing according to the requirement which relates to the concept of software-defined devices. The processed data is again stored to be sent to the server in the Send state. An example of an algorithm used to process data for detecting hammer hits is shown in algorithm 1.
4. **Send** - This state is used to send the processed sensor data to the server using the UDP protocol. It checks for the WiFi connection to be established. If connected, the data is sent. If not, the WiFi is turned on, a timer is set and the state changes to "Record". After the timer expires, the state changes to "Send" and the buffer of processed sensor data is sent (WiFi should be connected within that duration). The timeout helps to not miss the current sensor data while waiting for the WiFi connection.

4.4 Data Flow

An FSM-based design is proposed and implemented for the detection of hammer hits and movement on the machine, as shown in Fig. 3.

We considered 4 types of buffer for our code implementation (described in Fig. 2) depicting the relationship between the software(implemented states) and the hardware parts(buffers).

1. **Sensor Buffer** - The raw data from the MPU9250 IMU sensors on our device consists of 3-dimensional directional acceleration, gyroscope and magnetometer data (x, y, and z). The type of data that we will use depends on the requirement and the corresponding algorithm. For instance, for the hammer hit detection, we will be utilizing only the acceleration data. This buffer represents the source for the data flowing through states.
2. **Raw Data Buffer** - This is the buffer where the raw data collected from the sensor buffer is stored. It acts as the sink for the Record state and as a source for the Process state.
3. **Processed Data Buffer** - This is the buffer where the processed sensor data is stored. The buffer consists of the processed value and timestamp. It acts as a sink for the Process state and as a source for the Send state.

4. **Server** - The server buffer refers to the local server or optional cloud to which we will be sending the processed data. It acts as a sink for the overall data flow. We conducted our tests by sending the processed data to a local server through the UDP protocol.

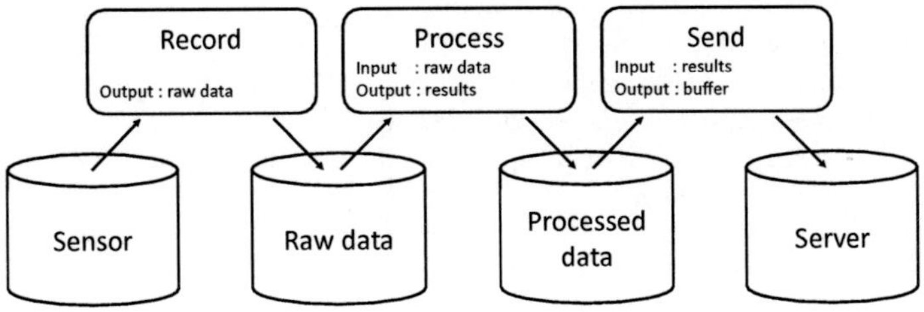

Fig. 2. Data Flow Through States

4.5 Software-Defined Event Recognition

By changing the state transitions and the process algorithm, we define the device. The raw data buffer is usually a 3D buffer (for 3 directional sensor inputs), while the processed data buffer(containing the output value) is a 1D buffer. This, in turn, reduces the size of data to be sent to the server. As mentioned before, the transition between the states and the flow can be inferred from Fig. 3.

Algorithm 1 presents the pseudo-code for detecting the event of hammer hits. Also, the processing part to detect movement and state filter events is referred from [15].

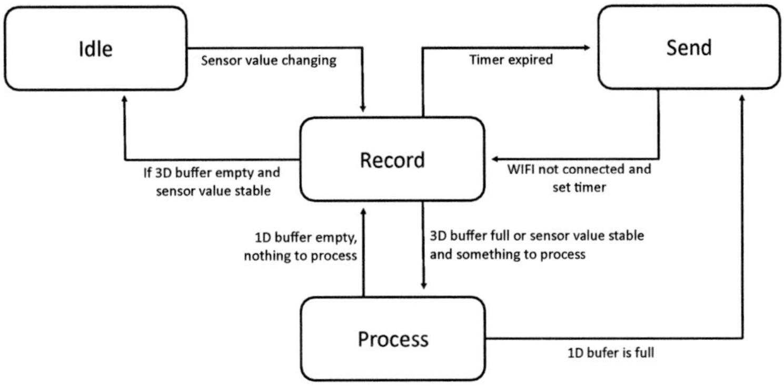

Fig. 3. State Transition Diagram

Algorithm 1: Process data to detect hammer hits

```
Hit intensity detection (sensordata);
Input  : acceleration data from sensor for x-, y- and z-axis
Output: gradated hammer hit intensity
winSize = 62
sqRoot = getAverageSquareRoot(sensordata)
window[ ] = Hanning(winSize)
windowedSlice[ ] = sqRoot * window[ ]
fftOutput[ ] = computeFFT(windowedSlice)
#Next we extract positive freq components
if (winSize%2) == 0 then
    posFreq = winSize/2
else
    posFreq = (winSize + 1)/2
end
for i = 1 to posFreq do
    windowedSlicePositive[i] = fftOutput[i]
end
normalizeResult[ ] = normalize(windowedSlicePositive[ ])
#Cut everything below -35dB
if normalizeResult[ ] < pow(10, −3.5) then
    normalizeResult[ ] = pow(10, −3.5)
end
#Compute power matrix by taking log.
power[ ] = 20 * log(normalizeResult[ ])
powerSum = sum(power[ ])
powerMean = mean(power[ ])
powerSD = deviation(power[ ])
band1 = powerMean − powerSD
band2 = powerMean + powerSD
band3 = powerMean − 2 * powerSD
band4 = powerMean + 2 * powerSD #and so on compute bands.
#Assign the values in this manner.
if (powerSum >= band1&&powerSum <= band2) then
    value = 1
else if (powerSum < band1&&powerSum >= band3)||||(powerSum <=
 band4&&powerSum > band2) then
    value = 2
else
    value = next value ...
end
return value
```

5 Results and Conclusion

In this paper, we set requirements for an IoT device in a product assembly context. As commercially available devices do not satisfy our requirements, we built our own solution and compared it with commercially available devices. It shows that our prototype device can compete with all researched IoT devices available in terms of size and weight. In fact, as the TI BLE Sensortag is not sold in Germany (where this research is conducted) due to legislative issues, our prototype presents the smallest and lightest inertial measurement IoT device available for us when considering programmability. Regarding reprogrammability, our device is on par with the most complete developer solution on the market (Bosch XDK). The main disadvantage compared to other devices is the limited battery capacity and lower versatility regarding environmental sensors.

6 Future Work

The used prototype device is not fully optimized in terms of size and energy usage, but for ease of manufacturing. Size savings of 50% are possible with a more integrated design. Also, WiFi connectivity was chosen as it is easier to develop and can be forwarded by standard WiFi routers. To enable more energy-efficient wireless communication, a switch to Bluetooth 4 is planned for the next design iteration, followed by Bluetooth 5 as the next step to enable longer range communication. At the moment, the device reconfiguration happens with a wired programming connection, which is planned to be switched to a wireless option. Also, more algorithms for event recognition will be developed and the existing ones optimized.

References

1. Kärcher, S., Cuk, E., Denner, T., Görzig, D., Günther, L.C., Hansmersmann, A., Riexinger, G.., Bauernhansl, T.: Sensor-driven analysis of manual assembly systems. Procedia CIRP (2018). https://doi.org/10.1016/j.procir.2018.03.241
2. Lotter, B., Wiendahl, H.-P.: Montage in der industriellen Produktion. Springer Verlag, Heidelberg (2006). https://doi.org/10.1007/3-540-36669-5
3. Koren, Y.: The Global Manufacturing Revolution. Wiley, Hoboken (2010). https://doi.org/10.1002/9780470618813
4. Varadinek, B., Indenhuck, M., Surowiecki, E.: Rechtliche Anforderungen an den Datenschutz bei adaptiven Arbeitsassistenzsystemen. Dortmund Berlin Dresden (2018). https://doi.org/10.21934/baua:bericht20180820
5. Ioan, U., Nicoleta-Cristina, G., Gheorghita, G.V.: An IoT architecture for things from industrial environment. In: 2014 10th International Conference on Communications (COMM). IEEE
6. Atzori, L., Iera, A., Morabito, G.: The Internet of Things: a survey. Comput. Netw., 2787–2805 (2010). https://doi.org/10.1016/j.comnet.2010.05.010
7. Boyes, H., Hallaq, B., Cunningham, J., Watson, T.: The Industrial Internet of Things (IIoT): an analysis framework. Comput. Ind., 1–12 (2018). https://doi.org/10.1016/j.compind.2018.04.015
8. Ferdoush, S., Li, X.: Wireless sensor network system design using Raspberry Pi and Arduino for environmental monitoring applications. Procedia Comput. Sci. **34**, 103–110 (2014). https://doi.org/10.1016/j.procs.2014.07.059

9. Robert Bosch GmbH: 06/14/19, XDK Cross Domain Development Kit. http://xdk.bosch-connectivity.com
10. Texas Instruments: 06/14/19, Hardware kits & Boards, Manuel Silverio-Fern. http://www.ti.com/de-de/design-resources/embedded-development/hardware-kits-boards.html
11. SparkFun Electronics: 06/14/19, Product Categories, Manuel Silverio-Fern. https://www.sparkfun.com/categories
12. Silverio-Fernández, M., Renukappa, S., Suresh, S.: What is a smart device? - A conceptualisation within the paradigm of the Internet of Things. Springer International (2018). https://doi.org/10.1186/s40327-018-0063-8
13. de Arriba-Pérez, F., Caeiro-Rodríguez, M., Santos-Gago, J.M.: Collection and processing of data from wrist wearable devices in heterogeneous and multiple-user scenarios. Sensors (Basel) (2016). https://doi.org/10.3390/s16091538
14. IDG Research Services: Studie Internet Of Things 2019 (German). https://www.q-loud.de/idg_iot_studie_2019
15. Cuk, E., Chaparro, V.: Methodology for optimizing manufacturing machines with IoT. In: 2018 IEEE International Conference on Internet of Things and Intelligence System (IOTAIS). https://doi.org/10.1109/IOTAIS.2018.8600907

Free-Obstacle Path Finding for Assistant Robot Based on Image Skeletonization

Fernando Martínez Santa[1(✉)], Santiago Orjuela Rivera[2], and Mario Arbulú[2]

[1] Universidad Distrital Francisco José de Caldas, Bogotá, Colombia
fmartinezs@udistrital.edu.co
[2] Corporación Unificada Nacional de educación superior CUN, Bogotá, Colombia
{santiago.orjuela,mario_arbulu}@cun.edu.co

Abstract. This document explores an autonomous navigation approach based on the generation of free-obstacle navigation diagrams by means of using the skeletonization technique and other image processing algorithms. These diagrams are turned into a path by means of a vector of points and performed by a hybrid robotic platform, composed of a humanoid robot and one with omni-directional locomotion. Four different configurations (origin, obstacles and destination) were studied in order to test the algorithm performance, computing the mean squared error and the average of relative error for each point in the path. For all of the tested configurations, the proposed algorithm chooses the shortest path through the navigation diagram. The final point and mean point-by-point relative errors were obtained for each of the 4 proposed configurations, where the average values over all of them were 1.11% for the final point relative error and 1.23% for the mean relative error, which validates the use of the algorithm for the autonomous generation of navigation paths for the hybrid robotic platform under study.

Keywords: Free obstacle path · Image skeletonization · Assistant robot · Hybrid robotic platform · Path planning · Navigation.

1 Introduction

Currently, robotics is highly focused on the human-machine interaction, one of the most important issues is that the robots can replace humans in specific tasks where both have to interact each other, for example being inside a house and doing chores. During the last years some human-like robotic platforms have been released to the broad public, so they have been used for researching as well. One of the most used platform is NAO^{TM}, which has been used for many projects like [1,12,16,19], this one basically consists of a human-like biped robot, which is so versatile, but as well as almost all biped robots the autonomy is limited, mainly due to the fact of moving using legs (power consumption higher than

Supported by CUN and Universidad Distrital.

D. F. Cortes Tobar et al. (Eds.): AETA 2019, LNEE 685, pp. 165–174, 2021.
https://doi.org/10.1007/978-3-030-53021-1_17

wheeled robots), for that reason some other developers have taken the decision of going back to wheels as locomotion medium, an example of that is [8]. In IDECUN research group at CUN university, a project named *Assistant robot for offices and homes* has been being developed since 2014 along with ARMOS research group at Universidad Distrital. This project is based on the union of two different robotic platforms: NAO^{TM} and Robotino^{TM}. The main idea is to join them as only one robot [2], Robotino as the main locomotion medium [11,20] and NAO as the manipulation and human-robot interface. Specifically, for this hybrid robot, a navigation approach based on the complex root location technique has been implemented [3,10], using an external camera that is able to take photos of the entire navigation scene. That approach used mainly image processing techniques previous to the implementation of the root location on the complex plan. On the other hand, in this paper another navigation approach is explored, this one is inspired on Voronoi diagrams, which has been one of the most used techniques for path planning and navigation for mobile single [4,5,9,17] and multiple robots [6,7,18]. This diagram is calculated by means of using image skeletonization technique, which has been previously explored by the authors [15].

2 Methodology

The navigation approach consists of using as main external sensor a fish-eye camera located on the scene, in order to cover it all. This camera automatically turns the image into a rectangular representation before sending it to the PC. After obtaining them, the images are pre-processed to obtain the obstacles position as well as the starting and ending points. Then the Voronoi diagram is computed through obtaining the skeleton of the obstacles image. Finally, the calculated path is sent to Robotino to run it. The error is computed by means of processing images of the scene while running.

The proposed assistance robot is composed of a humanoid robot NAO from the French company Aldebaran, and the wheeled robot Robotino from the German company FESTO which has omni-directional locomotion. Also, the proposed system has an external IP camera that provides global images of the environment in which the robot moves, a WIFI router that allows the communication of all the devices and a master computer that controls all the processes and functions of the system. The navigation process begins with the human machine interaction where an operator starting the process, immediately a global image of the scene is taken by the external camera and sent to the master computer. Afterwards, the area or object to which the robot must go is identified, by means of running the image pre-processing. Then, the main processing of images for the generation of the path is performed. After calculating the path to be performed, the execution command is given to Robotino platform which after reaching the destination point, notifies this fact to the system, in order to command NAO to perform its manipulation function or Robotino to correct its position. After accomplishing this function, the robot must be available again to perform the next task or designated function.

2.1 Image Pre-processing

The installed fish-eye camera gives a scene photo turned into rectangular format from its radial default representation, it offers an entire visualization of the navigation environment. Afterwards, a region of interest (ROI) is defined, reducing the image size as shown in Fig. 1(a).

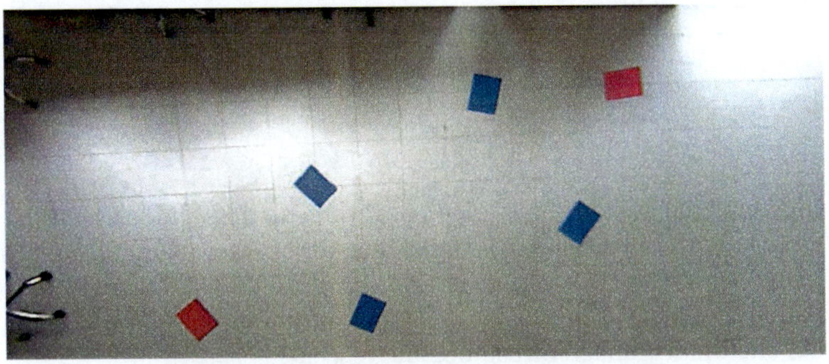

(a) ROI defined inside the navigation scene.

(b) scene without illumination pattern.

Fig. 1. First steps of image pre-processing.

Starting from the ROI image, a reduction of its illumination pattern is implemented in order to reduce the effect of the environment illumination on the rest of image processing algorithms to be used later. Instead of using the regular RGB representation to reach such reduction, the image was turned into HSV, making it easy to separate the lightness from the color. Then, only the S *(Saturation)* and V *(Value)* are blurred using 2D convolution with a matrix C defined as a disc of 30 pixels of diameter, obtaining the blurred matrices S_b and V_b as shown in the Eqs. 1 and 2.

$$S_b[n,m] = S[n,m] * C[n,m] \tag{1}$$

$$V_b[n,m] = V[n,m] * C[n,m] \tag{2}$$

After that, the blurred matrices (an approximation of the illumination pattern [10]), are subtracted from the original ones in order to reduce eliminate the illumination pattern. This is performed by means of using the Eqs. 3 and 4, where k is a constant defined by the average illumination value of whole image. Finally the matrices S_r and V_r with a reduced illumination pattern are obtained. Then the image is turned back into RGB using the original H matrix, S_r and V_r, as shown in Fig. 1(b).

$$S_r = S - S_b \cdot k \quad (3)$$

$$V_r = V - V_b \cdot k \quad (4)$$

For this implementation, both obstacles, the starting and the ending points are represented with different colors, light blue for obstacles and magenta for the points. Using again the HSV representation, the H matrix is thresholded using a single value in order to detect the starting and ending points (see Fig. 2(a)) and obtain the binary image of the obstacles scene as shown in Fig. 2(b).

(a) Starting and ending points.

(b) Binary image of the scene obstacles.

Fig. 2. First steps of image pre-processing.

2.2 Free Space Middle-Line Diagram

Voronoi diagrams obtain the middle line between near points in a plan [13], in this case instead of using the central point of the obstacles, the complete shape of those ones was used. For doing this, the image skeletonization technique was implemented on the obstacles image (see Fig. 2(b)) with a white background and black obstacles.

Image Skeletonization. Image skeletonization. Image skeletonization technique consists of doing an iterative controlled image erosion operation over a binary image [14], where each border pixel is eroded while this is not a selectable pixel for erosion of a near border. Finally, the skeleton or middle line of the object is obtained. In this case the background (in white) is eroded from the border of the obstacles to the middle point between it and the nearest object, globally obtaining a navigable network able to run without probability of any collision. The result of the implementation of this algorithm is shown in Fig. 3. https://es.overleaf.com/project/5b982a3d256fe851a4201eb1

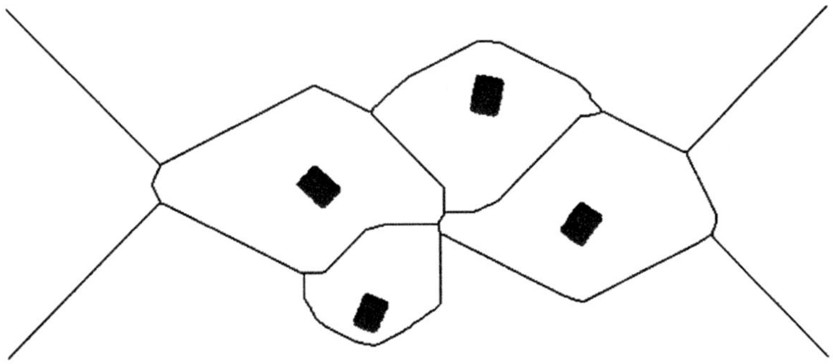

Fig. 3. Skeleton of the obstacles image.

Post-processing. After obtaining the skeleton of the obstacles image, it is necessary to choose a path from the starting point to the ending point, passing through the skeleton (middle points between objects and borders) [10]. First, the diagonal lines of the skeleton that takes to the corner of the scene are eliminated by means of end-points detection as shown in Fig. 4(a). Second, the origin and destination points are appended to the skeleton with a line to the nearest point in each case (see Fig. 4(b)). Then, the shortest path is computed by means of using the $A*$ algorithm as shown in Fig. 4(c). Finally, a decimation process is run on the obtained path in order to reduce the points to be sent to the robot (see Fig. 4(d)).

3 Results

The execution of the path generated by the skeletonization algorithm is carried out by sending the movement commands in distance units [mm] via WIFI to Robotino. These commands are generated by the used coordinate system on the artificial vision algorithms and converted to real magnitudes. The control of the execution is performed by proportional controllers using Robotino encoders, thus having a reading of the real distance traveled by the robot after performing the execution of each point in the path. Likewise, while carrying out the execution of the calculated path, a photographic record is taken to find the error between the generated path and the performed one, producing an image to compare both. With this data, both the final position error and the point-by-point mean error are calculated in the two coordinates over both paths. For test purposes, 8 different configurations were proposed for the calculation of paths using Voronoi diagrams generated by the proposed image skeletonization approach, of which 4 are presented in this document (see Fig. 5).

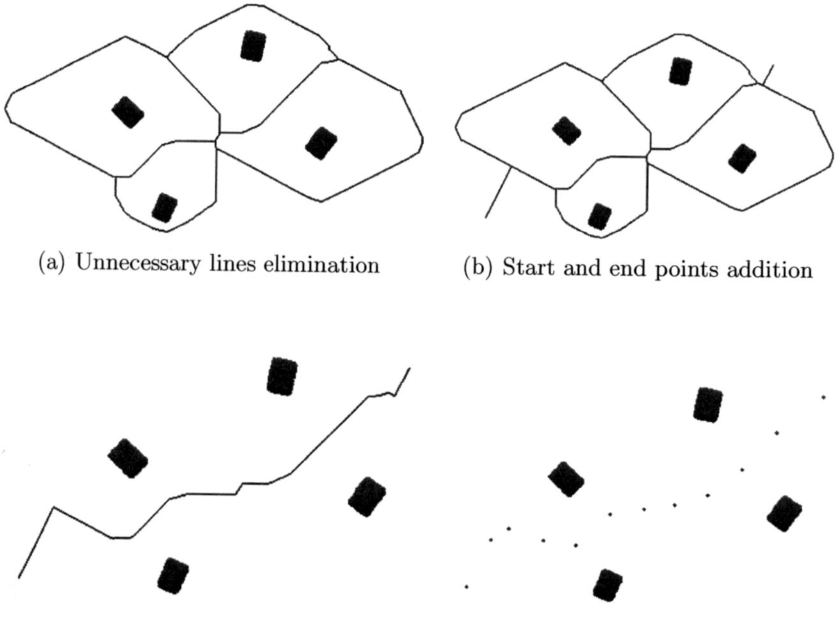

(a) Unnecessary lines elimination (b) Start and end points addition

(c) Path selection (d) Points reduction

Fig. 4. Path selection from the skeleton.

The proposed path generation algorithm always chooses and performs the best path based on the skeleton of the navigable space image, which in spite of not being the most efficient one in terms of geometry is the shortest one on the Voronoi diagram. The computed paths are shown over the real performed paths

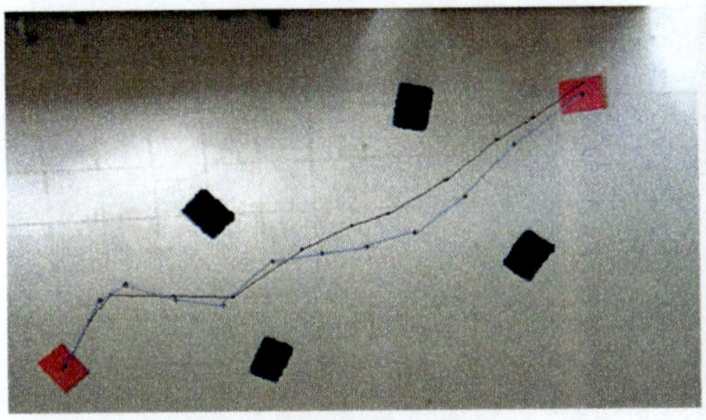

(a) Configuration 1

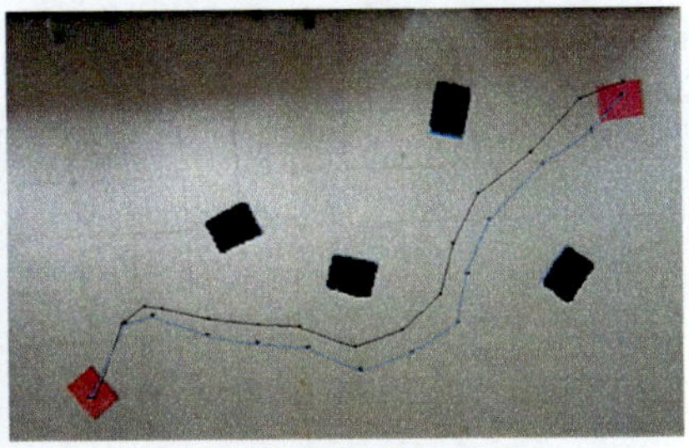

(b) Configuration 2

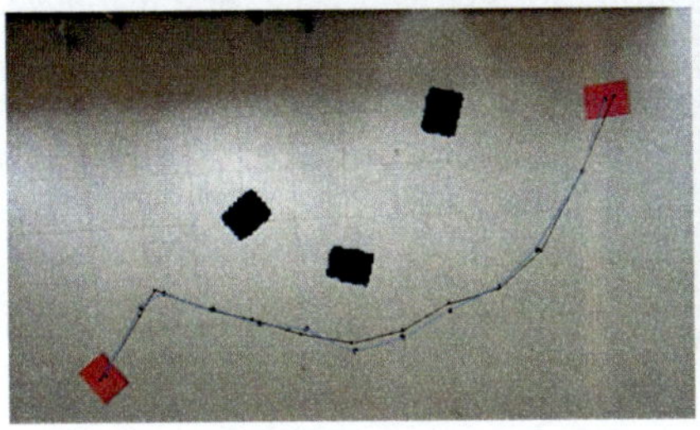

(c) Configuration 3

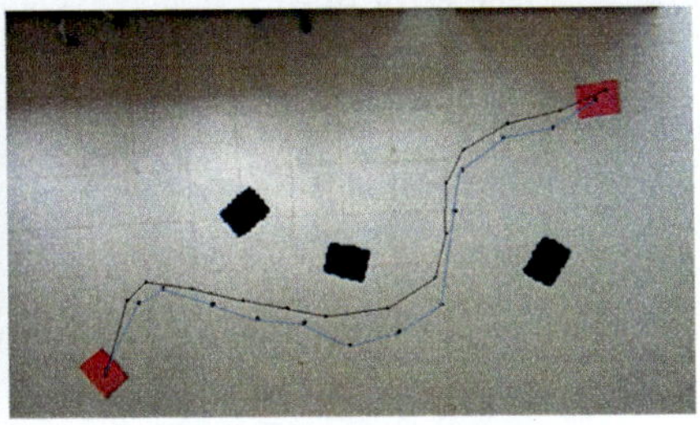

(d) Configuration 4

Fig. 5. Calculated paths (blue) vs performed paths (black).

for 4 of the performance testing configurations used for evaluating this navigation approach, as shown in Fig. 5. Having the computed and the real performed points of each path, it was possible to find the correspondent errors per coordinate, first of all the relative total error in x was found applying the Eq. 5, where only the final resultant point (x_f, y_f) was used, and the mean relative error over all path points (for x coordinate) applying the Eq. 6. In the same way, errors in y and the average error between x and y were computed as shown in Tables 1 and 2.

$$Ex_t = \frac{||\Delta x| - |\Delta v||}{\sqrt{(x_f - x_i)^2 + (y_f - y_i)^2}} \tag{5}$$

$$Ex_m = \frac{\frac{\sum_{j=1}^{n} ||\Delta x_j| - |\Delta v_j||}{n}}{\sqrt{(x_f - x_i)^2 + (y_f - y_i)^2}} \tag{6}$$

4 Conclusion

With the results of the generated and performed paths, it is evident that the path selection algorithm always chooses the shortest path, taking into account the network paths generated in the image skeleton, producing the necessary vectors for its execution. The generated paths can have turns with acute angles, also generating abrupt trajectory changes that make difficult a fluid navigation of the robotic platform. Those turns reduce their acuity with the decimation

Table 1. Final point error by configuration (relative to the total distance).

Configuration	Error X [%]	Error Y [%]	Av. error [%]
1	0,89	0,01	0,45
2	3,21	0,02	1,61
3	0,89	0,89	0,89
4	1,16	1,79	1,48

Table 2. Mean error of each point by configuration (relative to the total distance).

Configuration	Mean error X [%]	Mean error Y [%]	Av. error [%]
1	2,04	1,77	1,91
2	1,95	1,05	1,5
3	1,11	0,7	0,91
4	1,19	0,92	1,1

process applied to the path and the point conversion according to the image size, providing security since the paths are generated in the middle points between the obstacles. It must be taken into account that for a reliable execution, the distance between the trace of the path and the edge of the obstacle must be greater than the radius of the Robotino robotic platform (185 mm) with a tolerance (20 mm). It is important to know that the computed errors depend on the path shape, the stability of the communication between the PC and Robotino, a possible wheel sliding, and the data collected by the Robotino odometer, since these factors affect the correct displacement and execution of the commands given by the algorithm. It is important to highlight that under proper operating conditions, without interference in the signals and with certain paths it is possible to obtain final-point relative errors as low as 0,45% and point-by-point mean errors from 0,91% in the execution of the path. For these proposed test configurations, it was possible to notice that error in the X coordinate is always greater than in the Y coordinate, which is due to that in all of them the displacements are shorter in X, making the robot more susceptible to the inertia effects in this coordinate. Finally, the obtained data validate the use of the proposed approach as a practical path planning technique for mobile robots in known environments, and it is totally useful in the proposed application of the robot with a Robotino-based locomotion.

5 Future Work

In future work, other methods of generating curves for trajectories should be tested, such as Spline and Nurbs curves. These can provide a smoother navigation to the platform, avoiding damage to the humanoid robot mechanically coupled to Robotino due to inertia effects. It is also necessary to think of alternative control methods to have information about the movement of the robot, considering methods of PID control, Fuzzy and Neuro-fuzzy, as well as IPS reference systems, among others. A performance comparison between the path planning method proposed in this document and the previously proposed method will also be carried out [10], validating in this way the use of one algorithm or another for navigation based on the response of the scenario. Afterwards, the integration of all interaction, manipulation and artificial vision systems must be carried out in order to achieve a functional assistance platform.

Acknowledgements. Authors thank to Electronic Research Laboratory at CUN university for the simulations and practical tests carried out there, also to IDECUN (CUN) and ARMOS (Universidad Distrital) research groups for the given support.

References

1. Aguilera-Castro, D., Neira-Cárcamo, M., Aguilera-Carrasco, C., Vera-Quiroga, L.: Stairs recognition using stereo vision-based algorithm in NAO robot. In: 2017 CHILEAN Conference on Electrical, Electronics Engineering, Information and Communication Technologies (CHILECON), pp. 1–6. IEEE (2017)
2. Arbulú Saavedra, M., Martínez Santa, F., Ferro, R.: Asistencia Robotizada En Entornos Humanos. In: CIMM V Congreso de Ingeniería Mecatrónica. Cartagena, Colombia (2015)
3. Arbulú Saavedra, M.R., Martínez Santa, F., Montiel Ariza, H.: Metodología para el uso de la técnica de localización de raíces en la planeación de rutas para robots móviles, November 2015. https://doi.org/10.14483/udistrital.jour.tecnura.2015.4.a04. http://revistas.udistrital.edu.co/ojs/index.php/Tecnura/article/view/9545
4. Arslan, O., Koditschek, D.E.: Voronoi-based coverage control of heterogeneous disk-shaped robots. In: 2016 IEEE International Conference on Robotics and Automation (ICRA), pp. 4259–4266. IEEE (2016)
5. Bayar, G.: Development of a Voronoi diagram based tree trunk detection system for mobile robots used in agricultural applications. Ind. Robot Int. J. **44**(4), 521–531 (2017)
6. Breitenmoser, A., Schwager, M., Metzger, J.C., Siegwart, R., Rus, D.: Voronoi coverage of non-convex environments with a group of networked robots. In: 2010 IEEE International Conference on Robotics and Automation (ICRA), pp. 4982–4989. IEEE (2010)
7. Herrero, D., Martínez, H.: Range-only fuzzy Voronoi-enhanced localization of mobile robots in wireless sensor networks. Robotica **30**(7), 1063–1077 (2012)
8. Lafaye, J., Gouaillier, D., Wieber, P.B.: Linear model predictive control of the locomotion of Pepper, a humanoid robot with omnidirectional wheels. In: 2014 14th IEEE-RAS International Conference on Humanoid Robots (Humanoids), pp. 336–341. IEEE (2014)
9. Martínez Santa, F., Martínez Sarmiento, F.H., Jacinto Gómez, E.: Using the delaunay triangulation and voronoi diagrams for navigation in observable environments. Revista Tecnura **18**, 81 (2014). https://doi.org/10.14483/udistrital.jour.tecnura.2014.SE1.a06. http://revistas.udistrital.edu.co/ojs/index.php/Tecnura/article/view/8159
10. Martínez Santa, F., Orjuela Rivera, S., Arbulú Saavedra, M.: Global navigation approach for assistant robot. Tecnura **21**(51), 105–117 (2017)

11. Mbede, J.B., Melingui, A., Zobo, B.E., Merzouki, R., Bouamama, B.O.: zSlices based type-2 fuzzy motion control for autonomous robotino mobile robot. In: Proceedings of 2012 IEEE/ASME 8th IEEE/ASME International Conference on Mechatronic and Embedded Systems and Applications. pp. 63–68. IEEE, July 2012. https://doi.org/10.1109/MESA.2012.6275538. http://ieeexplore.ieee.org/lpdocs/epic03/wrapper.htm?arnumber=6275538
12. Montalvo López, M.E.: Programación por demostración del robot Aldebaran NAO H25 de la Universidad Politécnica Salesiana para la enseñanza y preservación de tradiciones, leyendas y expresiones orales del Ecuador (2017)
13. Saeedi, S., Paull, L., Trentini, M., Seto, M., Li, H.: Efficient map merging using a probabilistic generalized Voronoi diagram. In: 2012 IEEE/RSJ International Conference on Intelligent Robots and Systems (IROS), pp. 4419–4424. IEEE (2012)
14. Sangeetha, V., Vaithiyanathan, V., Sivagami, R., Divyalakshmi, K., Sundar, K.J.A., Ahmed, M.I.: Skeletonisation approaches for determining paths in an image: a Review. Indian J. Sci. Technol. **8**(35), (2015)
15. Santa, F.M., Gomez, E.J., Ariza, H.M.: Global path planning for mobile robots using image skeletonization. Indian J. Sci. Technol. **10**(14) (2017)
16. Shamsuddin, S., Yussof, H., Ismail, L., Hanapiah, F.A., Mohamed, S., Piah, H.A., Zahari, N.I.: Initial response of autistic children in human-robot interaction therapy with humanoid robot NAO. In: 2012 IEEE 8th International Colloquium on Signal Processing and its Applications (CSPA), pp. 188–193. IEEE (2012)
17. Vachhani, L., Mahindrakar, A.D., Sridharan, K.: Mobile robot navigation through a hardware-efficient implementation for control-law-based construction of Generalized Voronoi Diagram. IEEE/ASME Trans. Mechatron. **16**(6), 1083–1095 (2011). https://doi.org/10.1109/TMECH.2010.2076825
18. Wei, H.X., Mao, Q., Guan, Y., Li, Y.D.: A centroidal Voronoi tessellation based intelligent control algorithm for the self-assembly path planning of swarm robots. Expert Syst. Appl. **85**, 261–269 (2017)
19. Yun, S.K., Goswami, A.: Hardware experiments of humanoid robot safe fall using Aldebaran Nao. In: 2012 IEEE International Conference on Robotics and Automation (ICRA), pp. 71–78. IEEE (2012)
20. Zambrano Pérez, V.D.: Implementación de algoritmos de determinación de rutas para el robotino\textregistered de festo. Tesis maestr\'ia en diseño producción y automatización industrial (fiee), Escuela Politécnica Naciona EPN l Ecuador (2015). http://bibdigital.epn.edu.ec/handle/15000/11042

Neural Networks Identification of Eleven Types of Faults in High Voltage Transmission Lines

Laura Bautista F.[1], Cesar Valencia N.[1](✉), Gerson Portilla F.[1], Alfredo Sanabria[2], Carlos Pinto[2], Hernando González A.[3], Carlos Arizmendi P.[3], and David Orjuela C.[4]

[1] Mechatronics Engineering Faculty, Universidad Santo Tomás, Bucaramanga, Colombia
{laura.bautista,cesar.valencia,gerson.portilla}@ustabuca.edu.co

[2] Business International Faculty, Universidad Santo Tomás, Bucaramanga, Colombia
{alfredoenrique.sanabria,carlos.pinto}@ustabuca.edu.co

[3] Mechatronics Engineering Prog., Universidad Autonoma de Bucaramanga, Bucaramanga, Colombia
{hgonzalez7,carizmendi}@unab.edu.co

[4] School of Medicine and Health Sciences, Universidad del Rosario, Bogota, Colombia
alvaro.orjuela@urosario.edu.co

Abstract. In power transmission systems faults returning leaving them offline. This problem generates an economic impact on the interested parties, partly because in certain cases transmission lines protections act in a delayed manner or because the data processing generated by electrical protections tends to be a tedious. Artificial intelligence personnel have implemented a number of methods aimed to provide solutions for detection, classification and localization of said faults. In this work, a multilayer neural network capable of performing the process of classifying 11 types of faults in power transmission lines was implemented. As a result, a graphical interface allows users to intuitively visualize the faults.

Keywords: Neural networks · Power transmission · Confusion matrix · Multilayer perceptrons · GUI

1 Introduction

Within economic sectors of a country, one of the most importants is the energy sector. In Colombia, this sector stands out by the inclusion of generation, transmission, distribution, and commercialization of electric power that allows to provide a service in rural and urban sectors. Therefore, faults or interruptions that occur in their transmission or distribution medium would cause a reduction in transmission capacity and would affect final users.

A clear example of that is the exact moment at which any type of failure occurs within a power transmission line. It is inspected, data is collected by its electrical protections in order to solve these situations. This task becomes tedious taking into account the amount of data that is collected by the electrical protections. However, this allows to determine the type of failure, its severity and in this way apply reconnection protocols.

In addition to affecting the economy, it must be borne in mind that if the protection devices do not act in time, expenses would be generated since repairs could be needed

D. F. Cortes Tobar et al. (Eds.): AETA 2019, LNEE 685, pp. 175–184, 2021.
https://doi.org/10.1007/978-3-030-53021-1_18

in the faulty line, and execution time required for the repair also must be taken into account.

In order to help reduce the time it takes for the operator to identify the 11 symmetrical and asymmetrical faults that can occur in electric transmission lines, it is proposed to generate a support tool based on the identification properties that neural networks have. In addition, the proposed system should allow the determination of the section where the fault occurred, but not its exact location.

2 Literary Review

Faults in electrical systems are a constant problem. So different techniques have been implemented for detection, classification and location. A clear example of the use of alternative methods that meet this purpose can be found in [1] where it is proposed an algorithm capable of detecting and classifying high and low impedance faults using a diffuse inference system adapted to a network. The network inputs are the phase values of the current and voltage.

On the other hand, in [2] is proposed the development of an expert system that in real time processes the acquired information and helps specialists to identify faults origin. This system works exclusively on the RALZ protection. [3] uses a retro-propagation neural network for the classification of faults. The output of its network gives a binary code with which it defines the type of fault, the input information to the network (voltage and current of the phases) is obtained from the transmission model of the line. In [4] the authors use the Wavelet transform to perform the sampling of the voltage and current signals at both ends of the transmission line. The coefficients of the signals are compared with limit values to detect and classify the faults. On the other hand, the location it is done with the application of a neural network. [5] uses the current signals of each phase for the identification of the fault in the phase and the zero sequence current to determine the detection of the ground. In addition, the current signals are processed using the Wavelet transform. The system is flexible: when more precise information processing is required, the working windows are low and the sampling regions are shorter if high frequency information is required. In [6] the authors use the technique Fast Discrete S-Transform (FDST). Such scheme has been used in cases of the following failures: failure resistance, start angles and significant noise in the signals. The FDST manages to apply a frequency scaling, a band pass filter and interpolation techniques that lead to the processing of the algorithm. It low computing power and can be executed in real time. In [7], a MATLAB-implemented neural network with retro-propagation for the detection and classification of faults in the electric power transmission line. In other words, it takes the values for the three phases (current and voltage) as input the system. In [8] an adaptive auto-recurrent wavelet neural network is proposed used as an identifier for the fault and from the discriminant analysis and support vector machines implemented identifying the location of the faults. Using a signal based on components such as the identifier input, while a signal based on the system is used as the output, in order to combine fault diagnosis methods based on components and systems. Moreover, in [8] it is proposed the use of an adaptive auto-recurrent wavelet neural network as an identifier of the fault and from the discriminant analysis and support vector machines implemented identify the location of the faults, using a signal based on components such as the identifier input, while a signal based on the system is used as the output; in order to combine fault diagnosis methods based on components and systems. Finally, [9] proposes a neuro-diffuse technique for detection, classification and localization. The phases' impedances, current and voltage are the inputs variables. The technique consists of three phases [10]: fuzzification, membership functions, and defuzzification. After these three phases are obtained, binary values are extracted to be processed in a simulation of the design of a fault controller based on fuzzy logic, performed in SIMULINK, in order to verify the values RMS and identify the failure that has been presented.

3 Methodology

A training of a multilayer perceptron network [11] is required to identify the faults presented in the transmission lines based on a database obtained through the electrical modeling of 5 sections of an electric transmission line and the simulation of 11 types of faults. The process began with determining the sets that would be used in the training, validation and testing of multilayer perceptron neural networks, analyzing the data generated by the simulation of the section's electric circuits. At the moment of the acquisition [12] of the data, the format in which they are delivered must be taken into account, since the relays and electrical protections deliver the information in a comtrade format and for the purposes of this implementation, the data is needed in a format compatible with the MATLAB software (.mat).

Next, the generation of a support tool using multilayer perceptron neural networks was required to identify the type of failure through input variables that could be values of: current, voltage, voltage and current angles, taken one or two cycles after the fault took place, depending on the configuration, 4 configurations in total were performed.

Networks [13] with 1 and 2 hidden layers were trained with variations in their number of neurons in the following way: for a hidden layer 5, 10, 15, 20, 25 and 30 neurons and for two hidden layers 5–10, 10–10, 15–10, 15–15, 15–20 and 20–30 neurons.

Once the training tests of the neural networks were carried out, the results obtained were organized in a confusion matrix, where the output values of the network are compared with the real values, which allowed to establish which types of failures present greater complexity at the time of identification.

Finally, the algorithm [14] was implemented in a MATLAB GUI, which allowed the use of the support tool [15] in an intuitive way for operators of transmission lines.

4 Results

Network's performance validation allowed to determine which one was the one that best behaved before the identification of faults for each section and configuration (input data).

600 networks were trained for each configuration and the network with the smallest error in each section was selected as shown below

4.1 P1 – P2 Section

See Tables 1 and 2.

Table 1. Error results for the P1 – P2 section.

P1 – P2 section				
Conf.	1 layer		2 layers	
	Number of neurons	Error	Number of neurons	Error
1	15	2.51×10^{-3}	15–15	1.18×10^{-6}
2	15	6.83×10^{-4}	15–10	1.49×10^{-6}
3	30	4.75×10^{-2}	20–30	0.005972
4	10	8.06×10^{-2}	20–30	0.03791

Table 2. Percentages of classification according to thresholds, P1 – P2 section.

Threshold, P1 – P2 section				
0.4	0.45	0.5	0.55	0.6
96.98	97.08	97.05	96.98	96.88

4.2 P2 – P3 Section

See Tables 3 and 4.

Table 3. Error results for the P2 – P3 section.

P2 – P3 section				
Conf.	1 layer		2 layers	
	Number of neurons	Error	Number of neurons	Error
1	10	3.299×10^{-5}	10–10	1.662×10^{-6}
2	25	1.231×10^{-3}	15–10	3.572×10^{-11}
3	15	2.727×10^{-3}	15–10	4.769×10^{-10}
4	10	7.225×10^{-5}	15–15	4.710×10^{-7}

Table 4. Percentages of classification according to thresholds, P2 – P3 section.

Threshold P2 – P3 section				
0.4	0.45	0.5	0.55	0.6
99.48	99.5	99.51	99.5	99.46

4.3 P3 – P4 Section

See Tables 5 and 6.

Table 5. Error results for the P3 – P4 section.

P3- P4 section				
Conf	1 layer		2 layers	
	Number of neurons	Error	Number of neurons	Error
1	15	1.744×10^{-4}	15–10	1.815×10^{-13}
2	15	2.566×10^{-4}	15–10	1.493×10^{-14}
3	25	4.065×10^{-3}	15–10	1.889×10^{-9}
4	10	2.024×10^{-4}	15–10	4.768×10^{-12}

Table 6. Percentages of classification according to thresholds, P3 – P4 section.

Threshold P3 – P4 section				
0.4	0.45	0.5	0.55	0.6
99.81	99.82	99.81	99.81	99.80

4.4 P4 – P5 Section

See Tables 7 and 8.

Table 7. Error results for the P4 – P5 section.

P4 – P5 section				
Conf.	1 layer		2 layers	
	Number of neurons	Error	Number of neurons	Error
1	10	2.725×10^{-5}	5–10	2.575×10^{-7}
2	10	8.905×10^{-5}	15–10	5.247×10^{-8}
3	10	2.583×10^{-5}	15–10	3.426×10^{-7}
4	10	2.580×10^{-6}	15–10	5.752×10^{-8}

Table 8. Percentages of classification according to thresholds, P4 – P5 section.

Threshold P4 – P5 section				
0.4	0.45	0.5	0.55	0.6
98.73	98.80	98.81	98.75	98.67

4.5 P5 – P6 Section

See Tables 9 and 10.

Table 9. Error results for the P5 – P6 section.

P5 – P6 section				
Conf.	1 layer		2 layers	
	Number of neurons	Error	Number of neurons	Error
1	30	2.042×10^{-3}	10–10	2.600×10^{-6}
2	30	2.635×10^{-3}	15–20	1.176×10^{-3}
3	15	3.881×10^{-3}	15–10	3.103×10^{-5}
4	30	2.523×10^{-3}	15–15	2.798×10^{-2}

Table 10. Percentages of classification according to thresholds, P5 – P6 section.

Threshold P5 – P6 section				
0.4	0.45	0.5	0.55	0.6
98.95	99.01	99.04	99.03	98.94

For the analyzed sections the configuration that presented the best results in general was configuration 2 and the response threshold applied to the outputs of the neural network to encode the responses in binary and be able to make the comparison with the true outputs was 0.5. On the other hand, the parameters for the retraining of the neural networks were established in 2 hidden layers having the first 15 neurons and the second layer 10 neurons, according to the tests performed.

For the selection of the network implemented in the graphical interface, 1000 networks were trained. The training was carried out by series of 200 networks (5 series). In these groups, the initialization of weights and bias was carried out randomly, as was done in the previous tests.

The adopted selection methodology for the selection of the final network to be implemented in the GUI is described in Fig. 1, because it was observed that the network that presented the smallest error did not generate the best classification percentage.

The pseudocode for the GUI is presented in Fig. 2.

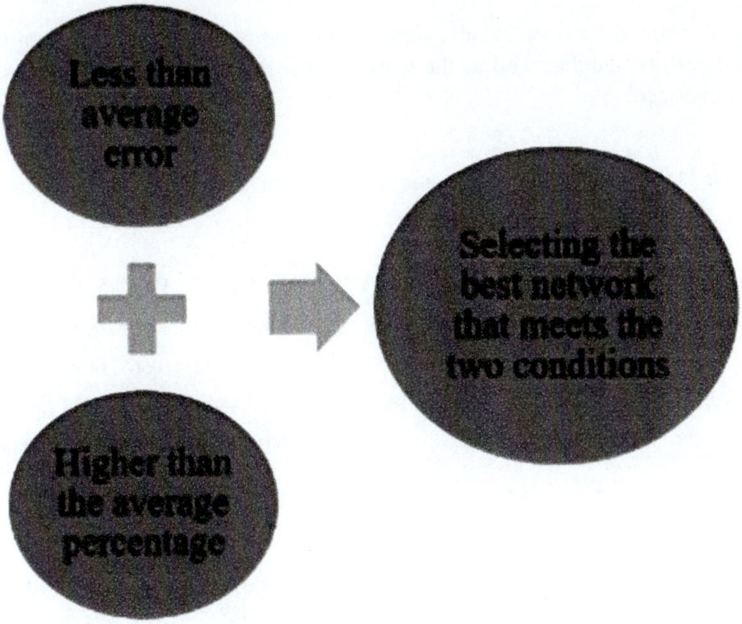

Fig. 1. Selection characteristics for the best network.

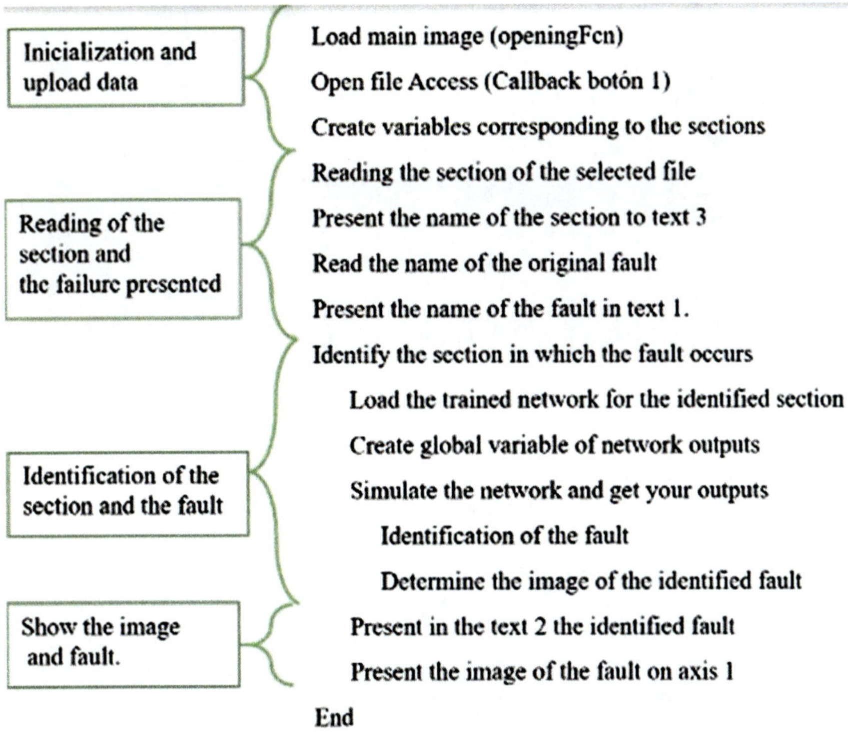

Fig. 2. GUI pseudocode

Table 11 show the values of the average neural networks in percentage of classification and performance, as well as the values of the final network implemented in the graphical interface.

Table 11. Percentage of classification and error of the average and final networks

		Section				
		P1 –P2	P2 – P3	P3 – P4	P4 – P5	P5 – P6
Average network	%	97,04674	99,48939	99,8028409	98,952991	99,09593
	Error	0,665706	0,667448	0,66737892	0,6671609	0,666629
Final network	%	98,48484848	100	99,8484848	99,9242424	99,92424
	Error	0,665609063	0,665874	0,66582732	0,66624164	0,666619

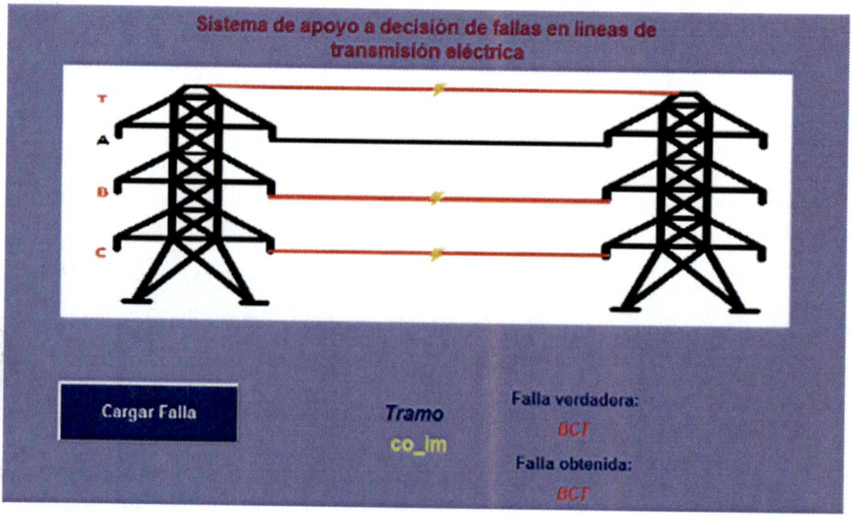

Fig. 3. GUI screenshot

The GUI has a main image where two transmission towers are presented that correspond to the ends of the analyzed section. The image, as shown in Fig. 3, displays the following information. The lines involved in the fault are represented in red and with a lightning symbol, as well as the name of the line also in red text. A button ("*cargar falla*" in Spanish) allows loading the file to be processed. A yellow text shows the section involved. Finally, two sets of text show the type of failure, that is, the failure presented, and the failure classified by the network. In this way, the user can easily interpret the results obtained by the decision support system.

5 Conclusion

The databases provided previously by the simulation of the circuits were analyzed, so it was possible to determine how balanced they were. Within the three sets that are considered essential (training, validation and testing), there are records for the 11 types of faults, that is, there are simulations that include variations in resistance, location and angle values. In this way, it is guaranteed that during the training and testing process of the network, complete information is available to allow adequate learning and better generalization when benchmark tests are performed.

A support tool was created based on a multilayer neural network capable of identifying the failure presented by the user. It was observed that configuration 2, in comparison to the other configurations, provided better classification results since the inputs of the neural network under this configuration are average values in the second cycle right after the voltage and current failure that allow to describe the network in the best way. The behavior of the fault and instantaneous values, a cycle after the failure, of angles that offer precision in terms of the opening time of the electrical protections, which leads to obtain a percentage of classification greater than 98% in all the sections.

The combination and selection of parameters, in addition to the re-training of the networks in established tests, generates variation in the results of identification of the faults, that is to say, the networks that were trained with a hidden layer obtained values of identification of faults smaller than the networks with two hidden layers. This type of response can be explained by observing the concept of an increase in the degrees of freedom for two hidden layers (taking the problem to a greater dimension), since during the training process a larger weight configuration is available that allows to improve the options of adjustment of the network.

An algorithm was created that allows coding the numerical outputs of the neural network in binary values "0" or "1" by using the threshold. Subsequently, and according to the previous classification, the faults are organized in such a way that the responses of the network can be compared with respect to the real values. In this way, the matrix of confusion is created allowing an observation of the faults classified wrongly. The confusion matrices show that ABC and ABCT failures are the ones that present the greatest complexity in their identification. This is due to the similarity between these faults since the level of coincidence is in the 3 lines (ABC). Thus, the use of networks with two hidden layers allowed to obtain better results not only for the two types of failures mentioned but in general terms for the 11 types of faults.

A GUI was created in the MATLAB software capable of intuitively approaching the user with the support tool. For this purpose, it was considered that said interface included graphic and textual interpretive spaces that present information about the sections and lines involved in the failure. In this way, the user should not interpret extensive data sets and can also select the records to be identified in a simple way in which the interface itself performs the processes of indexing, evaluation and presentation of responses.

The adopted methodology for the selection of network parameters depended on the lowest performance value in the tests performed. However, at the time of selection of parameters for the final network, modifications were made to it, since a change was observed between the error generated by the network and the percentage of classification of the network. That is, the networks that presented better performance showed in turn a low classification percentage that was reflected in the number of failures successfully classified. Therefore, a new methodology was adopted based on the selection of a network that had an error value lower than the average of the retrained networks and a higher value than the average of the classification percentage of the faults, ensuring that the network does not slow down the processes and is effective.

References

1. Calderon, J., Zapata, G., Ovalle, D.: Algoritmo neuro-difuso para la detección y clasificación de fallas en líneas de transmisión eléctrica usando ANFIS. Avances en sistemas e informática **4**, 101–108 (2007)
2. Degl'Innocentti, A., Salcedo, J., Crijenko, E.H.: Diagnóstico de fallas en líneas en tiempo real a través de un sistema inteligente. CIER, pp. 78–86 (2011)
3. Tayeb, E.B.: Faults detection in power systems using artificial neural network. Am. J. Eng. Res. (AJER) **2**, 69–75 (2013)
4. Shaik, A.G., Pulipaka, R.R.V.: A new wavelet based fault detection, classification and location in transmission lines. **2**, 35–40 (2015)
5. Yavav, A., Swetapadma, A.: A novel transmission line relaying scheme for fault detection and classification using wavelet transform and linear discriminant analysis. ASEJ **6**, 199–209 (2015)
6. Krishnanand, K., Dash, P., Naeem, M.: Detection, classification, and location of faults in power transmission lines. Int. J. Electr. Power Energy Syst. **67**, 76–86 (2015)
7. Jamil, M., Sharma, S.K., Singh, R.: Fault detection and classification in electrical power transmission systems using artificial neural network. SpringerPlus **4**(1), 1–13 (2015)
8. Torabi, N., Naghavi, F., Toliyat, H.A.: Real-time fault isolation in multiphase multilevel NPC converters using active semi-supervised fuzzy clustering algorithm with pairwise constraints. In: 2017 IEEE International Applied Power Electronics Conference and Exposition (APEC), pp. 570–577 (2017)
9. Babayomi, O., Oluseyi, P., Ofodile, N.A., Keku, G.: Fault diagnosis in an extra-high voltage power line. In: PowerAfrica, pp. 311–316 (2017)
10. Salazar, J.R.: Sistemas de tierra en líneas de trasnmisión. Xalapa, Veracruz (2010)
11. Valencia, C.H., Quijano, S.N.: Modelo de optimización en la gestión de inventarios mediante algoritmos genéticos. Iteckne **8**(2), 156–162 (2011). ISSN 2339-3483
12. Grainger, J.J., Stevenson, W.D.: Análisis de sistemas de potencia. McGraw-Hill, México (1996)
13. Haykin, S.: Neural Networks a Comprehensive Foundation. Pearson, Hamilton (2001)
14. Wilamowski, B.M., Irwin, J.D., Yu, H.: Levenberg-Marquardt Training, Intelligent Systems, pp. 12-1–12-12. CRC Press Editorial Taylor & Francis Group (2011)
15. Orjuela, A.D., Posada, H.F., Valencia, C.H., Mendoza, L.: On the use of neuroevolutive methods as support tools for diagnosing appendicitis and tuberculosis. In: WEA 2018. Communications in Computer and Information Science, vol. 915. Springer, Cham (2018)

Design of a Model for the Implementation of an Access Control to a WIFI Network

Fabian Blanco Garrido(✉), Wilfred Torres Cerón(✉), Fredys A. Simanca H.(✉), Pablo E. Carreño H.(✉), and Mauricio Alonso Moncada(✉)

Universidad Libre, Bogotá, Colombia
{fabian.blancog,wilfred.torresc,fredysa.simancah,
pabloe.carrenoh,mauricio.alonso}@unilibre.edu.co

Abstract. The usage of technology is moving ahead at a staggering speed, managing to increase its use in daily life, both at a business and personal level, which entails the need to protect the most important assets of the user, as well as to increase the mechanisms that are necessary to safeguard the information. In this article, we detail the design of a Model for the Implementation of Access Control to WIFI networks, which was done by means of a methodology composed of four phases: research, documentation, practice, and feedback. These were used to perform the different tests and validations, and thus adjust the configurations and performance. The model provides infrastructure managers a solution for the management of the access to the WIF resource, since the configuration that is proposed can be assigned in the same SSID to several network segments. The practice was carried out in a controlled manner and over a specific segment of the network. In the testing process and according to the methodology that was used, computer security policies were defined. This allowed to determine a more secure access control to WIFI networks, it is therefore a solution to the acquired need of mobility; likewise, the integration of the RADIUS and WIFI network service provides a complete framework for access control, and the most interesting thing is that this service can be executed on most operating systems almost free of charge.

Keywords: Radius · WIFI · SSID · Network segment · Authentication and authorization

1 Introduction

Information security tends to protect the most important assets of a company, but users in general do not conceptualize the risks they face in the use of computer tools and technologies that are available to guarantee confidentiality, integrity, and availability of information [1].

This should start from the people and the strategic leadership of the company management, allowing to write and inform in a natural language, the technical guidelines of the processes and technology that the entity supports (software, hardware, and information).

D. F. Cortes Tobar et al. (Eds.): AETA 2019, LNEE 685, pp. 185–192, 2021.
https://doi.org/10.1007/978-3-030-53021-1_19

Likewise, companies that seek to solve a functional need by the investment in technology, such as the use of their missionary applications, but that must also comply with the IT security guidelines. On the other hand, the WIFI service has the challenge of safeguarding the information of the user who uses the service [2].

It is worth highlighting the high degree of vulnerability that this service has [3]; the method of sharing the connection code to users, reduces the management burden for the network administrator, since he or she does not have to carry out many authentication controls when a new user arrives, as the configuration of the WIFI infrastructure is not modified; in spite of this advantage, the risk for information assets is clear, because unauthorized persons can enter the WIFI service and violate the pillars of IT security [4, 5].

The security of the information must start from the people and the strategic leadership of the management of the company, allowing to write and inform in a natural language, the technical guidelines of the processes and technology that the entity supports (software, hardware, and information), whose objective is to promote the use and exploitation of information and communication technologies in order to consolidate a competitive company with proactive and innovative users, who generate value in an environment of digital trust [6]. In the case of companies, the WIFI network must be part of its infrastructure, reason why the use of the network must be subject to the information security policies that are established and current regulations [7].

In fact, companies seek to solve a functional need for the use of their mission applications through technological investments, but in the same way, they must comply with the guidelines set out in the IT security principles of each company; since the WIFI service has the challenge of safeguarding the information of the user and of those who use the service [8, 9], pursuing the creation of conditions for a reliable use in the digital environment, by means of an approach based on risk management, preserving the confidentiality, integrity, and availability of the information.

Currently, the companies do not have a control and audit system for the connection to the WIFI network; there is no control on who is connecting to the network or what the user is doing with the service. In addition to this, it is not possible to keep statistics or apply the access control policies that the entity has.

This research allowed us to design a system to control the access to the WIFI network of a company, to authenticate a user, to grant the user permissions to access the different services, and to manage an event log showing the activities that the user is performing with this service.

It is then intended to disclose the main flaws at the security level, and design a model so that the WIFI service can be configured regardless of the manufacturer's brand, and be integrated into an access system that allows the user to be authenticated, giving the user an authorization profile, keeping track of the user's activity on the network and the services that are accessed. In this way, the organization will have better control and will be able to comply with the information security and privacy policies that must be handled, and thus respond quickly and effectively to possible internal attacks.

In order to validate the model, we performed a pilot with an SSID, which was incorporated with a Windows Server 2012 server with the RADIUS service, so each computer that was configured received the connection, authorization, and event registration policies from the server, framed with the information security and privacy policies of the organization and the good practices of the manufacturer for this technology.

1.1 State of the Art

When talking about a computer network, it is important to look at the past in order to understand not only the concept but also the necessity that promoted its existence. It is not hard to remember that any work in which sharing information was required had to be done manually with the help of typing machines or the printing press, so that several volumes of the documents could be printed at a time; also, transporting this information required to have means such as an airplane or a vehicle to travel and deliver the information, making it slow to deliver the message.

With the above, the world was moving very slowly and its technological advances were not very impressive at the level of communications, but all this changed in the Cold War, when the United States worked on a project called ARPANET whose objective was the connection of two devices that were not in the same place in order to exchange information. This war project gave the origin of the real concept of a computer network.

Basically, when talking about a computer network it can be defined as it is stated by [10] "the issue is sharing resources and the goal is that all programs, equipment, and especially data are available to anyone in the network, regardless of the physical location of the resource or the user". One example could be that of a group of office employees who share a printer. None of the individuals really need a private printer, on the other hand, a high-volume network printer is cheaper, faster, and easier to maintain than an extensive collection of individual printers.

It could be defined that a computer network is an interconnection of several end and/or intermediate devices where the main objective is to exchange information; it could also be said that a computer network is to extend local services so that other devices can consult it or use it.

Continuing with the technological advances at the level of computer networks, the improvement in speeds and in applications was important, but it still followed the limitations of a desk, a cable, and a personal computer.

Likewise, the need for mobility was just coming into being, a necessity to be constantly connected to the information in a safe way and regardless of the location. This is when in the late 90s, the first wireless network was made possible by Nokia, who gave the same connection benefits as if the device was connected by a cable, but with the subsequent challenge of improving the speed and integrity of the information.

After that, this first network was standardized with the IEEE 802.11 and its name was changed to WIFI ALLIANCE, the idea of the new project of WIFI technology was its popularization and standardization so that any corporate or personal device could use this service.

The work on the possible threats in wireless networks and the security protocols created for the WLAN networks presented by [11] carries out a comparative analysis of the WEP, 802.11i, WPA, and WPA2 protocols, in relation to the authentication method and the encryption technique, concluding with the selection of security protocols for the WLAN network applied to Universidad de Cienfuegos.

Security analyzes have also been conducted in wireless networks of SMEs by [12], who carries out a preliminary study in order to determine from a sample, analyzing: a) the level of vulnerability, b) the characteristics that they have, and c) the features of the network equipment. With this study, it would be possible to evaluate which are the technical and operational improvements that should be used in order to mitigate unauthorized access.

Wireless networks are currently being used with more frequency, and this is the case of the control applications for industrial environments. In this review article, the advantages and disadvantages between wireless and wired networks are presented, where a comparative analysis of the most used protocols in wireless networks for the control of industrial processes in real time is performed, showing the investigations that are related to the subject [13].

2 Materials and Methods

This section describes the materials and study methods that were used for the design of a model for the implementation of an access control to a WIFI network, detailing the methodology and design used for the realization of this model and its subsequent validation. This served to perform the different practices and experiments with this technology and to adjust configurations and performance.

2.1 Design of the Model

The basis for the design of the model was the compilation of the information that would allow the study of the problematic situation, in which companies require the implementation of access control to the WIFI network.

The phases that were carried out for the design of the model were: a) Research, b) Documentation, c) Practice, and d) Feedback.

Phase 1. Research

The basis of this phase was the consultation of bibliographic material, such as: books, documents, articles, and other publications that were available on WIFI technology. Afterwards, this theoretical basis was useful to carry out different practices and experiments, and thus be able to adjust configurations and performance.

Phase 2. Documentation

In this phase, we documented everything related to the use of WIFI standards, servers, and security policies, such as: IEEE protocols, information security and privacy policies, and the access system.

Phase 3. Practice

For this phase, we designed the structure of the WIFI networks with the corresponding logical topology of WIFI connection. Figure 1 shows the implemented topology, the idea of a data center where the critical equipment of the entity would be guarded and located was proposed, one of them is a new generation firewall that has IPS modules, content filter, application control, firewall, and VPN module, there is also a Core switch that is the most critical communication equipment that the entity would have, since this would connect all the floors of the company and also perform the communication between the different branches.

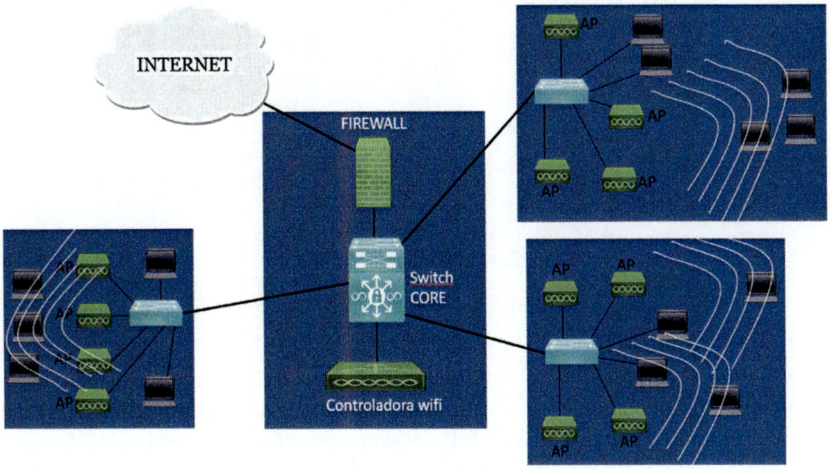

Fig. 1. Logical topology of WIFI connection

Finally, we have the WIFI controller, which is the centralized management of the Access Point. From this controller we can manage and deliver all the features, groups, SSID, and other factors that help users get connected.

For the configuration, two SSD **Enrolpro** and **Enrolvis** were created, in two different VLANs, with different addressing and working only with a PSK entry code, the RADIUS service was enabled on a server with Windows 2012 operating system and the network devices with the service of RADIUS.

Phase 4. Feedback

The base of this phase is the validation of the model where we must take into account:

a) Initially, we must take into account that the user must be part of the administrative group that is defined to perform access control and must also have the option to use NPS.
b) IP addressing according to the corresponding VLAN.
c) The logs that RADIUS has evidence that it is executing the profiles that are required.

The model for the implementation of access control to the WIFI network is shown in Fig. 2. Implementation model.

As shown in Fig. 2, the model for the implementation of access control to the WIFI network is composed of four phases, to be taken into account: 1) Information security policy, 2) Network access policy, 3) Network architecture, and 4) Configuration of Radius services on servers.

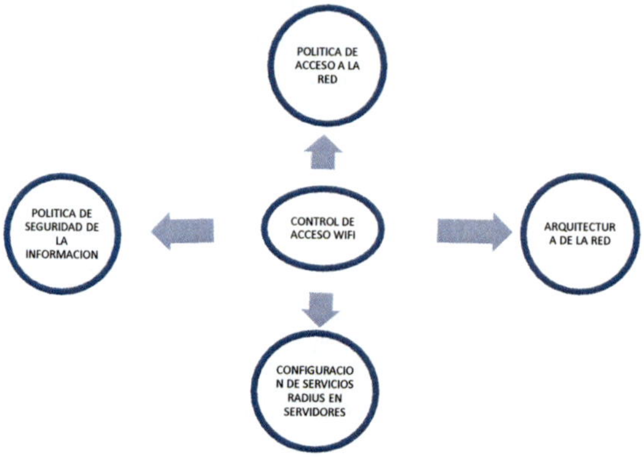

Fig. 2. Implementation model

Phase 1. Information security policy.
An information security policy should define the objective, direction, principles, and basic rules for the management of information security (taken from ISO 27001), this should be declared by the directives of the organization. Counting with an information security and privacy policy for organizations is to have the guide to action at a personal and professional level for each of the actors that interact with the organization.

Phase 2. Network access policy.
In the network access policy, it is necessary to identify the elements with which to authenticate and identify all users and devices in order to allow access or the use of logical or physical resources.

Phase 3. Network architecture.
In the network architecture, it is necessary to establish the topology to be implemented, taking into account the active devices of the network, the data center or critical servers of the organization, such as the firewall, content filter, application control, firewall, and VPN module, and the Core switch that is the most critical communication equipment that every organization has, since this is the main connectivity equipment.

Phase 4. Configuration of Radius services on servers.
In the configuration, it is necessary to define the operating system that is going to be used for the configuration of the services and network devices with the Radius service for integration with the WIFI network services.

3 Results and Discussion

It should be noted that wireless Internet networks are still under development for higher transmission speeds, in order to obtain technologies that improve connectivity, reliability, and security [14], however, good service and security depend on the proper use of the

WIFI network, since it is necessary, as indicated in the methodology, to implement an IT security policy in each situation depending on who is going to use the WIFI network.

There are platforms that allow both detecting the access points that are available in a certain place, and injecting false traffic in the middle of the encrypted communications, passing through breaking down the encryption of communications within the range of the device that is carrying the tool. All these sets of tools are used to give the administrator an idea of how strong or weak their WIFI installations are at the IT security level. That is to say, they are the basis for auditing the installations at the security level, with the WIFI network as the primary tool.

Organizations already have the WIFI service in their information assets, and it is considered one of their most critical assets; there are even many organizations that structure cabling only for critical equipment and for what is strictly necessary, such as printers, servers, switches, etc.

With the RADIUS solution, the organization can obtain one of the greatest advantages: the AAA (Authentication, Authorization, and Registration). With these components, it is possible to have a Wi-Fi network that complies with the security policies.

The integration of the RADIUS service and WIFI provides a complete framework for access control, and the most attractive part is that this service can be executed on most operating systems at almost no cost.

The integration of RADIUS into the infrastructure gives a granularity in assigning user groups authorization, achieving that when getting connected, the user could leave it in his or her own segment with a single SSID.

4 Conclusions

The WIFI Network is a solution to the acquired need of mobility, this technology allows to have the information anywhere and provides acceptable security standards.

The evolution of this technology has been rising in a dizzying way, and equipment and services are bing increasingly developed to work in it, even for cellular phones it is a must to have this technology activated.

The characteristics of both protocols and speed have the challenge and are very close to having the same conditions, of sending information by structured cabling.

Undoubtedly, the advance of the WIFI network is a globalization tool that has allowed worldwide connection in all areas, therefore, its good use, the education and addressing of the user is an important factor for the advancement of information safety and security control.

References

1. Zhuo, L., Xiang, L., Wenye, W., Cliff, W.: Review and evaluation of security threats on the communication networks in the smart grid. In: 2010 IEEE Military Communications Conference, MILCOM 2010, San Jose, CA, United States (2010)
2. Vinjosh Reddy, S., Sai Ramani, K., Rijutha, K., Mohammad Ali, S., Pradeep Reddy, C.: Wireless hacking-a WiFi hack by cracking WEP. In: 2010 2nd International Conference on Education Technology and Computer, Shanghai (2010)

3. Spaulding, J., Krauss, A., Srinivasan, A.: Exploring an open WiFi detection vulnerability as a malware attack vector on iOS devices. In: 2012 7th International Conference on Malicious and Unwanted Software, Fajardo, PR, United States (2012)
4. Aminanto, M.E., Kim, K.: Detecting impersonation attack in WiFi networks using deep learning approach, vol. 10144, pp. 136–147. Springer, Cham (2016)
5. Mojica San, J.F., Salcedo Parra, O.J., Acosta López, A.: Intrusion prevention model for WiFi networks, vol. 11251, pp. 66–73. Springer, Cham (2018)
6. MINTIC: Guía para la Implementación de Seguridad de la Información en una MIPYME. MinTIC, Bogotá (2016)
7. Simanca H., F.A., Blanco, F., Triana, E.: Las Redes MESH II. Universidad Libre, Bogotá (2018)
8. Chang, X., Wang, T., Rodríguez, R., Zhang, Z.: Modeling and analysis of high availability techniques in a virtualized system. Comput. J. **61**(2), 180–198 (2018)
9. Simanca H., F.A., Blanco, F., Triana, E.: Las redes MESH. Universidad Libre, Bogotá (2018)
10. Tanenbaum, S., Wetherall, D.J.: Redes de Computadoras. Pearson, México (2012)
11. González Paz, A., Beltrán Casanova, D., Fuentes Gari, E.R.: Propuesta de protocolos de seguridad para la red inalámbrica local de la universidad de cienfuegos. Revista Universidad y Sociedad **8**(4), 130–137 (2016)
12. Hernández Mendoza, M., Rodríguez Vidal, L.M., Aguilar Almanza, M.: Análisis de seguridad en redes inalámbricas de las MiPyME y propuesta de mejora. Revista Iberoamericana de Producción Académica y Gestión Educativa **7** (2016)
13. Monsalve Posada, J.F., Arias Londoño, A.: Redes inalámbricas para el control de procesos. Revista Politécnica **9**(17), 83–93 (2013)
14. Blanco, F., Ruiz, D.: Formulación de una Metodología para Diseñar e Implementar redes MESH como alternativa de solución para redes comunitarias o rurales. In: Congreso Iberoamericano de Ciencia, Tecnología y Educación, Buenos Aires (2014)
15. Cantabella Sabater, M.: Modelos y herramientas para la representación y análisis de datos en LMS para enseñanzas universitarias. Universidad Católica de Murcia, Murcia (2018)

Performance Study of Monocrystalline and Polycrystalline Solar PV Modules in Tropical Environments

Yimy Edisson García Vera[1(✉)], Oscar Daniel Díaz Castillo[2], Luz Ángela Cruz Pardo[3], and Luisa Fernanda Sanabria Pérez[3]

[1] Universidad de San Buenaventura - Bogotá, Carrera 8h No. 172 - 20, Bogotá D.C., Colombia
yegarcia@usbbog.edu.co

[2] Unipanamericana Compensar Institución Universitaria, Av Calle 32 No. 17 - 30, Bogotá D.C., Colombia
odiaz@ieee.org

[3] Universidad de Cundinamarca, Diagonal 18 No. 20-29, Fusagasuga, Colombia
langelacruz@ucundinamarca.edu.co, nanda1929@hotmail.com

Abstract. Photovoltaic solar technology has emerged as a sustainable and environmental friendly source of energy, capable of mitigate climate change caused by the intensive use of fossil fuels, in order to keep up with the current world's demand for energy. Current research effort is focused on the behavior of photovoltaic modules under real working conditions.

Different photovoltaic module technologies like monocrystalline and polycrystalline were tested under the same operating conditions: autonomous systems with comparable energy conversion capabilities, same electric load and similar data acquisition systems, operating under the same weather conditions (on Fusagasugá, Colombia) in a synchronous fashion over a three month period.

Variables like temperature, humidity, solar irradiance, power and I-V operation point were taken to estimate the conversion efficiency for each module related to those environment parameters. This paper presents comparison results between monocrystalline and polycrystalline technologies regarding operating parameters and environmental conditions (closely related to tropic environments), in order to determine the recommended photovoltaic solar technology to implement in a certain energy generation context.

1 Introduction

Nowadays up to 6.6 billions of the world's total population have access to electricity, however, about 1 billion people do not have it or have limited access [1]. A large part of this energy is obtained from fossil fuels such as gas, coal and oil, which are responsible for greenhouse gas emissions and global warming. Many countries are concern about reducing their dependence on fossil fuels, so many of them have opted for use of renewable energies, to ensure their energy supply and

D. F. Cortes Tobar et al. (Eds.): AETA 2019, LNEE 685, pp. 193–203, 2021.
https://doi.org/10.1007/978-3-030-53021-1_20

also to contribute to the reduction of gas emissions responsible of the greenhouse effect, which in turn contribute to climate change mitigation [2].

Photovoltaic (PV) solar energy has grown vertiginously recently, due to its reduded cost and increased efficiency. Nowadays PV solar energy has become a viable option for energy production, with added benefits from its low gas emissions level and extended lifetime (about 20 to 25 years). Recently it was included as a viable energy for energy supply in countries such as Colombia, where due to its geographical position presents great potential for this type of renewable energy [3].

Solar cells are devices that convert photons into DC electric power; they are based on thin films based on silicon, generating voltages in small ranges between 0.5 and 0.8 volts. Solar cells efficiency vary with manufacturing technology, polycrystalline and monocrystalline silicon are the most common type, with efficiencies about 13% and 17% respectively. Solar modules are manufactured and tested based on standard tests conditions (STC): solar irradiance of $1000\,\mathrm{W/m^2}$, ambient temperature of 25 °C, and an air mass of 1.5. However these conditions are not met since real conditions depend on the geographical location where they are installed and their efficiency can be affected by weather and climate conditions [4].

This paper evaluates the energy performance of two PV module technologies widely used in solar energy installations in Colombia, also commercially available in the Colombian market, such as monocrystalline and polycrystalline silicon. Other technologies such as amorphous silicon, thin-film, CdTe, CIGS or organic was not included in the study due to their lack of availability in the Colombian market. The variables evaluated are temperature and humidity. Tests were conducted for about 3 months, at different hours of the day at the University of Cundinamarca in Fusagasugá headquarters.

2 Background

Performance of PV cells largely depends on the climatic conditions to which they are subjected, mainly due that they are tested under laboratory environments (using STC). However these conditions differ from the real ones at the installation site, which causes performance deficiencies due to variations in environmental conditions such as irradiation, temperature and humidity [5]. Therefore knowledge and understanding of the performance of the PV module in the operating conditions of the installation site is of great importance for the correct selection of the solar module technology.

Different studies have been carried out to determine how atmospheric conditions affect PV modules. Perko et al. [6] presented a comparative analysis between solar modules, and concluded that thin film is the most efficient technology compared to monocrystalline and polycrystalline. Amin et al. [7] included a comparison of more than 3 solar cell technologies and study the operation of PV systems under different climatic conditions with polycrystalline, monocrystalline, amorphous silicon and CIS(Copper, Indium, Selenium) modules; this analysis conducted in Malaysia concludes that for this latitude the CIS cells had better performance. Other research presented by Bashir et al. [8] compared three technologies: monocrystalline, polycrystalline and amorphous silicon, concluding that the first two show better efficiency with high irradiance, while the amorphous silicon cells have a higher performance ratio with low irradiance due to its better sunlight absorption capacity.

Temperature is an important parameter in the performance of a PV cell, as shown by several investigations [9–11]. Higher temperatures produces a decrease in the open circuit voltage, which causes lower performance on the solar cells. Sansui et al. [12] studied the effect of the ambient temperature on the performance of PV modules in tropical regions, concluded that a proportionality relation exists between the ambient temperature and the power output of a PV module. Other important factors that affect the operation of PV modules such as ambient temperature and dust accumulation are analyzed in [13,14]; in these works the importance of maintenance in desert areas is highlighted. Dust and particules accumulation can affect the performance of solar modules. The wind is another factor considered in research as presented by Mattei et al. [15]; which present some temperature models with energy balances taking into account the ambient temperature, irradiation and wind speed.

Many studies focused on the behavior of solar modules are done taking into account parameters such as irradiation, temperature and humidity. Mustapha et al. [16], compared polycrystalline and monocrystalline technologies, the results indicated a direct relationship between the current generated by the modules and the irradiation and an independence of the temperature. However there is a decrease in the voltage due to an increase in temperature. Other studies such as the one carried out by Taşçıoğlu et al. [17] show a better performance of monocrystalline polycrystalline modules for a latitude like of Turkey.

The work carried out by Kalu et al. [18] is a project that used the Pvsyst simulation software, and performed a comparative analysis of different photovoltaic technologies. Polycrystalline, monocrystalline and thin film types were chosen for the modeling and simulation tests, with meteorological data used from the NASA database; the results show that for the geographical location of installation the best technology benefits was the polycrystalline.

There are other works carried out in Colombia where most common technologies found in the market are compared. The thesis proposed by Barbosa Janett [19] did a comparative study between monocrystalline and polycrystalline technologies. However no conclusive results on which of the two technologies had better energy performance, except for the similar behavior for open circuit conditions. On the other hand, the thesis done by Martinez Raúl [20] an efficiency study of PV modules was done in Bogotá, taking into account climatic variations such as temperature, humidity and irradiance; which found a better performance of monocrystalline modules. However, it is found that its performance is below that specified by the manufacturer.

The previous works related to the performance of solar modules of commercial technologies such as polycrystalline and monocrystalline have served as reference for this study and enhance the importance of knowing the operation of PV modules in outdoor conditions. However, many of the previous investigations only compare how solar modules work under open circuit and short circuit conditions that are common laboratory tests. It is necessary to analyze the performance of solar cells in real conditions, that is, under climatic conditions of the site or installation site and under load variations that can affect the output voltage.

3 Materials and Methods

3.1 Experimental Design

Two solar modules of similar characteristics have been prepared, whose specifications can be seen in Table 1.

Table 1. Comparison of the PV modules used

Characteristics	Polycrystalline	Monocrystalline
Peak power (W)	20	20
Maximum power current (A)	1.111	1.16
Maximum power voltage (V)	18	17.2
Open circuit voltage (V)	22.50	21.50
Short circuit current (A)	1.202	1.25
Dimensions (mm)	440 × 350 × 25	510 × 290 × 25

The theoretical efficiency of each PV module is calculated by Eq. 1, Where $P_{\max}$ is the maximum output power, G is the standard irradiance value ($1000\,\mathrm{W/m^2}$) and A is the module area in m^2. Using Eq. 1, calculated module efficiency for each technology is 13.52% for monocrystalline and 12.98% for polycrystalline.

$$\eta = \frac{P_{\max}}{G \times A} \times 100\% \tag{1}$$

By comparing the modules areas, a bigger efficiency of the monocrystalline module is observed for practically the same irradiation capture area (0.148 vs 0.154 m^2). Figure 1 shows a view of the installation site. For an optimal capture, the coordinates was taken into account (4 ° 20' 14.1" N, 74 ° 22' 17.8" W), south orientation was used, choosing a 15 ° inclination angle as recommended by the UPME.

Two low power autonomous PV systems were used for this work, one for each PV technology. Each system has its charge regulator, battery, and variable DC loads. For the dimensioning and autonomy, the IDEAM radiation database was taken into account [21], the system was designed with the worst radiation month method, which for the case of Fusagasugá is November and December since for both months there is a lower average radiation [22].

Fig. 1. Panoramic view of the system

Figure 2 shows a schematic view showing both the photovoltaic modules and the charge regulators, the storage system that are 12V/12Ah Netion type batteries, this is a dry maintenance-free deep cycle battery with high performance and electrical resistance against excessive discharges. The charge controllers are of PWM type with a current range of 10 A at 12 V. The test loads used are LED variable DC switching lamps, each with power variation between 6 W up to 20 W for each module.

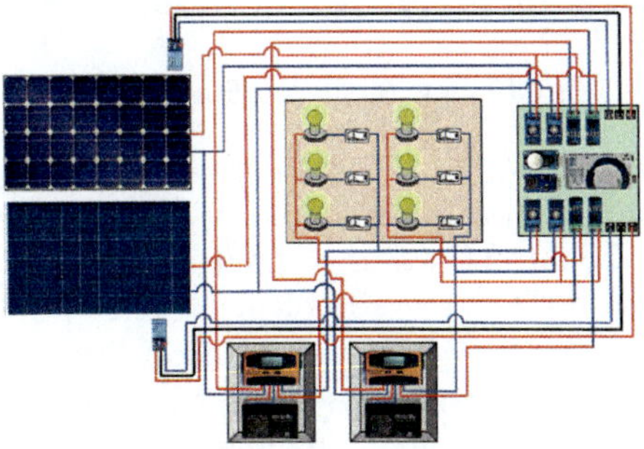

Fig. 2. Schematic view of the system

The system was designed with the following criteria: assuming one day autonomy, 25% of oversize design, irradiance of 4.6 kW/m^2, and choosing the month of worst radiation. The radiation table for the site is obtained from the SWERA tool [23]. Equation 2 calculates the system power output, where F_s is the oversize design factor, Consumption is the energy demanded by the load and the Irradiance is the one calculated based on the geographic location.

$$P = \frac{F_s \times \text{Consumption}}{\text{Irradiance}} \tag{2}$$

From the Eq. 2 and taking into account that the system power output must be 20 W, clearing the consumption gives 73.6 Wh per day. The 25% safety factor of compensates for the batteries and other components losses. The minimum battery capacity is set with Eq. 3, where Consumption is the calculated energy from Eq. 2, Days refers to the system time autonomy (one day), PD refers to the battery discharge depth and V is the generator voltage. Using Eq. 3, calculated minimum battery capacity is 7.7 Ah, so a commercial value of 10 Ah has been selected.

$$C_{\text{bat}} = \frac{\text{Consumption} \times \text{Days}}{PD \times V} \tag{3}$$

3.2 Data Acquisition System (DAQ)

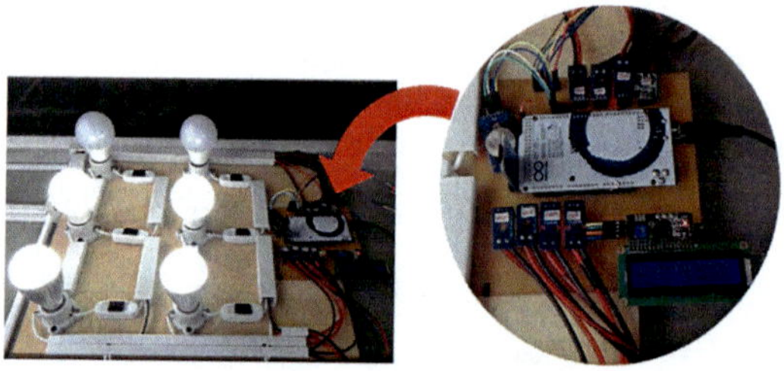

Fig. 3. View of the working system

To measure the temperature and humidity variables for each modules, as well as the voltages and currents (thus estimating power) supplied to the load, a data acquisition system (DAQ) was implemented [24], as shown in the Fig. 3.

This DAQ is based on an Arduino Mega 2560 R3 card, with four voltage sensor modules, four ACS712 current sensors and 2 temperature and humidity sensors based on the DHT11 module. These sensors measure currents and voltages in both the module and the load, with temperature and humidity data taken from both PV modules.

Additionally, a DS3231 real time clock (RTC) was used to improve time measurements, as well as a micro SD module for data storage. Irradiance and ambient temperature were taken manually. Data were acquired during March, April and May of 2018, with an 8 h on average with sunny and cloudy climatic conditions. Table 2 summarizes the instruments used.

4 Results and Discussion

Both systems were analyzed taking into account environmental data, like temperature and humidity. This environmental data was provided by IDEAM [25]. Average temperature in Fusagasugá was obtained with values between 15.2 °C and 21.8 °C. Air temperature remained at an average of 18.11 °C. Relative humidity data can be seen in Fig. 4, having an average of 77%, with two days of low humidity on March 15 and April 19 with a value of 19.98% and 18.48% respectively, and the highest humidity on April 14 with 97.80%.

Table 2. Overview of the measurement modules of the data acquisition system (DAQ)

Module	Specifications	View
Controller	Arduino Mega 2560 R3	
Temperature and humidity sensor	Model DHT11 Temperature range: 0 - 50 °C humidity range: 20 - 50 %)	
Voltage sensor	Voltage range: 0 - 25 V resolution: 4.9 mV	
Current sensor	Model ACS712 Current range: 0 - 20 A	
Solar Power Meter	Model SM206 range: 1 - 3999 W/m^2	

Temperature variaton of each module compared with ambient temperature differ greatly due to the heat generated by the photovoltaic process as explained in [8], as shown in the Fig. 5. Most of the radiation that reach a solar cell is not converted to electricity but heat, thereby increasing the module's own temperature [24].

Some studies show the inverse relationship between the relative humidity and the efficiency of the photovoltaic modules [26], the increase of humidity levels decreases the current and the output voltage of the modules and consequently the power output, due that the water vapor particles produce absorption, reflection (diffuse and non-diffuse), refraction and collision of solar radiation, which result in the reduction of solar irradiance [27]. Figure 6 shows how power output decreases when the humidity levels increase.

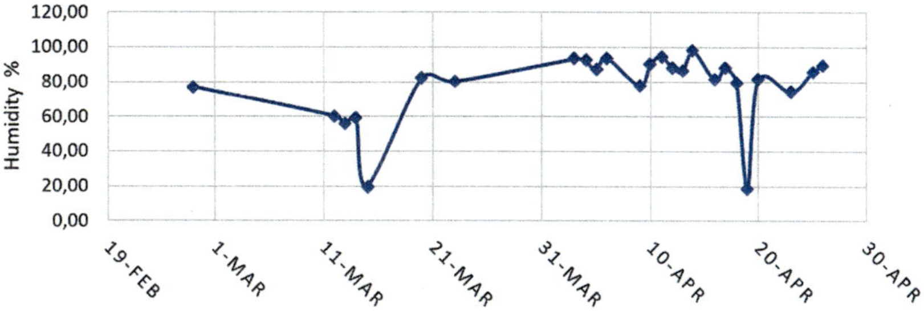

Fig. 4. Average humidity measured in Fusagasugá

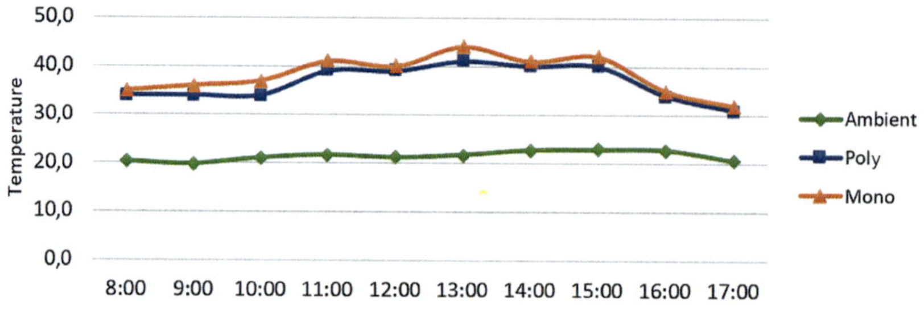

Fig. 5. Temperature comparison for each module (on average day)

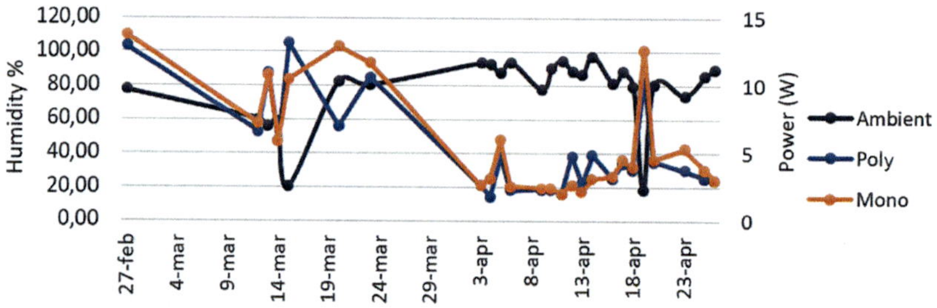

Fig. 6. Humidity and power output for each PV module

Figure 7 shows a simulation result done in Pvsyst, in order to compare the irradiance levels and power output with the real measurements done in the experiment site.

Figure 8 shows power output compared with irradiance along an average day, which shows that monocrystalline technology has better performance, mainly in noon hours, with peaks at 11:30 am, where the monocrystalline module outputs 14 W compared with 12 W delivered by polycrystalline module.

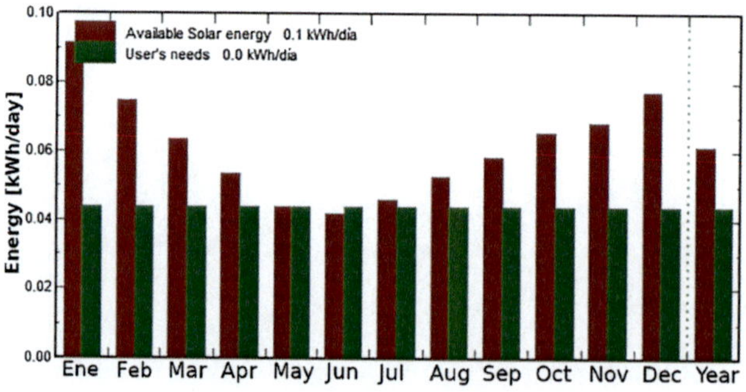

Fig. 7. Simulation using Pvsyst

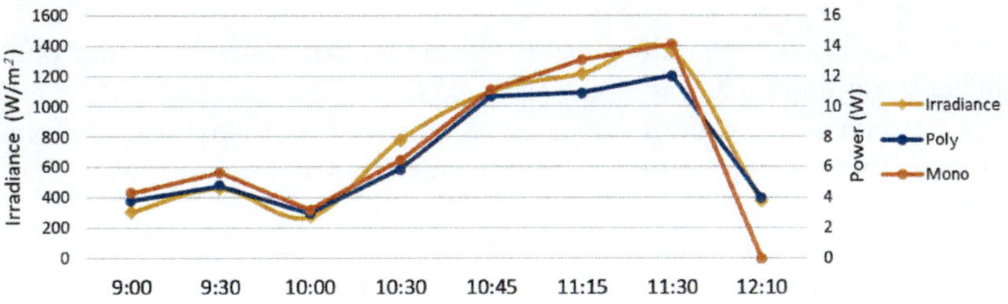

Fig. 8. Irradiance vs power output

To verify power output of both modules, a test with a 7 W load was done, where the monocrystalline module performs better than polycrystalline. The results are shown in Fig. 9.

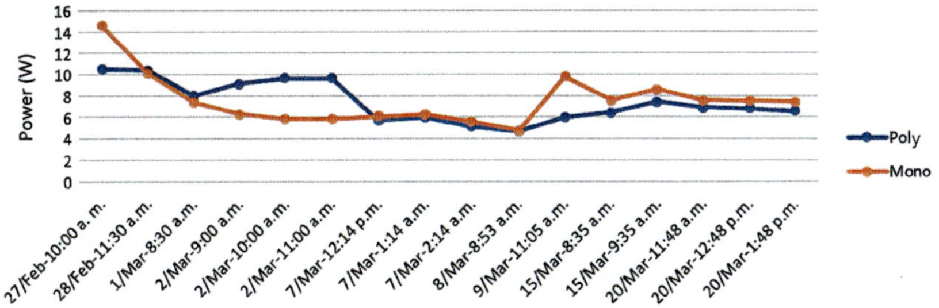

Fig. 9. Power output with load

5 Conclusions

This paper show a comparative study of the performance of two photovoltaic modules (monocrystalline and polycrystalline), under climatic conditions of Fusagasugá, supported by two autonomous photovoltaic systems with similar load characteristics. Both modules were almost always above the ambient temperature due to the heat produced during the photovoltaic process. However, temperature dropped in the afternoon hours in the same way as the power output. This did not show significant losses due to temperature.

It was shown that relative humidity has an influence on the performance of the photovoltaic modules, due to reflective and refractive factors that affect irradiance levels which results in a decrease in power output. The results show an inverse correlation between relative humidity and current supplied by each PV module.

Both technologies shown similar characteristics regarding irradiance and temperature variations; it is more notorious by increasing the load connected to the two systems. Although the difference between the two systems can not only be verified in the event of load variations (it could have been done under short circuit and open circuit test conditions done in laboratory conditions), the experimental setup includes load variation measurements to take account of real working conditions.

Regarding load variations the monocrystalline module had better performance under load variations, with a difference close to 2 W compared to the polycrystalline module. This difference is more noticeable when the load was 14 W. It is inferred that the small differences in both technologies had its effects increased under high power loads.

This results can help predict the output characteristics of the autonomous photovoltaic system, under environmental variations, which will allow improving designs in technical, operational and financial terms. However, it is recommended to perform extender tests for longer periods of time in similar latitudes of geographic areas of interest, with more detailed measurements, particularly measuring variations of solar irradiance that allow more conclusive results regarding this parameter.

References

1. IEA: Energy Access Outlook 2017. Technical report (2017). Cites: ieaEnergyAccessOutlook2017
2. IEA: IEA sees great potential for solar, providing up to a quarter of world electricity by 2050 — EMIS, May 2010
3. UPME: Integración de las energías renovables no convencionales en Colombia (2015)
4. Fuentes, M., Nofuentes, G., Aguilera, J., Talavera, D.L., Castro, M.: Application and validation of algebraic methods to predict the behaviour of crystalline silicon PV modules in Mediterranean climates. Sol. Energy **81**(11), 1396–1408 (2007)
5. Villar, P.D.: Energía solar fotovoltaica. Roble (2010)
6. Perko, J., Žnidarec, M., Topić, D.: Comparative analysis of electricity production from different technologies of PV modules. In: 10th International Conference on Sustainable Energy and Environmental Protection (2017)
7. Amin, N., Lung, C.W., Sopian, K.: A practical field study of various solar cells on their performance in Malaysia. Renew. Energy **34**(8), 1939–1946 (2009)
8. Bashir, M.A., Ali, H.M., Khalil, S., Ali, M., Siddiqui, A.M.: Comparison of performance measurements of photovoltaic modules during winter months in Taxila, Pakistan. Int. J. Photoenergy **2014** (2014)
9. Singh, P., Singh, S.N., Lal, M., Husain, M.: Temperature dependence of I-V characteristics and performance parameters of silicon solar cell. Sol. Energy Mater. Sol. Cells **92**(12), 1611–1616 (2008)
10. Rehman, S., El-Amin, I.: Performance evaluation of an off-grid photovoltaic system in Saudi Arabia. Energy **46**(1), 451–458 (2012)
11. Wen, C., Fu, C., Tang, J., Liu, D., Hu, S., Xing, Z.: The influence of environment temperatures on single crystalline and polycrystalline silicon solar cell performance. Sci. China Phys. Mech. Astron. **55**(2), 235–241 (2012)

12. Sanusi, Y.K., Fajinmi, G.R., Babatunde, E.B.: Effects of ambient temperature on the performance of a photovoltaic solar system in a tropical area. Pac. J. Sci. Technol. **12**(2), 176–180 (2011)
13. Mani, M., Pillai, R.: Impact of dust on solar photovoltaic (PV) performance: research status, challenges and recommendations. Renew. Sustain. Energy Rev. **14**, 3124–3131 (2010)
14. Jiang, H., Lu, L., Sun, K.: Experimental investigation of the impact of airborne dust deposition on the performance of solar photovoltaic (PV) modules. Atmos. Environ. **45**, 4299–4304 (2011)
15. Mattei, M., Notton, G., Cristofari, C., Muselli, M., Poggi, P.: Calculation of the polycrystalline PV module temperature using a simple method of energy balance. Renew. Energy **31**, 553–567 (2006)
16. Mustapha, I., Dikwa, M.K., Musa, B.U., Abbagana, M.: Performance evaluation of polycrystalline solar photovoltaic module in weather conditions of Maiduguri, Nigeria. Arid Zone J. Eng. Technol. Environ. **9**, 69–81 (2013)
17. Taşçıoğlu, A., Taşkın, O., Vardar, A.: A power case study for monocrystalline and polycrystalline solar panels in Bursa City, Turkey. Int. J. Photoenergy (2016)
18. Kalu, C., Ezenugu, I.A., Umoren, A.M.: Comparative study of performance of three different photovoltaic technologies. Math. Softw. Eng. **2**, 19–29 (2016)
19. Barbosa Urbano, J.: Estudio comparativo entre variables fotovoltaicas de dos sistemas de paneles solares (monocristalino y policristalino) en Bogotá. Ph.D. thesis, Universidad Sergio Arboleda, May 2013
20. Martínez Bernal, R.: Estimación de la eficiencia de un sistema de energía eléctrica basado en paneles solares considerando variables ambientales en la ciudad de Bogotá. Ph.D. thesis, Universidad Santo Tomás (2017)
21. IDEAM: Atlas Interactivo - Radiación IDEAM (2018)
22. NASA: NASA POWER — Prediction Of Worldwide Energy Resources (2018)
23. OpenEI: Solar and Wind Energy Resource Assessment (SWERA) — Open Energy Information (2018)
24. Soumia, B., Nallapaneni, M.K., Ali, T.: Data acquisition system: on the solar photovoltaic module and weather parameters monitoring. Procedia Comput. Sci. **132**, 873–879 (2018)
25. IDEAM: Atlas Interactivo - Climatológico - IDEAM (2018)
26. Kazem, H.A., Chaichan, M.T., Al-shezawi, I.M., Al-saidi, H.S., Al-rubkhi, H.S., Al-sinani, K., Al-waeli, A.H.A.: Effect of Humidity on the PV Performance in Oman (2012)
27. Amajama, J., Oku, D.E.: Effect of relative humidity on photovoltaic panels' output and solar illuminance/intensity. J. Sci. Eng. Res. (IJSEAS) **3**(4), 126–130 (2016)

An Identification of Bioimpedance Spectroscopy Parameters by the Least Squares Method

Stepan Ozana[1(✉)], Michal Prauzek[1], Jaromir Konecny[1], and Aleksandra Kawala-Sterniuk[2]

[1] Department of Cybernetics and Biomedical Engineering, VSB-Technical University of Ostrava, Ostrava-Poruba, Czech Republic
stepan.ozana@vsb.cz

[2] Faculty of Electrical Engineering, Automatic Control and Informatics, Opole University of Technology, Opole, Poland
http://www.vsb.cz

Abstract. This contribution deals with identification of bioimpedance spectroscopy parameters for implementation on embedded platforms. The paper's goal is to present method which is suitable for implementation in microcontrollers or field programmable gate arrays. The paper brings brief description of the state of the art and it reviews an equivalent electrical model and the least squares method as background of the work. In experimental part, the author presenting an experimental setup for data acquisition for electrical model and an identification approach including the implementation details of calculation of inverse matrix. In the result section the comparison between real and identified parameters is presenting and the results are evaluated in terms of accuracy. The proposed method can be considered as precise identification for the particular electrical model. The paper conclusion includes discussion about method accuracy and brief outline for future work.

Keywords: Bioimpedance · Identification · Least squares method · Spectroscopy

1 Introduction

Analysis of bioelectrical impedance becomes standard diagnostic method in general health-care and it is also nowadays applicable in many home-care or wearable electronic devices. Unlike other diagnostic methods (X-ray etc.) this approach works without ionizing radiation, which has negative effects on a human body. Recent progress in consumer electronic development leads to implementation of bioimpedance analysis to personal weights to extend or substitute body mass index (BMI) as major indication of body composition [8]. A basic principle implemented in consumer electronic is based on impedance analysis in one or more frequencies [6].

D. F. Cortes Tobar et al. (Eds.): AETA 2019, LNEE 685, pp. 204–213, 2021.
https://doi.org/10.1007/978-3-030-53021-1_21

The separate frequency analysis of bioelectrical impedance is valid and precise method but there are several disadvantages. The major downside lies in incapability to measure bioimpedance in multiple frequencies at the same time. This fact can lead to error in multi-frequency functional analysis such as impedance cardiography [9] or electrical impedance tomography [2]. Therefore there is significant need for real-time method that can evaluate bioimpedance on multiple frequencies in the same time.

This research paper present method for bioimpedance analysis by electric current impulse. Such approach can allow to calculate bioimpedance values on several frequencies in the same time and it can lead to multi-frequency impedance tomography [5]. In this contribution, the authors also present implementation principle of proposed approach as real-time implementation for microcontroller (MCU) or field-programmable gate array (FPGA).

This contribution is organized into five sections. Background section introduces a principle of bioimpedance analysis, a skin equivalent electrical model and a least squares method. Experimental section describes experiment data and proposed identification method including implementation perspective. The results are then discuses in Sect. 5. The final section brings major conclusions and outlines possible directions for future work.

2 Background

2.1 Skin Equivalent Electrical Model and Bioimpedance Analysis

The goal of bioimpedance analysis is to obtain impedance component in various frequencies. In general, a skin can be considered as conductive material with particular distribution of resistivity and capacitance in a serio-parallel architecture.

In this contribution we are using particular electrical model for human body electrical characteristics description. Such model was published in various references such as [3,7] or [1].

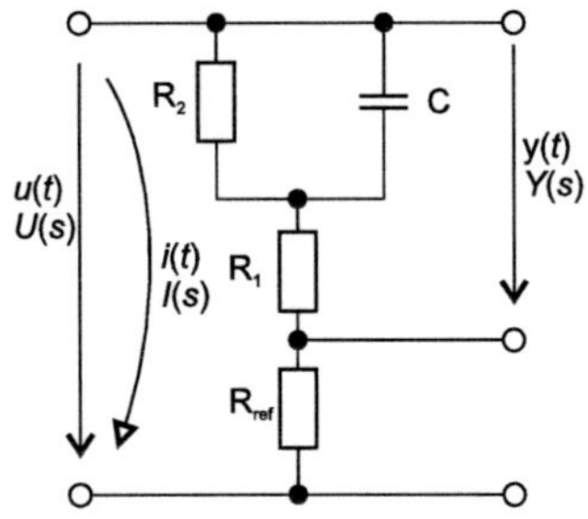

Fig. 1. Skin equivalent electrical model

The electric characteristics of biopotential electrodes are generally nonlinear and depends on the electrical current at their surface. The model assumes the low potential and low current and thus we are representing the biopotential electrode by linear model.

The idealized linear electrical model is depicted in Fig. 1. In this circuit R_2 and C are components that represent the impedance associated with the electrode-electrolyte interface and polarization at this interface. R_1 is the series resistance associated with interfacial effects and the resistance of the electrode materials themselves [7].

In this experiment we are using data from testing setup (see Fig. 2). This setup includes an impulse generator, a source of electrical current to guarantee a current limitation for safety reason, the electrical equivalent model and an osciloskope.

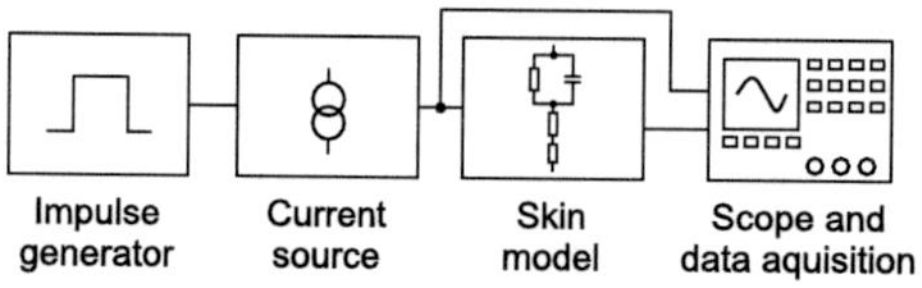

Fig. 2. Data acquisition experimental setup for evaluation of equivalent electrical skin model

The impulse generator triggers current source which drives electrical model byl current $i(t)$ introduced in Fig. 1. This situation results in voltage response in the circuit's input and output. Such responses are acquired by scope and digitalized data are stored on target's format.

2.2 Least Squares Method

The Least Squares method is a standard approach in regression analysis to approximate the solution of overdetermined systems, i.e., sets of equations in which there are more equations than unknowns. "Least squares" means that the overall solution minimizes the sum of the squares of the residuals made in the results of every single equation.

Let's suppose overdetermined set of linear algebraic equations

$$\boldsymbol{A} \cdot \boldsymbol{u} = \boldsymbol{b} + \boldsymbol{e} \tag{1}$$

where $\boldsymbol{A}$ is $m \times n$ matrix ($m >> n$), $\boldsymbol{b}$ is m-sized right side vector and $\boldsymbol{u}$ is n-sized unknown vector. Error vector $\boldsymbol{e}$ is a random signal with following properties:

$$\mathcal{E}\left\{e\right\} = 0 \tag{2}$$

$$cov\left\{e\right\} = \mathcal{E}\left\{\boldsymbol{e} \cdot \boldsymbol{e}^T\right\} = \sigma^2 \cdot \boldsymbol{I} \tag{3}$$

The Least Squares method finds unknown vector $\boldsymbol{u}^*$ according relation

$$\begin{aligned} \boldsymbol{u}^* &= \arg\min_{\boldsymbol{u}} \left\{\boldsymbol{e}^T \cdot \boldsymbol{e}\right\} \\ &= \arg\min_{\boldsymbol{u}} \left\{(\boldsymbol{A} \cdot \boldsymbol{u} - \boldsymbol{b})^T (\boldsymbol{A} \cdot \boldsymbol{u} - \boldsymbol{b})\right\} = \arg\min_{\boldsymbol{u}} || \, \boldsymbol{A} \cdot \boldsymbol{u} - \boldsymbol{b} \, ||_2^2 \end{aligned} \tag{4}$$

Minimizing this criterion, the best unbiased linear estimation is obtained:

$$\boldsymbol{u}^* = (\boldsymbol{A}^T \cdot \boldsymbol{A})^{-1} \cdot \boldsymbol{A}^T \cdot \boldsymbol{b} \tag{5}$$

Full derivation of the Least Squares method can be found in [4].

3 Proposed Identification Approach

There are several steps in presented approach and procedure composition is documented in Fig. 3. As the smoothing filter is beyond the scope of this paper, the identification procedure supposes smoothed data, as commented in the Conclusion.

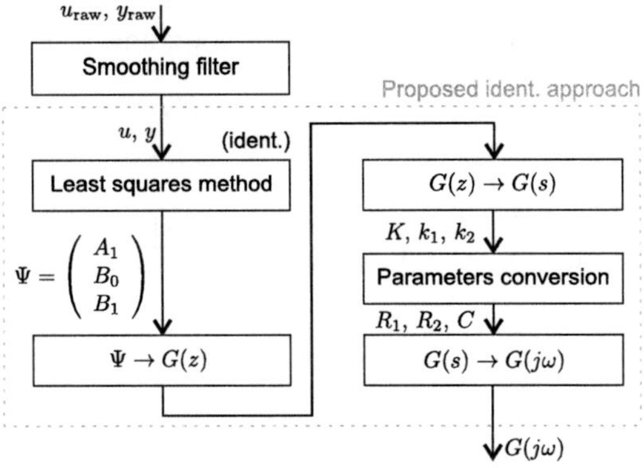

Fig. 3. Proposed identification approach block scheme

Firstly, supposing electrical circuit depicted in Fig. 1, $\mathcal{L}$-transform $U(s)$ of input voltage signal $u(t)$ is calculated as follows:

$$U(s) = \left(\frac{R_2}{R_2 C s + 1} + R_1 + R_{\text{ref}} \right) \cdot I(s) \tag{6}$$

Therefore, $\mathcal{L}$-transform $I(s)$ of current $i(t)$ is expressed as follows:

$$I(s) = \frac{U(s)}{\frac{R_2}{R_2 C s+1} + R_1 + R_{\text{ref}}} \tag{7}$$

Then the $\mathcal{L}$-transform $Y(s)$ of the output voltage $y(t)$ is calculated as follows:

$$Y(s) = \left(\frac{R_2}{R_2 C s + 1} + R_1 \right) \cdot I(s)$$

$$= \left(\frac{R_2}{R_2 C s + 1} + R_1 \right) \frac{U(s)}{\left(\frac{R_2}{R_2 C s+1} + R_1 + R_{\text{ref}} \right)} \tag{8}$$

$\mathcal{L}$-transfer function of the circuit defining relation between input and output signals is as follows:

$$G(s) = \frac{Y(s)}{U(s)} = \frac{\frac{R_2}{R_2Cs+1} + R_1}{\frac{R_2}{R_2Cs+1} + R_1 + R_{\text{ref}}} \tag{9}$$

This can be simplified into the following form containing time constants:

$$G(s) = \frac{R_2 + R_1}{R_2 + R_1 + R_{\text{ref}}} \cdot \frac{\frac{R_1R_2C}{R_2+R_1}s + 1}{\frac{R_1R_2C+R_{\text{ref}}R_2C}{R_2+R_1+R_{\text{ref}}}s + 1} = K\frac{k_1s+1}{k_2s+1} \tag{10}$$

where K, k_1 and k_2 are auxiliary parameters related to the elements R_1, R_2, C. For the identification purposes, the continuous-time transfer function is discretized with the use of Backward difference approximation:

$$G(z) = \frac{Y(z)}{U(z)} = G(s)|_{s=\frac{z-1}{Tz}} = \frac{B_0 + B_1z^{-1}}{1 + A_1z^{-1}} \tag{11}$$

where B_0, B_1 and A_1 are auxiliary parameters corresponding to vector of parameters $\boldsymbol{\Psi}$ used in the identification procedure. $\mathcal{Z}$-transform of the output can be expressed as the following equation:

$$Y(z)(1 + A_1z^{-1}) = U(z)(B_0 + B_1z^{-1}) \tag{12}$$

Time-domain equivalent of (12) is as follows:

$$y[k] + A_1y[k-1] = B_0u[k] + B_1u[k-1] \tag{13}$$

Based on (13) it is possible to express a recursive formula for obtaining current sample of output signal $y[k]$:

$$y[k] = -A_1y[k-1] + B_0u[k] + B_1u[k-1] \tag{14}$$

Real measured data are distorted by a noise signal $e[k]$. Using output error model, we obtain the following set of equations:

$$y[k] = -A_1y[k-1] + B_0u[k] + B_1u[k-1] + e[k] \tag{15}$$

$$y[k+1] = -A_1y[k] + B_0u[k+1] + B_1u[k] + e[k+1] \tag{16}$$

$$y[k+2] = -A_1y[k+1] + B_0u[k+2] + B_1u[k+1] + e[k+2] \tag{17}$$

$$\begin{aligned} y[k+m-1] = &-A_1y[k+m-2] + B_0u[k+m-1] \\ &+B_1u[k+m-2] + e[k+m-1] \end{aligned} \tag{18}$$

This set can be expressed in a matrix form as follows:

$$\boldsymbol{y} = \boldsymbol{Z}^T \cdot \boldsymbol{\Psi} + \boldsymbol{e} \tag{19}$$

where

$$\boldsymbol{y} = \begin{pmatrix} y[k] \\ y[k+1] \\ \cdots \\ y[k+m-1] \end{pmatrix} \tag{20}$$

$$\boldsymbol{e} = \begin{pmatrix} e[k] \\ e[k+1] \\ \cdots \\ e[k+m-1] \end{pmatrix} \tag{21}$$

$$\boldsymbol{\Psi} = \begin{pmatrix} A_1 \\ B_0 \\ B_1 \end{pmatrix} \tag{22}$$

$$\boldsymbol{Z}^T = \begin{pmatrix} -y[k-1] & u[k] & u[k-1] \\ -y[k] & u[k+1] & u[k] \\ -y[k+1] & u[k+2] & u[k+1] \\ \cdots & \cdots & \cdots \\ -y[k+m-2] & u[k+m-1] & u[k+m-2] \end{pmatrix} \tag{23}$$

Unknown vector of parameters $\boldsymbol{\Psi}$ can be calculated according (5), obtaining

$$\boldsymbol{\Psi} = \left(\boldsymbol{Z} \cdot \boldsymbol{Z}^T\right)^{-1} \cdot \boldsymbol{Z} \cdot \boldsymbol{y} \tag{24}$$

Once identified parameters $\boldsymbol{\Psi}$ are calculated, a discrete transfer function $G(z)$ is fully determined. Continuous transfer function $G(s)$ can be obtained by a Tustin approximation:

$$G(s) = G(z)|_{z=\frac{1+T/2}{1-T/2}} \tag{25}$$

Unknown parameters of elements R_1, R_2 and C can be computed by solving a set of algebraic equations:

$$\frac{R_2 + R_1}{R_2 + R_1 + R_{\text{ref}}} = K \tag{26}$$

$$\frac{R_1 \cdot R_2 \cdot C}{R_2 + R_1} = k_1 \tag{27}$$

$$\frac{R_1 \cdot R_2 \cdot C + R_{\text{ref}} \cdot R_2 \cdot C}{R_2 + R_1 + R_{\text{ref}}} = k_2 \tag{28}$$

Solving this set, we obtain identified parameters R_1, R_2 and C as follows:

$$R_1 = \frac{K \cdot R_{\text{ref}} \cdot k_1}{k_2 - K \cdot k_1} \tag{29}$$

$$R_2 = \frac{K \cdot R_{\text{ref}} \cdot k_1 - K \cdot R_{\text{ref}} \cdot k_2}{(K-1)(k_2 - K \cdot k_1)} \tag{30}$$

$$C = \frac{(k_2 - K \cdot k_1)^2}{K \cdot R_{\text{ref}} \cdot k_1 - K \cdot R_{\text{ref}} \cdot k_2} \tag{31}$$

Finally, the searched complex impedance of the circuit can be obtained via substitution $s = j\omega$ into (10):

$$G(j\omega) = \frac{R_2 + R_1}{R_2 + R_1 + R_{\mathrm{ref}}} \cdot \frac{\frac{R_1 R_2 C}{R_2+R_1} j\omega + 1}{\frac{R_1 R_2 C + R_{\mathrm{ref}} R_2 C}{R_2+R_1+R_{\mathrm{ref}}} j\omega + 1} \tag{32}$$

The most computationally extensive part is the calculation of inverse matrix which affects total computational costs of the identification approach.

In case of implementation of the algorithm in an embedded target such as MCU or FPGA where there are no sophisticated symbolic functions at disposal, the inverse matrix has to be computed with the use of simple mathematic operations. For this purposes, the authors have successfully tested one of the approaches based on simple loops containing multiplications and summations applied for each vector's element, see Algorithm 1. Therefore the effective implementation of this procedure on MCU or FPGA can lead to very fast bioimpedance representation for functional real-time analysis.

```
function b = matrix_inversion(a);
[r,c] = size(a); %input matrix dimensions;
b = eye(r); %identity matrix;
for j = 1 : r do
    for k = 1 : r do
        s = a(j,k); a(j,k) = a(j,k); a(j,k) = s;
        s = b(j,k); b(j,k) = b(j,k); b(j,k) = s;
    end
    t = 1/a(j,j);
    for k = 1 : r do
        a(j,k) = t * a(j,k);
        b(j,k) = t * b(j,k);
    end
    for L = 1 : r do
        if L = j then
            t = -a(L,j);
            for k = 1 : r do
                a(L,k) = a(L,k) + t * a(j,k);
                b(L,k) = b(L,k) + t * b(j,k);
            end
        end
    end
end
```

Algorithm 1: Pseudo-code for computation of an inverse matrix on an embedded target.

4 Results

The most important key result of the identification procedure proposed in this paper is obtaining of the unknown identified parameters corresponding to the

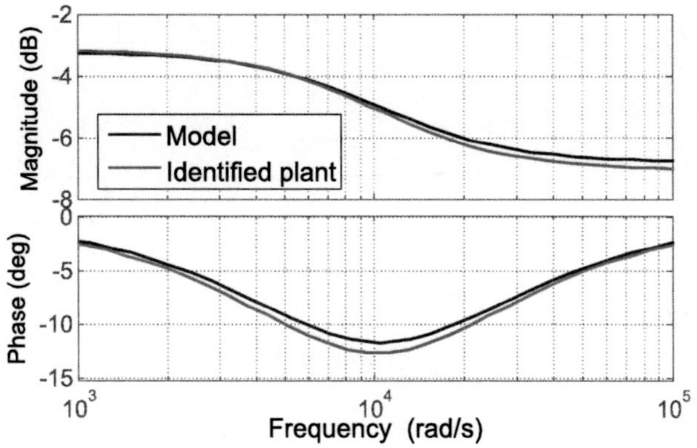

Fig. 4. Comparison in terms of the Bode diagram

R and C elements according Fig. 1. Comparison between original values used in a testing measurement chain according Fig. 1 and their identified values computed with the use of the proposed identification approach worked out as follows:

Original value of $R_1 = 120\Omega$, identified value $R_1 = 126,1\Omega$. Simlilarly, original value of $R_2 = 220\Omega$, identified value $R_2 = 205,5\Omega$. Finally, original value of $C = 1\mu F$, identified value $C = 0,9936\mu F$.

In terms of a relative error concerning the values of the parameters, there is a 5,08% relative error in R_1, 6,59% relative error in R_2 and 0,64% relative error in C. Having used this values for computation of complex impedance, the following expression is obtained:

$$G(j\omega) = 0.6885 \cdot \frac{7.7653 \cdot 10^{-5} j\omega + 1}{1.1706 \cdot 10^{-4} j\omega + 1} \tag{33}$$

Figure 4 shows a comparison of the Bode diagram between original plant, referred to as the model, and identified plant described by Eq. (33).

Table 1. Comparison between original and identified values

ω (rad·s^{-1})	Δ_{mag} (dB)	Δ_{ph} (deg)	ω (rad·s^{-1})	Δ_{mag} (dB)	Δ_{ph} (deg)
10^3	0.0627	−0.2357	$2 \cdot 10^4$	−0.1918	−0.7172
$2 \cdot 10^3$	0.0507	−0.4525	$3 \cdot 10^4$	−0.2153	−0.5217
$3 \cdot 10^3$	0.0325	−0.6361	$4 \cdot 10^4$	−0.2246	−0.4041
$4 \cdot 10^3$	0.0107	−0.7795	$5 \cdot 10^4$	−0.2291	−0.3283
$5 \cdot 10^3$	−0.0127	−0.8825	10^5	−0.2353	−0.1676
10^4	−0.1115	−0.9890			

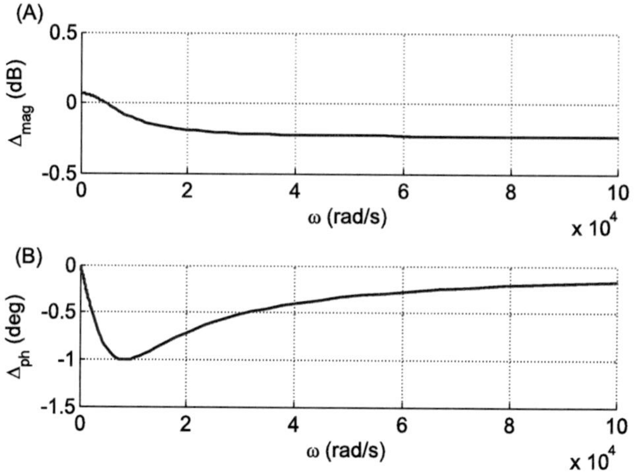

Fig. 5. (A) Absolute error in magnitude, (B) Absolute error in phase

It is obvious that each frequency is distracted with a different error in magnitude and phase. Table 1 summarizes the absolute errors in magnitude and phase for chosen frequency points in the most interested part of a frequency interval.

Graphical dependence between the frequency and magnitude absolute error is shown in Fig. 5A. Similarly, dependence between the frequency and phase absolute error is shown in Fig. 5B.

It is obvious that there exists a certain frequency where absolute error in magnitude is zero. Also, there exists a certain frequency where the absolute error in phase is maximal possible.

5 Conclusion and Future Work

The proposed identification approach in this paper is based on the Least Squares method in connection with continuous to discrete and discrete to continuous conversion carried out by standard bidirectional mathematical substitutions between $\mathcal{Z}$ and $\mathcal{L}$. It uses a measurement chain supposing prefiltering of the data. Simulation experiments have shown that prefiltering affects identification accuracy significantly. The results described in this paper currently suppose filtering based on Matlab *smooth* function which represents a software implementation of a simple filter that smooths the response data in column vector using a moving average filter. In a basic calling syntax, it uses the input data vector and so called span parameter representing number of data points for calculating the smoothed value. Currently, 20 points is considered as the span parameter so far. Although the accuracy achieved with the use of this kind of filter is sufficient enough, replacement of this filter with a sophisticated filter is planned for future work. Savitzky-Golay filter, median filter, local regression smoothing or a fractional filter are appropriate candidates that might increase the quality of identification procedure considerably.

Another issue related to the quality of identification procedure is the sampling period. The results achieved in the described experiment show that the current sampling period of 2 μs is a reasonable compromise between quality of the results and requirements on processor performance for the purposes of bioimpedance analysis.

It is also planned to analyze possibilities of using either more on-off pulses instead of a single pulse, or using a different type of a deterministic signal measured over a bit longer time interval to help the Least Squares method to get the most of its efficiency.

Generally, the most important key plan is porting of the proposes identification approach to an embedded target (MCU) to automate the identification process.

Acknowledgement. This research was funded by the European Regional Development Fund in the Research Centre of Advanced Mechatronic Systems project, grant number CZ.02.1.01/0.0/0.0/16_019/0000867 within the Operational Programme Research, Development and Education.

This work was supported by the project SP2019/107, "Development of algorithms and systems for control, measurement and safety applications V" of Student Grant System, VSB-TU Ostrava.

References

1. Ansede Pena, A.: A feasibility study of the suitability of an AD5933-based spectrometer for EBI applications, January 2009
2. Bayford, R.: Bioimpedance tomography (electrical impedance tomography). Annu. Rev. Biomed. Eng. **8**, 63–91 (2006)
3. Birlea, S., Corley, G., Birlea, M., Breen, P., Quondamatteo, F., ÓLaighin, G.: Detecting electroporation by assessing the time constants in the exponential response of human skin to voltage controlled impulse electrical stimulation, vol. 2009, pp. 1355–1358, September 2009
4. Boyd, S., Vandenberghe, L.: Introduction to Applied Linear Algebra: Vectors, Matrices, and Least Squares. Cambridge University Press, Cambridge (2018)
5. Michalikova, M., Abed, R., Prauzek, M., Koziorek, J.: Image reconstruction in electrical impedance tomography using neural network. In: Proceedings of the 7th Cairo International Biomedical Engineering Conference, CIBEC 2014, pp. 39–42 (2015)
6. Michalikova, M., Prauzek, M.: A hybrid device for electrical impedance tomography and bioelectrical impedance spectroscopy measurement. In: Canadian Conference on Electrical and Computer Engineering (2014)
7. Neuman, M.R.: Medical devices and systems. In: Biopotential Electrodes, pp. 47-1–47-13. The Biomedical Engineering Handbook: Third Edition, CRC Press is an imprint of Taylor & Francis Group (2006)
8. Prentice, A., Jebb, S.: Beyond body mass index. Obes. Rev. **2**, 141–147 (2001)
9. Van De Water, J., Miller, T., Vogel, R., Mount, B., Dalton, M.: Impedance cardiography the next vital sign technology? Chest **123**(6), 2028–2033 (2003)

Implementation of Smoothing Filtering Methods for the Purpose of Trajectory Improvement of Single and Triple Inverted Pendulums

Aleksandra Kawala-Sterniuk[1], Zdenek Slanina[2], and Stepan Ozana[2](✉)

[1] Faculty of Electrical Engineering, Automatic Control and Informatics, Opole University of Technology, Opole, Poland
kawala84@gmail.com

[2] FEECS, Department of Cybernetics and Biomedical Engineering, VSB – Technical University Ostrava, Ostrava, Czech Republic
{zdenek.slanina,stepan.ozana}@vsb.cz

Abstract. In this paper smoothing filters were applied in order to improve inverted pendulum's movement trajectory. The filtering was implemented in order to remove some signals' artifacts. The authors tested various classical filters such as Savitzky-Golay and median filters. The authors decided to test the data from two types of inverted pendulums – single and triple.

Keywords: Inverted pendulum · Control · Filtering

1 Introduction

The inverted pendulum is a classical problem in dynamics and control theory and is widely used in interalia education or as a reference for testing control algorithms. This is because it has the most balanced control and the lack of driving control system [1,2].

There are many methods that allow control over an inverted pendulum, such as classical control methods and methods based on the application of machine learning theory. Rocket control systems, robotics, construction cranes are the main areas of application of these methods, but they are most used in stabilizing cranes in a shipyard [2,3]. It can also be applied as a model for the most balanced control and the total lack of the driving control system. It's poor stability made it a target for numerous experiments as an example of non-linear system [1]. As it is one of the most popular non-linear systems – it may be successfully applied for testing purposes in research and education. The system is non-linear mostly due to inter alia gravitational effects. Another aspect of using inverted pendulum is strongly related with the problem of its position's stabilization and balancing of the pendulum during swinging up [20].

D. F. Cortes Tobar et al. (Eds.): AETA 2019, LNEE 685, pp. 214–223, 2021.
https://doi.org/10.1007/978-3-030-53021-1_22

A typical inverted pendulum system has two types of equilibria – stable and unstable. Stable equilibrium refers to a state similar to a pendulum suspended down. In the absence of any holding force, this system will return to its original state. Stable equilibrium does not need any control and is much simpler task management. Unstable equilibrium corresponds to the state in which the pendulum is located strictly at the top, requiring some force to maintain this position [3]. The system of the inverse pendulum (fixed, for example, on the motor shaft) can be applied in robotics when moving humanoid robots. With the help of engines, the angle of position of the component parts of the robot is changed, which makes it possible to maintain its equilibrium point and prevents the robot from falling. Also, this system is used to stabilize the position of rocket launcher [3,12].

This paper is mostly concentrated on implementation of the smooth filters for the purpose of disturbances removal in pendulum's trajectory. The study is at initial stage thus only basic single- and triple-inverted pendulums was applied. The author's of the hereof paper are currently working on application of the proposed solution for double-inverted pendulum. It is also planned to implement fractional filters, which have some smoothing features.

2 Inverted Pendulum

The inverted pendulum has been studied for decades and seems to be very thoroughly researched. This is because it revealed a number of dynamic phenomena especially in area of stabilisation of unstable equilibrium position. It is also one of the most versatile and popular non-linear system [18,19].

Under-actuated systems and their control have been a challenging problem especially because of some associated with it restrictions of the non actuated variables. As it was mentioned already before – such systems have been very thoroughly examined but their exact control remain a problem [5,20]. The inverted pendulum as one of the most popular nonlinear under actuated systems is till now used in linear and non-linear control as a benchmark sample. The system is highly non-linear mainly because the gravitational effects and centripetal forces and in certain configurations it may lack stability. To such becomes problem with balancing of the vertical pendulum in its upper position and the overall stabilization around this position [10].

2.1 Single-Inverted Pendulum

One of the main goals of this study was filtering of the inverted pendulum's trajectory in order to improve its stability. Some studied applied inter alia Kalman filters in order to cope with inter alia quantization noise, which has a very bad influence on state space controller [17].

The single-inverted pendulum (LQR – Linear Quadratic Regulator) is a very basic system unlike the double-inverted pendulum – one of the most popular systems applied in mechatronics for the purpose of control and estimation strategies study [16,17].

The perfect (impossible) case shows no noise interference. Real life however is different. The applied single-inverted pendulum (see: Fig. 1 and Fig. 2) consists of inverter line, actuator (DC motor) with incremental sensor, pendulum arm with incremental sensor.

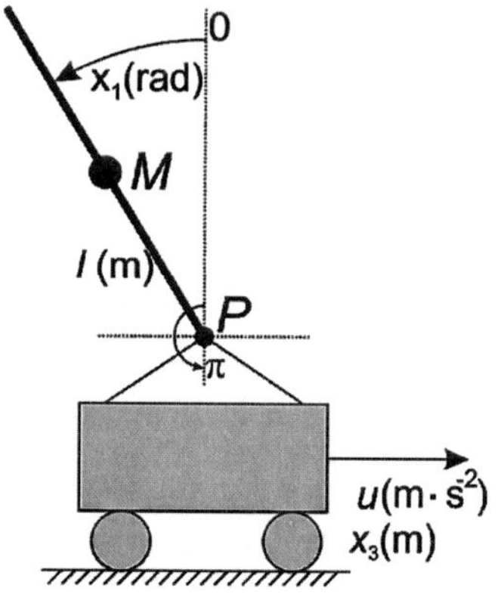

Fig. 1. Simplified scheme of the applied single-inverted pendulum.

Fig. 2. Implemented device.

The model (Fig. 1) consists of:

- x_1, x_2 – angle, angular velocity of the arm;
- x_3, x_4 – position, carriage speed;
- u – truck's acceleration;
- g, l, b – parameters (gravity acceleration, distance $|MP|$, friction coefficient).

Two-Degrees Control Structure (2-DOF) was illustrated in Fig. 3:

- **x***(t) – reference trajectory;
- u*(t) – reference control;
- **K**(t) – state controller.

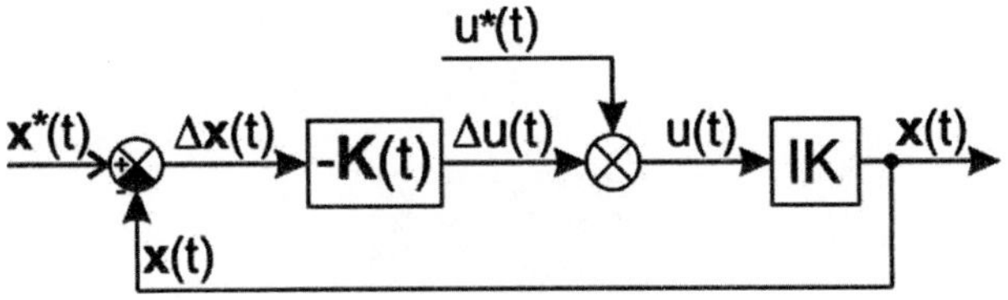

Fig. 3. Two-degrees control structure – scheme.

2.2 Triple-Inverted Pendulum

A typical triple link inverted pendulum is dynamic but highly unstable system [7,18]. It works that way that the lowest hinge is free for rotation and the torques of the upper two hinges are manipulated not only to stabilize the pendulum but also to control its attitude [9].

While using a model – Lagrange equation can be applied, where the r is for the motion displacement of the horizontal guide (Fig. 4) the θ_1, θ_2 and θ_3 are the angles of the lower, middle and upper pendulums relative to the vertical direction [1].

In Fig. 4 a simplified, classic model of a triple-inverted pendulum was presented Su.

It is possible to see three links $l^{(1)}$, $l^{(2)}$, $l^{(3)}$ with their mass and length: m_1, l_1, m_2, l_2 and m_3, l_3. The employed coordinates are $x(t)$, $\theta_1(t)$, $\theta_2(t)$ and $\theta_3(t)$, where x means the distance between the pendulum's base and the point on the track – x_0. The vector $x(t)$ means the time derivatives of the four coordinates as follows [1,4]:

$$x(t) = [x(t)\theta_1(t)\theta_2(t)\theta_3(t)x(t)\theta_1(t)\theta_2(t)\theta_3(t)]^T. \tag{1}$$

3 Research Methodology

For this study purposes the authors applied Savitzky-Golay filter with the following parameters: 19 (order) and 211 (frame length) and the Median Filter of the 30-th order. The results were satisfying. As mentioned above – Savitzky-Golay is a digital polynomial filter (or a least smoothing filter) [26].

In perfect, ideal case – there is no noise reference. Everything looks smooth. This is why models so frequently differ from reality [16]. Stabilisation's

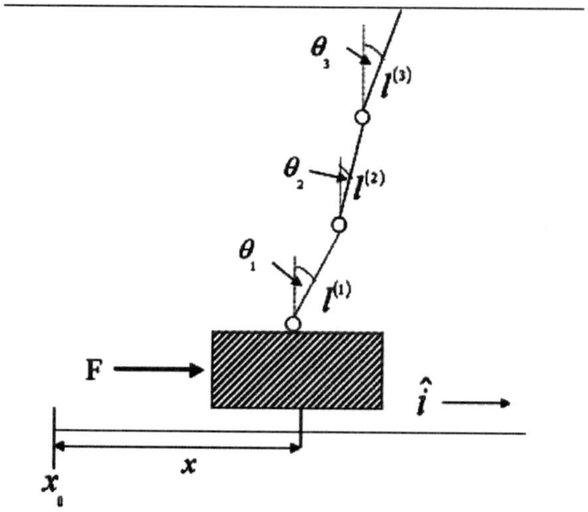

Fig. 4. Basic scheme of a typical triple-inverted pendulum [1,4].

approaches of such systems have been carried for decades [18]. The data presented in Figs. 5, 6 and 7 (x_2 and x_4) was a real data, coming from real inverted pendulum. Basic, median filter proved to be more efficient then Savizky-Golay filter.

It is visible, that some trajectories' smoothing and filtering was needed, however while the results from the data x_2 were very promising – while the same filters applied on x_4 gave different results. Further investigation is currently conducted by the authors of this paper.

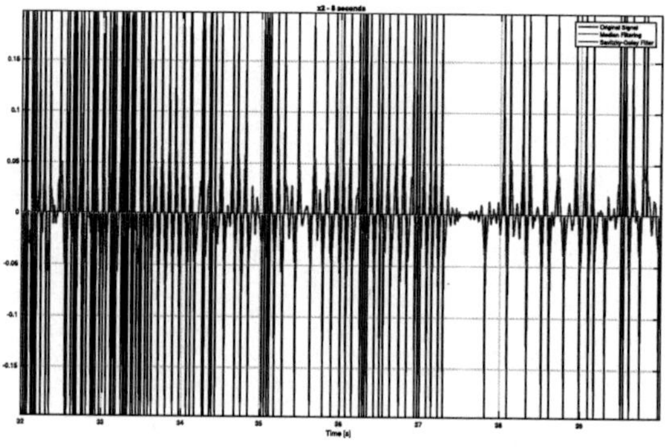

Fig. 5. x_2 – real signal, 8 s.

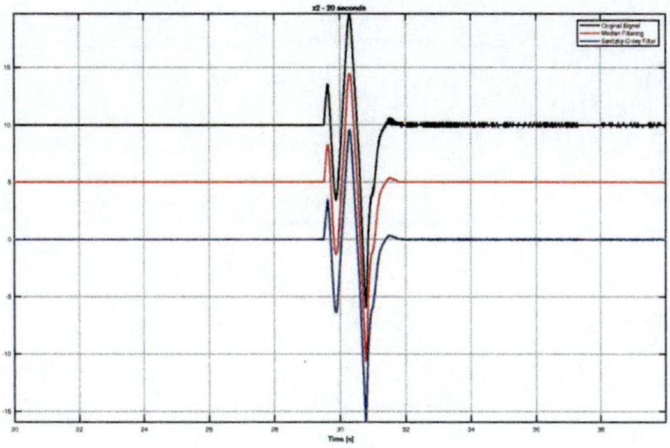

Fig. 6. x_2 – real signal, 20 s.

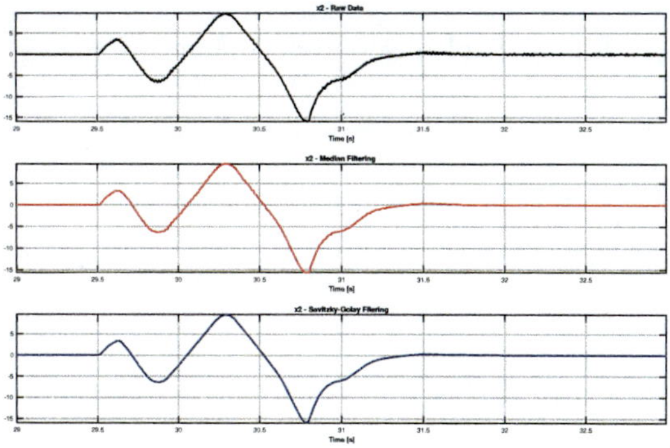

Fig. 7. x_2 – real signal, 4 s.

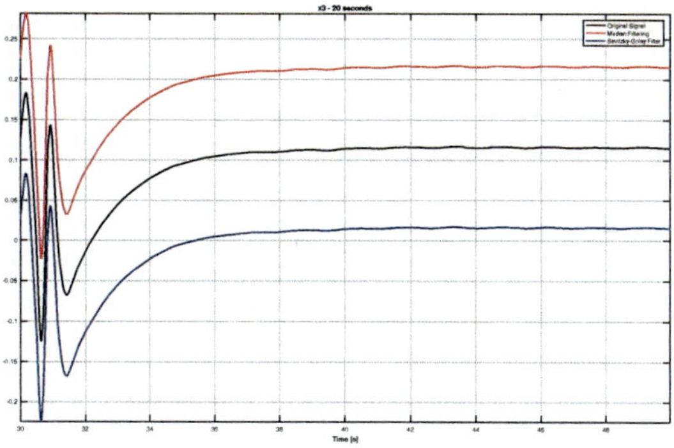

Fig. 8. x_3 – simulated signal, 20 s.

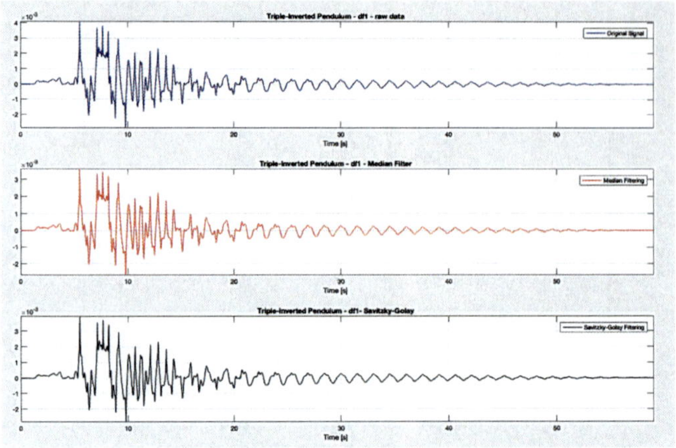

Fig. 9. Triple-inverted pendulum – $df1$.

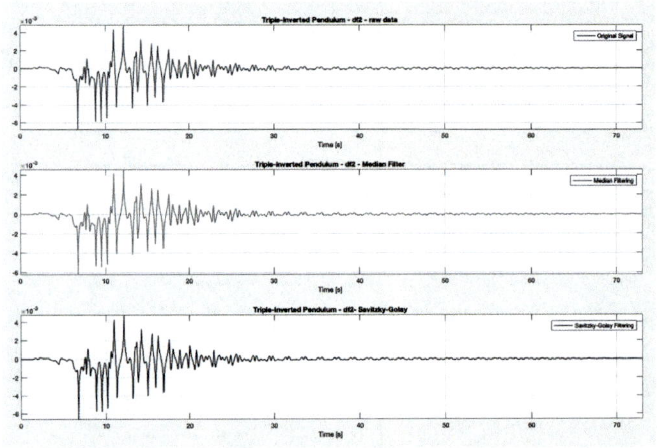

Fig. 10. Triple-inverted pendulum – $df2$.

Simulation is always different from the reality, no matter how hard we try to make is as close to it as possible. Data x_1 and x_3 were simulated. It is easy to observe that there is no interference in the simulated pendulum's trajectory (see: Fig. 8). Therefore adding extra filtering had no point, but was performed.

Same filters, based on promising results with the single-inverted pendulum wee applied to the triple-inverted pendulum (Fig. 9, 10 and 11).

It is very disappointing that smoothing filters successfully applied to biomedical data or to the single-inverted pendulum did not work properly on triple-inverted pendulum data. No visible change is noticed. The authors of the hereof paper are currently working on this topic.

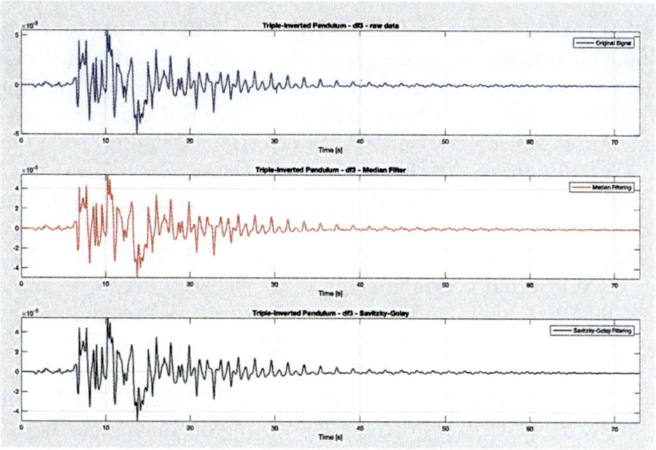

Fig. 11. Triple-inverted pendulum – *df*3.

4 Future Works

The study is still at initial stage as the authors of this paper for tests purposes only decided to apply basic inverted pendulum. The authors are currently working on application of the proposed filtering for double- and triple-inverted pendulums. It is also planned to implement fractional filters, which also have some smoothing features.

Why the next step involves double-inverted pendulum? Because it is the one of the most popular mechatronic systems and is successfully applied for control and estimation strategies. Dynamics of such pendulum is complex and provides a system perfect for design and development of new control methodologies [16]. It could be successfully applied for various important areas such as holding a balance of a robot or to design more stable walking robot [16,18].

Further research plan include also tests on double-inverted pendulums, or on rotary inverted pundulum systems (RIPS) [6,8].

Acknowledgment. This research was funded by the European Regional Development Fund in the Research Centre of Advanced Mechatronic Systems project, grant number CZ.02.1.01/0.0/0.0/16_019/0000867 within the Operational Programme Research, Development and Education.

This work was supported by the project SP2019/107, "Development of algorithms and systems for control, measurement and safety applications V" of Student Grant System, VSB-TU Ostrava.

References

1. Huang, X., Wen, F., Wei, Z.: Optimization of triple inverted pendulum control process based on motion vision. EURASIP J. Image Video Process. **73** (2018)

2. Siebert, W.M.C.: Circuits, Signals, and Systems. MIT Press, Cambridge (1986)
3. Druzhinina, O.V.: Modelirovanie i postroenie algoritma stabilizacii perevernutogo majatnika. Dinamika slozhnyh sistem – XXI vek, vol. 4 (2012). (in Russian)
4. Su, H., Woodham, C.A.: On the uncontrollable damped triple inverted pendulum. J. Comput. Appl. Math. **151**(2), 425–443 (2003)
5. Xianmin, Ch., Rongrong, Y., Kang, H., Shengchao, Z., Hao, S., Ke, S.: Linear motor driven double inverted pendulum: a novel mechanical design as a testbed for control algorithms. Simul. Model. Pract. Theory **81**, 31–50 (2018)
6. Hazem, Z.B., Fotuhi, M.J., Bingul, Z.: Comparison of friction estimation models for rotary triple inverted pendulum. Int. J. Mech. Eng. Robot. Res. **8**(1) (2019)
7. Hussein, M.T.: CAD design and control of triple inverted-pendulums system. Iraqi J. Mech. Mater. Eng. **2018**(3), 482–497 (2018)
8. Lopes, J.M., Moreira, L., Pinheiro, C., Sanz-Merodio, D., Figueiredo, J., Santos, C.P., Garcia, E.: Three-link inverted pendulum for human balance analysis: a preliminary study. In: IEEE 6th Portuguese Meeting on Bioengineering (ENBENG), pp. 1–4. IEEE (2019)
9. Furut, K., Ochiai, T., Ono, N.: Attitude control of a triple inverted pendulum. Int. J. Control **39**(6), 1351–1365 (1984)
10. Sun, L., Kong, H., Liu, C., Bi, L.: Overview of the control of the inverted pendulum system. J. Mach. Tool Hydraulics **7** (2008)
11. Farhang-Boroujeny, B.: Adaptive Filters: Theory and Applications. Wiley, Hoboken (2013)
12. Bobobekov, K.M.: The model of the inverted pendulum: special cases. In: Collection of Scientific Works of the Novosibirsk State Technical University, vol. 3, no. 81, pp. 21–42. Novosibirsk State Technical University (2015)
13. da Costa, V.L.R., Schettino, H.V., Camponogara, A., de Campos, F.P., Ribeiro, M.V.: Digital filters for clustered-OFDM-based PLC systems: design and implementation. Digit. Signal Proc. **70**, 166–177 (2017)
14. Tibdewal, M.N., Mahadevappa, M., Kumar Ray, A., Malokar, M., Dey, H.R.: Power line and ocular artifact denoising from EEG using notch filter and wavelet transform. In: 3rd International Conference on Computing for Sustainable Global Development (INDIACom), pp. 1654–1659. IEEE (2016)
15. Diniz, P.S., Da Silva, E.A., Netto, S.L.: Digital Signal Processing: System Analysis and Design. Cambridge University Press, Cambridge (2010)
16. Xu, B., Lyu, Y., Gadsden, Y.S.: Estimation and Control of a Double-Inverted Pendulum (2018)
17. Kloppelt, C., Meyer, D.: Comparison of different methods for encoder speed signal filtering exemplified by an inverted pendulum. In: 19th International Conference on Research and Education in Mechatronics (REM), pp. 1–6. IEEE (2018)
18. Masrom, M.F., Ghani, N.M., Jamin, N.F., Razali, N.A.A.: Stabilization control of a two-wheeled triple link inverted pendulum system with disturbance rejection. In: 10th National Technical Seminar on Underwater System Technology 2018, pp. 151–159. Springer, Cham (2019)
19. Kovalchuk, V.V.: Lyapunov's stability theory for a triple inverted pendulum with a follower forces. Application of Mathematics in Computer Sciences, Maths in the Modern Technical University, VI International Scientific-Practical Conference Kyiv 2017, p. 8. National Technical University of Ukraine (2018)
20. Ramirez-Neria, M., Gao, Z., Sira-Ramirez, H., Garrido-Moctezuma, R., Luviano-Juarez, A.: Trajectory tracking for an inverted pendulum on a cart: an active disturbance rejection control approach. In: 2018 Annual American Control Conference (ACC), pp. 4881–4886. IEEE (2019)

21. Zieliński, T.P.: Cyfrowe przetwarzanie sygnalow. Od teorii do zastosowan. Helion (2007)
22. Bhushana Rao, K.Ch., Krishna, B.T.: Comparative Analysis of Integer and Non-integer order Savitzky-Golay Digital Filters. IEEE (2017)
23. Liu, J.G.: Smoothing filter-based intensity modulation: a spectral preserve image fusion technique for improving spatial details. Int. J. Remote Sens. **21**(18), 3461–3472 (2018)
24. Einicke, G.A.: Smoothing, Filtering and Prediction: Estimating the Past, Present and Future (2012)
25. Baranowski, J., Bauer, W., Zagorowska, M., Piatek, P.: On digital realizations of non-integer order filters. Circ. Syst. Signal Process. **35**(6), 2083–2107 (2016)
26. Jahani, S., Setarehdan, S.K., Boas, D.A., Yucel, M.A.: Motion artifact detection and correction in functional near-infrared spectroscopy: a new hybrid method based on spline interpolation method and savitzky-golay filtering. Neurophotonics **5**(1), 015003 (2018)
27. Pander, T.: EEG signal improvement with cascaded filter based on OWA operator. Signal Image Video Process. **13**(6), 1165–1171 (2019)

Methods for the Characterization of the Variability of Solar and Wind Resource

C. Ojeda Avila, S. Salamanca Forero, and M. Bueno-López(✉)

Department of Electrical Engineering, Universidad de la Salle, Bogotá, Colombia
{cojeda17,ssalamanca05,maxbueno}@unisalle.edu.co

Abstract. In this paper, we propose the estimation of the electrical power that is possible generate using as primary source the energy from solar and wind resources, considering their variability. We use models from the specialized bibliography with the purpose of presenting a comparative study, and the data base is obtained from a meteorological station locate in Bogota, Colombia. For the solar variation, we consider especially two models, the linear regression by Angstrom-Prescott and the analysis of Bird and Hulstrom. In the estimation of wind energy, we used the Weibull model and the distribution of Rayleigh. Finally we present some results for each case using real data.

Keywords: Solar energy · Wind energy · Variability · Electric power

1 Introduction

In pursuit of environmental protection and the need to reduce the production of greenhouse gases, the humanity has progressed in eliminating the carbonization of economies, starting with large CO2 production centers, one of which is the Electricity generation [2]. Renewable energies, especially wind and solar resources, are expected to be able to dispatch a considerable amount of energy, but one of the challenges is the variability of the primary resource and the uncertainty of availability when it comes to producing electricity. As the existence of these resources depends on the climate, it is not easy to predict when the resource will be available [9].

Based on the need to have a prognosis to reduce the uncertainty about availability resources, the analysis of these data, extract from them significant patterns of behavior and make predictions which are vital to the improvement of the efficiency and the reduction of the costs of the services and for the decision making of the companies involved [5], as well the dispatch and re-dispatch, are achieved advantages when considering the variability of the resource since there is the possibility of doing a better planning and it would bring greater energy security for the country in question [3].

In the realization of this paper, we will consider the variability of the solar and wind resource, for this purpose a bibliographical review about models and distributions used or can be applied to reduce the uncertainty as to the availability

D. F. Cortes Tobar et al. (Eds.): AETA 2019, LNEE 685, pp. 224–233, 2021.
https://doi.org/10.1007/978-3-030-53021-1_23

of these resources and to estimate a generation potential is carried out. The data that have been used in the models are obtained from the meteorological station owned by the University of La Salle, located in the city of Bogota-Colombia.

2 Framework

2.1 Solar Energy

Solar energy is one of the natural resources that can be most exploited in terms of electricity generation today, since this source of energy is obtained directly from the sun, also called in the technical field solar energy Photovoltaics [6,7].

Capturing the sunlight has complications since the area of incidence of the rays of the sun is extensive, likewise the sun does not deliver the same amount of light or irradiation in the same period of time determined since it depends on factors such as the hour in the day, seasons in the year and the clarity or cloudiness of the sky [8].

Solar Geometry Decline: The declination corresponds to the angle between the equatorial plane and the line that joins the Earth and the sun. This can be calculated as:

$$\delta = 24,45 \cdot sen\left(2\pi \frac{d_n + 284}{365}\right) \tag{1}$$

Where d_n is the day of interest.

Also, it is important to consider the **Solar centil angle**. That term refers to the relative position of the sun with respect to the local normal surface, this is calculated as:

$$cos\,(\theta_{ZS}) = cos\,(\varphi) \cdot cos\,(\delta) \cdot cos\,(w) + sin\,(\varphi) \cdot sin\,(\delta) \tag{2}$$

Where φ is the latitude of the area, δ the solar declination and w the time angle.

Other factor that we consider is the **extraterrestrial solar radiation**, it determines a maximum theoretical value of the available solar energy incident on a surface is determined by:

$$I_o = I_{CS} \cdot E_o \cdot cos\,(\theta_o) \tag{3}$$

Where E_o is the eccentricity correction factor calculated as:

$$E_o = 1 + 0,033 \cdot cos\left(\frac{2\pi \cdot dn}{365}\right) \tag{4}$$

Radiation Estimation: The terrestrial solar radiation that affects the earth is affected by different factors in the atmosphere such as gases and aerosols, for the estimation of the total radiation incident on a surface, it is necessary to take into account the amount of radiation Diffused, reflected, global and direct, thus there are measuring devices such as the Pyranometer to obtain the available energy values incident on a horizontal surface [4].

2.2 Wind Energy

Wind energy is the kinetic energy that the wind contains, it can be used to be transformed into mechanical energy, hydraulic or electrical Wind energy is converted into electric energy by means of turbines. They can be classified by the orientation of their blades, currently the most used are the horizontal axis, its main components are: The rotor or wind collection system, gearbox or multiplier and electric generator.

To obtain the power values obtained from the wind, values such as density of dry air, wind speed and wind turbine area to be used must be taken into consideration, this expression is given by:

$$P = \frac{1}{2}\rho.A.v^3 \tag{5}$$

Where P is power in Watts, ρ is density of dry air, A is area from wind turbine, v is the wind speed.

3 Estimation Methods

3.1 Wind Estimation

Weibull Distribution: The distribution of Weibull composes a model that generalizes the exponential model and is defined by the following parameters: the threshold parameter or location parameter indicates the origin of times. The shape parameter is defined in terms of the exponent of the potential function that determines the model's failure reason. By applying the distribution of Weibull to the case of wind energy, it is necessary that the parameters of threshold and form will depend on the distribution and the data, being the parameter of form who will define the asymmetry of the distribution, in addition the parameter of scale are the average wind velocity data.

The construction of a Weibull distribution can be made from a histogram or from the values obtained when making a descriptive statistic of the data, the distribution is presented in (6) [6]:

$$p(v) = (\frac{k}{c})(\frac{v}{c})^{k-1}\exp{(-\frac{v}{c})}^k, \tag{6}$$

where k is the form factor, c the scale factor at (m/s) and v the velocity of the wind in (m/s) [6]. To find the numerical value of the parameters from least squares, we must define data from the histogram, its class (vi), relative frequency (f_i) and cumulative relative (F_i), to find new values $(y_i and x_i)$ that will help to find the Distribution parameters. To find y_i and x_i, we use (7) and (8) [1]:

$$y_i = \ln{(\ln{(1 - F_i)})} \tag{7}$$

$$x_i = \ln{(v_i)} \tag{8}$$

From (7) and (8) we get (9) and (10)

$$B = k = \frac{\sum_{i=1}^{n} X_i Y_i - \sum_{i=1}^{n} X_i \sum_{i=1}^{n} Y_i}{N \sum_{i=1}^{n} X_i^2 - (\sum_{i=1}^{n} X_i)^2} \tag{9}$$

$$A = \frac{\sum_{i=1}^{n} X_i^2 \sum_{i=1}^{n} Y_i - \sum_{i=1}^{n} X_i \sum_{i=1}^{n} X_i Y_i}{\sum_{i=1}^{n} X_i^2 - (\sum_{i=1}^{n} X_i)^2} \tag{10}$$

In this way the parameters for the distribution of Weibull are:

$$A = -k \ln c \tag{11}$$

So the parameter c is:

$$c = \exp -(\frac{B}{A})\frac{m}{s} \tag{12}$$

c corresponds to the speed obtained from the m/s distribution.

Rayleigh Distribution: The functional form of the Rayleigh distribution is [1]:

$$f(x) = \frac{2x}{b^2} \exp -\frac{x^2}{b^2} \rightarrow x \geq 0 \tag{13}$$

The function $f(x)$ represents the probability that the wind velocity x is in an interval between x and $x + dx$. The area under $f(x)$ is the unit.

$$\int_0^\infty f(x)dx = 1 \tag{14}$$

The distribution function of Rayleigh is a particular case of the Weibull for $k = 2$. The Rayleigh distribution function is expressed in terms of the average value x of the velocity as we show in (15)

$$f(x) = \frac{\pi x}{2x^2} e^{(-\frac{\pi}{4}(\frac{x}{x^2})^2)} \tag{15}$$

3.2 Solar Estimation

Linear Regression by Angstrom-Precot: This method seeks the correlation between the global solar radiation and the hours of sunstroke, this study was developed by Angstrom (1924) and it was subsequently modified by Prescott (1940), determining that the most simple relationship that links the Solar radiation with hours of solar brightness can be expressed as:

$$\left(\frac{\bar{H}}{\bar{H}_o}\right) = a + b\left(\frac{\bar{n}}{\bar{N}}\right) \tag{16}$$

Where $\bar{H}$ is the global radiation, H_o is the alien irradiation, $\bar{n}$ is the number of hours of sunshine effective brightness and $\bar{N}$es the number of hours of astronomical brightness possible to be recorded. The term $\left(\frac{\bar{H}}{\bar{H}_o}\right)$ is the Clarity index K_t and $\left(\frac{\bar{n}}{\bar{N}}\right)$ is the relative Holofonía, on the other hand A and B represent the relationship between the global solar radiation and the hours of Heat stroke [7].

Analysis of Bird and Hulstrom: From the direct radiation analysis that Bird and Hulstrom formulated, direct radiation can be determined on a horizontal surface considering different indices of cloudiness [5], as described in the Eq. (17).

$$I_{DH} = [0,9662 \cdot C \cdot \tau_{CTA}] \cdot sinA, \tag{17}$$

where C is the solar constant, A is the solar altitude in degrees, 0.9662 is a constant that represents the correction factor that adjusts the wavelengths from 0.3 to 3μm of the solar spectrum, and τ_{CTA} is the Atmospheric transmittance coefficient.

The atmospheric transmittance coefficient is calculated from the transmittance emitted by dispersion due to air molecules, gases, water vapour and aerosols. On the other hand, during the calculation and evaluation of the radiation it is necessary to know the multiple reflection between the soil and the atmosphere, taking into account meteorological data obtained from observations, so that the diffuse radiation on a surface Horizontal depends on the clarity of the sky, K_d.

$$I_{dH} = C * K_d * senA \tag{18}$$

Diffuse radiation is the sum of three different contributions: diffuse radiation due to the existence of air molecules, diffuse radiation due to the existence of dust particles (aerosols) and diffuse radiation by multiple reflection between the soil and the Atmosphere, expressed in (19).

$$I_{dH} = I_{dr} + I_{da} + I_{dm} \tag{19}$$

4 Results

4.1 Solar Energy Results

We considered the angle of inclination, which changes depending on the time of on a specific day of the year, this is given according to the Eq. (1). From the average irradiation data obtained from the station since the 80's, we can get the values of extraterrestrial solar radiation, hours of solar brightness and number of astronomical hours are obtained, also according the Eq. (18), the values of clarity index K_t and relative holophony (n/N) are obtained and presented in Table 1.

The linear regression is performed between the data obtained in the Table 2 in order to know the annual clarity index in Bogota (Fig. 1).

Once obtained, the diffuse fraction and the horizontal diffuse radiation are calculated in order to know the final power received by the solar panel, Table 3 shows the results.

4.2 Wind Energy Results

The variation of the wind speed is shown in the Fig. 2.

Table 1. Solar energy data

Month	Bogota irradiation (Wh/m^2)	Extraterrestrial radiation (Wh/m^2)	Shine hours (h)	Number of astronomical hours (N)	Index of clarity (Kt)	Relative holophony (n/N)
Junary	4681,9	9538,4	5,9	11,8	0,49	0,50
February	4312,7	10035	5,3	11,9	0,43	0,45
March	4322,2	10415,23	4,4	12	0,41	0,37
April	3716,7	10390	3,5	12,1	0,36	0,29
May	3506	10050	3,5	12,2	0,35	0,29
June	3658,9	9791,2	3,9	12,2	0,37	0,32
July	3917,3	9874,2	4,3	12,2	0,40	0,35
August	4168,2	10186	4,4	12,1	0,41	0,36
September	3947,8	10341	4,1	12	0,38	0,34
October	3961	10093	3,8	11,9	0,39	0,32
November	4017,7	9613,3	4,2	11,8	0,42	0,36
December	4241,4	9336,3	5,1	11,8	0,45	0,43

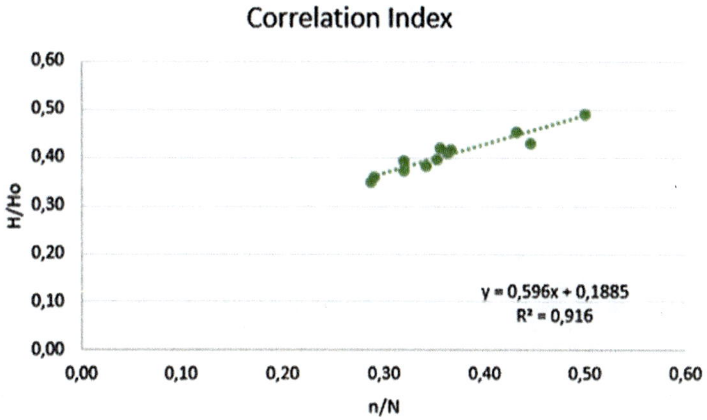

Fig. 1. Linear regression

Table 2. Coefficients of Angstrom and prescott

Coefficients of Angstrom and prescott	
a	b
0,188	0,596

When we apply the estimation methods for wind energy, we have 144 data per day, which can generate a 24 h window to consider its variability, in the distribution of Weibull as mentioned before takes the value of the scale parameter or the velocity of the wind in m/s, for Rayleigh is used the graph, where the

Table 3. Diffuse fraction, horizontal diffuse radiation and panel's power

Diffuse fraction	Horizontal diffuse radiation (Wh/m^2)	Panel (Wh/m^2)
0,45	2085,05	2596,85
0,51	2218,30	2094,40
0,53	2295,36	2026,84
0,60	2214,33	1502,37
0,61	2123,91	1382,09
0,58	2113,85	1545,05
0,55	2161,20	1756,10
0,54	2240,80	1927,40
0,57	2244,75	1703,05
0,56	2204,42	1756,58
0,53	2120,29	1897,41
0,49	2064,08	2177,32

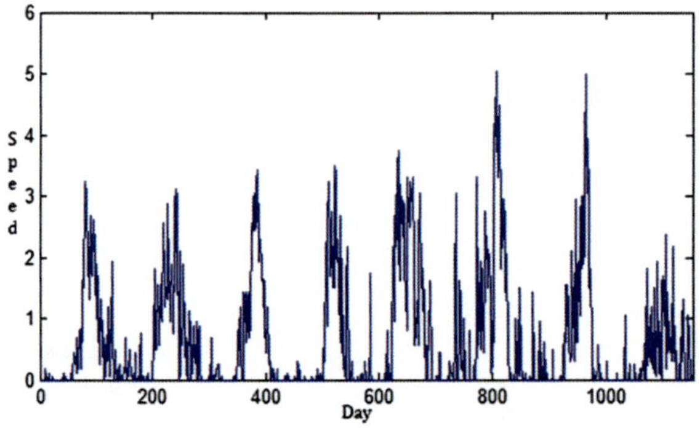

Fig. 2. Wind variation

Xmax point, is the most likely will be the speed, which results in the following result for the first 8 days that are considered from January 9, 2019. Table 4 shows the comparison between Weibull and Rayleigh.

After making the settings for each of the functions, the probability distributions (Weibull and Raileigh) are obtained per day for each method, an example of these are in the Fig. 3. Taking this into account, the variation in the generation is presented from day 09/01/2019 to 31/01/2019 taking into account the values of velocities obtained from each of the methods (Fig. 4):

Table 4. Weibull and Raileigh comparison

Day	T[°C]	Weibull		Rayleigh	
		m/s	W	m/s	W
09/01/2019	14,66	0,64	103,61	0,49	45,47
10/01/2019	14,21	0,82	217,49	0,56	68,18
11/01/2019	14,86	0,63	97,26	0,49	46,45
12/01/2019	14,35	0,59	82,04	0,41	26,32
13/01/2019	14,3	0,86	248,92	0,76	170,45
14/01/2019	14,43	0,73	150,89	0,80	198,43
15/01/2019	14,68	0,61	88,09	0,49	47,69
16/01/2019	14,55	0,65	106,72	0,29	9,977

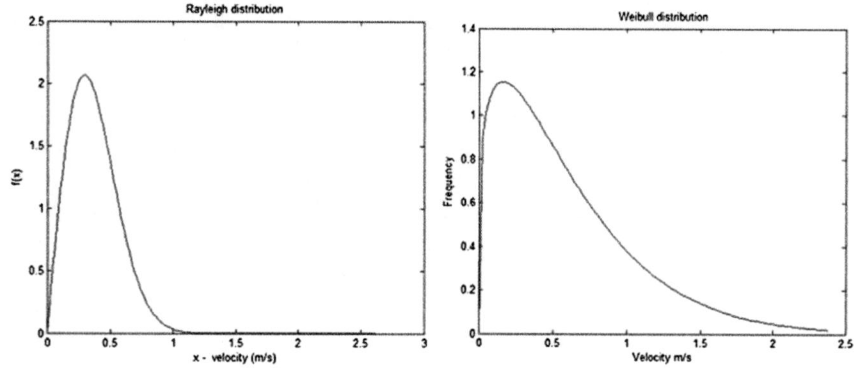

Fig. 3. Rayleigh and Weibull distribution

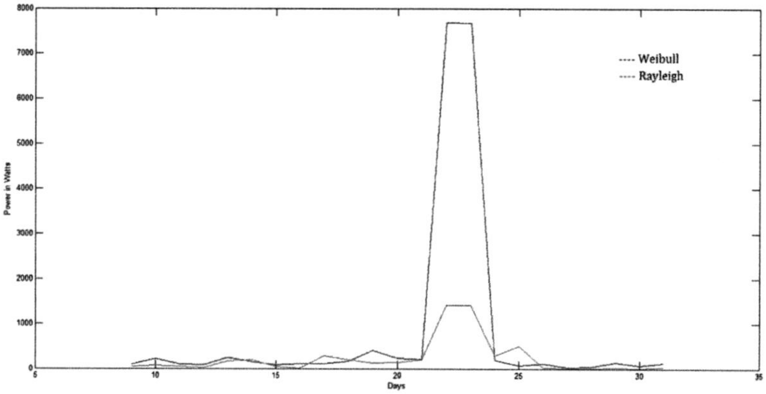

Fig. 4. Power calculated

5 Discussion and Conclusions

For extraterrestrial solar radiation is necessary to take into account the angle of inclination of the sun as shown in the Eq. (1) where the 24.45^o of inclination of the earth is not exceeded with respect to the line of Ecuador depending of the day of the year, in the same way the hours of astronomical brightness, shown in Table 1, do not find a significant variation, so it is possible to say that the astronomical hours can retake like 12 h for every month.

On the other hand, the clarity index shown in Table 3 does not exceed 0.5, so it can be said that the days in Bogota are mainly cloudy, thus reducing the probability of having a greater total global radiation on the surface. When performing the linear regression of Angstrom and Prescott is obtained that in index of clarity K_T on average of the year will be 0748 where most days will be clear. According to the data obtained, the total power received by solar panels located on the horizontal will be on average of 2600 Wh/m^2 during the first three months of the year and no more than 2000 Wh/m^2 during the remainder of the year.

Analyzing the results for the distribution of Weibull in which are taken the 144 data per day and conform to the method, the largest power is almost 8 kW with 2, 7 m/s, this being the speed that occurs more frequently on day 22/01/2019 taking into account a distribution as the Fig. 3. For the application of Rayleigh being a punctual case of the exponential distribution is used only an average value, without necessarily being this speed which occurs more frequently, in this case, no maximum value obtained with the function is equal to the Introduced with a maximum power value close to 1.5 kW with 1.6 m/s for day 22/01/2019.

By establishing a comparison between the results obtained for the distribution of Weibull and the distribution of Rayleigh at least in the velocity values, in most cases they are theoretically close, their difference is about 0.15 m/s Normally, except on the 22nd and 23rd of January where the greatest differences are present, up to 1.2 m/s in terms of power, the differences are presented in the same scales.

In conclusion, the implementation of solar energy in Bogotá is not entirely effective, since the days are mostly cloudy causing the majority of the incident solar radiation to disperse before reaching the surface.

The distribution of Weibull is considered better compared to that of Rayleigh, since it takes into account a greater amount of data and the speed that is encountered with the application of this distribution has a higher frequency of occurrence than that found with Rayleigh.

When considering the variability of the wind resource for this specific zone it is seen that the speeds are very low and therefore for a larger production it is recommended to use vertical axis wind turbines.

Acknowledgment. This paper is part of the project number 111077657914 and contract number 031-2018, funded by the Colombian Administrative Department of Science, Technology and Innovation (COLCIENCIAS) and developed by the ICE3

Research Group at Universidad Tecnologica de Pereira (UTP) and CALPOSALLE Group at Universidad de La Salle.

References

1. Allachi, M.O.: Predicción de energía eólica utilizando técnicas de aprendizaje automático. Master's thesis, Universidd Carlos III de Madrid, Madrid, Spain, March 2015
2. Bhattacharya, P., Bhattacharjee, R.: A study on weibull distribution for estimating the parameters. Wind Eng. **33**(5), 469–476 (2009)
3. Inácio, C.O., Borges, C.L.T.: Stochastic model for generation of high-resolution irradiance data and estimation of power output of photovoltaic plants. IEEE Trans. Sustain. Energy **9**(2), 952–960 (2018)
4. Maymon, G.: Some important statistical distributions (chapter 2). In: Maymon, G. (ed.) Stochastic Crack Propagation, pp. 9–18. Academic Press (2018). http://www.sciencedirect.com/science/article/pii/B9780128141915000024
5. Burgos, M.P., Aldana, S., Rodríguez, D.J.: Análisis del recurso energético eólico para la ciudad de bogotá dc para los meses de diciembre y enero, colombia. Avances. Investigacion e Ingenieria **12**(1), 1–7 (2015)
6. Pascual, A.F.: La predicción de energías renovables: oportunidades big data para la energía eólica. Technical report 1, Universidad Autonoma de Madrid, Madrid, Spain, July 2016
7. Salazar, G.A., Hernandez, L.A., Saravia, L.R., Romero, G.G.: Determinación de los coeficientes de la relación de angstrom-prescott para la ciudad de salta (argentina) a partir de datos tomados durante un año. Avances en Energías Renovables y Medio Ambiente **11**(1), 17–24 (2007)
8. Vanegas, M., Churio, O., Valencia, G., Ospino, A.: Calculo dela radiación total, directa y difusa a través de la transmisibilidad atmosférica en los departamentos del Cesar. Guajira y Magdalena **38**(07), 1–18 (2017)
9. Wan, C., Zhao, J., Song, Y., Xu, Z., Lin, J., Hu, Z.: Photovoltaic and solar power forecasting for smart grid energy management. CSEE J. Power Energy Syst. **1**(4), 38–46 (2015)

Auto-Tuning PID Based on Extremum Seeking Algorithm for an Industrial Application

A. Arias-Patiño[1], A. Zapata-Lombana[1], F. Salazar-Caceres[1], and M. Bueno-López[2](✉)

[1] Department of Automation, Engineering, Universidad de la Salle, Bogotá, Colombia
{aarias04,angieyzapata51,jfsalazar}@unisalle.edu.co
[2] Department of Electrical Engineering, Universidad de la Salle, Bogotá, Colombia
maxbueno@unisalle.edu.co

Abstract. The PID controller implementation on the industry is an active area of study, however, according to the technique used is possible not having optimal and good performance due to the difficulties in the process identification and parameter tuning of the system since there is no full information of the industrial process. In last years PID auto-tuning methodologies have been developed which let synthesize efficient polynomial controllers because the parameters, generally, are optimal in some sense. In this paper is showed a PID auto-tuning through Extremum Seeking algorithm (ES) in a flow control plant implemented with blocks programming in a programmable logic controller (PLC), this block system acquires input signals obtained by the plant sensors, in this case, a flow sensor, consequently a control signal is sent to the variable frequency drive of the pump system, the objective control is to maintain the flow constantly through the system in the desired reference. In the following section the block diagram of the controlled system is presented, the auto-tuning ES algorithm, the simulation results, and the real-time implementation on the industrial process, finally, some conclusions and final implementation comment considering some future work is developed.

Keywords: Auto-tuning · Extremum seeking algoritm · PID Control · Programmable logic controller

1 Introduction

Nowadays, the proportional - integral - derivative (PID) controller continuous being widely used, almost 95% of the applications since ease of implementation and good performance, moreover the proportional action adjusts the controller output according to the error signal, the integral action makes the steady-state error equals to zero and the derivative action anticipates the future action of the system, applying these elements there exist wide industrial applications which

D. F. Cortes Tobar et al. (Eds.): AETA 2019, LNEE 685, pp. 234–243, 2021.
https://doi.org/10.1007/978-3-030-53021-1_24

it is sufficient due to the effectiveness and good performance although in several cases there is no optimal performance by these controllers since bad tuning methodology in the loop control connected to the dynamical system is performed, mainly because full information of the system is required and there exist a lack of it in general [1].

This is the reason that in industrial applications have been developed methods which let to auto-tune the controllers in a certain configuration in closed loop[2], obtaining the parameters which let to guarantee optimal performance, moreover, the system identification plays a major role in this auto-tuning PID process[3].

The great advantage of the tuning methodologies is the independent of the human actions under disturbances, also uncertainties of the industrial environments. The tuning controllers follow two steps, first, system identification and second, adjusting the controller parameters. There exist several methods with different features like complexity, efficiency and design, among them is possible to find, Ziegle-Nichols, Kappa-Tau, IMC-PID adjust, Fractional PID, Successive Approximation (SAM-PID), Artificial Neural Networks (ANN), Fuzzy Particle Swarm Optimization (F-PSO), Iterative Feedback Tuning (IFT) and Extremum Seeking.

The Ziegler-Nichols tuning is commonly found in the industry, which is an off-line methodology to determine the PID parameters, in general, is made in the open-loop system, therefore there exists a chance to obtain an unstable system when the controller is implemented [4].

There exist other methodologies less conventional like the authors implemented in [5], the presented work considers a continuous algorithm to adjust the controller parameters before transient response of the system, based on neuro-dynamic programming, applying fuzzy rules to evaluate the likelihood of the system performance adjusting an evaluation signal which is used in the parameter adjusting process. Another common way to cope with this tuning problem is combining different techniques [6], using on-line intelligent algorithms using a combination of fuzzy logic and heuristic optimization like PSO. The IFT has the advantage to apply a performance criterion, which is based on the closed-loop system response, therefore, it is not constrained only with the plant model. On each iteration the system converges to a steady point according to the selected design criteria, the rate of change is controlled by the design, the objective functions might be varying to adjust the controller performance [7].

Extremum seeking is a model-free real-time optimization method minimizing a cost function, which quantifies the controller performance. Moreover, there is no model required which is an iterative form continuously modifies the cost function parameters, which are the controller gains minimizing in a local o global sense the system output [2]. As a control technique, ES lets to establish this optimization problem as a control problem according to the advantages mentioned and disturbance rejections properties [8]. The ES algorithm includes a sinusoidal perturbation in the actuation signal, and it is used to estimate the gradient of the objective function **J** minimized or maximized. Generally, the objective function is determined through sensor measurements of the system, although it depends on the internal dynamics and the input signal choose. In ES, the control variable **u** is the actuation signal or the parameters set which describes the control behavior, like frequency or gains in the PID controller [9]. The control scheme proposed in [9] applied in this work is observed on the Fig. 1:

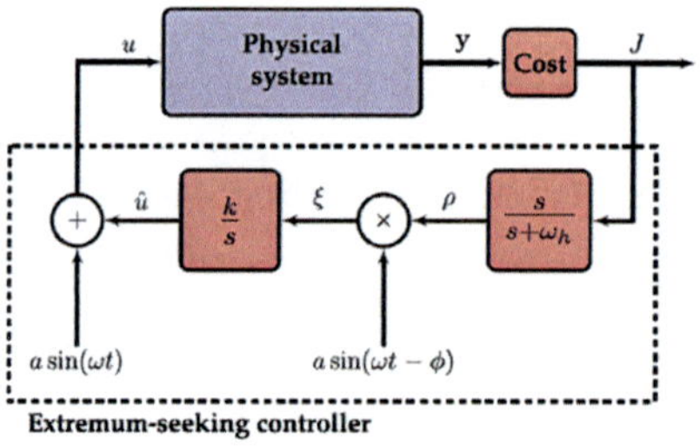

Fig. 1. Extremum seeking control scheme [9]

On the other way around, considering the implementation of this methodology looking for observe the system behavior, an industrial application was implemented in a flow control industrial process, including PLC programming, Human-Machine Interface (HMI), which integrate the whole hydraulic system, the sensor and control system in a SCADA (supervisory control and data acquisition), including industrial communication protocols, selecting the flow variable of this system. This measure was used in the ES algorithm and the results have been reported on this document [10].

2 Control System Loop and Auto-Tuning Algorithm

2.1 System Identification Method

Taking into account the signals obtained, and applying the ***First Order plus Dead Time (FOPDT)***, whose transfer function is defined as follow:

$$Gp(s) = \frac{k_p e^{-t_m s}}{\tau s + 1} \tag{1}$$

The transfer function representing the system identified is:

$$Gp = e^{-2.8s} \cdot \frac{0.02041}{5.796s + 1} \tag{2}$$

To implement the ES algorithm, first, a PI controller is selected with initial gains defined by the Root Locus method, adjusting the loop considering a canonical second-order polynomial, according to this the controller takes the form of:

$$C(s) = \frac{k(s + a)}{s} \tag{3}$$

The implementation generates the following controller parameters (k y a), replace them on the controller structure (3) before,

$$C(s) = \frac{28.62s + 45.5}{s} \tag{4}$$

2.2 Control Scheme

The Fig. 2, shows the control structure applied on the auto-tuning ES, in the first part the system is excited with a Heaviside function, then the controller is applied with the system in closed-loop form, afterward the cost function is defined with respect to the response, according to the steps this signal is the input for the ES algorithm which iteratively generates an output with the new controller parameters Kp and Ti.

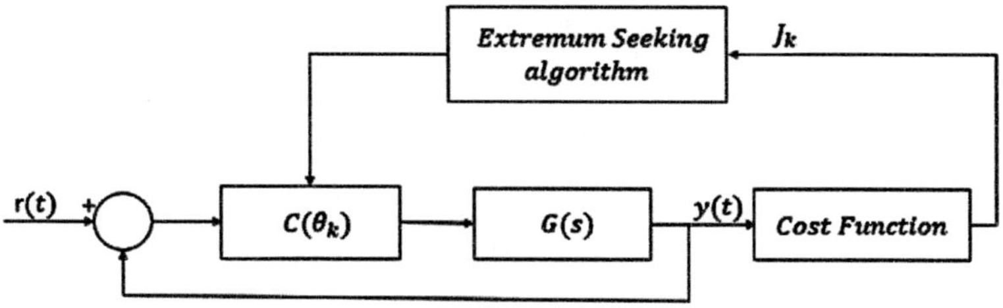

Fig. 2. System identification block diagram

Where:

$$C(\theta_k) = Kp_k(1 + \frac{1}{Ti_k s})$$

The function $C(s)$ is found in the Fig. 2, it can be shown in a detailed way in the Fig. 3, where the flow control loop is completely defined, based on the identification performed in the Sect. 2.

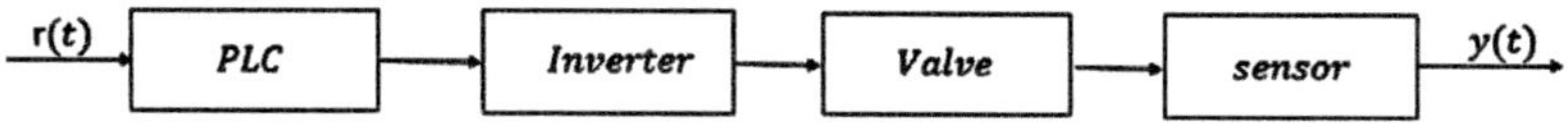

Fig. 3. Actuation and sensor systems

To implement the ES algorithm, identification is required which was performed on the flow control industrial plant, the control action takes the water pumped from reservoirs and transport the fluid into the pipe system maintaining the flow constant in the whole operation. This hydraulic system consists of two reservoirs (TK01 y TK02) and the auxiliar one (TK03), for each one the charge and discharge valves are included. The pump transport the fluid into the system, the instrumentation system keeps the information required, showing these signals in the HMI and the SCADA system. The system corresponds to that depicted in Fig. 4 and the Table 1 describes the whole elements in detail.

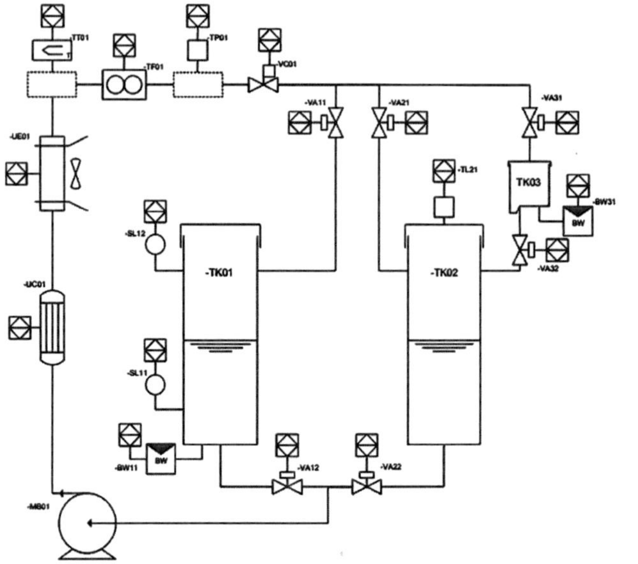

Fig. 4. Plant diagram used to implement the ES

2.3 Extremum Seeking Algorithm

Initially a cost function is defined, which produces the constants to be evaluated and improved while the code iterates, then we enter the initial constants, obtained by means of Root Locus, and finally we use the function transfer of the system, which has previously been applied PADE, because the identification was obtained with dead time.

Then the ES parameters such as frequencies, period, amplitudes and gains are entered; that for the case of the implementation that we make, being a PI controller, we enter vectors of two positions, one position for the values of the proportional constant and another for the integral constant and then the filter parameters.

Once the necessary parameters have been entered, the constants in the **for** cycle begin to be evaluated, until the best constant of that iteration is delivered, then these constants are introduced to the PI controller, and the error is calculated. This is done in such a way the current error minus the previous error is less than a certain parameter, previously determined, and at the end, the constants obtained are the optimal and the best for this controller.

The Extremum Seeking algorithm is described in Algorithm 1.

Algorithm 1. Extremum Seeking algorithm

```
function J(u, t)                                  ▷ Cost Function for u and t
    u ← valor inicial
    y0 ← J(u, 0)
    Gp ← transferfunction
                                                  ▷ Extremum Seeking Control Parameters
    frec ← sample frequency
    dt ← 1/frec
    T ← Total period of simulation (in seconds)
                                                  ▷ Perturbation Parameters
    A ← amplitude
    ω ← 20π
    φ ← 0
    K ← integration gain
                                                  ▷ High Pass Filter
    order ← 1
    freq ← in Hz
    [b, a] ← butter(order, freq · dt · 2, 'high')
    ys ← zeros(1, order + 1) + y0
    hpf ← zeros(1, order + 1)

    uhat ← u
    for i ← 1 to T/dt do
        t ← (i − 1) · dt
        time(i) ← t
        yvals(i) ← J(u, t)

        for k ← 1 to order do
            ys(k) ← ys(k + 1)
            hpf(k) ← hpf(k + 1)
        end

        ys(order + 1) ← yvals(i)
        hpfn ← 0

        for k ← 1 to order + 1 do
            hpfn(k) ← hpfn + b(k) · ys(order + 2 − k)
        end

        for k ← 2 to order + 1 do
            hpfn(k) ← hpfn − a(k) · hpf(order + 2 − k)
        end

        hpfn ← hpfn/a(1)
        hpf(order + 1) ← hpfn

        χ ← hpfn · sin(ω · t + φ)
        uhat ← uhat + χ · K · dt
        u ← uhat + A · sin(ω · t + φ)
        uhats(i) ← uhat
        uvals(i) ← u
    end
```

Table 1. Devices used in Fig. 4

List of elements	
BW11	Load cell
BW31	Load cell
MB01	Hydraulic pump
SL11	Level switch
SL12	Level switch
TF01	Flow transmitter
TK01	Process tank
TK02	Process tank
TK03	Process tank
TL21	Level transmitter
TP01	Pressure transmitter
TT01	Temperature transmitter
UC01	Heating unit
UE01	Cooling unit
VA11	Filling valve TK01
VA12	Drain valve TK01
VA21	Filling valve TK02
VA22	Drain valve TK02
VA31	Filling valve TK03
VA32	Drain valve TK03
VC01	Proportional control valve

3 Implementation and Results

From the code developed in Sect. 2.3 to obtain the best constants for the PI control, through the Extremum Seeking, the implementation is performed in a process control plant.

The final controller obtained employing this methodology, whose constants were improved through the iterations, compared to the initially used (Eq. 4) was as follows:

$$C(s) = \frac{144.2s + 57.42}{s} \tag{5}$$

Subsequently, these constants were entered into a block developed in the TIA Portal, in which, in addition to the constants, we have the value of the reference of the flow to reach and some parameters from the sensor and pump, as follows (Fig. 5).

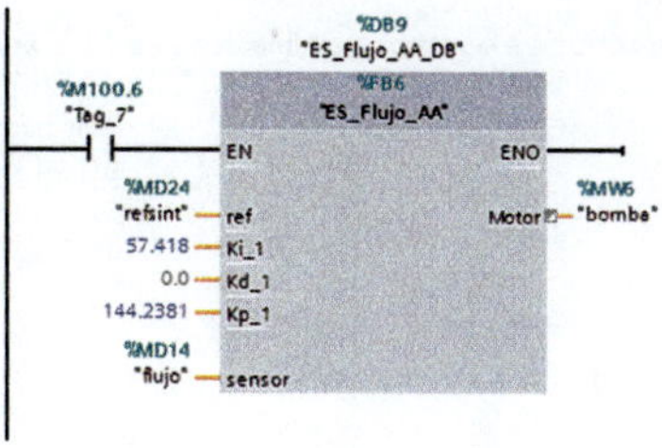

Fig. 5. Function PI TIA Portal

In the Fig. 6, it is possible to see the behavior of the initial controller whose constants were obtained employing the Root locus ($q_1(t)$) and the behavior obtained with Extremum Seeking ($q_2(t)$).

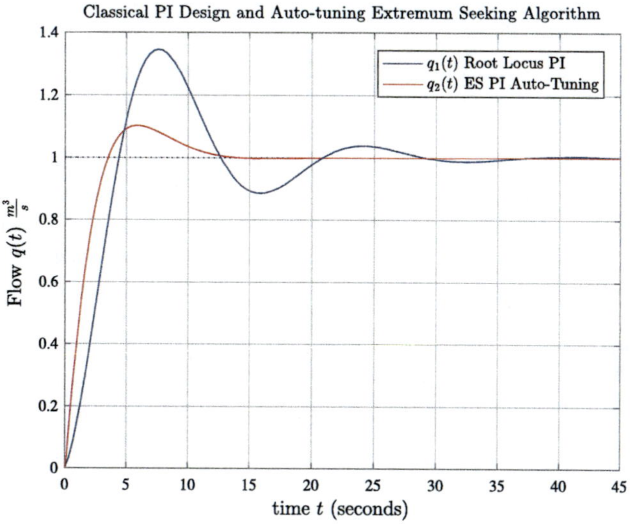

Fig. 6. Output of the controllers using as input one unit step function

We can see that the plant with the controller tuned using Extremum Seeking (Blue line) has a remarkably fast response and with little oscillation concerning the reference (Orange line).

4 Discussion

In Fig. 7 we can see the operation of the physical plant with the controller tuned through the Extremum Seeking, its response is fast and has an acceptable overshoot, the system perfectly controls the flow variable despite the disturbances

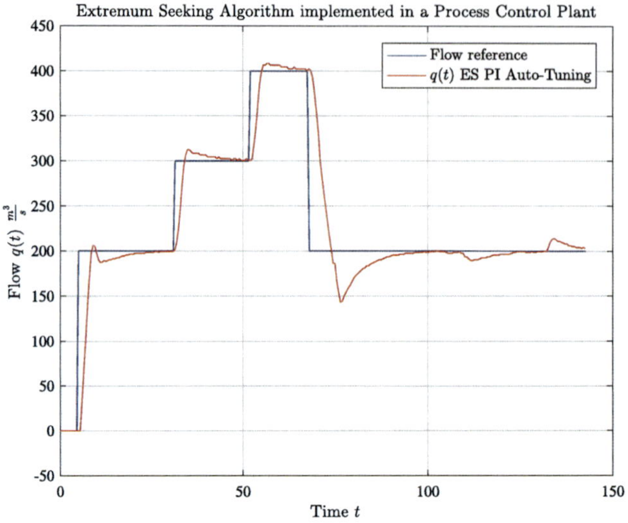

Fig. 7. Controller behavior at the plant

that arise in the system. It is possible to use this method as an offline auto-tuning, because having a stable initial controller, then we can simulate the behavior of the system using a unit step, with that response the error is calculated and the algorithm finds in a first iteration a few constants, which are entered back into the controller and continues to iterate until it finds the best constants according to the criteria previously established.

Finally, the last constants obtained are implemented as a PI controller in the plant.

5 Conclusions

The behavior of the plant with Extremum Seeking is very efficient in terms of flow control and its behavior is noticeably better than of a conventional controller. It facilitates the tuning of any controller with the algorithm since the initial constants don't need to be perfect, the only restriction is that they do not produce a loss of stability.

To develop the algorithm in Matlab to the Extremum Seeking facilitates the implementation for any plant, only by changing the transfer function in the code and the design criteria, such as initial conditions of the controller, amplitude, frequency and some basic parameter is possible to get a new answer.

The output of the final controller get by ES is very susceptible to small changes made to the filter design constants, such as to the algorithm gains.

In addition, the algorithm can function as an adaptive control through its discretization, to be implemented in any systems, which would further improve the behavior, because it modifies its response due to changes in the dynamic system, i.e. if there are disturbances during operation, it would cause changes in the system parameters however the system can continue to have an optimal operation. On the other hand, it is also possible to implement other self-tuning methodologies, such as PSO or IFT, to make a comparison of the effectiveness and behavior of them in complex systems.

Acknowledgment. This paper is part of project number 111077657914 and contract number 031-2018, funded by the Colombian Administrative Department of Science, Technology, and Innovation (COLCIENCIAS) and developed by the ICE3 Research Group at Universidad Tecnologica de Pereira (UTP) and CALPOSALLE Group at Universidad de La Salle.

References

1. Duque-Marín, A., Lopez, J.A., Navas, A.F.: Auto-tuning of a PID controller implemented in a PLC using swarm intelligence. Prospectiva **15**(1), 35–41 (2017)
2. Killingsworth, N.J., Krstic, M.: PID tuning using extremum seeking: online, model-free performance optimization. IEEE Control Syst. Mag. **26**(1), 70–79 (2006)
3. Yu, C.C.: Autotuning of PID Controllers: A Relay Feedback Approach. Springer, Heidelberg (2006)
4. Gade, S.S., Shendage, S.B., Uplane, M.D.: On line auto tuning of PID controller using successive approximation method. In: 2010 International Conference on Recent Trends in Information, Telecommunication and Computing, pp. 277–280, March 2010
5. Seidi Khorramabadi, S., Bakhshai, A.: Critic-based self-tuning PI structure for active and reactive power control of VSCs in microgrid systems. IEEE Trans. Smart Grid **6**(1), 92–103 (2015)
6. Bevrani, H., Habibi, F., Babahajyani, P., Watanabe, M., Mitani, Y.: Intelligent frequency control in an AC microgrid: online PSO-based fuzzy tuning approach. IEEE Trans. Smart Grid **3**(4), 1935–1944 (2012)
7. Hjalmarsson, H., Gevers, M., Gunnarsson, S., Lequin, O.: Iterative feedback tuning: theory and applications. IEEE Control Syst. Mag. **18**(4), 26–41 (1998)
8. Dochain, D., Perrier, M., Guay, M.: Extremum seeking control and its application to process and reaction systems: a survey. Math. Comput. Simul. **82**(3), 369–380 (2011). 6th Vienna International Conference on Mathematical Modelling
9. Brunton, S.L., Kutz, J.N.: Data-driven Science and Engineering: Machine Learning, Dynamical Systems, and Control. Cambridge University Press, Cambridge (2019)
10. ANITCO: Manual de Operación y Mantenimiento Unidad de Mantenimiento en Automatización (UEA), April 2015

Development of a Swine Health Monitoring System Based on Bio-Metric Sensors

Sebastian Rodriguez(✉), Carolina Chaves, and Alejandro Quiroga

Fundacion Universitaria Agraria de Colombia - Uniagraria, Bogota, Colombia
{Rodriguez.Jhoan,Chaves.Nydia}@uniagraria.edu.co,
Quiroga.Jaime@uniagraria.es.co
https://www.uniagraria.edu.co/

Abstract. The article presents the mechatronic design and construction of a bio-metric sensor for swine physiological variables reading and transmission, the design incorporates a temperature sensor and optic pulse heart beat sensor managed by a micro-controller on a wireless collar array. A receiver under a network structure of sensors is able to detect and storage data from unsupervised, freely-moving pigs to correctly evaluate the animal condition. The sensor system was tested under real conditions to check the toughness and endurance of the structure and the effects on the animal behaviour. Data transmission was successfully maintain under rough conditions where both the animal and instrumentation integrity was ensured.

Keywords: Biometrics · Animal health · Swine stress · Animal sensors · Sensor network · Wireless communications

1 Introduction

The size of global animal production, driven by dietary social evolution, results on high volume production of specific breeds of domesticated animal for food production [1]. As swine makes up for a large portion of protein production, it is crucial to characterize the impact of farming techniques since the animal can be reared in different locations and conditions before they are eventually slaughtered [2].

Precision farming tendencies allow professionals to make data-driven decisions to improve not only production indices but also animal well-fare conditions [3]. Precision Livestock Farming (PLF) is a way of managing a farm through the monitoring and recording of automated, real-time measurements of animal production, breeding, health and well-being [4]. Precision livestock farming has also been defined as the real-time monitoring technologies aimed at managing the smallest manageable production unit's temporal variability, known as "the per animal approach" [5].

D. F. Cortes Tobar et al. (Eds.): AETA 2019, LNEE 685, pp. 244–251, 2021.
https://doi.org/10.1007/978-3-030-53021-1_25

Biological research benefits from the technology of biome-tric sensors by monitoring of animal welfare in stressful environments [6]. The productive performance, and by extension, the well-ware of animal is adversely affected by extremes in the environment, such as temperature and crowding; in the same way the freedom of movement increases the risk of spreading livestock diseases [7], resulting in the contamination of meat products [8].

2 Swine Bio-Metric Variables

Swine rearing requires an specific control over external variables to ensure an optimal grow and development of the pig, the principal variables that affect the health of the animal are heart rate and temperature since reflect the impact of environmental conditions on the animal and how this can cope with them, animal stress detection can also be infer from this variables to predict and avoid negative effects of pathological conditions such as Swine Stress Syndrome which can culminate in the sudden dead of the animal [9].

3 Design of the Wearable Bio-Metric Sensor Transmitter

The sensor-transmitter system include 3 modules, see Fig. 3 which strap together around the neck of the pig, see Fig. 1, an Ear-Clip adapter for the heart rate sensor attached to the collar completes the assembly. The bio-metric sensor system is completely wireless with an autonomy of 6 hours between charges.

The battery module encloses the main power supply which can accommodate both 4-AA batteries pack or a Lithium polymer battery 7.2 v 2-cells pack, see Fig. 2.

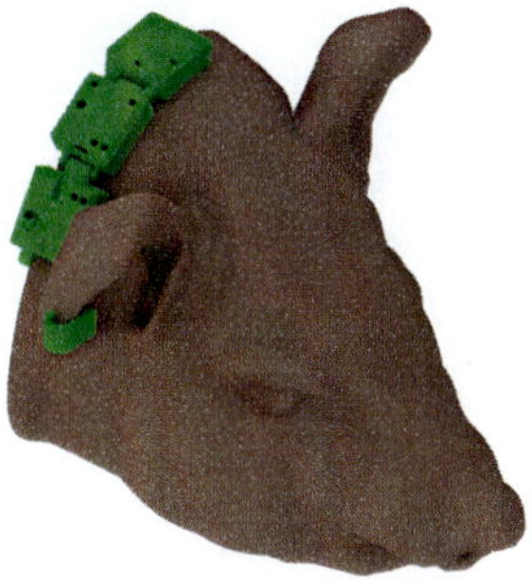

Fig. 1. Concept design of a pig wearing the bio-metric sensor transmitter.

The sensor module includes the temperature sensor probe in contact with the pig skin, the sensor is placed away from both the battery and micro-controller

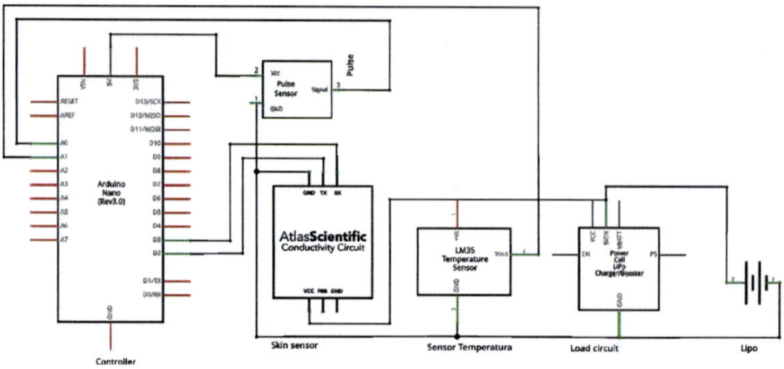

Fig. 2. Circuit diagram of the main components in the collar.

case to ensure no power-diffusive heat source would alter the temperature reading. The collar is assembled using the 3 principal modules linked using balsa wood pins and looped around the pig neck using a strap and buckle belt.

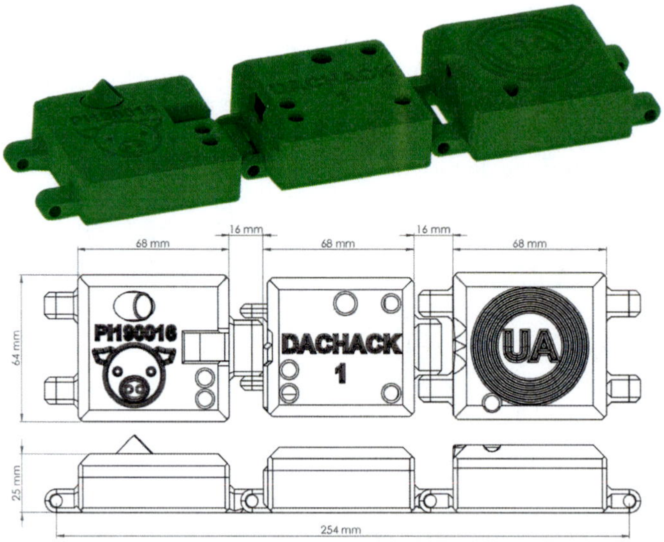

Fig. 3. Design of the Collar array. From left to right: Sensor module, Micro-controller module and Battery module.

3.1 Ear-Clip

The heart rate is measured using an optic analog Pulse Sensor sensible to the changes in light absorption caused by the expansion and contraction of blood

Fig. 4. The Ear-Clip features a opening clip to install the sensor effortless.

vessels during circulation, see Fig. 5. The change in volume caused by the expansion of blood vessels is detected by illuminating the skin with the light from a light-emitting diode (LED) and then measuring the amount of light either transmitted or reflected to a photodiode. The heart rate analog signal is computed on the collar array and only the calculated beats per minute (bpm) variable is transmitted to the receptor for later analysis. The ear-clip, Fig. 4, was chosen to take the heart rate measures since the skin is thinner and the optic receptor analog range can be adjusted to correctly identify peaks and valleys of photoplethysmogram.

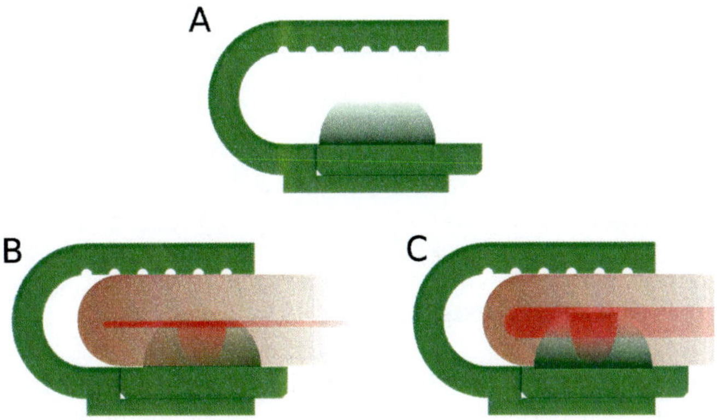

Fig. 5. Pulse sensor function principle on the Ear-Clip. A) Constant light emission. B) Low level reflection from blood vessels. C) High level reflection from blood vessels.

3.2 Receptor

The receptor uses a single a 2.4 GHz ISM band single chip radio transceiver operating ad −18 dB; a ATmega328 micro-controller constantly listen to one or multiple radio channels (frequencies) at 1 Mbps. The case design, see Fig. 6, embody a mount hook to secure the receptor to a high base in order to cover a wider area.

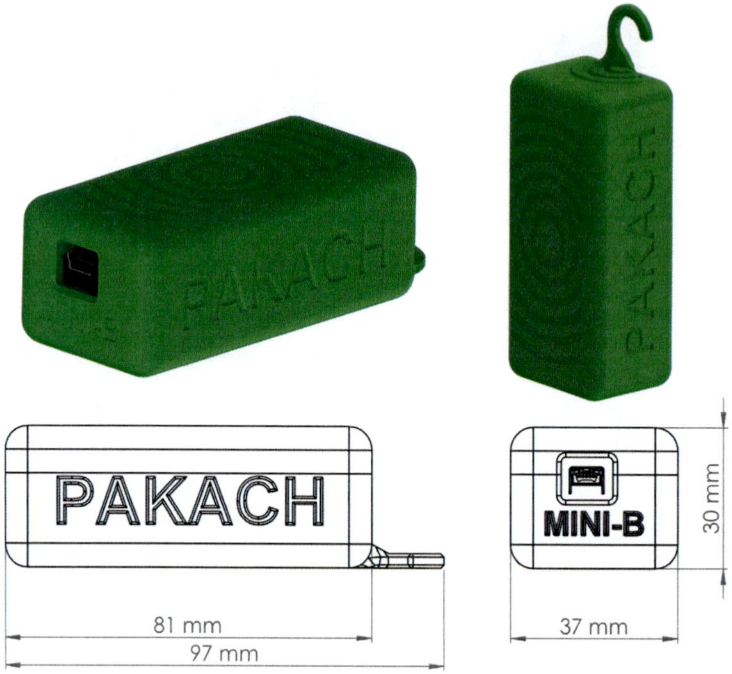

Fig. 6. Receptor design featuring a mini-USB port and a mounting hook.

The receptor connects to a personal computer using a mini-USB to USB cable and the data can be acquired through serial protocol communication at 57600 bits per second.

4 Sensor Network Design

An underlying problem in recording biome-metric signals from freely-moving animals over extended periods of time is the large volume of data that can be collected. Since both heart rate and temperature are variable that do not change rapidly a transmission protocol its created to send a data chain every time a heart beat its detected, about 1 to 2 packages per minute.

The network system allows the insertion of multiple transmitters collecting data from different specimens on the area; the Multiple Transmitters Single Receiver structure, shown in Fig. 7, represents a configuration on which each sensor system transmits its data on separate channels, in this case a single receptor scan a predefined band listening for any data package "message".

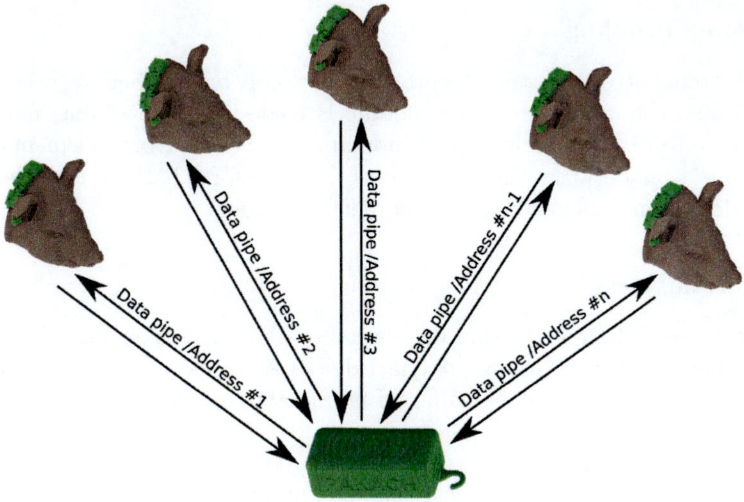

Fig. 7. Receptor...

5 Fabrication

The collar and receptor where constructed and sealed for a real life test, see Fig. 8 run to test the acquisition and transmission of data.

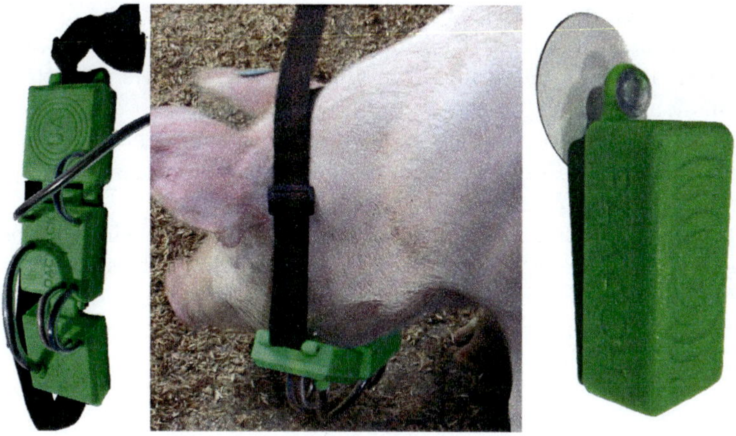

Fig. 8. Collar array strapped around a pig's neck.

5.1 Water Proofing

Both the transmitter and the receptor are enclosed on by their respective 3D printed cases, the circuits on each module is coated with low-taint, non-toxic silicone to ensure a watertight solid bonding of the electronic elements. The cable connection between the micro-controller, sensors, battery and Ear-Clip are also driven and enclosed by transparent PVC hoses.

5.2 Harmlessness

The material and design selection was driven by the concern on animal well-fare and harmlessness, the 3D printed polylactide (PLA) cases of the collar array and the Ear-Clip are bio-degradable [10] and no metallic material is in contact with the animal.

Although the mounting and removal of the collar array must be done by a professional withing a time no longer than 6 hours, the collar array has an safety measure where the pins keeping the modules together are made of uninsulated wood balsa, which under long exposure to humidity start to decompose ensuring that any disregarded animal wearing the collar can free itself avoiding any harm due to long exposure to the collar materials.

6 Conclusions

- The multiple module array of the sensor collar correctly adjusted to the size and shape of young and mature pigs with no observable affectation to the animal behavior; since the straps were not tighten to avoid any discomfort on the animal that could alter the heart rate or temperature readings, the collar rotated around the neck facing down but this didn't interfered with the data transmission.
- No significant damage was observed over the collar or the Ear-Clip after a couple of hours of testing even though the pigs lay over the assembly, bite it and rolled up over puddles in their enclosure.
- The sensor was able to read peak values of heart rate while it was being strapped around the pigs neck, this value correspond to the stress induced by the fear of the animal during handling; the process of installing and removing the sensor should be done as fast as possible to avoid creating adverse situations that could affect the health of the animal.

References

1. Neethirajan, S., Tuteja, S.K., Huang, S.T., Kelton, D.: Recent advancement in biosensors technology for animal and livestock health management. Biosens. Bioelectron. **98**, 398–407 (2017). https://doi.org/10.1016/j.bios.2017.07.015
2. Tullo, E., Finzi, A., Guarino, M.: Review: environmental impact of livestock farming and precision livestock farming as a mitigation strategy. Sci. Total Environ. **650**, 2751–2760 (2019). https://doi.org/10.1016/j.scitotenv.2018.10.018

3. Villeneuve, É., Akle, A.A., Merlo, C., Masson, D., Terrasson, G., Llaria, A.: Decision support in precision sheep farming. IFAC-PapersOnLine **51**(34), 236–241 (2019). https://doi.org/10.1016/j.ifacol.2019.01.048
4. Ndour, A., Loison, R., Gourlot, J.-P., Ba, K.S., Clouvel, P.: Biotechnologie, agronomie, société et environnement. BASE Biotechnol. Agron. Soc. Environ. **21**(1), 22–35 (2017). http://popups.ulg.ac.be/1780-4507/index.php?id=13496
5. Føre, M., Frank, K., Norton, T., Svendsen, E., Alfredsen, J.A., Dempster, T., Eguiraun, H., Watson, W., Stahl, A., Sunde, L.M., Schellewald, C.: Precision fish farming: a new framework to improve production in aquaculture. Biosyst. Eng. **173**, 176–193 (2018). https://doi.org/10.1016/j.biosystemseng.2017.10.014
6. Awasthi, A., Awasthi, A., Riordan, D., Walsh, J.: Non-invasive sensor technology for the development of a dairy cattle health monitoring system. Computers **5**(4), 23 (2016). https://doi.org/10.3390/computers5040023
7. Kumar, S., Singh, S.K.: Monitoring of pet animal in smart cities using animal biometrics. Future Gener. Comput. Syst. **83**, 553–563 (2018). https://doi.org/10.1016/j.future.2016.12.006
8. Awad, A.I.: From classical methods to animal biometrics: a review on cattle identification and tracking. Comput. Electron. Agric. **123**, 423–435 (2016). https://doi.org/10.1016/j.compag.2016.03.014
9. McCauley, I., Matthews, B., Nugent, L., Mather, A., Simons, J.: Wired pigs: ad-hoc wireless sensor networks in studies of animal welfare. In: Second IEEE Workshop on Embedded Networked Sensors, EmNetS-II, vol. 2005, pp. 29–36. IEEE (2005). https://doi.org/10.1109/EMNETS.2005.1469096
10. Raj, S.A., Muthukumaran, E., Jayakrishna, K.: A case study of 3D printed PLA and its mechanical properties. Mater. Today: Proc. **5**(5), 11219–11226 (2018). https://doi.org/10.1016/j.matpr.2018.01.146

Comparison of SST Topologies Suitable for Energy Applications

Tomáš Košťál[1](✉), Pavel Kobrle[1], Jakub Zedník[1], Jiří Pavelka[1], and Xiaofeng Yang[2]

[1] Department of Electric Drives and Traction, Czech Technical University in Prague, Technická 2, 166 27 Prague, Czech Republic
{kostatom,kobrlpav,zednijak,pavelka}@fel.cvut.cz
[2] School of Electrical Engineering, Beijing Jiaotong University, No. 3 Shangyuan Cun, Beijing 100044, China
xfyang@bjtu.edu.cn

Abstract. This paper presents an assessment of Solid State Transformer (SST) topologies from the usability in energy applications point of view. SSTs are nowadays at the center of researchers' attention mainly due to reducing size and weight of traction transformers in railway vehicles. Our contribution aims to use the SST as a cornerstone of an Energy Router that should replace a typical distribution substation with a classical transformer. After a general overview of SST topologies classification an overview of design guidelines for key SST components (ferrite cores for medium frequency (MF) transformers, semiconductors and high voltage capacitors) is presented for two topologies that seem to be promising for energy applications: classical three-stage SST and two-stage SST with a low voltage DC link and galvanic insulation in the first stage. These two topologies are then compared from an economical point of view based on acquisition cost of key components in order to evaluate whether the technology has matured to be used commercially.

Keywords: SST · Solid State Transformer · Energy router · Power converter design

1 Introduction

Power distribution grids are facing several significant challenges nowadays, such as accommodation of small unstable decentralized power sources and energy storage systems. One of the possible solutions is to deploy a device called Energy router. It is a replacement for a classical electromagnetic distribution transformer. Energy router is a complex device which does not have a unified structure because it is still under extensive research by many research teams around the globe [1]. The key part of Energy router is a power electronic module that typically consists of a so-called Solid State Transformer (SST) that allows controlling the power flow and the quality of electric energy in the distribution network. SST must in principle have lower efficiency than a classic transformer

D. F. Cortes Tobar et al. (Eds.): AETA 2019, LNEE 685, pp. 252–261, 2021.
https://doi.org/10.1007/978-3-030-53021-1_26

(e.g. 96.3% in case of a two-stage SST compared to 98.7% of classic transformer both for rated power of 1 MVA as stated in [1]) but it has the advantage of sophisticated control. SST can replace a classic transformer in applications where the size of the transformer should be decreased. This is not the case of power grid applications because size is not a crucial parameter there but in smart grids the control of power flow is an essential property. In the following sections, the general design procedure and comparison of two possible topologies suitable for SST is made and evaluated from an economical point of view. First, a basic summary of SST topologies will be presented.

Various topologies of SSTs have been proposed hence the situation can be rather confusing for an SST designer. Moreover the terminology has not yet stabilized in this dynamic field of research. In general, the main classification of SSTs is based on the number of stages that the SST comprises. Most authors agree on distinguishing three main "blocks" of an SST: medium voltage (MV, "primary", "grid", "EWAN", "source grid", "front-end") side, low voltage (LV, "secondary", "microgrid", "ELAN", "load grid", "back-end") side and galvanic insulation. Based on whether these three main blocks consist of dedicated converters or whether some of these functions are merged, three-stage SSTs, two-stage SSTs and single-stage SSTs can be distinguished [2–8]. Within the two-stage SST topologies, several sub-types can be differentiated based on whether there is a low voltage or medium voltage link between the stages [2, 3, 9]. Among other sources, [2] distinguishes four different types of SST, as shown in Fig. 1. Among single-stage SSTs and three-stage SSTs, there are two types of two-stage SSTs that differ in location of galvanic insulation (either in first or in second stage).

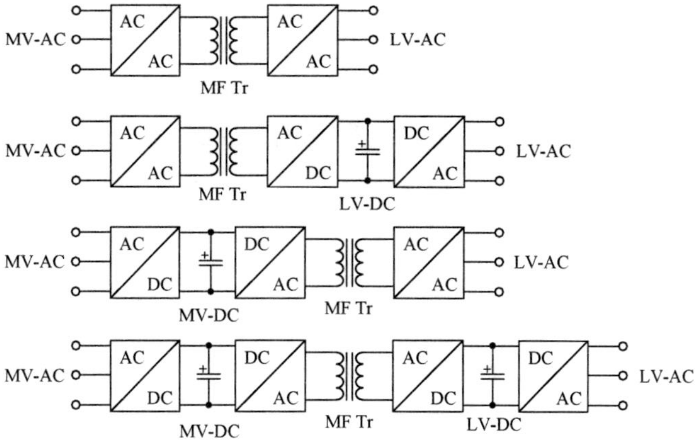

Fig. 1. Classification of SSTs based on stages as proposed by [2].

Topologies Used for Comparison

In the present paper we aim to compare the complexity of selected topologies and also their financial point. Our goal is to assess suitable topologies to be used as an SST in an Energy Router that should stand in place of $S_n = 630$ kVA distribution transformer on 20 kV supply grid voltage. 630 kVA and 20 kV are typical apparent power and voltage values of transformers used in distribution grids in many countries. Moreover we consider this apparent power value as an appropriate power of an Energy Router which could be connected to a microgrid with decentralised power sources and energy storage systems. For the purpose of our paper we did not pursue the single-stage topologies because we do not find them flexible enough to be scaled for a 630 kVA substation. Similarly we did not consider two-stage topologies with MV or high voltage (HV) DC link, as we think that the key advantage of two-stage topologies in field of high power applications (such as an Energy Router which should feed a microgrid) is the omission of the HV DC link which needs more expensive and bulkier capacitors than a low voltage DC link. We also did not take into account the LV-AC inverter as it is same in all of the cases. We consider the following two topologies as the most promising ones. First is a three-stage SST with a modular multilevel converter (MMC) on the "grid" side, middle stage with insulating transformers in input serial output parallel (ISOP) modular Dual Active Bridge (DAB) configuration and a three phase inverter on the "microgrid" side (Fig. 2).

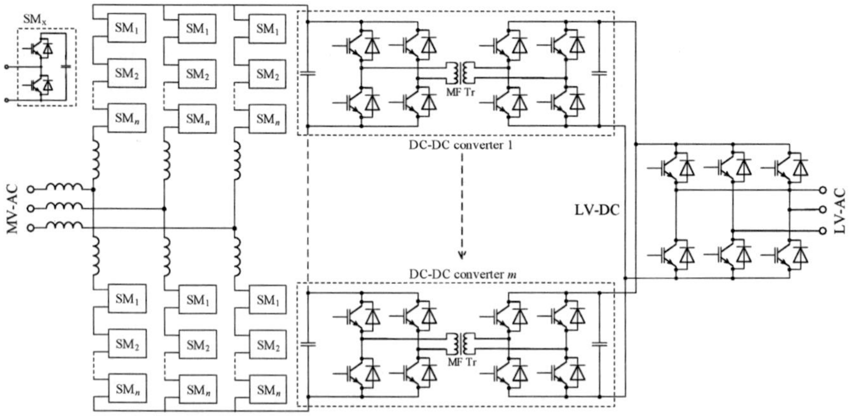

Fig. 2. Topology of a three-stage SST used for comparison in this paper [10].

Second is a two-stage SST based on the idea presented in [9] (Fig. 3). This topology uses a cascade of modules (labeled as "converter" in Fig. 3) connected in series for each phase (with star connection of phases). Each module consists of DAB connected to the full bridge. Outputs of the modules are connected in parallel, comprising a low voltage DC link (and the LV-DC bus as well).

2 General Design Guidelines for Key Components of SSTs

As mentioned above the rated apparent power for an SST is $S_\mathrm{n} = 630$ kVA and its rated MV phase-to-phase voltage $U_{\mathrm{MVRMS}} = 20$ kV. It is assumed for the design that this voltage in MV network can fluctuate within the range of $\pm 15\%$.

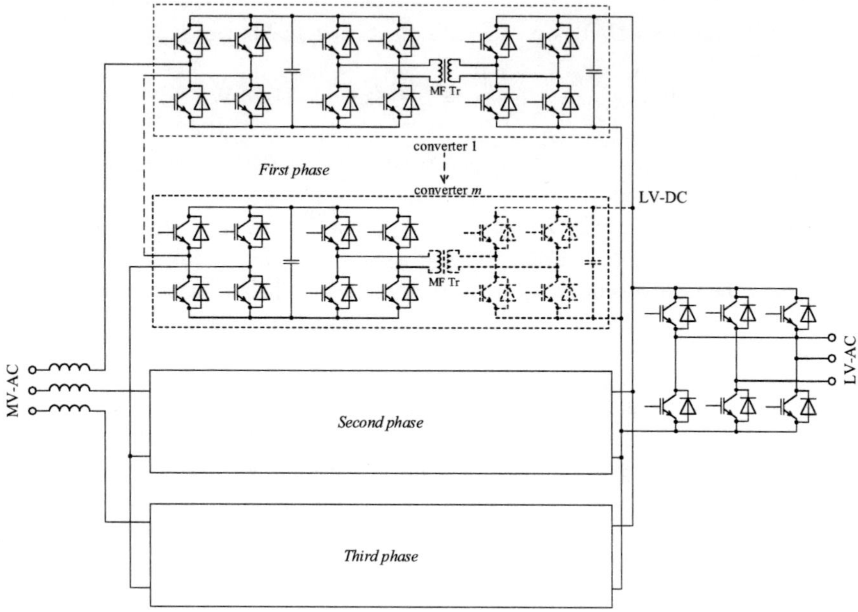

Fig. 3. Topology of a two-stage SST used for comparison in this paper.

2.1 General Design of a Three-Stage SST for a Particular Rated Apparent Power S_n and Resultant Parameters of Required Semiconductors

DC voltage U_{MVDC} required for PWM modulation of the MV side without injection of harmonics is

$$U_{MVDC} = 1.15 \cdot \sqrt{2} \cdot U_{MVRMS} = 1.15 \cdot \sqrt{2} \cdot 20\ \text{kV} = 32.6\ \text{kV} \tag{1}$$

When the third harmonic injection is used, then the required voltage U_{MVDCr} is lower (1.15×) and the needed DC voltage (which is the voltage per MMC arm) is

$$U_{MVDCr} = \frac{U_{MVDC}}{1.15} = \frac{32.6}{1.15}\ \text{kV} = 28.3\ \text{kV} \tag{2}$$

Maximal phase current $I_{MVphasemax}$ by the minimal phase voltage (15% voltage decrease respected with the coefficient k_f) is

$$I_{MVphasemax} = \frac{S_n}{\sqrt{3} \cdot k_f \cdot U_{MVRMS}} = \frac{630}{\sqrt{3} \cdot 0.85 \cdot 20}\ \text{A} = 21.4\ \text{A} \tag{3}$$

The number of submodules in the MMC arm is determined by the properties of the IGBTs that are used. The most common value of voltage U_{CE} of the IGBTs on the market is 1.2 kV. Therefore the maximal DC voltage of one capacitor will be $U_{DCcap} =$ 1 kV. The number of capacitors N_C (submodules) in MMC per arm is 29. A three phase MMC has 6 arms (Fig. 2) so the total number of submodules is 174. A suitable IGBT half bridge has to have $U_{CE} = 1.2$ kV and $I_C = 22$ A.

All DC/DC modules (DAB) operate with identical active power P_{one_module}.

$$P_{one\_module} = \frac{P_n}{N_{cap}} = \frac{630}{29}\ \text{kW} = 21.7\ \text{kW} \tag{4}$$

Input currents $I_{DCinput}$ of all modules (DAB) in the DC/DC converter are identical and their value is 21.7 A (as the voltage per module is 1 kV).

The number of modules (DAB) $N_{DC/DCmodules}$ in a DC/DC converter is the same as number of submodules in the MMC: $N_{DC/DCmodules} = 29$. Output currents $I_{DCoutput}$ of all modules in the DC/DC converter are also identical and their value is

$$I_{DCoutput} = I_{DCinput} \cdot p_T = 21.7 \cdot \frac{1000}{700}\ \text{A} = 31\ \text{A} \tag{5}$$

where p_T is the transformation ratio of the transformer. Each DC/DC module contains two full bridges with parameters $U_{CE} = 1.2$ kV, $I_C = 32$ A. Both sides of DC/DC modules will use the same full bridges for improving economy of scale and also for the fact that the lowest current of 1.2 kV modules is still higher than this value. Then the number of full bridges is twice the number of modules $N_{DC/DCmodules}$ which means 58 full bridges. The number of transformers is 29 (m = 29 in Fig. 2, one per each module).

2.2 General Design of a Two-Stage SST for a Particular Rated Apparent Power S_n and Resultant Parameters of Required Semiconductors

Maximal phase-to-zero voltage in MV network respecting the voltage fluctuation is

$$U_{MVmax} = \frac{1.15 \cdot \sqrt{2}}{\sqrt{3} \cdot U_{MVRMS}} = \frac{1.15 \cdot \sqrt{2}}{\sqrt{3} \cdot 20}\ \text{kV} = 18.8\ \text{kV} \tag{6}$$

When the third harmonic injection into the phase-to-zero voltage is used then the required voltage U_{MVmaxr} is lower (1.15×), so the voltage U_{MVmaxr} is then 16.35 kV. As a suitable voltage for one module, 1 kV has been chosen again, so the number of AC/DC modules in one phase is 17 (the closest higher integer number; m = 17 in Fig. 3).

Maximal phase current $I_{MVphasemaxRMS}$ flows when the phase voltage is minimal (0.85 of rated value as described above). This current has a sinusoidal waveform so its amplitude is

$$I_{MVphasemax} = \sqrt{2} \cdot \frac{S_n}{\sqrt{3} \cdot k_f \cdot U_{MVRMS}} = \sqrt{2} \cdot \frac{630}{\sqrt{3} \cdot 0.85 \cdot 20}\ \text{A} = 30.3\ \text{A} \tag{7}$$

Voltages of capacitors are constant in the DC side between the two full bridges and their value is equal to 1 kV. Therefore inner DC/DC converters in all modules operate with identical active power P_{one_module}.

$$P_{one\_module} = \frac{P_n}{N_{modules\_phase} \cdot 3} = \frac{630}{17 \cdot 3}\ \text{kW} = 12.35\ \text{kW} \tag{8}$$

Input currents of all modules DABs are identical and their value is 12.35 A. Output currents I_{DCout_module} of all modules are also identical and their value is

$$I_{DCout\_module} = I_{DCin\_module} \cdot p_T = 12.35 \cdot \frac{1000}{700}\ \text{A} = 17.7\ \text{A} \tag{9}$$

Each DC/DC module contains two full bridges with parameters $U_{CE} = 1.2$ kV, $I_C = 18$ A. Then the total number of full bridges is 153 as there are 3 bridges per module. Each module contains one MF transformer which means 51 transformers altogether.

3 Design of the Transformers

The fact that it uses higher frequency (MF) for the transformer is a key property of an SST. Thus the magnetic circuit of the transformer should be fabricated from ferrite material. Rated apparent power of transformer can be described as follows:

$$S_n = 2 \cdot B_{max} \cdot f_s \cdot \sigma_{Cu1} \cdot K_{Cufill} \cdot S_m \cdot S_{Cureal} \quad (10)$$

where
σ_{Cu1} is current density in the winding conductor [A·mm^{-2}]
S_m is cross section area of the magnetic circuit [m^2]
S_{Cureal} is cross section of conductor [m^2]
B_{max} is the maximal value of magnetic flux density in magnetic circuit [T]
f_s is supply frequency (switching frequency of the full bridges) [s^{-1}]
K_{Cufill} is the fill coefficient.

Recommended flux density for ferrite material is about 0.2 T. The following parameters in (10) can be estimated or selected: $B_{max} = 0.2$ T, $f_s = 10$ kHz, $\sigma_{Cu1} = 2$ A·mm^{-2}, $K_{Cufill} = 0.35$.

We obtain the following relation between the transformer apparent power S_n and the product ($S_m \cdot S_{Cureal}$) when the selected parameters are substituted in (10):

$$S_n = 2 \cdot 0.2 \cdot 10^4 \cdot 2 \cdot 0.35 \cdot S_m \cdot S_{Cureal} = 0.0028 \cdot S_m \cdot S_{Cureal} \quad (11)$$

To minimize the skin effect in a conductor at medium frequencies, copper strands must be used. Parameters of strand wires and ferromagnetic cores can be obtained from datasheets. For our three-stage SST the value of product $S_m \cdot S_{Cureal}$ is 7.64 m^2·mm^2 while for the two-stage it is 3.82 m^2·mm^2. The largest ferrite core from standard production of ferrite materials manufacturers, the Kaschke type 126/20, has the value 4.94 m^2·mm^2. This means that it is not suitable for the three-stage SSTs. However, more cores can be used next to each other to increase the magnetic circuit cross-section. In this case doubling the value is sufficient so there will be in total 4 U pieces for one three-stage SST MF transformer and two U pieces for the two-stage SST MF transformer. The total number of MF transformers required by each topology is stated in Sects. 2.1 and 2.2 respectively.

4 Determination of Capacitors' Values

In the case of three-stage SSTs there are three types of capacitors: in submodules of MMC, in the MV-DC link and on outputs of the galvanic insulation modules (DAB). In the case of two-stage SST there are two types of capacitors: one type is in the DC link and another type is on the outputs of the galvanic insulation modules (DAB), similarly to the previous case. Concerning the capacitor rated DC voltage we selected 1% voltage ripple of the capacitors' voltage in steady state operation as a main design criterion. A procedure of calculation of the optimal capacity has been proposed in our paper which is currently under the review process and its more detailed description would exceed the scope of this paper.

4.1 Design of Capacitors in Submodules of the MMC in Three-Stage SST

For the application in a particular SST, we use the following per unit values for calculation: grid voltage $u_G = 1$, grid current $i_G = 1$, average voltage across one capacitor in a module $u_{c(AV)} = 1$, voltage ripple $\Delta u_{cD} = 0.01$, number of submodules $N_{SM} = 29$. The base grid phase-to-phase voltage U_{MVRMS} is 20 kV, the base current I_N is 18.2 A ($I_{MVphasemax} \cdot k_f$), base phase impedance Z_N is 635.5 Ω (calculated from the U_{MVRMS} and I_N). The resulting formula for the capacitor designated as C_1 is as follows:

$$C_1 = \frac{N_{SM} \cdot u_G \cdot i_G}{\omega \cdot 2 \cdot 3 \cdot u_{c(AV)} \cdot \Delta u_{cD} \cdot Z_N} = \frac{29 \cdot 1 \cdot 1}{2 \cdot \pi \cdot 50 \cdot 2 \cdot 3 \cdot 1 \cdot 0.01 \cdot 634.5} \mathrm{F} = 2\,\mathrm{mF} \quad (12)$$

Recommended capacitor rated voltage should be twice as high as the operational voltage. The main parameters of these capacitors are then 2 mF at 2 kV. However, such a capacitor could not be obtained from standard distributors at the time of writing of this paper which was an important criterion for our comparison. As a solution, four 510 μF capacitors (the largest available) can be connected in parallel, so for 174 submodules, 696 capacitors are needed.

4.2 Design of Capacitors for MV DC Link in Three-Stage SST

For clarity, we designated the capacitor in MV-DC link as C_2. In (13) $t_n = \pi/6$ (time in period power transfer of the DAB is maximal), $f_s = 10$ kHz (switching frequency of the DAB), $\Delta u_c = 0.01 \cdot U_{cn}$ ($U_{cn} = 1$ kV being the nominal capacitor voltage to be maintained), $I_n = 21.7$ A. Capacity of C_2 can be calculated:

$$C_2 = \frac{I_n}{\Delta u_c} \cdot \frac{t_n}{f_s \cdot 2\pi}(1 - \frac{t_n}{2\pi}) = \frac{21.7}{10} \cdot \frac{\pi/6}{10^4 \cdot 2\pi}(1 - \frac{\pi/6}{2\pi})\,\mathrm{F} = 17\,\mu\mathrm{F} \quad (13)$$

The closest available capacity 22 μF rated voltage should be 2 kV.

4.3 Design of Capacitors on the Output of the Galvanic Insulation Modules (ISOP DAB) in Three-Stage SST

This capacitor is expected to behave similarly to the way described in Sect. 4.2, so Eq. (13) can be used for calculation. For the purpose of this work we designate the capacitor on the output of the galvanic insulation module in the case of a three-stage SST solution as C_3. Based on this we can calculate that capacitors with a capacity of 34 μF suitable for voltage of at least 1.4 kV would be needed.

4.4 Design of Capacitors for the Two-Stage SST

Two-stage SSTs have two types of capacitors – in the DC link within the particular modules (for the purpose of this paper designated as C_4) and on the output of the galvanic insulation module (DAB) (designated as C_5) which creates the LV-DC link between the stages of the SST. Both of these capacitors can be calculated as shown in Sect. 4.2 with the use of Eq. (13). Based on this calculation the suitable capacitors are: $C_4 = 10\ \mu F$ suitable for at least 2 kV and $C_5 = 22\ \mu F$ at 1.4 kV

5 Economical Comparison of Selected Topologies

Table 1 shows an overview of prices of key components of both two and three-stage SSTs. According to market research, IGBT H-modules and half bridges for rated voltage 1.2 kV have minimal nominal currents 40 A. Unit prices were stated from prices of lots of 100 (in the case of semiconductors) or 25 pieces (in the case of capacitors) from distributor Mouser. For both types of SSTs, the same components can be used. Ferrite cores are from manufacturing company Kaschke KG as other manufacturers on the market (such as TDK) do not offer such large cores in a series production. Prices of Kaschke cores

Table 1. Economical comparison of key components of the two selected topologies.

Component	Count	Type	Unit price [€]	Total price for required amount [€]	Total price of key components [€]
Two-stage SST					
H modules	**153**	Microsemi APT GLQ40H120T16	39	7344	12 145
Half bridge	–	–	–	–	
Capacitors C_4	**51**	Kemet PP film C44PXGR5220RASK	35	1785	
Capacitors C_5	**51**	Kemet PP film C44PXGR5220RASK	35	1785	
U cores	**114**	Kaschke 126/20	10.8	1231	
Three-stage SST					
H modules	**58**	Microsemi APT GLQ40H120T16	48	3362	115 680
Half bridge	**174**	Infineon BSM 50GB120DN2	58	10 092	
Capacitors C_1	**696**	B25620B1517K983	142	98 832	
Capacitors C_2	**29**	C44PXGR5220RASK	35	1015	
Capacitors C_3	**29**	B25620B1406K981	39	1131	
U cores	**116**	Kaschke 126/20	10.8	1252	

were stated based on an enquiry for a lot of 100 pieces. The number of cores is actually the number of particular U pieces where two pieces form a complete core of an MF transformer in the case of two-stage SST and four pieces (two and two in parallel next to each other) in the case of a three-stage SST. All prices are without VAT.

6 Conclusion

Discussion about the benefits and drawbacks of both topologies could be very long, however due to the constrained extent of this paper; only limited conclusions can be presented. Economical comparison, as well as the comparison of complexity of the selected topologies, confirms that the two-stage SST with LV-DC link is less complex for construction and its crucial components are much cheaper as well. Although the two-stage topology requires a greater number of transformers (51 compared to 29 in the case of a three-stage SST), the overall number of ferrite cores is almost identical (114 and 116 respectively) because the rated power of the transformer is almost double in the case of a three-stage SST and thus needs a larger magnetic circuit.

A three-stage SST contains more types of semiconductors and also larger quantities of them. A similar situation arises with capacitors. In the case of a two-stage SST, the same type of capacitor can be used for both cases. In the case of a three-stage SST, three different types need to be purchased. In the case of a three-stage SST, capacitor type C_1 is the most expensive item by far. This is because a capacitor with capacity 2 mF with rated voltage 2 kV should be used (this large capacity needs to be used as the MMC behaves like a large capacitive divider), which is however not available from standard distributors. To solve this issue, four capacitors with 510 μF in parallel combination need to be deployed, which makes the total number of 696 comparatively expensive capacitors. Based on our enquiry at a Czech company TrafoCZ, a typical distribution transformer with rated power $S_n = 630$ kVA for 20 kV network would cost around 9500 € (without VAT). This is not so far from 12 000 €, which is the approximate price of key components of the two-stage SST, which may come as a surprise. The overall manufacturing cost of an SST would be of course much higher but still it gives us hope that one day we will see the SST in an Energy Router application in place of currently used distribution transformers.

Acknowledgment. This material is based on the work within the Czech–Chinese common project supported by the Ministry of Education, Youth and Sports of the Czech Republic under the grant for INTER-EXCELLENCE, project No. LTACH-17001 and the National Key Research and Development Program of China No. 2016YFE0131700.

References

1. Kolar, J.W., Huber, J.E.: Solid-state transformers - key design challenges, applicability, and future concepts. In: 17th International Conference on Power Electronics and Motion Control (PEMC 2016), Varna (2016, unpublished)

2. Sen, S., Zhang, L., Zhao, X., Lei, Y., Huang, A.Q., Zhu, Q., Song, X.: Medium voltage single-stage dual active bridge based solid state transformer (DABSST). In: 2018 20th European Conference on Power Electronics and Applications (EPE'18 ECCE-Europe), Riga, pp. P.1–P.10 (2018)
3. Bhaskar, R., Agarwal, V.: Dual PID loop controller for HF link inverter in two-stage SST. In: 2016 IEEE 7th Power India International Conference (PIICON), Bikaner, pp. 1–4 (2016)
4. Shojaei, A., Joós, G.: A topology for three-stage solid state transformer. In: 2013 IEEE Power & Energy Society General Meeting, Vancouver, BC, pp. 1–5 (2013)
5. Liu, Y., Wang, W., Liu, Y., Abu-Rub, H., Li, Y.: Control of single-stage AC-AC solid state transformer for power exchange between grids. In: 2016 IEEE 11th Conference on Industrial Electronics and Applications (ICIEA), Hefei, pp. 892–896 (2016)
6. Zhu, Q., Wang, L., Huang, A.Q., Booth, K., Zhang, L.: 7.2-kV single-stage solid-state transformer based on the current-fed series resonant converter and 15-kV SiC Mosfets. IEEE Trans. Power Electron. **34**(2), 1099–1112 (2019)
7. Yao, T., Leonard, I., Ayyanar, R., Steurer, M.: Single-phase three-stage SST modeling using RTDS for controller hardware-in-the-loop application. In: 2015 IEEE Energy Conversion Congress and Exposition (ECCE), Montreal, QC, pp. 2302–2309 (2015)
8. Kolar, J.W., Oritz, G.: Solid-state-transformers: key components of future transaction and smart grid systems. In: Proceedings of the International Power Electronics Conference ECCE Asia (IPEC 2014), Hiroshima (2014)
9. Costa, L.F., Hoffmann, F., Buticchi, G., Liserre, M.: Comparative analysis of MAB DC-DC converters configurations in modular smart transformer. In: 2017 IEEE 8th International Symposium on Power Electronics for Distributed Generation Systems (PEDG), Florianopolis, pp. 1–8 (2017)
10. Kobrle, P., Košťál, T., Zedník, J., Pavelka, J., Yang, X.: Task of energy router in smart grids. In: Proceedings of the 2018 10th International Conference on Electronics Computers and Artificial Intelligence (ECAI), Iasi, pp. 1–6 (2018)

Neural Network Prediction and Decision Making System for Investment Assets

Cesar Valencia N.[1](✉), Alfredo Sanabria[2], Hernando González A.[3], Carlos Arizmendi P.[3], and David Orjuela C.[4]

[1] Mechatronics Engineering Faculty, Universidad Santo Tomás, Bucaramanga, Colombia
cesar.valencia@ustabuca.edu.co
[2] Bussines International Faculty, Universidad Santo Tomás, Bucaramanga, Colombia
alfredoenrique.sanabria@ustabuca.edu.co
[3] Mechatronics Engineering Prog., Universidad Autonoma de Bucaramanga, Bucaramanga, Colombia
{hgonzalez7,carizmendi}@unab.edu.co
[4] School of Medicine and Health Sciences, Universidad del Rosario, Bogota, Colombia
alvaro.orjuela@urosario.edu.co

Abstract. The problem is to test the weak hypothesis of efficient markets through three neural networks that can predict the trends of investment assets such as: The Dow Jones, gold and Euro dollar, according to theories of technical analysis to automate positions of both long and short investment in the Spot market. With regard to forecasting time series, multiple approaches have been tested, through statistical models such as [1–3], where forecasts are made from different information sources with characteristics differentiated (sasonality, tendency, periodicity), however, other actors have begun to gain strength by getting the first places in international competitions, this is the case of Neural Networks, in works published as [4–6] the results have shown that this type of model offers a real opportunity to work with time series of different characteristics.

Keywords: Neural network · Decision making · Invesment · Assests · Prediction

1 Introduction

Technical analysis is the study of market movements, mainly through the use of charts with the purpose of forecasting future price trends [7].

This type of analysis based on the theory of Dow Jones in 1882, states that the trends and prices of the past can be repeated in the future, is a cyclical analysis that is based on the psychology of the investor, that by observing the same pattern in the past infers that it will repeat itself in a cyclical way when awakening the same feelings in the investors, fear, euphoria, security etc.

According to [7] within the technical analysis there are 2 subsets called technicians and Chartists. The first one is more statistical and quantitative analysis.

The justification for this research is the high volatility presented in the financial markets and its importance in predicting trends in the most relevant assets in the world

D. F. Cortes Tobar et al. (Eds.): AETA 2019, LNEE 685, pp. 262–272, 2021.
https://doi.org/10.1007/978-3-030-53021-1_27

market, the Dow Jones, which is the world benchmark for measuring the economy. The strongest pair of currencies traded in the forex market is the euro dollar and commodities refuge in times of crisis is gold, through the theories of technical analysis and the construction of a system that allows us to analyze and make decisions in the Spot market in a period of high continuity and difficult operation as a result of sudden changes in prices.

The most recent theoretical developments on the stock markets were made based on the hypothesis of efficient markets emerged in the 70s in the Chicago school, postulated by Eugene Fama Nobel Prize in economics 2013, where he defined efficiency in markets as: "Where the prices of the assets reflect all the information available in the market" Fama 1970. According to this maxim, the market price of the assets is equal to their theoretical value, they are negotiated at equilibrium prices making it impossible for there to be assets over valued or undervalued, to the market to reflect all the possible information, prices would have random behavior, its prediction is not possible and you can not obtain higher yields than the market because the prices contain all the available information.

1.1 Dow Jones

Dow theory was born in the late nineteenth century, and its bases are presented in the 5 postulates of Dow in which it is emphasized that the market has a cyclical component and was the origin of Chartism, exposed by [8] (Fig. 1):

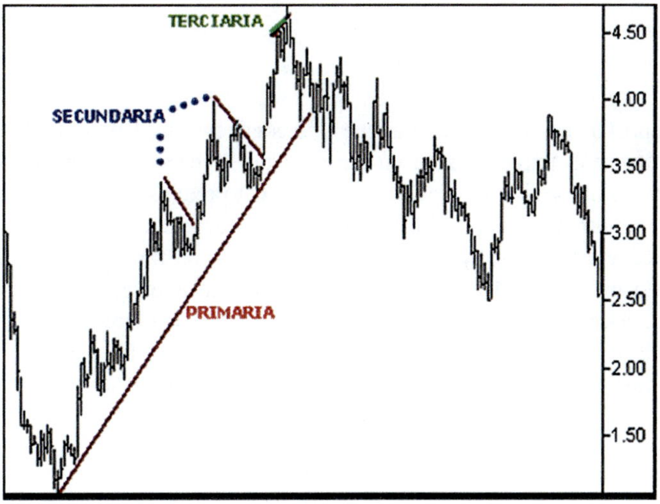

Fig. 1. Dow Jones theory

- The indices discount everything.
- The market has three trends.
- The main trends have three phases.

- The stockings must be confirmed between them.

Raphl Elliot after the financial crisis of 1929, begins an exhaustive study on the behavior of prices and begins to observe certain patterns of trends that he calls waves, identifiable, concludes that these can be repeated in market cycles, in five waves bullish and after these there would be a bearish trend (Fig. 2).

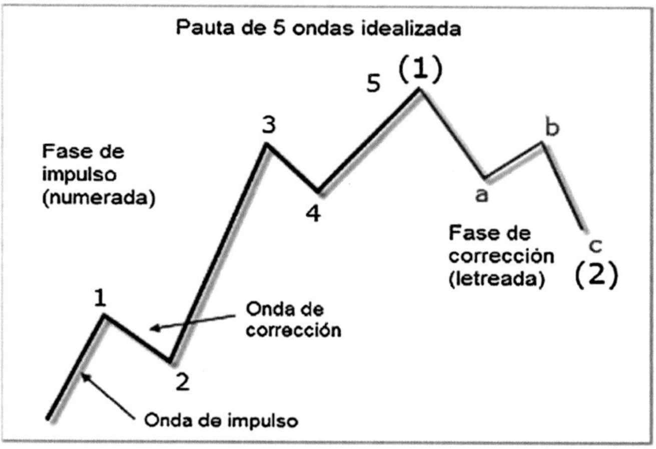

Fig. 2. Elliot waves

1.2 Dow Jones

There are approximate studies such as the one conducted by [9] entitled: "Technical analysis of financial markets based on artificial intelligence techniques." Where the author proposes, the design of a neural network that predicts the behavior of a robot according to the current situation of the market and its neuronal learning will be according to the historical events faced by the robot, according to [14] through technical analysis of the RSI, Stochastic, Macd, and Williams indicators, create investment portfolios with signals of purchase and sale of the Euro, IGBC, S & P500 and Dow Jones to maximize profitability by calculating a profitability index to this portfolio named by dynamic authors. Making the bench mark, with the creation of a portfolio of these securities

Table 1. Cost effectiveness.

	RENTABILIDAD A DICLEMBRE DE 20009											
	IGBC			Euro			S&P500			Dow Jones		
	Dinámica	Estática	Var%	Dinámica	Estática	Var%	Dinámica	Estática	Var%	Dinámica	Estática	Var%
RSI	152,73	108,67	41%	133,11	119,32	12%	121,34	90,25	34%	129,72	96,13	35%
MACD	142,80	108,67	31%	123,95	119,32	4%	102,10	90,25	13%	109,19	96,13	14%
Estocástico	120,15	108,67	11%	116,63	119,32	−2%	120,73	90,25	34%	119,40	96,13	24%
Williams	107,56	108,67	−1%	113,35	119,32	−5%	84,86	90,25	−6%	121,10	96,13	26%

without using technical analysis, calculating a profitability index called by static authors. Where they concluded that the dynamic index was several times higher than the static one according to each indicator, maximizing the profitability of this portfolio (Table 1).

In [9] Design a neural network to forecast the Japanese yen, including relevant economic fundamental indicators, concluding that the inclusion of the fundamental indicators decreased the prediction errors for the next 3 weeks.

In [11] he analyzes the obtaining in the general index of the Madrid Stock Exchange of profitability by means of: Technical analysis, neural networks, genetic algorithms, genetic programming and analogous occurrences.

In [12] they constructed a neural network Back propagation by contrasting it with an Arima and Garch econometric model using the Colombian stock index, concluding the high predictability of the artificial intelligence system vs the econometric model.

In [13] the authors propose by means of the technical analysis of the indicators: Macd, Rsi and moving averages Japanese candlesticks and Bollinger Bands, in the Euro/Dollar, using the Metatrader Negotiation platform.

This oscillator called in English Relative Strength Index created by mechanical engineer J. Welles Wilder Jr., published in his research New concepts in technical trading systems and measures the forces of supply and demand to determine if prices are over-bought (expensive) or over-sold (cheap). It is calculated used Eq. 1.

$$IFR = 100 - \frac{100}{1 + FR} \tag{1}$$

The most used period is 14 sessions that may be days or weeks at a shorter period, the more sensitive the indicator and the wider the amplitude and the opposite happens if the sessions are increased the indicator is softened and the amplitude is smaller. This indicator has values between 0 and 100 and an average line of 50. According to Welles Wilder, if the indicator exceeds the limit of 70, it is considered that prices are over bought and if it falls below 30, they are over-sold, the average line of 50 represents indecision for operators.

The work of [13] mentions that a value of about 80 purchase and one over sale of 20 will give more security to the analysis in 14 sessions. MACD (Moving Average Convergence Divergence): this indicator belongs to the family of oscillators created in 1979 by the administrator of investment funds Gerard Apple, is composed: by two lines one called Macd. which is the difference between two moving averages of 12 and 26 sessions and a signal line with an exponential average of 9 sessions. OVB (On-Balance Volumen): The volume indicator reflects the transactions made in the market, and precedes the price according to Charles Dow volume must confirm the trend, otherwise there is a high probability of presenting a market divergence.

1.3 Neural Networks Forecasting

A neural network, as humans do base on acquired knowledge generate solutions to different problems, the network takes as a reference the resolved problems in order to make decisions and perform data classification. They base their operation on the extraction of data from tables, experimental data or databases that are then used as inputs (voltage, current, system variables) of the network and output the established signals (boolean variables, speed, temperature, etc.)

Multilayer networks are based on sets of single-layer sensory units and were created to solve various problems with higher difficulty with supervised training, are constituted by the entrance of the layer, one or more hidden layers and the output layer.

Figure 3 shows the organization of a neural network that has 4 input signals, two hidden layers and three output signals. The input layer lets pass the information required for further processing in the hidden layers, finally, the output layer that allows the classification of the information, the final result will be delivered by the output signals.

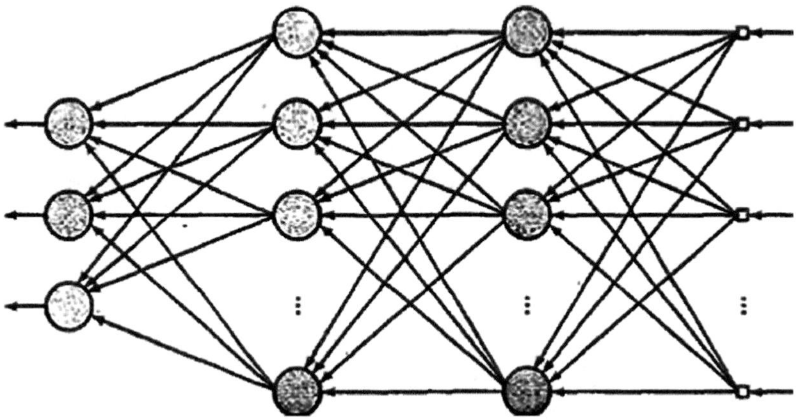

Fig. 3. MLP network

In order to forecast a time series using neural networks, it is necessary to take into account the temporal relationship that exists between the data for periods of time, this means that constant time slots must be maintained in the values presented as inputs, in order to achieve This effect uses a sample window that can vary in size and that would involve the historical ones to be considered to make the forecast.

For forecasting processes of time series, two procedures are used, the first one that makes the forecast in a single iteration for the necessary values, the second procedure is multi-step where a cyclical forecast is made where the predicted values can enter or not in the forecast sliding window like a Fig. 4.

2 Forescasting Results

The database available to perform the forecast tests consists of 1325 records that comprise the time period from September 3, 2012 to September 29, 2017, to have different perceptions of the forecast results were generated 5 sub bases of data, which are organized according to the size of the test set, for this case are 5, 10, 15, 20 and 45 records.

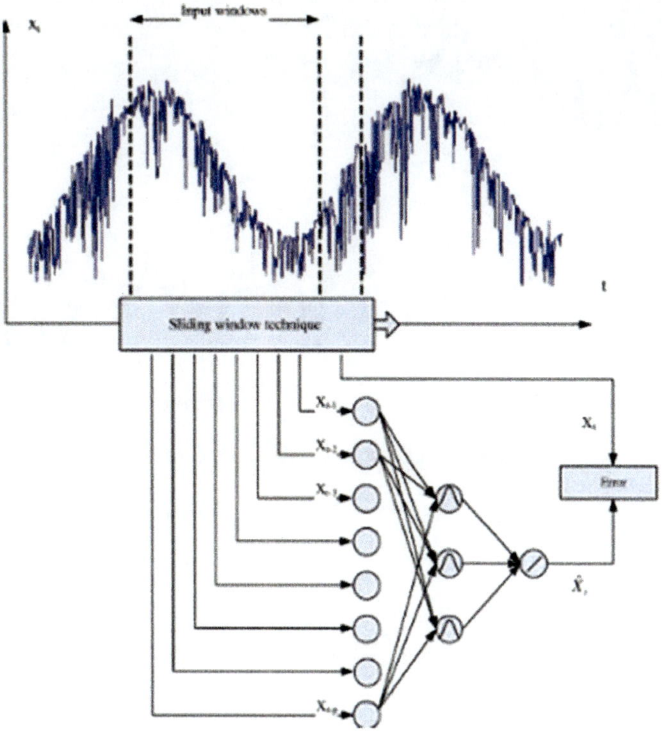

Fig. 4. Sliding window technique MLP network

2.1 Euro/Dolar Database

The Eurodollar indicator refers to the ratio of dollars deposited in the foreign currency, that is, those that are outside the banking system of the United States, this means that it does not contain any close relationship with the euro but with European banks where they are deposited.

In Figs. 5, 6 and 7, it is possible to notice the behavior of the network response for the euro/dollar database with a test window of 45 records, in Table 2 the error metrics for the data are presented. sub databases used in this series (Table 4).

The results presented are from the test set, that is, outside the sample.

2.2 Gold Database

Being one of the most precious metals since many centuries, the negotiations based on this are characterized by being very desirable since their value remains stable despite moment of movement in the financial market, has been known as a "quotable asset" and denominated in some parts like "refuge".

In Figs. 8, 9 and 10, it is possible to notice the behavior of the network response for the gold database with a test window of 10 records, in Table 3 the error metrics for the data are presented. sub databases used in this series.

The results presented are from the test set, that is, outside the sample.

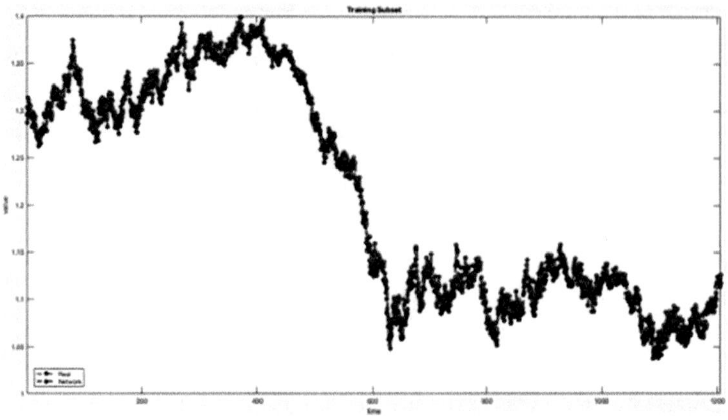

Fig. 5. Training subset result (45 records)

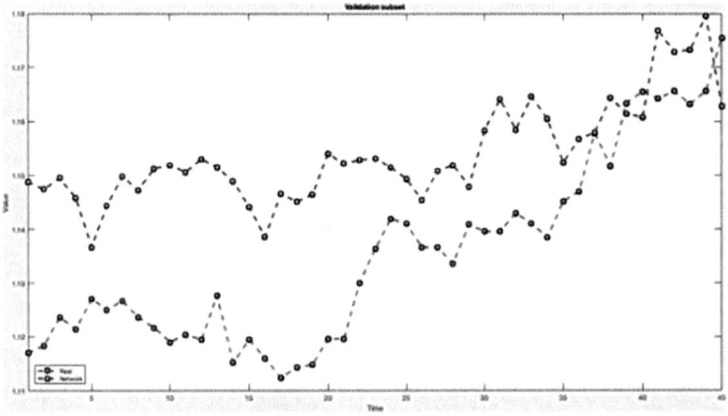

Fig. 6. Validation subset result (45 records)

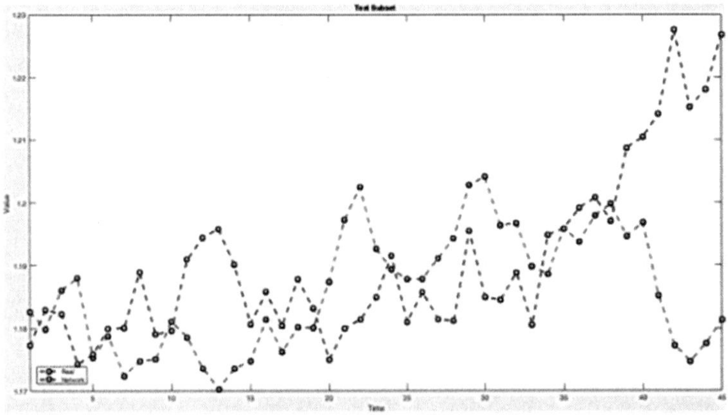

Fig. 7. Test subset result (45 records)

Table 2. Euro/Dollar results.

	5	10	15	20	45
MAPE	1.3209	1.7613	**0.6963**	0.8113	0.9633
RMSE	0.0176	0.0245	**0.0100**	0.0121	0.0142

Table 3. Gold results.

	5	10	15	20	45
MAPE	0.8985	0.8615	1.3969	**0.6111**	0.7903
RMSE	0.0125	0.0128	0.0198	**0.0092**	0.0117

Table 4. Dow Jones results.

	5	10	15	20	45
MAPE	**1.3816**	3.5482	1.4679	1.8002	3.5104
RMSE	**0.0193**	0.0475	0.0196	0.0230	0.0470

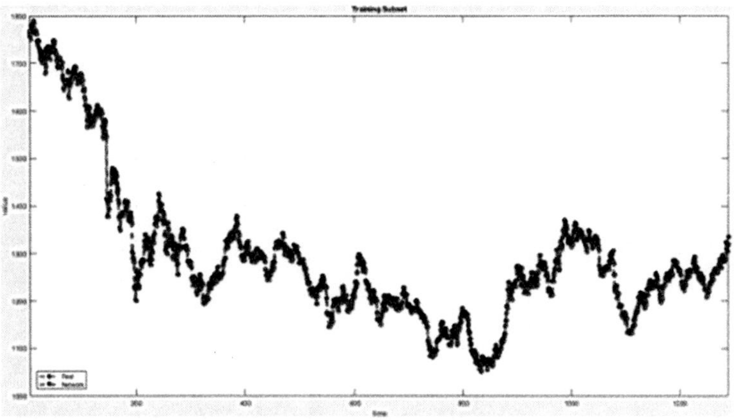

Fig. 8. Training subset result (10 records)

2.3 Dow Jones Database

Denominated as a stock market index, it is composed of 30 of the most important shares of all the listed companies in the Nasdaq and the New York Stock Exchange, however, transportation and public services actions are not included, without being the stock exchange index. of New York, is one of the most used to monitor the behavior of this, also stands out because it is one of the oldest indices used to determine the behavior of important actions.

It is known as a balanced index, which represents the movement of the most important stocks in the New York Stock Exchange.

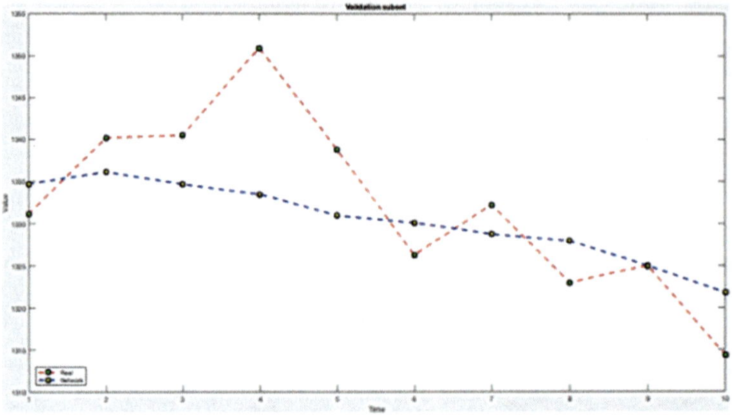

Fig. 9. Validation subset result (10 records)

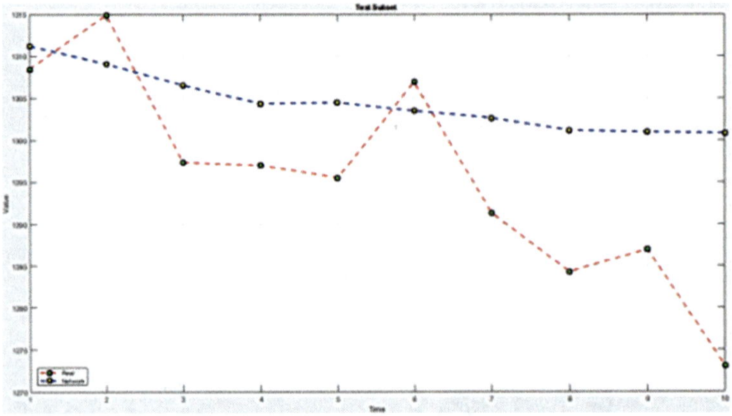

Fig. 10. Test subset result (10 records)

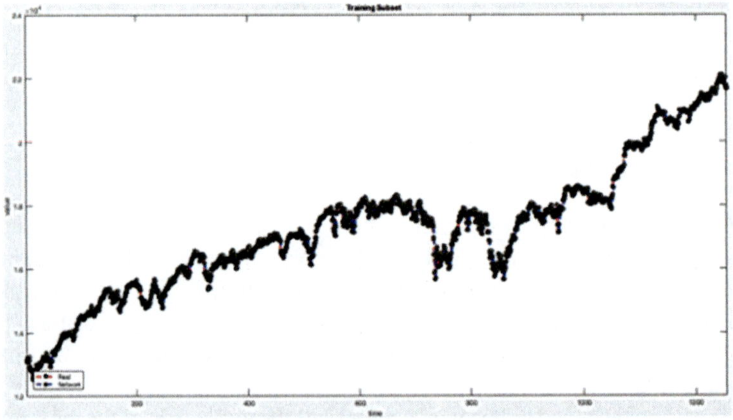

Fig. 11. Training subset result (15 records)

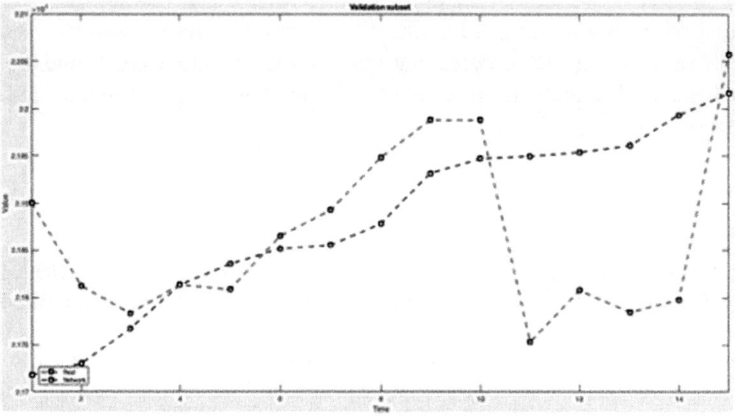

Fig. 12. Validation subset result (15 records)

In Figs. 11, 12 and 13, it is possible to notice the behavior of the network response for the DowJones database with a test window of 15 records, in Table 3 the error metrics for the data are presented. sub databases used in this series.

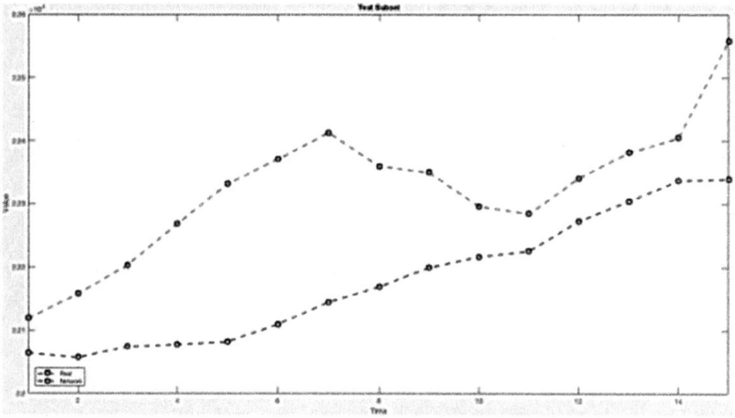

Fig. 13. Test subset result (15 records)

3 Conclusions

For the tests carried out on the three intelligent systems, multiple configurations of their most relevant parameters were tested, presenting several databases and still yielding the same results presented in Sect. 2; this is mainly due to the non-stationary properties of the myoelectric signals making the classification more complex despite being two gestures only, and forcing to use robust intelligent systems.

The results presented show the efficiency that RNA handles in the portable embedded system, which lends itself to multiple applications in various areas. The success of the device was to train the neural network in a separate computer, since the computational expense that requires a training is high, and to program in the embedded system only the structure of the neural network leaving space to the synaptic weights and the biases, since which are the data that is acquired from training, thus making the embedded system suitable for processing the RNA.

References

1. Singh, S., Taylor, J.H.: Statistical analysis of environment Canada's wind speed data. In: 2012 25th IEEE Canadian Conference on Electrical & Computer Engineering (CCECE), Montreal, QC, pp. 1–5 (2012)
2. Asami, A., Yamada, T., Saika, Y.: Probabilistic inference of environmental factors via time series analysis using mean-field theory of ising model. In: 2013 13th International Conference on Control, Automation and Systems (ICCAS), Gwangju, pp. 1209–1212 (2013)
3. Shenoy, D.S., Gorinevsky, D.: Stochastic optimization of power market forecast using non-parametric regression models. In: 2015 IEEE Power & Energy Society General Meeting, Denver, CO, pp. 1–5 (2015)
4. Valencia, C.H., Quijano, S.N.: Modelo de optimización en la gestión de inventarios mediante algoritmos genéticos. Iteckne **8**(2), 156–162 (2011). ISSN 2339-3483
5. Cherif, A., Cardot, H., Boné, R.: SOM time series clustering and prediction with recurrent neural networks. Neurocomputing **74**(11), 1936–1944 (2011)
6. Valencia, C.H., Vellasco, M.M.B.R., Figueiredo, K.T.: Trajectory tracking control using echo state networks for the CoroBot's Arm. In: Kim, J.H., Matson, E., Myung, H., Xu, P., Karray, F. (eds.) Robot Intelligence Technology and Applications 2. Advances in Intelligent Systems and Computing, vol. 274. Springer, Cham (2014)
7. Murphy, K.: Executive Compensation. Marshal School of Business (1999)
8. Murphy, J.: Análisis técnico de los mercados financieros. Gestión, Barcelona (2003)
9. Pina, A.: Análisis técnico de mercados financieros basado en técnicas de inteligencia artificial (2014)
10. Villa, F., Muñoz, F., Henao, W.: Pronóstico de las tasas de cambio. Una aplicación al Yen Japonés mediante redes neuronales artificiales. Scientia et technica (2006)
11. M González Martel, C.: Nuevas perspectivas del análisis técnico de los mercados bursátiles mediante el aprendizaje automático: aplicaciones al índice general de la bolsa de Madrid (2003)
12. Cruz, E.A., Restrepo, J.H., Varela, P.M.: Pronóstico del índice general de la Bolsa de Valores de Colombia usando redes neuronales. Scientia et technica (2009)
13. Castellanos Vargas, O.E., Jaramillo Jaramillo, J.M.: Cuantificación de riesgo en estrategias de análisis técnico del mercado de divisas usando redes neuronales (2007)
14. Roncancio Millán, C.A., Valenzuela Reinoso, A.F.: Desarrollo de un modelo de Trading algorítmico para índices bursátiles y divisas. Retrieved (2010)

Movement Control System for a Transradial Prosthesis Using Myoelectric Signals

John Bermeo-Calderon[1], Marco A. Velasco[1], José L. Rojas[1], Jesus Villarreal-Lopez[1](✉), and Eduard Galvis Resrepo[2]

[1] Universidad Santo Tomas, Cra 9 # 51-11, Bogotá, Colombia
jesusvillareal@usantotomas.edu.co
[2] Universidad EAN, Cl 79 # 11-45, Bogotá, Colombia

Abstract. The design of a robotic prosthetic hand, based on the control of myoelectric signals is a multidisciplinary task involving areas such as medicine, mechanics and electronics. The implementation of a robotic prosthetic hand could help restore the daily activities of people with amputation, improving their self-esteem and quality of life. This article is mainly focus on the development of a transradial prosthetic hand motion control system. The system uses myoelectric signals processing, for a series of preset gestures that are detected in real-time. Experimental tests of pattern recognition show that the signals have variable amplitudes and frequencies, for instance, the signal strength and the phenomenon of muscle fatigue vary depending on the subject. Hand movement recognition is achieved using EMG feature extraction technique such as Average Amplitude Change, Root Mean Square, Mean Absolute Value, among others. The actual results show that the implemented control system can be effective in performing six gestures or movements on a prototype robotic hand prosthesis.

Keywords: Electromyography · Myoelectric signal (EMG) · Handheld prostheses · Finite state machines

1 Introduction

Although, it is difficult to quantify accurately, people in the world who requires prostheses to replace a missing limb are estimated at millions. In 2015, at least 3 million people in the world were estimated to have had amputations [1]. In the United States, in 2012, the number of people with loss of a higher limb was close to 100,000, for instance [2]. Particularly, in Colombia, according to the National Health System database, the number of people needing senior limb prostheses was around 11,000 in 2018 [3].

Both vascular diseases and diabetes are two causes of disability from partial or total limb amputation are. There are an estimated 170 million people in the world suffering from diabetes and this number is expected to double by 2030. Moreover, work-related injuries and specially the internal armed conflicts result in a significant number of cases of amputees in the Colombian context [4].

The prostheses are the most used devices to replace the limbs of amputees. However, in the Colombian context, these devices are no technological enough to control the motion

D. F. Cortes Tobar et al. (Eds.): AETA 2019, LNEE 685, pp. 273–282, 2021.
https://doi.org/10.1007/978-3-030-53021-1_28

in an automatic way, due mainly to the application of control systems in prostheses are usually done at the research level. The implementation in actual prosthetic systems are scarce due to the high cost of the automatic control system which involve high accuracy data acquisition and motorized actuators [5].

Electromyography is the study of electrical signals that come from muscle activity, such activity is measured through muscle fibers, for instance, the potential for action that occurs once there is synaptic transmission of the binding neuromuscular [7]. The electromyogram (EMG) can be recorded in three levels: at the level of a muscle fiber, at the level of a single motor unit and at the level of the whole muscle. The entire muscle signal register is called surface EMG and can be measured from the surface of the muscle, or also from the surface of the skin. A surface EMG records the activity of multiple motor units.

Surface EMGs are mainly used in behavioral studies, physical therapy and sports medicine evaluations, as well as in the detection of myoelectric signals to allow the control of prostheses. EMG signals can be used to control a prosthesis using different schemes: On/Off control, Proportional, Direct, Finite State Machine Control, Pattern Recognition, Control Posture and Regression Control [8]. Within these schemes, the finite state machine control has functions that make it convenient to implement low-cost prosthetic myoelectric control: acquisition of process data to be controlled through digital inputs and similar decision-making, decision-making based on scheduled rules, intervention on external devices through digital and analog outputs, communication with external systems and low cost of implementation and modification of actions [9].

This work presents the implementation of a myoelectric control system in a top limb prosthetic system, which use of the 8 channel-sensor Myo™ Gesture Control Armband device. The system uses the MATLAB® environment to monitoring and digital processing of electric signals. The whole system aims to improve the quality of life of patients with upper limb amputation and considering hands have very important functions for the development of communicative, recreational and work activities. In the proposed system, the movement control is done by an electromechanical system, which has the advantage of the absence of complex interfaces, allowing a simple signal processing which involve low computational burden [6]. The paper is organized as follows, Sect. 2 describes the technical details related to EMG signal acquisition the proposed detection method; Sect. 3 presents the results; and Sects. 4 and 5 discuss the results and presents the conclusions.

2 Materials and Methods

2.1 Myoelectric Signal Acquisition System

A myoelectric signal measurement (EMG) device, called Myo™ Gesture Control Armband, was used for the signal acquisition system, which was developed by the ThalmicLabs™, the technical specifications are shown at the beginning and were obtained from [7, 10].

Myo™ Gesture Control Armband

Abrief description of the Myo™ Gesture Control Armband device will then be made, which was used for the collection of the electromyography (EMG) signal. In addition to the description of the procedure carried out by the device at the time of detecting any signal, the elements that make up the device, the main uses, such information was collected from [7, 10].

The Myo™ Armband Gesture Control is a portable device capable of reading the muscle activity of the forearm, can grant contactless control with different devices gestures and hand movements. Is an electromyography (EMG) device that includes a set of sensors and a processor connected to the EMG sensor set.

Data Acquisition

In this subsection the data was obtained, using the Device Myo™ Gesture Control Armband and the MATLAB software®, the Myo™ was placed on the forearm of the test subject, for this case the Author of this document, as shown in Fig. 1, in order to obtain the signals of the muscles involved in the movement of the hand among which are extender muscles and flexors located in the area of the forearm.

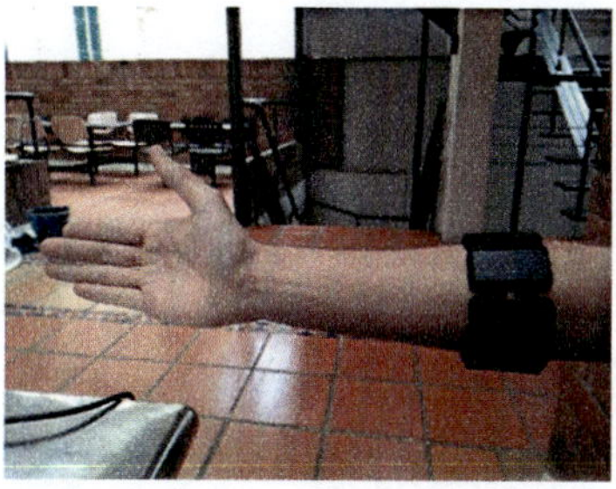

Fig. 1. Location of the Myo [Authors]

The Myo™ band device has 8 signal channels corresponding to the 8 EMG sensors, which take the signals simultaneously, for the monitoring and subsequent treatment of the signals the device was connected to MATLAB®, which generated a record of the signals for each of the channels. From an open source interface available on the internet [11].

After having successfully configured the interface the GUI (user interface) was carried out, which comes by default in the MyoMex package; that interface facilitates the visualization of the activity of each of the variable is measured by the IMU sensor, but shows the signals recorded by the 8 EMG channels in a single graph, in Fig. 2 you can see the environment of the user interface that comes by default in MyoMex.

In Fig. 2, the user interface shows the different data that the Myo can transmit, including orientation, gyroscope, accelerometer, a display of the device's position, signals EMG with its 8 channels and an image of the gesture that is being presented now. Because the development of the experiment requires the development of a GUI (graphical user interface) that can move the activity of the EMG signal channels individually and is view simultaneously. For the development of the necessary GUI, some of the code used for the development of the GUI shown in Fig. 2 was considered, by titling the way

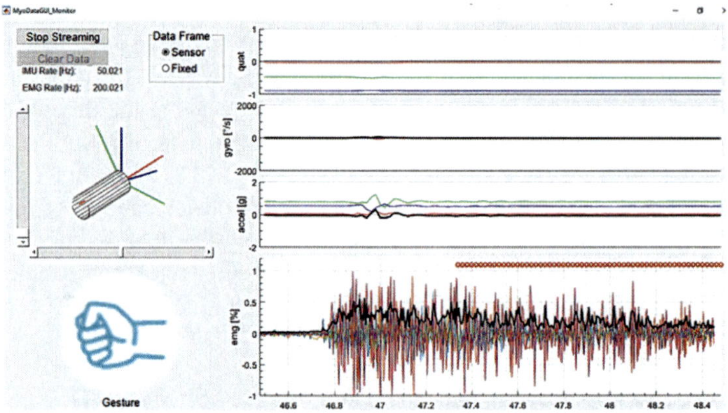

Fig. 2. User interface [Authors]

it calls the variables and, in the structures, it uses for setting the sampling time. A user interface display adapted for the experimentation phase can be seen in Fig. 3.

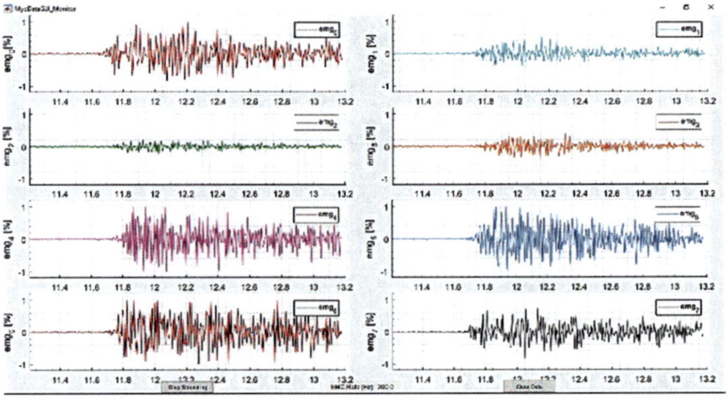

Fig. 3. User interface developed for the experimental phase [Authors]

In Fig. 3 you can see the user interface that was developed to have a greater perception of the activity of the sensors, while performing a certain movement.

Implementation of the Prosthetic Control System

The prototype prosthesis model was taken from a model developed for the InMoov initiative developed to create an anthropomorphic android [12], said model was downloaded and arranged to be printed. The printing was developed on a Zortrax-branded 3D printer ®, making use of the material developed by the company of the same name known as Z-Ultrat®; within the most popular printing parameters important has a nozzle size of 0.4 mm and a layer size of 0.14 mm. In addition, to establish density with a honeycomb arrangement and a density of 70%, because for the parts high resistance was required when it was used. For the movement of the fingers is done by using tension ropes, which for that matter was used fishing nylon of 100 lb. For actuators, the use of 5 MG995 servos was required to tighten the cables to extend or flex the fingers. Among the specifications most important are found are metal gear servos, because this allows the output torque to be approximately 12 kg at its maximum load and about 9 kg on average. For development cards or microcontroller an Arduino MEGA board was used® to make use of 2 of its serial ports (one for communication with Matlab® and one for communication with the servo control card). The servo control card used was a Lyxmotion SSC-32U connect to 32 servos continuously with a resolution of one microsecond. In addition, provide two-way connection in case you require feedback of the position of the servos.

In schematic shown in Fig. 4 shows in a simplified way the way in which the operation of the prototype was performed. As part of the generation of gestures in the forearm is shown, the recognition and comparison of the gestures based on the description shown in Sect. 2.1, when recognizing the gesture this is recorded in Matlab® by the Execution of the GUI created for data collection which were ad-friendly to track the gestures captured by the Myo™ Gesture Control Armband collection data is by using Bluetooth™. Subsequently, during the execution of the GUI, a connection is made between Matlab® and Arduino®, which is made by serial port at a rate of 9600 bits/s, while Matlab® performs the detection of the gesture desired this sends a numeric value per serial port to the microcontroller so that the microcontroller would subsequently send back via serial port the motion commands by using the SSC-32U servo control card with which is operated. Finally, the control card sends via PWM the movement signals to the servos performing these motor movements based on the commands set for each gesture. An image of the elements used for the operation of the prosthesis can be seen in Fig. 5.

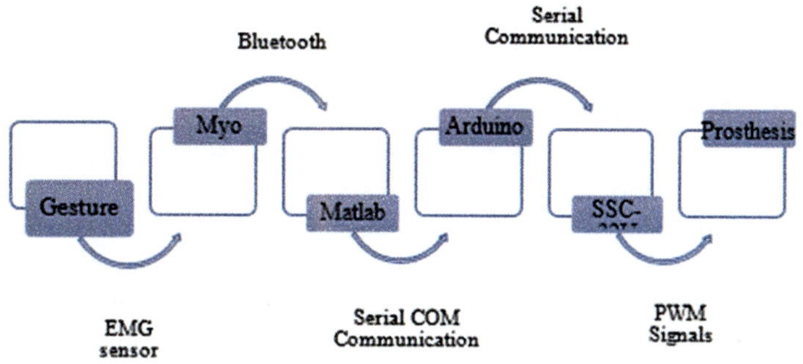

Fig. 4. Schema of the information flow for the operation of the prosthesis prototype [Authors]

Experimental Development

To obtain the samples of the movements of each of the fingers and thus establish the characteristics that can be used for the generation of the parameters used for the subsequent prosthesis control was developed by: Establish the Connection of the Myo to MATLAB, the test subject should start with a relaxed muscle posture, so it is necessary that the user has the arm extended downwards. Execute the user interface shown in Fig. 3, the test subject develops some of the movements shown in Fig. 6, for approximately 5 s repeatedly. After 5 s of movement the test subject must return to the resting state to

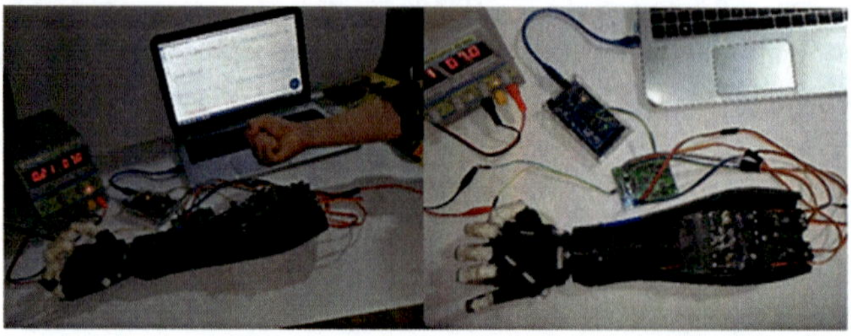

Fig. 5. Elements used for the drive and prototype of prosthetics developed [Authors]

stabilize the reading of the EMG signals, the readings obtained from steps 4 and 5 are treated for direct form while the user interface is running to perform signal processing simultaneously and are recorded in an Excel sheet.

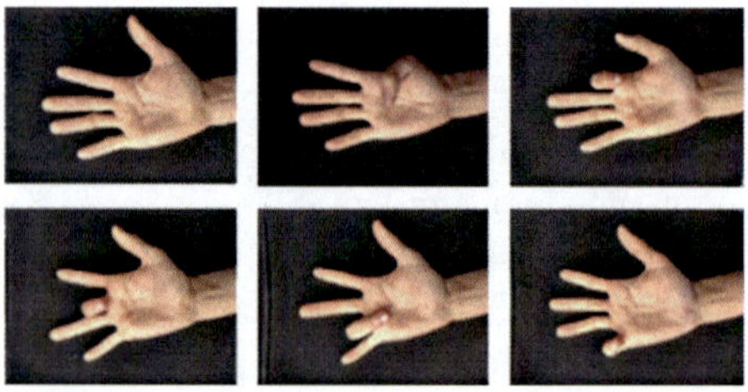

Fig. 6. From left to right: resting position, thumb movement, index movement, middle finger movement, ring movement and little finger movement [Authors].

3 Results and Analysis

After the data collection of the EMGs proceeds with the processing of the signal, for the processing of the signals are made use of characteristics for the processing of these, these characteristics were selected from the bibliography revision EMG signal processing the characteristics listed below were taken from [13]. Among the main ones are ACC (Average Amplitude Change), RMS (Root Mean Square), iEMG (Integrated EMG), MAV (Mean Absolute Value), SSI (Simple Square Integral), VAR (Variance of EMG) [13–17]. After the selection of the characteristics, these characteristics are extracted from the signals obtained from the experimental phase, this extraction was carried out using the MATLAB® software; the registration of the features for the middle finger will be displayed below.

Figure 7 shows one of the characteristics obtained to obtain a description of the activity of the sensors during the generation of the movement of the middle finger. For the specific case of use of the MAV feature and the middle finger the largest activities are presented channels 1-5-8, while the pinky presents activity in the channels 1-5-7-8; this generation of activity contributes to that during reading you can monitor these channels to generate the movement of the motor assigned to the movement of the proposed prosthesis. This strategy is repeated with the other characteristics, as well as the fingers involved in the movements. In addition, as future work, Machine Learning algorithms will be used for some machine learning algorithm to generate descriptors to assign movement to the respective engines. Based on it limiting the use of Machine Learning algorithms it was decided to use the gestures generated by default to carry out the operation of the prosthesis. This drive was carried out through the interconnection of the physical model, making use of the GUI developed for data collection. In addition, the use of the Arduino MEGA microcontroller. The information flow is displayed later.

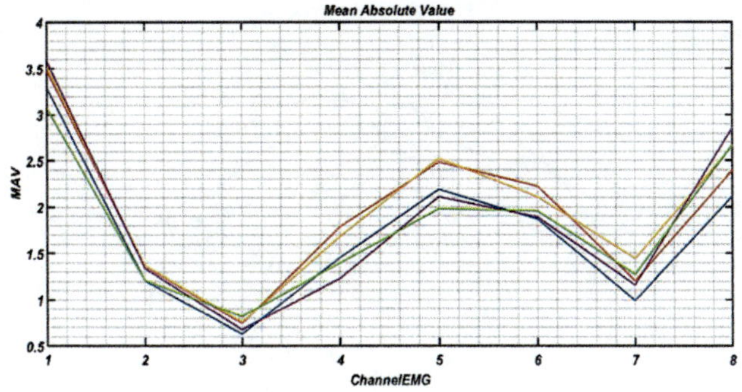

(a)

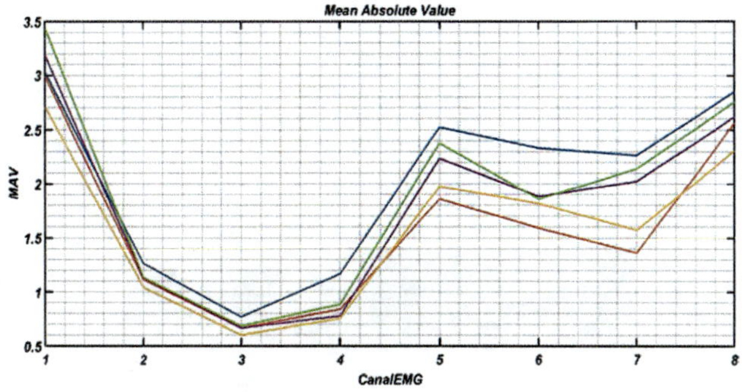

(b)

Fig. 7. MAV (mean absolute value) for the middle (a) and pinky finger (b) [Authors]

Due to the need for the use of the Myo™ Gesture Control Armband device for the operation of the prosthesis it was necessary to use the different gestures identified by default of the device. Among those are: (Thumb to Pinky, Fingers Spread, Wave In, Wave Out and Fist). The recorded information is sent to the Arduino MEGA to define the drive sequences of the engines. Because the default gestures of the Myo™ Gesture Control Armband were used the control implemented was finite state machine control, because the gestures are predefined as states and transition between states is also predefined from inputs [8]. With the help of this control strategy, the sequences that can be executed were determined by detecting the input gestures provided by the Myo™, the finite state machine diagram is presented below.

Figure 8 refers to the finite state machine control diagram used for the drive of the prosthesis based on the acceptance state presented during motor drive. Among the states present are: fist position (e_0), position to open hand (e_1), aiming position (e_2), clamp position (e_3) and tripod position (e_4); being those states influenced by the variables of transition: fist (g_0), FingersSpread (g_1), WaveIn (g_2), Wave Out (g_3) and Thumb To Pinky (g_4), which allow the prosthesis to perform some of the most commonly used movements in other related works [18], [19]. The different movements made by the prototype prosthesis are observed in the Fig. 9.

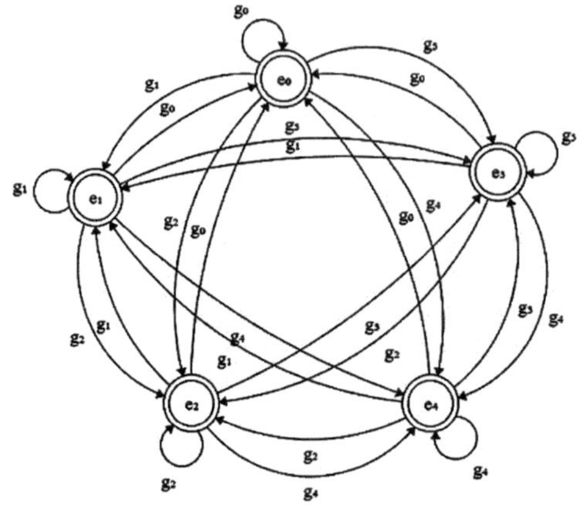

Fig. 8. Diagram for the machine control of finite states used in the prototype [Authors]

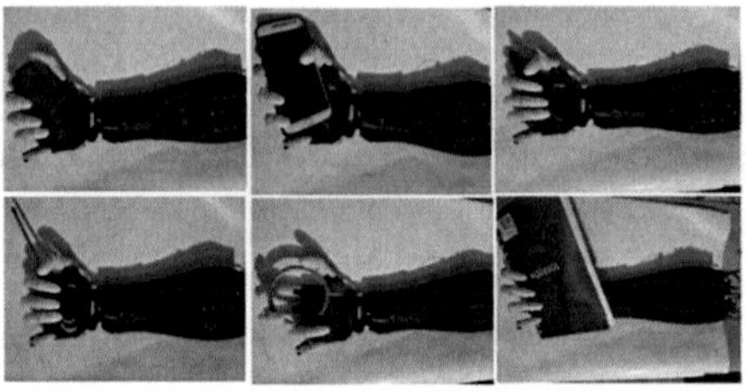

Fig. 9. Different types of grips available in the prototype [Authors]

4 Conclusions

For the generation of strategies for measuring myoelectric signals, it is necessary to have a basic knowledge about the muscle activity that you want to study. EMG signals have unstable magnitudes, because they vary from person to person, showing different reactions in each of them. The signals depend on the effort applied to the muscle and each muscle has different types of fatigue, causing the signals to weaken. Considering the information obtained from the characteristics of the Obtained EMG signals, a parameter can be temporarily estimated to differentiate the signals that are generated from the individual movements of each one of your fingers. In this way it has tools for the characterization of the signals and their subsequent applicability in the control of hand prostheses. Due to the complexity of estimating classifiers, it is concluded that the use of Machine Learning algorithms is necessary to develop the learning technique in the control of the prosthesis. Its data processing capacity and robustness at the time classifiers are generated, allows to have a large amount of input data. For the use of finite-state machine-based control strategies, it is concluded that such a strategy is useful when a fixed number of properly identified positions is possessed; but is useless when it is necessary to establish multiple functions from set positions, because the state change occurs from the detection of the EMG signal until the position is identified and related as a valid variable. It is also possible to solve this difficulty by implementing approaches related to pattern Machine Learning.

References

1. Slade, P., Akhtar, A., Nguyen, M., Bretl, T.: Tact: design and performance of an open-source, affordable, myoelectric prosthetic hand. In: Proceedings - IEEE International Conference on Robotics and Automation, vol. 2015-June, no. June, pp. 6451–6456 (2015)
2. Kyberd, P.J., Losier, Y.G., Fougner, A., Stavdahl, O., Parker, P.A.: Control of upper limb prostheses: terminology and proportional myoelectric control - a review. IEEE Trans. Neural Syst. Rehabil. Eng. **20**(5), 663–677 (2012)
3. Ruiz, Y.: Fabrilab: Print Quality of Life in 3D. The Spectator, p. 1 (2018)
4. Ocampo, M.L., Henao, L.M., Lorena, V.: Member amputation lower: functional changes, immobilization and physical activity. Univ. of the Rosary. Rehabilitation Human Rights, vol. 42, pp. 1–26 (2010)
5. Prosthetics: Hand Prosthetics Myoelectrics I Prosthetic (2015). http://protesica.com.co/myoelectricas/. Accessed 24 Apr 2017
6. L6pez, N.M., From Diego, N., Hernández, R., Pérez, E., Ensinck, G., Valentinuzzi, M.E.: Customized device for pediatric upper limb rehabilitation in obstetric brachial palsy. Am. J. Phys. Med. Rehabil. **93**(3), 263–266 (2014)
7. Morán Medina, Y.S.: Classifier based on neural networks of the state of movement of the forearm using surface EMG signals. Monterrey Institute of Technology and Higher Studies (2016)
8. Geethanjali, P.: Myoelectric control of prosthetic hands: state-of-the-art review. Med. Devices Evid. Res. **9**, 247–255 (2016)
9. Alavi, M., Aliaga, S., Murga, M.: Finite state machines. Enlighten Yourself **8**(1), 41–57 (2016)
10. ThalmicLabs: MYO 2013 (2018). https://www.myo.com/techspecs
11. Tomaszewski, M.: Home - Mark Tomaszewski. http://www.mark-toma.com/view/. Accessed 03 Jul 2019
12. Langevin, G.: "InMoov" open-source 3D printed life-size robot (2012). http://inmoov.fr/. Accessed 22 Mar 2017
13. Phinyomark, A., Phukpattaranont, P., Limsakul, C.: Feature reduction and selection for EMG signal classification. Expert Syst. Appl. **39**(8), 7420–7431 (2012)
14. Tkach, D., Huang, H., Kuiken, T.A.: Study of stability of time-domain features for electromyographic pattern recognition. J. Neuroeng. Rehabil. **7**, 21 (2010)
15. Phinyomark, A., Quaine, F., Charbonnier, S., Serviere, C., Tarpin-Bernard, F., Laurillau, Y.: EMG feature evaluation for improving myoelectric pattern recognition robustness. Expert Syst. Appl. **40**(12), 4832–4840 (2013)

16. Phinyomark, A., Phukpattaranont, P., Limsakul, C.: Fractal analysis features for weak and single-channel upper-limb EMG signals. Expert Syst. Appl. **39**(12), 11156–11163 (2012)
17. Asghari Oskoei, M., Hu, H.: Myoelectric control systems—a survey. Biomed. Signal Process. Control **2**(4), 275–294 (2007)
18. Robert, M., Henderson, A.: Book review 'counter intelligence transforming KMT intelligence on Taiwan'. Int. J. Intell. CounterIntelligence **15**(2), 275–317 (2002)
19. Kyberd, P.J., et al.: MARCUS: a two degree of freedom hand prosthesis with hierarchical grip control. IEEE Trans. Rehabil. Eng. **3**(1), 70–76 (1995)

System for Analysis of Human Gait Using Inertial Sensors

Diego Fernando Saavedra Lozano(✉) and Javier Ferney Castillo Garcia(✉)

Grupo de Invertigación en Electrónica Industrial y Ambiental-GIEIAM, Universidad Santiago de Cali, Cali, Colombia
{diego.saavedra00,javier.castillo00}@usc.edu.co

Abstract. Gait laboratories using image processing are the standardized method to analyze human gait, but these are expensive, require restrictive workspaces, and post-processing times. On the other hand, inertial sensors are cheaper and reach high performance with lower computational cost in its implementation. In this work, a wireless system was implemented for the analysis of human gait using inertial sensors. The inertial sensor was chosen according to its specifications, cost, and pursuant to the characteristics of the human gait. In total, seven inertial sensors were used, arranged on the pelvis, thighs, legs and feet. A calibration algorithm was implemented to adjust the sensor angles with those of their respective body segment based on a known posture. The data capture system was verified comparing the angle of flexion-extension of the knee with the angle obtained by the artificial vision system from the sagittal plane and comparing it with bands of normality. As a result, it was possible to implement a low-cost inertial motion capture system for the analysis of human gait, with its respective graphical interface to visualize the orientation, flexion-extension angles for each segment and joint of the lower limbs. It is concluded that is possible to develop a complete tool for the kinematic analysis of human gait with the information of the artificial vision system or an anthropometric model.

Keywords: Capture system · Human gait · Inertial sensors

1 Introduction

In this article, first we introduce the characteristics of human gait, methods of measurement and antecedents. In the materials and methods section, several inertial sensors are compared to select the best one, according to its cost-benefit, then the prototype of the capture system is shown, the positioning of the sensors, calibration methods and calculation of the kinematics. In the results and discussion section, the obtained gait parameters and the graphic interface implemented are exposed, the limitations of the capture system are also analyzed, and a verification is made using an artificial vision system and the data obtained from knee flexion-extension with a normal band. Finally, in the conclusions section, we talk about the results obtained and possible future improvements for the development of a tool for the analysis of human gait that will have applications for rehabilitation.

D. F. Cortes Tobar et al. (Eds.): AETA 2019, LNEE 685, pp. 283–292, 2021.
https://doi.org/10.1007/978-3-030-53021-1_29

The standard method for the analysis of human gait is performed in specialized laboratories, where a high-speed multi-camera system is used in conjunction with the use of force plates, to capture the three-dimensional data of a person's gait and the reaction forces. Although this method is very accurate, it needs a considerably large workspace and high-speed graphics signal processing devices, being an expensive method due to its complexity and pre-calibration experiment requirements [1]. The analysis using multi-cameras of high speed is limited to laboratories and is not easily applicable to study the gait in daily activities, or to follow the process of rehabilitation of an individual person [2].

Laboratories of human gait are difficult to apply in daily life, a subjective analysis is usually carried out for diagnosis or rehabilitation, which consists of a specialist doctor observing and evaluating the quality of a patient's gait. Walking can be affected by various pathologies, such as neurological, muscular, skeletal, antalgic postures, general weakening, psychological or psychiatric problems [3], so a subjective analysis it is problematic, due to its poor accuracy and qualitative nature, having a negative effect on the diagnosis and monitoring of the evolution of rehabilitation processes [4]. According to the World Health Organization, it is estimated that 15% of the world's population are disabled [5]. In addition, neurological diseases such as cerebrovascular accidents represent 60% of the causes of gait difficulties [6]. The rehabilitation of the march requires long periods of recovery, and without proper treatment, can have social consequences with economic and health implications [7]. Rehabilitation therapies with appropriate systems could improve it efficacy, because the progress of a patient in rehabilitation, can be evaluated quantitatively by comparing their walking parameters with the walking parameters of a healthy subject [8].

In Colombia, 6.3% of people with disabilities, 29.5% have limitations to move or walk [9]. The armed conflict in Colombia continues increasing the number of people in conditions of disability. Alterations in gait drastically reduce the quality of life, affecting the affected person's independence and ability to [11]. The effectiveness of a treatment depends on determining the pathophysiological causes and acting on dysfunctions and disabilities, it is necessary to have easily accessible tools that are objective and allow accurate and precise quantification of the state of the patient's progress, as well as the evolution of it in rehabilitation [12].

Nowadays, with the increasing interest and application areas of human gait measurement, different systems have been developed for its measurement, some based on the use of camera arrays, inertial sensor arrays, ultrasound sensors, transducers, etc. Inertial sensors stand out with the advantage of being more economical and simple systems, as they do not require large work spaces or large computational expenses, being applicable to follow the process of evolution of a patient even outside the laboratory [4].

1.1 Gait Analysis

The human gait includes a sequence of coordinated movements and relays that allow displacement [13]. The study of these movements is known as gait analysis [14]. The planes for the gait analysis divide the human body, frontal plane oriented vertically, dividing the body in the anterior and posterior zone; the sagittal plane that is oriented vertically and orthogonal to the frontal plane, dividing the body into right and left; the transverse plane is oriented horizontally, dividing the body into the upper and lower zone [15].

Gait phases divides the walk into eight movements, in this way the different movements of the segments of the lower limbs and the angles of the joints can be identified. The walking cycle is divided into initial contact, loading response, mid stance, terminal stance, pre-swing, initial swing, mid swing and terminal swing [14]. When the foot is in contact with the ground it is known as the stance period and in the opposite case it is known as the swinging or oscillating period. Normal gait has an average cadence of 110 steps per minute [16], meaning that a walking cycle has an average frequency of 0.9 Hz. According to [17] the highest harmonics are in the trajectories of the toe and the heel, 99.7% of the signal power being contained below 6 Hz.

2 Materials and Methods

With the development and miniaturization of inertial sensors, due to its use in electronic devices such as smartphones and videogames, its cost has been reduced. Several inertial sensors were compared in the market, such as the GY-25 that delivers the orientation angles, the GY-91 that delivers the values of each sensor, accelerometer, gyroscope and magnetometer with its raw value, the BNO055 that performs a fusion of the three sensors through a 32-bit MCU, the x-IMU, although it does not have the highest resolution of the compared sensors, implements a AHRS filter and has a high sampling rate, USB and Bluetooth connection, as well as other features, but at the same time its extra features increase its price. Finally, the InertialCube4 has a very high precision, implementing a Kalman filter, but with a very high price. The summary information of the comparison of the inertial sensors is shown in Table 1.

Table 1. Comparation of inertial sensors.

Name	Accuracy (°)	Fs (Hz)	Price (US)	Size (mm)	Consumption (mA)
GY-25	1	100	5	11.5 × 15.5 × 2	15
GY-91	Without fusion	157	10	25 × 1 × 3 m	6
BNO055	2.5	100	35	20 × 27 × 4	12.3
x-IMU	1	512	320	33 × 42 × 10	150
InertiaCube4	1 yaw, 0.25 pitch and roll	200	995	36 × 27.7 × 13.8	40

Some inertial sensor selection criteria that were considered for decision making were in the first instance, an adequate relationship between accuracy and price, that the sensor included sensory fusion, because certain advanced knowledge is required to implement a Kalman filter correctly, that the sampling frequency was greater than the Nyquist frequency of the gait (12 Hz), that the size and weight were lower compared to other options and its consumption was not greater than 30 mA, in order to have sufficient system autonomy.

The GY-91 counts with higher resolutions than the BNO055 sensor, but it does not merge the data, it delivers the raw values of the accelerometer, gyroscope and magnetometer. The inertial Cube4 is a bit larger than the other sensors, and its consumption exceeds the limit by 10 mA, but with the advantage of having the best accuracy of all the sensors compared. The factor to discard this sensor was its high price. Likewise, the x-IMU was discarded because it does not have such a high accuracy and has unnecessary features that increase its price. The BNO055 was discarded because its price was still high, because the system must have seven inertial sensors for each lower extremity and its accuracy was lower than the compared sensors. The advantage of this sensor is that it has a magnetometer, which reduces the errors caused by the effect of the drift present in the yaw angle.

Finally, we chose the GY-25 sensor, which has an accuracy and enough resolution, compared to other sensors, has the smallest size and weight, its sampling frequency is higher than necessary, its price is better considering that it performs the fusion of the data and its current consumption is the least. The disadvantage of this sensor is that it does not have a magnetometer, although with the limitations of this work it was not necessary because it was proposed to use only inertial sensors, which only have an accelerometer and gyroscope (IMU) and not inertial sensors in set with magnetometers (IMMU). Like other sensors that perform fusion (such as BNO055), it is not specified what type of fusion algorithm the GY-25 uses, since it is proprietary algorithms of the manufacturer.

The datasheet did not have all the characteristics of the GY-25 sensor, for that reason a high-torque DH-03 servomotor was used to calculate some extra characteristic of the sensor, DH-03 servomotor has a control precision of 0.32° and angular velocity of 1.0 s by 60° to 12 V, It was verified that the accuracy and precision of GY-25 were of 1° and in addition tests of repeatability and linearity were made for the angle of yaw, roll and pitch. The repeatability test consisted in varying the angle from 0° to 10° in 0.5 s intervals for one hour and the linearity test in varying the angle between −45° and 45° in 8 s intervals at constant speed for one hour. It was found that the yaw angle error was 2.9°/min, the angle of roll and pitch have no drift error and its repeatability is 1°. The linearity error for the yaw angle was $\pm 1.1\%$, the pitch error of $\pm 1.0\%$ and the roll error of $\pm 1.0\%$ (for a scale value of 90°). No hysteresis was observed at any of the orientation angles.

As it was required to connect seven inertial sensors to hardware ports, two Arduinos (Mega) were used, it has four hardware serial ports, and for the communication system the nRF24L01 module was used. The design of the capture system is shown in Fig. 1. The capture system communicates wirelessly with another 16-bit MCU (Arduino Nano), which also has an nRF24L01 module and sends all the sensors data serially to the computer.

The capture system wirelessly sends the values of the angles of each sensor through two nodes (each MCU is a node) of a two-to-one network, the receiver is connected to another MCU which is responsible for sending serially the information of all the sensors to the computer. In addition, the time in which the data is captured is also sent using the MCU time to avoid changes in time due to serial communication or processing.

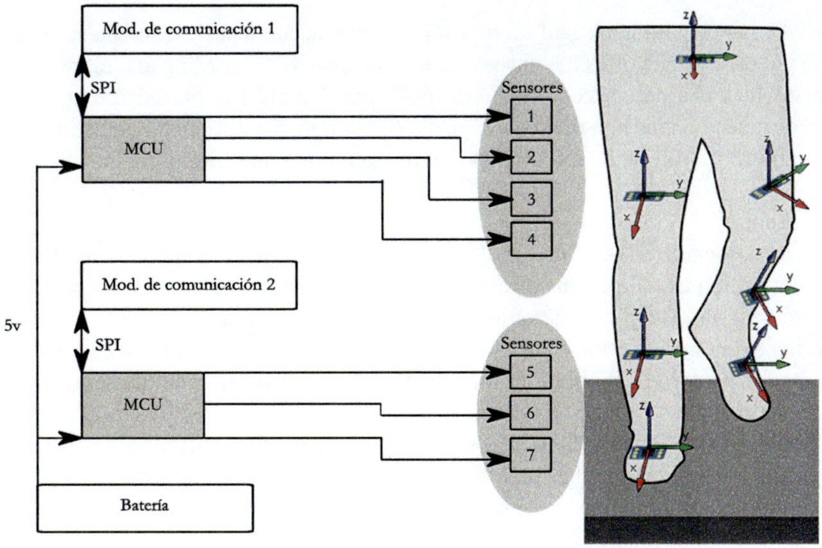

Fig. 1. Capture system and positioning of the sensors.

Computer programming was done in Python since it has many libraries to process data, implement serial communication, perform interfaces and graphics, having a broad community and versatility when developing a program, as well as having a good learning curve compared to other languages such as C/C++ or Java. Although Python is not the programming language with faster execution time, the working frequency is 100 Hz, so it is not necessary to use a compiled language.

The program of the interface works with two threads, the first thread orientation of the capture system and time in which data is captured. The second thread generates an interface that is constantly updated as the angles of the capture system arrive, and allows to visualize the angles of orientation and flexion-extension of the selected segment in an organized manner, choosing by means of the selection buttons and check boxes, in the interface is also shown a two-dimensional kinematic model of the sagittal plane, in addition the second thread is also responsible for keeping all information in a comma separated vector file (CSV) of the data obtained by the system capture when closing the program. The thread of the serial communication works at 1 ms and the thread of the interface works at 200 ms, so that desynchronization does not occur when graphing the data. The system was not restricted to some operating interval in real time, so it is an interactive system.

Different positions of the sensors in the body were tested, but due to the gimbal lock, by the use of the Euler angles that are delivered by the GY-25 (it does not calculate quaternions), the sensor has an impact at 90° between the angles roll and pitch, it was decided to put the sensors perpendicular to the segment of the corresponding limb in the case of the hip, femur and tibia. In the case of the foot the sensor is parallel to it, as seen in Fig. 1. In this way it is avoided to reach the cardan block, decreasing the error of the drift in the yaw angle.

Elastic bands with hook-and-loop fasteners are placed on each segment, so they are easy to put on and take off. The sensors are attached to each according to their respective number, which is ordered according to the positions that must be placed, the first sensor should be placed on the hip, the second on the right thigh, the third in the right leg, fourth in the right foot, fifth in the left thigh, sixth in the left leg and seventh in the left foot.

Calibration

It should be considered that the GY-25 oversees making the data merge with a microcontroller, so it must correct the errors of the inertial sensors such as deviation, scaling, misalignment and stability. Therefore, it processes of calibration should be respected, according to the specifications when turning on the GY-25, the sensor must be still during 3000 ms for its correct calibration, for which it is necessary to feed the sensor before being able to adhere it to the body.

To calibrate the sensors, the sensors are aligned facing the same place in an approximate zero orientation for the three angles, the capture system is turned on and 3 s are waiting before mounting the sensors on the body. Then the method to calibrate the angles of orientation, so that the axes of the inertial sensor are aligned with those of their respective segment, is that the person with the capture system put must stand upright supported on a wall. Once that position is assumed, the interface is started, and after five seconds the Software automatically adjusts the yaw, pitch and roll angles of each sensor to match those of the previously described position (by means of an arithmetic sum and obtaining the difference of the current angle and angle of the posture). In this way, the values of the orientation angles of each inertial sensor are aligned with those of its respective segment.

Artificial Vision System

The artificial vision system that was used works with the OpenCV library and a webcam, allowing the identification of color markers through the hue, saturation and value (HSV) color model. From the sagittal plane, placing colored markers (non-reflective) on the joints, you can follow the hip, the head of the left femur (sagittal plane), knees and ankles. So that each joint can be located according to their X and Y coordinates in pixels or with their equivalence in centimeters, with an accuracy of 2 mm with the camera used.

Criterion of Inclusion of Tests

The inclusion criteria established that for the tests performed the subject was 22 years old or older, that the subject would not have had any type of accident that affect their mobility. Three tests were done with different people endorsed by the ethics committee of the Universidad Santiago de Cali. The initial tests, with only the capture system were made walking at the normal speed of the person, in a straight line for five meters, when reaching the end, the subject turned and returned to the starting position continuously for thirty seconds. The tests with the system of capture and system of artificial vision were made in a treadmill at speed of 0.25 m/s for thirty seconds. Five data tests were taken in which a subject had the inertial capture system and the artificial vision system with colored markers arranged in the greater trochanter of the left leg, knees and ankles (from the left sagittal plane).

3 Results

By using the capture system with inertial sensors, the orientation of the segments of the lower limbs corresponding to the hip, thighs, legs and feet was obtained in time. In addition, with the artificial vision system, the positions (in two dimensions) of the markers in time were acquired. This information is used to calculate parameters such as stride speed, step length, stride length, cadence, step angle, gait phases, orientation of the body segments and joint angle.

The interface was made in Matplotlib, see Fig. 2, in this one you can select which segment of the extremities you want to observe, using the buttons in the lower part it can be choose between the hip, right thigh, leg right, right foot, left thigh, left leg and left foot. Once the limb has been selected, the check boxes in the lower left is using to select what orientation values are plotted as the angle of yaw, pitch, roll, and angle of flexion. The horizontal axis shows the time in milliseconds of an interval of 18 s and in the vertical axis the values of the angles in degrees, with a variable interval according to the maximum value of the angles plotted. The two-dimensional kinematic model calculated in 200 ms intervals with constant values for the yaw angle was included in the upper right corner. In addition, in the lower left you have additional options such as zoom, save and move the graph.

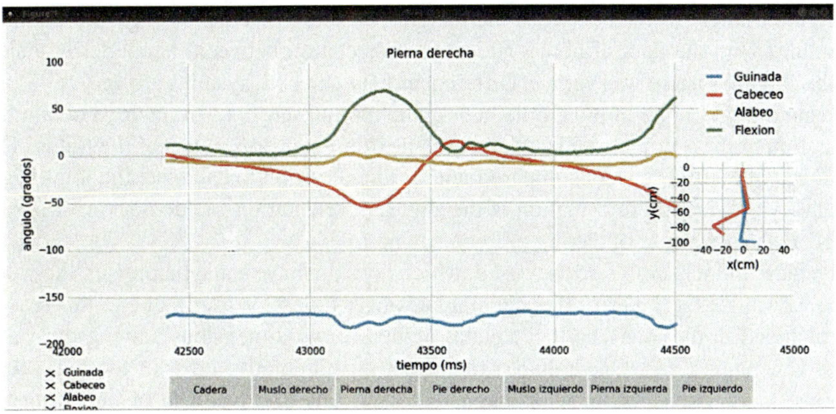

Fig. 2. Graphic interface of the capture system.

Through the inertial sensors we can obtain the orientation of the body segments corresponding to the hip, thighs, legs and ankles. In addition, with the information of the orientation of the segments of the body can calculate the angle between joints, such as knee angle, which was compared with the angle calculated by artificial vision and trigonometry, both angles are shown in the Fig. 3. It should be considered that the joint angles obtained with the information of the inertial sensors are more accurate than that calculated by artificial vision with a camera, because the vision system may have errors due to the perspective given that the movement of the human march is in three dimensions.

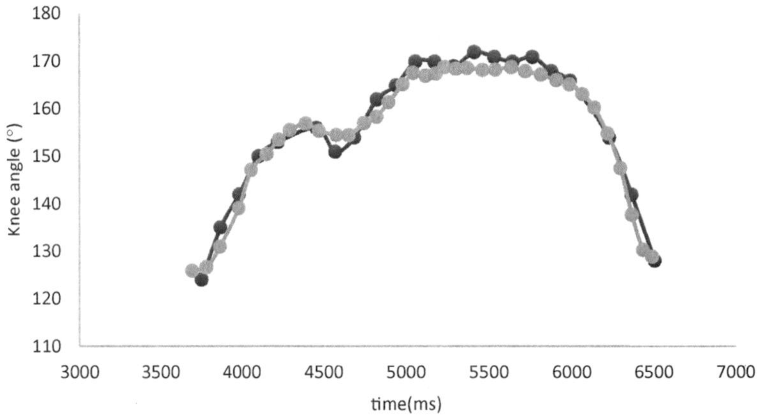

Fig. 3. Knee angle comparison using inertial sensors and artificial vision.

In Fig. 3, the knee angle comparison is made using the capture system with inertial sensors and the angle obtained by artificial vision from the sagittal plane. To calculate the angle of the knee, the inclination angles of the tibia and the femur are subtracted. Through vision, the angle of the knee is estimated mathematically using the law of cosine due to the triangle formed between the markers placed on the head of the femur, knee and ankle. With the data of both systems the correlation between the angle by vision and the capture system was 98.5%. Differences between both systems are very few, thus verifying that the angle of pitch obtained by the inertial sensors is reliable to obtain the orientation of the segments and calculate the flexion-extension angles of the joints. The advantage of inertial sensors compared with artificial vision to calculate the joint angle is that they provide full information of the angles of orientation and do not vary with the perspective (in the case of having only one camera).

Because the sensor GY-25 was used, which does not have a magnetometer, the value of the yaw angle has an error that accumulates over time (2.9°/min), due to this reason was included in the interface the skeletal model in two dimensions leaving the value of the yaw angle fixed. One solution to correct the cumulative error of the gyroscope is to use inertial sensors set with magnetometers. The disadvantage of obtaining the orientation of the inertial sensors with magnetometer is that the angles can vary in the presence of magnetic disturbances in the environment. Another solution for short distances and walking in the same direction, it's to generate a trend line and adjust the data eliminating the error caused by the drift, the disadvantage of this solution is that it requires post-processing, losing the possibility of implementing a system in real time.

The results of the flexion-extension angle for three subjects were compared with the normality bands [18], which provide a set of reference data of normal walking values of a healthy person. In Fig. 4, the normality bands were compared for knee extension flexion angle, with the data obtained with the capture system by three different healthy people, as can be seen, the subjects of the sample are contained within the bands of normality.

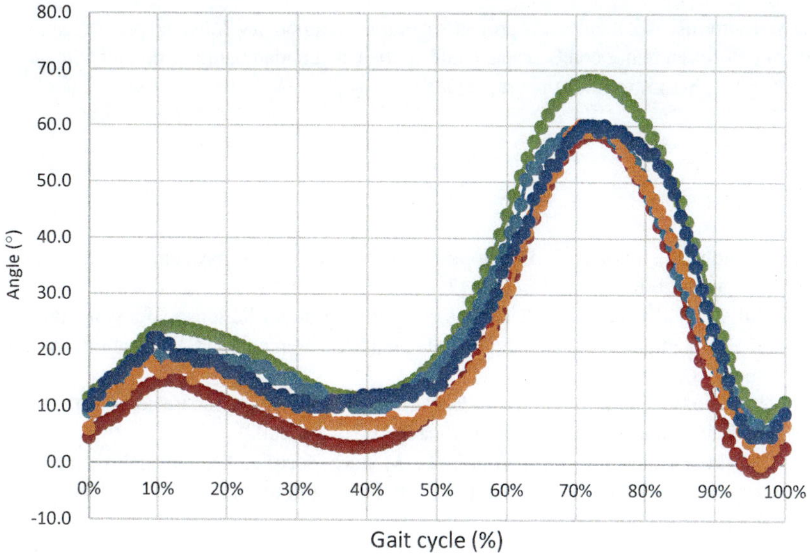

Fig. 4. Bands of normality of knee flexion-extension angle in conjunction with the tests performed.

4 Conclusions

It was verified that the flexion-extension angle calculated with the orientations of the inertial capture system has an error less than 2° with the angle calculated by the artificial vision system from the sagittal plane. The advantage of the inertial sensors versus the artificial vision system is that they also offer the yaw and roll angles, so it is possible to obtain more information for the analysis of the movement of the gait and they don't have errors due to the perspective. Also, the data obtained from the capture system were compared with normal bands and the individuals analyzed were in the normal range. As future work, the inertial motion capture system must be validated in a laboratory for human gait analysis.

Gait phases can be found automatically with trigger signals using a kinematic model, so the length of each lower limb segment most be feeding back with an anthropometric model or artificial vision system. Thus, the inertial system can work independently for applications of tracking the gait outside the laboratory. This will allow having a more versatile system by not restricting the analysis of the human march to a closed space.

Without magnetometer you can't calculate the direct kinematics in real time, due to the integral error that accumulates in the yaw angle, so if you want to build a better kinematic model of the lower extremities in real time you should choose to add a magnetometer, except for the legs (because the yaw angle is the same as that the corresponding thigh). However, for short time measurements, the variation of the yaw angle is 2.9°/min, so it must be analyzed if this error can be tolerated, or it can be reduced with post-processing techniques if subject walk in linear trajectories.

Acknowledgments. We thank to the project "Plataforma Robótica y Sensorial para Rehabilitación Cognitiva y Física en Niños con Discapacidad", of the Universidad Santiago de Cali with the filing code: 829-621118-135, for the support provided.

References

1. Liu, T., Inoue, Y., Shibata, K.: Development of a wearable sensor system for quantitative gait analysis. Meas. J. Int. Meas. Confed. **42**(7), 978–988 (2009)
2. Gastaldi, L., et al.: Technical challenges using magneto-inertial sensors for gait analysis. In: 2016 IEEE International Symposium on Medical Measurements and Applications (MeMeA), pp. 1–6 (2016)
3. Villa Moreno, A., Gutiérrez Gutiérrez, E., Pérez Moreno, J.C.: Consideraciones para el análisis de la marcha humana. Rev. Ing. Biomédica **2**(3), 16–26 (2008)
4. Muro-de-la-Herran, A., García-Zapirain, B., Méndez-Zorrilla, A.: Gait analysis methods: an overview of wearable and non-wearable systems, highlighting clinical applications. Sens. (Switz.) **14**(2), 3362–3394 (2014)
5. Observatorio Nacional de Discapacidad: Cómo van las estadísticas en discapacidad (2016). http://ondiscapacidad.minsalud.gov.co/Paginas/Home.aspx. Accessed 15 Feb 2017
6. Stolze, H., et al.: Prevalence of gait disorders in hospitalized neurological patients. Mov. Disord. **20**(1), 89–94 (2005)
7. Bureau, M., Eizmendi, G., Olaiz, E., Zabaleta, H., Medina, J., Pérez, M.: Diseño de un nuevo exoesqueleto para neuro-rehabilitación basado en detección de intención. In: II Congreso Internacional sobre Domótica, Robótica y Teleasistencia para todos (2007)
8. Zeni, J.A., Higginson, J.S.: Differences in gait parameters between healthy subjects and persons with moderate and severe knee osteoarthritis: a result of altered walking speed. Clin. Biomech. **24**(4), 372–378 (2009)
9. DANE: Boletín Censo General: Discapacidad-Colombia (2005). http://www.dane.gov.co/censo/files/boletines/discapacidad.pdf. Accessed 15 Feb 2017
10. Así Vamos en Salud: Seguimiento al sector salud en Colombia (2013). http://www.asivamosensalud.org/indicadores/estado-de-salud/poblacion-con-discapacidad-georeferenciado. Accessed 15 Feb 2017
11. Willems, P., Schepens, B., Detrembleur, C.: Marcha normal. EMC - Kinesiterapia - Med. Física **33**(2), 1–29 (2012)
12. Chaler Vilaseca, J., Garreta Figuera, R., Müller, B.: Técnicas instrumentales de diagnóstico y evaluación en rehabilitación: estudio de la marcha. Rehabilitacion **39**(6), 305–314 (2005)
13. Mariana Haro, D.: Laboratorio de análisis de marcha y movimiento. Rev. Médica Clínica Las Condes **25**(2), 237–247 (2014)
14. Tao, W., Liu, T., Zheng, R., Feng, H.: Gait analysis using wearable sensors. Sensors **12**(2), 2255–2283 (2012)
15. González Mejia, S., Ramírez Scarpetta, J.M., Avella Rodríguez, E.J.: Control techniques for the balance of a robot biped: a state of the art. Tecnura **19**(43), 139–162 (2015)
16. Carmen, M.: Cinesiologia de la marcha humana normal (2006)
17. Winter, D.A.: Biomechanics and Motor Control of Human Movement, 4th edn. David A. Winter (cloth) 1. Human mechanics. 2. Motor ability. 3. Kinesiology. I. Title. QP303.W59, vol. 7 (2009)
18. Bovi, G., Rabuffetti, M., Mazzoleni, P., Ferrarin, M.: A multiple-task gait analysis approach: kinematic, kinetic and EMG reference data for healthy young and adult subjects. Gait Posture **33**(1), 6–13 (2011)

Monte Carlo Sensitivity Analysis of Biomass to the Input Parameters of a Microalgal Culture Model

Gianfranco Mazzanti[1(✉)], Sangregorio Soto Viyils[2], Claudia L. Garzón-Castro[2(✉)], and John A. Cortés-Romero[3]

[1] Dalhousie University, F2215-1360 Barrington Street, Halifax, NS B3H 4R2, Canada
mazzanti@dal.ca

[2] Engineering Faculty, CAPSAB Research Group, Universidad de La Sabana, Campus Universitario del Puente del Común, Km 7 Autopista Norte de Bogotá, Chía, Cundinamarca, Colombia
claudia.garzon@unisabana.edu.co

[3] Universidad Nacional de Colombia, Av. Carrera 30 No. 45-03, Edif. 411 Of. 203, Bogotá, Colombia

Abstract. Microalgal cultures are developing into a promising ecofriendly technology for a host of applications. In this paper we investigate the response of the biomass *(X_{SS})*, computed from a photo-bioreactor (PBR) algal growth numerical model, to its nine input variables. We explored this response using a Monte Carlo (MC) sensitivity analysis (SA) and Spearman's Rho concordance coefficient method (MC-Rho). The model represented a continuous culture of the green microalgae *Chlorella vulgaris*. The photo-bioreactor was simulated using MATLAB as a steady state operation. Four important control variables were selected to represent different states of operation: dilution rate *(D)*, incident light intensity *(I_m)*, hydrogen ion potential *(pH)*, and partial pressure of CO_2 (P_{CO_2}). For each test point, a statistical array of values was generated of all the input variables with a 2% relative standard deviation. The output values were computed using the model, and then correlated with the input values. The model output turned out to be very sensitive to the four input variables that directly affect the carbon uptake and retention: dilution rate *(D)*, partial pressure of CO_2 (P_{CO_2}), mass conversion yield *(Y_r)*, and gas-liquid mass transfer coefficient *($k_L a$)*, in that order. The magnitude of the output relative standard deviation decreased as the output biomass *(X_{SS})* increased. However, the ratio of the output standard deviation to the input standard deviation was strongly non-linear and essentially unpredictable outside of a MC simulation.

Keywords: Biomass production · Continuous culture · Microalgae · Monte Carlo · Sensitivity analysis · Spearman's Rho

1 Introduction

This paper reports on the use of algorithms based on the stochastic Monte Carlo (MC) method, to analyze the sensitivity of the output to the variation of the input variables of

D. F. Cortes Tobar et al. (Eds.): AETA 2019, LNEE 685, pp. 293–302, 2021.
https://doi.org/10.1007/978-3-030-53021-1_30

a model microalgal culture in a photo-bioreactor (PBR). In recent decades the need for environmentally friendly technologies has become paramount. A promising branch of these technologies is the cultivation of microalgae [1]. They are used for industrial production of high value compounds and as promising alternative for carbon dioxide (CO_2) mitigation. To be sustainable, biomass production in continuous cultures [2–4] requires automation. Supply streams, agitation, aeration, and other operating variables must be controlled to respond to random disturbances of the bioprocess. Therefore, the variables that have the greatest influence need to be identified. The variables reported to be controlled in published works are usually chosen from previous literature reports [5–8], *e.g.* oxygen supply, light intensity, *pH* and temperature. A better method to determine *which variables have the greatest influence on biomass productivity in a continuous culture* is the Sensitivity Analysis (SA). SA provides quantification of the relative importance of the impact that each input variable has on the value of a chosen output variable.

Thus, the aim of the present study was to identify the hierarchy of the input variables, and their standard deviations, with respect to their impact on the total biomass concentration using a growth model of the green microalga *Chlorella vulgaris.*

2 Growth Model

The growth model of the green microalga *Chlorella vulgaris* used here is described in detail in references [9, 10]. This model combines two sub-models: the Monod model for the effect of light, and the Contois model for the limitation by a substrate. The model incorporates the combined influence of several operational parameters on the total biomass: dilution rate (*D*), incident light intensity (I_{in}), hydrogen ion potential (*pH*), and carbon dioxide partial pressure $\left(P_{CO_2}\right)$. Other parameters included are discussed below.

The effective volume (9.6 L) of the PBR is assumed constant (equal inlet and outlet volumetric flow rates), and under perfectly stirred conditions. For this study we reduced the model to its steady state conditions at 25 °C. Under these assumptions, the change rate of the cell number density (*X(t)*) is expressed by (1).

$$\frac{dX(t)}{dt} = \mu \cdot X(t) - D \cdot X(t) = 0 \tag{1}$$

Where, μ is the specific growth rate (h^{-1}), X(t) is the cell number of microalgae per unit culture volume or biomass (=X_{SS} under steady state, in 10^9·cell·L^{-1}), and *D* (h^{-1}) is the dilution ratio, *i.e.* the inflow rate of medium per culture volume in the PBR. Under steady state they do not change with time. The steady state equations are (2), (3) and (4).

$$\frac{I_{in} \cdot \left(1 - C_1 \cdot X_{SS}^{C_2}\right) \cdot A_r}{V \cdot X(t)} - E = 0 \tag{2}$$

$$\left(\frac{\mu_{max} \cdot E}{K_E + E}\right) \cdot \left(\frac{[TIC]}{K_{CL} \cdot X_{SS} + [TIC]}\right) - D = 0 \tag{3}$$

$$\frac{-D \cdot X_{SS}}{Y_r} - D[TIC] + k_L a \left(\frac{P_{CO_2}}{H} - \frac{[TIC]}{1 + \frac{K_1}{[H^+]} + \frac{K_1 \cdot K_2}{[H^+]^2}}\right) = 0 \tag{4}$$

Where, $[TIC]$ is the concentration of Total Inorganic Carbon (TIC) (mol·L^{-1}) in the liquid of the PBR, Yr is the mass conversion yield (ratio of the biomass produced to the amount of TIC consumed, 10^9·cell·L^{-1}·mol^{-1}) and k_La is the overall gas-liquid mass transfer coefficient of carbon dioxide (h^{-1}).

Furthermore, H is Henry's constant (atm·L·mol^{-1}) for carbon dioxide in the culture medium. The maximal specific growth rate is μ_{max} (h^{-1}), K_E is the half saturation constant of the light intensity accessible per cell (μE·s^{-1}·10^9·cell^{-1}), and K_{CL} is the $[TIC]$ limitation constant in (mmol·10^9·cell^{-1}). Additionally, E is the available light intensity (μE·s^{-1}·10^9·cell^{-1}), A_r is the reactor illuminated area (m^2), V is the PBR volume (L), C_1 and C_2 are constants which determine the light absorption by the algae for the photo-bioreactor geometry. The constants of the chemical equilibrium $CO_2 \leftrightarrow HCO_3^-$ and $HCO_3^- \leftrightarrow CO_3^{2-}$, K_1 and K_2 at 25 °C correspond to values of $pK_1 = 6.35$ and $pK_2 = 10.33$.

Bioprocess parameters were experimentally determined [9] at 25 °C. They are summarized in Table 1 along with the nominal values for the simulations, as reported by [10].

Table 1. Model parameters of *Chlorella vulgaris* used in the simulation.

Parameter	Unit	Value
PBR geometry		
C_1		0.49
C_2		−0.92
A_r	m^2	0.31
V	L	9.6
$[TIC]$ properties		
k_La	h^{-1}	1.36
pH		6
H	atm·L·mol^{-1}	29
Operating conditions		
I_{in}	μE·s^{-1}·m^{-2}	90
P_{CO_2}	atm	0.05
μ_{max}	h^{-1}	1.068
K_E	μE·s^{-1}·10^9·cell^{-1}	0.0817
K_{CL}	mmol·10^9·cell^{-1}	0.0038
Y_r	10^9·cell·L^{-1}·mol_TIC^{-1}	4.353

3 Sensitivity Analysis Methodology

To quantify the response of the model output to its inputs, large sets of input variables behaving statistically were fed to the model. The probability distribution and the standard deviations of X_{SS} were computed for each subset. The distribution applied to each parameter's value was lognormal, with a standard deviation $\sigma_m = 0.02 \cdot Par_m$. In this context, the value $\sigma_{rin} = 0.02$ is the relative standard deviation ratio. The same standard deviation ratio was used for all input parameters, regardless of whether they were tested in a target range or left at the nominal value.

Global SA (GSA) measures the contribution of a group of input variables to the output, expressed as the non-dimensional function,

$$S_i = \frac{\partial \text{Output}}{\partial \text{Input}} \cdot \left(\frac{\text{Input}}{\text{Output}} \right) \tag{5}$$

GSA is computed in terms of total sensitive indices, T, calculated from S_i.

$$\mathrm{T} = \sum\nolimits_{\mathrm{i=n}}^{\mathrm{n}} \mathrm{S_i} \tag{6}$$

The procedure for GSA through MC included the following steps [11]: 1) selection the input variables to be tested, 2) setting a large enough range for each selected variable value to cover reasonable variations, 3) generation of a series of independent random numbers with known distribution within the range selected, 4) MC sampling: 4.1) calculation of the output corresponding to each combination of input variables, and 4.2) evaluation of the GSA for each variable through correlation coefficients.

The most widely used methods are Spearman's rank-based (Spearman's Rho) and Kendall's concordance coefficient (Kendall's Tau). Furthermore, both Spearman Rho and Kendall Tau correlations combine high efficiency with a bounded and smooth influence function [12]. In this paper we only used Spearman Rho, since it was proportional to Tau.

Further numerical tests were conducted to observe the effect of the magnitude of the standard deviation of the most influential processing parameters on the standard deviation of the output biomass. These tests were conducted by exploring input standard deviations $\sigma_{rin} = 0.01, 0.02$ and 0.03 for selected operational conditions.

4 Results and Discussion

4.1 Selection of Preliminary Exploratory Parameters

The growth model process parameters and biological parameters. The process parameters can be controlled, whereas the biological parameters are intrinsic of the microorganism. The sensitivity indices were computed for the five processing parameters (D, I_{in}, pH, P_{CO_2}, $k_L a$) and for four biological parameters (μ_{max}, Y_r, K_E, K_{CL}) at a temperature of 25 °C. A separate paper (in preparation) will study the effects of temperature.

Using the model, values of steady state biomass (X_{SS}) were computed as a function of a subset of test-value inputs D, I_{in}, pH, and P_{CO_2}. They were chosen due to their assumed importance to the growth [10, 13].

The nominal value of D was 0.0934 with range of operation between 0.05 and 0.21 h^{-1}. The nominal value of I_{in} was 90 $\mu E \cdot s^{-1} \cdot m^{-2}$ with range of operation between 90 and 300 $\mu E \cdot s^{-1} \cdot m^{-2}$. The nominal pH was 6, with a range from 5 to 7. The nominal P_{CO_2} was 0.05 atm with a range from 0.01 to 0.07.

4.2 Evaluation of the Impact of the Test Process Variables and Biological Parameters on the Steady State Biomass (X_{SS}) Output

The values of X_{SS} are plotted in Fig. 1. as a function of test values of (a) D, *(b) pH*, (c) I_{in}, and (d) P_{CO_2}. The dotted black lines represent the value of X_{SS} for the values tested within each range. The marker 'clouds' correspond to the statistical effects on X_{SS} from the parameter's standard deviations of 2%, applied to generate the MC realizations.

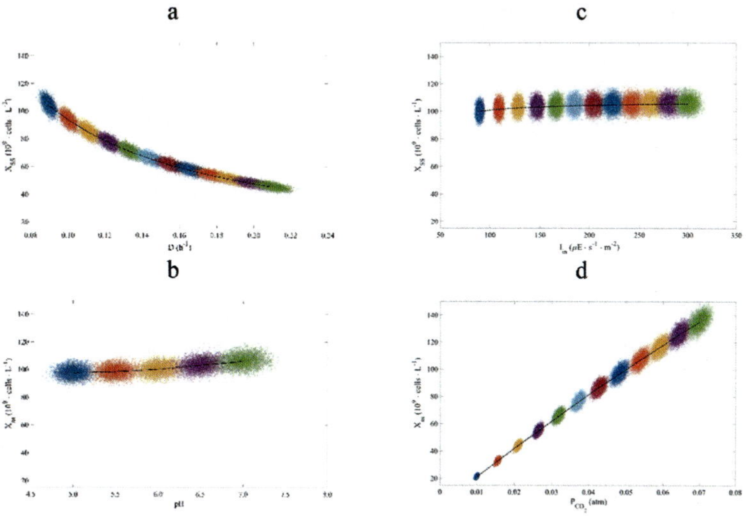

Fig. 1. Steady state biomass (X_{SS}) as a function of test values of D, I_{in}, pH, P_{CO_2}, represented by dotted lines. The statistical effects of the five processing parameters (D, I_{in}, pH, P_{CO_2}, k_La) and the four biological parameters (μ_{max}, Y_r, K_E, K_{CL}) are seen in the 'clouds' surrounding each test point. Each test point included 5000 statistical realizations.

The values of X_{SS} decrease with D (panel a) and increase with the other parameters (panels b, c, and d). These general trends have different ranges and curvatures. To highlight the non-constant slopes of these curves, we have plotted the values of S_i from (5) in Fig. 2. Note the different scale of the two y axes.

The magnitudes of the sensitivity index are strongly nonlinear with changes in pH. Changes in the other three parameters, though non-linear as well, have smoother curves. D and P_{CO_2} have high sensitivity indices, whereas I_{in} displays the smallest values. These sensitivity indices are not constant, even when the variables are changed one by one.

To further evaluate the impact that the input parameters have on the output, we used the data generated by the MC procedure. The value of T for each variable was

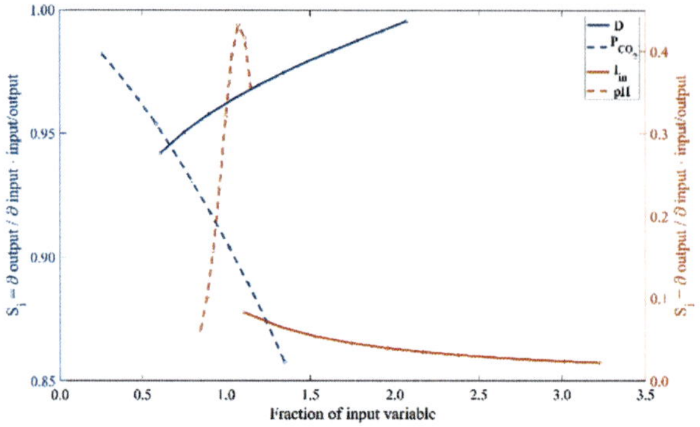

Fig. 2. Sensitivity indices (S_i) computed by (5) for (X_{SS}) as function of target parameters D, I_{in}, pH, P_{CO_2}. The values of the parameters in the x-axis are presented as a nondimensional ratio between the parameter and its nominal value.

computed as the correlation index Rho, using all the outputs produced by the random sets of inputs. These Spearman Rho values are plotted in Fig. 3. The Rho indices were computed from 5000 realizations at each parameter test value in triplicate. The total number of realizations per parameter was 60000. The correlation Rho values did not show a trend with respect to the range of the targeted variables. Thus, the averages of three runs are shown in Fig. 3, together with their standard deviations. The correlation Rho values are almost identical, regardless the variable chosen to be the explored over its range, while the other input parameters were varying around their nominal values. Recall that the standard deviation applied to all input variables was 2% of the nominal or tested value.

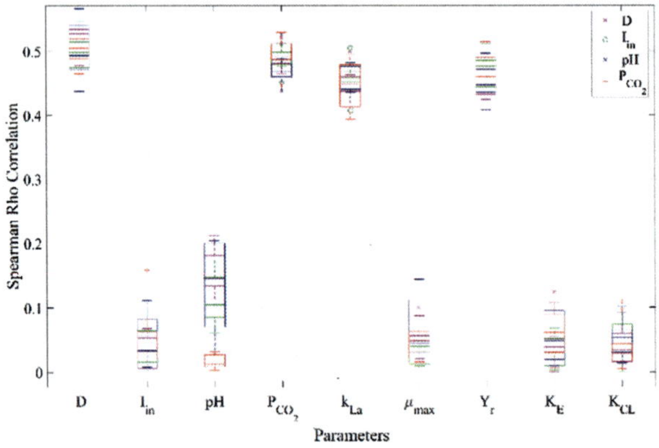

Fig. 3. Spearman Rho correlation coefficients (to the steady state biomass X_{SS}) for each one of the five processing parameters ($D, I_{in}, pH, P_{CO_2}, k_La$) and the four biological parameters (μ_{max}, Y_r, K_E, K_{CL}). This plot summarizes results from three sets of 5000 realizations for each test point of each preliminary parameter.

The Rho values divide the input variables into two groups. A high impact group that includes D, P_{CO_2}, Y_r, and k_La, and a low impact group that includes I_{in}, pH, K_{CL}, K_E, and μ_{max}. the variables of the high impact group have a strong, direct effect on the capacity to retain carbon, *i.e.* to increase the biomass. The low impact group parameters have in common an indirect effect on the output, both physically and in the mathematical model. It seems surprising that I_{in} is in the low high impact group. However, the operational conditions had been optimized at the limit where an increase in light intensity is strongly compensated by the microalgae attenuation of light penetration (self-shadowing).

4.3 Effect of the Input Standard Deviations on the Standard Deviations of the Steady State Biomass (X_{SS}) Output

The next step is to evaluate the effect that the standard deviation of the input parameters has on the standard deviation of the output biomass.

The two-dimensional probability distributions of X_{SS} for 12 test D values are plotted in Fig. 4 using a colormap to represent the value of the probability (10^4 realizations/D, with all parameters varying statistically). The distribution applied to the input variables was lognormal, with a relative standard deviation ratio $\sigma_{rin} = \sigma_{in}/Par_m = 0.02$ where Par_m is the test or nominal value of the parameter, *e.g.* D or K_{CL}. The same standard deviation ratio was used for all input parameters, regardless of whether they were tested in a target range or left at the nominal value. These data sets are very useful when planning a control strategy, since one has both the uncertainty of the tracked variable and of the control variable. However, for the sensitivity study, we focused on the effect that a 'commanded' value of a process variable has on the overall output. Thus, we computed the relative standard deviations of the probability distribution of the output, as illustrated in Fig. 5 by the orange histograms. Further experimental studies are necessary to determine the actual standard deviations of each particular parameter.

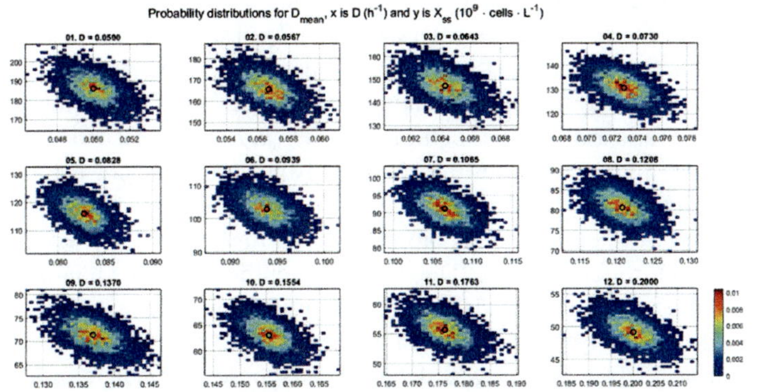

Fig. 4. Probability distribution (0 to 0.01 in the color bar) of X_{SS} for 12 targeted D values (10^4 realizations/D value, with all parameters behaving statistically). The test value and nominal response are indicated by the black circles.

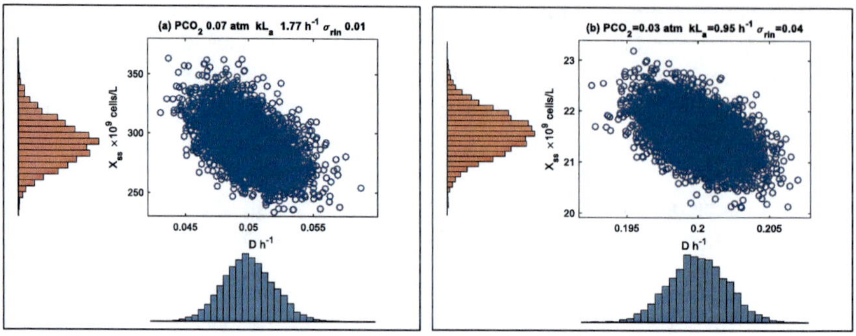

Fig. 5. Probability distribution of X_{SS} (orange histograms) for 2 targeted D values (5000 realizations with all parameters behaving statistically). (a) $P_{CO_2} = 0.07$ atm, $k_L a = 1.77$ h^{-1}, and $\sigma_{rin} = 0.01$, and (b) $P_{CO_2} = 0.03$ atm, $k_L a = 0.95$ h^{-1}, and $\sigma_{rin} = 0.04$. Note the different scales for each plot.

To evaluate the effect of the magnitude of the input standard deviations, the output relative standard deviations σ_{rout} were computed via MC for three levels of σ_{rin}: 0.01, 0.02 and 0.04. The σ_{rout} were computed from the distributions plotted as orange histograms in Fig. 5 $\sigma_{rout}/\sigma_{rin}$. The values of σ_{rout} were computed for low, medium and high biomass conditions produced by combinations of the input parameters P_{CO_2} and $k_L a$ at levels of (0.03, 0.95), (0.05, 1.36), and (0.07, 1.77). The levels of D were 0.05, 0.10 and 0.20 h^{-1}.

The response of the relative output standard deviation σ_{out} to the relative input standard deviation σ_{rin} is quite complex, as seen in Fig. 6. In this figure, we have plotted the ratio $\sigma_{rout}/\sigma_{rin}$ for two extreme conditions of operation and their midpoint. These conditions of operation can produce the smallest and the largest biomass, within the practical operational range of this bioreactor. In Fig. 6, panel (a) corresponds to process input parameters (D, P_{CO2}, $k_L a$) that produce the smallest biomass output (X_{SS}). Panel (b) corresponds to process input parameters (D, P_{CO2}, $k_L a$) at their nominal values, where the biomass output (X_{SS}) is at a mid-value. Finally, panel (c) corresponds to process input parameters (D, P_{CO2}, $k_L a$) for the largest biomass output (X_{SS}). There is a general trend of reduction of the ratio as the values of the pair (P_{CO2}, $k_L a$) increase, and larger biomass (X_{SS}) is obtained. This response, however, is strongly nonlinear and not-monotonic. This creates a practical impossibility for prediction of the output standard deviation that could be used in control development. This bioprocess will therefore easily deviate from its target if a basic controller such as PID is used. The output responds nonlinearly to the inputs, and on top of that the standard deviation of that output is extremely nonlinear to the standard deviation of the input. An active disturbance rejection controller will perform much better [7], since it will define the response parameters in a short time frame ad without needing an exact model of the main response nor its standard deviation response.

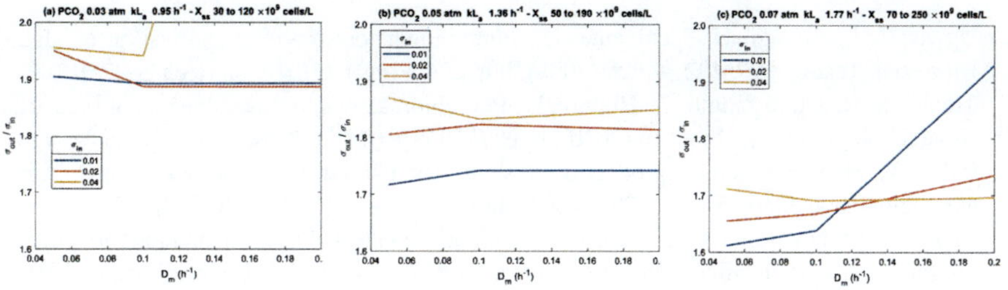

Fig. 6. Ratio of output over input relative standard deviations σ_{out}/σ_{in}, plotted as a function of three values of the dilution ratio D (h^{-1}). Each plot corresponds to a combination of process input parameters (P_{CO_2}, $k_L a$) indicated in the panel title.

5 Conclusions and Perspectives

The analysis of the sensitivity of the output to five processing parameters (D, I_{in}, pH, P_{CO_2}, $k_L a$) and four biological parameters (μ_{max}, Y_r, K_E, K_{CL}) allowed the successful ranking of the input variables. The analysis was conducted through a combined Monte Carlo method and Spearman's Rho correlation (MC-Rho) of the biomass at steady state (X_{SS}) (model of *Chlorella vulgaris*). It revealed that the input variables with the strongest impact were D, P_{CO_2}, Y_r, and $k_L a$. This led to the conclusion that they are of high impact because they directly affect the carbon uptake and retention, which in hindsight makes much sense.

Not surprisingly, the output has a nonlinear response to the input variables. Moreover, the standard deviation of the output has a very strong and hardly predictable dependence on the standard deviations of the input variables.

Process control strategies must respond to deviations from the targeted biomass by specifying, for instance, a target D value. The output biomass for this value of dilution rate, however, is not deterministic. The biomass is instead described by the probability distributions illustrated in this paper. The actual values of the standard deviations need, however, to be determined experimentally. Other models and PBR types can be explored using this methodology. As indicated in the text the effects of other important parameters such and temperature need to be explored as well.

Acknowledgements. The authors thank Ministerio de Ciencia, Tecnología e Innovación Minciencias for funding of the Project "*Diseño e implementación de una estrategia de Control por Rechazo Activo de Perturbaciones de orden fraccionario aplicado al crecimiento de microalgas*". The support and facilities provided by Universidad de La Sabana, Dalhousie University and Universidad Nacional de Colombia are gratefully recognized. Mr. Sangregorio thanks the Engineering PhD Program of the Universidad de La Sabana for the "Carlos Jordana" Scholarship awarded. Dr. Mazzanti acknowledges partial funding from Natural Sciences and Engineering Research Council of Canada (NSERC), Gobernación de Cundinamarca, and Colombia BIO.

References

1. Dominic, V., Murali, S., Nisha, M.: Phycoremediation Efficiency of three micro algae *Chlorella vulgaris, Synechocystis salina and Gloeocapsa gelatinosa* **XVI**(1), 138–146 (2009)

2. Ugwu, C.U., Aoyagi, H., Uchiyama, H.: Photobioreactors for mass cultivation of algae. Bioresour. Technol. **99**, 4021–4028 (2008). https://doi.org/10.1016/j.biortech.2007.01.046
3. Tredici, M.R., Chini Zittelli, G., Rodolfi, L.: Photobioreactors. In: Encyclopedia of Industrial Biotechnology. Wiley, Hoboken (2010). https://doi.org/10.1002/9780470054581.eib479
4. Pulz, O.: Photobioreactors: production systems for phototrophic microorganisms. Appl. Microbiol. Biotechnol. **57**, 287–293 (2001). https://doi.org/10.1007/s002530100702
5. Mairet, F., Muñoz-Tamayo, R., Bernard, O.: Adaptive control of light attenuation for optimizing microalgae production. J. Process Control **30**, 117–124 (2015). https://doi.org/10.1016/j.jprocont.2015.03.010
6. Raso, S., Van Genugten, B., Vermuë, M., Wijffels, R.H.: Effect of oxygen concentration on the growth of *Nannochloropsis* sp. at low light intensity. J. Appl. Phycol. **24**, 863–871 (2012). https://doi.org/10.1007/s10811-011-9706-z
7. Sousa, C., Compadre, A., Vermuë, M.H., Wijffels, R.H.: Effect of oxygen at low and high light intensities on the growth of *Neochloris oleoabundans*. Algal Res. **2**(2), 122–126 (2013). https://doi.org/10.1016/j.algal.2013.01.007
8. Costache, T.A., Gabriel Acién Fernández, F., Morales, M.M., Fernández-Sevilla, J.M., Stamatin, I., Molina, E.: Comprehensive model of microalgae photosynthesis rate as a function of culture conditions in photobioreactors. Appl. Microbiol. Biotechnol. **97**(17), 7627–7637 (2013). https://doi.org/10.1007/s00253-013-5035-2
9. Filali, R., Tebbani, S., Dumur, D., Isambert, A., Pareau, D., Lopes, F.: Growth modeling of the green microalga *Chlorella vulgaris* in an air-lift photobioreactor. In: IFAC Proceedings, pp. 10603–10608 (2011). https://doi.org/10.3182/20110828-6-IT-1002.01955
10. Garzón-Castro, C.L., Delgado-Aguilera, E., Cortés-Romero, J.A., Tello, E., Mazzanti, G.: Performance of an active disturbance rejection control on a simulated continuous microalgae photobioreactor. Comput. Chem. Eng. **117**, 129–144 (2018). https://doi.org/10.1016/J.COMPCHEMENG.2018.06.006
11. Sobol, I.: Global sensitivity indices for nonlinear mathematical models and their Monte Carlo estimates. Math. Comput. Simul. **55**, 271–280 (2001)
12. Croux, C., Dehon, C.: Influence functions of the Spearman and Kendall correlation measures. Stat. Methods Appt. **19**, 497–515 (2010). https://doi.org/10.1007/s10260-010-0142-z
13. Mayo, A.W.: Effects of temperature and pH on the kinetic growth of unialga *Chlorella vulgaris* cultures containing bacteria. Water Environ. Res. **69**(1), 64–72 (1997). https://doi.org/10.2175/106143097x125191

Characterization of People with Type II Diabetes Using Electrical Bioimpedance

Luis Carlos Rodríguez Timaná(✉) and Javier Ferney Castillo García(✉)

Grupo de Investigación en Electrónica Industrial y Ambiental – GIEIAM, Universidad Santiago de Cali, Cali, Colombia
{luis.rodriguez11,javier.castillo00}@usc.edu.co

Abstract. Diabetes is a disease that causes the death of a person every seven seconds around the world and is also expensive. In 2014, it invested 600 billion dollars to be treated worldwide, that is why the need arises develop technological projects that allow the analysis of specific patterns in people suffering from this disease, in order to detect the pathology in a non-invasive manner and reduce costs, for which an electrical bioelectrical impedance analyzer was developed for the analysis of diabetes. The integrated AD5933 was used as a bioimpedance signal acquisition device and a beaglebone black development platform to process said data. Bioelectrical impedance data were taken from 5 healthy people and 3 people with the pathology. The data were processed by mathematical methods such as linearization with least squares and correlation, which allowed us to find parameters to differentiate between the signals of people with diabetes and people without diabetes. It was determined that people with diabetes have a curve that relates their bioimpedance in a range of frequencies from 10 kHz to 80 kHz, a curve that presented a high correlation to a power function of the form *aXb*. It was observed that the people who presented values of coefficients (*a*) greater than 38000 and exponents (*b*) less than −0.659 were people with diabetes, this in turn allowed to find the equation of a line that separates the two populations $W0 + 1W * a + W2 * b = 0$.

Keywords: Electrical bioimpedance · Type II diabetes · AD5933

1 Introduction

Diabetes is a disease that without being treated properly can cause death. The World Health Organization estimated that in the world about 347 million people had diabetes in 2013 and it was also estimated that in 2012, 1.5 million people died from this disease, it is estimated that by the year 2030, diabetes will be the seventh cause of mortality [1]. Rodríguez and Plata 2015 explain that type II diabetes mellitus due to the complications it can have, due to its chronic nature and the means required for its treatment, is considered a highly expensive disease. Evidence obtained in recent years shows that an early diagnosis and good control of diabetes reduce the progress of chronic complications of the disease, such as retinopathy, nephropathy, neuropathy and death, and that, at the same time, improves the quality of life of these people [2].

D. F. Cortes Tobar et al. (Eds.): AETA 2019, LNEE 685, pp. 303–318, 2021.
https://doi.org/10.1007/978-3-030-53021-1_31

The newspaper El País, based on estimates made by the DANE, ensures that in Valle del Cauca there are 2,685,865 adults, between 20 and 60 years on average, which means that at least 112,806 Valle del Cauca in this same age range presents symptomatic symptoms associated with diabetes, which is considered a global pandemic related to the production of insulin in the human body. DANE estimates that for every 100 thousand inhabitants, 19 from this disease in the Valle del Cauca, according to the figures, the municipalities with the highest mortality from this disease are Andalusia, Buga, Yotoco, Pradera, Seville, Palmira, Guacarí, Caicednia, Roldanillo and Riofrío [3].

The search for methods that are quick, precise, easily accessible and inexpensive to analyze the probability that a person has diabetes is a latent need. In Colombia, between 7 and 8 percent of adults can have diabetes, which means that we would be talking about 7 to 8 people out of every 100, over 30 years old, suffering from it. These statistics correspond to type II diabetes, which, in more than a third of those who have it, tend to develop chronic complications of the kidneys (renal failure) due to lack of control of their blood glucose levels, explains Pablo Aschner Montoya, endocrinologist, clinical epidemiologist and deputy director of the Colombian Diabetes Association [4].

As with the previous statistics, there are many more, both for Colombia and for the rest of the world. That is why if there is a rapid, accurate and non-invasive method to determine whether or not a person has diabetes, the global statistics would change, since the disease could be controlled and treated in time, in addition to monitoring this would be much simpler and even less expensive. There are different methods to measure blood sugar levels, currently the most used method is invasive, and it is glucometry. This method uses a device called glucometer which submits a drop of blood from the patient to an enzyme called glucose oxidase and then apply a small voltage to this compound causing an electric current to be generated which is measured by the device that calculates and delivers the glucose value of the blood drop. This method can become very expensive if it is used for a long time since the glucose oxidase enzyme comes in strips which have only one use as well as the needles used to obtain the blood.

There are other methods such as photoplethysmography which is non-invasive. [5] used photoplethysmography as a technique to apply infrared light to the skin and detect sugar levels in it depending on the absorption rate of infrared radiation from the blood, since the greater the absorption, the greater the level of sugar, but the method has a problem and is that it requires a sensor for heart rate detection, which is expensive. [6] made an electronic device that allows to relate glucose levels with the impedance response of the human body. To do this, they used the electric bioimpedance method and the conventional glucometry method to find a relationship with the reading of these two devices, obtaining positive results, but requiring for their study, invasive tests.

Knowing then that this disease causes the death to people around the world, that it is expensive, in some cases difficult to access for treatment and that there are different researches and developments that try to study diabetes in a non-invasive way but they are expensive and of high complexity This article presents the development of a team which is an alternative method to study type II diabetes non-invasively, painlessly and at a low cost. The equipment is based on the study of the electrical bioimpedance of the human body. A pair of electrodes, an AD5933 bioimpedance analysis chip and a beagle bone black development platform are used to do the study and determine if a person presumably suffers from this pathology. This equipment does not require a calibration with the person's sugar levels.

The research project and the data obtained, are governed by an Ethics Committee of the Faculty of Engineering of the Santiago de Cali University.

2 Theoretical Framework

2.1 Diabetes

Guisán explains that diabetes is a disease that causes an alteration and a serious disorder of the organic metabolism. These metabolic alterations of diabetes are due to insufficient absolute or relative availability of the hormone called insulin, which is produced by the pancreas. Insulin is responsible for regulating the amount of glucose in the blood. Within diabetes, there are several fundamental groups or types of diabetes, the most common are diabetes mellitus type 1 and diabetes mellitus types II. The portal of the Ministry of Health explains that diabetes mellitus type 1, can affect people of any age, but is more in children and youth, usually diagnosed before 30 or 40 years of age, affects more to men in the first half of life and to women in the second. People who suffer from this disease, their pancreas, is unable to produce enough insulin, so they need insulin injections to control blood glucose levels and thus be able to live. Basically, your pancreas does not produce enough insulin. On the other hand, type 2 diabetes mellitus affects between 90–95% of people who have diabetes. It is the most common diabetes in adults and the elderly, although there have been cases in which this type of diabetes is diagnosed in children and young people, but the most common age of onset occurs after 30 or 40 years. The most serious aspect of this disease is that it is diagnosed after many years of suffering, due to personal carelessness, not showing symptoms or even lack of good medical control [7].

Diabetes mellitus type 2 is characterized because although your pancreas produces insulin, sometimes it does not produce it in the right amounts (sometimes less or sometimes more) and the body is not able to adequately administer this insulin. Inheritance is a basic factor in the transmission of diabetes, age, sex and dietary transgressions have a very limited pathogenic value. The hereditary factor has a great importance. Obesity is one of the most notorious predisposing factors. Although not all obese people suffer from or become diabetic, statistics clearly show that excess body weight favors the onset of diabetes. Another factor also closely related to diabetes is nutrition and this is clearly influenced by social position. Figure 1 shows the symptoms that type II diabetes can produce.

According to the IDA (International Diabetes Federation) (2015), in 2013 Colombia was the second country with the highest number of diabetics in South and Central America. The numbers leave no doubt about the concern generated by this problem. And is that in this region (does not include Mexico) there are 24.1 million adults diagnosed with type II diabetes. In 2013, 2.1 million people were diagnosed, this makes Colombia the second country with the most cases of type II diabetics, after Brazil. By 2035, the number of 24.1 million diabetics in Central and South America is expected to increase by almost 60% to 38.5 million of type II diabetes [8].

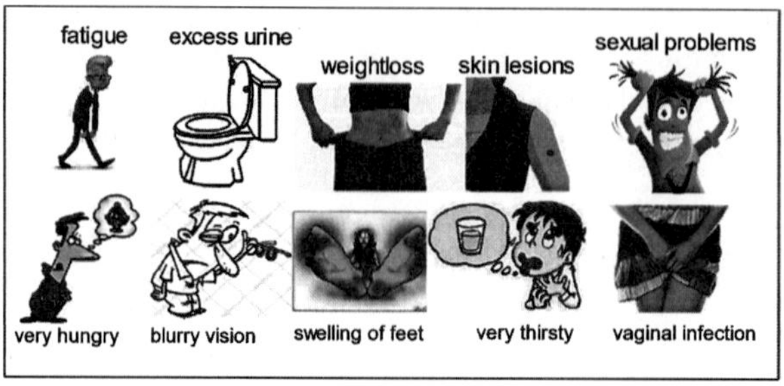

Fig. 1. Symptoms that some patients with type II diabetes may have. Source: Own.

More recent studies of the International Federation of Diabetes (2015), show that in 2015 in Colombia, there were more than 3 million cases of diabetes. In that same year, the international federation, announced the ages of the Colombian people who were suffering from diabetes at the time. These prevalence data are presented in Fig. 2, where the dotted line refers to the distribution of the prevalence of diabetes by age across the world, the red line refers to the distribution in Colombia and the black line refers to the distribution in Central and South America (SACA). According to the North American Diabetes Association, the non-early detection of type II diabetes, in the United States, produces 40% of cases of kidney failure, is the leading cause of new cases of blindness in adults and produces 20% of the cases. diabetic retinopathy and 10% proteinuria [8].

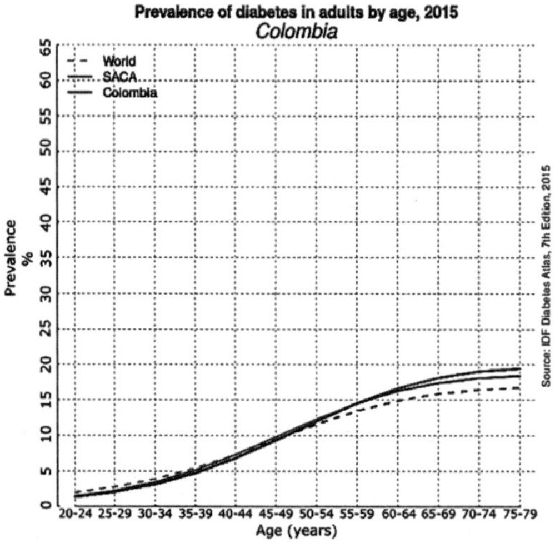

Fig. 2. Prevalence of diabetes in adults for the year 2015. Source: International Diabetes Federation.

The International Federation of Diabetes (2015), says that about 4.6 million people between 20 and 79 years died of diabetes in 2011, accounting for 8.2% of the worldwide mortality from any cause in people of this age group. This number of deaths is similar in magnitude to the sum of deaths due to several infectious diseases that are among the main priorities of public health and is equivalent to one death every seven seconds. 48% of deaths of diabetic origin occur in people under 60 years.

2.2 Electric Bioimpedance

Bioimpedance is a non-invasive, painless and effective method that is based on applying a small current through the body to measure the ability of the body to transport a quantity of electrical energy, which determines the impedance of the body; This is called electrical bioimpedance. In simpler words, a small current is applied to a human body at a strategic point to be measured and at another extreme, the voltage is measured, to calculate the impedance.

Figure 3 is the representation of the electrical impedance response at different frequencies to a person. The dotted curve is the representation closest to the impedance curve. This representation is given by a potential curve. The mathematical model of the potential function is presented in Eq. 1.

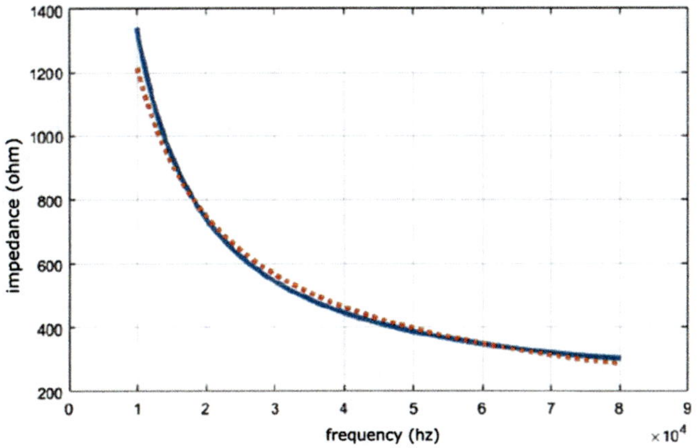

Fig. 3. Impedance curve of a person. Source: Own.

The representation obtained in Fig. 3 is a response characteristic of the model referred to the human body when applying an electric current to different frequencies. By varying the frequency, that said response varies, which is the impedance of the body. To give a relationship to human tissue with an electrical circuit, the Cole-Cole circuit appears. This circuit allows to understand why at different frequencies, different impedance responses are obtained. In Fig. 4, the electrical circuit that simulates human tissue is observed.

From Fig. 3, the solid line represents the impedance curve and the dotted line represents the trend of the impedance curve.

$$Y = aX^b \quad (1)$$

where a is the coefficient and b is the exponent.

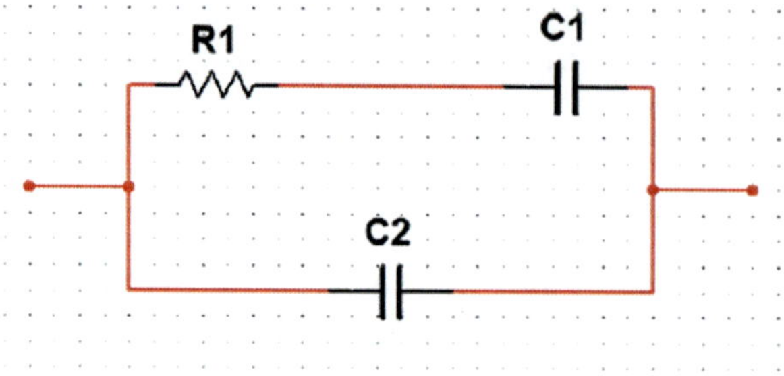

Fig. 4. A Cole-Cole circuit. Source: Own.

For low frequencies, the loading and unloading process is slow enough for the two capacitors to be charged, the total capacity would be C1 + C2. For high frequencies, capacitor C1 can not be charged and discharged at the rate of variation of the applied voltage, it is as if it were not, thus being the equivalent capacitance C2.

There are many applications in which bioimpedance is involved, but the most common has to do with determining the body composition of the human body, this study allows to establish the percentage of fat mass, fat-free mass and excess or lack of water in our body. In the case of body composition, there are many devices that allow these analyzes to be made. There are very few studies that exist on the application that is given to bioimpedance in diabetes, either for characterization or monitoring. The study of body composition helps in the treatment of chronic kidney diseases and, to know how much a patient can absorb x-rays. That is why bioimpedance can have many applications, everything depends on the study to which it is submitted [9].

2.3 Electrical Safety in Patients

The tissues of the human body allow the conduction of electricity between two points of the body that come into contact with an electric field, presenting an electrical resistance that varies according to various factors (example, contact area, skin moisture, frequency of the current, physical characteristics of each person, this implies that electrical currents will be produced between the two contact points. The passage of current through the body can cause damage if its magnitude is large enough (or long-lasting) to stimulate The nervous system or large muscle mass Two types of electric shock are identified The macro shock occurs when the current flows through a wide area of the skin, passing through the heart when passing from one part of the body to another, the micro shock occurs when the current It flows through a small area of the skin and there are electrodes or catheters directly connected to the heart, allowing the circulation of I take you for the myocardium [10].

In Table 1, we observe the currents that are allowed by the ICONTEC standard to be applied in people. It should be explained that the leakage current to ground, is the current that flows from the power supply through the insulation to the protective conductor of ground or pole to ground as it is also called. The leakage current of the enclosure is the current that flows from one part of the enclosure to ground, to another part of the enclosure (there is no circulation by the patient). The patient leakage current is the current that flows from the applicable parts to ground through the patient and the auxiliary current to the patient is the one that flows through it, but this current does not leave the elements of the applicable part.

Table 1. Example of the maximum values of currents allowed by the IEC 60601 standard for type B and BF equipment in direct current.

Measured current	Normal condition (mA)	Condition of first defect (mA)
Electrical current from leak to ground	0.5	1
Leakage current of the envelope	0.1	0.5
Electric leakage current of patient	0.01	0.05
Auxiliary electric current of patient	0.01	0.04

3 Materials and Methods

3.1 Materials

Integrated AD5933. The AD5933 is a high-precision impedance analyzer that combines a frequency generator with a 12-bit analog-to-digital converter (ADC) and 1 MSPS. It can generate up to 100 kHz frequencies and with an impedance measuring range of 100 Ω up to 10 MΩ. The received signal is sampled by the ADC and processed by an integrated digital signal processor that performs the Fourier transform. Through I2C communication, the integrated returns the real (R) and imaginary (I) value of the signal already processed [11]. This device was used to generate the sweep of frequencies and obtain the bioimpedance signal of each person. This integrated needs a calibration stage, this is explained in the methodology section.

Beagle Bone Black. It is a free hardware card that has a 1 GHz ARM Cortex-A8 CPU, 3D graphic accelerator, a pair of 32-bit SS RISC PRU, 512 MB of RAM and 2 GB of internal storage, plus a microSD slot. It has a USB connection, Ethernet, micro-HDMI output and two 46-pin connectors. This device was used to process the impedance data that was delivered by the AD5933 [12].

Other Devices. General purpose electrodes were used to make the interface between the integrated AD5933 and the human body. These are ideal, since they have a good adherence to the skin, have a low probability of generating infections or similar problems, are economical and have a low impedance which allows obtaining the bioimpedance signals with little noise.

3.2 Methods

The following mathematical methods were used to characterize the bioimpedance signals obtained from people.

Linearization Using the least Squares Method. Linearization is a numerical analysis technique in which we try to find the equation of the line that best represents a set of data. The equation to which we want to apply this method is to the potential equation shown in Eq. 1. The best-known method to find the values of (a) and (b) is called least squares linearization, which consists of submitting the system at different values of (X) taking the outgoing values corresponding to the dependent variable (Y), thus allowing obtaining the values of (a) and (b) to complete the equation [13]. The least squares can be used to linearize any type of function, so that the power function is linear, the following steps must be performed:

Natural logarithm must be applied to Eq. 1, the result of this is observed in Eq. 2.

$$\ln(Y) = \ln(a) + b\ln(X) \tag{2}$$

The parameters of both equations are equalized, getting so:

$$Y = \ln\ Y; \tag{3}$$

$$a = \ln\ a; \tag{4}$$

$$b = b; \tag{5}$$

$$X = \ln\ X \tag{6}$$

Finally, solving the following system of equations, we can find the values of a and b, which are necessary to find the equation that will allow linearizing the signal obtained from bioimpedance and thus be able to find the best characteristics of it. This system of equations is presented in Eq. 7 and 8.

$$\sum \log Y_i = n\ \ln\ a + b \sum \ln\ X_i \tag{7}$$

$$\sum \ln\ X_i * \log Y_i = \ln\ a \sum \ln\ X_i + b \sum (\ln X_i)^2 \tag{8}$$

Correlation. The correlation allows finding the maximum relationship between strength and direction between two statistical variables. To do this you must start by applying Eq. 9, where you can find the error e that there may be when linearizing the equation.

$$\sum\nolimits_{i=1}^{n} e_i^2 = \sum\nolimits_{i=1}^{n} (y_i - \widehat{y_i})^2 = \sum\nolimits_{i=1}^{n} (y_i - f(x_i))^2, \tag{9}$$

where x_i is the dependent variable, and y_i is the independent variable.

The sum of the squares of the divided residues on the number of samples (n) is the variance of variables also known as residual variance, in Eq. 10 and 11 it can be understood in a better way.

$$\frac{\sum_i^n e_i^2}{n} = \frac{\sum_i^n (e^i - \bar{e})}{n} = S_e^2 \tag{10}$$

$$e_i = y_i - \widehat{y_i} \Rightarrow y_i = \widehat{y_i} + e_i \Rightarrow S_y^2 = S_{\hat{y}}^2 + S_e^2, \tag{11}$$

where S_e^2 is the variance of the error, S_y^2 is the variance of the output y_i.

The main problem is to find the values where the residual variance defines whether it is a good or bad adjustment, which generates a question where the correlation coefficient R^2 solves the above. This is explained in Eq. 12.

$$R^2 = \frac{S_{\hat{y}}^2}{S_y^2} = 1 - \frac{S_e^2}{S_Y^2} \tag{12}$$

With the above equations we can find the correlation between two functions [11], this data allows us to understand how good was the signal taken to the patient and thus avoid signals of noise.

LDA Linear Discriminant Analysis. This method allows the exploration of learning machines, pattern recognition and statistical analysis to find the maximum separation between two or more classes of objects, by searching for linear combinations of characteristics. Equation 13 shows the function of a linear discriminant, a vector of input characteristics $X = (x_1, \ldots, x_d)^T$ where an output $Y(x)$ is generated for each x of an *xth* class.

$$Y(x) = w^T.x + w_0 = \sum w_i.x_i + w_0, \tag{13}$$

where $w = (w, \ldots, w_d)^T$ is the vector of weights and w0 is the off-set also known as the threshold.

The appropriate value for vector w is calculated by maximizing Fisher's separability mean. When you have two classes J is defined by Eq. 14.

$$J(\varpi) = \frac{(\mu_{y1} - \mu_{y2})^2}{P_1\sigma_{Y1}^2 + P_2\sigma_{Y2}^2}, \tag{14}$$

where μ_{y1} and μ_{y2} is the mean for each class, and σ_{Y1}^2 and σ_{Y2}^2 is the variance for each class, P_1 and P_2 refer to the a priori probability of each class.

In Eq. 15 the maximum argument of Fisher's separability is shown.

$$w^* = \arg\, max_w J(w) = \left(P_1 \textstyle\sum_{X_1} + P_2 \sum_{X_2}\right)^{-1} (\mu_{X1} - \mu_{X2}) \tag{15}$$

where μ_{X1} and μ_{X2} is the average for classes C1 and C2. In the future a new entry x is assigned to a class C1 or C2.

The decision of $y = 0$, is the lumbral where the error can be generated by not being able to decide which class belongs as shown in Eq. 16 [14].

$$X \in \begin{cases} C1 & , if \ y > 0 \\ C2 & , if \ y < 0 \\ limite\ de\ desici\acute{o}n & , if \ y = 0 \end{cases} \tag{16}$$

4 Methodology

Figure 5 presents the flow diagram of the algorithm implemented for bioimpedance equipment for characterization of people with type II diabetes.

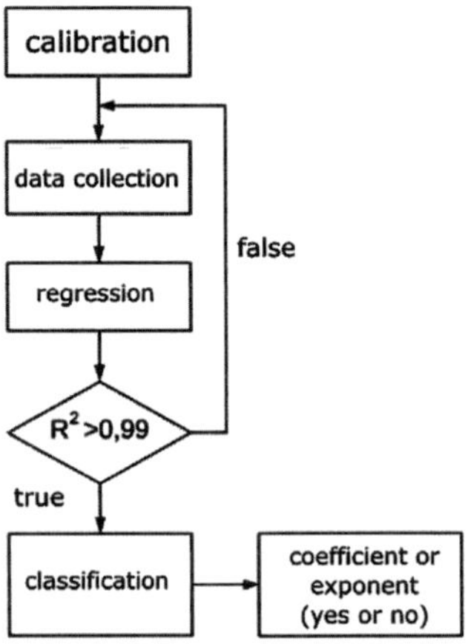

Fig. 5. Flow diagram of bioimpedance equipment. Source: Own.

At the time of data collection, there were different inclusion and exclusion criteria, this to avoid factors that alter blood glucose, such as alcohol consumption constantly or women in gestation period or in their menstrual cycle. Considering this, the ages of the people who would serve in the project were also defined. Tests were conducted on people aged 30–75 years (age ranges of people with type II diabetes) to organize and classify the characteristics of each population group.

The integrated AD5933 requires a calibration stage before the bioimpedance tap can be made. For this purpose, a resistor must be placed on the V_{ref} and V_{out} pins of the integrated one, which allows having a reference of the magnitude of the impedance to be measured. This same resistance must be placed in the output of the integrated, being this removed at the time of making the bioimpedance. For this, different resistance values were used, concluding that the best representation of the received signal was obtained with a resistance of 510 Ω with a tolerance of 1%.

Having the module properly calibrated, it is passed to the **data collection** where a frequency sweep is made that starts at 10 KHz up to 80 KHz, making an increment of every 250 Hz, where as a result the different values are obtained by frequency. This frequency range was determined because at low frequencies the current signal applied to the body travels through the membranes of the cells, without these being traversed and at high frequencies, the signal cuts the membrane, thus allowing to cross the cell.

In the **regression** stage, a mathematical model was obtained that allowed us to determine which function resembled the frequency versus magnitude graph of the bioimpedance. As a result, the power function was the one with the highest correlation. If the correlation is greater than 0.99, it indicates that the input data is valid and can continue to the classification stage.

In the **classification** stage, linear discriminate analysis LDA (Linear Discriminant Analysis) was used to generate a mathematical model that will use the values of coefficient and exponent of the linearized bioimpedance signals, to perform the training process and thus be able to classify the two populations and determine which person has type II diabetes or not. With the model already trained, a sample can be made to a patient and entered into the system to validate in which class it is.

With the card and electrodes ready, the first data collection was performed on a patient diagnosed medically with type II diabetes. For this, the following protocol was carried out:

Protocol for Testing. After identifying the inclusion/exclusion criteria, the following recommendations must be fulfilled:

- You must not have any metallic object on your body.
- You must not have a period of more than 5 h without having consumed any food.
- You should not consume foods with a high percentage of sugars (chocolates).
- You must not have any electronic device with you.
- You must not have done physical activity in the 2 h prior to the test.
- Not having presented any symptoms of excessive fluid loss.
- Not having ingested drinks with a high alcohol content.

Procedure for the Test. Is the next:

- The patient must be seated with the palms of their hands facing upwards and placed on their legs.
- Clean the area of the forearm of the left hand, to remove dead skin cells to improve the contact interface and ensure good adherence for the placement of the electrodes.

Fulfilling the above requirements, we proceed to take the first shot of bioimpedance to a person, the process takes 3 min. After that the glucometry test is performed on the person to have one more parameter of analysis and study. Equipment under normal operating conditions does not require glucose measurement.

In Fig. 6, the general scheme of the constructed equipment is observed. The computer is optional since the bioimpedance team is the one that performs the processing of the impedance data obtained, but it also has the possibility of such data being sent by I2C communication so that another team can process or analyze them in a different way.

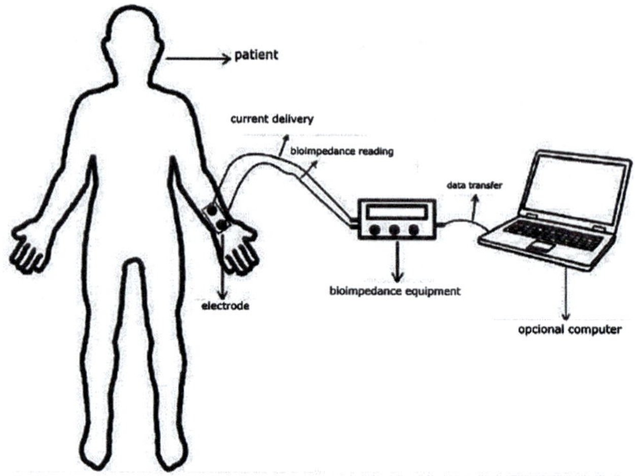

Fig. 6. General scheme of the electrical bioimpedance equipment. Source: Own.

5 Results and Discussion

Figure 7 shows the curve of the amplitude of the bioimpedance of a healthy person and dotted the bioimpedance curve of a person with type II diabetes.

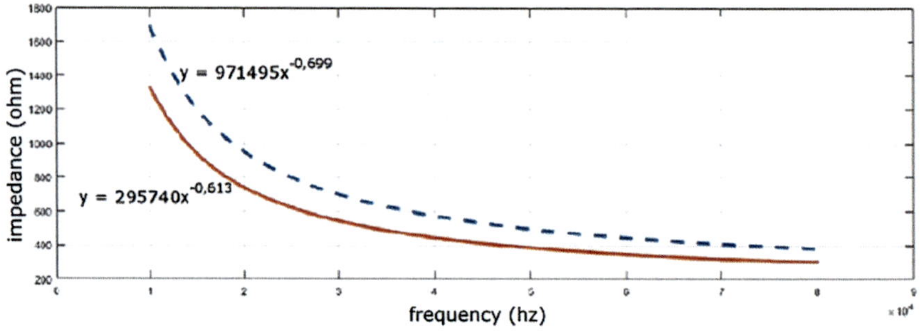

Fig. 7. Representation of populations of diabetic people and healthy people. Source: Own.

In the curve of the person diagnosed with type II diabetes, it is evident that the impedance value is greater than that of the healthy person since his impedance response starts from 1.7 KΩ and the healthy person starts from 1.35 KΩ. Both curves present in their mathematical representation a different exponent value and that is why the signals decay at different Ω/Hz ratio but as the frequency increases, these curves tend to join.

From Fig. 7, the solid line represents the impedance curve of a healthy person and the dotted line represents the impedance curve of a person with diabetes.

The physiological explanation of these results is given by the physical characteristics of healthy people and diabetics. People with type II diabetes have a greater number of glucose cells in their blood compared to healthy people. These cells are composed of a cellular membrane which, seen electrically, functions as a condenser. As the signal applied by the team is synodal, if the frequency is low, the signal will only travel through the intracellular and extracellular fluid, obtaining a resistive (real) response, but if the frequency is high, the signal will pass through the cell membrane and you will get a response of reactance impedance and resistance (imaginary more real). In Fig. 8A, it is observed a representation of how the current signal at low frequencies surrounds the cell membranes but never traverses them, whereas in Fig. 8B, the signal with high frequency traverses the cells completely.

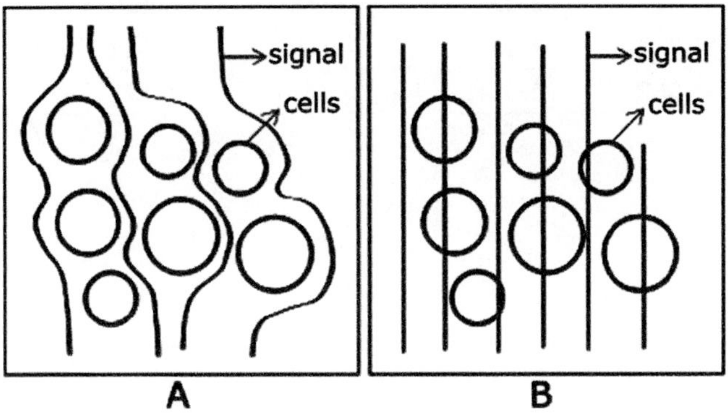

Fig. 8. Signal of current at low and high frequencies by cellular tissue. Source: Own.

That's why people who have type II diabetes have a different response to people who do not have it, since the same signal applied to the healthy must go through diabetic a greater amount of intracellular fluids, extracellular fluids and get different values of reactance due to the fact of traversing a greater number of cell membranes.

Table 2 shows a relationship of coefficients, exponents and correlation, obtained from the potential curve representative of the bioimpedance signals of some users. The people who present (*) are people with a medical diagnosis of type II diabetes. It is evident that their coefficient values are different to the users that do not have the pathology.

Graphing the values of coefficient and exponent of Table 2, evident differences of the populations of diabetic and healthy people are observed. In Fig. 9, said representation is shown. It is possible to say, that in the figure more shots are evidenced by each of the six people, this to have a better representation.

From Table 2, it was observed that a person who is not diagnosed medically as a diabetic (USER 5), presented values of coefficient and exponent corresponding to people who are diabetic. This user presents some of the symptoms seen in Fig. 1 and is currently in the medical process to determine if he has diabetes.

From Fig. 9, the tables represent the population of healthy people and the diamonds represent the population of people with diabetes. The line is the separation of $W0 + W(1) * x(1) + W(2) * x2 = 0$.

Table 2. Different parameters of the power functions representative of the bioimpedance curves of different people.

USUARIO	Coefficient (a)	Exponent (b)	Correlation (R^2)
USER 1 (*)	971495	−0,699	0,9947
USER 2 (*)	501394	−0,667	0,9907
USER 3	315195	−0,617	0,9915
USER 4	235367	−0,581	0,9859
USER 5	731528	−0,695	0,9925
USER 6	293641	−0,615	0,9879

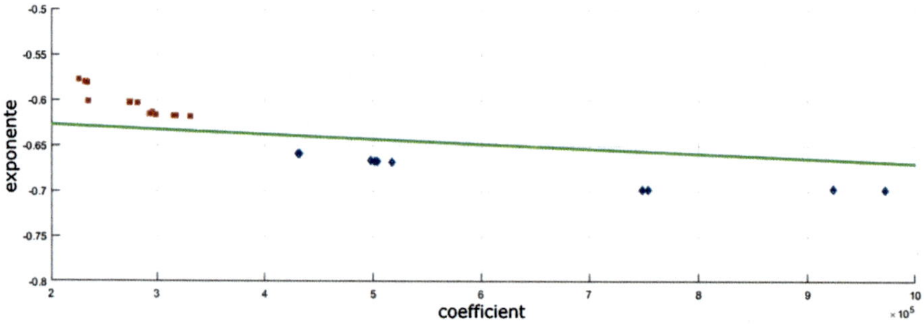

Fig. 9. Representation of populations of diabetic people and healthy people. Source: Own.

6 Conclusions

The electrical bioimpedance team using mathematical methods such as least squares and correlation allows to find patterns with which a person with type II diabetes can be characterized. Unlike the conventional invasive method, this equipment uses a technique that prevents the person to be punctured to obtain a drop of blood, thus preventing pain and possible infections. It is an alternative method that without being costly unlike other developments, can help to study this pathology and contribute to the study of it.

With the team It was determined that people who have diabetes have a curve that relates their bioimpedance with respect to different frequencies, a curve that showed a similarity to a potential function, it was evaluated that the people who had a higher coefficient of 38,000 and exponent less than −0,659 were people with diabetes.

It is important to know in studies like this, the physiological part of the human body since this research was conducted in order to find different characteristics that would allow differentiating using a non-invasive method a person with type II diabetes and a healthy person. Since type II diabetes is the one that generates many insulin cells in the body but is unable to absorb glucose, then a different amount of these cells and sugars will be found in the body of a diabetic person in the body of a healthy person. which allows the applied current to travel different paths and can be diversified between the two study populations.

This device requires at least 63 target and 63 control people, all this having a statistical power of at least 80% and a level of confidence of 95% to think later about being able to validate the system, but with the people studied and the given classification to perform the study, the desired differences were found.

With the results obtained, the use of this equipment can be projected to follow the treatments or to demonstrate the evolution of this pathology. As future work the study is projected to perform a continuous glucose measurement equipment - non-invasive using electric bioimpedance.

References

1. Alvero Cruz, J.R., Correas Gomez, L., Ronconi, M., Fernandez Vazquez, R., Porta, J.: La bioimpedancia electrica como metodo de estimacion de la composicion corporal: normas prácticas de utilizacion. Rev. Andal. Med. Deporte **4**(4), 167–174 (2011). https://doi.org/10.1016/j.ramd.2015.05.004
2. Analog Devices: AD5933: 1 MSPS, 12-Bit Impedance Converter, Network Analyzer. Datasheet, 40 (2013). http://www.analog.com/en/rfif-components/direct-digital-synthesis-dds/ad5933/products/product.html
3. Augusto, C., Estupiñan, S., Helena, L., Chaparro, F.: De Dispositivos Médicos (n.d.)
4. Coley, G.: Beaglbone Black System Reference Manual (2014)
5. Correlación, T.R.Y.: Tema 4: regresión y correlación, vol. 4, no. 1, pp. 1–15 (n.d.)
6. Dispositivos, S.D.E., Edici, P.: Guía para clasificación de dispositivos médicos según riesgo (2016)
7. El, M., La, X.: Monitorización 3 M (n.d.)
8. General, H., Gonz, M.G.: La diabetes mellitus y su detección temprana, vol. 5, pp. 5–7 (2002)
9. Guisán, S.: Gran enciclopedia de la ciencia y de la técnica. In: Océano, 4th edn., pp. 740–742 (1998)
10. Gómez-Cadenas, J.J.: Ajuste por Mínimos Cuadrados, 15 (2005). http://benasque.org/benasque/2005tae/2005tae-talks/232s5.pdf
11. Kamat, D.K., Bagul, D., Patil, P.M.: Blood glucose measurement using bioimpedance technique. Adv. in Electron. **2014**, 1–5 (2014). https://doi.org/10.1155/2014/406257
12. Lopez Lopez, C., et al.: Medicina del Deporte **3**(1), 170–178 (2011). https://doi.org/10.1016/S1888-7546(14)70058-9
13. Masoomia, R.: Enhancing LDA-based Discrimination of left and right hand motor imagery: outperforming the winner of BCI competition 2, p. 395. IEEE (2015)
14. Ministerio de Salud y Protección Social: Analisis de la Situacion de Salud (ASIS) Colombia 2015. Instituto Nacional de Salud, p. 175 (2015)
15. Ortega, M.: Diabetes, un dulce enemigo de la vida moderna (2012). http://www.portafolio.co/tendencias/diabetes-dulce-enemigo-vida-moderna-93734
16. Pais, E.: Más de 112 mil personas en el Valle del Cauca sufren de diabetes (2013)
17. Paul, B., Manuel, M.P., Alex, Z.C.: Design and development of non invasive glucose measurement system. In: 2012 1st International Symposium on Physics and Technology of Sensors (ISPTS-1), pp. 43–46 (2012). https://doi.org/10.1109/ISPTS.2012.6260873
18. Rodríguez, L., Plata, G.: La calidad de vida percibida en pacientes diabéticos tipo 2. Revista Javeriana (2015). Revistas.javeriana.edu.co

19. Salazar Gómez, A.J., Cuervo Ramírez, D.K.: Protocolo de pruebas de seguridad eléctrica para equipos electromédicos: caso de estudio de equipos de telemedicina. Revista de Ingeniería, Universidad de Los Andes, pp. 27–32 (2013). https://doi.org/10.16924/riua.v0i38.87
20. Consejería de Salud: Portal de la Consejería de Salud (2012)
21. Día Mundial de la Salud: Diabetes (2016)
22. Yang, Y., Zhang, W., Sun, Q.: Design and preliminary test of a palm bio-impedance spectroscopy measurement system for biometric authentication. In: 2014 International Symposium on Computer, Consumer and Control, pp. 824–827, February 2014. https://doi.org/10.1109/IS3C.2014.218

Embedded System for Electrical Load Characterization Based on Artificial Neuronal Networks in the Management of Electrical Demand in a Domotic System

Kevin Andrés Suaza Cano(✉), Ángel Stiven Angulo Gamboa, and Javier Ferney Castillo Garcia

Grupo de Investigación en Electrónica Industrial y Ambiental-GIEIAM, Universidad Santiago de Cali, Cali, Colombia
{kevin.suaza00,angel.angulo00,javier.castillo00}@usc.edu.co

Abstract. A load characterization system was developed in an embedded platform, in order to identify electrical devices used in the home. For the characterization process, the most representative electrical parameters of the different loads were defined, which were used in the training of an artificial neural network implemented in an embedded platform with a network topology with the best performance in terms of computational resources, time of execution and percentage of error. The network topologic had two hidden layers each one with 10 neurons. With the characterization of electrical charges, an intelligent home automation system could be created which can generate savings of up to 23% compared to traditional home automation systems or 69% savings compared to a home without any automation or system control. The proposed demand management system can actively manage the loads due to the knowledge of the elements connected to the network, identifying periods of low consumption which can be related to charging processes completed in mobile phones, laptops or standby mode for televisions. The identification of charges facilitates the implementation of management schemes and control of electric charges.

Keywords: Home automation · Characterization of electric charges · Artificial Neural Network · Learning machines · Embedded system

1 Introduction

Currently the energy consumption in home is high, this consequence of a very low energy efficiency, which translates into a high impact on the environment, this makes necessary to have a greater degree of control over the electrical devices used in the home. In the beginning, domotics was restricted to being a tool at the hand of the user, providing the ability to make some adjustments on the house like turning lights on or off, or setting timers, however, the technology has advanced very quickly and domotics has been assigned more complex tasks over the years.

D. F. Cortes Tobar et al. (Eds.): AETA 2019, LNEE 685, pp. 319–327, 2021.
https://doi.org/10.1007/978-3-030-53021-1_32

In the current market you can find different advances in terms of safety, energy and comfort, like intelligent cameras that brings the opportunity of access their images in real time from anywhere in the world, another of the great aspects in which domotics has evolved in terms of security, is being able to control the doors, windows and blinds of the home through an Internet connection [1, 2]. One of the most important features of home automation and one of its main characteristics is the ability to control different appliances in the home that directly influence energy consumption, in areas such as air conditioning and lighting, all this in order to generate energy savings and have a better use of the resources [3, 4]. One of the major trends in terms of home automation is the use of personal assistants such as Google Home, or Amazon Alexa, this type of devices have the ability to recognize voice commands from the user and receive direct commands from it, although one of its main features is the ability to control different devices on home [5, 6].

The objective of this paper is present the implementation of an artificial neural network in an embedded system for the characterization of electrical loads for the management of electrical demand in a domotic system, where one of its main characteristics is the versatility in the implementation of different network topologies and the definition of the electrical parameters that make it possible to model or characterize the most common electrical loads in a home.

2 State of the Art

Studies related to domotic systems based on demand management or load connection/disconnection control systems are presented. Widely used systems share similar characteristics such as hardware definition for communication protocols, load control system (contactors or solid state), temporization for the connection/disconnection of the different electrical devices.

In the first instance [7] addresses the problem of excessive consumption of electrical resources in the home and raises the need to implement automation systems on a massive scale.

The most basic aspect of home automation is the ability to control one or more aspects of the home, such as turning on/off lights or devices connected to the electrical network, [8] and to be able to provide information on basic aspects such as temperature, humidity [9] and in some cases more detailed information such as light intensity, motion detection and magnetic fields to detect the opening of doors or windows [10]. In the research work of [11] the implementation of a traditional domotics system using KNX technology is carried out in a 67 m^2 house. This system, although reliable, becomes obsolete when compared to technological proposals such as the one carried out by [12]. In its research work, the implementation of a decentralized on/off system is carried out, using ESP8266 modules connected through a wireless network, which allows a greater level of flexibility and ease of implementation, all this together with a lower total cost of the system.

One of the areas where domotics has greater capacity for evolution, is in automatic decision making, providing the domotics the ability to decide which electrical devices are connected to the electrical grid, allows a more precise control over the electricity consumption of the home. Different approaches to this idea have been made in the literature and proposals have been found that have the capacity to generate electricity savings, [13] in this work, he carries out a system of scheduling of electrical charges, with which he manages to make the appropriate use of the photovoltaic resource that a home has, making use to a lesser extent of the home electrical network, however, a variety of works have been found in which the objective is to carry out a system that can classify electrical charges on the basis of their voltage and current signals, [14] makes an approach to this idea, making use of Artificial Neural Networks achieves the identification of fundamental characteristics in a radio signal, for its part, [15] develop a system with the ability to correctly identify a cooler using an ANN trained solely from the power data of the device, [16] makes a similar system, however in its development are used the general consumption data of the house and with the use of the transformed Hilbert-Huang achieves a correct characterization of elements with high energy consumption such as the washing machine, clothes dryer and refrigerator.

The proposed home automation system will characterize and identify electrical loads, which has the ability to identify an appliance and its mode of operation (full load or low consumption). The characterization uses electrical parameters that model the different loads from their characteristics in power, harmonic distortion among others. The demand management uses the information of the appliances connected to the network and establishes their mode of operation, those in low consumption are disconnected after their identification. The ecosystem is made up of outlet modules, switch modules and a master module where the actions that depend on the identified appliance are defined. The activation of the loads will be carried out with solid state power electronic devices.

3 Materials and Methods/Methodology

3.1 Embedded System

An embedded system was development, which allowed the acquisition of voltage and current data of each one of the most common household appliances, and then these data were used to extract the electrical parameters that allowed differences between the several appliances and their states of consumption.

Hardware. The selection of the platform carries out to meet the needs of the development phase and that could be used as a final product. The requirements were raised, an embedded system that can have readings on voltage and alternating current (AC), that counts on different means of wireless communication, for its later implementation in a domotic system, that count with great capacities of memory and a high power of compute for the implementation of the system of characterization. Having as main objective a low-cost platform, easy in the implementation of a wireless communication network with a high level of processing and sufficient memory capacity. The development platform selected was the ESP32, given that its low price, has a powerful dual core processor Xtensa LX6 32 Bits at 240 MHz, 4 MB flash memory for the program and 540 KB of integrated RAM, WI-FI b/g/n and Bluetooth 4.2 Low Energy integrated and the ability to perform analog voltage reading with a resolution of 12 Bits. Other platforms with similar characteristics have the disadvantage of being more expensive, making it impossible to develop a low-cost system.

In the data acquisition phase, voltage and electrical current samples were taken from the most common household appliances in the home. To perform this task, circuits were implemented to couple these physical magnitudes to the working ranges of the ESP32 platform, and a storage system was implemented through an external SD memory connected to one of the SPI modules available on the development platform.

The last requirement of the embedded system, was to provide it with the ability to control on and off on the load that was connected to it, for this was designed and implemented a control system, using a triac BTA20 semiconductor device.

Software. The first stage of the project consisted of collecting voltage and current data from the most commonly used appliances in the home, to fulfill this purpose the necessary hardware was developed and an algorithm was developed, which was intended to perform the storage of this information in an SD memory using parameters previously specified, these parameters are, have the ability to generate a text file for each appliance analyzed, the data was taken at a fixed frequency of 1 kHz.

Data Acquisition. The acquisition of voltage and current data of the most commonly used household appliances and their characteristics as model, manufacturers and type are presented in the Table 1.

Table 1. Average and standard deviation of the electrical features used.

Feature/device	D (W)	Fp	P (W)	Phi (°)	P_{Max} (W)	Q (VAR)	S (VA)	TDH	σ_n^2 (A^2)	I_{RMS} (A)
Cell	0,62 (0.05)	0,60 (0.05)	8,17 (3.38)	51,49 (4.33)	73,02 (21.9)	10,16 (3.43)	13,20 (4.76)	0,20 (0.12)	0,01 (0.01)	0,11 (0.04)
Iron	0,49 (0.44)	0,48 (0.45)	346,9 (362)	50,97 (37.7)	822,3 (748)	91,10 (118)	377,9 (362)	0,41 (0.49)	26,83 (28.1)	3,73 (3.59)
Computer	0,85 (0.05)	0,85 (0.02)	26,58 (2.80)	30,76 (4.77)	90,48 (14.0)	15,81 (2.16)	31,14 (2.51)	0,05 (0.13)	0,06 (0.01)	0,25 (0.02)
Blender	0,42 (0.04)	0,41 (0.05)	61,36 (10.9)	64,94 (3.03)	410,8 (150)	132,1 (6.03)	146,2 (9.39)	0,15 (0.11)	1,95 (0.23)	1,39 (0.08)
Laptop	0,63 (0.05)	0,61 (0.05)	30,59 (4.38)	50,81 (3.99)	261,2 (20.9)	37,94 (1.34)	49,22 (3.10)	0,18 (0.07)	0,19 (0.05)	0,44 (0.05)
TV	0,76 (0.20)	0,75 (0.21)	39,26 (20.4)	34,28 (21.7)	171,5 (18.4)	21,07 (7.72)	47,97 (13.3)	0,12 (0.14)	0,20 (0.10)	0,43 (0.12)
Low consumption	0,19 (0.10)	0,16 (0.09)	3,90 (12.5)	78,88 (6.99)	54,71 (62.8)	10,17 (21.7)	11,12 (25.1)	0,53 (0.67)	0,06 (0.25)	0,10 (0.23)

3.2 Electrical Characteristics

Taking as information the voltage and current data collected from household electrical devices, a correct differentiation must be made between the different types of devices recorded, but just analyzing this information is not possible to give a correct analysis, it is necessary to have electrical features from these signals that provide information, allowing the classification of different household devices. The following features were used: Instantaneous Power (P), Maximum Power (P_{Max}), Distorted Power (D), Power Factor (Fp), Reactive Power (Q), Apparent Power (S), Offset Angle (ϕ), Total Harmonic Distortion (TDH), Current Variance (σ_n^2), Root Mean Square (I_{RMS}) [17].

3.3 Learning Machines

The learning machines are computational techniques used to develop prediction algorithms based on data samples, there are different classification techniques which are divided into two classes, unsupervised learning machines and supervised learning machines [18].

In the training process of all classification methods used, (Cross-Validation) was performed, defined by [19] As the process of making a random separation of the input data and their corresponding outputs to make use of part of the data for system training and with the remaining data to perform its validation. In the implementation three types of learning machines were analyzed, which will present ease in their development in an embedded system, the types of machines used were: Neural networks, decision trees and K nearest neighbors.

An artificial Neural Network (ANN) is a mathematical abstraction of the process of communication and information generation of physical neural networks. One type of implementation of artificial neural networks widely used are the multilayer perceptron MLP (Multi-Layer Perceptron), this are an evolution of the single-layer neural network, its operation is based on the use of a series of hidden layers, which is ideal for solving non-linear problems. [20] a MLP is formed by at least 3 layers, the first is the input layer, this is where the data is received to be used by the ANN, the second layer is part of the group of hidden layers being necessary to have at least one layer of that type and finally has the output layer which delivers the data generated by the ANN [21]. Equation 1 is a mathematical representation of a neural network, where x are the inputs to the neuron, w is the matrix of synaptic weights and b is the value of the tracks [22].

$$y = f(x \cdot w + b) \tag{1}$$

3.4 Implementation of the Load Characterization System on the Embedded Platform

The implementation of the load characterization system consists of providing the embedded system with the ability to take voltage and current data to perform the electrical feature extraction, making use of these electrical features in the process of generate a result through the use of the ANN previously trained, the training is done on a desktop computer. When the training process ends, the information of the synaptic weight matrix, bias value, network topology and characteristic normalization values are transferred.

To execute the neural network, a series of for cycles are used to realize the calculation of the output of each one of the neurons, in total 3 for cycles was used, the first one to go through all the layers, the second one used to go through all the neurons on a layer and the last one to do the calculations for all the inputs on a neuron. The output of each one of the neurons is evaluated on the activation function.

4 Results and Discussion

The Table 1 shows the average values of each of the features of all household appliances, the value in brackets corresponds to the standard deviation of the value of each feature.

A Decision Tree classifier was trained because this algorithm has a low computational cost. It was implemented using conditionals and basic mathematical operations. Cross validation results in an error rate of 23.4%. The decision tree is implemented in the embedded system and its operation is verified, obtaining as a result, that the system does not have the capacity to carry out the correct characterization of the electrical appliances that present strong variations in their electrical features, due to the fact that this classifier uses the definition of thresholds, therefore it does not have the capacity to identify correctly elements that produce high levels of noise and cannot use features that have a high level of variation such as reactive power and current variance.

A k-Neighbor closer KNN classifier was trained. This classifier is a much more robust system than the decision tree and had an average error rate of 1.1538% and a standard deviation of 0.7278%, however, its implementation requires 38 KB of volatile memory and very long execution times which makes it impossible to implement in an embedded system.

The ANN training process was carried out and a list of values of synaptic weights, biases, values for data normalization and activation functions was generated. Different network topologies were used, which implied different capacities in the size of bytes to use number of neurons to use, the Table 2 shows the results of the different trained topologies. The topology with two hidden layers, each with 10 neurons, provides an average error level of 2.051% and a standard deviation of 1.2385%, without greatly compromising system memory and execution times. An adjustment was made to the KNN classifier, approximating the value of volatile memory that consumes to the values of use of the artificial neural network, this in order to appreciate under equal conditions which classifier presents a higher performance, it was found that the KNN presents error levels higher than the artificial neural network, this information is shown in the Table 3.

Table 2. Tests performed on different neural network topologies. Source Authors.

Topology	Neurons	Average error	Size (bytes)	Standard deviation	Time (us)
10 4	14	3,71	836	3,27	215
10 8	18	7,05	1044	7,93	277
10 10	20	2,05	1148	1,23	310
10 4 4	18	13,20	936	28,53	266
10 4 8	22	7,17	1048	7,38	326

Table 3. Comparison between different classifiers.

Classifier	Average error	Standard deviation	Size (bytes)
Decision tree	23.4%	1.32	100
KNN	1.15%	0.72	34320
KNN (adjusted)	83.8%	18.65	1232
Neuronal network	2.0%	1.23	1148

The artificial neural network implemented in the embedded system made it possible to evaluate the signal establishment times of the connected appliances, the average times was 4.5 ± 2 s.

Table 4. Result of the simulation of monthly electricity consumption in a home without a home automation system, with a traditional control system and an intelligent system with load characterization.

First name	Consumption (KW/h)	STDBAY (KW/h)	Consumption in normal state KW/month (CO2-Kg)	Traditional home automation consumption KW/month (CO2-Kg)	Consumption load characterization KW/month (CO2-Kg)
Cellphone	0,037125	0,004	194,94 (72,1275)	114,75 (42,455)	53,964 (19,965)
Griddle	1,2	0	207,36 (76,72)	0	0
Computer	0,065	0,005	367,2 (135,86)	367,2 (135,86)	281,16 (104,03)
Blender	0,55	0	39,6 (14,65)	0	0
Portable	0,12	0,004	535,68 (198,2)	276,48 (102,295)	259,344 (95,955)
TV	0,1575	0,006	1412,64 (522,675)	732,24 (270,93)	286,632 (106,055)
Monitoring system	0,0007425	825×10^{-11}	0 (0)	0,5346 (0.2)	0,26730297 (0.1)

The demand management proposed for use in an intelligent home automation system, with the ability to recognize that appliances are connected to the system and to exercise control actions that ensure low electricity consumption, without making major effects on the user, requires a correct characterization and identification of the electrical loads connected to the system. To evaluate the proposed system, a simulation is carried out based on the time that the devices remain connected to the electrical network and what would be the consumption without load management actions what we call normal consumption, a load management based on traditional domotic systems and load management, where the characterization of the household appliances is used as a basis for intelligent demand management. The results of this simulation and the corresponding reduction in terms of carbon dioxide production are presented in Table 4. Simulated systems include the electrical consumption of demand monitoring and management devices.

5 Conclusions

The implementation of mathematical and statistical tools, such as learning machines for the process of characterization of electrical loads require an adequate definition and selection of technical and economic specifications for implementation in embedded systems or low-cost development systems.

An algorithm was implemented to adjust artificial neural networks of up to 50 neurons in a low-cost embedded system to classify an error rate of 2%, which enables the management of electric demand in a domotic system with the capacity to characterize loads with high levels of energy savings.

Traditional domotic systems base their operation on actions taken completely by the user without any type of autonomous decision making. The characterization of electrical loads offers the opportunity to develop intelligent domotic systems that execute advanced control actions on a house.

Acknowledgments. This research was supported by the Santiago de Cali University (Universidad Santiago de Cali). We thank our colleagues from the university who provided insight and expertise that greatly assisted the research.

References

1. Siswanto, A., Katuk, N., Ku-Mahamud, K.R.: Biometric fingerprint architecture for home security system. In: IACE (2016)
2. Taryudi, Adriano, D.B., Ciptoning Budi, W.A.: IoT-based integrated home security and monitoring system. IOP Conf. Ser.: J. Phys. **1140**, 012006 (2018)
3. Guerard, G., Levy, L.-N., Pousseur, H.: Multi-agent model for domotics and smart houses. Scitepress (2018)
4. Escobar Gallardo, E., Villazón, A.: Sistema de Monitoreo Energético y Control Domótico Basado en Tecnología "Internet De Las Cosas". Invest. Desarrollo **18**, 103–116 (2018)
5. Ciabattoni, L., Ferracuti, F., Ippoliti, G., Longhi, S., Turri, G.: IoT based indoor personal comfort levels monitoring. In: ICCE (2016)
6. Ammari, T., Kaye, J., Tsai, J.Y., Bentley, F.: Music, search, and IoT: how people (really) use voice assistants. ACM Trans. Comput.-Hum. Interact. **29**, 1–28 (2019)
7. Hassanpour, V., Rajabi, S., Shayan, Z., Hafezi, Z., Arefi, M.M.: Low-cost home automation using arduino and modbus protocol. In: ICCIA (2017)

8. Ovalles, F.O., Bolivar, A.E., Rodriguez, A.J.: Use of an embedded system with WiFi technology for domotic. IOP Publishing (2018)
9. Dobrescu, L.: Domotic embedded system. In: ECAI (2014)
10. Banu, B., Merlin, S., Suhitha, S.: Domotics using labview. Int. J. Res. **5**, 1–7 (2018)
11. Morón, C., Payán, A., García, A., Bosquet, F.: Domotics project housing block. In: MDPI (2016)
12. Garcia, V.H., Vega, N.: Low power sensor node applied to domotic. Springer (2018)
13. Moya Ch., F.D., Da Silva, L.C.P., Lopez A.J.C.: A mathematical model for the optimal scheduling of smart home electrical loads. WSEAS Trans. Power Syst. **13**, 300–310 (2018)
14. Toro Betancur, V.: Algoritmo de Estimación de Parámetros y Modulación de Una Señal Recibida Por Un SDR. EAFIT (2017)
15. Bouazzaoui Cherraqi, E., Maach, A.: Load signatures identification based on real power fluctuations. Springer (2018)
16. García Ortiz, M.: Adaptación y Aplicación de la Transformada de Hilbert-Huang a Sistemas Eléctricos: Aplicaciones en el Estudio de la Gestión de la Demanda y Caracterización de Transitorios. Springer (2016)
17. Bollen, M.H.J., Gu, I.Y.H.: Signal processing of power quality disturbances. IEEE Press Series on Power Engineering, Danvers (2006)
18. Witten, I.H., Eibe, F., Hall, M.A., Pal, C.J.: Data Mining: Practical Machine Learning Tools and Techniques. Elsevier, Cambridge (2017)
19. Zhang, Y., Yang, Y.: Cross-validation for selecting a model selection procedure. J. Econ. **187**, 95–112 (2015)
20. Chen, X., Liu, G., Shi, J., Xu, J., Xu, B.: Distilled binary neural network for monaural speech separation. In: IJCNN (2018)
21. Zhang, H., Wang, Z., Liu, D.: A comprehensive review of stability analysis of continuous-time recurrent neural networks. IEEE Trans. Neural Netw. Learn. Syst. **25**, 1229–1262 (2014)
22. Khosrojerdi, S., Vakili, M., Kalhor, K.: Thermal conductivity modeling of graphene nanoplatelets/deionized water nanofluid by MLP neural network and theoretical modeling using experimental results. Int. Commun. Heat Mass Transf. **74**, 11–17 (2016)

Virtual Reality Interface for Assist in Programming of Tasks of a Robotic Manipulator

Daniel Santiago Rodríguez Hoyos(✉), José Antonio Tumialán Borja, and Hugo Fernando Velasco Peña

Universidad de la Salle, Bogotá D.C 111711, Colombia
{Danielsrodriguez09,jtumialan,hfvelascop}@unisalle.edu.co

Abstract. The creation of interfaces that contribute to the interaction between a programmer and an industrial robot are based on implementing adequate procedures that facilitate people without knowledge of robotics and programming of robotic manipulators to generate tasks or activities that the robotic arm executes. In this article, it describes the creation of a programming interface using virtual reality, as an immersion tool in the work environment of the yaskawa motoman HP20D robot. This virtual environment was developed from 3D models, later exported to the Unity 3D video game engine, in this interface the functions are programmed that allow the user to interact with the robot according to the movements and trajectories performed by the user in the work environment. A MYO Armband sensor is used to operate the position and orientation of the end effector of the robot within the environment and the Oculus Riff DK2 virtual reality glasses are used to visualize the entire space, with a first-person perspective. As a result, a computational tool is obtained that allows generating trajectories, recording them, simulating them and generating the script or programming code to be implement in the robotic manipulator controller.

Keywords: Robotics · Human-robot interaction · Virtual reality · Programming

1 Introduction

The technology evolves with the purpose of facilitating human activity, as well as the fact of providing independence of previous knowledge in the actions developed in a computer, when referencing in the academy there are advances like the one that describes an embedded system, signal reader electromyography and inertial measurements for the control of visible tasks in a computer, as is the movement of the cursor, its main objective to help people with disabilities [1]. Related to this project that involves the same nature of signals, unlike that will be applied in control tasks, such as the movement of the final effector of a robotic manipulator located in a virtual environment.

The peripherals that are used conventionally in the construction of a human computer interface are mouse and keyboard, there is the need to alternating this technology by one that allows to give versatility to the user in the environment, tools that are suited

D. F. Cortes Tobar et al. (Eds.): AETA 2019, LNEE 685, pp. 328–335, 2021.
https://doi.org/10.1007/978-3-030-53021-1_33

to the tasks objective of the application. Therefore, it is important to resort to projects that are based on the use of other peripherals; as it was done with the 3D manipulation application, which using smartphones called mobile devices, allows rapidity to interact unambiguously with 3D objects, the practical thing is that the user simply has to perform a simple and intuitive gesture with two fingers and rotate the handheld device to perform manipulations on 3D objects [2].

The motivation to develop a tool is given by its interface and the purpose of facilitating the interaction between a user programmer and a manipulator robot in the task of creating trajectories that will be executed by the robot, one of the aids to implement for the elaboration of this project, 3D models are applied in virtual reality technology that has been welcomed by many sectors, among them the academic and the industrial. An example of this is the developed application, a 3D virtual environment for robotics and the purpose of the project to evaluate robotic and virtual reality techniques, implementing a simulator in robotic education [3].

2 Methods

As a project process and functional model of an automatic system, the following methodology is presented, which complies with the objectives set out in this research, and builds from modules like virtual environment, peripherals for user interaction. Starting with the description of the interface with which the user has an interaction, for which stipulates a development request described individually below.

2.1 3D Virtual Environment Modeling

The need to create a virtual environment is based on allowing the user to make critical errors such as the collision of the robot with objects during the programming of a trajectory.

The interface and virtual environment are created from 3D models, including the Yaskawa Motoman HP20D robot who is the second main actor within the interface against the programmer user. To create the virtual model of the work area, the laboratory of robotics of La Salle University is taken as a reference from the plans, a virtual representation of it is constructed through the software Sweet Home 3D and them implemented in Unity 3D (See Fig. 1).

Unity 3D is a platform for video game development, it provides essential features as a tool for this project:

- Rendering of 3D Objects, as interactive environments
- Compilation of virtual reality applications
- Kinematic calculation of objects within the environment
- Custom logic programming in C# and JavaScript.

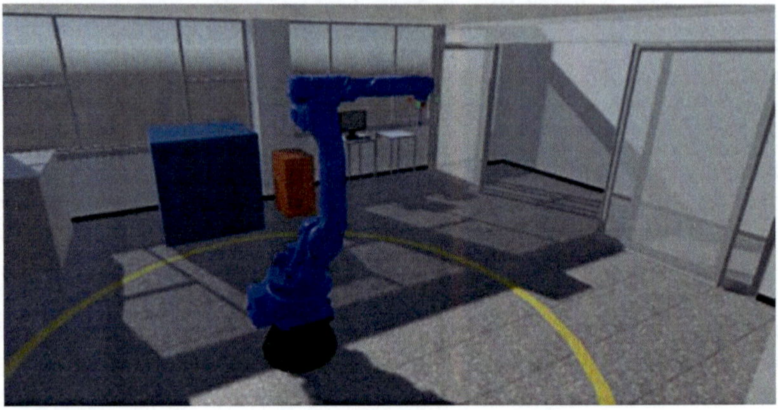

Fig. 1. Virtual Unity 3D workspace.

2.2 Implementation of Peripherals

The implementation of virtual reality glasses occurs naturally, thanks to the Unity3d video game engine. Therefore it is important to emphasize the implementation of the MYO armband.

This device is a band of Thalmic Labs, usable in the right or left forearm, used for the control of gestures with hands and movement control; allows you to connect to devices such as cell phones, computers and many more with bluetooth connection. As a device for controlling hand gestures, Thalmic Labs characterized 4 generalized movements for use on any forearm and identified at the time of performing the calibration of the device on the user [7] (See Fig. 2).

Fig. 2. Gestures recognized by MYO Armband

The gestures are used as interface functionality, giving a specific action to 2 of these gestures.

- Make a fist: Activation of movement control on the end effector of the robot.
- Wave Right: Registration and storage of the current point of the final effector.

Once the movement control is activated, it is necessary to obtain the reference for the movement of the virtual arm which the user observes in order to relate the movement of the user's arm to that of the robot, giving the user control over the final effector. To solve the control of the target point, the compass of the device is used, which transmits the relative rotation information of the three axes of the MYO Armband and will be used to control the rotation of the forearm point of the 3D model of the arm (See Fig. 3).

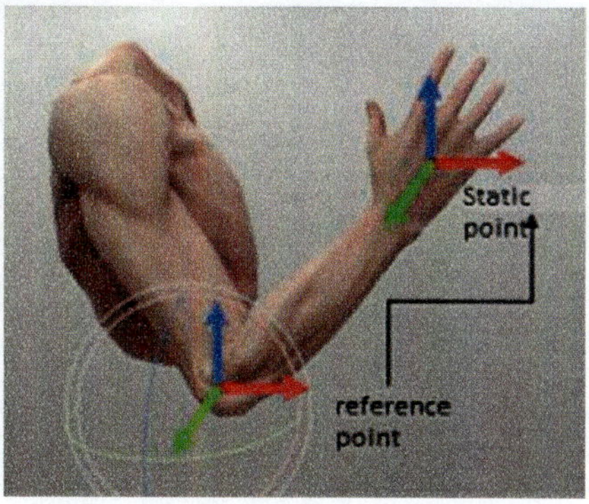

Fig. 3. Dynamic of virtual upper extremity

The orientation of the MYO Armband applied to the forearm is obtained through a programming script.

2.3 Implementation of Inverse Kinematics

One of the important and core aspects of the application is the movement animation of the robot, with this the user will observe the behavior of the robot according to the points on which the robot is positioned and trajectories that it is expected to perform. Mentioned previously is the robot's kinematics which is implemented in a programming script which is executed continuously, and varies the values of each articulation.

From the observed coordinate system (See Fig. 4), The orientation of the final effector is determined, reflected in the virtual environment with the orientation gizmo located in the end effector of the robot.

Its kinematics is constructed as a C# script in Unity 3D that is executed in a cyclical way each visualization frame, obtained its mathematical calculation of [6]. The need to find the values in degrees of each joint according to the objective point of the final effector is supplied from the following algorithm constructed to give a solution to the inverse kinematics of the robot.

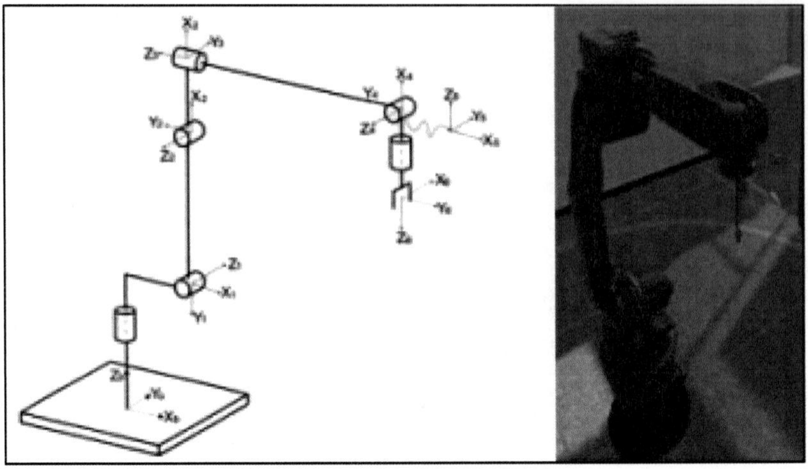

Fig. 4. Robot coordinate system. (a) Coordinates of articulations, (b) Gizmo of orientation in Robot.

Algorithm 1 Implementation of inverse kinematics

Require: Homogeneous transformation matrix Gameobject "Sphere Target bot" T.

Ensure: Array of angles corresponding to the joints of the robot $Q[\,]$.

```
Initialize Q[ ] en null
Assign rotations matrix Rbe
Assign Positions vector Pbe
Assign Positions vector Pbaux = T • (0, 0, −lh, 1)^T
Calculation of the angle S = Atan2(Pbaux(2), Pbaux(1))
Calculation of Pw (Mariño, 2015) donde Pw = ( sqrt(Paux(1)^2 + Paux(2)^2) − L1x ; 0 ; Paux − L1z )
Calculation of a1 donde a1 = sqrt(l3^2 + l4^2)
if sqrt(Pw(1)^2 + Pw(3)^2) > (L2 + a1) then
    return Q[ ] Array of angles HP20D.
end if
Carry out rotations in Rb3 = RotZ(S) * RotY(L) * RotY(−U) * RotZ(PI/2) * RotX(PI);
R3e = Rb3' • Rbe.
Calculation of the angle L (PI/2) − θ
Calculation of the angle U γ + θ2
Calculation of the angle R Atan2(−R3e(1,3), R3e(3,3))
Calculation of the angle B Atan2(R3e[1,2], sqrt(R3e(2,1)^2 + R3e(2,2)^2))
Calculation of the angle T Atan2(−R3e(2,1), R3e(2,2]));
return Q[ ] Array of angles HP20D
```

3 Results

By integrating the virtual environment developed with the peripherals used by the user, the main result is an interface for a software application that implements virtual reality technology, its main function is to be a programming tool for the Yaskawa robot. motoman HP20D, as well as providing ease to people without deep knowledge in programming of manipulators (See Fig. 5).

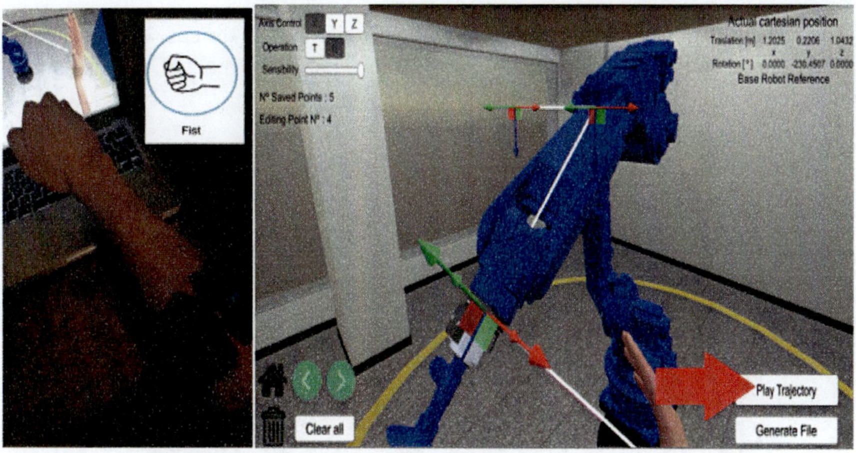

Fig. 5. User in interaction with the programming tool

And as previously described, the hand gestures of the user allow to activate the operation on the end effector and the storage of desired points, these are the fist of the hand and the opening of the palm out determined by code. Respectively.

The procedure to evaluate the programming tool is create a task with a tool on the final effector at the virtual environment. It was observed in the tool that a successful linear trajectory is constructed in suitable conditions to be evaluated in the real robot and to be able to make a comparison (See Fig. 6).

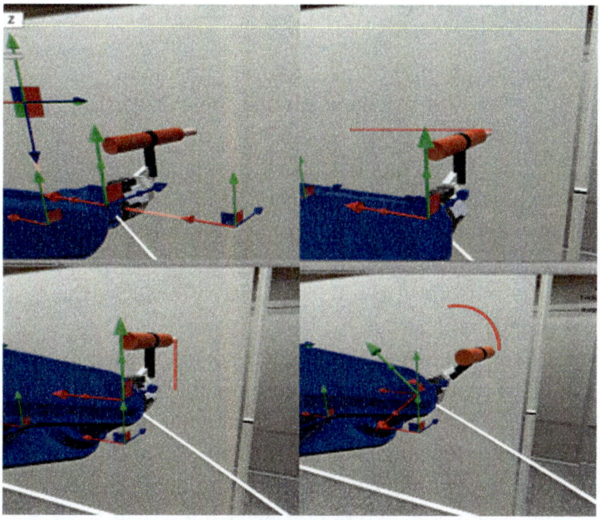

Fig. 6. Performing task of robot with a marker tool.

Then it have to play the same trajectory with the real robot. In the case of the trajectory executed in real robot it is concluded that it fulfills what was expected, it must enter to evaluate the details of the position errors (See Fig. 7).

Fig. 7. Evaluating the created task in the real robot and their work environment.

4 Conclusions and Future Work

The integration of the virtual environment and the positioning tool was successfully carried out. Although the result of characterizing the MYO Armband sensor was not as expected, an alternative interaction to the keyboard and mouse of a computer is achieved, with an unconventional tool that provides an interface with additional features in the construction of a concept of user immersion in the virtual reality environment. The usual techniques to define the tasks of a robot are to write machine code interpreted by the controller of the robot, but with this, we seek to contribute to the theory that the method of teaching the trajectory to robot in a way guided by the gestures of a user programmer could be quickly and flexible [4].

The developed interface manages to shorten robot manipulation procedures, in relation to the preparation of trajectories that are programmed point-to-point in the robot's teach pendant, compared with the naturalness of moving the end effector of the robot with gestures and rotating movements of the robot. Hand and forearm respectively. In projects that want to give continuity to this research, apply in new developments the alternative of creating trajectories with curvilinear interpolation, as well as including the same task in the generation of the machine code for its later execution in the real robot. Also, open the possibility of directly connecting the application to the control of the robot online and approach real-time execution.

Associating this development with the educational approach, the results presented with this interface can be related to projects that seek to give importance to the application 11 of virtual reality in the teaching of robotics as shown in the work in [5], which an application concludes a high potential as a learning and practice tool. This scope can be applied in the same way in this project, since the virtual environments give the flexibility of implementation and execution.

References

1. Ancans, A., Rozentals, A., Nesenbergs, K., Greitans, M.: Inertial sensors and muscle electrical signals in human-computer interaction. In: 2017 6th International Conference on Information and Communication Technology and Accessibility (ICTA), Muscat, pp. 1–6 (2017)
2. Tseng, P.H., Hung, S.H., Chiang, P.Y.: EZ-manipulator: designing a mobile, fast, and ambiguity-free 3D manipulation interface using smartphones. Comput. Vis. Media **4**, 139 (2018)
3. dos Santos, M.C.C., Sangalli, V.A., Pinho, M.S.: Evaluating the use of virtual reality on professional robotics education. In: 2017 IEEE 41st Annual Computer Software and Applications Conference (COMPSAC), Turin, pp. 448–455 (2017)
4. Cousins, M., Yang, C., Chen, J., He, W., Ju, Z.: Development of a mixed reality based interface for human robot interaction, In: 2017 International Conference on Machine Learning and Cybernetics (ICMLC), Ningbo, China, pp. 27–34 (2017)
5. Crespo, R., García, R., Quiroz, S.: Virtual reality simulator for robotics learning. In: 2015 International Conference on Interactive Collaborative and Blended Learning (ICBL), Mexico City, pp. 61–65 (2015)
6. Mariño, D.: Interfaz natural para la programación de un robot manipulador a través de un Kinect. Universidad de la Salle, pp. 35–40 (2015)
7. ThalmicLabs, Myo gesture control armband (2018). www.myo.com/. Accessed 14 Apr 2019

Glucose Control for T1D Patients Based on Interval Models

Fabian León-Vargas[1(✉)], Maira García-Jaramillo[2], Andrés Molano[1], Hernán De Battista[3], and Fabricio Garelli[3]

[1] Facultad de Ing. Mecánica, Electrónica y Biomédica, Universidad Antonio Nariño, Bogotá, Colombia
{fabianleon,andres.molano}@uan.edu.co

[2] Facultad de Ingeniería, Universidad EAN, Bogotá, Colombia
magarcia@universidadean.edu.co

[3] Facultad de Ingeniería, Universidad Nacional de La Plata, La Plata, Argentina
{deba,fabricio}@ing.unlp.edu.ar

Abstract. In this paper, a new glucose control strategy for patients with type 1 diabetes based on the dynamic simulation of an interval model is presented. Particularly, a safety scheme based on an interval insulin-on-board estimation model for artificial pancreas systems is designed, which includes a combination of soft and hard insulin-on-board constraints that can be implemented as part of an automatic control strategy. This control strategy is tested in 10 virtual type 1 diabetes patients from the UVa/Padova simulator approved by the Food and Drug Administration agency of the United States, showing a satisfactory performance in comparison to the same control system without the safety scheme. The proposed approach avoids episodes of mild and severe hypoglycemia under demanding assessment scenarios.

Keywords: Artificial pancreas · Glucose control · Diabetes · Interval simulation · Insulin-on-board

1 Introduction

Diabetes is a disease characterized by high blood glucose levels due to the non-generation of insulin by the pancreas (type 1 diabetes) or inadequate use of it (type 2 diabetes). In type 1 diabetes (T1D), the treatment consists of supplying exogenous insulin analogues to the body. A slow-acting insulin allows steady-state glycemic control in the system, while a fast-acting insulin is used to counteract the transient effect of meals. However, unlike traditional insulin therapy, based on multiple daily injections of fast and slow insulin analogs, intensive insulin therapy, based on the continuous supply of rapid-acting insulin through an infusion pump, has been shown to achieve better glycemic control [1]. Similarly, the treatment for T1D has benefited from continuous glucose monitors (CGM), which, unlike traditional glucometers, provide an estimate of the blood

D. F. Cortes Tobar et al. (Eds.): AETA 2019, LNEE 685, pp. 336–344, 2021.
https://doi.org/10.1007/978-3-030-53021-1_34

glucose concentration continuously; which allows to observe the behavior of glucose in real time and make decisions accordingly.

In the last decade, the treatment has focused on the development of control systems that use the CGM measurements to automatically calculate the amount of insulin to be delivered continuously to the patient through the insulin pump. This development is known as artificial pancreas (AP).

One of the main problems that still persist in the AP development is how to deal with the variability present in the glucose-insulin system of T1D patients. This variability can be due to multiple factors such as exercise, stress, alcohol consumption, etc., which induces changes in the dynamics of the food absorption subsystem, as well as in the absorption and action of insulin.

In this paper, a safety control strategy to deal with the variability associate with the insulin-on-board (IOB) estimation through an interval model considering parametric uncertainty, is presented. A combined scheme of soft and hard constraints limiting the performance of the main AP controller according to some safety IOB thresholds, was designed. In this work, the interval safety layer was conditioned to one PID-like controller widely used in AP implementations, and evaluated in a glucose-insulin dynamics simulator approved as a substitute for clinical trials in animals by the Food and Drug Administration (FDA) agency of the United States.

2 Interval IOB Model

In general terms, an IOB model allows to estimate how much insulin still to act in the body, which depends on the patient's glucose-insulin dynamics. Having an IOB estimation can help to prevent an excess of insulin and therefore hypoglycemia events [2]. However, the IOB estimate requires determining the duration-of-insulin-action (DIA) of the patient, which is not usually considered by physicians and can lead to poor glycemic control of the disease [3].

The IOB model to use in this work was presented by León-Vargas et al. in 2013 [2]. In order to consider uncertainty related to the DIA in this IOB model, Modal interval analysis (MIA) was applied here. It is a mathematical approach to go beyond the limitations of classic intervals in terms of their structural, algebraic and logical features [4]. One of the main problems in interval computations is the existence of multiple instances of the same variable in the expression to be evaluated, leading to overestimation of the result. MIA reduces the impact of this problem as each interval function to be evaluated is automatically analyzed and put, if possible, in its optimal form, yielding an exact computation of the range [5]. Basics of MIA theory and some applications related to diabetes can be found in [4,6,7].

Here, uncertainty in the DIA is considered through the K_{DIA} parameter. Applying the theorem of coercion to optimality and considering the definition of a dual operator as $Dual([a_1, a_2]) = [a_2, a_1]$, the rational computation of the interval IOB model can be performed from Eq. (1).

$$\begin{aligned} & C_1(t+1) = C_1(t)(1 - \Delta t K_{DIA}) + \Delta t u(t) \\ & C_{12}(t+1) = C_{12}(t)(1 - \Delta t K_{DIA}) + \Delta t u(t) + \Delta t Dual(K_{DIA}) C_1(t) \\ & C_2(t+1) = C_{12}(t+1) - Dual(C_1(t+1)) \\ & IOB = C_1(t) + C_2(t) \end{aligned} \tag{1}$$

subject to $\Delta t < 1/C_2(t)$, where $C_1(t)$ and $C_2(t)$ are compartments, $u(t)$ is the insulin dose, and $C_{12}(t) = C_1(t) + C_2(t)$ is a new state created to avoid under and overestimation of the model.

As a result, upper and lower bounds that define an envelope for all possible values of IOB, are obtained (Fig. 1). This envelope is determined by the set of trajectories involved in the simulation from all possible responses taking values inside the interval of variation.

3 Control Scheme

Figure 1 shows an example of an interval IOB estimation, represented by the upper (IOB_u) and lower band (IOB_l). The envelope width is dynamic and is related to the uncertainty associated with the K_{DIA} parameter of the model. In previous works, a single restriction imposed on the IOB value ($\overline{IOB}$) was used as a threshold to reduce the risk of hypoglycaemia [8]. In this work, a threshold for each IOB estimation band, whose effect is to restrict the action of the main controller in a soft and a hard way, is considered.

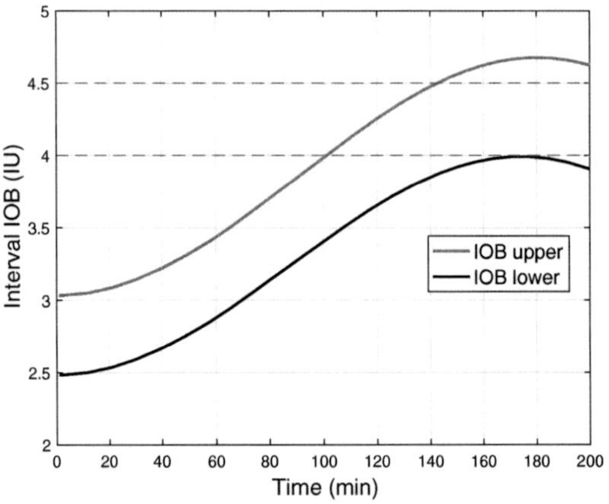

Fig. 1. Interval IOB model estimation and the IOB threshold to each band.

Both, the upper (IOB_u) and lower (IOB_l) band of the interval IOB envelope are used here to establish a soft and hard performance limitation, respectively, on the main AP controller considering a corresponding IOB threshold. The soft/hard performance limitation basis is detailed in Sect. 3.1.

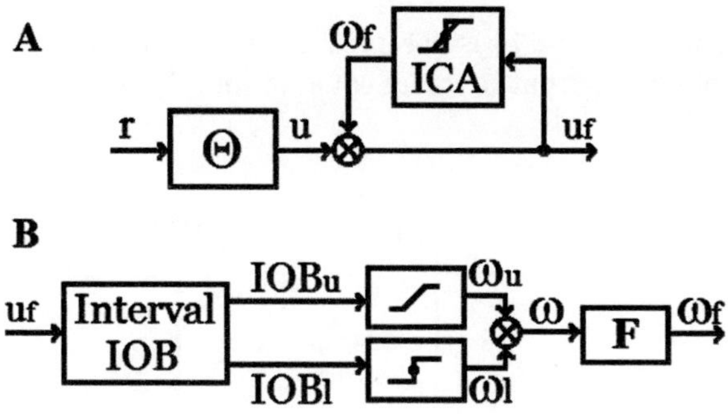

Fig. 2. A: General scheme of the proposed control system. B: Interval constraint algorithm (ICA) scheme.

Figure 2 (A) shows the concept of the proposed control, where Θ represents the main controller of the AP system, r is the set-point, u is the output of the main controller, ω_f is the conditioning signal of the algorithm proposed, and u_f is the modulated control action that is finally sent to the insulin pump. Figure 2 (B) shows the internal components of the proposed interval constraint algorithm (ICA).

3.1 Constraints Basics

Soft constraint scheme is related to the upper band of the interval IOB envelope. Equations (2) and (3) represent the relationship between IOB_u and ω_u of the scheme shown in Fig. 2 (B).

$$\omega_u(i) = \begin{cases} 1 & \sigma_u(i) < 0 \\ max\,(1 - \sigma_u, 0) & otherwise \end{cases} \tag{2}$$

where

$$\sigma_u(i) = IOB_u(i) - \overline{IOB_u} \tag{3}$$

Dynamics between ω_u and σ_u is observed in Fig. 3 (A). As $IOB_u(i)$ moves away above the $\overline{IOB_u}$ threshold, the modulation over the amount of insulin delivered to the patient becomes greater, i.e., insulin command decreases inversely proportional to σ_u.

Hard constraint scheme is related to the lower band of the interval IOB envelope. Equations (4) and (5) represent the relationship between IOB_l and ω_l of the scheme shown in Fig. 2 (B).

$$\omega_l(i) = \begin{cases} 1 & \sigma_l(i) < 0 \\ 0 & otherwise \end{cases} \tag{4}$$

where

$$\sigma_l(i) = IOB_l(i) - \overline{IOB_l} \tag{5}$$

Dynamics between ω_l and σ_l is observed in Fig. 3 (B). Once $IOB_l(i)$ reaches the $\overline{IOB_l}$ threshold, the output becomes equal to zero, i.e, insulin is not delivered to the patient.

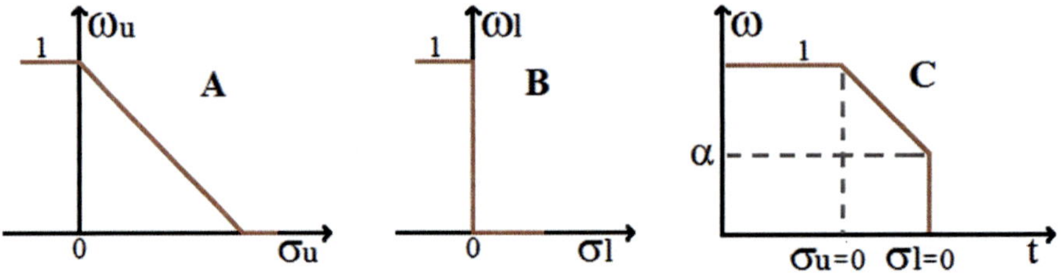

Fig. 3. A: relationship between ω_u and σ_u. B: relationship between ω_l and σ_l. C: example of the relationship between ω and both σ_u and σ_l in the soft-hard constraint scheme over time for the case shown in Fig. 1.

3.2 Soft-Hard Constraints Combination

In this scheme, a modulation between the outputs of both constraints (ω_u and ω_l) presents a progressive conditioning of the main control action as the interval IOB envelope moves above the corresponding threshold. Therefore, the combined result of both constraints on a particular case such as that shown in Fig. 1 would have the result shown in Fig. 3 (C).

Equation (6) represents the combination of the soft and hard constraints as a function of the upper and lower bands of the interval IOB envelope and the corresponding IOB threshold.

$$\omega(i) = \begin{cases} 1 & \sigma_u < 0 \\ max\,(1 - \sigma_u, 0) & \sigma_l < 0 \\ 0 & otherwise \end{cases} \tag{6}$$

where σ_u and σ_l are obtained from Eqs. (3) and (5), respectively, while $\overline{IOB}_u$ and $\overline{IOB}_l$ are computed using Eq. (7).

$$\begin{aligned} \overline{IOB}_l &= IOB_b + \tfrac{CHO}{I2C} \\ \overline{IOB}_u &= (1 + \beta_u)\,\overline{IOB}_l \end{aligned} \tag{7}$$

where IOB_b is the corresponding IOB value from the basal insulin, $I2C$ is the subject-specific insulin-to-carbohydrate ratio, and CHO is a tuning parameter related to the carbohydrates ingested. The β_u parameter is related to the percent of uncertainty defined to the K_{DIA} parameter of the interval IOB model.

It is important to note that, in Fig. 3 (C), α corresponds to the value that ω takes when the hard constraint works. This value depends both on the width of the IOB interval envelope (which is dynamic), and on $\beta_u \overline{IOB_l}$, and can be expressed from Eq. (3), (5) and (7) as shows Eq. (8).

$$\alpha = \sigma_u - \sigma_l = IOB_u(i) - IOB_l(i) - \beta_u \overline{IOB_l} \tag{8}$$

Considering that the conditioning signal ω may present a discontinuous behavior due to the switching caused by the hard constraint, a first order filter is added to the output of the ICA scheme (block F in Fig. 2 (B)) using Eq. (9), where τ^{-1} is related to the filter cut-off frequency. The smoothed signal ω_f is then used to modulate the final control action (Fig. 2 (A)).

$$\omega_f\,(i+1) = \left(1 - \tau^{-1}\right)\omega_f\,(i) + \tau^{-1}\omega(i) \tag{9}$$

3.3 Main Control Algorithm

In this work, the main control algorithm corresponds to a PID-like controller widely implemented in AP trials, which considers an insulin feedback (IF) component in its design. Eq. (10) presents the overall definition about this PID-IF controller. However, the reader can find it in depth in [9].

$$\begin{aligned} u(t) &= u_c(t) - \gamma I_p(t) \\ u_c(t) &= k_p \left[e(t) + \tfrac{1}{\tau_I} \int e(t)dt + \tau_D \tfrac{de(t)}{dt} \right] \end{aligned} \tag{10}$$

where $e(t) = r - y_g(t)$ is the glucose tracking error, and $y_g\,(t)$ is the current measurement from the CGM system. The controller gain (k_p) is adjusted according to the total daily dose of insulin I_{TDD} and the factor 1500 related to the '1500-rule', Eq. (11).

$$k_p = \frac{60}{1500}\frac{I_{TDD}}{\tau_D}\left(1 + \frac{\gamma}{K_{cl}}\right) \tag{11}$$

4 Results

The control system comprises the combined scheme of soft-hard constraints conditioned to the main controller presented in Sect. 3.3, whose integral action was replaced by the basal insulin specific to each virtual patient [2]. The γ and τ_D parameters were set to 5/6 and 90 min respectively. The CHO parameter was set to 40 g for the corresponding IOB thresholds. A set-point of 110 mg/dL was fixed to each of the 10 virtual patients from the UVa-Padova simulator approved by the FDA. Reference [10] is suggested to the reader to fully understand the features and constraints related to this simulator commonly used in pre-clinical tests.

4.1 Test Scenario

The test scenario corresponds to one simulation day for each virtual patient, which includes three meals totaling 200 g of carbohydrates and distributed as follows: 60 g at 8:00, 80 g at 13:00 and 60 g at 18:00. Uncertainty of 20% ($\beta_u = 0.2$) is assumed in the IOB estimation, taking a baseline of K_{DIA} equal to $0.0122\,\mathrm{s}^{-1}$ in the interval model, which corresponds to the most common value found on virtual patients from the UVa-Padova simulator [2].

The same scenario was assessed under two performance conditions: a nominal gain (Eq. (11)) and an aggressive gain (60% greater) in the main controller, and also comparing the inclusion or not of the ICA scheme proposed in this work.

4.2 Performance Obtained

Table 1 shows the performance results of the test scenario considered. # Hypos corresponds to the number of hypoglycemia events below 70 mg/dL for at least 15 min, Excursion is related to the postprandial glucose excursion in mg/dL, Mean glucose in mg/dL, A+B and C+D zones indicates the number of patients whose daily glucose lies in those zones of the Control-Variability Grid Analysis (CVGA) plot, see Fig. 4, and finally % Time Hypo and % Time Hyper are related to the time percent below 70 mg/dL and above 180 mg/dL, respectively. Figure 5 shows the performance obtained for one subject.

Table 1. Resulting performance metrics.

	# Hypos	Excursion	Mean glucose	A+B zones	C+D zones	% Time Hypo	% Time Hyper
Nominal gain							
PD-IF	10	51.9 ± 5.3	120 ± 9.4	7	3	7.1 ± 11	11 ± 4
PD-IF with ICA	0	61.1 ± 6.3	135 ± 7.4	10	0	0 ± 0	15 ± 5.9
Aggressive gain							
PD-IF	28	37.7 ± 9.5	101 ± 7.9	0	10	24 ± 11	5 ± 2.8
PD-IF with ICA	0	54.8 ± 7.1	128 ± 7.8	10	0	0 ± 0	11 ± 6.4

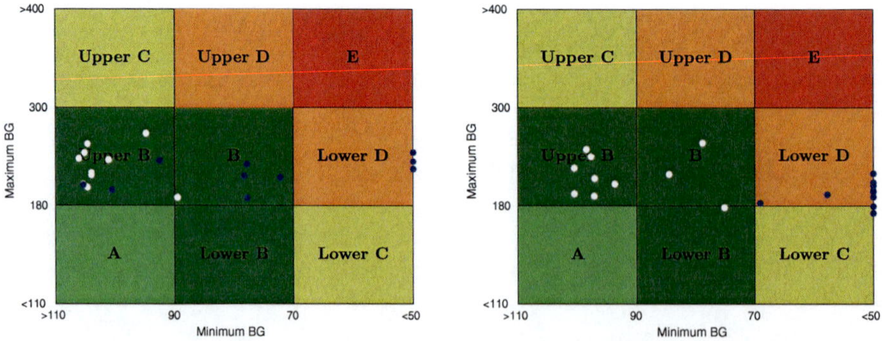

Fig. 4. Left: Nominal gain. Right: Aggressive gain. White and blue dots correspond to results of the main controller with and without the ICA scheme, respectively.

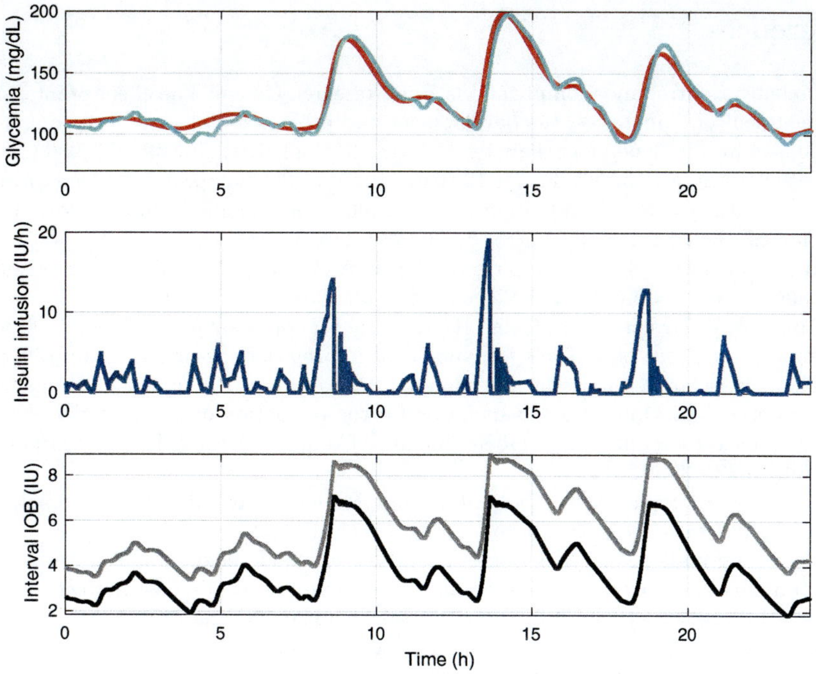

Fig. 5. Performance obtained using the aggressive gain of the PD-IF with ICA on subject #10. Blood glucose concentration and glucose measurements from the CGM system are shown in red and cyan, respectively.

5 Conclusion

A new safety scheme to reduce the risk of hypoglycemia without detracting performance of the main controller was designed. It is based on an interval IOB model, whose value allows to limit the insulin delivered on a full closed-loop control system under two constraints conditions. Results show that, in comparison with the main controller without the ICA scheme, no hypoglycemia events were found in any virtual patient assessed, even when the main controller gain was substantially more aggressive than its nominal value.

Acknowledgments. F. León-Vargas and A. Molano were supported by Antonio Nariño University Project #2018222 and Colciencias Project #110180763081. M. García-Jaramillo was supported by Universidad EAN Project #TO_P0_0518. H. De Battista and F. Garelli were supported by Agencia Nacional PICT 2014-2394, 2016-2258, CONICET (PIP 2015-0837), Universidad Nacional de La Plata (I216) of Argentina.

References

1. Diabetes Control and Complications Trial Research Group: The effect of intensive treatment of diabetes on the development and progression of long-term complications in insulin-dependent diabetes mellitus. N. Engl. J. Med. **329**, 977–986 (1993)
2. León-Vargas, F., Garelli, F., De Battista, H., Vehí, J.: Postprandial blood glucose control using a hybrid adaptive PD controller with insulin-on-board limitation. Biomed. Signal Process. Control **8**, 724–732 (2013)
3. Walsh, J., Roberts, R.: Pumping Insulin: Everything You Need for Success on a Smart Insulin Pump. Torrey Pines Press, San Diego (2006)
4. Sainz, M.A., Armengol, J., Calm, R., Herrero, P., Jorba Jorba, L., Vehi, J.: Modal Interval Analysis: New Tools for Numerical Information. Springer, Cham (2013)
5. Calm, R., García-Jaramillo, M., Bondia, J., Sainz, M.A., Vehí, J.: Comparison of interval and Monte Carlo simulation for the prediction of postprandial glucose under uncertainty in type 1 diabetes mellitus. Comput. Methods Programs Biomed. **104**(3), 325–332 (2011)
6. Gardeñes, E., Sainz, M.A., Jorba, L., Calm, R., Estela, R., Mielgo, H., Trepat, A.: Modal intervals. Reliable Comput. **7**(2), 77–111 (2001)
7. García-Jaramillo, M.: Prediction of postprandial blood glucose under intra-patient variability and uncertainty and its use in the design of insulin dosing strategies for type 1 diabetic patients. Ph.D. thesis, Universitat de Girona (2011)
8. León-Vargas, F., Garelli, F., De Battista, H., Vehí, J.: Postprandial response improvement via safety layer in closed-loopblood glucose controllers. Biomed. Signal Process. Control **16**, 80–87 (2015)
9. Palerm, C.: Physiologic insulin delivery with insulin feedback: a control systems perspective. Comput. Methods Programs Biomed. **102**(2), 130–137 (2011)
10. Man, C.D., Micheletto, F., Lv, D., Breton, M., Kovatchev, B., Cobelli, C.: The UVA/PADOVA type 1 diabetes simulator: new features. J. Diabetes Sci. Technol. **8**(1), 26–34 (2014)

Inverse Reinforcement Learning Application for Discrete and Continuous Environments

Yeison Suarez, Carolina Higuera(✉), and Edgar Camilo Camacho

Universidad Santo Tomás, Bogotá, Colombia
{yeisonsuarez,edgarcamacho}@usantotomas.edu.co,
c.higuera10@uniandes.edu.co

Abstract. We show the application of inverse reinforcement learning (IRL) in discrete and continuous environments based on the apprenticeship focus. The objective is to learn a mathematical definition of a task based on trajectories made by an expert agent. To achieve this, there must be features functions that in some way describe abilities that the agent can learn. Therefore, the description of a task can be formulated as a linear combination of those functions. This learning method was applied in Open-AI gym environments to show the learning process of a task using demonstrations.

Keywords: Inverse reinforcement learning · Reinforcement learning · Quadratic programming

1 Introduction

Reinforcement Learning (RL) is one of the best known algorithms for autonomous learning. It uses the agent's positive and negative experiences to learn how to behave in order to solve a defined task. However, the information of what is the task must be included in the rewards given to the agent.

Although RL has shown successful results, it highly depends on an established reward function. But, in some cases, designing the reward function can become more complex than the learning itself [1]. On the other hand, if we have an expert agent, it is possible to exploit its knowledge to infer the reward through an Inverse Reinforcement Learning (IRL).

Learning the reward function instead of the policy directly has some advantages, as shown in [7]. First, with the reward function it is possible to understand the purpose of the expert agent. Second, the reward is adapted to disturbances and is transferable to other environments, while with policies the agent can learn actions that are not relevant to the task, for instance, errors in the demonstrations. Third, the policy learned after inferring the reward function allows knowledge transfer for more complex tasks.

D. F. Cortes Tobar et al. (Eds.): AETA 2019, LNEE 685, pp. 345–355, 2021.
https://doi.org/10.1007/978-3-030-53021-1_35

2 Theoretical Background

2.1 Reinforcement Learning

The RL problem can be described as a Markov Decision Process (MDP), where for each time step t, the agent receives the state of its environment s_t and takes an actions in order to learn how to perform a task. As feedback, the agent receives a numeric reward r_{t+1} and moves to a new state s_{t+1} [8].

The agent should learn the best way to perform the task, which is reflected in the rewards that it gets every time it makes an action. For this reason, the agent has to learn to maximize the rewards in the long-term. To do this, the agent maps, with the function $Q(s_t, a_t)$, the expected discounted utility of taking action a_t in state s_t if he behaves optimally thereafter. From the Q-value, the agent can generate a policy $\pi_t(s, a)$ that indicates the probability of applying action a_t when it is in state s_t that assures the maximum long-term reward.

Q-learning is widely use as RL algorithm. It updates the Q-values according to greedy actions, without taking into account the current policy, which may include exploratory actions. The updating rule for each Q-value is:

$$Q(s_t, a_t) = (1 - \alpha)Q(s_t, a_t) + \alpha[r + \gamma \max_{a \in A} Q(s_{t+1}, a)]$$

Where:

- r: reward at time step t ($r \in \mathbb{R}$)
- α: learning rate at time step t ($0 \leq \alpha \leq 1$)
- γ: discounting rate ($0 \leq \gamma < 1$).

2.2 Inverse Reinforcement Learning

According to Abbeel and Ng in [1], since the RL field is based on the fact that the reward function is the most clearly, robustly and transferable definition of the task, it makes sense to consider an RL approach in which the reward function is what has to be learned. The problem of deriving the reward function from observations is known as Inverse Reinforcement Learning - IRL [6].

IRL assumes that an expert already optimized a reward function $R(t)$ that is unknown for the learner agent, which can be expressed as a linear combination of known features ϕ_i describing possible subtasks or skills to be learned:

$$r(s) = w_1\phi_1(s) + w_2\phi_2(s) + \cdots + w_n\phi_n(s) = \boldsymbol{w}^{\boldsymbol{T}} \boldsymbol{\phi}(\boldsymbol{s})$$

Because the expert knows the truly reward function $\boldsymbol{w}^{*^T} \boldsymbol{\phi}(\boldsymbol{s})$, with $\|\boldsymbol{w}^*\|_1 \leq 1$ for rewards bounded by 1, the utility expectation of the expert policy is:

$$\begin{aligned} U^{\pi_E}(s) &= \mathbf{E}\left[\sum_{t=0}^{\infty} \gamma^t R(s) \middle| \pi_E\right] = \mathbf{E}\left[\sum_{t=0}^{\infty} \gamma^t \boldsymbol{w}^{*^T} \phi(\boldsymbol{s}) \middle| \pi_E\right] \\ &= \boldsymbol{w}^{*^T} \mathbf{E}\left[\sum_{t=0}^{\infty} \gamma^t \phi(\boldsymbol{s}) \middle| \pi_E\right] = \boldsymbol{w}^{*^T} \mu(\pi_E) \end{aligned}$$

Where, $\mu(\pi_E)$ is known as *feature expectations* given policy π_E. For the case of the learning agent, its utility can be express as:

$$U^{\tilde{\pi}}(s) = \boldsymbol{w^T}\mu(\tilde{\pi})$$

The objective of IRL is to find values for the weights $w_1, w_2, \ldots, w_n$ such that the utility of the policy learned by the agent $\tilde{\pi}$ is close to the utility of the expert policy π_E:

$$\|\mu(\tilde{\pi}) - \mu(\pi_E)\|_2 \leq \epsilon$$

This means, IRL is seeking for a reward function $R = \boldsymbol{w^T}\boldsymbol{\phi(s)}$, for which the expert policy performs better than the learner by a margin: $U^{\pi_E}(s) \geq U^{\tilde{\pi}}(s) + \Delta$. In terms of feature expectations:

$$|\boldsymbol{w^T}\mu(\tilde{\pi}) - \boldsymbol{w^*}^{\boldsymbol{T}}\mu(\pi_E)| \leq \Delta$$

And, due to we want to approximate $\mathbf{w}$ to $\mathbf{w}^*$:

$$\mathbf{w^T}|\mu(\tilde{\pi}) - \mu(\pi_E)| \leq \Delta \qquad (1)$$

Because we are looking for a margin Δ, the IRL problem is similar to the one solved in Support Vector Machines (SVMs), for finding the maximum margin hyperplane separating two set of points [1]. This similarity can be view taking the expert feature expectations as class $+1$ and the others as class -1, as show in Fig. 1.

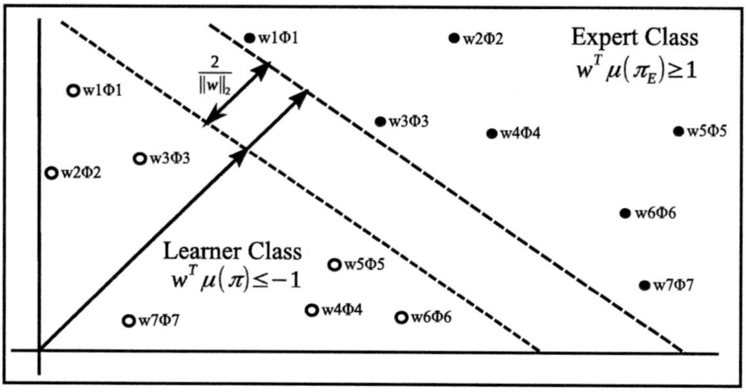

Fig. 1. IRL problem as SVM classification

The distance of the vector from the origin to the class $+1$ hyperplane is $1/\|\mathbf{w}\|_2$ and the distance of the vector from the origin to the class -1 hyperplane is $-1/\|\mathbf{w}\|_2$. Therefore, the margin size is:

$$\frac{1}{\|\mathbf{w}\|_2} - \frac{-1}{\|\mathbf{w}\|_2} = \frac{2}{\|\mathbf{w}\|_2}$$

To maximize the margin, we have to minimize the weights, resulting the following quadratic programming problem:

$$\min \frac{1}{2}\|\mathbf{w}\|_2^2$$

Regarding the constraints, they are related with Eq. 1. Due to the expert policy has to do it better that the learner by a margin:

$$\mathbf{w}^{\mathbf{T}}|\mu(\tilde{\pi}) - \mu(\pi_E)| \leq \frac{2}{\|\mathbf{w}\|_2}$$

And, because bounded rewards by 1, we have $\|\mathbf{w}\|_2 \leq \|\mathbf{w}\|_1 \leq 1$.
The complete quadratic programming problem is formulated as:

$$\begin{aligned} \underset{\mathbf{w}}{\text{minimize}} \quad & \frac{1}{2}\|\mathbf{w}\|_2^2 \\ \text{subject to} \quad & \mathbf{w}^{\mathbf{T}}\left[\mu(\pi_E) - \mu(\tilde{\pi})\right] \geq -2 \end{aligned}$$

In Algorithm 1 we summarize the process for approximating $\mathbf{w}$ to $\mathbf{w}^*$ through quadratic programming and the feature expectations from both learner and expert.

Algorithm 1. weightsUpdate

Result: $\mathbf{w}$ and optimizer status:= {feasible, infeasible}

Requirements: *Learner* $\mu(\tilde{\pi})$ and *Expert* $\mu(\pi_E)$ feature expectations, convex optimization solver;

ObjectiveFunction $= \min \frac{1}{2}\|\mathbf{w}\|_2^2$;

Constraints $= \mathbf{w}^{\mathbf{T}}\left[\mu(\pi_E) - \mu(\tilde{\pi})\right] \geq -2$;

$\mathbf{w}, \mathit{status}$ =Optimizer.solve(ObjectiveFunction, Constraints);

3 Framework: A Discrete Environment Problem

We used the Open-AI gym suite due to it provides easy to use environments to test reinforcement learning algorithms. Specifically, we applied IRL to the Taxi-v2 task, which was introduced in [5] to illustrate some issues in hierarchical RL. The goal in the task is to pick-up and drop off a passenger in a specific location.

The environment is a 5×5 grid with five walls, five possible positions for the passenger (R, G, B, Y and inside taxi) and four possible positions for the destination (R, G, B, Y). The taxi, which is the learner agent, has 6 movements: up, down, right, left, pickup passenger and drop-off passenger. The total number of states for the environment is 500, divided in taxi row (5), taxi column (5), passenger location (5) and destination (4).

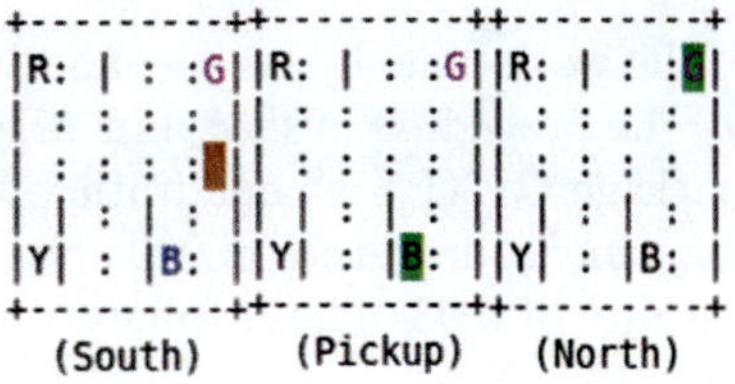

Fig. 2. Taxi-v2 Environment

The episode starts with a random location for the agent (yellow if empty, green if full), and ends when the agent carries the passenger (blue) from his location until the desired destination (magenta). Figure 2 shows the graphical interface of the environment.

The gym environment gives a reward of -1 for every time step. When the agent pickup correctly the passenger from his location to destination it gives a reward of $+20$ and, for every pickup and drop-off illegally, the reward is -10.

In order to apply IRL to learn the implicit reward function, we extracted 20 trajectories doing a manual control of the taxi. Furthermore, we proposed two types of convex functions as features, based on the distance between the taxi to the passenger when the taxi is free, the distance between the taxi to the destination when the taxi has a passenger and the time the route takes.

Features Option 1

Passenger: $$\phi_1 = 1 - \left[\frac{1}{2}\exp\left(\frac{-(taxi_x - pass_x)^2}{2}\right) + \frac{1}{2}\exp\left(\frac{-(taxi_y - pass_y)^2}{2}\right)\right]$$

Destination: $$\phi_2 = 1 - \left[\frac{1}{2}\exp\left(\frac{-(taxi_x - dest_x)^2}{2}\right) + \frac{1}{2}\exp\left(\frac{-(taxi_y - dest_y)^2}{2}\right)\right]$$

Time spent: $$\phi_3 = 1 - \exp\left(\frac{-steps^2}{2}\right)$$

Features Option 2

Passenger: $$\phi_1 = 1 - \text{manhattanDistance}\,(taxi, pass)^2$$

Destination: $$\phi_2 = 1 - \text{manhattanDistance}\,(taxi, dest)^2$$

Time spent: $$\phi_3 = -steps^2$$

To run IRL, we follow the work done by Abbeel and Ng in [1] called *Apprenticeship learning via IRL*. The procedure is shown in Algorithm 2. We code it in Python and we use the package CVXPY as quadratic optimization solver [2,4]. We train the IRL for the taxi environment for 60000 episodes, with a discounting rate of 0,99 and a learning rate of 0,03. Those values were set experimentally.

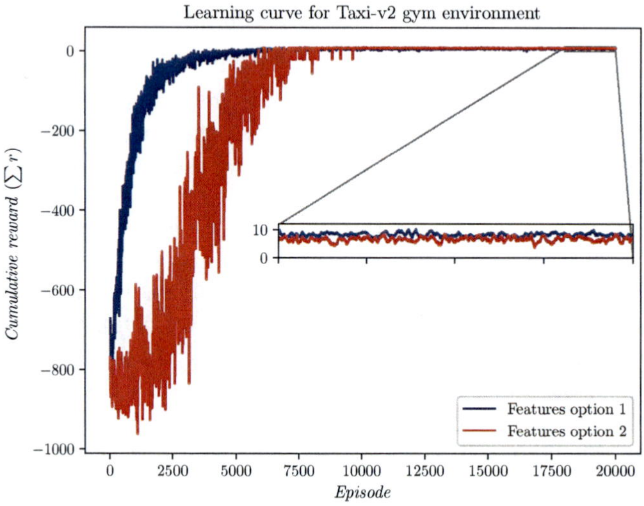

Fig. 3. Learning curve for the Taxi-v2 gym environment

Figure 3 shows the cumulative reward during each episode, given by the gym environment during the IRL training. It shows that with features option 2 the apprenticeship is slower in comparison to features option 1. However, both options converge to the same maximum cumulative reward. This means that both make the agent to learn policies with the same long-term utility.

In addition, the weights given by the IRL algorithm to each feature in the reward function are:

Features Option 1 $r(s) = -3,908 \times 10^{-3}\phi_1(s) - 3,908 \times 10^{-3}\phi_2(s) - 3,908 \times 10^{-3}\phi_3(s)$
Features Option 2 $r(s) = -1,65 \times 10^{-3}\phi_1(s) - 35,1 \times 10^{-3}\phi_2(s) + 11,4 \times 10^{-3}\phi_3(s)$

Algorithm 2. IRL for discrete environment

Requirements: expert trajectories, features $\boldsymbol{\phi}$, discount factor γ, learning rate α, exploration rate ϵ, episodes;
Initialize weights $\mathbf{w}$;
Calculate feature expectations $\mu(\pi_E)$ from expert trajectories;
for *episode in episodes* **do**
  Initialize gym environment;
  Get current state s_t;
  while *not done* **do**
    Choose action a from policy $\tilde{\pi}$ derived from tabular $Q(s,)$ with ϵ-*greedy* exploration;
    Take action a and observe s_{t+1} ;
    Calculate reward $r = \mathbf{w}^\mathbf{T}\boldsymbol{\phi}(\boldsymbol{s_{t+1}})$;
    Update policy with Q-Learning:
    $Q(s_t, a_t) = (1-\alpha)Q(s_t, a_t) + \alpha[r + \gamma \max_{a' \in A} Q(s_{t+1}, a')]$;
    Update $s_t \leftarrow s_{t+1}$;
    if *is time to update weights* **then**
      Save and append learner trajectory;
      Calculate feature expectations $\mu(\tilde{\pi})$ from learner trajectories;
      do
        $\mathbf{w}$, *status* = weightsUpdate($\mu(\tilde{\pi})$, $\mu(\pi_E)$);
        if *status=infeasible* **then**
          Remove last learner feature expectation;
      while *status=infeasible*;
    Check environment done;
  end
end

Notice that the definition for features option 1 rewards the agent if the taxi moves away from the passenger and the destination. However, the sign of the weights makes the inferred reward function to have the same purpose of the expert agent, which is to move closer to both, passenger and destination. At testing, the average cumulative reward of the learner is 7.4, while the expert wins, in average, 8.25 in each demonstration. Figure 4 shows followed by the expert and the learner agent with features option 1 in the same environment. Although both agents follows different trajectories, they receive the same reward at the end of the episode, which is 9, +20 for legal drop-off and −11 for the eleven steps.

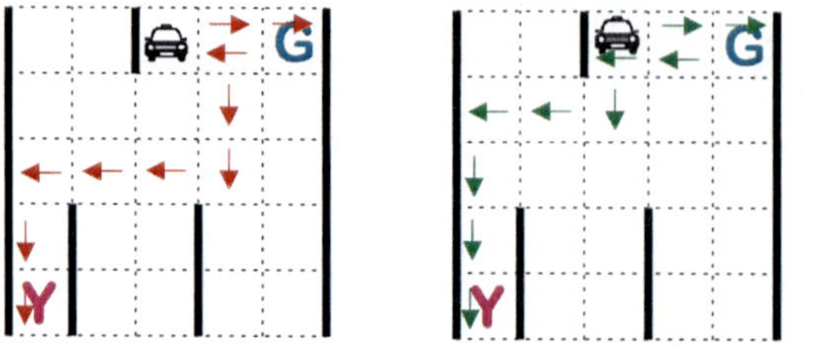

Fig. 4. Policies followed by expert and learner agents in he Taxi-v2 environment

4 Framework: A Continuous Environment Problem

The cartpole [3] is a classic control problem in the Open-AI gym suite. It consists of a pole that is attached by a joint to a cart. The cart can move without friction while the pendulum remains upright. The episode ends when the pole angle exceeds ±12° or the cart moves more than ±2.4 units from the center.

The state is conformed by four variables, as shown in Table 1. Because the ranges for each observation are continuous, we have to run Q-learning inside the IRL algorithm with a neural network instead of the tabular form. Also, the agent can apply two actions: 0 to push the car to the left and 1 to the right.

Table 1. States in the cartpole problem [3]

	Cart position	Cart velocity	Pole angle	Pole velocity at tip
Min	−4,8	$-\infty$	−24°	$-\infty$
Max	+4,8	∞	+24°	∞

The expert trajectories and utilities were extracted from a policy π_E trained with neural Q-Learning. To infer the expert task, we proposed four features, one for each observation in the state.

Cart Position: $\phi_1(s) = \exp\left(-\frac{pos^2}{2\times 1{,}5^2}\right)$ Pole Angle: $\phi_3(s) = \exp\left(-\frac{angle^2}{2\times 6^2}\right)$

Cart Velocity: $\phi_2(s) = \exp\left(-\frac{vel^2}{2\times 0{,}5^2}\right)$ Pole Velocity at Tip: $\phi_4(s) = \exp\left(-\frac{tip^2}{2\times 0{,}5^2}\right)$

The IRL procedure for a continuous problem is presented in Algorithm 3 and 4. Our Q neural network has two layers with 24 units and relu activations. Because we want to approximate the Q-values for each pair (s, a), we are facing a regression problem. Therefore, our loss function was Mean Squared Error. In order to help the neural network to remember earlier trajectories, we use a memory buffer with 50000 samples and a batch size of 64.

Algorithm 3. IRL for discrete environment

Requirements: expert trajectories, features ϕ, discount factor γ, learning rate α, exploration rate ϵ, episodes;
Initialize weights $\mathbf{w}$;
Initialize Q neural network and memory;
Calculate feature expectations $\mu(\pi_E)$ from expert trajectories;
for *episode in episodes* **do**
  Initialize gym environment;
  Get current state s_t;
  while *not done* **do**
    Choose action a from policy $\tilde{\pi}$ derived from tabular $Q(s,)$ with ϵ-*greedy* exploration;
    Take action a and observe s_{t+1} ;
    Calculate reward $r = \mathbf{w}^{\mathbf{T}} \boldsymbol{\phi}(\boldsymbol{s_{t+1}})$;
    Add trajectory step to memory (s_t, a_t, r, s_{t+1});
    Call experienceReplay() to update policy $\tilde{\pi}$;
    Update $s_t \leftarrow s_{t+1}$;
    if *is time to update weights* **then**
      Save and append learner trajectory;
      Calculate feature expectations $\mu(\tilde{\pi})$ from learner trajectories;
      do
        $\mathbf{w}, status$ = weightsUpdate($\mu(\tilde{\pi})$, $\mu(\pi_E)$);
        if *status=infeasible* **then**
          Remove last learner feature expectation;
      while *status=infeasible*;
    Check environment done;
  end
end

Algorithm 4. experienceReplay

Requirements: Batch size, Memory;
if *Memory size* $\geq$ *Batch size* **then**
  for $(s_t, a_t, s_{t+1}, r_t, done)$ *in random sample of memory* **do**
    if *done* **then**
      $QValue = r_t$;
    else
      $QValue = r_t + \gamma *$ modelPredict(s_{t+1});
    end
    QValues[a_t]=QValue;
    Fit model with new QValues;
  end
end

Figure 5 shows the evolution during training of the agent to keep the pole upright, satisfying the environment constraints. The training returns the weights for each feature. The inferred reward function is:

$$r(s) = -988.583\text{e}{-6}\,[\phi_1(s) + \phi_2(s) + \phi_3(s) + \phi_4(s)]$$

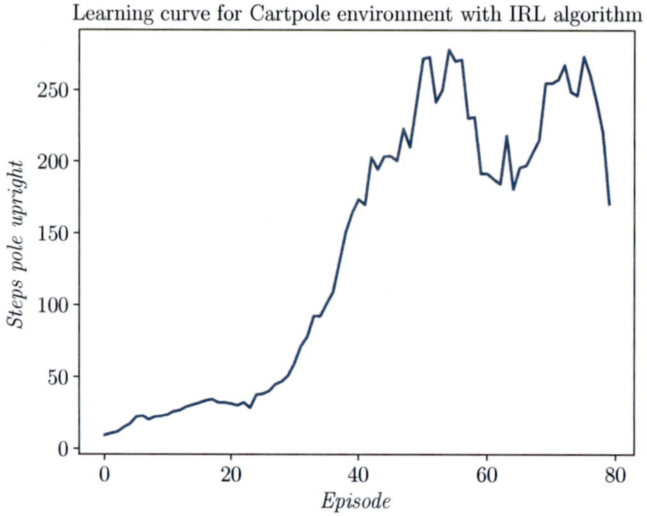

Fig. 5. Learning curve for the Cartpole-v1 environment

A final remark with the continuous environment is that in some cases during training, when the agent almost know how to perform the task, the random samples of the memory result in very similar states. So, when carrying out the optimization process with neuronal network, the system forgets earlier stages in the trajectory, causing a fall in the learning process and damaging the value of the weights that were obtained in previous episodes.

5 Conclusions

It was show that through the expert behavior on a certain task, an agent can learn it without necessarily following the same trajectory. For this, we need to define features that are significant for the reward function of the task. The weights of those features can be found establishing the problem as a SVM, looking for the maximum margin that separates the hyper plane from the expert and learner classes. We tested the IRL algorithm with discrete and continuous environments. We found that the learner accomplishes the task demonstrated by the expert, although following a different trajectory. However the utility gained by both agents were the same.

References

1. Abbeel, P., Ng, A.Y.: Apprenticeship learning via inverse reinforcement learning. In: Proceedings of the Twenty-First International Conference on Machine Learning, ICML 2004, New York, NY, USA (2004)
2. Agrawal, A., Verschueren, R., Diamond, S., Boyd, S.: A rewriting system for convex optimization problems. J. Control Decis. **5**(1), 42–60 (2018)
3. Barto, A.G., Sutton, R.S., Anderson, C.W.: Neuronlike adaptive elements that can solve difficult learning control problems. In: Artificial Neural Networks, pp. 81–93. IEEE Press, Piscataway (1990). http://dl.acm.org/citation.cfm?id=104134.104143
4. Diamond, S., Boyd, S.: CVXPY: a Python-embedded modeling language for convex optimization. J. Mach. Learn. Res. **17**(83), 1–5 (2016)
5. Dietterich, T.G.: Hierarchical reinforcement learning with the MAXQ value function decomposition. CoRR cs.LG/9905014 (1999). http://arxiv.org/abs/cs.LG/9905014
6. Ng, A.Y., Russell, S.J.: Algorithms for inverse reinforcement learning. In: Proceedings of the Seventeenth International Conference on Machine Learning, ICML 2000, pp. 663–670. Morgan Kaufmann Publishers Inc., San Francisco (2000). http://dl.acm.org/citation.cfm?id=645529.657801
7. Piot, B., Geist, M., Pietquin, O.: Bridging the gap between imitation learning and inverse reinforcement learning. IEEE Trans. Neural Netw. Learn. Syst. **28**, 1814–1826 (2017)
8. Sutton, R.S., Barto, A.G.: Reinforcement Learning: An Introduction, 2nd edn. The MIT Press, Cambridge (1998)

Self-Tuning Control Based on Generalized Minimum Variance Criterion Under the Sliding Mode Concept for Linear and a Class of Nonlinear Systems

Anna Patete[1(✉)] and Katsuhisa Furuta[2]

[1] Universidad de Los Andes, Mérida 5101, Venezuela
apatete@ula.ve
[2] Tokyo Denki University, Tokyo, Japan
http://webdelprofesor.ula.ve/ingenieria/apatete/index.html

Abstract. The stability of the implicit self-tuning algorithm based on the generalized minimum variance criterion under the sliding mode concept is a research line that has been studied for several years, mostly focusing on theorems definition and their demonstrations, to assure overall stability of the close-loop system and reference tracking by the output signal. The studies have been carrying out for linear systems and some class of nonlinear systems, with parametric uncertainties. In this paper a generalized algorithm is presented, named Self-Tuning Control based on Generalized Minimum Variance Criterion under the Sliding Mode Concept (STC-GMVC-SMC) for Linear and a class of Nonlinear Systems. The algorithm summarizes the research work done, for more than one decade, for linear and some type of nonlinear systems and the results has been compiled in a ten steps easy to follow algorithm.

Keywords: Control system · Discrete-time · Generalized minimum variance · Self-tuning control · Sliding mode control

1 Introduction

The generalized minimum variance criterion (GMVC) under the sliding mode concept (SMC) is a research line studied by several researchers, including [1] and [2]. The stability of the implicit self-tuning control (STC) based on the GMVC under the SMC is a relative new research line studied mostly by focusing the research on demonstrating, using a Lyapunov function, the overall stability of the close-loop system, which is one of the most important issue of self-tuning algorithms. The new STC has been studied for discrete-time single input single output (SISO) linear models [3], auto-regressive with exogenous input (ARX) linear models [4], contemporaneous autoregressive-moving average (CARMA) linear models [4], a defined type of time-varying models [5], and for multi input

D. F. Cortes Tobar et al. (Eds.): AETA 2019, LNEE 685, pp. 356–364, 2021.
https://doi.org/10.1007/978-3-030-53021-1_36

multi output (MIMO) linear models in [6,7]. Letter on, the proposed STC for some type of bilinear models was studied in [8–10], and recently for a class of nonlinear models [11,12].

The proposed algorithm in this paper is a ten steps generalized algorithm, named Self-Tuning Control based on Generalized Minimum Variance Criterion under the Sliding Mode Concept (STC-GMVC-SMC) for Linear and a class of Nonlinear Systems, that works for discrete-time SISO linear models defined as in [3–6], for some type of bilinear models as in [10], and for a class of nonlinear models defined as in [11].

This paper is organized as follows: Sect. 2 presents the basic notion of the self-tuning control design for the linear system case; in Sect. 3, in analogy with the linear case, basic notion of the self-tuning control design for a class of nonlinear system case is given; then Sect. 4 presents the proposed generalized algorithm, STC-GMVC-SMC for Linear and a class of Nonlinear Systems. Some remarks conclude the paper.

2 STC-GMVC-SMC for Linear Systems

This section compiles the results given in [3–5], then this section considers single-input single-output (SISO) discrete-time linear systems. The nominal model structure is given as in (1):

$$A^0(z^{-1})y_k = z^{-d}B^0(z^{-1})u_k \tag{1}$$

where polynomials $A^0(z^{-1})$ and $B^0(z^{-1})$ have not common factors. The plant with parametric uncertainties is given as follows (2):

$$A(z^{-1})y_k = z^{-d}B(z^{-1})u_k \tag{2}$$

with polynomials $A(z^{-1})$ and $B(z^{-1})$ defined as

$$\begin{aligned} A(z^{-1}) &= 1 + a_1 z^{-1} + a_2 z^{-2} + ... + a_n z^{-n} \\ B(z^{-1}) &= b_0 + b_1 z^{-1} + b_2 z^{-2} + ... + b_m z^{-m}, b_0 \neq 0 \end{aligned}$$

where $a_i \in [\underline{a}_i \ \bar{a}_i]$ and $b_j \in [\underline{b}_j \ \bar{b}_j]$ for $i = 1, 2, \ldots, n$ and $j = 0, 1, \ldots, m$ respectively.

For the linear model n, m and d are considered known. Lower and upper bounds ($\underline{a}_i$, $\underline{b}_j$, $\bar{a}_i$, $\bar{b}_j$, for $i = 1, 2, \ldots, n$ and $j = 0, 1, \ldots, m$) of plant parameters in (2) are known.

The control objective of the generalized minimum variance criterion under the sliding mode concept (GMVC-SMC) for discrete-time models is to minimize the variance of the controlled variable s_{k+d}, which is defined in the deterministic case as

$$s_{k+d} = C(z^{-1})(y_{k+d} - r_{k+d}) + Q(z^{-1})u_k \tag{3}$$

The polynomials

$$C(z^{-1}) = 1 + c_1 z^{-1} + c_2 z^{-2} + ... + c_n z^{-n}, \tag{4}$$

$$Q(z^{-1}) = q_0(1 - z^{-1}) \tag{5}$$

are to be designed, r_k is the reference signal, and the error signal e_k is defined as: $e_k = y_k - r_k$. The idea comes from the discrete time sliding mode control [1,2]. Equation (3) is rewritten as in (6):

$$s_{k+d} = G(z^{-1})u_k + F(z^{-1})y_k - C(z^{-1})r_{k+d} \tag{6}$$

where the polynomial $G(z^{-1})$ is defined as:

$$G(z^{-1}) = E(z^{-1})B^0(z^{-1}) + Q(z^{-1}), \tag{7}$$

and polynomials $E(z^{-1})$ and $F(z^{-1})$ satisfy the Diophantine equality,

$$C(z^{-1}) = A^0(z^{-1})E(z^{-1}) + z^{-d}F(z^{-1}) \tag{8}$$

then the minimum generalized variance control input needed to vanish s_{k+d} in (6) is given by:

$$u_k = -G(z^{-1})^{-1}[F(z^{-1})y_k - C(z^{-1})r_{k+d}] \tag{9}$$

where, the polynomials $C(z^{-1})$ and $Q(z^{-1})$ are chosen to make the control system satisfy

$$T^0(z^{-1}) = A^0(z^{-1})Q(z^{-1}) + B^0(z^{-1})C(z^{-1}) \tag{10}$$

belong to the open unit disk, and the polynomials (Q, C), (A^0, C), and (B^0, Q) have no common zeros outside of the unit disk [13].

The reference signal r_k is considered bounded, i.e $|r_k| < m_r$ for all k, where m_r is a positive constant, polynomials $C(z^{-1})$ and $T^0(z^{-1})$ must be Schur. Let the recursive self-tuning be defined by (11) y (12):

$$\hat{\theta}_k = \hat{\theta}_{k-1} + \frac{\Gamma_{k-1}\phi_{k-d}}{\mu + \phi_{k-d}^T \Gamma_{k-1}\phi_{k-d}}[s_k + C(z^{-1})r_k - \phi_{k-d}^T\hat{\theta}_{k-1}], \tag{11}$$

$$\Gamma_k = \frac{1}{\mu}(\Gamma_{k-1} - \frac{\Gamma_{k-1}\phi_{k-d}\phi_{k-d}^T\Gamma_{k-1}}{\mu + \phi_{k-d}^T\Gamma_{k-1}\phi_{k-d}}) \tag{12}$$

where $0 < \mu \leq 1$. Then the minimum generalized variance control input needed to vanish s_{k+d} in (6) is given by (13):

$$u_k = -\hat{G}(z^{-1})^{-1}[\hat{F}(z^{-1})y_k - C(z^{-1})r_{k+d}] \tag{13}$$

where polynomials $\hat{F}(z^{-1})$ (14) and $\hat{G}(z^{-1})$ (15) are inicialy given by the nominal values of $F(z^{-1})$ and $G(z^{-1})$ computed using the Diophantine equality (8) and using (7). And the signal vector ϕ_k, the parameter vector θ and its estimate $\hat{\theta}$ are defined in (16), (17), (18) respectively.

$$\hat{F}(z^{-1}) = \hat{f}_0 + \hat{f}_1 z^{-1} + ... + \hat{f}_{n-1}z^{-n+1}, \tag{14}$$

$$\hat{G}(z^{-1}) = \hat{g}_0 + \hat{g}_1 z^{-1} + ... + \hat{g}_{m+d-1}z^{-(m+d-1)}, \tag{15}$$

$$\phi_k^T = [y_k, y_{k-1}, ..., y_{k-n+1}, u_k, ..., u_{k-m-d+1}], \tag{16}$$

$$\theta^T = [f_0, f_1, ..., f_{n-1}, g_0, ..., g_{m+d-1}], \tag{17}$$

$$\hat{\theta}_k^T = [\hat{f}_0, \hat{f}_1, ..., \hat{f}_{n-1}, \hat{g}_0, ..., \hat{g}_{m+d-1}]. \tag{18}$$

For the case of linear auto-regressive with exogenous input (ARX) SISO system, the model with parametric uncertainties is given by (19):

$$A(z^{-1})y_k = z^{-d}B(z^{-1})u_k + \xi_k \tag{19}$$

where ξ_k represents the disturbance (white noise) signal with media 0. In this case the generalized minimum variance criterion is considered taking the expectation of the disturbance signal, i.e. $E\{\xi_k\}$. For the contemporaneous autoregressive-moving average (CARMA) model (20) with parametric uncertainties:

$$A(z^{-1})y_k = z^{-d}B(z^{-1})u_k + D^0(z^{-1})\xi_k \tag{20}$$

where polynomial $D^0(z^{-1})$ is defined as in (21)

$$D^0(z^{-1}) = 1 + d_1^0 z^{-1} + d_2^0 z^{-2} + ... + d_l^0 z^{-l}, \tag{21}$$

$D^0(z^{-1})$ is considered known and must be Schur. In this case the signals y_k and u_k are defined as filtered signals of the form: $y_k^{'} = \frac{1}{D^0(z^{-1})}y_k$, and $u_k^{'} = \frac{1}{D^0(z^{-1})}u_k$. Then (20) is rewritten as show in (22), which has the same structure as the ARX model given in (19).

$$A(z^{-1})y_k^{'} = z^{-d}B(z^{-1})u_k^{'} + \xi_k \tag{22}$$

For the linear SISO time-variant case, the model for time-varying parameters is defined as in (23):

$$\theta_k = \theta_{k-1} + \eta_k \tag{23}$$

where η_k is a signal with media 0. In this case $0 < \mu < 1$ should be chosen.

3 STC-GMVC-SMC for a Class of Nonlinear Systems

In analogy to the previous linear case, a generalized minimum variance control based on the concept of discrete-time sliding mode control, is presented for a class of nonlinear systems. In general, the class of nonlinear systems [11] is defined as a SISO time-invariant model with the following structure:

$$A^0(z,q)y_k = B^0(z,q)u_k \tag{24}$$

where $A^0(z,q)$ and $B^0(z,q)$ are polynomials of the form

$$A^0(z,q) = 1 + a_1(q)z^{-1} + a_2(q)z^{-2} + ... + a_n(q)z^{-n},$$
$$B^0(z,q) = b_0(q) + b_1(q)z^{-1} + b_2(q)z^{-2} + ... + b_m(q)z^{-m}.$$

with q being a function defined as in (25),

$$q = h(y_k) = h_{11} + h_{12}y_{k+n-1} + h_{13}y_{k+n-1}^2 + ... + h_{1n}y_{k+n-1}^{n-1} \tag{25}$$

when (25) is substituted in (24), a nonlinear model with polynomial structure is obtained, as in (26)

$$y_{k+n} + a_1(q)y_{k+n-1} + a_2(q)y_{k+n-2} + ... + a_n(q)y_k = b(q)u_k \tag{26}$$

If

$$\begin{aligned} a_1(q) &= a_{11} + a_{12}y_{k+n-1} + ... + a_{1n}y_{k+n-1}^{n-1}, \\ a_2(q) &= a_{21} + a_{22}y_{k+n-1} + ... + a_{2n}y_{k+n-2}^{n-1}, \\ &\vdots \\ a_n(q) &= a_{n1} + a_{n2}y_{k+n-1} + ... + a_{nn}y_k^{n-1}. \\ b(q) &= b_0 + b_1y_k + ... + b_ny_k^n, \end{aligned}$$

then, writing (26) in the present time k and grouping, the following model (27) is obtained

$$\begin{aligned} &A^0(z^{-1})y_k + \Delta_{y_k^2}(z^{-1})y_k^2 + ... + \Delta_{y_k^n}(z^{-1})y_k^n \\ &= z^{-d}B^0(z^{-1})u_k + \Gamma_{y_ku_k}(z^{-1})y_ku_k + \Gamma_{y_k^nu_k}(z^{-1})y_k^nu_k \end{aligned} \tag{27}$$

with,

$$\begin{aligned} A^0(z^{-1}) &= 1 + a_{11}z^{-1} + ... + a_{n1}z^{-n} \\ \Delta_{y_k^2}(z^{-1}) &= a_{12}z^{-1} + ... + a_{n2}z^{-n} \\ &\vdots \\ \Delta_{y_k^n}(z^{-1}) &= a_{1n}z^{-1} + ... + a_{nn}z^{-n} \\ B^0(z^{-1}) &= b_0 \\ \Gamma_{y_ku_k}(z^{-1}) &= b_1 \\ &\vdots \\ \Gamma_{y_k^nu_k}(z^{-1}) &= b_n \end{aligned}$$

Polynomials in (27) have not common factors, and n, m and d are known. Then, the generalized minimum variance control input required to vanish s_{k+d}, defined same as in (3), using the nominal nonlinear model (27) is given by:

$$u_k = -\frac{F(z^{-1})y_k + P_{y_k^2}(z^{-1})y_{k+d}^2 + ... + P_{y_k^n}(z^{-1})y_k^{n+d} - C(z^{-1})r_{k+d}}{G(z^{-1}) + P_{y_ku_k}(z^{-1})y_k + ... + P_{y_k^nu_k}(z^{-1})y_k^n} \tag{28}$$

where

$$P_{y_k^2}(z^{-1}) = -E(z^{-1})\Delta_{y_k^2}(z^{-1}),$$
$$\vdots$$
$$P_{y_k^n}(z^{-1}) = -E(z^{-1})\Delta_{y_k^n}(z^{-1}),$$
$$P_{y_k^2 u_k}(z^{-1}) = E(z^{-1})\Gamma_{y_k u_k}(z^{-1})$$
$$\vdots$$
$$P_{y_k^n u_k}(z^{-1}) = E(z^{-1})\Gamma_{y_k^n u_k}(z^{-1})$$

When the plant parameters in (27) are not accurately known and parameter identification should be performed, the following STC-GMVC-SMC (Self-Tuning Control based on Generalized Minimum Variance Criterion under the Sliding Mode Concept) for Nonlinear Systems algorithm is given. In analogy with the STC-GMVC-SMC for Linear Systems, the nonlinear model with parametric uncertainties is (29)

$$A(z^{-1})y_k + \Delta_{y_k^2}(z^{-1})y_k^2 + ... + \Delta_{y_k^n}(z^{-1})y_k^n \quad (29)$$
$$= z^{-d}B(z^{-1})u_k + \Gamma_{y_k u_k}(z^{-1})y_k u_k + \Gamma_{y_k^n u_k}(z^{-1})y_k^n u_k$$

Then, the controller uses identified parameters as follows

$$u_k = -\frac{\hat{F}(z^{-1})y_k + \hat{P}_{y_k^2}(z^{-1})y_{k+d}^2 + ... + \hat{P}_{y_k^n}(z^{-1})y_k^{n+d} - C(z^{-1})r_{k+d}}{\hat{G}(z^{-1}) + \hat{P}_{y_k u_k}(z^{-1})y_k + ... + \hat{P}_{y_k^n u_k}(z^{-1})y_k^n} \quad (30)$$

Using the recursive Eqs. (11) and (12), where

$$\phi_k^T = [y_k, ..., y_{k-n+1}, y_k^2, ..., y_{k-n-d+1}^2, ..., y_k^n, ..., y_{k-n-d+1}^n, u_k, ..., u_{k-d+1},$$
$$y_k u_k, ..., y_{k-n-d+1}u_{k-d+1}, y^2 u_k, ..., y_{k-d+1}^2 u_{k-d+1}, y^n u_k, ..., y_{k-d+1}^n u_{k-d+1}]$$
$$\theta^T = [f_0, ..., f_{n-1}, P_{y^2 0}, ..., P_{y^2 n+d-1}, ..., P_{y_k^n 0}, ...,$$
$$P_{y_{n+d-1}^n}, g_0, ..., g_{d-1}, P_{y_k u_k 0}, ..., P_{y_k u_k d-1}, P_{y_k^2 u_k 0}, ..., P_{y_k^2 u_k d-1}]$$
$$\hat{\theta}_k^T = [\hat{f}_0, ..., \hat{f}_{n-1}, \hat{P}_{y^2 0}, ..., \hat{P}_{y^2 n+d-1}, ..., \hat{P}_{y^n 0}, ...,$$
$$\hat{P}_{y^n n+d-1}, \hat{g}_0, ..., \hat{g}_{d-1}, \hat{P}_{y_k u_k 0}, ..., \hat{P}_{y_k u_k d-1}, \hat{P}_{y_k^2 u_k 0}, ..., \hat{P}_{y_k^2 u_k d-1}]$$

4 STC-GMVC-SMC Algorithm for Linear and a Class of Nonlinear Systems

In this section a generalized algorithm is proposed. The generalized algorithm is based on the results presented for the different linear model cases, where for each separate case its closed-loop stability has been proved it (see [3–6]). The generalized algorithm proposed in this paper works for the class of nonlinear systems described in previous section, where the overall closed-loop stability was proved it in [11]. And for the particular case when $a(q) = a_0$ and $b(q) = b_0 + b_1 y_k$ the case is the same as for bilinear, first order, systems considered in [10].

Theorem 1 (Patete-Furuta). *STC-GMVC-SMC for Linear and a class of Nonlinear Systems. Given a positive definite matrix Γ_0 and the initial parameters vector $\hat{\theta}_0$, if the estimate $\hat{\theta}_k$ of the estimated controller ((13) or (30)) satisfies the recursive Eqs. (11) and (12), then the closed-loop system is stable and the output signal follows the reference signal in steady state.*

Algoritm: STC-GMVC-SMC for Linear and a class of Nonlinear Systems

1. Given the real model structure.
2. Compute the nominal model (1) for linear case or (27) for nonlinear case.
3. Design $C(z^{-1})$ as in (4) to be Schur.
4. Design $Q(z^{-1})$ as in (5), such that $E_\xi\{T^0(z^{-1})\}$ be Schur.
5. Compute $E(z^{-1})$ and $F(z^{-1})$, using the Diophantine equality (8).
6. Compute $G(z^{-1})$ as in (7).
7. Compute the nominal control law u_k (9) for linear case or (28) for nonlinear case.
8. In the linear case, choose the forgetting factor μ, as it is the case: $\mu = 1$ for time-invariant model or $0 < \mu < 1$ for time-variant model.
9. Compute the recursive equations given by (11) and (12), each sampling time.
10. Compute the self-tuning control law u_k (13) for linear case or (30) for non-linear case, each sampling time.

The proposed algorithm also works for MIMO linear systems defined as in [6], considering the respective dimension for MIMO systems.

4.1 Simulated Example

Using the proposed algorithm for a first order nonlinear system then, Step 1: Given the real model structure,

$$A(z^{-1})y_k + \Delta_{y_k^2}(z^{-1})y_k^2 = z^{-d}(B(z^{-1})u_k + \Gamma_{y_k u_k}(z^{-1})y_k u_k) \tag{31}$$

where $A(z^{-1}) = 1 + a_{11}z^{-1}$, $\Delta_{y_k^2}(z^{-1}) = a_{12}z^{-1}$, $B(z^{-1}) = b_0$, $\Gamma_{y_k u_k}(z^{-1}) = b_1$, and $d = 1$. With parametric uncertainties: $a_{11} \in [-0.8; -0.2]$; $a_{12} \in [0.1; 0.5]$; $b_0 \in [0.1; 0.5]$; $b_1 \in [0.1; 0.9]$. Then the following real model is considered

$$(1 - 0.3z^{-1})y_k + 0.2z^{-1}y_k^2 = z^{-1}(0.4u_k + 0.7y_k u_k) \tag{32}$$

Step 2: Compute the nominal model (27), using the parametric uncertainties intervals.

$$(1 - 0.5z^{-1})y_k + 0.3z^{-1}y_k^2 = z^{-d}(0.3u_k + 0.5y_k u_k) \tag{33}$$

Step 3: In this case $C(z^{-1})$ is designed as: $C(z^{-1}) = 1 + 0.75z^{-1}$, and Step 4: $Q(z^{-1}) = 20(1 - z^{-1})$. Step 5: From the Diophantine equality (8), $E(z^{-1}) = 1$ and $F(z^{-1}) = 1.25$. Then, Step 6: $G(z^{-1}) = 20.3 - 20z^{-1}$. Step 7: Thus the nominal control law u_k (28) is

$$u_k = -\frac{1.25y_k - 0.3z^{-1}y_{k+1}^2 - (1 + 0.75z^{-1})r_{k+1}}{(20.3 - 20z^{-1}) + 0.5y_k} \tag{34}$$

Step 8: As the model is time-invariant, then $\mu = 1$. Step 9: Compute the recursive Eqs. (11) and (12), each sampling time, where $\phi_k^T = [y_k, y_k^2, u_k, u_{k-1}, y_k u_k]$, $\theta^T = [f_0, p_{y^2 0}, g_0, g_1, p_{y_k u_k 0}]$, $\hat{\theta}_k^T = [\hat{f}_0, \hat{P}_{y^2 0}, \hat{g}_0, \hat{g}_1, \hat{P}_{y_k u_k 0}]$. And s_k is computed as in (6):

$$s_k = (20.3 - 20z^{-1})u_{k-1} + 1.25y_{k-1} - (1 + 0.75z^{-1})r_k \tag{35}$$

Step 10: Finally the self-tuning control law u_k (30) computed each sampling time is given by,

$$u_k = -\frac{\hat{f}_0 y_k + \hat{P}_{y^2 0} z^{-1} y_{k+1}^2 - (c_0 + c_1 z^{-1}) r_{k+1}}{(\hat{g}_0 + \hat{g}_1 z^{-1}) + \hat{P}_{y_k u_k 0} y_k} \tag{36}$$

Figure 1 shows the output responses of the real model (32), when the GMVC-SMC is applied using the nominal control law (34) and when the proposed STC-GMVC-SMC algorithm is applied using the recursive control law (36). The GMVC-SMC is not able to control the system to the reference because of the presence of uncertainty parameters, on the contrary, the STC-GMVC-SMC algorithm is able to track the reference in steady-state, and the output dynamic presets good performance.

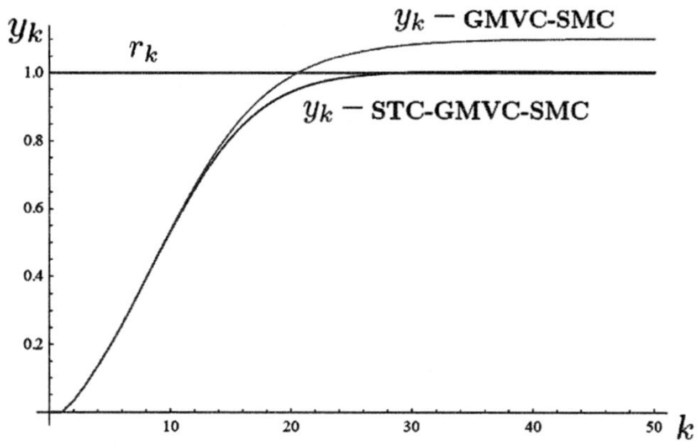

Fig. 1. Output responses of the real model (32).

5 Conclusions

In the aim to find a control law that works for a large group of real process, a generalized ten steps easy to follow, implicit self-tuning control algorithm for linear and a class of nonlinear systems was proposed in this paper. The algorithm may be applied to discrete-time SISO or MIMO linear models, to ARX or CARMA linear models, to a defined structure of time-varying linear models, to some type of bilinear models, and to a defined class of nonlinear models. This algorithm represents a more general implicit (or also called direct) self-tuning control, based on the generalized minimum variance control, using the discrete-time sliding mode control concept, for which the proposed algorithm was named Self-Tuning Control based on Generalized Minimum Variance Criterion under the Sliding Mode Concept (STC-GMVC-SMC) for Linear and a class of Nonlinear Systems. As a future work, the proposed algorithm is to be applied to some nonlinear real process to show its performance.

References

1. Furuta, K.: Sliding mode control of a discrete system. Syst. Control Lett. **14**(2), 145–152 (1990)
2. Furuta, K.: VSS type self-tuning control. IEEE Trans. Ind. Electron. **40**(1), 37–44 (1993)
3. Patete, A., Furuta, K., Tomizuka, M.: Stability of self-tuning control based on Lyapunov function. Int. J. Adapt. Control Signal Process. **22**(8), 795–810 (2008)
4. Patete, A., Furuta, K., Tomizuka, M.: Self-tuning control based on generalized minimum variance criterion for auto regressive models. Automatica **44**(8), 1970–1975 (2008). https://doi.org/10.1016/j.automatica.2007.11.008
5. Patete, A., Furuta, K.: Stability of implicit self-tuning controllers for a class of time-varying systems based on Lyapunov function. Rev. Téc. Ing. **34**(2), 148–155 (2011)
6. Sugiki, A., Ohata, A., Patete, A., Furuta, K.: Design of an implicit multivariable self-tuning controller. In: Proceedings of 47th IEEE Conference on Decision and Control, Mexico, pp. 1340–1345 (2008). https://doi.org/10.1109/CDC.2008.4739049
7. Furuta, K., Ohata, A., Sugiki, A.: Self-tuning control based on discrete-time sliding mode with applications. In: Proceedings of 18th IFAC World Congress, Italy, vol. 18, pp. 774–779 (2011)
8. Patete, A., Furuta, K., Ríos, M.: Self-tuning of bilinear systems based on generalized minimum variance criterion. In: XIII Congreso Latinoamericano de Control Automático y VI Congreso Venezolano de Automatización y Control, Venezuela, pp. 557–562 (2008)
9. Patete, A., Ríos, M., Gómez, C., Furuta, K.: Stability of a self-tuning control for an extended class of bilinear systems. Presented in XIV Congreso Latinoamericano de Control Automático 2010 y XIX Congreso de la Asociación Chilena de Control Automático 2010, Chile (2010)
10. Patete, A., Ríos, M., Gómez, C., Furuta, K.: Self-tuning control for an extended class of bilinear systems, case of study: nuclear fission model. Rev. Cienc. Ing. **35**, 13–22 (2014)
11. Patete, A., Velasco, M., Furuta, K.: Implicit self-tuning control for a class of nonlinear systems. Rev. Cienc. Ing. **38**(2), 131–140 (2017)
12. Patete, A., Velasco, M., Furuta, K.: Nonlinear implicit self-tuning control applied to a slider-crank mechanism. Rev. Cienc. Ing. **38**(1), 41–52 (2017)
13. Aström, K.J., Wittenmark, B.: Adaptive Control. Addison-Wesley, Reading (1989)

Calculation of the Depth of Field in Digital Camera Using Fuzzy Inference Systems

Luini Hurtado-Cortés(✉) and Juan González-Toro(✉)

Universidad Distrital Francisco José de Caldas, Bogotá, Colombia
llhurtadoc@udistrital.edu.co, juan2002manuel@yahoo.es

Abstract. This paper presents the development of Fuzzy Inference Systems of Mamdani and Sugeno to calculate the depth of field in a digital camera. A method is proposed to find the minimum distance from the object to the camera where the image has the greatest depth of field, which is determined by the focal length and the diaphragm aperture. To do this, membership functions for these three variables are designated. The linguistic rules in the form of IF-THEN to correlate the inputs (focal length and diaphragm aperture) with the output (minimal distance) are shown. The results of the Mamdani and Sugeno Fuzzy Inference Systems with graphs and contour surfaces are presented, and they were analyzed and compared with theoretical level curves. The proposed method avoids the use of complex algorithms that require a large memory capacity and extensive processing in a common digital camera. The Matlab® Fuzzy Logic Toolbox was used for this work.

Keywords: Depth of field · Focal length · Diaphragm aperture · Fuzzy Inference Systems

1 Introduction

A common digital camera does not allow manipulation in a separate way both the diaphragm aperture and the focal length (this last related with the zoom). In this kind of camera when the focal length increases (reduces), the diaphragm aperture reduces (increases). Other options are not allowed. In addition, the cameras have two types of zoom: optical zoom and digital zoom. The optical zoom is obtained by altering the focal length with movement of a lens system. In contrast, the digital zoom is obtained by amplifying the image taken, but this affects their quality. For this paper, only the optical zoom has been considered.

The diaphragm aperture, also known as f number, and the focal length affect the depth of field, which determines the range of distances where objects are clearly focused [1]. For a small diaphragm aperture, if the focal length does not change, a great depth of field is obtained.

The depth of field is related to the hyperfocal distance. The depth extends from half of hyperfocal distance to infinity. If the focal length and the f number are combined, a desired depth of field can be found. A method using fuzzy logic to find the minimum

D. F. Cortes Tobar et al. (Eds.): AETA 2019, LNEE 685, pp. 365–374, 2021.
https://doi.org/10.1007/978-3-030-53021-1_37

distance of the object where the image has a depth of field that extends to infinity, through the focal length and aperture system, is proposed.

A possible method is to provide a microprocessor in a digital camera to make calculations using equations and algorithms for these variables. In this situation, a large processing time is necessary.

Another usual method is storing the data, by means of a lookup table, in a Read Only Memory (ROM) incorporated in a microprocessor [2]. The advantage of this method is the reduction of processing time, but the disadvantage is that it requires a mechanism with a greater memory capacity. The above situations are not always available in a common digital camera. Because of this, the "depth of field calculators" are used, which are available on the web.

The alternative method proposed in this paper is the use of Fuzzy Inference Systems (FIS), a technique widely used by software and/or hardware for decision-making and automatic adjusting, using rules of the form IF-THEN. It avoids the use of equations and complex algorithms. The method minimizes both the memory size and the processing time. With these advantages, a "fuzzy depth calculator" can be included in a common digital camera, where it is possible manipulation in a separate way of focal length and diaphragm aperture.

There are some approaches to this technique, in [3] the design of a Mamdani fuzzy inference system to determine the zoom of a digital camera from the focal length, regardless the depth of field, is presented. However, in the simulations show that some improvements and extensions can be made to this focus.

In [4] an ANFIS approach has been developed for the detection of effective digital camera focal distance parameters with additional parameters, which help the measurement of image coordinates that provide the basis for three-dimensional modeling obtained from two-dimensional images. However, this work also does not consider the phenomenon of depth of field.

This paper is not concerned with the technical mechanisms to focus an image with great depth of field in a digital camera; rather, it focuses on the application of Fuzzy Inference Systems to calculate the desired depth of field.

The distance minimum from the camera to the object is calculated to achieve the greatest depth of field possible. This distance is determined by using the Mamdani FIS and Sugeno FIS with rules derived from sources such as tables, databases and photographers' opinions.

In several studies, the results of Mamdani FIS and Sugeno FIS are compared when applied to situations engineering, as the air conditioning systems [5] or cases of economy as the price variation [6]. This work makes the same comparison on the case of the depth of field in a digital camera.

The paper is organized as follows: Sect. 2 presents optical basic concepts related to the depth of field, as such the focal length and diaphragm aperture. Section 3 describes the fundamental concepts of FIS and Mamdani FIS and Sugeno FIS, and IF-THEN rules proposals. Section 4 shows the results as graphs and contour surfaces, which are compared with theoretical curves level for the depth of field. The conclusions are given in the last section.

2 The Depth of Field

Digital cameras have a lens system for varying the focal length, known as Afocal Zoom System [7]. The focal length is the distance between the lens and the focal plane of the camera where the sharp image is formed when the lens focuses to infinity [8]. In a digital camera, sensor CCD (Charge-Coupled Device) it's located in this plane.

The ratio of the minimum and maximum focal length available in the lens system is the maximum optical zoom [9]. In a common digital camera with focal lengths from 5.0 mm to 25.0 mm, the maximum zoom is 5X. The optical zoom image achieves a real approach to the object.

If a lens (camera) focuses the object, there is some distance in front (near limit) and behind (far limit) of the object, which the image will be acceptably sharp. This zone of sharpness is called depth of field [9].

Now, the hyperfocal distance is the point where the lens focuses to achieve an image reasonably sharp, that is, the greatest depth of field possible is obtained. The depth of field is extending from the middle of hyperfocal distance (near limit) to infinity (far limit) [10], this is illustrated in Fig. 1.

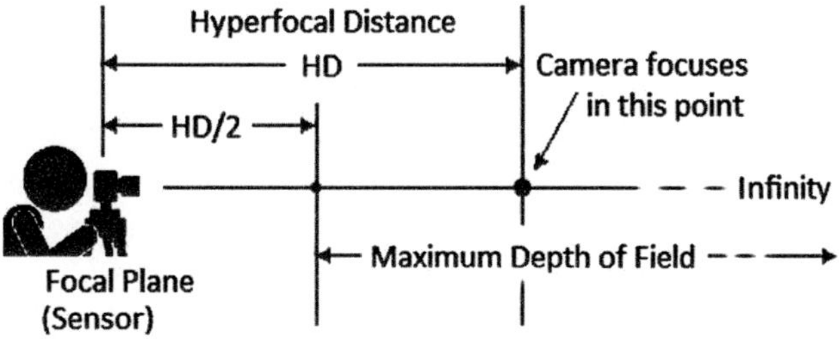

Fig. 1. Hyperfocal distance and depth of field.

Additionally, the diaphragm aperture affects depth of field. The *f number* is the ratio of the focal length and the diameter of the diaphragm aperture. The most common notation of *f* number is "*f/n*" (also known as *f* stop) with *n* real number [8]. When an *f* number is high (more closed aperture) a large depth of field is reached.

3 Fuzzy Inference Systems

3.1 Basic Concepts and Description of FIS

Fuzzy logic is a set of mathematical principles based on degrees of membership that allows modeling information through linguistic rules to relate qualitative variables instead of numeric variables. Such variables form sets whose elements have a membership gradual rather than total, giving their fuzzy nature variables. The membership degree varies from 0 to 1. A fuzzy subset has no specific boundaries and is defined as:

$$A = \left\{ \sum_{i} \mu_A(x_i)/x_i \right\} \tag{1}$$

With $0 \leq \mu_A(x_i) \leq 1$ and $x_i \in X$, where X is the universal set, x_i is an element of X, A is a fuzzy subset of X and $\mu_A(x_i)$ is a membership function of x_i in A [11]. The membership functions may have triangular, trapezoidal, Gaussian shapes, among others, and these are defined for input and output variables.

Fuzzy rules IF-THEN are used to formulate the conditional statements that comprise fuzzy logic. Such rules have the form: IF *"x is A" and "y is B"* THEN *"z is C"*, where "*A*, *B* and *C*" are linguistic variables defined by fuzzy sets, "*x*" and "*y*" are input variables and "*z*" is the output variable. The part "IF" (premise) contains the input variables and the part "THEN" (conclusion) includes output variables [12]. A Fuzzy Inference System (FIS) is the process of transformation from input variables to output variables using fuzzy sets. The FIS most commonly used are the Mamdani FIS and Sugeno FIS.

The inference process is defined as the way in which concrete outputs from fuzzy inputs are generated; it is the main difference of Mamdani FIS and Sugeno FIS. The Mamdani FIS uses methods of defuzzification because its consequents are fuzzy [13]. Some known methods to obtain quantitative outputs (crisp) are "center of area", "area bisector", "mean of maxima" and "maximum center". The Sugeno FIS uses weighted averages to calculate concrete outputs [14]. These outputs are linear or constant functions. Therefore, the Sugeno FIS is computationally faster because it avoids the defuzzification process and replaces it with weighted averages.

3.2 Application of FIS to Obtain the Depth of Field in Digital Cameras

The proposed method has a mechanism for calculating the distance from the camera to the object, when it is permitted to separately control the focal length and aperture diaphragm. This action is not available in common digital cameras.

The focal length and diaphragm aperture system (input variables "*x*" and "*y*") is initially discussed with the Mamdani FIS to determine the minimum distance (near limit) that must be the object (output variable "*z*") and obtain the greatest depth of field. The values for the focal length and diaphragm aperture (*f* number) are taken into ranges from 5.0 mm to 25.0 mm, and from *f/2.8* to *f/5.6*, respectively. The above values are typical for a common digital camera of 5X optical zoom.

Three membership functions of triangular shape are designated for each input. In the focal length, membership functions have labels as "Low, Normal and High" and are centered on 5 mm, 15 mm and 25 mm. The diaphragm aperture has membership functions' labels like "Wide, Moderate and Narrow". They are centered at *f/2.8*, *f/4* and *f/5.6*, respectively. Figure 2 and Fig. 3 illustrate the membership functions for input variables.

For the minimum distance (output variable), nine membership functions are defined, these are associated to three sets called "Near, Medium, Far" each subdivided into "less (−), middle (m) and more (+)" with values centered on 0.4 m, 0.6 m, 0.9 m, 4.0 m, 5.7 m, 8.0 m, 11.0 m, 16.0 m and 22.0 m. The above values were obtained from tables used by expert photographers [1, 8, 15]. Figure 4 show the membership function for distance.

The IF-THEN rules used in this paper are shown in Table 1.

Later, the same focal length and aperture system is analyzed with the Sugeno FIS. The description of the input variables is the same as the Mamdani FIS, but now to each output variable (minimum distance) is designated a linear function that generates a plane.

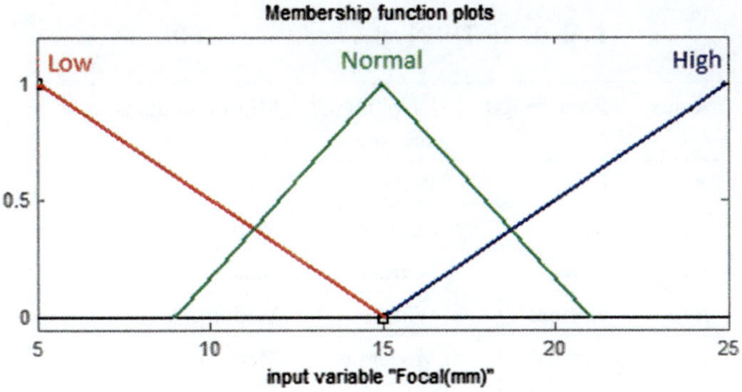

Fig. 2. Membership functions for focal length.

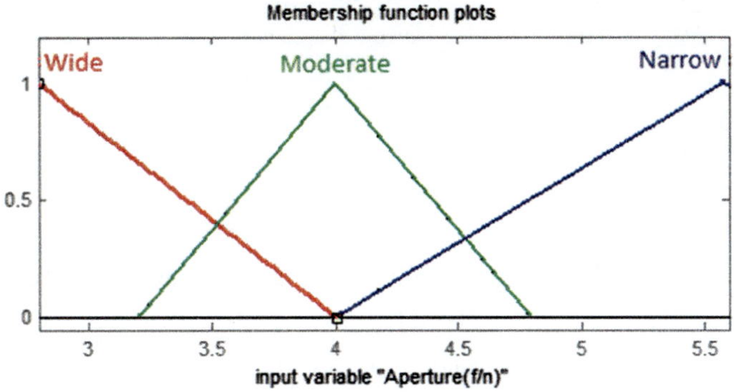

Fig. 3. Membership functions for diaphragm aperture.

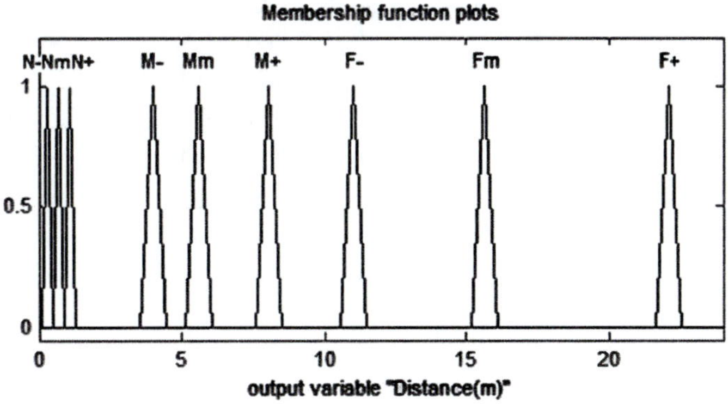

Fig. 4. Membership functions for distance.

The Sugeno FIS smooths the transitions between functions for contour surface [14]. To further smooth these transitions, it is given a slight inclination to the planes; these

Table 1. IF-THEN rules for Mamdani FIS.

Rules	Focal length	Diaphragm aperture	Minimum distance
R1	Low	Wide	Near +
R2	Low	Moderate	Near m
R3	Low	Narrow	Near −
R4	Normal	Wide	Medium +
R5	Normal	Moderate	Medium m
R6	Normal	Narrow	Medium −
R7	High	Wide	Far +
R8	High	Moderate	Far m
R9	High	Narrow	Far −

have equations of the form:

$$ax + by + c = z \tag{2}$$

For Sugeno FIS, the Table 2 is modified by adding the column showed in Table 1.

Table 2. Output variables for Mamdani FIS.

Minimum distance	Function
Near +	0.2x − 0.2y + 0.9
Near m	0.1x − 0.2y + 0.6
Near −	0.1x − 0.2y + 0.4
Medium +	0.1x − 0.3y + 0.8
Medium m	0.1x − 0.3y + 5.7
Medium −	0.1x − 0.3y + 4.0
Far +	0.0x − 0.2y + 22
Far m	0.0x − 0.2y + 16
Far −	0.0x − 0.3y + 11

4 Results and Discussion

The graphs obtained with Mamdani FIS and Sugeno FIS, are illustrated in Fig. 5 and Fig. 6.

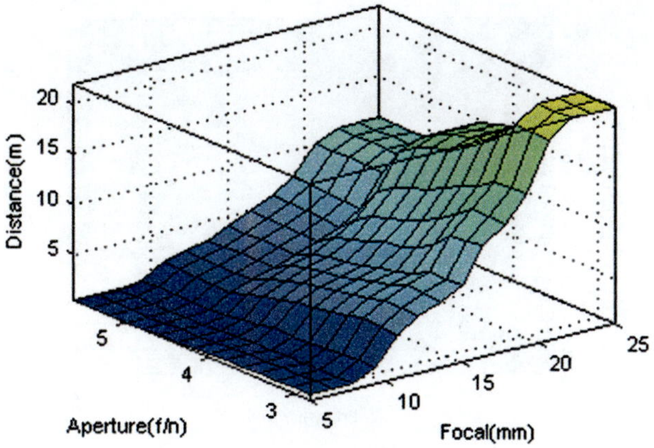

Fig. 5. Mamdani FIS surface.

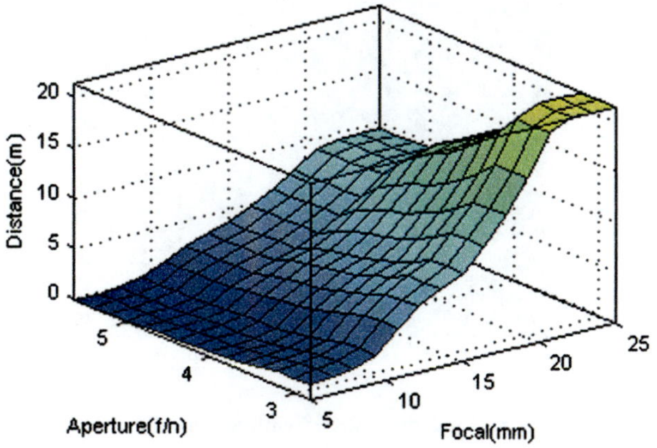

Fig. 6. Sugeno FIS surface.

The surfaces show that, for focal length and diaphragm aperture system, the Mamdani FIS and Sugeno FIS work similarly, however the second surface is smoother than the first, this means, it presents more moderate changes. This aim is reinforced by the subtle inclinations suggested for planes of output variables.

Subsequently, the contour surfaces (view "x" and "y") of the initial graphs in Fig. 5 and Fig. 6 are shown in Fig. 7 and Fig. 8, where "height or level" expresses the distance from the object, and they are compared with the curves theoretical level of Fig. 9.

In the contour surfaces, distances are designated with the following colors, according to Table 3.

Data, algorithms and equations are required to obtain the last graph [1, 13]. In Fig. 9 each contour has labels that express the minimum distance in meters. If storing this information in a camera is desired, it must have great memory capacity and processing, impractical option in a common digital camera.

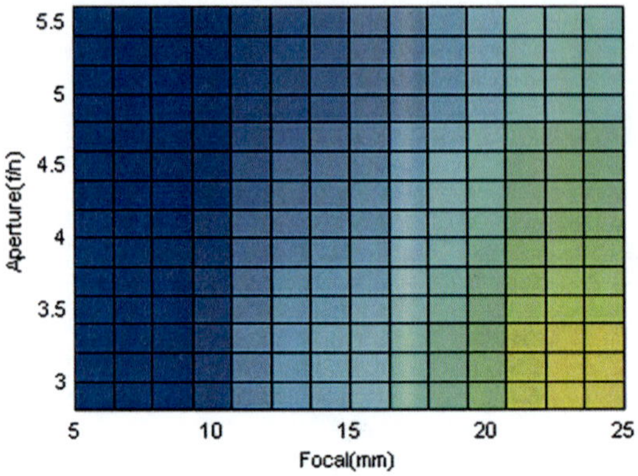

Fig. 7. View "*x*" and "*y*" with Mamdani FIS.

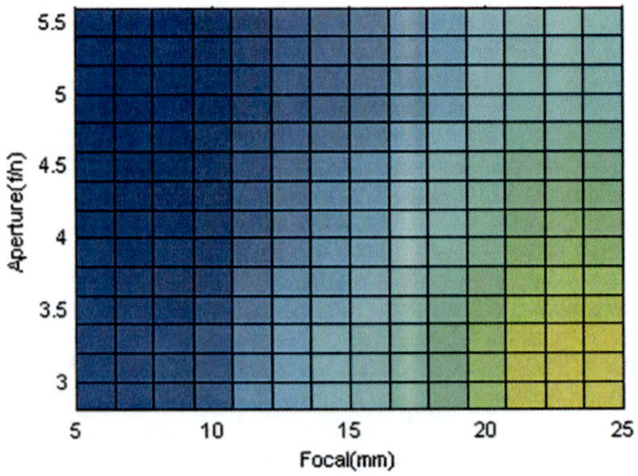

Fig. 8. View "*x*" and "*y*" with Sugeno FIS.

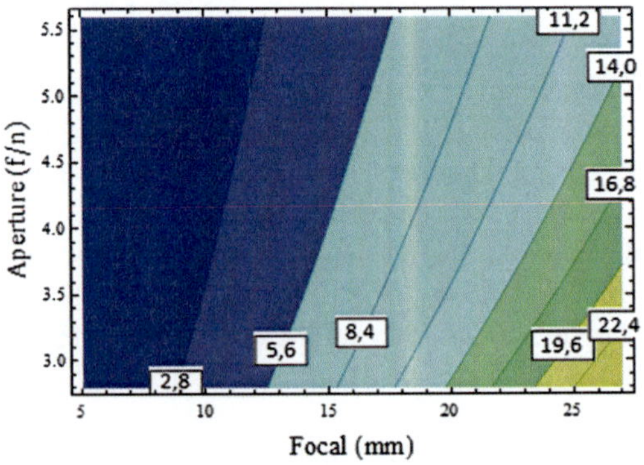

Fig. 9. Theoretical contours.

Table 3. Colors assigned to distances.

Colour	Minimum distance
Blue	Near –
	Near m
	Near +
	Medium –
Turquoise	Medium m
	Medium +
	Far –
Green	Far m
Yellow	Far +

The similarity is observed in the two graphs (Mamdani FIS and Sugeno FIS) with theoretical curves level. The blue region (near distances) dominates the three surfaces and this occurs especially in the smaller focal lengths area, independent of the diaphragm aperture.

The turquoise and green regions are narrow strips surrounding the yellow region and they manifest from areas of intermediate focal lengths and wide aperture (smaller *f/n*) until the areas of high focal lengths and smaller apertures (larger *f/n*). The yellow region (farther distances) occurs in areas of longer focal lengths and larger apertures.

To conclude the discussion of the results, Table 4 shows some numerical calculations for minimum distance, obtained from the Mamdani FIS, Sugeno FIS and "depth of field calculator" on the web [13]. This calculator considers other variables such as the circle of confusion and the CCD sensor size. A circle of confusion of 0.005 mm and a sensor 1/2.3 in. (6.2 mm × 4.6 mm) are used for calculations; these data are typical values in a common digital camera.

Table 4. Numerical calculations of minimum distance with Mamdani FIS, Sugeno FIS and Depth calculator.

	Mamdani FIS	Sugeno FIS	Depth calculator
10 mm, *f/3.6*	3.40 m	3.14 m	3.00 m
15 mm, *f/4.5*	4.96 m	5.02 m	5.00 m
20 mm, *f/5*	8.21 m	8.25 m	7.95 m
25 mm, *f/3.6*	17.8 m	17.5 m	17.55 m
25 mm, *f/3.2*	22.0 m	21.4 m	19.8 m

The calculations of the first row (blue region) show that the values of both FIS are slightly higher than the actual value of the calculator, a fact that is not a problem since at distances greater than the one actual (theoretical minimum distance), the depth of field sought is obtained. Something similar occurs in the third (turquoise region) and the last row (yellow region). The second (turquoise-blue border region) and the fourth row (green-yellow border region) have very similar values.

5 Conclusions

A method that calculates the minimum distance from the object where the image has greatest depth of field possible for a digital camera is proposed. This distance is determined by focal length and diaphragm aperture. To achieve this purpose, the Mamdani and Sugeno Fuzzy Inference Systems (FIS) were used in this calculation. The results obtained by contour surfaces (view "*x*" and "*y*") in both FIS are compared with the curves of theoretical level and the similarity and resemblance between them are evident, so does the calculations of some particular values. This demonstrates the effectiveness of the proposed method, since acceptably good results are obtained without resorting to complex algorithms; avoiding the use of procedures that require large memory capacity and extensive processing in a common digital camera.

References

1. Greenleaf, A.: Photographic Optics. The MacMillan Company, New York (1950)
2. Hoad, P., Illingworth, J.: Automatic control of camera pan, zoom and focus for improving object recognition. In: Fifth IEEE International Conference on Image Processing and its Applications, no. 410, pp. 291–295 (1995)
3. Elamvazuthi, I., Vasant, P., Webb, J.: The application of Mamdani fuzzy model for auto zoom function of a digital camera. IJCSIS Int. J. Comput. Sci. Inf. Secur. **6**(3), 244–249 (2009)
4. Ergün, B., Sahin, C., Kaplan, U.: ANFIS definition of focal length for zoom lens via fuzzy logic functions. In: Ramakrishnan, S. (ed.) Modern Fuzzy Control Systems and its applications, pp. 297–320. IntechOpen (2017)
5. Kaur, A.: Comparison of Mamdani-type and Sugeno-type fuzzy inference systems for air conditioning system. IJSCE Int. J. Soft Comput. Eng. **2**(2), 323–325 (2012)
6. Zaher, H., Kandil, A., Fahmy, R.: Comparison of Mamdani and Sugeno fuzzy inference systems for prediction (with application to prices of fund in Egypt). Br. J. Math. Comput. Sci. **4**(21), 3014–3022 (2014)
7. Zou, T., Tang, X., Song, B., Wang, J., Chen, J.: Robust feedback zoom tracking for digital video surveillance. Sensors **12**(6), 8073–8099 (2012)
8. Hecht, E.: Optics. Addison Wesley, San Francisco (2002)
9. Davis, P.: Photography. Wm. C. Brown Pub., Iowa (1972)
10. Shaw, J.: Landscape Photography. Amphoto Books, New York (1994)
11. Dubois, D., Prade, H.: Fuzzy Sets and Systems: Theory and Applications. Academic Press, New York (1980)
12. Zadeh, L.: Outline of a new approach to the analysis of complex systems and decision processes. IEEE Trans. Syst. Man Cybern. **3**(1), 28–44 (1973)
13. Mamdani, E., Assilian, S.: An experiment in linguistic synthesis with a fuzzy logic controller. Int. J. Man Mach. Stud. **7**(1), 1–13 (1975)
14. Sugeno, M.: Industrial Applications of Fuzzy Control. Elsevier Science Inc., New York (1985)
15. Cambridge in colour homepage. http://www.cambridgeincolour.com. Accessed 29 Oct 2018

Vision Based Upper Limbs Movement Recognition Using LSTM Neural Network

Andrea Rey, Alison Ruiz, Camilo Camacho(✉), and Carolina Higuera

Universidad Santo Tomás, Bogotá, Colombia
{andrearey,alisonruiz,edgarcamacho,carolinahiguera}@usantotomas.edu.co

Abstract. This paper presents the theoretical background and the implementation of a Long Short-Term Memory (LSTM) Neural Network architecture to recognize arm movements from video clips. The pose points (corresponding to the position of six body parts: shoulders, elbows and wrists) are extracted with a pre-trained Convolutional Pose Machine. Those points generate sequences over time with 66 (x, y) pairs, which are the input for a neural network, to classify them in 20 movement classes. Our architecture has 128 LSTM cells and presented 92.5% of accuracy on testing data and an execution time of around 6.64 ms.

Moreover, we present the methodology used to create our dataset, with 2400 samples of 20 different arms movements, recorded by 6 persons with different physical appearance in a controlled environment.

Keywords: Movement recognition · Pose recognition · Recurrent neural network · Long short-term memory neural network

1 Introduction

There are several situations where robotic assistants can improve the quality of life, as supporting elderly people and patients with mobility problems, or backing staff in their duties in offices, stores or medical centers. To achieve this, it is mandatory to develop an human-robot interaction that allows communication to be as fluid and natural as possible. These desired features are not only provided by oral interaction, but also by body language, for which it is necessary to identify correctly the movements that humans perform in front of the robot, so it can react correctly.

The movements or actions that humans perform during the interaction are not only based on the instantaneous position of their body parts, but also on the change of those positions. This fact implies that the information extracted from the video must be processed using a temporal point of view, analyzing more than the last captured frame. Therefore, a classification model with memory is needed, such as a Long-short Term Memory (LSTM) architecture. This approach is used to recognize arms movements from points in a pose estimation, provided by a Convolutional Pose Machine [3,4].

D. F. Cortes Tobar et al. (Eds.): AETA 2019, LNEE 685, pp. 375–383, 2021.
https://doi.org/10.1007/978-3-030-53021-1_38

This paper is organized as follows. In Sect. 2 we present related works about movement recognition. In Sect. 3 we provide a theoretical background and a basic implementation example. Following, in Sect. 4 is described the dataset creation process and the recording environment. Finally, Sect. 5 shows the performance of the model trained with our dataset.

2 Related Work

The pose estimation in each video frame is the first challenge found in the body movement recognition problem. The models that perform this task are known as Pose Machines [14]. The pose is the set of spacial points where body parts are detected. Some methods use Convolutional Neural Networks (CNNs) to generate body parts heatmaps from RGB images, and then post process them to extract the corresponding points [2–4]. Other processes involve RGB-D images, where is taken advantage of the depth information to improve the accuracy of the estimation [15].

The next step is to detect the movement or action from the sequences or changes in the pose points. Convolutional Neural Networks are used in [6,7] from 3D pose sequences. Likewise, Recurrent Neural Networks [17] and LSTM [9,10] take poses directly. In other cases, RAW data is used, as in [12,16]. In [11], spatial rank pooling is proposed to describe the evolution of poses. For cases of detecting particular movements, [13] presents a fall detector that uses a skeleton provided by a RGB-D camera, and a binary SVM as its classification core.

3 Architecture

Long-Short Term Memory (LSTM) is a special type of recurrent neural network, that allows to omit long-term dependencies. Its main feature is the ability to remember information during long periods of time. The architecture of a cell is shown in Fig. 1, where $x^{<t>}$ is the cell input in time t.

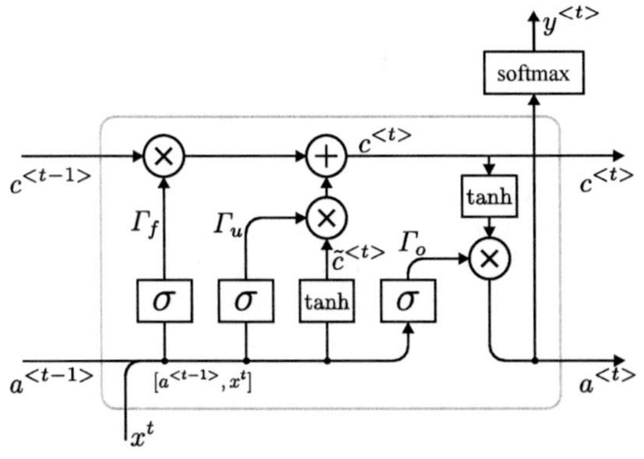

Fig. 1. LSTM cell architecture

$C^{<t>}$ is the **Cell State**, computed as:

$$c^{<t>} = (\Gamma_u * \tilde{c}^{<t>}) + (\Gamma_f * c^{<t-1>}) \tag{1}$$

Γ_f is the **Forget gate**, a value between 0 and 1. It is 0 to forget the previous cell state or 1 to keep it. This value is computed from the linear combination of the previous cell output and the new data, with parameters W_f as weights and b_f as bias, passed through a sigmoid activation function.

$$\Gamma_f = \sigma(W_f[a^{<t-1>}, x^{<t>}] + b_f) \tag{2}$$

$\tilde{c}^{<t>}$ is the **Candidate Cell State**, the new state of the cell according to weights W_c and bias b_c:

$$\tilde{c}^{<t>} = \tanh(W_c[a^{<t-1>}, x^t] + b_c) \tag{3}$$

Γ_u is the **Update gate**, a value between 0 and 1, and represents how much of the candidate cell state is going to be transferred to the current cell state:

$$\Gamma_u = \sigma(W_u[a^{<t-1>}, x^{<t>}] + b_u) \tag{4}$$

The cell's output is defined as:

$$a^{<t>} = \Gamma_o * \tanh(c^{<t>}) y^{<t>} \tag{5}$$

where Γ_o is the **Output Gate**:

$$\Gamma_o = \sigma(W_o[a^{<t-1}, x^{<t>}] + b_o) \tag{6}$$

Each cell can select how important is to forget the old information (through the *forget gate*) or to remember the current one (using the *update gate*). LSTM cells provide the capability to effectively associate memories and inputs over time to the network. This approach involves a dynamic structure over time and high prediction capacity [5]. In general, Recurrent Neural Networks (RNN) can be connected in different ways, depending on the application [8]. In the case of this work, it is needed to generate a unique output (class) from a sequence of data (movement), so it is used a *many to one* architecture, where it is only used the last output, as shown in Fig. 2.

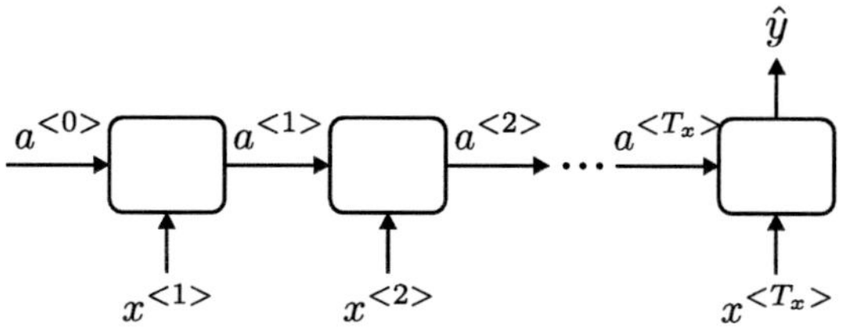

Fig. 2. *Many to one* architecture

4 Dataset

Our dataset includes 20 arms movements listed in Table 1. Figure 3 shows some examples. For the construction of the dataset, 6 people recorded 120 samples by movement, for a total of 2400 samples in a balanced dataset.

Table 1. Movements included in the dataset

Up-down	Down-up	Front-up	Up-front
Front-down	Down-front	Center-down	Down-center
Center-sides	Sides-center	Center-front	Front-center
Sides-front	Front-sides	Up-down	Down-up
Center-up	Up-center	Sides-up	Up-sides

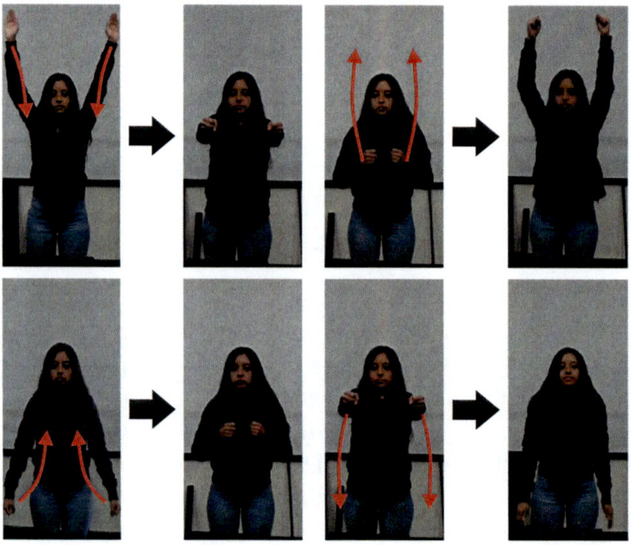

Fig. 3. Movements examples. From left to right and from top to bottom: Up-Front, Center-Up, Down-Center and Front-Down.

The following subsections present the methodology that we used for the creation of the dataset.

4.1 Pose Extraction

In order to get the joint positions and create the movement sequences, a Convolutional Pose Machine (CPM) was used [3,4], previously trained with the MPII

Human Pose Dataset [1]. The desired points are extracted from the maximums of the heat maps given by the convolutional network (one by each interest point). Then, points are associated between them, in order to determine which of them belong to the same person, according to the 2D trust maps and a set of vector fields of affinities of parts that encode the degree of affinities between parts (e.g. association between the neck and a shoulder, between a shoulder and an elbow, etc.), given by the CPM too. This analysis produces the set of 2D points corresponding to the body parts of the person in front of the camera, as shown in the Fig. 4.

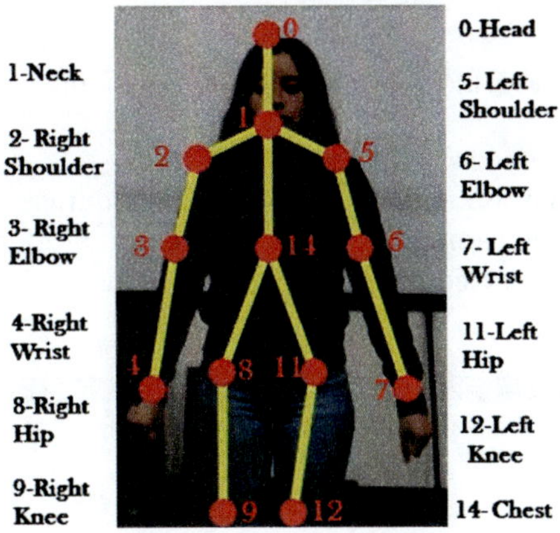

Fig. 4. Points extracted from the Pose Machine

The CPM provides 15 key points, however, Fig. 4 only presents 13, the 3 remaining correspond to feet. For the case of this work, only 7 points were used: chest, shoulders, elbows and wrists (points 2 to 7 and 14).

4.2 Recording Environment Setup

A camera was installed 2 m from the place where the person performs the movements and 1.2 m from the floor (Fig. 5), considering the height for all possible actors. The light conditions are the average for office environments. The camera is connected to a laptop that records the clips of each movement. The camera configuration was 30 FPS and 640 $\times$ 480 pixels resolution.

4.3 Graphical User Interface

In order to easily control the recording of the video clips, a Graphical User Interface (GUI) was developed, which is presented in Fig. 6. The GUI includes the following sections: 1. *Capture Parameters*, where FPS and resolution can be established; Sections 2, 3 and 4 *Dataset Information*, allow to manage people

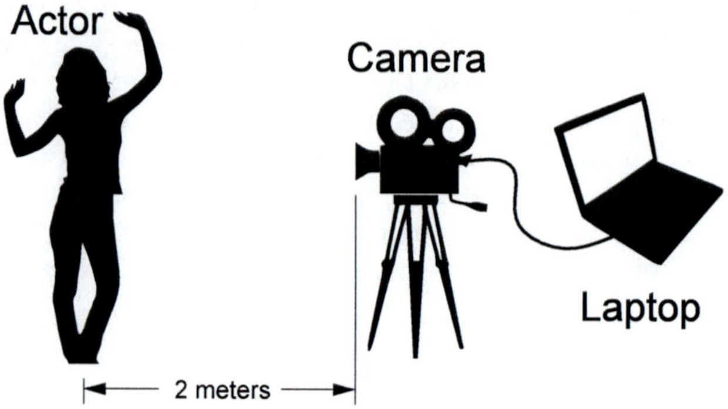

Fig. 5. Recording environment setup

and movements information and the path where the dataset is stored; Section 5. *Visualization*, where the camera image is shown; and section 6. *Record Control*, where the recording is started and stopped.

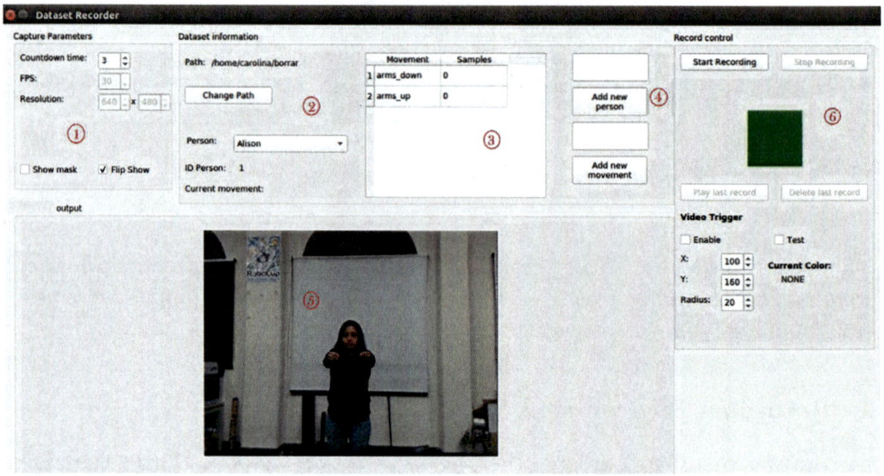

Fig. 6. Dataset recorder interface

4.4 Post Processing

Each video was processed, frame by frame, by the CPM, generating a sequence or sample of bidimensional (x, y) data for each body point. These sequences have 30 points per second (because of the 30 FPS frame rate of the captured videos), and the quantity of points depends on the length of each video. For that reason, the samples length was resized to the average duration of all video clips

(66 frames), and then were interpolated to 30 FPS again, getting sequences of the same length for the complete dataset.

In addition, bidimensional points were approximated to metric units, and then, they were centered around the chest point. After that, the chest point was discarded. In summary, each sample of the dataset has 6 points (shoulders, elbows and wrists) with the reference at the chest location, with 66 bidimensional (x, y) pair per point.

5 Results

A *many to one* architecture was implemented (recall Fig. 2). The input is the set of bidimensional points, 6 points $\times$ 2 dimensions = 12, so $x^{<t>} \in \mathbb{R}^{12}$, for a sequence length of 66. There are 20 possible classes, so $y^{<t>} \in \mathbb{R}^{20}$. We implemented 128 cells. Loss was computed using softmax with cross-entropy. Adam optimization was used for training with learning rate of 10^{-3}. Figure 7 presents the learning curve obtained with 70% of the dataset and 400 epochs. It can be noticed that, around 300 iterations, the training converged.

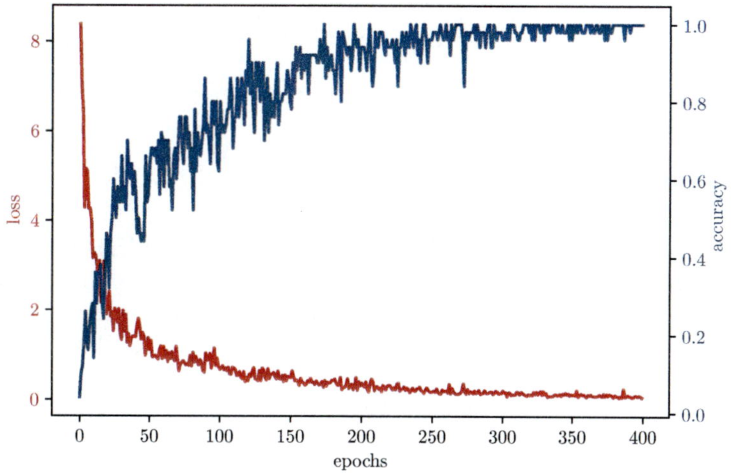

Fig. 7. Training behavior of LSTM neural network

The remaining 30% of the dataset was used for evaluation, obtaining an accuracy of 92.5%. The complete confusion matrix is presented in Fig. 8.

Network processing time was tested in CPU, with a laptop with the following specifications: Intel® Core^{TM} i5-5200U, 8 GB RAM, no dedicated GPU, Ubuntu 16.04; getting an average inference time of 6.64 ms per data sequence, appropriate for real-time applications.

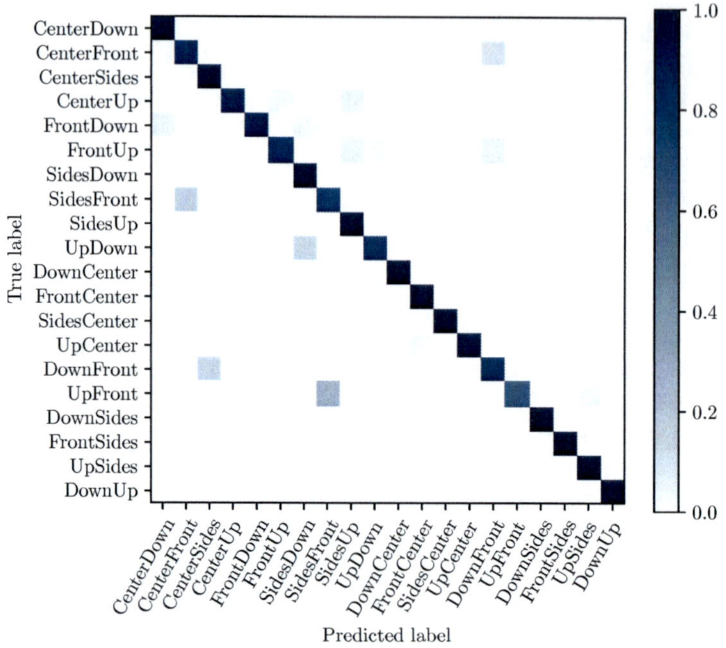

Fig. 8. Confusion matrix for our arms movement recognition problem

6 Conclusions

A Long-Short Term Memory (LSTM) neural network was implemented for recognition of movements from pose points in video clips, getting and accuracy of 92.5% on testing data. This result shows that a sequential model (neural networks with memory) are appropriate for movements recognition. The tested architecture has an average processing time of 6.64 ms per data sequence, which is fast enough to be used in real time applications.

Besides, our arms movements dataset was created recording video clips, with 2400 samples of 20 different movements, recorded by 6 people with different physical appearance in a controlled environment. In addition, the post processed data sequences were also generated, allowing the using of the spacial pose points instead of the RAW image. This is an useful material for research in machine learning and artificial vision, due to it allows inputs for data series classification.

References

1. Andriluka, M., Pishchulin, L., Gehler, P., Schiele, B.: 2D human pose estimation: new benchmark and state of the art analysis. In: IEEE Conference on Computer Vision and Pattern Recognition (CVPR), June 2014

2. Bulat, A., Tzimiropoulos, G.: Human pose estimation via convolutional part heatmap regression. CoRR abs/1609.01743 (2016). http://arxiv.org/abs/1609.01743
3. Cao, Z., Hidalgo, G., Simon, T., Wei, S.E., Sheikh, Y.: OpenPose: realtime multi-person 2D pose estimation using part affinity fields. arXiv preprint arXiv:1812.08008 (2018)
4. Cao, Z., Simon, T., Wei, S.E., Sheikh, Y.: Realtime multi-person 2D pose estimation using part affinity fields. In: CVPR (2017)
5. Hochreiter, S., Schmidhuber, J.: Long short-term memory. Neural Comput. **9**(8), 1735–1780 (1997). https://doi.org/10.1162/neco.1997.9.8.1735
6. Huang, Z., Wan, C., Probst, T., Gool, L.V.: Deep learning on lie groups for skeleton-based action recognition. CoRR abs/1612.05877 (2016). http://arxiv.org/abs/1612.05877
7. Ke, Q., Bennamoun, M., An, S., Sohel, F.A., Boussaïd, F.: A new representation of skeleton sequences for 3D action recognition. CoRR abs/1703.03492 (2017). http://arxiv.org/abs/1703.03492
8. Lipton, Z.C.: A critical review of recurrent neural networks for sequence learning. CoRR abs/1506.00019 (2015). http://arxiv.org/abs/1506.00019
9. Liu, J., Wang, G., Hu, P., Duan, L., Kot, A.C.: Global context-aware attention LSTM networks for 3D action recognition. In: 2017 IEEE Conference on Computer Vision and Pattern Recognition (CVPR), pp. 3671–3680, July 2017
10. Liu, J., Shahroudy, A., Xu, D., Wang, G.: Spatio-temporal LSTM with trust gates for 3D human action recognition. CoRR abs/1607.07043 (2016). http://arxiv.org/abs/1607.07043
11. Liu, M., Yuan, J.: Recognizing human actions as the evolution of pose estimation maps. In: 2018 IEEE/CVF Conference on Computer Vision and Pattern Recognition, pp. 1159–1168, June 2018
12. Oreifej, O., Liu, Z.: HON4D: histogram of oriented 4D normals for activity recognition from depth sequences. In: 2013 IEEE Conference on Computer Vision and Pattern Recognition, pp. 716–723, June 2013
13. Seredin, O.S., Kopylov, A.V., Huang, S.C., Rodionov, D.S.: A skeleton features-based fall detection using Microsoft Kinect V2 with one class-classifier outlier removal. In: ISPRS - International Archives of the Photogrammetry, Remote Sensing and Spatial Information Sciences, vol. XLII-2/W12, pp. 189–195 (2019). https://www.int-arch-photogramm-remote-sens-spatial-inf-sci.net/XLII-2-W12/189/2019/
14. Wei, S.-E., Ramakrishna, V., Kanade, T., Sheikh, Y.: Convolutional pose machines. CoRR abs/1602.00134v4 (2016). https://arxiv.org/pdf/1602.00134.pdf
15. Shotton, J., Fitzgibbon, A., Cook, M., Sharp, T., Finocchio, M., Moore, R., Kipman, A., Blake, A.: Real-time human pose recognition in parts from single depth images. In: CVPR 2011, pp. 1297–1304 (2011)
16. Yang, X., Tian, Y.: Super normal vector for activity recognition using depth sequences. In: 2014 IEEE Conference on Computer Vision and Pattern Recognition, pp. 804–811, June 2014
17. Du, Y., Wang, W., Wang, L.: Hierarchical recurrent neural network for skeleton based action recognition. In: 2015 IEEE Conference on Computer Vision and Pattern Recognition (CVPR), pp. 1110–1118, June 2015

Kinematic Model Analysis and ROS Control of Cable Driven Continuous Robot Manipulator

Hernando Leon-Rodriguez[1,2(✉)], Yefry Moncada[2], Simon Mosqueda[2], Cecilia Murrugarra[2], and Michael Canu[2]

[1] Electronics Engineering Department, El Bosque University, Bogota, Colombia
hefrainl@unbosque.edu.co

[2] Bioengineering Department, El Bosque University, Bogota, Colombia
{ymoncada,smosqueda}@unbosque.edu.co

Abstract. This paper describes the developed of cable continues redundant manipulator and its kinematic model analysis. The goal is to determine the accuracy of the experimental prototype with the constant curvature kinematic model. The experimental validation of the constant curvature kinematic model is evaluated to establish a relation between the mathematical model and continuum robot, and its feasibility of applying for any type of continuous redundant structure by computing the end effector trajectory based on the displacement of passive cables located along the structure. The continuous redundant prototype robot is composed by 2 segments with 7 modules sections powered by 6 driven cables where each segment is controlled with 3 lineal motors. The length of the segment is desired but it is also allows to add modular sections and connect them like serial chain. The radius of concave and/or convex curvature allows the robot manipulator to follow different trajectories and displacements in their workspace. The paper also describes the relevant components and the position feedback control implemented via ROS.

Keywords: Continuous redundant robot · End effector · Mathematical model · Cable robotic system

1 Introduction

The robotics researchers was mainly focused in automated manipulators for dangerous work environments or manufacturing processes to replace human operator, especially in those with executed routines tasks. Industrial robots have traditionally open links with connected joints directly actuated [1]. In addition, these types of robots have been proved to be useful in many industrial tasks, however, they have obvious weaknesses in constricted spaces due to the lack of degrees of freedom (D.O.F.), low level of flexibility and in the real world the environment and task most of them are unknown for rigid link manipulator [2].

Nowadays, some of these problems have been resolved through biologically inspired designs, which aim to reproduce the locomotion properties and adaptability of certain

D. F. Cortes Tobar et al. (Eds.): AETA 2019, LNEE 685, pp. 384–393, 2021.
https://doi.org/10.1007/978-3-030-53021-1_39

animal extremities such as octopus, or squid tentacles [3], snake-like structures [4], elephant trunks [5] and others. This type of robotics mechanisms are known as a continuum or hyper-redundant robots where contrasts from traditional manipulators in the sense that they are capable of possessing a high range of flexibility and more degrees of freedom enabling them to traverse precisely through almost any "space" and around many obstacles. I recent years, they have gained wide interest from the medical field [6], specifically, in conducting Minimal invasive surgery (MIS) [7]. And also, in industrial applications they are considered to be ideal for probing cramped and hazardous industrial environments like inspect nuclear reactors [8], as well as confined spaces [9], to search and rescue in disaster situation [10]; One of the best examples is been developed by a British company OC robotics [11].

The research of multiple DOF are been proposal a few year back by mimicking a snake's locomotion [12], followed by the work of Chirikjian and Burdick that extended on this approach of fitting the hyper-redundant robot to analytical desirable mathematical curves by establishing them as the product of a Bessel function with trigonometric models [13], moreover, they described the behaviour of the constant curving structure using the Frenet-Serret Frames equations [14]. Subsequently, the work of Walker [15], Zheng, Webster and Jones [16] allowed for the integration of these models, with the D-H parameters [17], the integral representation and a product of exponentials (POE) [18], to finally obtaining a reliable kinematic model derived to constant-curvature, which is capable to accurate describes the behaviour of any continuum robot and is one of the most well-known and used for modelling.

The classifications of type of robots by their defining feature are continuously curving back-bone or core. This backbone is known as continuous redundant and can be segmented and resembled to serial rigid-links, which together give the appearance of one continuous structure. The shape can be modified with diverse types of actuators, as example: pneumatic mechanism [19], tendon mechanism [20].

Most continuum redundant robots have a few actuators; these are also known under-actuated considering the number of DOFs produced. These under-actuated systems had a substantial simplicity on the control system, with some disadvantage in speed, strength and limited workspace due to the reduce dexterity. Nevertheless, these robots by their type of actuation and structure can be classified in three main categories, tendon-driven serpentine manipulators (TSM), Tendon-driven continuum manipulators (TCM) [21], and concentric tube manipulators (CTM) [18].

This work presents the description of technical characteristics in the mechanical design, motion, electronic system and measurements. The paper is based in the evaluation of the relation between the mathematical models of continuous robot and some experimental measuring in the end effector in planar trajectory [22]. This work was done by tracking the displacement of passive cables located along the structure, with the purpose of estimating the resulting shape of the planar trajectory. The further research work and experimental validation implemented of the continuous redundant robot is presented by analysing two segments in a formal description of the Mathematical model and validation.

2 Prototype Design

Figure 1 (left) is showing a render and prototype of the continuous redundant manipulator; it is composed by 2 segments with 7 sections each and free joints between them. The bases are made in MDF in order to locate the manipulator on top of them and underneath are located the 6 lineal actuators. Each segment is power by 3 lineal motors where its mathematical model and analysis is presented in the next section.

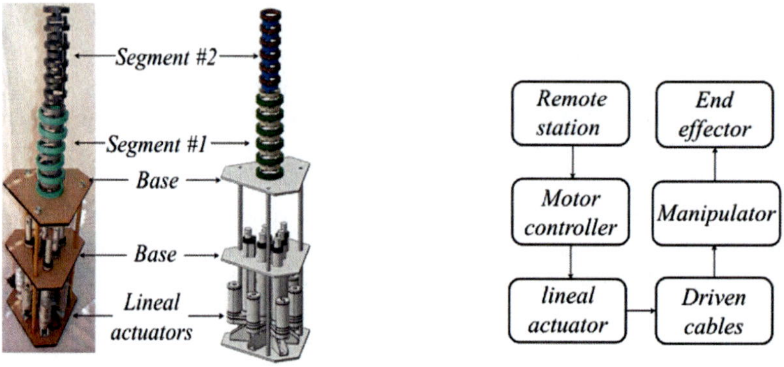

Fig. 1. (Left) Prototype platform of continuous redundant robot composed by: 2 segments, 14 sections, 6 cables (3 for each segment), and 6 lineal actuators. (Right) Block diagram of prototype platform for the continuous redundant manipulator

Figure 1 (right) is showing the block diagram of the experimental prototype controller from the remote stations (ROS computer) to end up in the end effector.

The Continuous redundant Manipulator system showing in Fig. 2 was implemented to position the end effector in a determined 3D location by executed the following steps. The ROS software located in the remote stations send a specific position command through the motor controllers. The motors using the motor control boards until the lengths of the tension cables are reached. When the end effector is close as possible to the theoretical lengths the systems stops by compared the final position of the actuator through of lineal sensor located in the stroke of the lineal motor.

The main components are listed as follow:

- **Processing central** by a remote station computer with ROS software.
- **Motor Control Boards** using six own developed amplifiers cards mastered with an Arduino card.
- **Lineal Actuators and Transmission Mechanics** by six ECO-WORTHY: stroke length 150 mm, rated at 5.7 mm/s, 1500 N pull and 1200 N push force and encoder.
- **Cables wire**, stainless-steel (316), with 0.28 m in length and 1 mm diameter.
- **Lineal sensors**, Contact transducer SoftPad sensor with internal variable resistance from 100 Ω to 10 kΩ, with 200 mm length and 5 V.
- **Manipulator** 2 segments were designed: the first segment is composed by 7 Polypropylene tube sections with 0.20 m in length and 0.14 m in diameter, with

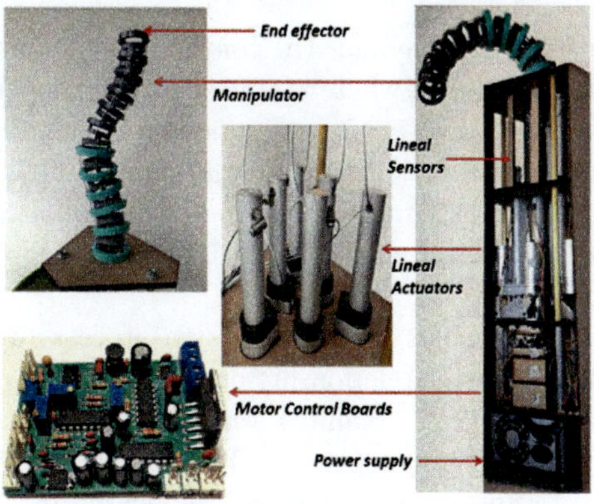

Fig. 2. Continuous redundant manipulator prototype and its mechanical architecture.

seven sections, free joints. The second segment is composed by 7 PVC tube sections with 0.22 m in length and 0.08 m in diameter, with seven sections, free joints.

- **End effector** finally locating point of manipulator.

The kinematic model used to obtain the necessary cable lengths to move the manipulator to a certain position was simulated with MATLAB, using the equations of the further section.

3 Analysis and Mathematical Model

This section descripts the manipulator shape and end-effector position by analysing the robust kinematic model based on constant curvature, describing the movements of the structure and workspace.

The classic modelling of an articulated robot system consists in compute the position and orientation of the end effector based in world reference coordinates; this are achieved by defining a robot with link or rigid elements and lineal or rotated joints. However, this mathematical method is not suitable to a system that has more degrees of freedom than those needed to carry out; as a result, not useful to continuous robots.

$$A = \begin{bmatrix} \cos^2(\phi)[\cos(ks-1)]+1 & \sin(\phi)\cos(\phi)[\cos(ks-1)] & -\cos(\phi)\sin(ks) & \frac{cos(\phi)[\cos(ks-1)]}{k} \\ \sin(\phi)\cos(\phi)[\cos(ks-1)] & \cos^2(\phi)[\cos(ks-1)]+\cos(ks) & -\sin(\phi)\sin(ks) & \frac{sin(\phi)[\cos(ks-1)]}{k} \\ \cos(\phi)\sin(ks) & \sin(\phi)\sin(ks) & \cos(ks) & \frac{\sin(ks)}{k} \\ 0 & 0 & 0 & 1 \end{bmatrix} \tag{1}$$

The kinematical model developed for continuous redundant manipulator [23] and [14], uses a substitutions on the Denavit-Hartenberg (D-H) transformation matrix, with the aim of obtain a model that defines a more accurate behaviour of translations and rotations with standard homogeneous transformation matrix A(θ, d), Eq. 1.

Further derivations is requires with different mathematical analysis, in which a series of substitutions are made on the D-H transformation matrix, in order to obtain a model that relates a more approximate behaviour. The conventional notation involving a homogeneous matrix is yielded by A(θ, d), where: θ represents the independent rotations of the joints and d the translations. However, for continuous robots, performing this analysis involves a homogeneous matrix for each section that makes up a segment, so authors such as Ian D. Walker, which associates the shape of the robot with a curve in space by the middle of the Frenet Serret equations [24], one can establish an equivalent to rotations and translations.

The Fig. 3 shows the variables for the first step process to obtaining a reliable mathematic model, is to establish a relationship between the D-H parameters θ, and the trunk parameters s, k, φ, as follow, obtaining a near approach to the continuum robot movements in terms of the cable length [16, 23]. The space transformations between the D-H parameters and s, k, φ and actuators length is shown in Fig. 4.

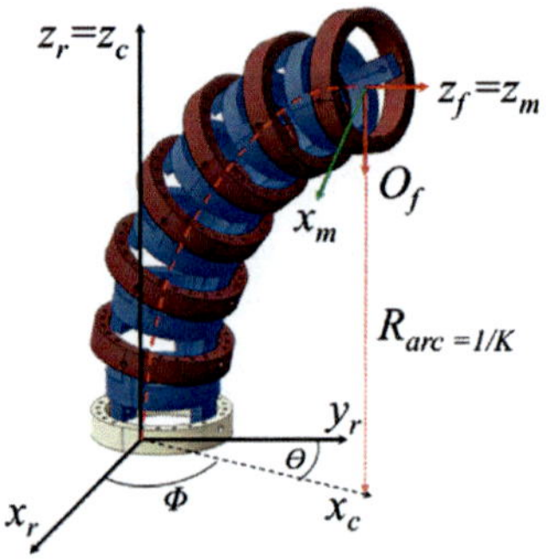

Fig. 3. Parameters for continuous redundant Robot Manipulator with free joint sections.

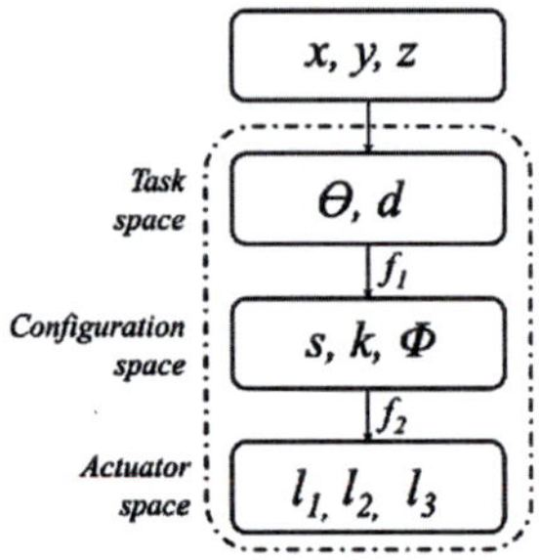

Fig. 4. Space transformations between the D-H parameters to actuations spaces s, k, φ, l_1, l_2, l_3.

According to the coordinate system and the geometric relationship with the arc parameters, the homogeneous matrix of the n segment can be calculated as Eq. 1, where the matrix of transformation A, allows us to know the position and orientation to perform an analysis of the bending of the continuum robot of the end effector and its 3D workspace. Figure 5 is showing the computer simulation.

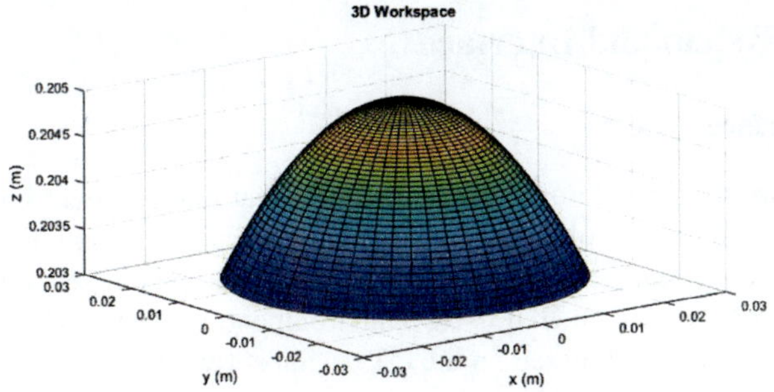

Fig. 5. Workspace of Continuum Robot Manipulator with 0.22 m length.

In Eq. 2, specify the space position of the end effector as follow:

$$\begin{bmatrix} P_x \\ P_y \\ P_z \end{bmatrix} = \begin{bmatrix} \frac{\cos(\phi)[\cos(ks-1)]}{k} \\ \frac{\sin(\phi)[\cos(ks-1)]}{k} \\ \frac{sin(ks)}{k} \end{bmatrix} \tag{2}$$

Through Eqs. 3, 4 and 5 can be computed to find the position of the end effector based on the arc parameters, these parameters can be controlled by adjusting the length of the tensor cables, as a result, it is necessary to derive the mapping from the linear actuators length to the arc parameters.

$$\phi = \tan^{-1}\left(\frac{\sqrt{3}}{3} \cdot \frac{l_3 + l_2 - 2l_1}{l_2 - l_3}\right) \tag{3}$$

$$k = 2\frac{\sqrt{l_1^2 + l_2^2 + l_3^2 - l_1 l_2 - l_2 l_3 - l_1 l_3}}{d(l_1 + l_2 + l_3)} \tag{4}$$

$$s = \frac{l_1 + l_2 + l_3}{3} \tag{5}$$

Finally to obtain the position of the end effector in relation to the cable lengths is necessarily to know the following parameters from a section of the manipulator; the length of one of its segments, the curvature of the section (k), the angle at which the curvature (φ) is found, the distance from the center of the manipulator to the location of the cables (s), and the number of segments in the manipulator section (n); with a combinations of all of these variables finally its obtain the Eq. 6.

$$\begin{bmatrix} l_1 \\ l_2 \\ l_3 \end{bmatrix} = \begin{bmatrix} s[1 - kd\sin(\phi)] \\ s\left[1 + kd\sin\left(\frac{\pi}{3} + \phi\right)\right] \\ s\left[1 - kd\cos\left(\frac{\pi}{6} + \phi\right)\right] \end{bmatrix} \tag{6}$$

4 Control System and Interface

4.1 User Interface

The Fig. 6 shows the graphic user interface (GUI) that allows direct interaction with electronic devices to be controlled through graphic icons and visual indicators, avoiding modifications of code in situations in which the changes implemented are punctual and frequent. In order to work with GUI in ROS environment, we had work with the "RQT" package; this is a framework based on the QT design software.

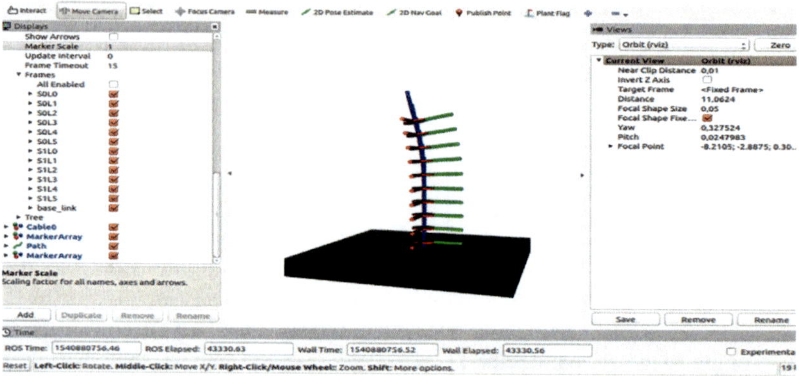

Fig. 6. Graphic user interface in ROS of the continuous redundant manipulator.

4.2 Control Algorithm

The system used a PID control algorithm in which the position signal delivered by a subsystem that is continuously monitoring to be able to move the actuator from an initial position to a desired position with less possible error. When the control algorithm is working with the control pack, the system takes one input position from virtual world, and then, reviews the status data from feedback sensors resulting of try to reach the target position. It uses a generic loopback feedback mechanism, usually a PID controller, to

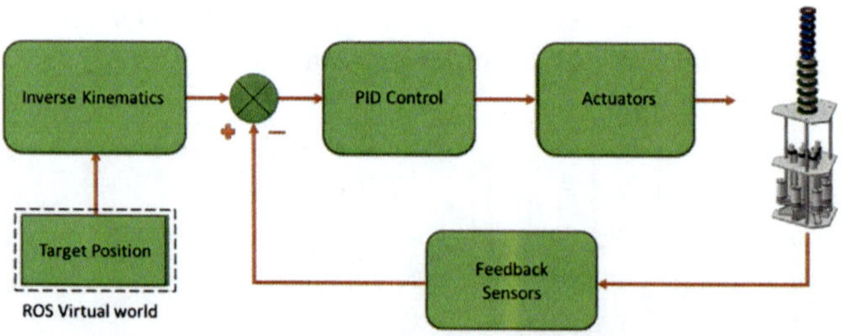

Fig. 7. Internal software architecture for the continuous redundant manipulator.

control the output, that is, the information sent to the actuators. To adapt the control to the robotic system, the control parameters must be calculated based on the electromechanical characteristics of the actuators. See Fig. 7.

The calculation algorithm is responsible for establishing the linearization and performs the conversion of the obtained variables by the feedback sensors of the robotic system; these numerical variables is interpreted by the control algorithm as distances, in order to obtain the actuator position as close as possible to the target position.

5 Results

The implementation of ROS as the main system resulted in a structured project in a modular manner, which allowed the implementation of some scripts that can model any type of continuous robot regardless of its size or number of sections or disks. Figure 6 showing the simulation of the continuous robot model in the RVIZ environment, in which the dynamics of an internal or vertebrate cable is responsible for carrying out the movement according to the trajectory that was established. In Addition, when it was implemented the constant curvature model was necessary to generate a different transformation to each virtual model discs to make-up the segments; as a result, when a variation was made in the angle curvature, it is reflected at each time in the virtual model and reproduced by the prototype showing in Fig. 8.

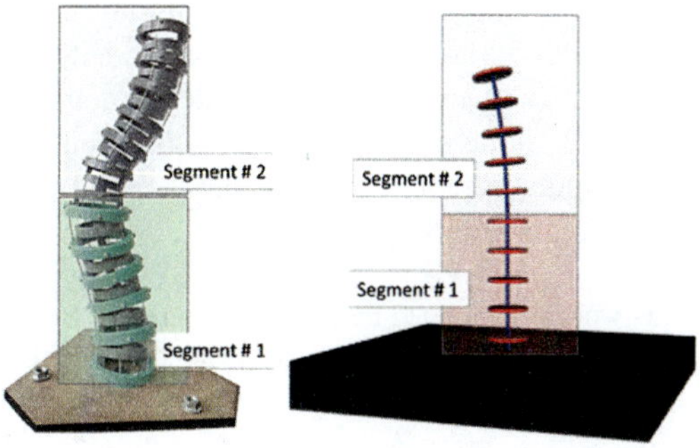

Fig. 8. Position and comparison between physical prototype (left) and simulation (right).

Figure 9 (top) is showing the data results generated by feedback sensors on a real tests. The contact sensor or SoftPot was necessary to calibrate with previous acquisition data; and the following results were obtained: when an end-to-end displacement was executed, this is obviously considered to the maximum values k for each tendon trajectory. The Fig. 9 (bottom) also is showing the values behaviour of configurations space parameters s, k, φ, performed by the prototype in one desired target positions.

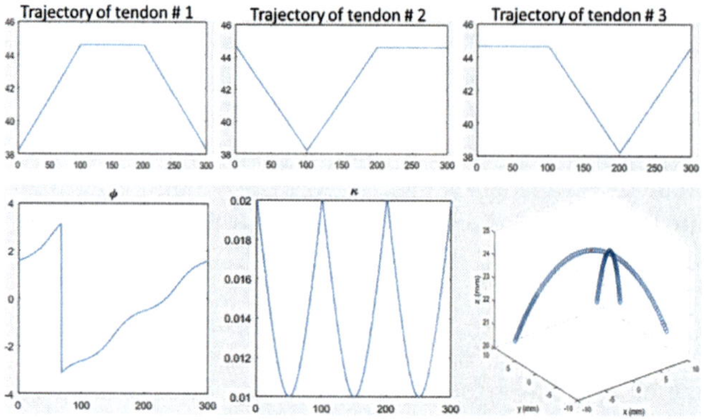

Fig. 9. (top) Tendon's trajectory actuated in segment #2; (bottom) graphic results of configurations space parameters s, k, φ, executed in segment #1.

6 Conclusions

This paper presents a further test manipulator experiments for two segments based in the mathematical model presented by [22] where the model in the space transformations between the D-H parameters and s, k, φ, l_1, l_2 and l_3 was implemented, simulated, controlled from ROS system and tested. Its model is the constant curvature kinematic used in most continuous robot manipulators. The modular continuous redundant manipulator proposed with 2 set of segments, whose length can be extended even further by adding sections to the segments (links) and, in turn with a new define configuration by extending the length of the manipulator. Some difficulties where observed in the executed prototype. The top segment manipulator requires one specific central chamber to go through the wire cables in order to minimize its error.

Acknowledgements. This work is supported by a grant from the El Bosque University and the Research Vice-rectory with the project number PCI-2017-8832. The authors also wish to thank the Electronic Engineering and Bioengineering departments of El Bosque University for all the support received.

References

1. Pires, N.: Robot manipulators and control systems. Industrial Robots Programming. Building Applications for the Factories of the Future (2007)
2. Li, C., Rahn, C.D.: Design of continuous backbone, cable-driven robots. J. Mech. Des. **124**(2), 265–271 (2002)
3. Laschi, C., Cianchetti, M., Mazzolai, B., Margheri, L., Follador, M., Dario, P.: Soft robot arm inspired by the octopus. Adv. Robot. **26**(7), 709–727 (2012)
4. Crespi, A., Badertscher, A., Guignard, A., Ijspeert, A.J.: AmphiBot I: an amphibious snake-like robot. Robot. Auton. Syst. **50**(4), 163–175 (2005)
5. Tiefeng, S., Libin, Z., Mingyu, D., Guanjun, B., Qinghua, Y.: Fruit harvesting continuum manipulator inspired by elephant trunk. Int. J. Agric. Biol. Eng. **8**(1), 57–63 (2015)

6. Burgner-Kahrs, J., Rucker, D.C., Choset, H.: Continuum robots for medical applications: a survey. IEEE Trans. Robot. **31**(6), 1261–1280 (2015)
7. Burgner, J., Rucker, D.C., Gilbert, H.B., Swaney, P.J., Russell, P.T., Weaver, K.D., Webster, R.J.: A telerobotic system for transnasal surgery. IEEE/ASME Trans. Mechatron. **19**(3), 996–1006 (2014)
8. Ma, S., Hirose, S., Yoshinada, H.: Development of a hyper-redundant multijoint manipulator for maintenance of nuclear reactors. Adv. Robot. **9**(3), 281–300 (1994)
9. Liu, S., Yang, Z., Zhu, Z., Han, L., Zhu, X., Xu, K.: Development of a dexterous continuum manipulator for exploration and inspection in confined spaces. Ind. Robot: Int. J. **43**(3), 284–295 (2016)
10. Meng, G.Z., Yuan, G.M., Liu, Z., Zhang, J.: Forward and inverse kinematic of continuum robot for search and rescue. Adv. Mater. Res. **712–715**, 2290–2295 (2013)
11. OC Robotics OC. LaserPipe Remote in-bore laser welding of industrial pipelines (2016)
12. Hirose, S.: Biologically inspired robots: snake-like locomotors and manipulators. Appl. Mech. Rev. **48**(3), B27–B27 (1995)
13. Chirikjian, G.S., Burdick, J.W.: The kinematics of hyper-redundant robot locomotion. IEEE Trans. Robot. Autom. **11**(6), 781–793 (1995)
14. Chirikjian, G.S., Burdick, J.W.: Kinematically optimal hyper-redundant manipulator configurations. IEEE Trans. Robot. Autom. **11**(6), 794–806 (1995)
15. Jones, B.A., Walker, I.D.: Kinematics for multisection continuum robots. IEEE Trans. Robot. **22**(1), 43–55 (2006)
16. Webster, R.J., Jones, B.A.: Design and kinematic modeling of constant curvature continuum robots: a review. Int. J. Robot. Res. **29**(13), 1661–1683 (2010)
17. Hannan, M.W., Walker, I.D.: Kinematics and the implementation of an elephant's trunk manipulator and other continuum style robots. J. Robot. Syst. **20**(2), 45–63 (2003)
18. Li, Z., Wu, L., Ren, H., Yu, H.: Kinematic comparison of surgical tendon-driven manipulators and concentric tube manipulators. Mech. Mach. Theory **107**, 148–165 (2017)
19. Mahl, T., Hildebrandt, A., Sawodny, O.: A variable curvature continuum kinematics for kinematic control of the bionic handling assistant. IEEE Trans. Robot. **30**(4), 1–15 (2014)
20. Li, Z., Ren, H., Chiu, P.W.Y., Du, R., Yu, H.: A novel constrained wire-driven flexible mechanism and its kinematic analysis. Mech. Mach. Theory **95**, 59–75 (2016)
21. Walker, I.D.: Continuous backbone continuum robot manipulators. ISRN Robot. **2013**, 1–19 (2013)
22. Mosqueda, S., Moncada, Y., Murrugarra, C., León-Rodriguez, H.: Constant curvature kinematic model analysis and experimental validation for tendon driven continuum manipulators. In: Proceedings of the 15th International Conference on Informatics in Control, Automation and Robotics (ICINCO 2018), Oporto, Portugal, 28th–31th July 2018, vol. 2, pp. 211–218 (2018). ISBN 978-989-758-321-6
23. Jones, B.A., McMahan, W., Walker, I.D.: Practical kinematics for real-time implementation of continuum robots. In: IEEE International Conference on Robotics and Automation, pp. 1840–1847 (2006)
24. DHB: Differential Geometry, Reading, vol. 1, no. 4 (2013)

Password Authentication Attacks at Scale

Junade Ali(✉) and Malgorzata Pikies

Cloudflare, London, UK
{junade,malgorzata}@cloudflare.com

Abstract. Whilst there is much literature on password authentication attacks, there has been little study of real online brute force attacks. This paper provides analysis of the most prolific attackers and provides analysis of distributed brute force attackers using unsupervised learning technique for clusterization. *Compromised credential checking* has recently emerged as an approach to improve password security by deterring users from using re-using breached passwords. Analysing an implementation on a large online gaming site, a 184% median increase in daily password change events in the first 30 days of rollout and an ongoing 117% increase in the following 20 days is observed.

Keywords: Password authentication · Authentication attacks · Distributed brute force attacks · Credential stuffing · *Compromised credential checking*

1 Introduction

Brute force attacks are a common source of malicious internet traffic, particularly for websites built on WordPress software [1], unfortunately there has been little empirical research into the success rate of these attacks or the characteristics of how such attacks are conducted.

Previous guidance for preventing brute force attacks included the use of rate-limiting the frequency of login requests from distinct IP Addresses [2]. Recent strategies have sought to distribute brute force attacks, such they are harder to detect [3]. Recently *compromised credential checking* services have emerged to allow website administrators to warn users when their password appears in a data breach to increase the difficulty of such attacks [4].

In this paper, we perform the first widescale evaluation of brute force attacks, including success rates. We perform such an analysis by reviewing attacks against WordPress websites by the most prolific attackers. Using unsupervised learning, we are able to cluster and identify distributed brute force attackers. Furthermore, by using data provided by a popular online gaming website, we show the impact of enabling *compromised credential checking* warning messages on users changing passwords and enabling Two Factor Authentication.

D. F. Cortes Tobar et al. (Eds.): AETA 2019, LNEE 685, pp. 394–403, 2021.
https://doi.org/10.1007/978-3-030-53021-1_40

2 Related Work

There have been multiple advancements in the field of brute force attacks. [3] demonstrates a how brute force attacks can be performed in a distributed and asynchronous fashion using botnets. Such attack strategies can fly under the radar of rate limiting tools, as attack traffic can appear to be small from each individual source. [5] describes a machine learning strategy for detecting distributed bruteforce attacks on the SSH protocol. Unfortunately there is no work in studying such attacks on websites, as opposed to SSH traffic. Additionally there is no literature studying attacks on large internet infrastructure networks as opposed to single networks with highly localised traffic.

[6] uses a simulated attack against a WordPress website using password dictionaries and the WPScan tool to simulate an attack. The paper outlines some sources of data for forensic evaluation and proceeds to mention some counter measures to prevent such an attack, including lockout measures, Two Factor Authentication and strong passwords. The paper also notes that it is possible for the WPScan to extract usernames from a WordPress installation, allowing such attacks to made easier. The work focuses on a simulated, toy model of attack; instead of passive analysis of network traffic. Whilst [1] provides some passive analysis of brute force attacks, the success rate of such attacks is not evaluated and traffic is constrained to a single university network.

[7] extensively considers a vast quantity of alternative authentication mechanisms to act as alternatives to passwords; despite a wide-ranging search, the authors fail find an alternative to password authentication on the basis that such mechanisms do not retain the full set of benefits that passwords offer. The authors claim that as a result of failing to consider a sufficiently wide range of real-world constraints, many academic proposals have failed to gain traction. [8] finds a gap between the technology available to limit password reuse and it's utilization. Even amongst a well-connected, well-educated technologically savvy demographic there was still trouble understanding the nature of attacks against reused passwords and the sampled users rarely utilized technology such as password managers to aid in limiting password reuse. Further, [8] demonstrated that password reuse becomes a greater problem as users acquire more accounts online, as more accounts can result in greater password reuse.

[4] provides empirical evaluation of *compromised credential checking* services, which allow for websites to check if a user's credentials are present in a data breach. The work provides a theoretical basis and an evaluation of popular *compromised credential checking* API services. The work does not cover the impact in user behaviour after such strategies have been implemented.

[9] evaluates changes in user behaviour after receiving password security notifications, they find that 88.41% of users who received the fear appeal later set unique passwords, whilst only 4.45% of users who did not receive a fear appeal would set a unique password. This study was based on 135 students from an information systems course and there has been no further work in literature to evaluate behaviour changes at scale in production websites.

3 WordPress Authentication Attack Data

3.1 Attack Model and Data Collection

WordPress uses a login script known as `/wp-login.php` to authenticate users to the administration area. Upon a HTTP POST request containing valid credentials, a HTTP 302 redirect is issued to direct the user to the administration panel or back to the site. If the login fails, a response with a HTTP 200 status code is returned with a message informing the user to reattempt the login. The WordPress login script also allows for other actions, such as password reset requests - but will operate as a login script when no query parameters are set in the URL.

Our interest is in the most prolific attackers, so we aim to select the most active IPs conducting these attacks. Data was collected from a 1% HTTP request sample of WordPress logins and obtained during a 7 week period from the beginning of May to mid-July 2019. We stored data from the top 500 most active IP Addresses each day and required each IP Address have at least 10 requests in the sample each day. Such data was acquired using a 1% sample of limited HTTP headers across the Cloudflare reverse proxy network, proxying over 16 million active websites as of June 2019. Such data did not contain any HTTP body information for privacy purposes.

In total, we captured WordPress login requests from 9053 unique IP addresses alongside the HTTP User Agent and Referrer headers. To exclude any non-WordPress websites from the analysis, we required that each site produced both a HTTP 200 and a HTTP 301 response code during the overall time period for `/wp-login.php`. This yielded 1290896 websites that received 166534363 login requests.

It is important to note that WordPress attack data in this paper is calculated from a 1% request sample. Accordingly, it is important to note that actual traffic volume may be up to 100 times higher.

3.2 Success Rate and Variety Ratio

Recall that WordPress yields a HTTP 302 status code on a successful login and a HTTP 200 on a failed login. For each IP a success rate was calculated as:

$$\epsilon_{tot} = \frac{N_{302}}{N_{200} + N_{302}}, \tag{1}$$

where N_{302} is the number of successful requests and N_{200} is the number of failed login attempts. The subscript *tot* stands for total, as the success rate was calculated using requests from the whole time period. There were 1634 IPs that had $\epsilon_{tot} = 0.0$ are analysed separately in Sect. 3.3.

Figure 1 shows the distribution of the success rate ϵ_{tot} with two distinct peaks.

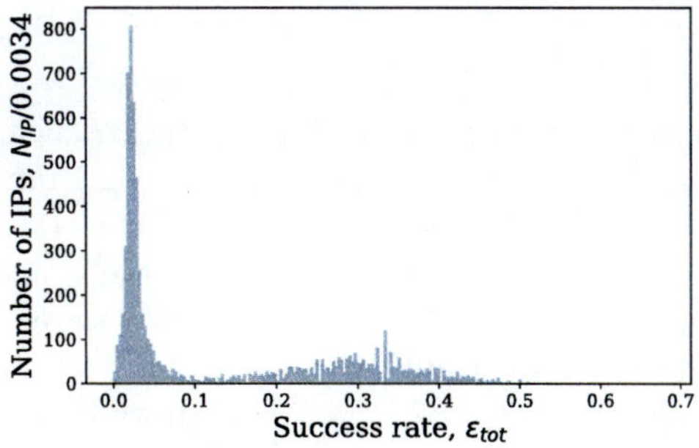

Fig. 1. Success rate ϵ_{tot} distribution. Horizontal axis corresponds to the number of unique IP numbers falling into a given bin (each bin size is 0.0034). The total number of IP is equal to 7 419.

Further investigation showed interesting clusters in two dimensional distributions of success rate with number of unique zones and a variety ratio of HTTP status codes (Fig. 2). We define the variety ratio as follows:

$$R_{\mathcal{V}} = \frac{N_{302} + N_{200}}{\sum_i N_i}, \quad (2)$$

where the denominator sums over all possible status response codes. Web server rate limiting may issue HTTP 403 or HTTP 429 status codes when blocking excess traffic, hence decreasing the variety ratio. The two plots from Fig. 2 show that the high success rate IPs had a tendency to attack the same websites and got more diverse response codes than their less successful counterparts.

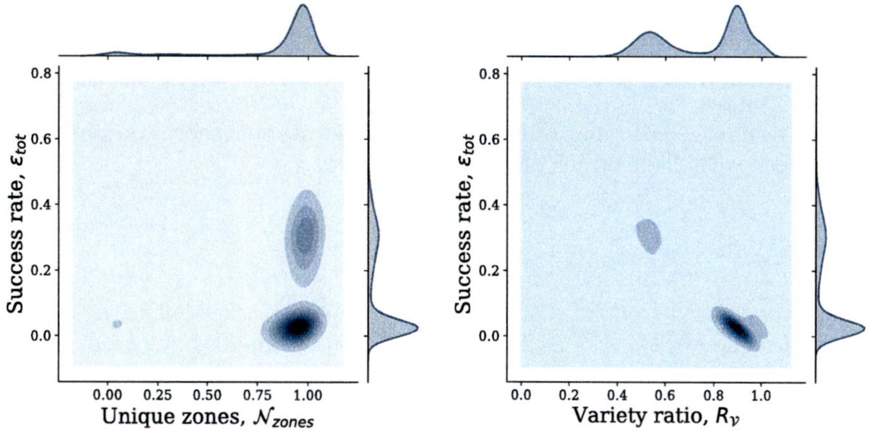

Fig. 2. Kernel density estimate plots of the success rate vs the normalised number of unique zones (left) and the success rate vs the variety ratio (right) of 7419 unique IPs. The number of unique zones is normalised to the total number of POST requests sent by a given IP.

3.3 Unsuccessful Attacks

We noticed an area of low volume brute force attacks, such attacks generally appear to be classical (high volume of requests coming from one IP). For this region, the median value of sampled requests by an IP is 47. Hence, in order to reduce the pollution of this data sample with legitimate traffic we further apply a conservative requirement for an IP to have at least 47 sampled requests (in the examined period of time). As a result, the number of sampled requests falling into this category is 1000318, from 820 unique IPs towards 16248 distinct zones. The majority of requests comes from networks (classified based on their ASNs using the MaxMind GeoIP2 database [10]) registered in United States, Turkey, Spain and France. The highest IP variety is seen in United States and Turkey.

10.70% had a user agent strings of `WPScan v3.5.3` (https://wpscan.org/) indicating usage of WPScan to perform attacks [11], in 2.79% of cases such user agent strings were not set at all. The second most common referrer path is empty (32.83% of requests), despite the fact that human users would submit their login request from a login form (with their browser therefore setting a referrer path). An interesting feature is that it is common to find user agent strings containing double slashes, for example `//wp-login.php` represented 1.09% of requests.

3.4 IPs Clusterisation

When the requirement of at least 47 sampled requests per IP was applied to this sample, we were left with 165515869 sampled requests, from 7072 unique IPs towards 1290699 distinct zones. Instead of applying rough requirements on the success rate to divide attackers from Fig. 1 we used an unsupervised learning method. We chose `sklearn`'s [12] implementation of the agglomerative hierarchical clustering [13] using the distributions defined in Sect. 3.2 as features. In this technique, each IP is initially considered as a unique cluster. Similar clusters are then merged together. The algorithm continues until the requested number of clusters (3, in our case) are created. We used the Euclidean distance as a metric and Ward's linkage criterion. The result of the algorithm is shown on Fig. 3. We now proceed with the analysis of the three clusters.

4 Analysis

The least populated group of IPs is in Cluster 0, these attackers made the highest volume of requests (88.8% of all requests in the data sample). In terms of unique zones, the least populated is Cluster 1, where large group of IPs sent the smallest amount of requests. The most populated cluster by unique IPs is Cluster 2, who attacked the highest number of unique zones. An additional data sample has been gathered for all IPs belonging to these clusters. The sample contained information about the number of POST (N_{POST}) and GET (N_{GET}) requests sent by a given IP towards a WordPress login page from the same period of time as the main data sample. Whilst a POST request processes a login, a GET

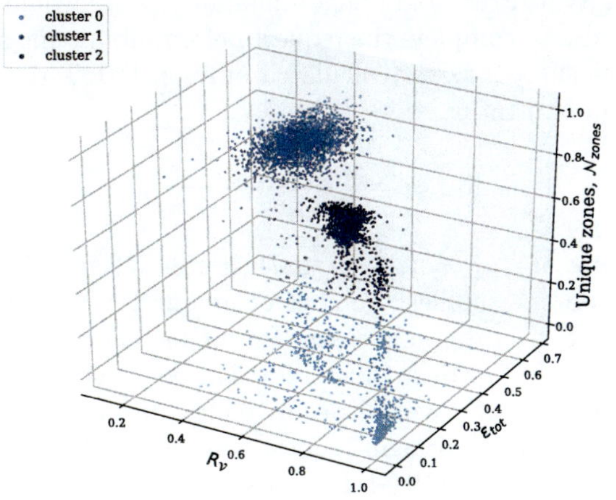

Fig. 3. Clusters obtained with the agglomerative hierarchical clustering technique.

Table 1. Clusters' summary. Zones may have occurrences in all clusters.

Feature	Cluster 0	Cluster 1	Cluster 2
IPs	879	2775	3418
ASNs	176	215	711
Zones	1128569	628322	1183511
Requests (1% sample)	146932324	1380871	17202674
ϵ_{tot} (mean)	4.91%	30.1%	2.5%
ϵ_{tot} (median)	3.55%	30.0%	2.27%
N_{req}(IP) (mean)	167158.5	497.6	5033.0

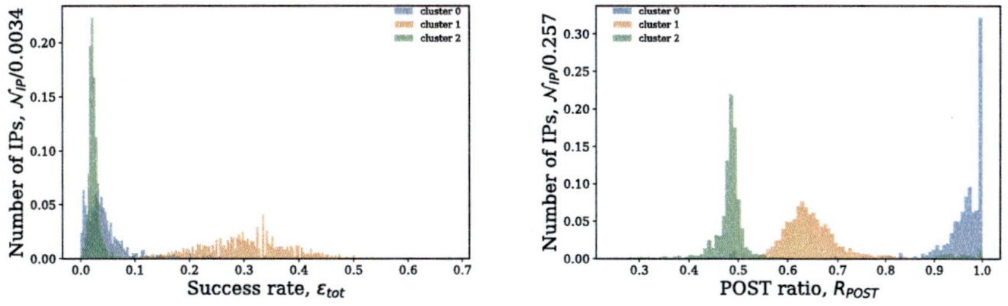

Fig. 4. Success rate ϵ_{tot} distributions and POST/GET requests ratio R_{POST} of the three clusters. Horizontal axis corresponds to the number of unique IP numbers falling into a given bin. Histograms for each cluster are normalised to unity.

request shows the WordPress login page - human visitors would need to make a GET request in order to complete the request before submitting a POST request. Figure 4 shows the success rates (calculated only on POST requests) combined with a distribution of a ratio:

$$R_{\text{POST}} = \frac{N_{\text{POST}}}{N_{\text{POST}} + N_{\text{GET}}}. \tag{3}$$

The highest ratio of R_{POST} belongs to Cluster 0, which had the highest number of requests from the smallest number of unique IPs. The highly successful attackers (30.1% success rate on average) from Cluster 1 had higher number of POST requests than GET, whilst Cluster 2 balanced their requests evenly. Interestingly, WordPress doesn't have any security measures requiring the login page be loaded before submitting a login request, through code inspection we found that WPScan [11] made HTTP GET requests to the login endpoint to follow any redirects (such as from HTTP to HTTPS) before conducting an attack. The median success rate for IPs in all 3 clusters was 3.49%, but IPs in Cluster 1 saw a dramatically higher success rate of 30.0%. Exact values are given in Table 1.

The three most common user agents among Cluster 0 are `Mozilla/5.0 (Windows NT 6.1; WOW64; rv:18.0) Gecko/20100101 Firefox/18.0` (75.0% of requests), `Mozilla/5.0 (Windows NT 10.0; Win64; x64) AppleWebKit/ 537.36 (KHTML, like Gecko) Chrome/63.0.3239.132 Safari/537.36` (12.3%) and `Mozilla/5.0 (Linux; U; Android 2.2) AppleWebKit/533.1 (KHTML , like Gecko) Version/4.0 Mobile Safari/533.1` (1.7%), where the total number of unique user agents is 3874391. Cluster 1, with very successful attackers is highly dominated by one user agent: `Mozilla/5.0 (Windows NT 6.1; WOW64) AppleWebKit/537.36 (KHTML, like Gecko) Chrome/58.0.3029.110 Safari /537.36 SE 2.X MetaSr 1.0` (99.5%). Similar behaviour is seen in cluster 2, where 98.0% of requests had `Mozilla/5.0 (X11; Ubuntu; Linux x86_64; rv:62.0) Gecko/20100101 Firefox/62.0` as user agent. For all clusters, the dominating among requests referrer path was empty, and the second most common was `/wp-login.php`. As with the unsuccessful attackers, referrer paths containing double slashes were present (such as `//wp-login.php`).

The majority of the requests coming from Cluster 0 belong to a network (Autonomous System) registered in Poland yet they are all originating from only 8 IPs, in contrast a majority of IP Addresses belonging to the cluster were assigned to the United States. The vast majority of requests from Cluster 1 come from China (99.45% of requests) representing 2762 of IPs used. Attackers from Cluster 2 are more spread among countries, with a domination of United States, but the majority of requests was coming from a French network.

5 The Effectiveness of Compromised Credential Checking

Many approaches to defending against authentication attacks have already been developed [2]. Recent *compromised credential checking* approaches rely on altering users behaviour such that they set credentials which are less vulnerable to

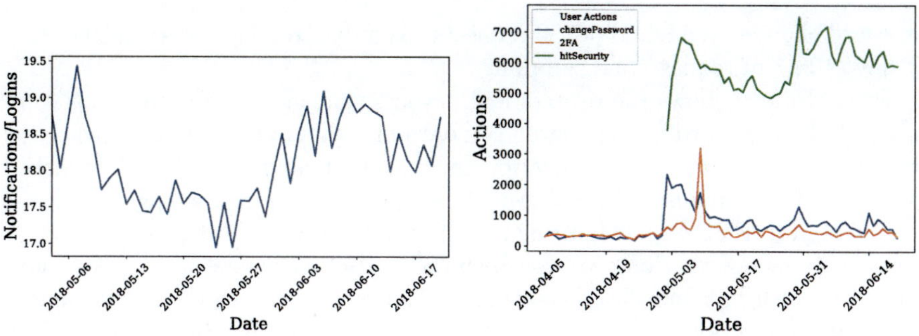

Fig. 5. Ratio indicating the number of login actions that results in the presentation of a security notification (left). User account changes by day (right).

Table 2. Clusters' summary. Zones may have occurrences in all clusters.

Date Range (Inclusive)	Median Password Changes	Median 2FA Enables
6th Apr. 2019 to 1st May 2019	310.5	359.5
2nd May 2019 to 31st May 2019	881.0	521.5
1st June 2019 to 20th June 2019	674.0	436.5

attack. In this section we will explore how effective such strategies are at adjusting user behaviour.

EVE Online is an online video game developed by CCP Games. To protect users, fear appeals were implemented to encourage users with breached credentials to change their password or enable Two Factor Authentication).

Passwords which users supply on their login pages are checked against a service called "HaveIBeenPwned" to validate if a password has been in a data breach. To protect the security of user passwords, password hashes are queried on the basis of k-anonymous buckets. Empirical analysis of *compromised credential checking* services and their protocols (including "HaveIBeenPwned") can be found in [4].

On login, users are presented with a message indicating they use a commonly set password and informing them they should change their password and enable Two Factor Authentication, with a call-to-action to continue to the game.

The website deployed such fear-appeals into production on 2nd May 2018. EVE Online shared raw data with us on the effectiveness of this approach containing day-by-day aggregate information of the total number of login sessions, those prompted with a security message and the amount of users changing their password or enabling Two Factor Authentication. The data supplied was from the period of the 6th April 2018 through to the 20th June 2018. It is important to note that such fear appeals were originally deployed for only the login page and not the registration page, meaning the data only represents fear appeals presented on a login. The data in total includes 1630627 login sessions, 52899 password change events and 36736 requests to enable to 2 Factor Authentication by day (100% of requests, no sampling).

Figure 5 (left) shows the ratio of login events that yielded a notification indicating that a password had appeared in a data breach. On the day of launch, 3794 of 19731 logins received a notification indicating their password had breached (just over 19% of logins). This value declined, however it increased again towards the end of May. EVE Online informed us they believed this was a result of a newly released a new registration process which resulted in more new user registrations (some of which had breached passwords). Figure 5 (right) shows the change in user behaviour upon receiving such fear appeals. The day before launch, on the 1st of May, there were 363 password change events and 452 users enabling Two Factor Authentication. The day of the launch, on the 2nd May, there were 2339 password change events and 633 users enabling 2 Factor Authentication. A month after notifications were rolled out, on the 2nd July, there were 698 password change events and 455 password change events.

There was a median 184% increase in password change events and a 45% in users enabling 2 Factor Authentication in the 30 days after rollout. From 30 days to 50 days after rollout, there continued to be a 117% increase in median password change events and a 21% increase in users enabling 2 Factor Authentication. Data is described in Table 2.

6 Conclusions

Using a simple unsupervised machine learning technique, we were able to separate attackers into 3 key clusters by analysing the ratio of response codes, success rate and unique zones attacked. Whilst the median success rate (per IP) for all attacks considered was 3.49%, we find a cluster of successful IPs that has achieved a median success rate of 30% per IP. The most successful cluster, Cluster 1, had 99.5% of requests containing the same distinctive user agent and 99.45% of requests coming from the same single country - making it highly likely such attacks were co-ordinated in some way. This highlights the importance of finding approaches to mitigate distributed brute force attacks. Further work is needed to study why some attackers are able to achieve such a dramatically higher success ratios than the median baseline.

We find that *compromised credential checking* provides promising opportunities improve password security through changes in user behaviour. Analysis of implementation on a large online gaming website found a 184% median increase in daily password change events in the first 30 days of rollout. Further work should attempt to study how this impact changes on other types of websites and user demographics.

References

1. Husák, M., Velan, P., Vykopal, J.: Security monitoring of http traffic using extended flows. In: 2015 10th International Conference on Availability, Reliability and Security, pp. 258–265. IEEE (2015)

2. Florêncio, D., Herley, C., Van Oorschot, P.C.: An administrator's guide to internet password research. In: 28th Large Installation System Administration Conference (LISA14), pp. 44–61 (2014)
3. Salamatian, S., et al.: Why botnets work: distributed brute-force attacks need no synchronization. IEEE Trans. Inf. Forensics Secur. **14**(9), 2288–2299 (2019)
4. Li, L., et al.: Protocols for checking compromised credentials. arXiv preprint arXiv:1905.13737 (2019)
5. Najafabadi, M.M., et al.: Machine learning for detecting brute force attacks at the network level. In: 2014 IEEE International Conference on Bioinformatics and Bioengineering, pp. 379–385. IEEE (2014)
6. Kyaw, A.K., Sioquim, F., Joseph, J.: Dictionary attack on WordPress: security and forensic analysis. In: 2015 Second International Conference on Information Security and Cyber Forensics (InfoSec), pp. 158–164. IEEE (2015)
7. Bonneau, J., et al.: The quest to replace passwords: a framework for comparative evaluation of web authentication schemes. In: 2012 IEEE Symposium on Security and Privacy (SP), pp. 553–567. IEEE (2012)
8. Gaw, S., Felten, E.W.: Password management strategies for online accounts. In: Proceedings of the Second Symposium on Usable Privacy and Security, pp. 44–55. ACM (2006)
9. Jenkins, J.L., et al.: Improving password cybersecurity through inexpensive and minimally invasive means: detecting and deterring password reuse through keystroke-dynamics monitoring and just-in-time fear appeals. Inf. Technol. Dev. **20**(2), 196–213 (2014). https://doi.org/10.1080/02681102.2013.814040
10. MaxMind, LLC: GeoIP2 (2019). https://www.maxmind.com/en/geoip2-country-database
11. WPScan. https://wpscan.org/. Accessed 04 July 2019
12. Pedregosa, F., et al.: Scikit-learn: machine learning in Python. J. Mach. Learn. Res. **12**, 2825–2830 (2011)
13. Ward Jr., J.H.: Hierarchical grouping to optimize an objective function. J. Am. Stat. Assoc. **58**(301), 236–244 (1963). https://doi.org/10.1080/01621459.1963.10500845

Power-Sharing Controller of Microgrid Based on Parallel Inverters Under the Droop Technique

Enrique Sanabria(✉), Alonso Chica, and Sergio Díaz

Departamento de Ingeniería Electrónica, Universidad Central, Calle 21#4-40, Bogotá, Colombia
{esanabriat,achical,sdiazc3}@ucentral.edu.co

Abstract. In low voltage electrical grids, small generation (smaller than 100 kVA) have taken importance in the last years with the growing emergence of alternative energy sources. Generally, some of these generators require power inverters to convert DC power into AC power. In this context, when the low voltage electrical grid has a load requirement, it is necessary to have generators to supply this load under agreements of power values delivered by each generator in a time frame. In this way, when it is required to use parallel inverters for the electrical energy generation, these must be connected under a power sharing technique that guarantees the stability of the system. Therefore, this work presents a power sharing controller of three levels. Simulation results are presented in Matlab/Simulink and experimental results of power-sharing control in grid-connected mode between two power inverters and the electrical grid, and in island mode between two inverters.

Keywords: Voltage inverter · Power sharing control · Droop control

1 Introduction

The adverse effects of climate change and the growing demand for energy have led to a growing concern about the efficient use of energy and the penetration of alternative energy sources. These aspects have driven the interest in distributed generation systems for their advantages over traditional systems in terms of reducing losses and penetration of renewable energies. In fact, there is a trend in the expansion of this type of systems based on renewable energy sources [1], so it can be said that the generation and consumption of energy has gone in this direction [2–4].

On the other hand, the use of renewable energy sources and their inherent intermittency also brings some difficulties in the operation of power systems, such as the need for storage systems, bidirectional flow of power and the eventual connection and disconnection with the traditional electric power distribution systems. In this aspect, the concept of microgrids [3, 5] emerges, defining them as local energy distribution systems that have the capacity to generate, store and consume electrical energy and must also be able to operate connected to the main grid (grid-connected mode) or independently (island mode).

These microgrids can be classified according to the voltage bus in DC (direct current), AC (alternating current) or mixed type. However, an AC microgrid can be composed

D. F. Cortes Tobar et al. (Eds.): AETA 2019, LNEE 685, pp. 404–413, 2021.
https://doi.org/10.1007/978-3-030-53021-1_41

of DC type energy sources, as is the case with most micro-turbines, wind power plants, fuel cells and photovoltaic cells. In this case, properly controlled voltage inverters are used to connect these energy sources to the microgrid.

Additionally, the correct operation of AC microgrids based on voltage inverters depends in good part on the correct power sharing. Regarding this, despite the fact that power-sharing control has been discussed in the literature for several decades, it has been mainly done by considering synchronous generators in grids with centralized generators. On the other hand, the power-sharing control in distributed generation systems based on voltage inverters has caused interest only until a decade ago.

This work shows the design and testing of the voltage inverter controller (internal level) and power-sharing control (primary level) of a system consisting of two voltage inverters connected in parallel and operating in grid-connected mode, and in island mode. Section 2 shows a summary of the power sharing technique based on Droop control. Section 3 shows the modeling and control of the power system. Section 4 shows the experimental results of the test prototype. Finally, the conclusions are shown.

2 Power Sharing by Droop Control

The Droop control technique allows controlling the power sharing of the demanded power by the loads between the inverters connected to the power system without using communications between them. This is done when each inverter emulates the behavior of a synchronous generator, that is, it reduces its frequency (speed of rotation) for increases in the active power it must deliver.

To understand the idea behind this technique, consider Fig. 1. This shows an inverter (represented by the AC voltage source) that imposes a voltage $E\angle\theta_E$ V and that through a line with impedance $Z_l\angle\theta$ Ω, is connected to an AC load bus whose voltage is $V_o\angle 0°$ V.

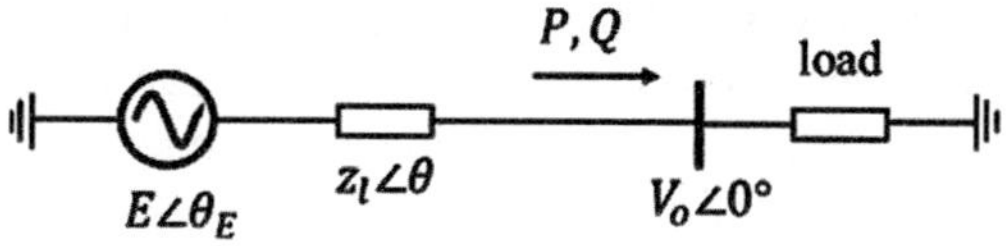

Fig. 1. Equivalent circuit of a distributed generator connected to a local load.

Considering that, the connection line is mainly inductive ($\theta_l \approx 90°$) and that the phase of the inverter voltage is very small ($\sin(\theta_E) \approx \theta_E$ and $\cos(\theta_E) \approx 0$), it can be shown that The active and reactive power delivered by the inverter can be calculated approximately as:

$$P \approx \frac{EV_o}{X}\theta_E. \tag{1}$$

$$Q \approx \frac{V_o(E - V_o)}{X}. \tag{2}$$

X is the reactance of the line.

From the previous expressions, it can be seen that the active power depends predominantly on the phase between the voltage of the inverter and the voltage on the load, while the reactive power is strongly influenced by the voltage of the inverter. Thus, since phase variations can be represented as variations in frequency, this concept is integrated into inverters connected in parallel using the Droop P/Q method [6]. Therefore, the frequency f and the amplitude E of the output voltage reference of the inverter can be expressed as:

$$\begin{aligned} f &= f_0 - mP, \\ E &= E_0 - nQ \end{aligned} \tag{3}$$

Where f_0 and E_0 are the frequency and voltage of the inverter when it without load. On the other hand, the coefficients m and n represent the inclination of the Droop curves [6], which for the case of the mainly inductive line can be chosen following the conventional Droop method [3, 8] as:

$$m = \frac{\Delta f}{P_{max}}. \tag{4}$$

$$n = \frac{\Delta E}{Q_{max}}. \tag{5}$$

Where Δf and ΔE are the maximum allowed deviations of frequency and amplitude of the output voltage of the inverter, respectively.

3 Modeling and Control

The control strategy for the power system is divided into two main parts: the control of the voltage of each inverter and the power-sharing control between the inverters. For the design of the controller of each inverter, its model is taken in average values. This can be obtained from the analysis of the circuit shown in Fig. 2, where it is shown, the inverter, the inductance (L_F) and capacitance of its filter (C_F) with its respective series resistance (R_F and R_{esr}), the inductive line (represented by L_C and R_C) used to connect it with the load and the load itself (R). In a similar way, the values of the components and the characteristics of the considered inverter are shown in Table 1. Thus, the representation in the state space of this circuit can be written as:

$$\frac{d}{dt}\begin{bmatrix} i_F \\ i_{AC} \\ v_C \end{bmatrix} = \mathbf{A_{inv}}\begin{bmatrix} i_F \\ i_{AC} \\ v_C \end{bmatrix} + \mathbf{B_{inv}}[V_I], \quad \begin{bmatrix} i_F \\ v_o \end{bmatrix} = \mathbf{C_{inv}}\begin{bmatrix} i_F \\ i_{AC} \\ v_C \end{bmatrix} + \mathbf{D_{inv}}[V_I] \tag{6}$$

Where:

$$\mathbf{A_{INV}} = \begin{bmatrix} \frac{-(R_F+R_{esr})}{L_F} & \frac{R_{esr}}{L_F} & \frac{-1}{L_F} \\ \frac{R_{esr}}{L_C} & \frac{-(R+R_{esr}+R_C)}{L_C} & \frac{1}{L_C} \\ \frac{1}{C_F} & \frac{-1}{C_F} & 0 \end{bmatrix}, \quad \mathbf{B_{INV}} = \begin{bmatrix} \frac{1}{L_F} \\ 0 \\ 0 \end{bmatrix}, \tag{7}$$
$$\mathbf{C_{INV}} = \begin{bmatrix} 1 & 0 & 0 \\ R_{esr} & -R_{esr} & 1 \end{bmatrix}, \quad \mathbf{D_{INV}} = [0]$$

Where V_I represents the average voltage imposed by the IGBts in each switching period, *vo* represents the voltage that would be measured on the capacitor C_F (including its series resistance) and R is a resistance used to represent the load of the inverter.

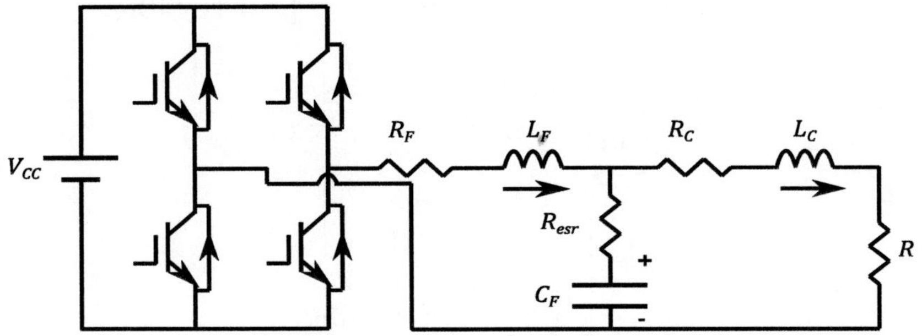

Fig. 2. Considered voltage inverter circuit.

Table 1. Characteristics of voltage inverters.

Parameter	Value	Parameter	Value
Switching frequency	10080 Hz	Inductance of the line (L_C)	500 μH
Fundamental frequency	60 Hz	Capacitor series resistance (R_{esr})	0.01 Ω
Rated voltage	120 V	Series resistance of the filter inductor (R_F)	1.3 Ω
Filter capacitance (C_F)	4.704 μF	Resistance line series (R_C)	0.143 Ω
Filter inductance (L_F)	2.787 mH	Load resistance (R)	29 Ω

In addition to the inverter model, the use of filters in the acquisition system is considered. Thus, since two filters are used to pass second-order drops to measure the voltage on the capacitor and the current of the filter inductor, the new state space can be written as:

$$\begin{aligned} \frac{d}{dt}X &= \mathbf{A_{TOT}}X + \mathbf{B_{TOT}}[V_I] \\ Y &= \mathbf{C_{TOT}}X + \mathbf{D_{TOT}}[V_I] \end{aligned} \tag{8}$$

Where:

$$\mathbf{A_{TOT}} = \begin{bmatrix} \begin{bmatrix} \mathbf{A_{INV}} & 0_{3x2} \\ \mathbf{B_F}\begin{bmatrix}0 & 1\end{bmatrix}\mathbf{C_{INV}} & \mathbf{A_F} \end{bmatrix} & 0_{5x2} \\ \mathbf{B_F}\begin{bmatrix}0 & 1\end{bmatrix}\mathbf{C_{INV}} & \mathbf{A_F} \end{bmatrix}, \quad \mathbf{B_{TOT}} = \begin{bmatrix} \begin{bmatrix} \mathbf{B_{INV}} \\ \mathbf{B_F}\begin{bmatrix}0 & 1\end{bmatrix}\mathbf{D_{INV}} \end{bmatrix} \\ \mathbf{B_F}\begin{bmatrix}0 & 1\end{bmatrix}\mathbf{D_{INV}} \end{bmatrix}$$

$$\mathbf{C_{TOT}} = \begin{bmatrix} \begin{bmatrix} \mathbf{C_{INV}} & 0_{2x2} \\ \mathbf{D_F}\begin{bmatrix}0 & 1\end{bmatrix}\mathbf{C_{INV}} & \mathbf{C_F} \end{bmatrix} & 0_{3x2} \\ \mathbf{D_F}\begin{bmatrix}1 & 0\end{bmatrix}\begin{bmatrix} \mathbf{C_{INV}} & 0_{2x2} \\ \mathbf{D_F}\begin{bmatrix}0 & 1\end{bmatrix}\mathbf{C_{INV}} & \mathbf{C_F} \end{bmatrix} & \mathbf{C_F} \end{bmatrix} \tag{9}$$

$$\mathbf{D_{TOT}} = \begin{bmatrix} \begin{bmatrix} D_{INV} \\ \mathbf{D_F}\begin{bmatrix}0 & 1\end{bmatrix}\mathbf{D_{INV}} \end{bmatrix} \\ \mathbf{D_F}\begin{bmatrix}1 & 0\end{bmatrix}\begin{bmatrix} D_{INV} \\ \mathbf{D_F}\begin{bmatrix}0 & 1\end{bmatrix}\mathbf{D_{INV}} \end{bmatrix} \end{bmatrix}$$

$$X = \begin{bmatrix} i_F \; i_{AC} \; v_C \; x_1 \; x_2 \; x_3 \; x_4 \end{bmatrix}^T$$

$$Y = \begin{bmatrix} i_F \; v_C \; v_{C_{Flt}} \; i_{F_{Flt}} \end{bmatrix}^T$$

Where the variables x_1, x_2, x_3 and x_4 correspond to the state variables of the two second order filters used for the conditioning of signals. Similarly, $v_{C_{Flt}}$ and $i_{F_{Flt}}$ correspond to the capacitor voltage and filter inductor current measured and filtered.

3.1 Current and Voltage Control of the Inverters

Based on the model obtained, the controllers are designed considering an internal current loop and an external voltage loop with a controller. Additionally, to facilitate the design of the current loop, the decoupling of the capacitor voltage was carried out, adding this to the manipulated variable of the inverter (V_I).

For the design of the current controller, it was considered that the zero crossing frequency of the system should be at most a decade below the inverter switching frequency of 10080 Hz [10]. Thus, 1 kHz is taken as the zero crossing frequency of this loop and a controller is designed to guarantee enough phase and gain margins. In this case a PI with an integral gain equal to $1.078 \bullet 10^4$ and a proportional gain of 17.1571 is used.

With the designed current loop controller and the inverter model with that closed loop, the voltage loop controller is designed, which must deliver the reference current (i_{ref}). For this, a zero crossing frequency is established at half the frequency used for the current loop, that is, 500 Hz. In this case, following a classical procedure [11] for the design of the PI, an integral gain is obtained equal to 15.12 and a proportional gain of 0.04813. Additionally, a resonant controller tuned to the fundamental frequency is used in the voltage loop to reduce the voltage error in steady state, so that the voltage loop controller remains:

$$H_{C\text{-}Voltage} = \frac{0.04813s + 15.12}{s^2 + 25.13s + 1421 \cdot 10^5} \tag{10}$$

3.2 Power-Sharing Control (Droop Curves)

The controller for the active power sharing is designed so that each inverter delivers a maximum active power of 530 W when the frequency is 59.9473 Hz and when it is 60 Hz it delivers 0 W. Thus, Δf is 0.0527 Hz and f_0 is 60 Hz, Therefore, the slope of the curve, of frequency vs active power, is calculated as:

$$m = \frac{0.0527\,\text{Hz}}{530\,\text{W}} = 0.0001\,\text{Hz/W} \tag{11}$$

For its part, the curve for the reactive power sharing is designed so that each inverter delivers 10 VAR when the amplitude is equal to 169.99859 V and that for 170.026 V the reactive power delivered is 0 VAR. Thus, ΔE is 0.00141 V; therefore, the slope of the curve, of amplitude versus reactive power, is calculated as:

$$n = \frac{0.00141\ \text{V}}{10\ \text{VAR}} = 0.001\ \text{V/VAR} \tag{12}$$

4 Simulation Results

The use of MATLAB/Simulink allowed building the model of two identical inverters that operate in island mode and grid-connected mode, validated by simulation.

4.1 Operation in Island Mode

For the simulation of the operation in island mode, it is taken into account that one of the inverters (#1) is powered by a resistive load that demands 530 W. On the other hand, initially the other inverter (#2) is not connected to the load; however, it operates in order to synchronize its output voltage with the voltage on the resistive load. After 0.8 s the inverter #2 receives an external command that enables its connection to the load and operates in parallel with the inverter #1. Thus, the load will demand each of the inverters 265 W.

The response of the active and reactive power of the two inverters is shown in Fig. 3, where the expected behavior for the active power is shown. On the other hand, it is seen that the reactive power of the inverter #1 is initially 19 VAR and when the system stabilizes with the inverters operating, it has a value of 4.9 VAR for each and with an adequate transient response at the moment of its interconnection.

4.2 Operation in Grid-Connected Mode

For the simulation of the operation of the inverters in grid-connected mode, an initial operating condition is taken into account where a resistive load of 64 Ω is only fed by the grid. The two inverters initially synchronize their output voltages with the voltage on the load and after 2 s are connected to the grid.

In this case, since the grid imposes the operating frequency at 60 Hz, the inverters do not deliver active power. Considering this, to force them to deliver power, the reference active power was adjusted at the tertiary level of each one.

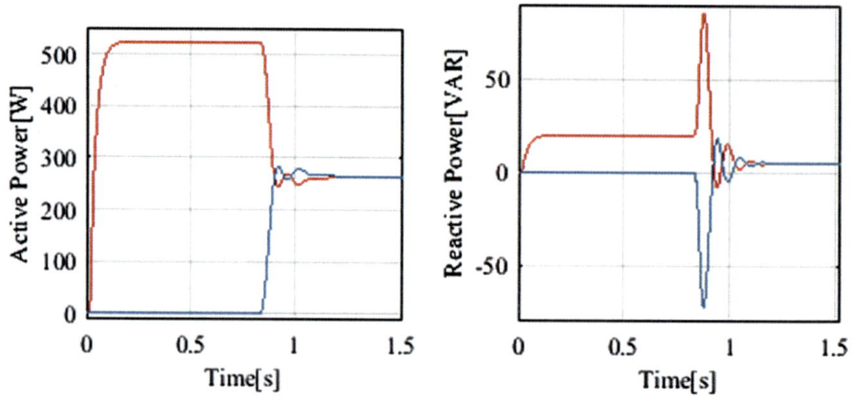

Fig. 3. Behavior of active and reactive power in island mode. Inverter #1 (red), Inverter #2 (blue).

The behavior of the active and reactive power of each inverter is shown in Fig. 4, where it can be seen that for 3.5 s the active power of the inverters is stabilized at 0 W. Subsequently, new transients are presented that are due to the variations of the adjustment points defined by the tertiary control, however, in all cases it is seen that the inverters operate in a stable manner and the power is adequately shared. Similar behaviors can be seen in the behavior of reactive power.

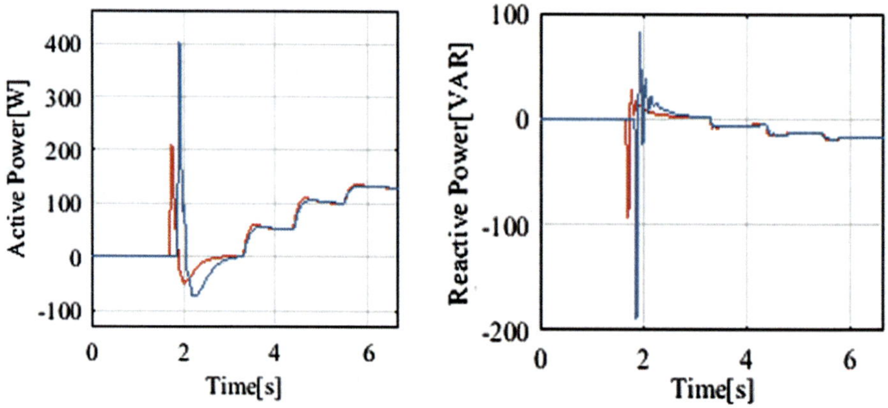

Fig. 4. Behavior of active and reactive power in grid connected mode. Inverter #1 (red), Inverter #2 (blue).

5 Experimental Results

The experimental results are obtained using the scheme of Fig. 5. The prototype is based on Semitech power modules from Semikron and the control law is implemented in a

DSC TMS 320F28335 microcontroller. The measurements presented here were obtained using the FLUKE 1730 power analyzer.

Below is the behavior of the active power of the inverters and the power grid for the operation in island mode and grid-connected mode.

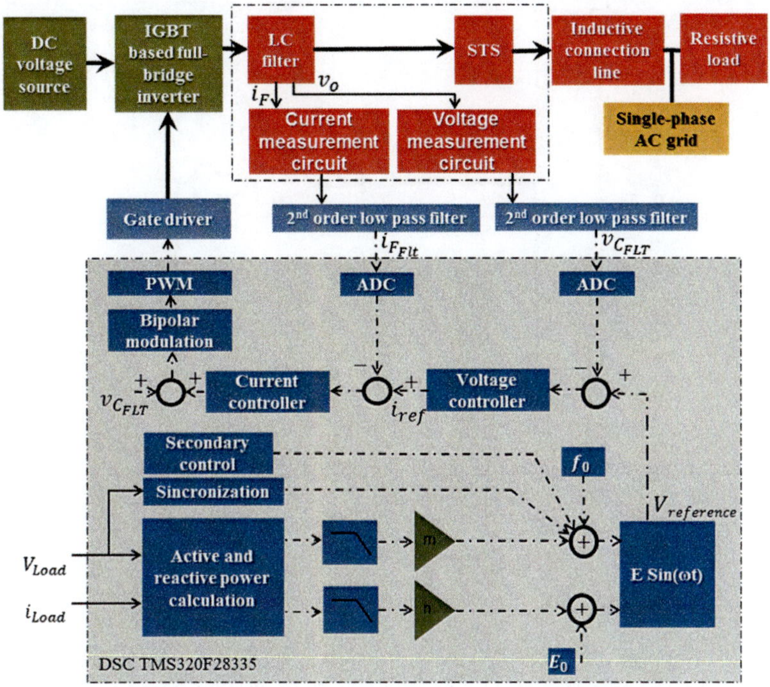

Fig. 5. Block diagram of the experimental prototype.

5.1 Operation in Island Mode

The results of the operation in island mode are shown in Fig. 6. For this test the two inverters are initially disconnected from the load. After 3 s the inverter #1 is connected to the load and after one minute, the second inverter is connected, thus power-sharing demanded equally between the two inverters. 4 min after the start of the test, to verify different active power-sharing values, the slopes of the frequency vs. power curves of the inverters are modified, forcing each inverter to deliver a different fraction of the total power demanded by the load.

5.2 Operation in Grid Connected Mode

The experimental results of the operation in grid-connected mode are shown in Fig. 7. In this test, initially the inverters are disconnected from the load and the electric grid supplies all the power demanded. Then, after 30 s, inverter #2 is connected and at 40 s the reference of active power is changed sequentially so that it delivers different power levels. The same procedure is followed with inverter #1 after one minute and 10 s from the start of the test.

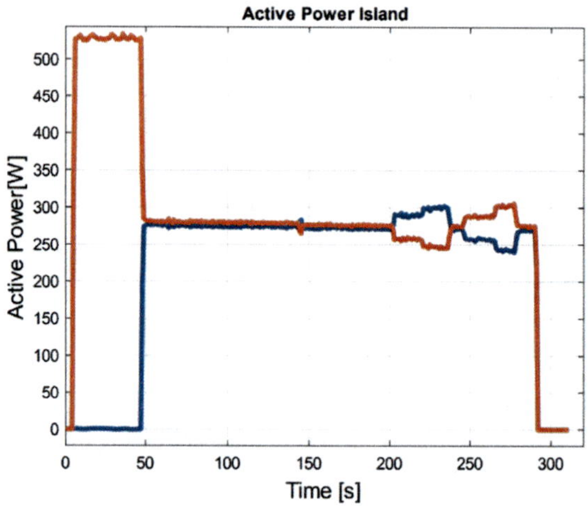

Fig. 6. Active power-sharing control in island mode. Inverter #1 (red), Inverter #2 (black).

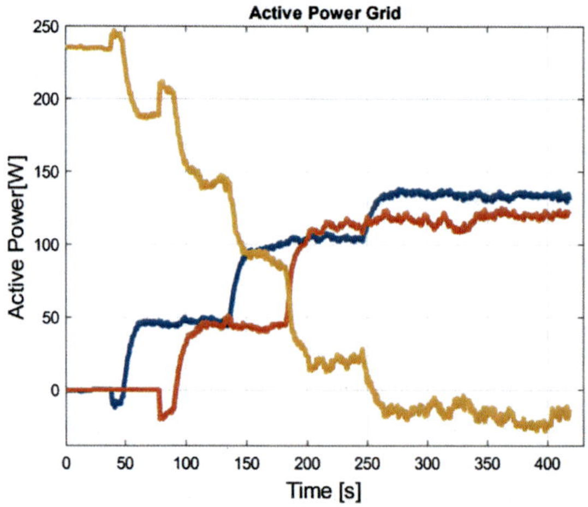

Fig. 7. Active power-sharing control in grid connected mode. Inverter #1 (red), Inverter #2 (black), Electrical Grid (blue).

6 Conclusions

This work presented a verification of the Droop technique for power-sharing among two VSI through simulation and experimental results. Furthermore, the dynamic model of a

VSI was presented by considering the use of second-order filters for current and voltage measurement. This addition allows considering the effect of the gain and phase shift associated with the measurement stage on the closed-loop inverter stability.

It was verified that both in island mode and in grid-connected mode, the frequency variation curves of the inverters correspond to the calculated active power. Likewise, the amplitude variation curves of the inverters correspond to the reactive power calculated. Also, the calculated coefficients for Droop curves can be used in both grid-connected mode and island mode.

Another important point that must be concluded in this project is the line impedance that the outputs of each inverter have, since they constitute the stability of the Droop model, because the load of each inverter must have a large inductive component in the case of impedance with inductive output.

References

1. OECD: Renewables information, p. 492. OECD Publishing, Paris (2018)
2. Vergara, P., Rey, J., Shaker, H., Guerrero, J., Jorgensen, B., da Silva, L.: Distributed strategy for optimal dispatch of unbalanced three-phase islanded microgrids. IEEE Trans. Smart Grid **10**, 3210–3225 (2019)
3. Guerrero, J., Vasquez, J., Matas, J., de Vicuna, L., Castilla, M.: Hierarchical control of droop-controlled AC and DC microgrids—a general approach toward standardization. IEEE Trans. Ind. Electron. **58**, 158–172 (2011)
4. Zhong, Q.: Robust droop controller for accurate proportional load sharing among inverters operated in parallel. IEEE Trans. Ind. Electron. **60**, 1281–1290 (2013)
5. Planas, E., Andreu, J., Gárate, J.I., De Alegría, I.M., Ibarra, E.: AC and DC technology in microgrids: a review. Renew. Sustain. Energy Rev. **43**, 726–749 (2015)
6. Bendib, A., Kherbachi, A., Kara, K., Chouder, A.: Droop controller based primary control scheme for parallel-connected single-phase inverters in islanded AC microgrid. In: 2017 5th International Conference on Electrical Engineering - Boumerdes (ICEE-B). IEEE, Boumerdes, pp 1–6 (2017)
7. Tayab, U.B., Roslan, M.A.B., Hwai, L.J., Kashif, M.: A review of droop control techniques for microgrid. Renew. Sustain. Energy Rev. **76**, 717–727 (2017)
8. Ortega, R.: Aplicación de técnicas de control para reducción de THDv en inversores monofásicos de generación de energía que operan en modo isla en el contexto de microrredes. Tesis Doctoral, Departamento de Ingeniería Electrónica, Universidad Politécnica de Valencia (2012)
9. Domínguez, C.: Control Jerárquico de Potencias en Microrredes Aisladas. Proyecto final de carrera. Universidad de Sevilla, Sevilla, España (2012)
10. Seguí, C.S.: Fundamentos básicos de la electrónica de potencia. Editorial de la UPC, Valencia (2002)
11. Erickson, R.: Fundamentals of power electronics. Kluwer Academic Publishers, Boston (2004)

Audio Scene Classification Based on Convolutional Neural Networks: An Evaluation of Multiple Features and Topologies in Short Time Segments

Juddy Y. Morales[1](✉), Juan D. Castillo[2], Brayan M. León[2], Roberto Ferro Escobar[2], and Andrés E. Gaona[2]

[1] Corporación Unificada Nacional de Educación Superior, Bogotá, D.C., Colombia
juddy.morales@cun.edu.co
[2] Universidad Distrital Francisco José de Caldas, Bogotá, D.C., Colombia

Abstract. The classification of acoustic scenes has been gaining importance in recent years, either because of the applications that it may have or because of the challenge of implementing a computational tool that allows the proper detection of complex and diverse sounds, such as those presented in real environments. In this work a convolutional neuronal network is implemented, trained with features such as mel-frequency cepstral coefficients, gamma tone y discrete Fourier transform, extracted to the sounds in segments of 10 s and 1 s. Cross validation is used with 80% of the data for training and 20% for validation with the DCASE2018 database evaluating the performance for different topologies of the neural network.

Keywords: Acoustic scene classification · Convolutional neural networks · Deep neural networks

1 Introduction

The task of Acoustic Scene Classification (ASC) consists of identifying environments by the sounds that characterize them. The basic idea is to extract information from audio signals and then enter it into a system capable of assigning a label suitable for the recording environment, which generally corresponds to urban environments. In 1997 MIT researchers created the first database of urban environments, this included classes such as voices, people, traffic, metro and others. For the classification, they implemented recurrent neural networks and close neighbors [1]. Since then, numerous techniques have been explored to try to solve the ASC problem [2]. The proposed solutions for the ASC task generally include artificial neural networks, such as Multilayer Perceptrons (MLP) [3, 4], Vector Support Machines (SVM) [5, 6], Deep Neural Networks (DNN) [6, 7] and Convolutional Neural Networks (CNN) [7, 8]. For the training and validation of these models, it is necessary to extract features, usually performed in short audio segments with a maximum overlap of 50%, which are then concatenated to form matrices that

D. F. Cortes Tobar et al. (Eds.): AETA 2019, LNEE 685, pp. 414–422, 2021.
https://doi.org/10.1007/978-3-030-53021-1_42

represent times equal to or greater than one second of audio. Commonly used features are the MFCC coefficients [3, 5, 7], gamma tones [5, 9], DFT [10], etc.

In the solutions proposed with MLP, topologies can be found where sets of neurons of the first hidden layer are trained with different features and their outputs are totally connected to a hidden back layer [4]. But similarly simple topologies are composed of few layers and neurons [3, 6] or complex topologies of DNN whose performance can be comparable to that obtained by CNN [3, 7].

On the other hand, SVMs have also been explored to solve the problem, in jobs such as [6] they train an SVM and obtain better performances than training MLP or CNN. Other researchers implement the SVM as a classification layer replacing the commonly used SoftMax layer, assuming that the generalization capacity is higher in the SVMs [5]. Currently convolutional networks are being widely used in ASC because they have demonstrated superior results to any other type of classifier, [11, 12]. Most of the works developed for ASC using convolutional networks are based on topologies that combine convolutional layers, activation functions, pooling layers, etc. [13, 14], also highlighting the importance of concatenating Dropout layers, which in most cases reduces the problem of over-adjustment [7, 15].

When analyzing the topologies, there are cases where the complexity is high for the ASC task, since they take as a reference the techniques used for the analysis of images, where a large number of layers and convolutional filters are implemented [8, 10, 15–18]. This represents high training times and can trigger over excessive adjustment in the data. It is also possible to find solutions in which few convolutional layers and filters are used [6, 19] and other cases where attempts are made to combine classifiers such as CNN and RNN in an attempt to improve performance [20–22].

2 Acoustic Scene Classification Framework

2.1 Database

The work will be developed with the TUT Acoustic Scenes 2018 database, used for the DCASE2018 challenge. This consists of ten acoustic scenes, as they are: airport, shopping center (inside), metro station, pedestrian street, public square, street with average traffic level, bus trip, underground subway ride, tram ride and urban park, as shown in Fig. 1. Each scene consists of 864 audio segments of 10 s sampled at 48 kHz. These were recorded in six important European cities: Barcelona, Helsinki, London, Paris, Stockholm and Vienna [19].

1	Airport
2	Bus
3	Metro
4	Metro-Station
5	Park
6	Public-Square
7	Shoppung-Mall
8	Street-Pedestrian
9	Street-Traffic
10	Tram

Fig. 1. Classes of scenarios included in the database

2.2 Feature Extraction

The sounds of urban environments present variations in long, medium and short time intervals (depending the nature of the scene). Therefore, the choice of the size of the segments and the overlap between them is fundamental for the performance of the system. If the analysis is carried out in long segments of time, it will be complex to extract generalizing features of the environment. On the other hand, if the segments are too short, the number of coefficients extracted and that must be analyzed is high and the complexity of the system increases, while the capacity for generalization is affected [16].

For the development of the work, each 10-s audio is divided into segments of 100 ms with overlap of 50%, to which 40 MFCC coefficients, 40 gamma tone coefficients and 100 DFT coefficients, the segments of the individual features are extracted they are concatenated to obtain a spectrogram that represents the energy of the frequency content of the signal as it varies over time.

For the extraction of MFCC coefficients and gamma tones, filters are applied in which the bandwidth increases while the frequency increases, similar to how the cochlea of mammals works. The filters used in MFCC are triangular [3], while for gamma tones filters are used based on the impulse response of a gamma envelope and a sinusoidal tone [23].

2.3 Convolutional Neural Network

The general CNN model is composed of alternating convolutional, pooling, activation and fully connected layers, which are sequentially connected [24]. The convolutional layers, which give the name to this type of networks, operate under an input vector that performs operations of convolutional type, based on a kernel for each filter. The activation function in most cases is ReLU (Rectified Linear Unit), since it makes the calculation of the partial derivatives easier and therefore the training time is reduced [25]. The max-pooling layers aim to reduce the dimensionality and maintain the invariance at the scale for the data set. Fully connected layers are concatenated to a layer called loss function, where it is common to use Softmax, which obtains a probability distribution in all classes giving an estimate to the output [25]. Additionally, Dropout layers can be implemented, which allow to reduce the overfitting of a network by eliminating weights in a random way, forcing the obtaining of precision even with loss of information. Another layer of importance is Batch Normalization (BN), which has some advantages with respect to the global normalization of the data, being more important that it reduces the change of covariance and also the dependence of the gradient to a previously established amount of data.

The implemented architecture is summarized in Fig. 2, has 1 input layer with a variable size of [100 × 20 × 1] when tests are performed with interval segments of one second, while the size is [100 × 199 × 1] when the tests were performed with segments of ten seconds. The network has 2 concatenated convolutional blocks, where a convolutional layer of [3 × 3] is implemented, followed by a layer of BN, the activation function ReLU, a layer of Dropout with probability of elimination of weights of 20%, later they are connected again a convolutional layer, BN and ReLU, followed by a Max Pooling layer of [2 × 2]. The number of convolutional filters is the same for each convolutional layer within each block. After the second convolutional block a Dropout layer is implemented, followed by a fully connected layer with 100 units, a new Dropout layer, to finally have a fully connected layer with 10 units corresponding to the number of classes, whose outputs are connected to the Softmax loss function.

	Layers	Size
1	Image Input	[100 x (20;199) x 1]
2-9	**Convolutional Block 1**	
10-17	**Convolutional Block 2**	
18	Dropout	20%
19	Fully Connected	100
20	ReLU	
21	Dropout	20%
22	Fully Connected	10
23	Softmax	

	Layers	Size
2	Convolution	[3 x 3]
3	Batch Normalization	
4	ReLU	
5	Dropout	20%
6	Convolution	[3 x 3]
7	Batch Normalization	
8	ReLU	
9	Max Pooling	[2 x 2]

Fig. 2. Convolutional Neural Network architecture

With the set of features obtained, 80% is taken for training and 20% for validation. For experiments, the number of convolutional filters in each convolutional block is varied. In the first block, the number of filters is varied by 16, 32, 64 and 128. For the second block, it is varied by 32, 64 and 128 filters. It is established that the second block must have a greater number of filters than the first block, so only combinations of filters of 16–32, 16–64, 16–128, 32–64, 32–128 and 64–128.

The training of the network is done on a server owned by the CECAD (Centro de Computación de Alto Desempeño) of the Francisco José de Caldas District University, which has a 16-core processor with a frequency of 2 GHz and 32 GB of RAM. The network was trained for a maximum of 25 times, with a constant learning rate of 0.001, under the ADAM (Adaptive Moment Estimation) optimizer for network training [25].

3 Experimental Results

Under the previously established, the first experiment was conducted with audio segments of 10 s, the results can be seen in Fig. 3 where the training and validation performance for the different input features is shown by varying the number of filters in each convolutional block. The best performance in network validation is obtained with Gamma coefficients, but the overfitting of training data can be evidenced, since the difference between training and validation performance is between 8 and 24%. In this sense, the lowest overfitting to the data is obtained with MFCC coefficients, but it should be considered that the validation performance is between 5 and 11% lower than that obtained with Gamma coefficients. On the other hand, although a greater number of DFT coefficients were extracted, the obtained yield is in some cases barely comparable to that achieved with the other input features.

The Fig. 4 is the confusion matrix for gamma-type input features extracted to 10-s audio segments, with 16 and 32 filters in the first and second convolutional blocks, respectively. It can be seen that the lowest yields are obtained by the classes metro station 50.3%, metro 52% and pedestrian street 52%. On the other hand, the highest yields are obtained for the shopping center 91.3% and park 86.1%.

When performing the experiments with audio segments of 1 s, it was observed that the performance in training and validation is approximately the same, so the over-adjustment is low, since the network has a greater number of data to train and can be more easily extract generalizing characteristics from each acoustic scene.

The Fig. 5 shows that the best performance for data validation is obtained with Gamma coefficients reaching 80%, hence the generalization capacity of these features can be evidenced since it exceeds the yields obtained with DFT and MFCC, which in all cases has the lowest yields.

The best results were obtained with Gamma coefficients, the topology of the network was configured with 32 and 64 filters in the first and second convolutional block, respectively. Figure 6 shows the confusion matrix for the validation data, where the overall performance reaches 81.6%. It can be seen that the lowest performance is obtained for the subway station class with 70.7%, for this class it has the highest number of false positives with the metro and pedestrian street classes.

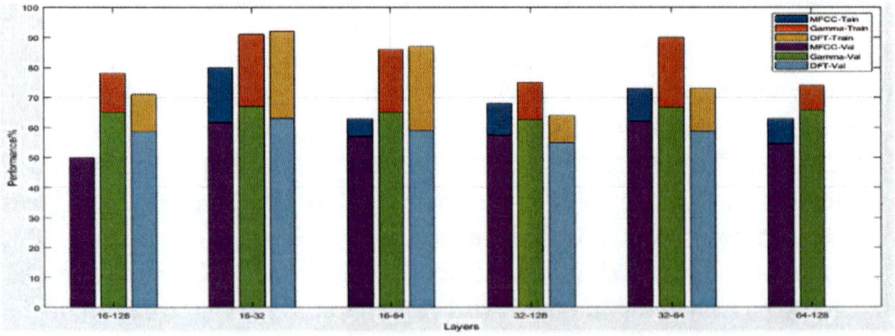

Fig. 3. Performance obtained with different filters and input features for 10-s audio segments

Fig. 4. Confusion matrix for CNN with 16 and 32 filters for the first and second convolutional block with gamma input features for 10-s segments.

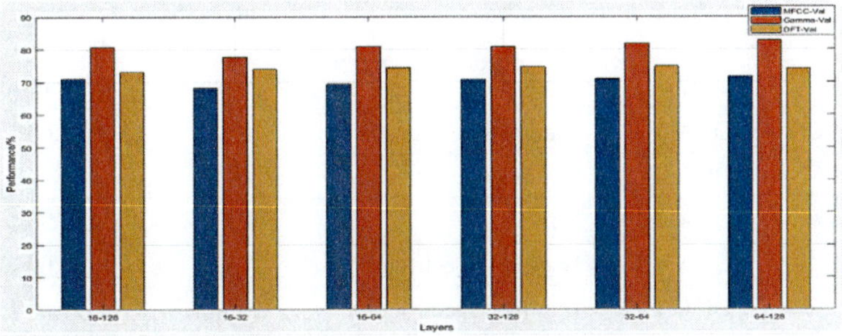

Fig. 5. Performance obtained with different filters and input features for 1-s audio segments

Table 1 summarizes the performance obtained by the best models implemented and some previous works validated with different databases. The implemented system with data length of 10 s exceeds by about 7% the baseline model that uses the DCASE2018 database.

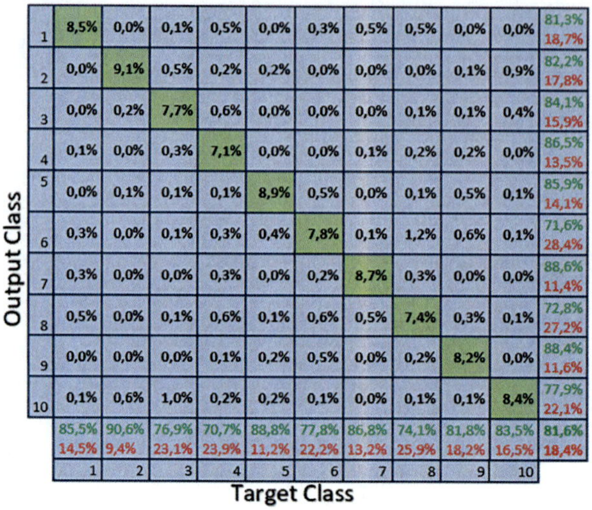

Fig. 6. Confusion matrix for CNN with 32 and 64 filters for the first and second convolutional block with gamma input features for 1-s segments

Table 1. Performance comparison on different databases

Author	Database	Features	Classifier	Performance
Proposed CNN	DCASE18	Gammatone 1 s	CNN	81,60%
Proposed CNN	DCASE18	Gammatone 10 s	CNN	66,90%
[3]	DCASE16	MFCC	MLP	82,90%
[6]	DCASE16	Log-mel Spectrogram	SVM	80,00%
[17]	DCASE17	Log-mel Spectrogram	CNN	91,80%
[7]	DCASE17	MFCC, mel energies	CNN	83,65%
[8]	DCASE18	Log-mel Spectrogram	CNN	76,90%
[22]	DCASE18	Log-mel Spectrogram	CNN	72,18%
Baseline [19]	DCASE18	Log-mel band energies	CNN	59,70%
[5]	LITIS	MFCC, Gammatone, Log-frequency filter bank	RNN + SVM	97,50%
[9]	LITIS	Gammatone	LTE + SVM	96,20%
[20]	LITIS	Log-mel Spectrogram	CNN + SVM	93,78%

4 Conclusions

Models trained with 1-s segments had better performance in both training and validation, since the neural network was able to obtain generalizing features more easily, almost totally avoiding overtraining. However, the variation in the performance of the different topologies is minimal, showing the limit achievable with the different features used. Even so, the Gamma coefficients were the features that best represented the input audios, obtaining better performances than for example the DFT coefficients that had a greater number of features, which allows us to think about the possibility of reducing the number of features used and verify to what extent it is possible to reduce the complexity of the system.

References

1. Sawhney, N.: Situational Awareness from Environmental Sounds (1997)
2. Barchiesi, D., Giannoulis, D., Stowell, D., Plumbley, M.D.: Acoustic scene classification: Classifying environments from the sounds they produce. IEEE Sig. Process. Mag. **32**(3), 16–34 (2015)
3. Prakruthi, U.S., Kiran, D., Ramasangu, H.: High performance neural network based acoustic scene classification. In: 2018 2nd International Conference on Inventive Systems and Control (ICISC), pp. 781–784. IEEE, Ukkadam (2018)
4. Fraile, R., Blanco-Martın, E., Gutiérrez-Arriola, J.M., Sáenz-Lechón, N., Osma-Ruiz, V.J.: Classification of Acoustic Scenes Based on Modulation Spectra and Position-Pitch Maps (2018)
5. Phan, H., Koch, P., Katzberg, F., Maass, M., Mazur, R., Mertins, A.: Audio Scene Classification with Deep Recurrent Neural Networks (2017)
6. Mafra, G., Duong, N., Ozerov, A., Pérez, P.: Acoustic scene classification: an evaluation of an extremely compact feature representation. In: 2016 Detection and Classification of Acoustic Scenes and Events. IEEE, Budapest (2016)
7. Khalid, H., Mazhar, H., Muhammad, G.: An improved acoustic scene classification method using convolutional neural networks (CNNs). Am. Sci. Res. J. Eng. Technol. Sci. **44**(1), 68–76 (2018)
8. Sakashita, Y., Aono, M.: Acoustic scene classification by ensemble of spectrograms based on adaptive temporal divisions (2018)
9. Phan, H., Hertel, L., Maass, M., Koch, P., Mertins, A.: Label Tree Embeddings for Acoustic Scene Classification (2016)
10. Han, Y., Lee, K.: Acoustic scene classification using convolutional neural network and multiple-width frequency-delta data augmentation (2016)
11. Mesaros, A., Heittola, T., Virtanen, T.: Acoustic scene classification: an overview of DCASE 2017 challenge entries. In: 2018 16th International Workshop on Acoustic Signal Enhancement (IWAENC), pp. 411–415. IEEE, Tokyo (2018)
12. Mesaros, A., Heittola, T., Benetos, E., Foster, P., Lagrange, M., Virtanen, T., Plumbley, M.D.: Detection and classification of acoustic scenes and events: outcome of the DCASE 2016 challenge. IEEE/ACM Trans. Audio Speech Lang. Process. **26**(2), 379–393 (2017)
13. Dang, A., Vu, T.H., Wang, J.C.: Acoustic scene classification using convolutional neural networks and multi-scale multi-feature extraction. In: 2018 IEEE International Conference on Consumer Electronics (ICCE). IEEE, Las Vegas (2018)
14. Valenti, M., Diment, A., Parascandolo, G., Squartini, S., Virtanen, T.: DCASE 2016 acoustic scene classification using convolutional neural networks. IEEE, Budapest (2016)
15. Zhang, Z., Xu, S., Cao, S., Zhang, S.: Deep convolutional neural network with mixup for environmental sound classification. In: Chinese Conference on Pattern Recognition and Computer Vision (PRCV). LNCS, Guangzhou, pp. 356–367 (2018)
16. Mariotti, O., Cord, M., Schwander, O.: Exploring Deep Vision Models for Acoustic Scene Classification (2018)

17. Han, Y., Park, J., Lee, K.: Convolutional Neural Networks with Binaural Representations and Background Subtraction for Acoustic Scene Classification (2017)
18. Yin, Y., Shah, R.R., Zimmermann, R.: Learning and fusing multimodal deep features for acoustic scene categorization. In: 2018 ACM Multimedia Conference on Multimedia Conference, pp. 1892–1900. ACM, (2018)
19. Mesaros, A., Heittola, T., Virtanen, T..: A multi-device dataset for urban acoustic scene classification (2018)
20. Phan, H., Chén, O.Y., Pham, L., Koch, P., De Vos, M., McLoughlin, I., Mertins, A.: Spatio-Temporal Attention Pooling for Audio Scene Classification (2019)
21. Ren, Z., Qian, K., Wang, Y., Zhang, Z., Pandit, V., Baird, A., Schuller, B.: Deep scalogram representations for acoustic scene classification. IEEE/CAA J. Automatica Sinica **5**(3), 662–669 (2018)
22. Phaye, S.S.R., Benetos, E., Wang, Y.: SubSpectralNet - Using Sub-Spectrogram Based Convolutional Neural Networks for Acoustic Scene Classification (2018)
23. Gammatone-like spectrograms. http://www.ee.columbia.edu/~dpwe/resources/matlab/gammatonegram/. Accessed 21 June 2019
24. Wang, Z., Liun, J.: A review of object detection based on convolutional neural network. In: 2017 36th Chinese Control Conference (CCC), pp. 11104–11109. IEEE (2017)
25. Indolia, S., Goswami, A.K., Mishra, S.P., Asopa, P.: Conceptual understanding of convolutional neural network-a deep learning approach. Procedia Comput. Sci. **132**, 679–688 (2018)

Microservices-Based Architecture for Resilient Cities Applications

Zeida Solarte(✉), Juan D. Gonzalez, Lyda Peña, and Oscar H. Mondragon

Universidad Autonoma de Occidente, Cali, Colombia
zsolarte@uao.edu.co

Abstract. The dramatic urban population growth worldwide brings several problems to cities. Urban centers must be prepared to face critical everyday situations and sudden risk events. Information systems play a fundamental role in achieving resilience in cities, allowing quick and effective responses to threatening events. This paper presents a software architecture based on reactive architectures and microservices concepts in order to facilitate applications development in the context of resilient cities.

Keywords: Resilient cities · Software architecture · Reactive architectures · Microservices

1 Introduction

The increase in the population of cities around the world and urban vulnerabilities have forced public administrators to include programs in their strategic plans with the aim of transforming their cities into resilient cities. Cali is the third largest city in Colombia, and some studies recognize it as the largest urban conglomerate in the country exposed to high seismic activities. Additionally, Cali is in constant flood risk, since it is located between the valleys of two rivers [5]. Additionally, other socio-economic factors, brings the need to improve the resilience capacities of the city.

To achieve the goal of turning a city into a resilient one, it is necessary to leverage integrated information systems that allow reacting to everyday events and emergency situations, facilitating decision-making. For this reason, we propose ResCity, a computational platform which aim is to help Cali to become a resilient city.

This paper presents the proposal of a software architecture for ResCity based on microservices that facilitates the process of applications development in the context of resilient cities. The remainder of this paper is structured as follows. Section 2 presents the main concepts related to the problem to be solved. Section 3 describes the work methodology for the development of the architecture proposal, which is presented in Sect. 4. Finally, Sect. 5 provides some conclusions.

D. F. Cortes Tobar et al. (Eds.): AETA 2019, LNEE 685, pp. 423–432, 2021.
https://doi.org/10.1007/978-3-030-53021-1_43

2 Background

2.1 Resilient Cities

The term resilience is used in a wide variety of fields. Although the context of the term may change, the concept of resilience is related to the capability of an element to return to a stable state after a disruption [1].

In 1973, Holling [12] defined the term resilience associated with ecological systems: "resilience determines the persistence of relationships within a system and is a measure of the ability of these systems to absorb changes of state variables, driving variables, and parameters, and still persist".

According to the United Nations Strategy for Disaster Reduction [8] a resilient city "has the capacity of resist, absorb, accommodate, adapt to, transform and recover from the effects of a hazard in a timely and efficient manner, including through the preservation and restoration of its essential basic structures and functions through risk management". On the other hand, the OECD [9] defined the resilient cities as "cities that have the ability to absorb, recover and prepare for future shocks (economic, environmental, social & institutional)", it also lists the 4 areas that drive resilience, which are economy, governance, society, and environment.

The Rockefeller foundation [10] establishes seven qualities that allow resilient cities to "withstand, respond to, and adapt more readily to shocks and stress", these qualities are: reflective, resourceful, robust, redundant, flexible, inclusive and integrated.

Improving city resilience is an important aspect, for cities like Cali, which are at high risk. On the other hand, the post-disaster recovery and reconstruction can provide an important opportunity to solve serious structural problems in the city and prevent future suffering for affected populations [11].

2.2 Reactive Architectures

In recent years applications and the ways to develop them have changed. New applications (e.g., reactive applications), should have milliseconds response times, 100% uptime, and handle data in the order of the Petabytes. A reactive application needs: react to events, react to load, react to failure and react to users [3].

The reactive architectures are those that satisfy the four principles of the *Reactive Manifesto* and let developers build reactive systems. Reactive Systems are easier to develop and change, more flexible, loosely-coupled, scalable, tolerant of failure and highly responsive [2]. The four principles defined by the Reactive Manifesto for reactive systems are responsiveness, resilience, elasticity, and message-driven communication.

2.3 Microservices

Microservices are the base of an architectural style that focuses on the development of a single application as a suite of small services, each running in its own process and communicating with lightweight mechanisms, often an HTTP resource API. These microservices are based on business capabilities and are implemented independently, can be written in different programming languages and use different data storage technologies [16].

Microservices were not invented, they emerged as a real-world pattern, guided by domain-driven design, continuous delivery, on-demand virtualization, infrastructure automation, small autonomous teams, and systems at scale. Many organizations have discovered that with the microservices architecture, they can deliver software faster and adopt newer technologies [17].

The use of microservices provides many and varied benefits, improving those generated by the distributed systems. Newman [17] lists these benefits: technology heterogeneity, resilience, scaling, ease of deployment, organization alignment, composability, and optimization for replaceability.

Although microservices have a concept similar to SOA (Service Oriented Architecture), it shows distinctive characteristics that blend into something unique and make it different from SOA itself: the size is comparatively smaller than a typical service, and the functionality related to the context is combined in a limited unique business capacity (bounded context) [6].

3 Methodology

Domain-driven Design (DDD) is a methodology that captures relevant knowledge of the domain through collaborative work between business experts and software engineers during the modeling stage [7]. According to Evans et al. [7], a domain model contains all the parts that are required to understand the domain. In order to facilitate implementation process, they propose the use of three patterns of model elements that allow expressing that implementation process: *Entities*, *Value Objects*, and *Services*.

Entities are objects that are characterized by having an identity in the system, which allows them to be identified and follow their thread of continuity, even though there are other entities with the same attributes. Entities have attributes, responsibilities, and relationships, which are defined according to their identity.

Value objects represent concepts that do not have (do not require) identity. Generally, they describe some characteristic of something, and for this reason, they only have attributes.

Services are operations that do not belong conceptually to any specific domain object and are defined in terms of what they can do. A service has three features: (1) the operation that defines it, which is related to the domain concept, but is not part of an entity or value object; (2) its interface is defined in terms of other elements of the domain, and (3) the operation has no status, so any client can use an instance of the service without considering other instances previously created. Services can be distributed in different layers of the architecture in domain services, application services and infrastructure services.

In recent years the DDD has gained relevance because it provides the basis for decomposing the domain in contexts, each of whom groups coherent domain concepts. These contexts correspond to microservices that provide the different capacities of the business [18].

4 Architecture

4.1 Functional and Non-Functional Requirements

The architecture design for ResCity platform was based on the functional and non-functional requirements defined during its initial stage. These requirements were collected during several meetings carried out with government offices of Cali city. Among the government entities involved in the requirements' collection are the office of resilience, the environment office, the risk management office, the ICTs office, and the education secretariat, among others. On the other hand, requirements related to the characteristics necessary for a system to be considered resilient were taken into account.

Functional requirements are related to the reception of data from different information sources and their management. Examples of those sources are sensors, social networks, events reported by citizens, and government databases. Among these requirements, we also consider those related to data analysis and information delivery through alerts, messages, and reports. Additionally, there are functional requirements that include the management of the platform components and the delivery of information related to its state.

Concerning non-functional requirements, those are related to the features of resilient systems, among which we considered tolerance to failures, interoperability, low coupling, speed, scalability, and flexibility.

4.2 Proposed Architecture

Our architecture design is founded on *reactive* and *microservices-based* architectures. Those architectures are designed to support resilience, which is a fundamental feature of ResCity platform. Additionally, those types of architectures comply with the features defined in the non-functional requirements. The proposed architecture is shown in Fig. 1a.

The API Gateway provides a single interface to the outside world and performs routing functions towards the different services. The API Gateway implements security functions related to the authentication and authorization, usage monitoring, and orchestration of own and third parties APIs.

The services implement the functionalities of the domain, which correspond to the functional requirements that are not related to the computational infrastructure. Each service exposes an API to facilitate communication with the API Gateway and with the other developed services. Each service contains three components which are SH (Self-Healing), LB (Load-Balancing) and AH (Auto-Scaling), as shown in Fig. 1b. The SH controls the instances life cycle and allows their automatic recovery, the LB allows the balancing between different instances in a transparent way when consuming a service, and the AH allows to perform the automatic scaling of the instances in accordance to configuration conditions.

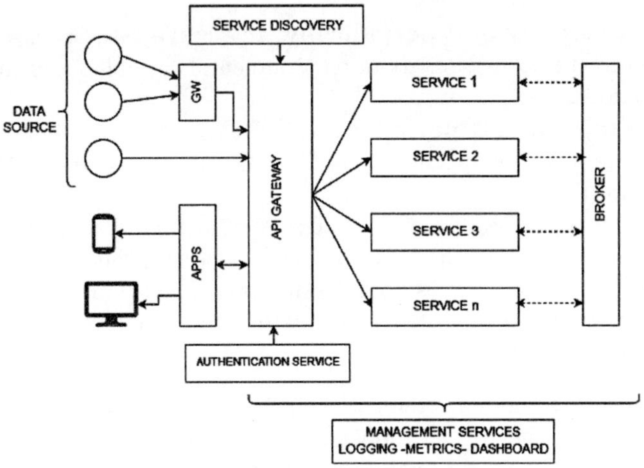

(a) General Architecture

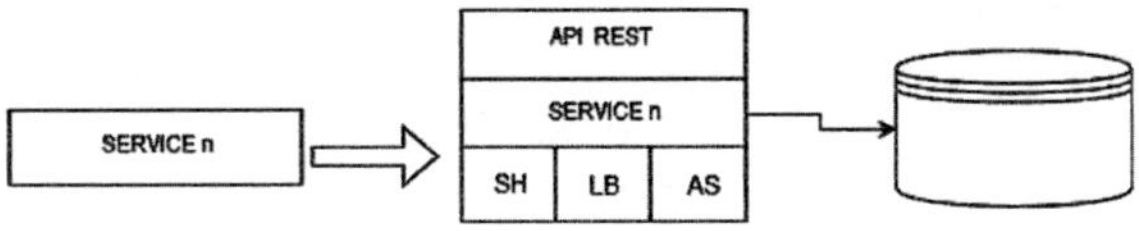

(b) Detailed Service

Fig. 1. ResCity platform architecture

The Broker implements another communication mechanism between the services, based on the publication/subscription scheme. The main feature of this component is that it speeds up the sending and receiving of messages, especially those related to information streams generated at high speed, and acts as a messaging queue, allowing the system to be more robust. The Authentication Service allows the authentication of external applications on the platform. It communicates directly with the API Gateway in order to grant access to the services. The management services component implements the functional requirements related to the computational infrastructure.

4.3 Domain-Driven Services Definition

After performing the analysis of the functional requirements for resCity, a platform which principal domain has to do with city resilience, we defined 3 higher-level contexts completely independent of each other. These contexts, which define the sequence of work of the platform are *information input*, *information processing and analysis*, and *information output*.

The platform input data can come from different sources: sensors located in different parts of the city, social networks, databases of government entities, citizen reports, events reported directly to the platform, citizens sensors, files of data, web pages, among others. These data sources can be classified into 3 types, according to the way in which information is obtained:

- **Data sent directly and periodically:** this is the case of sensors that are reporting direct information about environmental variables, and their sending is done periodically.
- **Direct data, but not periodic:** data of this type are events and citizen reports. Data sent are specifically related to some variable or situation, but are not periodic.
- **Data extracted through the analysis of information entered from other sources:** this is the case of social networks, databases, web pages, and data files. These sources do not send data directly or continuously. The information must be analyzed in order to extract relevant data and it is usually done by demand.

Considering that the management of each of these types of sources is done differently and that they do not depend on each other, we define each of them corresponding to a different domain context. In this way, we define the *Bounded Contexts* as *Sensors*, *Events*, and *Other sources.*

Regarding the processing and analysis context, there are requirements related to the analysis of information and workflows. Given the characteristics of each one of them, we defined 2 Bounded Context: *Analysis* and *Workflows.*

The information output context refers to the mechanisms for the delivery of relevant information to citizens and government entities, including messages, alerts, and reports. Messages and alerts can be grouped given the similarity in their handling in a single Bounded Context called *Notifications.* On the other hand, the reports are handled differently and represent a different Bounded Context called *Reports.*

Figure 2 depicts the designed Bounded Contexts and its relationships.

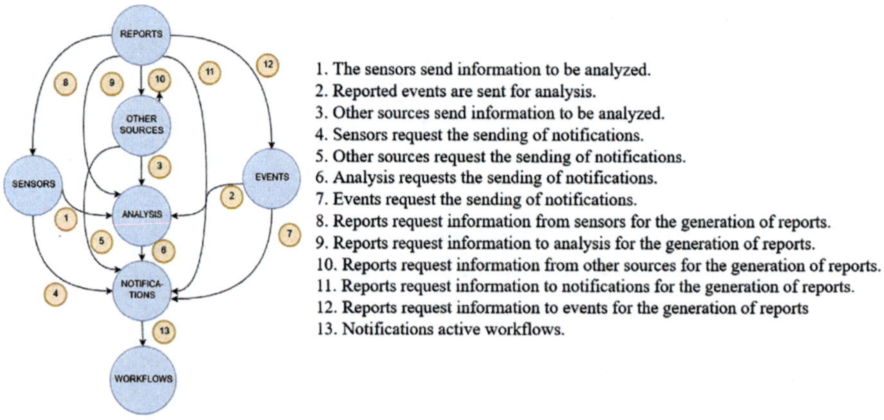

Fig. 2. Bounded contexts

4.4 Test Scenario

In order to validate the designed architecture, we developed an application for the constant monitoring of data generated in Twitter social network. The goal of this application is to find information that can alert about a fortuitous event. This scenario required the implementation of three Bounded Contexts: Other sources, Analysis, and Reports. As shown in Fig. 3.

Because of data coming from a social network, the use of the context Other Sources is required and we defined the *Twitter* entity within it. In order to perform the initial filtering of the data, the entity *Rules* is required. Given that the streaming data is sent directly to the analysis context and this is responsible for storing the results of the performed analyses, we considered it was not necessary to store it. In this context we also defined the entity *Variable*, which represents the variable that is going to be analyzed through the Tweets. Each entity has a repository to implement the requirements associated with their management (i.e., set, get, update, delete).

The context Analysis has the *Rule* entity, which contains the information related to the rules associated with the analysis that is required to be done on the received tweets. There is also the *Analysis* entity, which represents the types of analysis that can be performed. There is a repository available to implement the requirements associated with the management of the rules, which must be able to be created, consulted, modified and eliminated. Additionally, there is a repository for the storage of the results of the analyzes. Finally, this context has a *Data Analysis* service.

In the Reports context, we have the *Report* entity which represents the different types of reports that can be built. There is a repository for the management of the types of reports and there is a service for reports generation.

The Other sources and Analysis contexts communicate with each other through the Broker, which acts as a messaging queue. This is due to the high flow of data generated by social networks and to ensure that data are not lost and can be analyzed completely. The Reports context queries the results through the API offered by the Analysis context for that purpose. All data generated by the source (i.e., the Twitter API) pass through the API Gateway, as well as the data sent to the client application. Fig. 4 shows how the test scenario was deployed and the technologies used for the implementation of each component.

The API Gateway required by the architecture is supplemented by the *API Server Proxy* offered by Kubernetes Orchestration Framework [14]. Among Kubernetes main functions is to connect users outside the cluster with the services implemented within it. The Service Discovery is performed by Kubernetes *CoreDNS*, the Broker is implemented by Apache Kafka [13]. The Dashboard functionality is supplied by the Kubernetes *Dashboard*. The Logging System is covered by Elastic Stack services[4]. Kubernetes also allows implementing a metrics server.

As for the services components, the SH (Self-Healing) is implemented by the Kubernetes controllers, since these control the life cycle of the Pods, and keep the number of replicas at the desired value. The LB (Load-Balancing) is

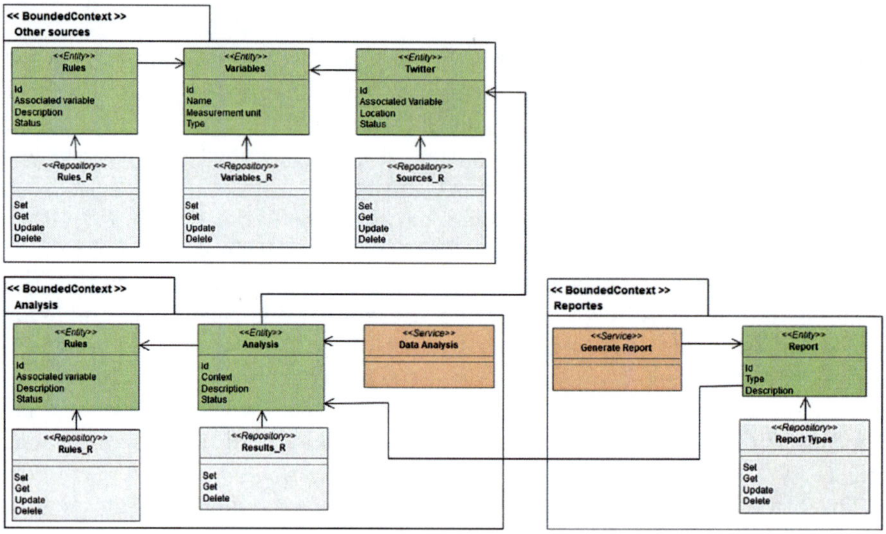

Fig. 3. Services included in the test scenario

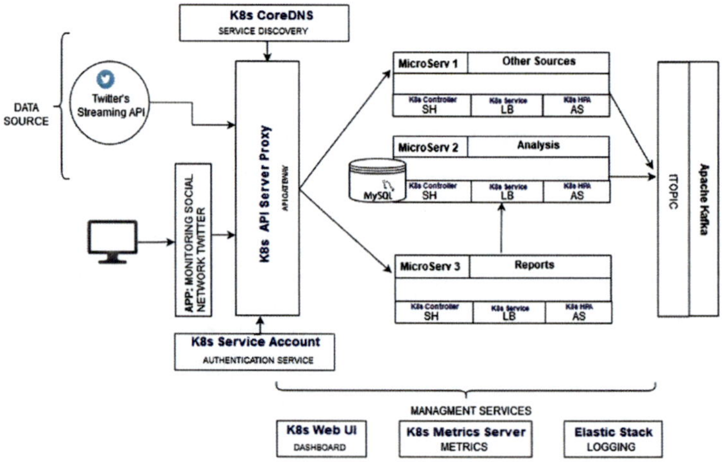

Fig. 4. Architecture with services for the test scenario

achieved when a Kubernetes service is created since the service performs load balancing between its instances. Finally, the Auto Scaling is achieved through the implementation of Horizontal Pod Autoscalers (HPAs).

Repositories or databases are implemented using persistent volumes which, as the name implies, are Kubernetes storage units where data persists in the event of a failure. This is necessary to keep the information due to the ephemeral nature of the instances and the possibility of autoscaling.

The detailed characteristics of the mentioned components can be found in Kubernetes documentation [14].

The external client that consumes the services consists of a web application that we developed in PHP using the Laravel framework [15] and JavaScript. The application shows in a dashboard the data obtained from the Report context. Output data correspond to the result of the analysis of the tweets, that is, the count of the hashtags that are being monitored. Figure 5 shows the application output. Results are presented through a table and a graph. The graph allows visualizing the behavior of the messages over time.

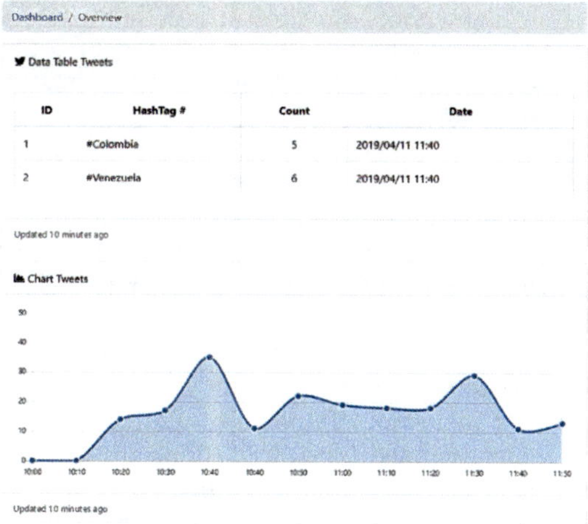

Fig. 5. Client application dashboard

5 Conclusions

Resilience is a characteristic that becomes more relevant every day in the urban context. Massive migration to cities, natural disasters and the changing socio-economic situation of populations require structures and procedures that allow them to respond efficiently and effectively to potentially catastrophic situations.

To improve the resilience capacity in cities, it is essential to have software systems that provide adequate and timely information for decision making at different levels in the urban context.

Applying the domain-driven design methodology, a reactive architecture based on microservices has been proposed. This architecture facilitates the development of specialized software for resilient cities since it considers essential aspects related to resilient capabilities.

In order to validate the proposed architecture, we explored the computational infrastructure that supports distributed data analysis applications and provides the required resilience capabilities. Kubernetes is a technology for the orchestration of containerized applications, which offers essential characteristics for the support of resilient applications, which makes it suitable for use in projects such as the one developed in this work.

References

1. Bhamra, R., Dani, S., Burnard, K.: Resilience: the concept, a literature review and future directions. Int. J. Prod. Res. **49**(18), 5375–5393 (2011)
2. Boner, J., Farley, D., Kuhn, R., Thompson, M.: The reactive manifesto (2014)
3. Bonér, J.: Reactive Microservices Architecture. O'Reilly Media, Inc., Sebastopol (2016)
4. B.V., E.: Elastic stack and product documentation. https://www.elastic.co/guide/index.html. Accessed 12 July 2019
5. Campos, A., Holm-Nielsen, N., Diaz, C., Rubiano, D., Costa, C.R., Ramirez, F., Dickson, E.: Analysis of Disaster Risk Management in Colombia. A Contribution to the Creation of Public Policies. World Bank, Washington, DC (2011)
6. Dragoni, N., Giallorenzo, S., Lafuente, A.L., Mazzara, M., Montesi, F., Mustafin, R., Safina, L.: Microservices: Yesterday, Today, and Tomorrow, pp. 195–216. Springer, Cham (2017)
7. Evans, E.: Domain-Driven Design: Tackling Complexity in the Heart of Software. Addison-Wesley Professional (2004)
8. U. O. for Disaster Risk Reduction. Terminology (2017)
9. O. for Econmic Co-operation and D. OECD. Resilient cities (2019)
10. The Rockefeller Foundation: City resilience framework (2014)
11. Gomes, P.J., Pena, L.A.: Urban resilience: a conceptual framework. Sustain. Cities Soc. **50** (2019)
12. Holling, C.S.: Resilience and stability of ecological systems. Annu. Rev. Ecol. Syst. **4**(1), 1–23 (1973)
13. Kafka, A.: Apache kafka: a distributed streaming platform. https://kafka.apache.org/. Accessed 11 July 2019
14. Kubernetes: Production-grade container orchestration. https://kubernetes.io/. Accessed 11 July 2019
15. Laravel: Laravel framework. https://laravel.com/. Accessed 11 July 2019
16. Lewis, J., Fowler, M.: Microservices. A definition of this new architectural term (2014)
17. Newman, S.: Building Microservices. O'Reilly Media, Inc., Sebastopol (2015)
18. Richards, M.: Microservices vs. service-oriented architecture (2015)

Optoelectronic Oscillator at S-Band and C-Band for 5G Telecommunications Purpose

Juan Fernando Coronel[1(✉)], C. Camilo Cano[2], Margarita Varón[2], Héctor Guarnizo[1], and Mónica Rico[3]

[1] Universidad El Bosque, Bogotá, Colombia
jfcoronelr@unal.edu.co
[2] Universidad Nacional de Colombia, Bogotá, Colombia
[3] Universidad Católica de Colombia, Bogotá, Colombia

Abstract. This article presents the implementation results for two optoelectronic oscillators at 2.2 GHz and 5.7 GHz; and the proof-of-concept of the integration of photonics components into a microwave communication link. The oscillators are characterized for frequency stability. The phase noise values achieved are −101.1 dBc/Hz at 10 kHz offset for the 2.2 GHz optoelectronic oscillator and −100.3 dBc/Hz at 10 kHz offset for the 5.7 GHz optoelectronic oscillator. The implemented microwave link is evaluated using the Error Vector Magnitude (EVM) for a 2.2 GHz transmission with a 4 symbols QPSK modulation.

Keywords: Optoelectronic oscillator · Signal reference · Phase noise · Error Vector Magnitude (EVM)

1 Introduction

The arrival of 5G technology promotes the development of networks with technology integration in order to accomplish the 5G network goals, such as latency reduction up to 1 ms, spectrum management enhancement as well as the user density network capability [1].

From the network infrastructure point of view, for 5G networks it is necessary to give a ubiquitous radio access to users at outdoor or indoor environments. This represents a challenge for the spectrum management in order to make better spectrum assignation. Some network operators have been testing frequency bands from 3 GHz up to 70 GHz for 5G applications [2, 3].

The 5G network architecture has as fundamental pillar the use of radio over fiber (RoF) technology in order to maximize the network bandwidth and reduce the latency from the radio access to the network core. For this reason, the development of an optoelectronic signal carrier distribution, based on optoelectronic oscillators is proposed [4]. The advantage of this technique is that it can be scaled from femtocells to microcells.

To assure radio access spectrum efficiency, it is necessary to optimize the bandwidth required for each user in order to accomplish the data rate required for 5G communication. In consequence, according to the Shannon formula for channel capacity (C),

D. F. Cortes Tobar et al. (Eds.): AETA 2019, LNEE 685, pp. 433–441, 2021.
https://doi.org/10.1007/978-3-030-53021-1_44

this depends on the signal-to-noise ratio (S/N) and the channel bandwidth (B) as shown in (1).

$$C = B \cdot log_2(1 + S/N) \tag{1}$$

From this equation, it is necessary to enhance the signal-to-noise ratio in order to take advantage of the channel assigned bandwidth. In consequence, when the baseband message is upconverted into microwave frequency, it is required to provide a high spectral purity signal reference to avoid phenomena such as the frequency drift that could lead to channel overlapping or carrier noise that lead to signal noise.

To face the challenge of carrier generation stability, optoelectronic oscillators (OEO) are proposed as an alternative to electronic oscillators to generate microwave frequency carriers with reduced noise and high spectral purity [5].

This article is focused on the implementation of an optoelectronic oscillator using a delay-line architecture to generate carriers at microwave frequencies for S and C-band that are of interest for 5G network access and carrier distribution to assure the network synchronization and the demonstration of the integration of the optoelectronic oscillator into a microwave wireless link.

The article presents in its second section a brief theoretical background of the opto-electronic oscillator operation, the third section presents the OEOs implementation and characterization through the spectrum analysis and the phase noise measurement, the fourth section describes integration of an OEO into a microwave transmission link and its performance, finally the fifth section presents the work conclusions.

2 The Optoelectronic Oscillator

The optoelectronic oscillator was first presented in 1995 by Yao and Maleki [5, 6] as a high spectral purity carrier generator. This hybrid architecture is shown in Fig. 1. It is composed of a laser source (Laser Diode), a Mach-Zender electro-optical amplitude

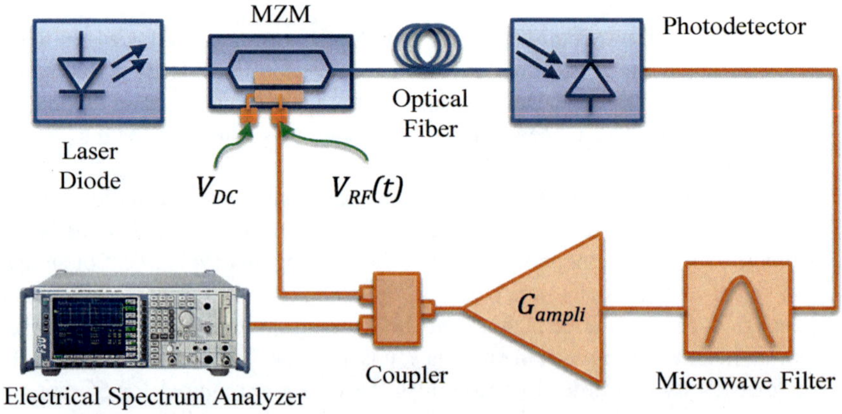

Fig. 1. Delay line based optoelectronic oscillator schematic configuration

modulator (EOM), an optical fiber delay line, a high-speed photodetector to converts the modulated and "delayed" light into electrical domain to be filtered by a microwave bandpass cavity filter.

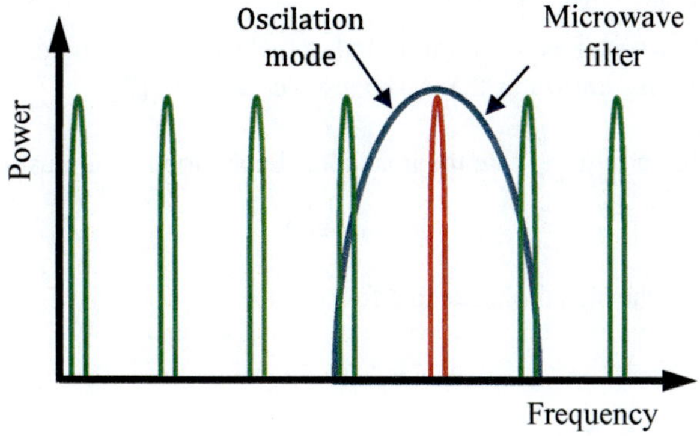

Fig. 2. Selection of the oscillation mode in the optoelectronic oscillator.

The purpose of the microwave bandpass filter is to select an oscillation mode from the frequency comb produced by the resonant modes of the optical fiber delay line, this is shown in Fig. 2. Then, the selected modes - it is possible to get several oscillating modes depending on the microwave filter bandwidth – are amplified with a low noise microwave amplifier and re-injected into the OEO loop. Finally, part of the loop signal is extracted through a microwave coupler. This signal corresponds to the generated carrier.

To get a deeper understanding of the operation principle of the optoelectronic oscillator, Fig. 3 is presented. From this figure, the components of the oscillator are presented with their respective input/output signals. From Fig. 3. It can be concluded that the open-loop optoelectronic oscillator output voltage (V_{OL}) is given by (2).

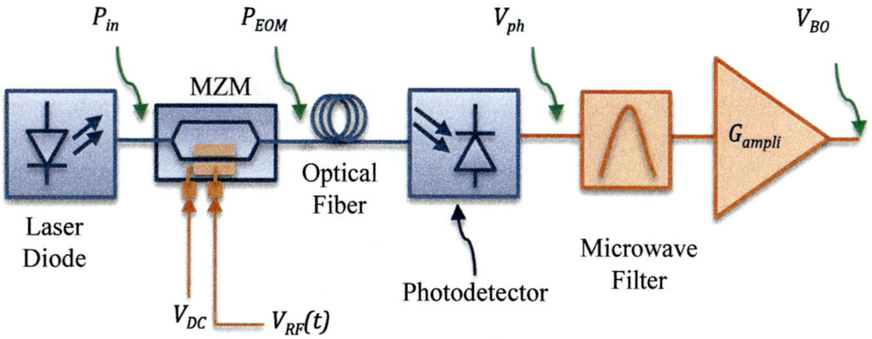

Fig. 3. Optoelectronic oscillator in open loop configuration.

$$\overline{V_{OL}} = \overline{G_A} * \overline{\alpha_E} * \bar{F} * R_{ph} * \bar{S} * \bar{A} * \overline{H_{mod}} * P_{opt} \quad (2)$$

Where, G_A is the voltage gain of the microwave amplifier, α_E represents the electrical losses of the loop, F represents the microwave filter insertion loss, R_{ph} is the photodetector impedance load (Ω), S is the photodetector responsivity (A/W), A is the optical fiber transfer function, H_{mod} is the electro-optical modulator transfer function and P_{opt} is the optical detected power (W) [6].

To obtain a self-sustained oscillation, the optoelectronic oscillator open loop must satisfy two conditions, known as the Barkhausen conditions [7]:

- The gain condition requires that the normalized open loop gain must be at least 1:

$$|V_{OL}| = 1 \tag{3}$$

- Phase condition: the signal phase must follow (4):

$$\omega_{RF}\tau + \phi_o = 2k\pi \tag{4}$$

Where τ is the open loop time delay, ω_{RF} is the oscillation angular frequency, k is an integer.

The oscillation frequency is given by (5), this expression considers the effects of the loop delay, the initial phase (ϕ_o), and the oscillation multiple accepted by the microwave bandpass filter:

$$f_{osc,k} = \frac{k}{\tau} - \frac{\phi_o}{2\pi\tau} \tag{5}$$

3 Implementation and Characterization

For this work, two delay line based optoelectronic oscillators are implemented at 2.2 GHz and 5.7 GHz. Both oscillators are built using a 500 m single mode optical fiber delay line with a 1550 nm DFB laser source (DFB Fitel FRL15CWD). The difference from

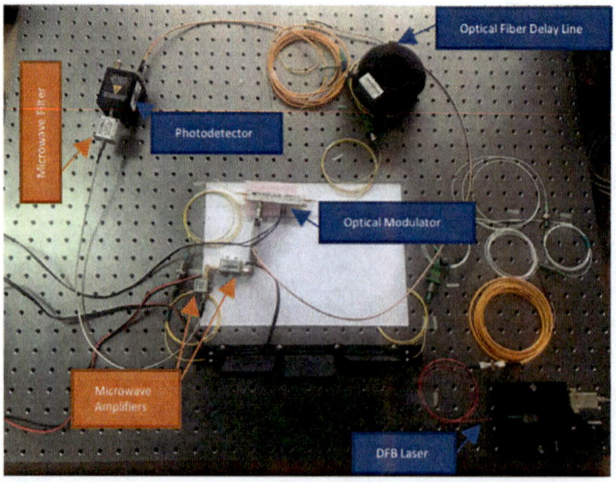

Fig. 4. Optoelectronic oscillator testbench

the 2.2 GHz to the 5.7 GHz oscillator is the microwave filter that is tuned at the oscillator nominal frequency and the microwave amplifiers that are changed for each frequency. Figure 4 presents the optoelectronic oscillator testbench. The elements used for each oscillator are summarized in Table 1.

Table 1. Optoelectronic Oscillators Components

Component	2.2 GHz OEO	5.8 GHz OEO
Photodiode	New Focus 1414	
Amplifier	Minicircuits ZX60-2522 M-S+	Minicircuits ZX60-24-S+
Filter	Minicircuits ZFBP-2400-S+	Minicircuits ZFBP-5800-S+

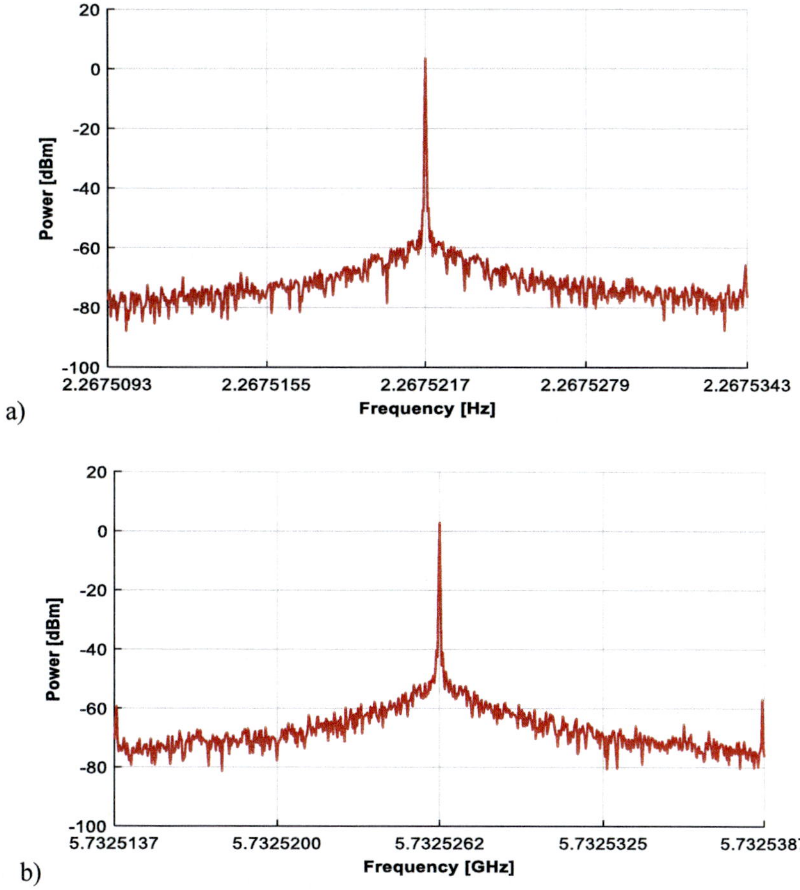

Fig. 5. Optoelectronic oscillator output spectrum at a) 2.2 GHz and b) 5.8 GHz.

The output carrier spectrum of the oscillators are presented in Fig. 5. For the 2.2 GHz OEO (Fig. 5a) and 5.7 GHz OEO (Fig. 5b). The carrier power for the 2.2 GHz OEO is 3.6 dBm and for the 5.7 GHz OEO is 2.8 dBm.

The implemented optoelectronic oscillators are characterized in terms of the frequency stability through the phase noise. This measurement is performed using a FSH8 Rohde and Schwarz electrical spectrum analyzer. Both phase noise curves are presented in Fig. 6. From these curves it can be inferred that frequency stability of the 2.2 GHz optoelectronic oscillator (black line) is better than the 5.7 GHz optoelectronic oscillator (red line), considering the lower phase noise value of the first one.

The phase noise values measured at 10 kHz offset from the carrier were, for the 2.2 GHz oscillator −101.1 dBc/Hz and for the 5.7 GHz optoelectronic oscillator −100.3 dBc/Hz.

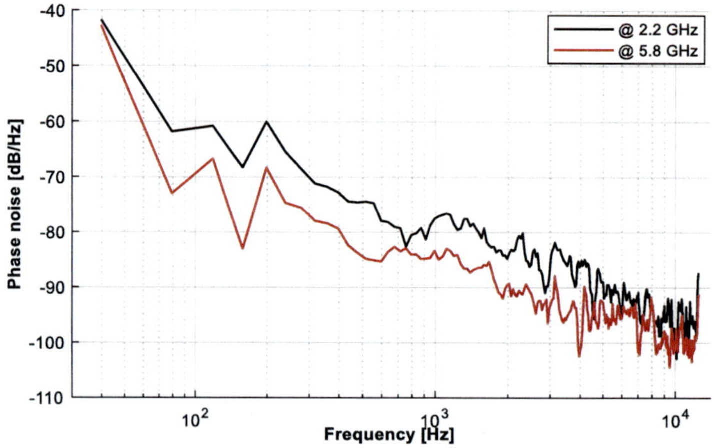

Fig. 6. Optoelectronic oscillator phase noise curves.

4 Microwave Link Demonstration Using a Optoelectronic Oscillator

To present a proof-of-concept of the integration of the optoelectronic oscillator feasibility in a microwave link, is described in this section. Figure 7 presents the schematic setup of the transmitter/receiver implemented system. The system is composed of a transmitter and receiver blocks synchronized using the carrier distributed from the 2.2 GHz optoelectronic oscillator implemented and characterized along the previous sections.

The transmitter and receiver blocks are composed of the following elements: A computer with a GNURadio software using a Universal Software Radio Peripheral (USRP) model B200-mini from National Instruments. To upconvert/downconvert the baseband signals, microwave passive mixers are used with a 2.2 GHz carrier of the local oscillator (optoelectronic oscillator). For transmission, the radio frequency mixer output is amplified and transmitted through an omnidirectional antenna. For reception, the received

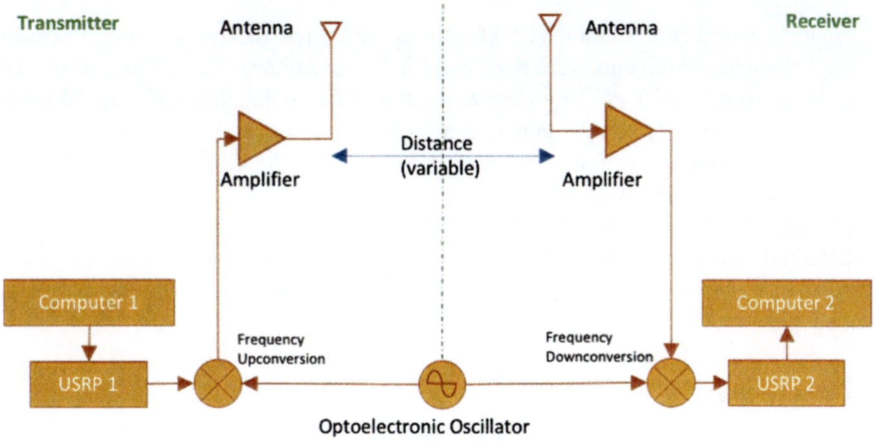

Fig. 7. Microwave link schematic setup.

signal is also amplified to compensate insertion losses of the passive mixer used at the receiver. The baseband received signal is demodulated at the receiver.

The transmitted message for this setup is a random sequence of numbers from 1 to 100 loop. This information is coded using 4 symbols and modulated using QPSK. To evaluate the performance of the microwave link using the optoelectronic oscillator, the received information will be processed to compute the Error Vector Magnitude (EVM) [8] for three different distances between the emitter and receiver (1.1, 2.6 and 4.6 m).

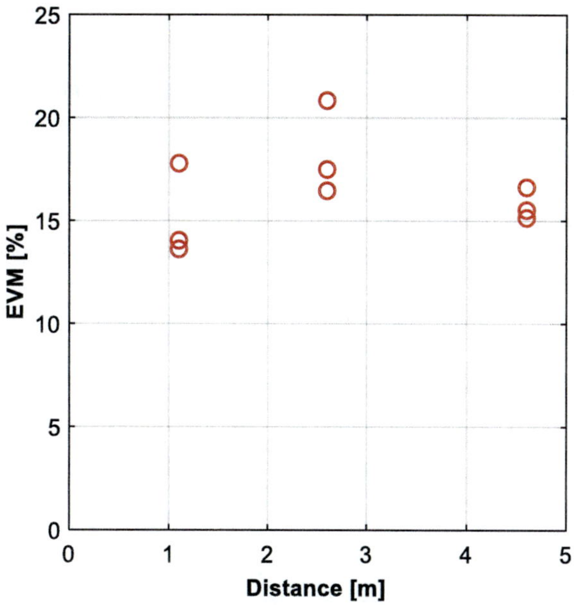

Fig. 8. Microwave link EVM vs antennas separation.

Figure 8 presents the computed EVM value vs the emitter-receiver distance. For each distance, three measurements were performed for each distance. The EVM average for these measurements is 16.4%. From Fig. 8, it can be inferred that the highest EVM value is 20.8% for 2.6 m between the antennas and the lowest is 13.6% for 1.1 m. If these results are observed by the antennas separation, for 1.1 m the average EVM is 15.2%, for 2.6 is 18.3% and for 4.6 is 15.8%.

The performed measurements do not lead to a direct correlation between the EVM and antennas separation. However, it is important to remark that the goal of this article is to demonstrate the feasibility of the microwave link implementation using photonic components.

5 Conclusion

In this work, two optoelectronic oscillators are implemented for 2.2 and 5.7 GHz using 500 m optical fiber delay line. These oscillators are characterized in terms of frequency stability through the phase noise measurement. The implemented optoelectronic oscillator phase noise measured values at 10 kHz offset are summarized in Table 2.

Table 2. Optoelectronic oscillators phase noise values measured at a 10 kHz offset.

Frequency [GHz]	Phase noise at 10 kHz offset [dBc]
2.2	−101.1
5.7	−100.3

According to the delay line based optoelectronic oscillator theory, the phase noise can be reduced by increasing the delay line length [9]. Tests with longer fibers are considered for this research. The development of the optoelectronic oscillators using stable laser sources and external modulators makes possible the generation of high-spectral purity carriers to be distributed through optical or electrical transmission lines to obtain a reliable synchronization source for 5G remote access points.

To demonstrate the feasibility of implementation of photonics components in a microwave link, the 2.2 GHz optoelectronic oscillator is used as local oscillator in a software defined radio-based system. From the link performance, measured through the EVM, it is demonstrated that is possible to establish a microwave frequency digital communication using an optoelectronic oscillator as signal reference achieving EVM average value of 16.4%.

From this proof-of-concept, the research team is motivated to consider higher frequency band implementations, specially, those bands considered for 5G communications at millimetric wavelengths.

Units Summary

Units used to express the optoelectronic oscillator and microwave link figures-of-merit measurements are summarized in Table 3.

Table 3. Optoelectronic oscillator figure-of-merit units.

Magnitude	Unit
Oscillator frequency	GHz
Length	m
Oscillator carrier power	dBm
Optical wavelength	nm
Phase noise	dBc/Hz
Error Vector Magnitude (EVM)	%

References

1. Blust, S., Buonomo, S.: Forging paths to 5G. ITU News Mag. **2**, 14–18 (2017)
2. GSMA Intelligence: Understanding 5G: perspectives on future technological advancements in mobile, December, pp. 3–15 (2014)
3. Lee, J., et al.: Spectrum for 5G: global status, challenges, and enabling technologies. IEEE Commun. Mag. **56**(3), 12–18 (2018)
4. Iezekiel, S.: Radio-over-fiber technology and devices for 5G: an overview. In: Broadband Access Communication Technologies X, vol. 9772, p. 97720A, February 2016
5. Maleki, L.: Sources: The optoelectronic oscillator. Nat. Photonics **5**(12), 728–730 (2011)
6. Yao, X.S., Maleki, L.: Opto-electronic oscillator and its applications. In: Microwave Photonics (MWP 1996), pp. 265–268 (1996)
7. Igor, B., Zdenko, B.: Barkhausen criterion and another necessary condition for steady state oscillations existence. In: 2013 23rd International Conference on Radioelektronika (RADIOELEKTRONIKA), no. 1, pp. 151–155 (2013)
8. Alkhalifah, O., Alrabiah, O., Ragheb, A., Esmail, M.A., Alshebeili, S.: Investigation and demonstration of 5G signal transmission over fiber/FSO/wireless links. In: 2017 International Conference on Electrical and Computing Technologies and Applications (ICECTA 2017), January, vol. 2018, pp. 1–4 (2018)
9. Chembo, Y.K., Volyanskiy, K., Larger, L., Rubiola, E., Colet, P.: Determination of phase noise spectra in optoelectronic microwave oscillators: a Langevin approach. IEEE J. Quantum Electron. **45**(2), 178–186 (2009)

Design and Implementation of a Remote Virtual Laboratory by Internet Applied to KUKA LBR IIWA 14 R820

Susan Juliet Martinez[1], Jose Guillermo Guarnizo[2(✉)], and Jonathan Avendano[3]

[1] Facultad de Ingeniería, Universidad ECCI, Bogotá, Colombia
susanj.martinezg@ecci.edu.co
[2] Facultad de Ingeniería Electrónica, Universidad Santo Tomás, Bogotá, Colombia
jose.guarnizo@usantotomas.edu.co
[3] Facultad de Ingeniería – Bioingeniería, Universidad El Bosque, Bogotá, Colombia
javendanop@unbosque.edu.co

Abstract. In the educational field, it is necessary to have the ability to experiment for learning process, that is where virtual laboratories allow having more freedom to do it, even when there are no physical resources. In this paper is presented a web-based virtual laboratory with the KUKA LBR IIWA 14R820 robot. A virtual robotic arm was used in V-REP simulator allowing to control the robotic arm using the direct kinematic model, directly controlling the position of each joint or through the inverse kinematic model that was developed using Denavit-Hatenberg methodology. For connection with the remote API through the web and design of user interface, LabVIEW 2016 was used with some external libraries. For connection with laboratory through the web, the user had to have a computer with Internet access and have LabVIEW Run-Time Engine 2016 plugin installed. In experimental tests, it was observed that the system was robust since it allows streaming to several users without affecting the work of who is managing the program. It was possible to verify that the performance of the laboratory is linked to the working capacity of the user's computer.

Keywords: Virtual laboratory · Industrial manipulator · Kinematic model

1 Introduction

Teaching in engineering requests laboratories to implement practical issues however traditional laboratories present different problems such as costly, limited resources or restricted physical access. One alternative is using virtual laboratories or remote laboratories [1], which allow the students to explore the experimentation process, problem resolution, data analysis and scientific implementation in a virtual environment, setting on screen results, using peripherical equipment the condition of the experiments.

D. F. Cortes Tobar et al. (Eds.): AETA 2019, LNEE 685, pp. 442–452, 2021.
https://doi.org/10.1007/978-3-030-53021-1_45

Different Works based on virtual laboratories have been presented. For example, in [2] is presented a virtual laboratory for digital signal processing in images, this work was validated through case studies and surveys to professors and students. In [3] is presented a study in the simulation of the AESOP robot (Automated Endoscopic System for Optimal Positions), allowing visualization of a "virtual surgery" practiced by a medical doctor. Other research is focused on remoted laboratories connected to the Internet which permit students to interact with components and recourses out of range. For example MIT Lab [4], which is an environment which provides general solutions to web applications development framework.

In [5] is presented the implementation of a virtual laboratory for bioengineering applications: Rephys. This is a web laboratory providing remote access to real experiments. In [6] is presented the design of a virtual platform called (Electronics and Optics e-Learning for Embedded Systems), that included virtual classrooms, forums, a virtual remote laboratory and an application designed using Moodle. In [7] is presented a virtual platform for the simulation and local and remote control of a SCARA robotic arm UV-CERMA for educational purposes. In [8] is presented a SCADA real time laboratory with remote access based on the web, which controlled an induction motor using real instruments. In [9] is developed a virtual platform for a remote application based on Web for monitoring and failure detection applied to A.C machine. In [10] is presented a remote-control system based on a web server for different purposes, through the use of PIC microcontrollers, and ethernet communication.

In this paper is presented the design and implementation of a remote virtual laboratory by Internet applied to KUKA LBR IIWA 14 R820 industrial manipulator. This project is composed of a virtual reality simulation environment, the remote application and a web user interface, involving direct and inverse kinematics model. The direct kinematic model is obtained first in order to calculate the final position of the end-effector with respect to the base of the robot and the joint´s angles. Subsequently, the inverse kinematic model is used to obtain the final position of the end-effector. For the simulation environment is used the robot simulator V-REP (Virtual Robot Experimentation Platform) [11], to simulate the behavior of the robot and the interaction with other robots or objects.

2 KUKA LBR IIWA 14 R820 Robot

This industrial manipulator weighs 29.9 kg, offering a payload of 14 kg. This robot is considered as a redundant manipulator with seven rotational joints, which form seven axes. In Fig. 1 is presented the coordinate reference system based on Denavit-Harteberg parameters, which values are presented in Table 1, where Lu, Lf y Lh correspond to upper section arm, forearm and the and link lengths, respectively [12]. Inverse kinematic model is presented in [12].

Table 1. Denavit-Harteberg parameters for KUKA LBR IIWA 14 R820 Robot.

i	αi	ai	di	Θi
1	90°	0	0	q1
2	90°	0	0	q2
3	90°	0	Lu	q3
4	90°	0	0	q4
5	90°	0	Lf	q5
6	90°	0	0	q6
7	0	0	Lh	q7

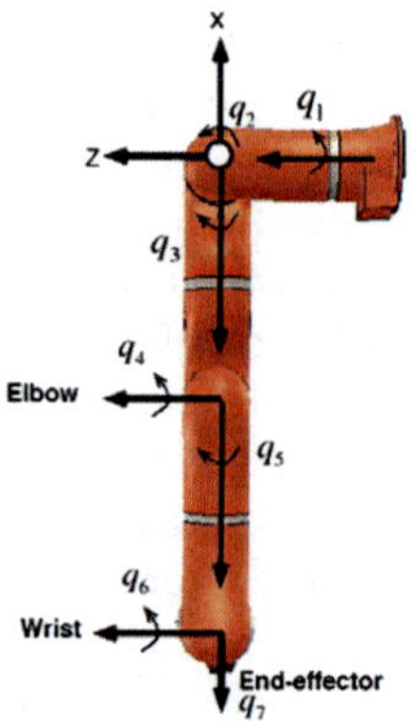

Fig. 1. Kuka robot reference system [12].

3 Simulation Environment

3.1 Specifications

The V-REP software was used for KUKA LBR IIWA 14 R820 simulation. The operation mode of V-REP working like a distributed control schema. All objects into the integrated development environment (IDE) can be controlled by embedded scripts, like a remote API. This script was developed using the LUA language. Figure 2 shows the flow diagram of the script developed. The script uses only ports connections and commands for remote API connections. For visualization remote API, it is included in the simulation an image sensor, with a maximum resolution of 512 × 512.

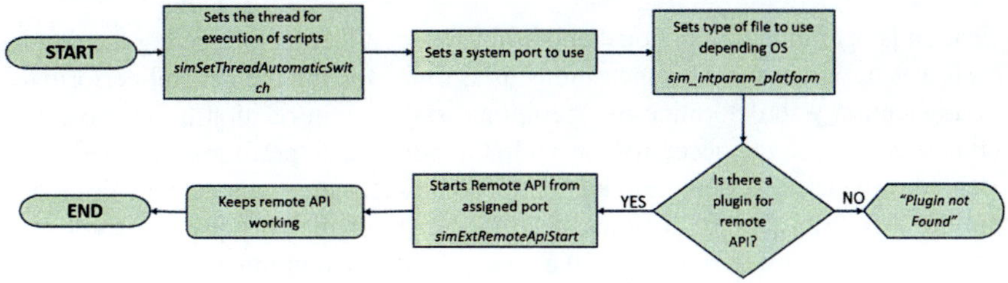

Fig. 2. Flow diagram of the script developed.

3.2 Remote API

The remote API was developed using LabVIEW software. LabVIEW is chosen due to it let us connect our simulation environment, developed in V-REP, through libraries. Additionally, the LabVIEW libraries let us make the user interface and a web connection. Figure 3 shows the Virtual Instrument (VI) diagram.

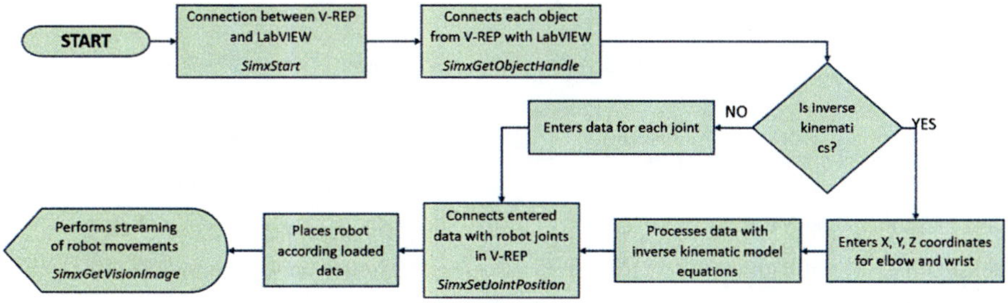

Fig. 3. VI diagram.

First of all, the algorithm developed begins the connection between LabVIEW and V-REP, using the Simx Library. In the second part, the algorithm connects the V-REP objects with LabVIEW objects. Next, the proposed algorithm catches the data that the user inputs. Here, inverse or forward kinematics is chosen. If the inverse kinematics is chosen, the program takes the tridimensional space ubication (x, y, and z-axes) entered by the user, the data entered by the user correspond to the location of the elbow and the wrist with the end effector. This data is sending to the V-REP to resolve the kinematics problem. Otherwise, if the selection is forward kinematics the program takes the angles in radians for each robot joint. In the fourth part, the algorithm sends the data that the user entered to the Simulation Environment in V-REP. Finally, for each robot joint, the spatial ubication was found in the simulation environment. These spatial ubications were sent to the LabVIEW user interface. The robot movements were visualized through real-time streaming video.

4 Assessment of the Application

For the assessment of the application three elements were taken into consideration: availability, performance and response latency. Availability is the measure in which an application is operational, functional and usable for completing or meeting the requirements it was designed for. This measure is used for analyzing the general performance of an application y to determine its operational statistics in relation to its capacity to function according to the necessity for it. On the other hand, performance is defined as the degree in which a system, application or component meets its designed functions within the given limitations, such as speed, accuracy or memory use. Lastly, latency refers to a time interval during which the response of an event must be executed. Time intervals establish a response window given by a start time (minimum latency) and a finish time (maximum latency). These can be specified as absolute times (e.g. time of the day) or compensations of an event that occurred at a specific time.

4.1 Experimental Setup

In order to assess performance, availability and latency on the developed application this latter was set up in a server. After that, the developed remote API was executed on 5 computers with the features mentioned on Table 2.

Table 2. Test computer features

Model	Storage	Memory	Processor	Operating system
Dell E7450	500 GB	16 GB	Core i7 - 2.6 GHz	Windows 7 SP1
Dell E5450	500 GB	4 GB	Core i5 - 2.5 GHz	Windows 7 SP1
Dell E6430	500 GB	4 GB	Core i5 - 2.5 GHz	Windows 7 SP1
Dell E6420	500 GB	4 GB	Core i5 - 2.5 GHz	Windows 7 SP1
Dell E6520	500 GB	2 GB	Core i5 - 2.5 GHz	Windows 7 SP1

In order to carry out the performance tests, it was necessary to install LabVIEW Run-Time Engine 2016, 32 bits version on each computer. It was noticed that the plugins of the browsers Google Chrome and Mozilla with their updated versions are not sup-ported by National Instruments, which is why it was necessary the use of Internet Explorer. All tests were performed in a LAN network.

Availability Assessment. For the purpose of availability assessment of the application, the computers described on Table 2 were connected to a LAN network in which a computer is in charge of executing the robot's simulation environment and the remaining 3 computers are connected remotely to the developed application. It was observed that these 3 computers are able to visualize at the same time the execution of the application in real time, but only one of the computers is able to control the robot.

The remote visualization from the front panels allows for the connection of multiple clients for visualizing the same front panel of the VI. However, only one client has control over that VI. According to the connection tests that were performed the client who has control of the simulation cannot be interrupted as long as they are controlling the application regardless of the number of clients that are connected. Another client can control the application only in the event of the original controller gives up control. By doing so, a failure on the execution of the VI due to the connection of multiple clients can be prevented.

Performance Assessment. To assess the performance of the application, LabVIEW offers a tool called Remote Manager Connection Panel, which allows managing the queue of clients, disconnecting a client and monitoring data traffic for each client. Figure 4 shows the transmission rate of data between a client having control of the VI and the application. For each time a data was sent, a change in the data traffic rate is observed. The average rate remained at 2.7 Mbps while at the time of sending and receiving data it remained between 3.0 and 4.0 Mbps. For the evaluation of the execution time of the simulation, different movement tests were carried out directly in the simulator with different data to know its response capacity.

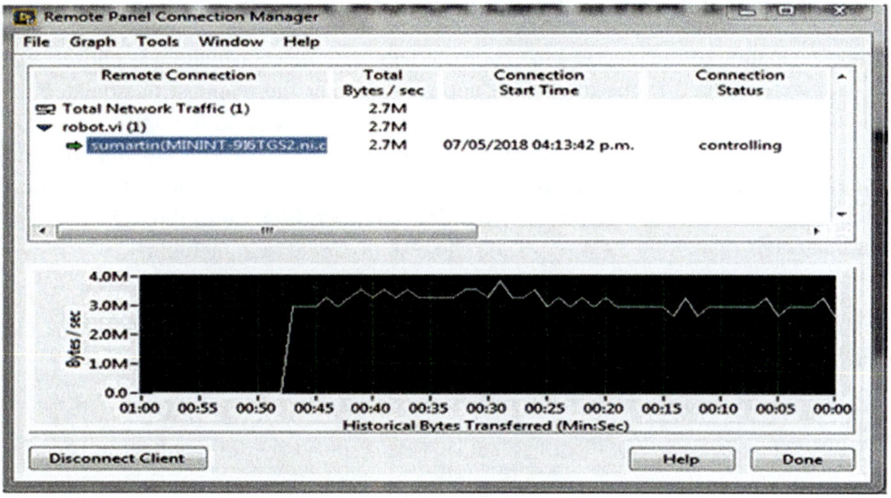

Fig. 4. Remote panel connection manager

The response times of the simulation obtained showed that the time taken by the robot to reach a desired position is approximately 50 ms. Subsequently, the equations of the inverse kinematic model were entered, and regard-less of the type of data that was generated when solving the equations, they always maintained the same execution times.

Evaluation of Response Latency. For the evaluation of the latency in the response, LabVIEW offers a tool that counts in milliseconds the time in which the client takes control of the VI and its interaction with the application. Connection tests were performed between the VI and a client computer, where the time in which the client was connected to the simulation was measured. Table 3 shows some values of the experimental tests carried out with the different computers.

Table 3. Response latency of test computers

Model	Response latency (ms)	Model	Response latency (ms)	Model	Response latency (ms)
Dell E7450	3.47	Dell E6430	10.2	Dell E6520	15.9
Dell E5450	5.34	Dell E6420	9.27		

It is noticeable that the application execution times are related to the computer model. The best time was obtained in the E7450 model, which indicates that the response times are related to the RAM memory of the client in which the remote API is executed.

4.2 Evaluation of the Inverse Kinematic Model

For the evaluation of the inverse kinematic model a specific coordinate was chosen and the data were entered in the developed application, later the responses calculated by the inverse kinematic model found, for each joint, are observed. Additionally, in V-REP the final position of the effector was observed. All the results were compared with the mathematical data calculated using equations and the ideal data of the final location of the effector. The tests were performed 5 times, one of those times through cartesian coordinate axis (X+, X−, Y+ Y− y Z+). The results can be seen in Table 4.

Table 4. Joint and position data registered in the application, mathematically calculated joint and position data.

	X+ (1,0,0)	X− (−1,0,0)	Y+ (0,1,0)	Y− (0,−1,0)	Z+ (0,0,1)	X+ (1,0,0)	X− (−1,0,0)	Y+ (0,1,0)	Y− (0,−1,0)	Z+ (0,0,1)
A	0	3.14	1.57	−1.57	0	0	3.14	1.57	−1.57	0
B	1.57	1.57	1.57	1.57	0	1.57	1.57	1.57	1.57	0
C	−3,14	2.03	−0.78	0.78	0	−3.14	2.09	−0.78	0.78	0
D	7.65	1.71	1.08	1.08	0	0	1.57	0	0	0
E	−3.14	−2.75	−1.57	1.57	0	−3.14	−3.14	−1.57	1.57	0
F	1.57	1.57	1.57	1.57	0	1.57	0	1.57	1.57	0
G	−1.57	−0.46	−0.78	0.78	3.14	−1.57	0	−0.78	0.78	3.14
x	0.82	−0.82	−0.11	0	0	1	−1	0	0	0
y	0.39	0.55	0.86	−0.44	0.42	0	0	1	−1	0
z	0.27	0.4	0.42	0.42	1.28	0	0	0	0	1

Figures 5, 6, 7, 8, 9 and 10 show the comparison of each position that the inverse kinematic model found against the mathematically calculated data. For each joint the comparison between the data found by the simulator and the final effector position is shown.

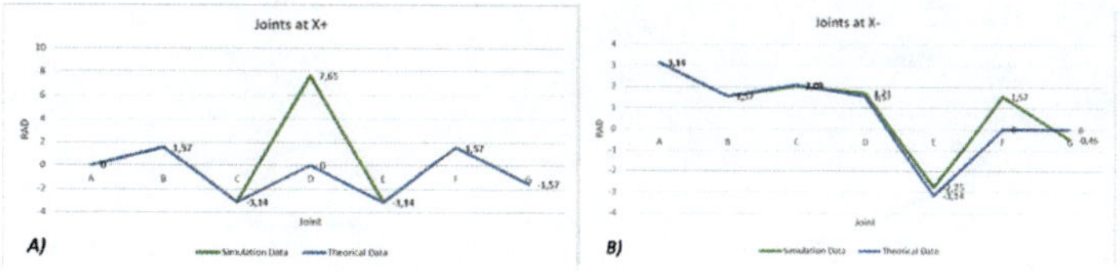

Fig. 5. Joints position. A) Joints at X+. B) Joints at X−.

As can be seen, A and B there are no discrepancies between the simulated data and the calculated data. These joints are exclusively responsible for the total direction of the arm, so it is to be expected that they will not show differences. On the other hand, joints C, D, E, F and G are responsible for positioning the final effector at a specific point. It should be noted that the anatomy of the robot, which is comparable to the anatomy of a human arm, allows the final wrist or effector to be fixed. However, the elbow can rotate freely without altering the position of the final effector, which explains the inconsistencies between the calculated data and the data obtained in those joints.

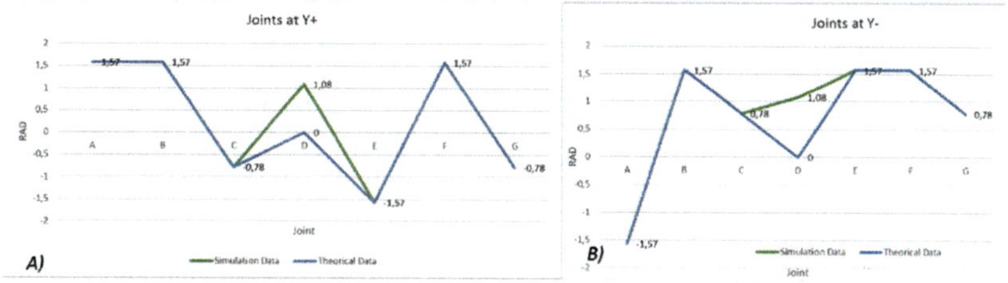

Fig. 6. Joints position. A) Joints at Y+. B) Joints at Y−.

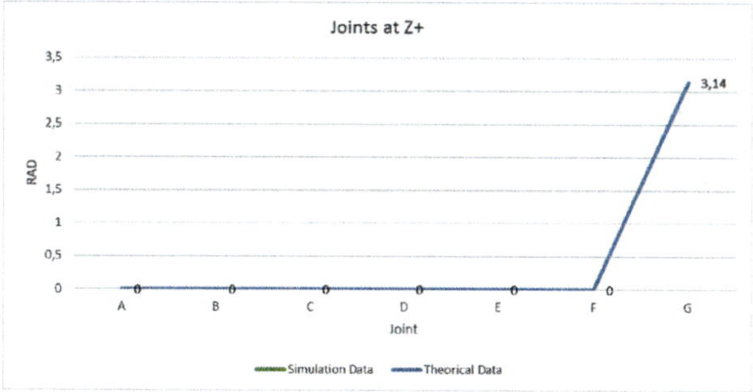

Fig. 7. Joints position at Z+

Table 5 shows the average of the Standard Deviations (SD) and the Mean Square Error (MSE) for each position. The experimental results of the application show an average MSE of 1.82 and a SD of 0.25.

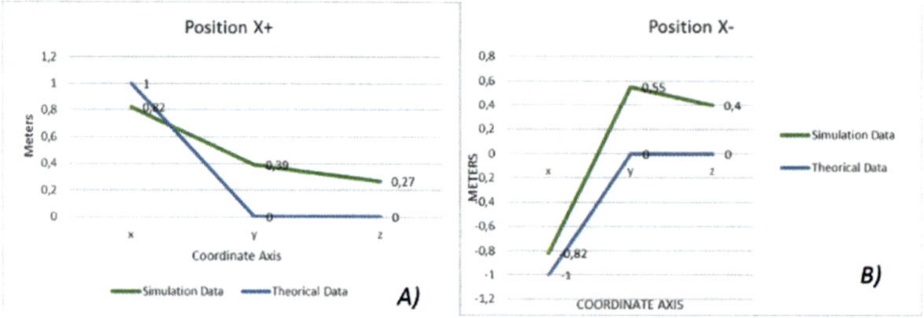

Fig. 8. End effector position at X. A) Position at X+. B) Position at X−.

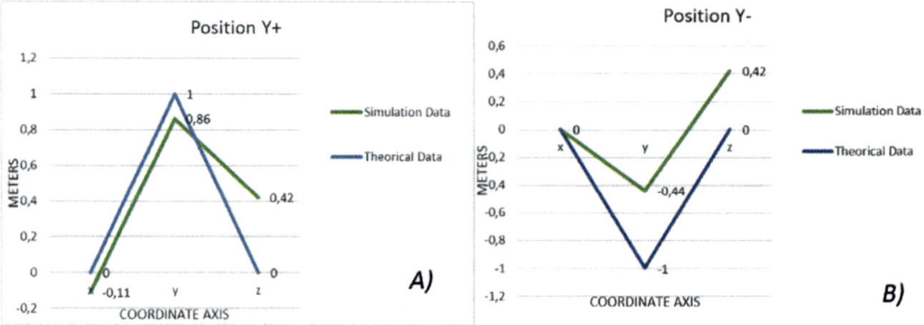

Fig. 9. End effector position at Y. A) Position at Y+. B) Position at Y−.

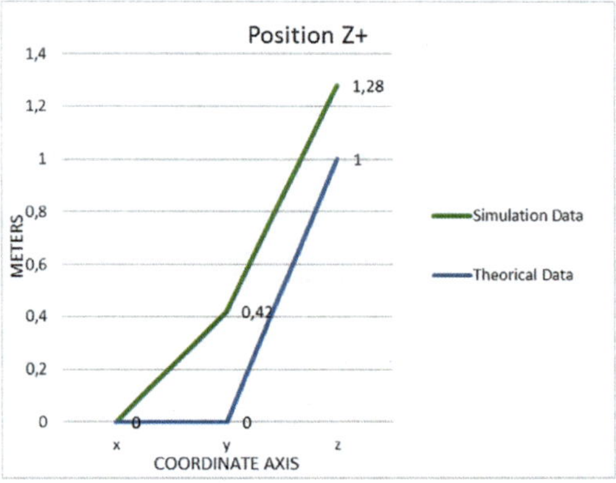

Fig. 10. End effect

Table 5. MSE and SD average.

Position	SD	MSE
X+	0,773	8,360
X−	0,265	0,407
Y+	0,109	0,167
Y−	0,109	0,167
Z+	0	0

5 Conclusions and Future Works

In this paper is developed a web-based virtual laboratory using V-REP simulator in order to programming different trajectories applied to KUKA LBR IIWA 14 R820 using the inverse kinematic model. Software LabVIEW is presented in the model through the use of remote APIs and external libraries. In accordance with the results obtained, it was observed that the performance of the system depended of the computer used by the user. It is also recommended using a server with high performance. Furthermore, in the validation of kinematic model, there are some differences among the trajectories obtained, analyzing results, it was found that given the differences between anatomy of the simulated robot and the ideal model, there are several mathematical responses for the same position of the final effector. In future works it is proposed to design a remote API using different programming environments in order to improve the performance and flexibility with other computer programs, to design and to implement a multi-user connection and visualization and testing this platform in a class of mechatronic and electronic students.

References

1. Pirker, J., Lesjak, I., Parger, M., Gütl, C.: An educational physics laboratory in mobile versus room scale virtual reality - a comparative study. Lect. Notes Netw. Syst. **22**, 1029–1043 (2018)
2. Castillo, A., Ortegon, J., Vazquez, J., Rivera, J.: Virtual laboratory for digital image processing. IEEE Latin Am. Trans. **12**(6), 1176–1181 (2014)
3. Tzagkas, D., Gketsis, Z.E., Stavrakakis, G., Zervakis, M.E., Hatzilias, P.V.: Simulation model of robotic arm for laparoscopic surgery. In: 2007 European Control Conference (ECC), Kos, Greece (2007)
4. Hardison, J.L., DeLong, K., Bailey, P.H., Harward, V.J.: Deploying interactive remote labs using the iLab Shared Architecture. In: 38th Annual Frontiers in Education Conference, Aratoga Springs, NY (2018)
5. Barros, C., Pinto Leão, C., Soares, F., Minas G., Machado, J.: Issues in remote laboratory developments for biomedical engineering education. In: 2013 International Conference on Interactive Collaborative Learning (ICL), Kazan (2013)
6. Sandid, H., Elhedda, W.: Remote laboratories across the mediterranean, EOLES project, a case study and a point of start. In: Information and Communication Technologies Innovation and Application (ICTIA), Sousse, Tunisia (2014)
7. Buitrago-Molina, J.T., Carvajal-Guerrero, J.S., Zapata-Castillo, C.A.: Virtual environment for local and remote. Tecno Lógicas **17**(32), 67–74 (2014)
8. Aydogmus, Z., Aydogmus, O.: A web-based remote access laboratory using SCADA. IEEE Trans. Educ. **52**(1), 126–132 (2009)
9. Yazidi, A., Henao, H., Capolino, G.-A., Betin, F., Filippetti, F.: A web-based remote laboratory for monitoring and diagnosis of AC electrical machines. IEEE Trans. Ind. Electron. **58**(10), 4950–4959 (2011)
10. Shalini, N., Siva Sakthi, S.: An embedded web server based remote motor control system using virtual panel. In: 2014 International Conference on Circuits, Power and Computing Technologies (ICCPCT 2014), Nagercoil (2014)

11. Rohmer, E., Singh, S.P.N., Freese, M.: V-REP: a versatile and scalable robot simulation framework. In: 2013 IEEE/RSJ International Conference on Intelligent Robots and Systems, Tokyo (2013)
12. Wang, Y., Artemiadis, P.: Closed-form inverse kinematic solution for anthropomorphic motion in redundant robot arms. Adv. Robot. Autom. **2**(3), 1–7 (2013)

Measurement System to Monitor the Presence of Patients in Nursing Home Beds Using FBG Sensors and a VCSEL Optical Source

Andrés Triana[1(✉)], Camilo Cano[2], Carlos Perilla[2], and Margarita Varón[2]

[1] Universidad Santo Tomás, Bogotá, Colombia
catrianai@unal.edu.co
[2] Universidad Nacional de Colombia, Carrera 45 # 26-85, Bogotá, Colombia

Abstract. This paper shows the results of a research intended to use optical fiber sensors to instrument patients' beds in a nursing home. It is proposed an interrogation methodology to detect the reflected wavelength of the sensors using a VCSEL laser as the optical source. The FBG sensor is packaged into a plastic composite and characterized. The results show that identification of the reflected central wavelength can be achieved with the proposed scheme and confirm the expected linear behavior of the sensors.

Keywords: Optical fiber sensor · FBG · Fiber Bragg gratings · Interrogation

1 Introduction

The elderly population is known to be growing not only in Colombia but also in the majority of countries in the world. On the other side, the number of professionals specialized in caring for this population is very poor in comparison. In a country like Colombia there are around 80 professional gerontologist to take care for a population of around 6 million people [1, 2]. Therefore the government has established regulations regarding specially the infrastructure quality in the nursing homes and the quality standards for the medical and nursing staff [3, 4].

Taking into account the above mentioned situation, the implementation of an effective monitoring and warning system could assist the process of patient's observation and caring by recording the sleep (time in bed) patterns and alerting whenever a sudden movement is presented in any patient's bed. Optical fiber technology is one of the best ways to implement a sensing network with the requirements described before due to the advantages of fast transmission, multiplexing capabilities and rapid growing of the market [5].

During the last decades, fiber optical sensors (FOS) have been widely studied and many monitoring systems have been proposed in several fields [6, 7]. Between the proposed schemes, fiber Bragg gratings devices (FBG) are one of the most robust and used sensing devices [8]. FBGs use the inherent advantages of the optical fiber such as

D. F. Cortes Tobar et al. (Eds.): AETA 2019, LNEE 685, pp. 453–459, 2021.
https://doi.org/10.1007/978-3-030-53021-1_46

immunity to electromagnetic interferences, light weight and low losses [8, 9], and can incorporate a great number of sensors in a single optical path.

This research is intended to present and discuss a measurement system using FBG optical sensors in order to detect sudden movements in nursing home beds and generate an alarm to alert the health professionals, accelerating the care process given to elderly population in this kind of institutions. The investigation validates the use of an interrogation setup based on the tuneability of a VCSEL laser under the operational range of the sensing devices. It also proposes to use a plastic composite packaging to protect the sensors, the characterization of the complete sensing device is presented here. From this point, it is possible to set a trigger signal to originate the alarm and integrate the measurement prototype with an information interface able to monitor and transfer the alarm signals when necessary.

2 Sensor Design

2.1 Fiber Bragg Grating Sensing System

Fiber Bragg gratings (FBGs) are optical devices embedded into the core of an optical fiber and basically consists of a periodic structure with alternating refractive index values. The FBG acts at a specific wavelength value determined by the spacing between the periodic gratings, basically, when light travels along the optical fiber a particular wavelength (known as the Bragg wavelength) gets reflected and therefore filtered in transmission as depicted in Fig. 1. The relationship between the Bragg wavelength (λ_B) and the periodic spacing of the grating (Λ) is given in Eq. 1 where η_{eff} is the effective refractive index of the optical fiber [9].

$$\lambda_B = 2\eta_{eff}\Lambda \tag{1}$$

Due to its filtering behavior, FBGs were originally used in optical communication applications, nevertheless, due to the direct relationship between the reflected wavelength λ_B and the grating spacing Λ, (i.e. the geometrical properties of the optical fiber modify the reflected wavelength) these devices were also proposed as sensing elements. The two physical variables directly involved with a shift of the Bragg wavelength are temperature and strain, any variation in the temperature surrounding the optical fiber or its elongation/stretching will result in a shift (positive or negative) of the reflected wavelength λ_B. Other variables affecting the optical fiber can be measured indirectly and have been widely researched (e.g. pressure, stress, acceleration, among others).

Having into account the working principle and the FBG sensing features described above, the measurement setup is completed with the optical interrogator, it typically involves a light source, a circulator to couple back-reflected light and a photo-detector or wavelength analyzer to detect the Bragg wavelength reflected by the sensors. Different systems and methodologies have been proposed to detect the reflected wavelength from any sensing FBG array [11–13], however, the interrogation approach proposed in this paper is a cost-effective setup, using a low-cost tunable VCSEL (Vertical-Cavity Surface-emitting Laser) as the optical source. VCSEL lasers' main advantages [12] over other optical sources are their cost, due to their integrated manufacturing process, and their low energy consumption.

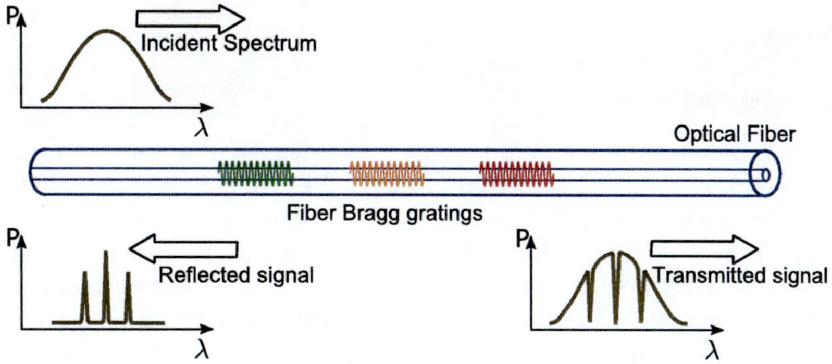

Fig. 1. Working principle of FBG devices, light from an optical source is sent through the optical fiber, the manufactured grating acts as a filter reflecting the light around the specific Bragg wavelength other wavelengths are completely transmitted.

2.2 Packaging of FBG Sensors

The design and arrangement of the sensing devices should be taken into consideration in order to find an appropriate configuration given a determined application assuring protection of the sensors and minimum intrusion with the environment. In this investigation, the FBG sensing device was embedded into a plastic composite of 45 × 15 × 3 mm (see the image in Fig. 6), the plastic packaging has been reported to be the less intrusive material and therefore presents a very low influence over the measured variables [1]. Each packaged sensor then is merged into a mat that can go beneath the patients' bed as illustrated in Fig. 2. From this figure, it can be seen that, as FBG sensors are point sensors, the distribution of various sensors can be distributed alongside the sensing mat. Nevertheless, since the application described here is intended to monitor the movement of the patients and prevent problems produced from falls a single FBG sensor installed in each patient's bed would be enough.

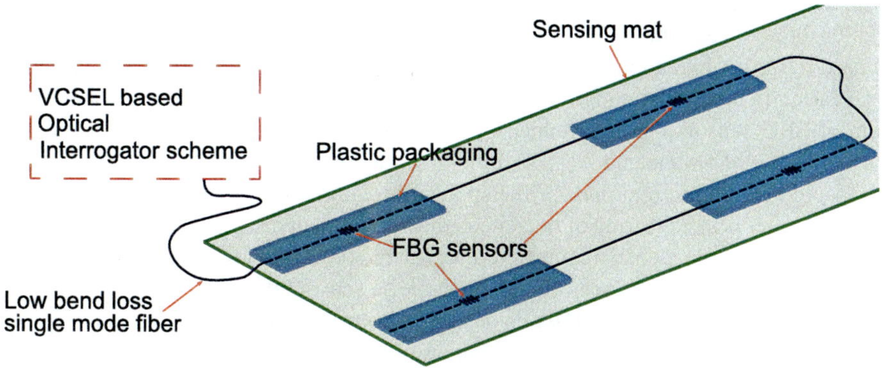

Fig. 2. Schematic representation of the packaged sensors inside the mat.

3 Results

3.1 Interrogating Setup

The characterization of the embedded sensor was carried out using the schematic representation shown in Fig. 3. As described before, a VCSEL emitting light @1550 nm is used to illuminate a single FBG sensor; once the specific wavelength gets reflected from each FBG sensor, it is conducted through a circulator to a photo-detector which is synchronized with the swept performed in the emitting VCSEL. In this way, according to the fineness of the emitted wavelength, the photodetected power points out the reflected wavelength from the sensors.

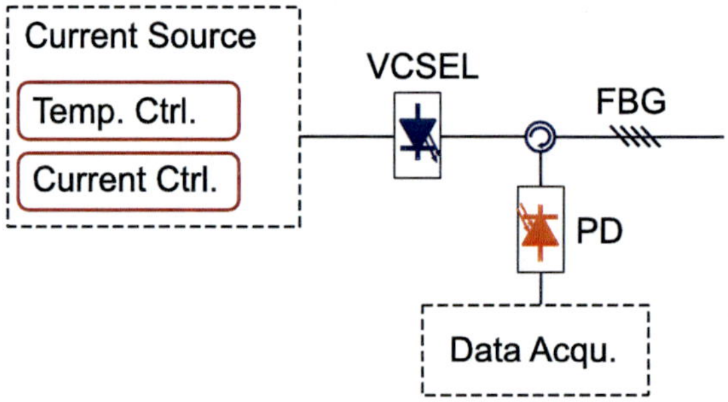

Fig. 3. Diagram of the interrogation setup used to illuminate the optical FBG sensors and retrieve their reflected wavelength. A VCSEL laser is tuned by sweeping its bias current, the reflected wavelength from the sensing devices is detected in the photo-detector (PD) and consequently the reflected wavelength can be tracked.

Using the setup depicted in Fig. 3. A range of wavelengths is scanned with the VCSEL laser. Since VCSELs are semiconductor devices, their thermo-electrical characteristics allow to induce a shift in the emitted wavelength [14] which means the device can be tuned in a range of about 4 nm as seen in Fig. 4. This figure shows as well the relationship between the VCSEL emission wavelength and its temperature, which can be used to further extend the range in which optical FBG sensors can be interrogated.

Bearing in mind the tuning range obtained from the VCSELs, the operational spectrum of FBG sensors can be scanned in order to find out the spectral position of the sensor or its Bragg wavelength at a specific time. In this process it is important to define the spectral resolution required for the application, a fine tuning of the VCSEL would exhibit a better peak detection of the reflected wavelength at the cost of incrementing the scanning time, while a coarse tuning would present some resolution problems allowing for a faster scanning of the operational range of the sensing device.

In Fig. 5 it is depicted the situation described above: a FBG device was interrogated using the proposed setup setting two different incremental steps for the bias current in the VCSEL. In Fig. 5a), the VCSEL bias current was swept in steps of 0.2 mA, while in Fig. 5b), the step increment was 0.05 mA. It is clear that a better scanning resolution, i.e. smaller incremental steps of the VCSEL bias current, would allow for more precision in determining the central wavelength measured by FBG sensors, while a scan with

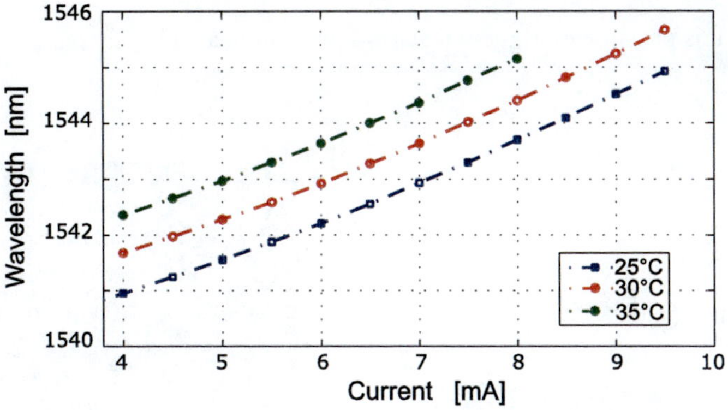

Fig. 4. Experimental tuning of the emission wavelength of a VCSEL laser. The VCSEL is tuned by sweeping its bias current while controlling its temperature. A linear shift is observed as the device temperature increases.

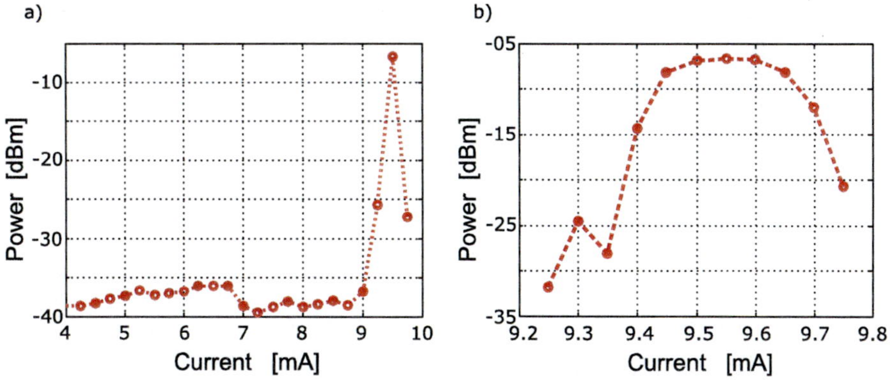

Fig. 5. Experimental readout from an interrogated FBG sensing device.

bigger incremental steps would be appropriate to initially find the operational range of the sensor. Therefore, the two scanning resolution values can work together to optimize the Bragg wavelength measurement.

3.2 Sensor Characterization

The embedded FBG sensor was characterized following the schematic representation setup depicted in Fig. 6a), a commercial optical interrogator was utilized to perform the experimental characterization, it acquires the reflected wavelength from a FBG sensor centered at 1540.4 nm. The packaged sensor, which was designed to be placed inside a mat in contact with the patient's bed, is supported from the two axial edges while pressure is applied to the sensing structure. Figure 6b) shows the experiment performed at the lab. and in Fig. 6c), the obtained characterization values are presented along with the linear approximation. The resulting characterization exhibits a highly linear behavior from which it is possible to trigger an alarm signal (indicated in the figure to be around 1537.5 nm).

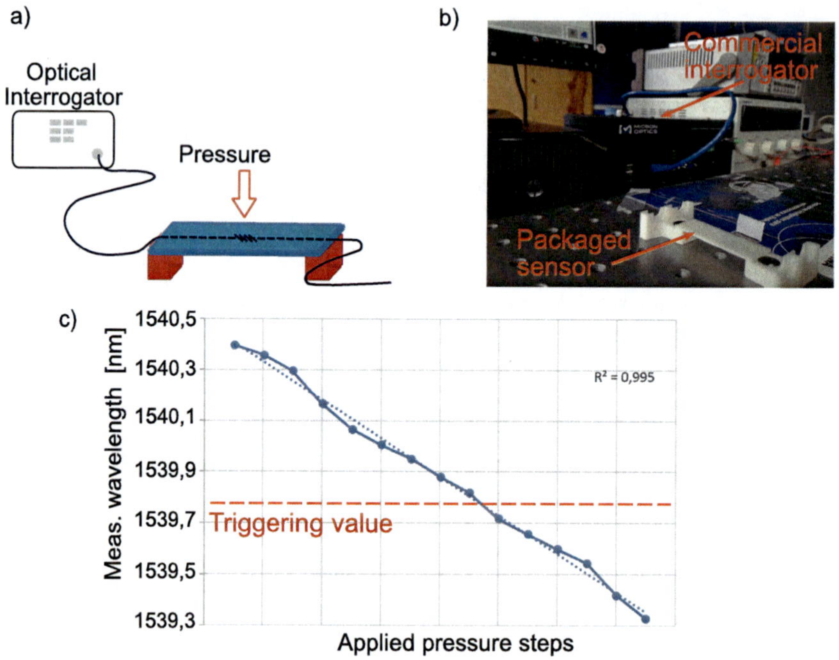

Fig. 6. Experimental characterization of the proposed sensor. In a) it is shown the schematic representation of the experiment: a commercial optical interrogator is used to retrieve the reflected wavelength from the sensor under test. b) shows the experimental setup implemented in the laboratory, and c) shows the resulting characterization of the sensor, the reflected wavelength is plotted as a function of the incremental pressure applied to the sensing structure.

4 Discussion

We have proposed the use of an optical measurement system to detect the presence of patients in their beds on a nursing home, the overall system comprehends the use of FBG sensors manufactured inside low bend-loss optical fiber, packaging of the sensing area using a plastic composite and embedding the sensors in a mat that can be placed beneath the patients' beds. The proposed interrogation system uses a VCSEL laser electrically tuned around the operational range of the sensors. The interrogation setup was demonstrated to be stable, obtaining a variation in wavelength of ~4 nm while driving its bias current from 4 to 9 mA approximately. Interestingly, the temperature (which is needed to be steady during the tuning process) can also be shifted to reach other wavelength values when necessary. It was also demonstrated how the resolution in the incremental steps can be used first to detect the position of the sensor, by using a coarse incremental in current, and then, using a fine tuning, the central wavelength of the sensor can be retrieved with higher accuracy.

The characterization of the FBG sensor was performed after the packaging of the optical fiber sensor into a plastic composite to enhance the protection of the sensing element. The packaged sensor demonstrated to follow a very good linear behavior, which makes it easy to set a wavelength value limit that triggers the alarm signal. The optoelectronics interrogation system is demonstrated to work; therefore, the future work involves the development of the software interface to register the measured signals and communicate the alarm to the health professionals whenever a sudden movement is detected.

References

1. Méndez, L.S.: Trabajo de cuidado y vejez: condiciones laborales, dinámicas organizacionales y devaluación social. Ediciones Uniandes, Bogotá D. C., Colombia (2014)
2. Borda, E.M.: Valoración geriátrica integral: ¿tienen acceso a ella los adultos mayores institucionalizados en colombia? reflexión desde lo encontrado en un hogar geriátrico. Rev. Asoc. Colomb. Gerontol. Geriatr 2258–2264 (2016)
3. Stuck, A.E.: Comprehensive geriatric assessment for older adults. Br. Med. J. 343 (2011)
4. Kan, A.V.: Task force on frailty assessment of older people in clinical practice. J. Nutr. Health Aging 29–37 (2008)
5. marketsandmarkets.com: Optical Sensing Market by Industry (Aerospace & Defense, Utilities, Oil & Gas, Medical, Construction, and Consumer Electronics), Application, Method, and Geography - Global Forecast to 2023. https://www.marketsandmarkets.com/Market-Reports/optical-sensing-market-197592599.html. Accessed 10 July 2019
6. Cusano, A., et al.: Fiber Bragg Grating Sensors: A Look Back. de Fiber Bragg Grating Sensors: Recent Advancements, Industrial Applications and Market Exploitation, Bentham Science Publishers Ltd. (2011)
7. Yan, P., et al.: High-speed FBG-based fiber sensor networks for semidistributed strain measurements. IEEE Photon. J. **5**(2), 7200507 (2013)
8. Cano, C., et al.: Reliability of strain and temperature measurements based on fiber bragg gratings sensors in the early age of concrete shrinkage. In: de Latin America Optics and Photonics Conference, Cancún (2014)
9. Hill, G., et al.: Fiber Bragg grating technology fundamentals and overview. IEEE J. Lightwave Technol. **15**, 1263–1276 (1997)
10. Erdogan, T.: Fiber grating spectra. IEEE J. Lightwave Technol. **15**, 1277–1294 (1997)
11. Kalli, K., Othonos, A.: Fiber Bragg Gratings. Fundamentals and Applications in Telecomunications and Sensing. Artech House, Norwood (1999)
12. Zhou, Z., et al.: Techniques of advanced FBG sensors: manufacturing, demodulation, encapsulation and their application in the structural health monitoring of bridges. Pac. Sci. Rev. **5**, 116–121 (2003)
13. Triana, A., et al.: A code division design strategy for multiplexing fiber bragg grating sensing networks. Sensors **17**(11), 2508 (2017)
14. Gilard, O., et al.: Optoelectronics in space: main qualification pitfalls and how to avoid them. In: de International Symposium on Reliability of Optoelectronics for Systems (2014)
15. Wu, L., et al.: Selection and characterization of packaged FBG sensors for offshore applications. Sensors **18**, 3963–3980 (2018)
16. Mogg, S., et al.: Temperature sensitivity of the threshold current of long-wavelength ingaasgaas vcsels with large gain-cavity detuning. IEEE J. Quantum Electron. **40**, 453–462 (2004)

Online Parameter Adjustment of an Active Disturbance Rejection Controller for a Robotic Manipulator via Simulated Annealing

Edwin Villarreal-López[1,2](✉), Horacio Coral-Enriquez[1], Sebastian Medina-Camacho[1], and Luini Hurtado-Cortés[2]

[1] Facultad de Ingeniería, Universidad de San Buenaventura Bogotá, Bogotá, Colombia
{evillareal,hcoral,smedina}@usbbog.edu.co
[2] Facultad de Ingeniería, Universidad Distrital Francisco José de Caldas, Bogotá, Colombia
llhurtado@udistrital.edu.co

Abstract. In this paper, the parameterization of active disturbance rejection joint controllers for the Mitsubishi RV-M1 robotic manipulator is addressed by the simulated annealing metaheuristic performed online. The proposed approach solves the problem of controller bandwidth and input gain selection due to nonlinearities and practical considerations by means of an optimization process. A cost function based on the integral square error is defined over a closed path for each joint. Then, the controllers parameters are modified according to the cooling schedule. Finally, experimental results show better performance than traditionally tuned active disturbance rejection controllers in a reduced number of iterations.

Keywords: Active disturbance rejection control · Simulated annealing · Robot control

1 Introduction

Modern control strategies for robotic manipulators usually relies on very close mathematical descriptions of the real system. However, despite of the multiple advantages of precise mathematical modeling in controller design, system identification process can be complex and time consuming. Besides, this model may not reflect the full range of parameter deviations and uncertainty related to functional characteristics due to changes in operating conditions [16] thus controller design must take these features into consideration. Active Disturbance Rejection Control (ADRC) paradigm [1] establishes a different approach for controller design where only essential information may be considered, and the leftover dynamics and disturbances are lumped together and obtained by means of an

D. F. Cortes Tobar et al. (Eds.): AETA 2019, LNEE 685, pp. 460–469, 2021.
https://doi.org/10.1007/978-3-030-53021-1_47

Extended State Observer (ESO). This term represents the unknown exogenous and endogenous system features and is injected into the control law in order to reject it providing robustness to the ADRC scheme.

Many researches have propose ADRC applications in numerous fields, such positioning control [14,15], aerospace applications [6,7,12], electric machinery [5,11] and wind turbines [3,8]. Although many ADRC variations, it usually consists on an input differentiator block which can efficiently improve the tracking performance, the Extended State Observer for system state and disturbance estimation, and a control law in charge of providing a stable tracking error dynamics while cancelling the effect of the lumped disturbance.

Due to the online disturbance estimation, the ADRC scheme is highly robust and require a few parameter adjustments referring to the controller, the observer and the input gain. The chosen values determine the estimation and tracking error dynamics and can be selected in terms of performance requirements [18]. Although theoretical relations between observer bandwidth and sampling time have been established [10], for implementations purposes there may be practical considerations related to signal to noise ratio and sampling intervals that difficulties the controller parameterization.

Previous works have focused on optimal ADRC design. In [13] an auto-tuning method based on noise quantification for ADRC parameterization is proposed and experimentally verified on a power converter. On the other hand, self-learning strategies have also been developed to cope with this problem. In [17] a least square support vector machine optimization method was applied to improve the performance of pneumatic servosystems through a training phase performed with the system model. In addition, in [4] it is detailed an adaptive scheme for permanent magnet synchronous motors with varying load inertia using a fuzzy inference system and a compensation gain self-tuned according to the identified variations. However, these strategies have not been applied for ADRC online parameter adjustment based on real systems performance. This problem is important for repetitive trajectory tracking under disturbances.

Robotic manipulators are usually used in highly structured environments for repetitive tasks. This fact can be exploited to improve the controller performance related to a cost function minimization. To ensure global convergence, this paper proposes an online optimization process based on simulated annealing for ADRC parameterization.

The reminder of this paper is organized as follows. Initially in Sect. 2 the system model of a Mitsubishi Movemaster RV-M1 robotic manipulator is presented. Then, the ADRC controller structure and the optimization strategy by means of simulated annealing are detailed in Sect. 3. In Sect. 4 the proposed method is applied for a repetitive trajectory tracking problem through an experimental setup. Finally, concluding remarks are given in Sect. 5.

2 Robotic Manipulator

The Mitsubishi Movemaster RV-M1 is a widely used robot in academic and industrial environments. Table shows its main technical specifications given by the manufacturer. It is a 5 degrees of freedom (DOF) manipulator with 1.2 kg of load capacity and brushed DC actuators for each joint. The dynamics description of this robot can be given by:

$$\tau = \mathbf{M}(\mathbf{q})\ddot{\mathbf{q}} + \mathbf{C}(\mathbf{q},\dot{\mathbf{q}})\dot{\mathbf{q}} + \mathbf{G}(\mathbf{q}) \tag{1}$$

where τ is the torque vector of each joint, $\mathbf{M}$ represents the inertia matrix, $\mathbf{C}$ lumps together centrifugal and Coriolis forces and $\mathbf{G}$ depends on gravitational torques. Additionally, system states are given by the joint angles:

$$\mathbf{q} = \begin{bmatrix}\theta_1 \; \theta_2 \; \theta_3 \; \theta_4 \; \theta_5\end{bmatrix}^T \tag{2}$$

Numeric parameters of $\mathbf{M}(\mathbf{q})$, $\mathbf{C}(\mathbf{q},\dot{\mathbf{q}})$ and $\mathbf{G}(\mathbf{q})$ can be found using identification techniques and detailed physical descriptions of each link. Now, including actuator dynamics given by their rotational inertia J, viscous and Coulomb frictions f_v and f_c, respectively, and k_m as the motors torque constant, the complete robot description is given by:

$$\mathbf{KU} = (\mathbf{M}(\mathbf{q}) + \mathbf{J})\,\ddot{\mathbf{q}} + (\mathbf{C}(\mathbf{q},\dot{\mathbf{q}}) + \mathbf{F_v})\,\dot{\mathbf{q}} + \mathbf{G}(\mathbf{q}) + \mathbf{F_c}\,(\dot{\mathbf{q}}) \tag{3}$$

where, $\mathbf{U}$ are the applied voltages vector, $\mathbf{K}$, $\mathbf{J}$, $\mathbf{F_v}$ and $\mathbf{F_c}$ are diagonal matrices with the motor torque constants, rotational inertia, viscous and Coulomb friction coefficients, respectively.

In order to perform the system decoupling, matrix $\mathbf{M}\,(\mathbf{q})$ can be decomposed into a constant term $\bar{\mathbf{M}}$ and configuration-depend terms depending on joint angles [9].

$$\mathbf{M}\,(\mathbf{q}) = \bar{\mathbf{M}} + \mathbf{\Delta M}\,(\mathbf{q}) \tag{4}$$

where $\bar{\mathbf{M}}$ is a diagonal matrix whose elements represent the average inertia of each link, and $\mathbf{\Delta M}\,(\mathbf{q})$ are the left over nonlinear dynamics. With this factorization, the system dynamics can be described by:

$$\ddot{\mathbf{q}} = \kappa\mathbf{U} \underbrace{- \left(\bar{\mathbf{M}} + \mathbf{J}\right)^{-1}\left[(\mathbf{M}\,(\mathbf{q}) + \mathbf{J})\,\ddot{\mathbf{q}} + (\mathbf{C}\,(\mathbf{q},\dot{\mathbf{q}}) + \mathbf{F_v})\,\dot{\mathbf{q}} + \mathbf{F_c}\,(\dot{\mathbf{q}}) + \mathbf{G}\,(\mathbf{q})\right]}_{-\varsigma} \tag{5}$$

where ζ is considered as the endogenous disturbance representing the nonlinear and coupling terms. We define the high frequency gain as:

$$\kappa = \left(\mathbf{K_m^{-1}}\left(\bar{\mathbf{M}} + \mathbf{J}\right)\right)^{-1} \tag{6}$$

Numeric estimation of the terms in the main diagonal of κ can be found considering the respective joint average inertia, motor torque constant and inertia[1].

[1] Numerical values for these terms can be found in the following link: https://drive.google.com/open?id=1u1eFxBjpM2IJ3iKp8Tasl7S7UeheMBf8.

3 Optimized Active Disturbance Rejection Control

3.1 Active Disturbance Rejection Control of the Robotic Manipulator

Matrices in Eq. 6 are diagonal, which allows decoupled controller design for each joint. thus, from Eqs. 5 and 6 we have:

$$\ddot{q}_i = \kappa_i u_i + \xi_i \tag{7}$$

where κ_i are the terms on the diagonal of the κ matrix and $\xi = \zeta_{end} + \zeta_{exo}$ denotes the lumped disturbance with endogenous component ζ_{end} and exogenous ζ_{exo}, representing the joint disturbance. Robotic manipulators are usually intended to work on constant or slow varying load displacements. According to this consideration, the disturbance internal model for the i-th joint is given by:

$$\frac{d\xi_i}{dt} \approx 0 \tag{8}$$

Although numerical approximations of κ_i can be found from 3, in this paper it will be chosen from the proposed optimization scheme.

ADRC schemes are usually composed by the ESO for state and disturbance estimation, an input differentiator and the control law for tracking purposes. Figure 1 shows the second order ADRC scheme. The corresponding control law is:

$$u_i = \frac{1}{\kappa_i}\left[\ddot{r}_i - k_{2i}(\hat{\dot{q}}_i - \dot{r}_i) - k_{1i}(\hat{q}_i - r_i) - \hat{\xi}_i\right] \tag{9}$$

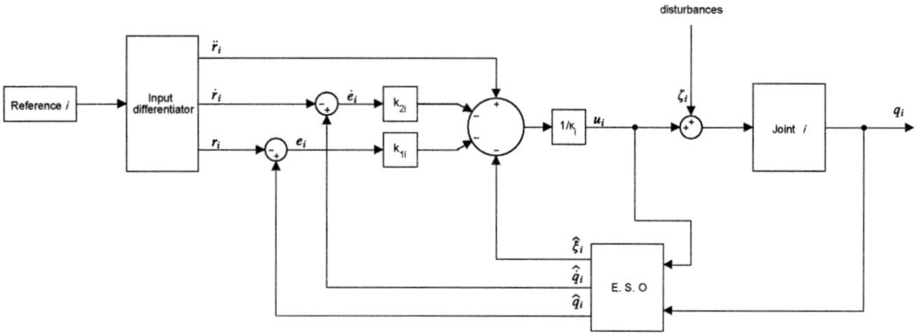

Fig. 1. Second order ADRC for joint angle tracking

3.2 Online ADRC Parameter Adjustment via Simulated Annealing

Simulated annealing is a stochastic search algorithm inspired by the controlled heating and cooling of metals, commonly applied to increase their strength. It is probability based technique for global optimization [2]. This iterative algorithm comprises the generation of a neighbour solution in the search space and an evaluation through a cost function. Improvements are immediately accepted, while worse solutions may be chosen according to a decreasing probability. These features allow the exploration of the search space and convergence to the global optimum.

Parameters to Estimate. According to the proposed control law shown in 9, if the tracking error for each joint is $e_i = q_i - r_i$, then its dynamics is given by:

$$\ddot{e}_i + k_{2i}\dot{e}_i + k_{1i}e = \xi_i - \hat{\xi}_i \tag{10}$$

choosing the controller bandwidth ω_{ci} and considering a low estimation error of disturbances, then in the Laplace domain we have:

$$(s^2 + k_{2i}s + k_{1i})e_i = (s + \omega_{ci})^2 \tag{11}$$

thus, we have that both controller gains can be selected from ω_{ci}, according to:

$$\begin{aligned} k_{2i} &= 2\omega_{ci} \\ k_{1i} &= {\omega_{ci}}^2 \end{aligned}$$

Ideally, the controller performance is directly related to ω_{ci}. However, for practical considerations, increasing this value causes performance degradation related to noise sensitivity and undesired oscillations. Therefore, this parameter will be selected through the optimization process.

Further, due to nonlinearities κ_i is not a constant value and depends on the current system state. Considering that ADRC performance is highly sensitive to this value, it will also be included in the global search.

Cost Function. Since repetitive trajectories are commonly used in industrial tasks for robot manipulators, the integral square tracking error (ISE) over a closed path C is selected as the fitness function in order to evaluate the performance of each set of the controller parameters.

$$ISE_i = \oint_C e_i(t)^2 dt \tag{12}$$

Optimization Algorithm. Algorithm 1 describes the simulated annealing heuristic implemented to perform the search of each of the selected parameters. After initialization, a neighbor solution is generated close to the current solution considering the step size η_k. If the new solution reduces the ISE then it is adopted, otherwise it may be accepted depending on the probability given by the temperature T_k. Finally, T_k and η_k are reduced according to the cooling profiles.

Algorithm 1: Simulated annealing search for ADRC parameterization

Input : Cooling schedule

$s_k \leftarrow s_0$;
$f_k \leftarrow f_0$;
$T_k \leftarrow T_{\max}$;
while $T_k > T_{min}$ **do**
  $T_k = T_{k-1} \times Cooling_1()$;
  $\eta_k = \eta_{k-1} \times Cooling_2()$;
  $s_k = s_{k-1} + (\eta_k \times sign(Uniform(0,1) - 0.5))$;
  $f_k = fitness(s_k)$;
  $\Delta = f_k - f_{k-1}$;
  if $\Delta > 0$ *or* $Uniform(0,1) < T_k$ **then**
    $s_k = s_{k-1}$;
end

4 Experimental Results

In order to evaluate the performance of the proposed parameter search, the experimental test setup shown of Fig. 2 was built. The control algorithm and the simulated annealing heuristic were implemented using Matlab-Simulink Real Time environment with a sampling frequency of 2.5 KHz. To ensure bounded derivatives for tracking purposes, a sinusoidal closed trajectory was selected for each joint. The complete path period of $T = 10$ s was also used as the online optimization sampling rate. Each of the 5-DOF ADRC controllers bandwidths ω_{ci} and κ_i gains were optimized simultaneously using the Algorithm 1 with exponential cooling and $\eta_\omega = 3$, $\eta_\kappa = 1$.

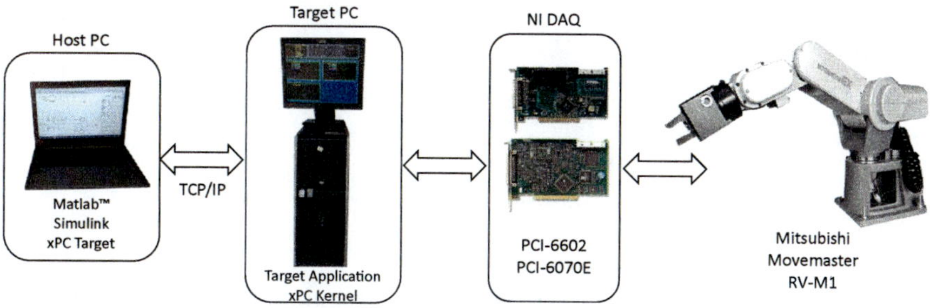

Fig. 2. Experimental test setup

To prove the usefulness of the proposed technique, gains with poor performance were selected at the cycle start. Optimization begins after the initialization phase at $t = 30$ s. Figure 3 top-left shows the ISE over for 500 s of operation on each DOF. After a transient at the beginning, the error starts decreasing rapidly to low values. This fact is shown in detail at the top-right where it can be seen small oscillations due to optimization process.

For comparison purposes, the same trajectory was applied to a sliding mode controller (SMC), a conventional tuned ADRC and a state space controller (SSC) for each joint. The evolution of the overall error for each controller is shown at bottom-left figure. The poorest performance is obtained by the SSC and the ADRC. Also, it can be seen a transient period for the proposed scheme in the first 100 s followed by a fast convergence. This fact is detailed at bottom-right figure by means of a detailed view of this performance measure after 40 cycles of optimization. It can be seen the stable ISE for the SMC while the proposed technique stills shows improvements in the selected cost function.

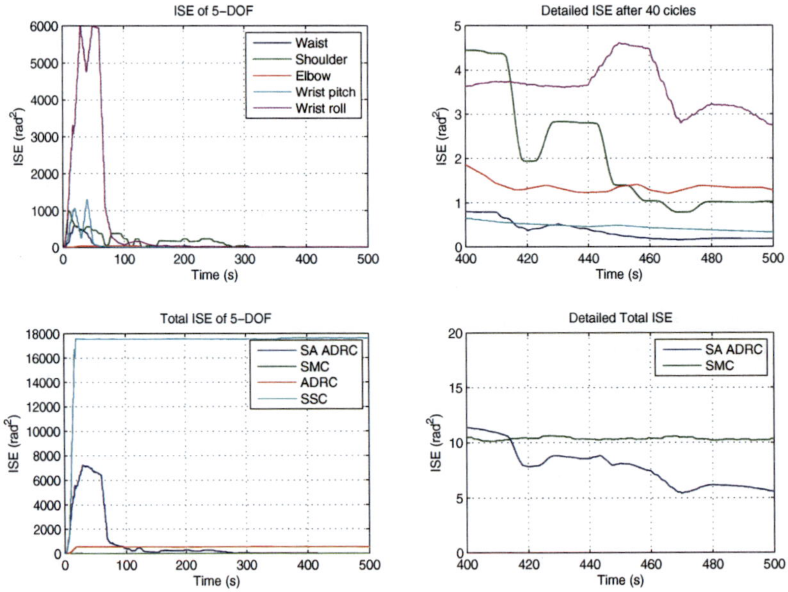

Fig. 3. Experimental results for ISE for each robot joint.

Figure 4 presents the parameters evolution through optimization. It displays the decreasing size of gains corrections steps, according to the selected cooling scheme.

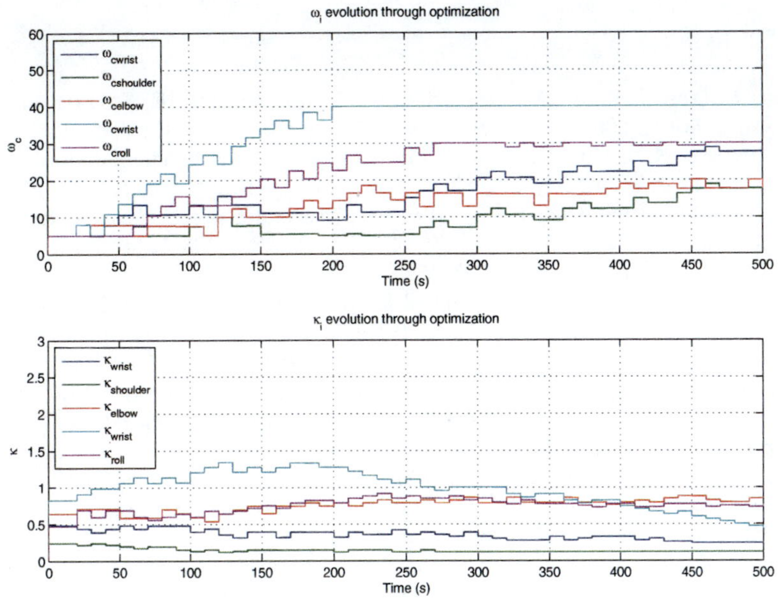

Fig. 4. Parameters evolution through optimization

5 Conclusions

In this paper, a self-tuning mechanism for ADRC controllers through simulated annealing is presented. A decoupled model for each joint is obtained based on the concept of disturbance estimation where the interactions between joints and external disturbances are lumped together and estimated online via the ESO. The tracking control law considers the disturbance estimation to reject its effects via cancellation. Controller bandwidth ω_c and high frequency gain κ are selected by means of online optimization to minimize the ISE. The simulated annealing parameters are selected to soften the steps on the search space while converges to the solution. To verify the effectiveness of the proposed approach, 5 ADRC controllers were implemented for each DOF of the Mitsubishi Movemaster RV-M1 robotic manipulator and tuned simultaneously. It is shown by means of the experimental setup the ability of the strategy to improve the tracking performance online on repetitive tasks.

Acknowledgments. The authors would like to thank the anonymous reviewers for their suggestions which contribute to improve the quality of the paper. This work was supported by the Universidad de San Buenaventura Bogotá, under research grant FI-014-005 as part of the Doctoral Research on Engineering of the first author at the Universidad Distrital Francisco José de Caldas.

References

1. Han, J.: From PID to active disturbance rejection control. IEEE Trans. Industr. Electron. **56**(3), 900–906 (2009). https://doi.org/10.1109/TIE.2008.2011621
2. Kirkpatrick, S., Gelatt Jr., C., Vecchi, M.: Optimization by simulated annealing. Science **220**(4598), 671–680 (1983). https://doi.org/10.1126/science.220.4598.671
3. Li, S., Li, J.: Output predictor-based active disturbance rejection control for a wind energy conversion system with PMSG. IEEE Access **5**, 5205–5214 (2017). https://doi.org/10.1109/ACCESS.2017.2681697
4. Li, S., Liu, Z.: Adaptive speed control for permanent-magnet synchronous motor system with variations of load inertia. IEEE Trans. Industr. Electron. **56**(8), 3050–3059 (2009). https://doi.org/10.1109/TIE.2009.2024655
5. Liu, X.Q., Hu, J.Q., Zhou, L.: Active disturbance rejection control of three-motor synchronous control system. Zhongguo Dianji Gongcheng Xuebao/Proc. Chin. Soc. Electr. Eng. **30**(12), 80–85 (2010)
6. Miao, J., Wang, S., Zhao, Z., Li, Y., Tomovic, M.: Spatial curvilinear path following control of underactuated AUV with multiple uncertainties. ISA Trans. **67**, 107–130 (2017). https://doi.org/10.1016/j.isatra.2016.12.005
7. Rida Mokhtari, M., Choukchou Braham, A., Cherki, B.: Extended state observer based control for coaxial-rotor UAV. ISA Trans. **61**, 1–14 (2016). https://doi.org/10.1016/j.isatra.2015.11.024
8. Rojas-Cubides, H., Cortés-Romero, J., Coral-Enriquez, H., Rojas-Cubides, H.: Sliding mode control assisted by GPI observers for tracking tasks of a nonlinear multivariable Twin-Rotor aerodynamical system. Control Eng. Pract. **88**, 1–15 (2019). https://doi.org/10.1016/j.conengprac.2019.04.002
9. Sciavicco, L., Siciliano, B.: Modelling and Control of Robot Manipulators, 2 edn. Advanced Textbooks in Control and Signal Processing. Springer, London (2000). https://www.springer.com/gp/book/9781852332211
10. Shi, S., Li, J., Zhao, S.: On design analysis of linear active disturbance rejection control for uncertain system. Int. J. Control Autom. **7**(3), 225–236 (2014)
11. Sira-Ramírez, H., Linares-Flores, J., García-Rodríguez, C., Contreras-Ordaz, M.: On the control of the permanent magnet synchronous motor: an active disturbance rejection control approach. IEEE Trans. Control Syst. Technol. **22**(5), 2056–2063 (2014). https://doi.org/10.1109/TCST.2014.2298238
12. Su, Z., Wang, H., Yao, P., Huang, Y., Qin, Y.: Back-stepping based anti-disturbance flight controller with preview methodology for autonomous aerial refueling. Aerosp. Sci. Technol. **61**, 95–108 (2017). https://doi.org/10.1016/j.ast.2016.11.028
13. Sun, B., Gao, Z.: A DSP-based active disturbance rejection control design for a 1-kW H-bridge DC-DC power converter. IEEE Trans. Industr. Electron. **52**(5), 1271–1277 (2005). https://doi.org/10.1109/TIE.2005.855679
14. Sun, G., Ren, X., Li, D.: Neural active disturbance rejection output control of multimotor servomechanism. IEEE Trans. Control Syst. Technol. **23**(2), 746–753 (2015). https://doi.org/10.1109/TCST.2014.2336595
15. Tang, H., Li, Y.: Development and active disturbance rejection control of a compliant micro-/nanopositioning piezostage with dual mode. IEEE Trans. Industr. Electron. **61**(3), 1475–1492 (2014). https://doi.org/10.1109/TIE.2013.2258305
16. Villarreal-López, E.: A new scheme for multiple fault detection and isolation for rotational mechatronic systems through analytical redundancy and adaptive filtering. DYNA **86**(209), 40–48 (2019). https://doi.org/10.15446/dyna.v86n209.73363

17. Zhao, L., Zhang, B., Yang, H., Li, Q.: Optimized linear active disturbance rejection control for pneumatic servo systems via least squares support vector machine. Neurocomputing **242**, 178–186 (2017). https://doi.org/10.1016/j.neucom.2017.02.069
18. Gao, Z.: Scaling and bandwidth-parameterization based controller tuning. In: Proceedings of the 2003 American Control Conference, vol. 6, pp. 4989–4996 (2003). https://doi.org/10.1109/ACC.2003.1242516

Methodology, Based on the Correlation and the Discrete Wavelet Transform to Debug and Correct the Misalignment Signal Amplitude, A-Scan, for Images by Time of Flight Diffraction, D-Scan

Jairo Alejandro Rodríguez Martínez(✉)

MEM Research Group (Electronic Modeling Monitoring), Electronic Engineering Faculty, Santo Tomás University, Cra. 9 no. 51-11, Bogotá 110231, Colombia
jairorodriguez@usantotomas.edu.co

Abstract. This article presents a non-invasive ultrasonic methodology for assessing the quality of welding in metal joints, but mentioning their most frequent defects. Subsequently, the concept of amplitude scanning, A-Scan, is ex plained as well as the ultrasonic imaging technique known as TOFD time of flight diffraction in D-Scan mode is detailed. Then, the problem regarding the misalignment of the A-Scan signals and their distorting effect on the TOFD images is raised, and in like manner the implementation of an algorithm based on the correlation that allows for this problem to be eliminated is described. On the other hand, due to the non-stationary of the ultrasound signals, signal processing techniques are used to eliminate the noise present in these signals, Wavelet analysis and synthesis is used for this purpose in the domains of continuous and discrete time, which finally lead to multiresolution filtering applied to the A-Scan signals.

Keywords: Welded joint · A-Scan · D-Scan · Correlation · Wavelet · Multiresolution

1 Introduction

Welding defects are one of the main causes of structural weakening, in ship hulls, airplane wings, in metal seats, in embedded reinforcements in concrete slabs, etc., and they occur due to different circumstances when carrying out the welding process, generating discontinuities that are unacceptable from the perspective of officially established codes or standards. In a welded joint, defects or discontinuities may occur due to reasons such as improper material selection, improper welding procedures, or improper preparation of the parts to be welded, among others. Discontinuities that usually occur in welded joints are the lack of penetration at the root of the weld; lack of fusion between the weld metal and the base metal; inclusion of slag, which consists of non-metallic solids contained in the weld metal and the base metal; porosity, which represents gases trapped

D. F. Cortes Tobar et al. (Eds.): AETA 2019, LNEE 685, pp. 470–481, 2021.
https://doi.org/10.1007/978-3-030-53021-1_48

during the solidification of the weld metal; cracks, or linear rupture in the deposited metal; undermining, or a crack in the base metal adjacent to the root. Previous articles have presented the design and construction of a system for a non-invasive assessing of welded joints using the ultrasonic imaging technique known as flight time diffraction, TOFD [1, 2], as well as a statistical methodology to correct a type of distortion that affects those images, known as misalignment, that is caused by the way the welded joint is explored [3]. As the statistical procedure described there involves several steps, that fact causes the process to slow down, so that in this paper a solution to this difficulty is proposed by means of a simpler algorithm, which will be described later. The other problem to solve is the elimination of noise in the ultrasound signals, but since these are non-stationary, the signal is processed using Wavelets, which will also be explained further on.

2 Background

2.1 Time of Flight Diffraction

In the technique known as time of flight diffraction, TOFD [4], involves using two ultrasound transducers, a *T* transmitter, and an *R* receiver and located between these two the welding region to be explored. Such transducers are designed to deploy a wide angular acoustic field, so that, when the transmitter sends a wave, it propagates through four trajectories. The first, the so-called direct or lateral wave, which transits parallel to the surface of the metal on which the transducers are located; the fourth, called background echo due to the reflection of the wave at the lower border of the metal under examination, and limited by these are the two remaining ones, which are formed as a consequence of the interaction of the acoustic wave front with the edges of a discontinuity in the welding, which translates into a wave phenomenon known as diffraction (see Fig. 1).

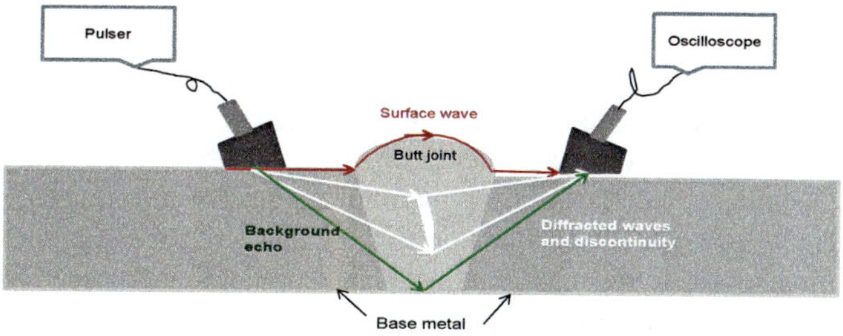

Fig. 1. Illustration of the TOFD technique. The transducer on the left is activated by a source of pulsating voltage, so that the acoustic field it generates is displayed in the form of: a lateral wave (red color), waves diffracted by the discontinuity (white color), and the background echo (green color). These four waves arrive at the transducer on the right, to be later captured by an oscilloscope enabled as an analog-digital converter.

Figure 1 illustrates one of the imaging techniques used for the detection of discontinuities in welded joints, which is based on the diffraction experienced at the front of an ultrasonic wave when its propagation path interacts with the edges of a discontinuity, so that if this did not exist, then diffraction would not occur. These concepts are clarified in the following images.

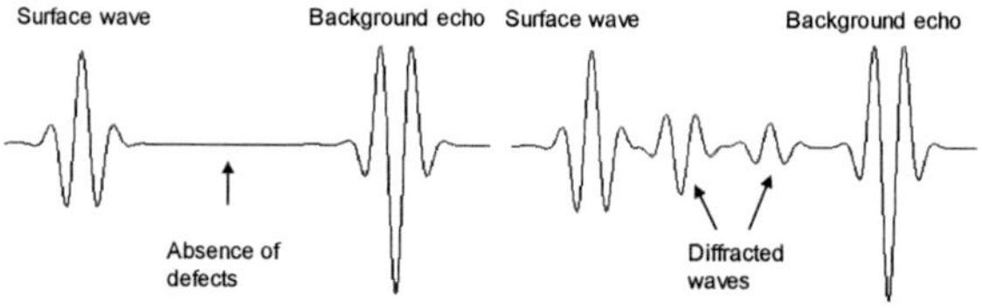

Fig. 2. Waveform of the signals displayed on the oscilloscope. The left image shows a non-diffracted wave, while in the image on the right a diffracted wave is observed.

The waveforms shown in Fig. 2, which represent voltage amplitude signals as a function of transit time, are called amplitude scanning, A-Scan, but observing the image on the right in the same figure, we can infer that each of the four signals has a pulse shape and that each one invests different transit time, due to which they arrive at the receiver at different instants and phase changes. The first pulse represents the direct or lateral wave, the second and third pulse represents the diffraction produced by the upper and lower edge of the discontinuity, and the fourth pulse is the reflection of the wave in the lower wall of the base metal.

Returning to Fig. 1, which shows the experimental arrangement to perform the exploration of a test tube (the base metal) using the TOFD technique, in particular, the disposition of transducers to send and receive information, as well as the welded joint two metallic pieces, but containing a discontinuity, then the procedure initiates by performing a parallel sweeping to the welded joint the pair of transducers displacing small spatial intervals. Now, each time the transducers advance a certain distance, the receiver transducer sends the captured waves to an oscilloscope where they are displayed in the form of A-Scan signals, such as those shown in Fig. 2, and which contain data referring to the presence or absence of dis-continuities in the welded joint. Once the exploration of the welded joint is completed, the oscilloscope sends them to a computer, where enough of them are stored in order to process them using computer procedures that construct a two-dimensional map, called D-Scan, a type of image that it is constructed by moving the transducers in a perpendicular direction to the transverse beam, and that therefore contains information regarding a longitudinal plane parallel to the welded joint, in line with each small advance of the pair of transducers.

2.2 Distortion of TOFD Images

An image, D-Scan, like that of Fig. 3, is the result that provides an algorithm that groups together a series of scans, A-Scan, but that visualizes the upper panorama of said grouping by assigning gray tones proportional to the relative amplitude of said signal for each position of the pair of transducers [4], finally exhibiting an image, similar to an X-ray, of the interior lateral of the welded joint [2, 3].

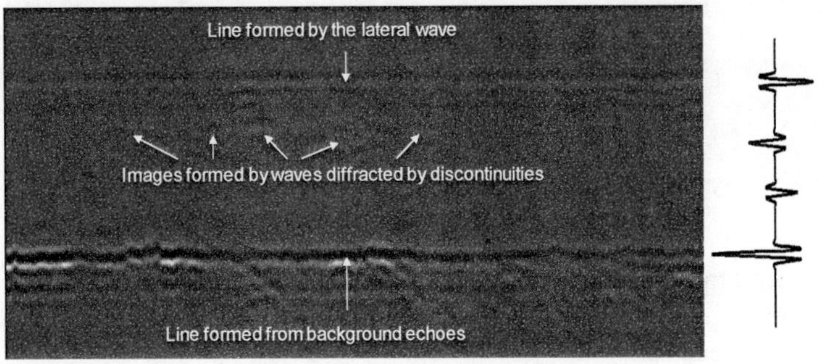

Fig. 3. D-Scan image typical the TOFD technique. For clarity, an A-Scan wave has been added to the right side, in order to visualize the relationship between both

The region of interest in the images of the TOFD technique (see Fig. 3), is that found between the diffracted waves, because in that area the defects or discontinuities that generate them are found; however, due to the liquid that is used as an acoustic coupler between the piece to be examined and the transducers, or due to possible roughness on the surface of the piece being scanned, or because the displacement of the pair of transducers is not performed with rectilinear movement, there is a distortion in the image, called misalignment, which can lead to diagnostic errors by an inspector [2].

In order to correct the problem of the zigzag of the transducers, described in the previous paragraph, an algorithm has been implemented based on the measurement of the lag time caused by the misalignment of the A-Scan signals (see Fig. 7), for which the statistical technique is used based on cross-correlation.

2.3 Calculation of Delay Time

In a generic way, if at a certain point in space there is a source that emits waves, the delay time is the time interval needed to arrive at other places in space [5]. However; for the case under study a pair of transducers, transmitter-receiver, is available so that for a certain position of the pair the emitter is activated by the voltage source to send an ultrasonic pulse scanner to the region of the welded joint the metallic pieces, after which an A-Scan signal is formed that is received by the receptor. The procedure is repeated by moving the pair to another position, but it may happen that the duo deviates from the trajectory due to any of the reasons explained in Sect. 3, so that the A-Scan signal arriving at the detector will do so with a certain temporary lag with respect to the preceding one, and consequently misalignment occurs.

Denoting $s(t)$ to the signal A-Scan and $x(t)$ to the one initially captured by the receiver, then $y(t)$ will be the one received for the next position of the pair of transducers, events that can be modeled by means of the expressions,

$$\begin{aligned} x(t) &= s(t) + v_1(t) \\ y(t) &= s(t-\tau) + v_2(t) \end{aligned} \quad (1)$$

Where τ is the lag time, v_1 and v_2 additive noise signals that statistically can be considered as zero mean, stationary, and not correlated, neither between them, nor with $s(t)$.

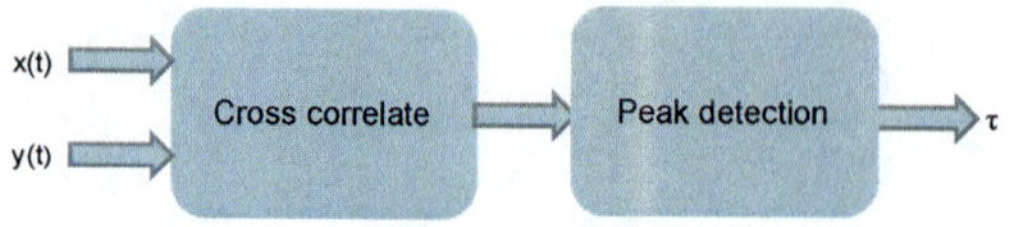

Fig. 4. System diagram for calculating the delay time between two signals

Figure 4 illustrates the method used to estimate the lag time (or delay), as the maximum value that provides calculation of the continuous-time cross-correlation, between *x(t)* and *y(t)*, by means of,

$$r_{xy}(\tau) = \int_{-\infty}^{\infty} x(t)y(t+\tau)dt \tag{2}$$

or in discrete time [5] given by the expression,

$$r_{xy}(L) = \sum_{n=-\infty}^{\infty} x(n)y(n+L) \tag{3}$$

where, L represents the delay sample, or lag [5].

It is illustrative to observe that if in the expression for the cross correlation in continuous time (2) the signals (1) are replaced, it is obtained that the maximum value of the correlation is expected for the delay corresponding to $\tau = \mathrm{D}$; that is, when the signals are aligned.

However, an examination of Fig. 2 would suggest that the lag times between signals without defect, or between signals with and without defect, would not be distinguished, but that difficulty is settled, since the maximum value provided by the cross correlation depends fundamentally both on the lateral waves and the bottom echo, since these will always have the greatest amplitude.

Next, it will be explained how to use the lag time between the A-Scan signals in order to align them, and thus eliminate that distortion of the D-Scan images.

The first A-Scan wave is chosen as the reference signal, and then the cross-correlation and the corresponding offsets are calculated, between the remaining A-Scan signals and the reference signal. With this stored data, the maximum correlation values and the moments of lag in which it occurs are subsequently determined.

Finally, and taking into account the aforementioned, that the A-Scan signals are aligned for the lag times, the D-Scan images are constructed, but now free of distortion.

A simulation will be described below to illustrate the use of the cross correlation algorithm applied to deliberately misaligned artificial signals.

The experiments and simulations reported in this paper were carried out taking into account the transducer information supplied by the manufacturer. Thus, based on this data (see Fig. 5), both the waveform in the time domain and the frequency response are demonstrated.

The images in Fig. 5 and Fig. 6 were made considering that the vibration of the surface of a transducer is possible to be modeled as a sinusoid oscillating at the resonance frequency F_0, modulated by a Gaussian envelope [4].

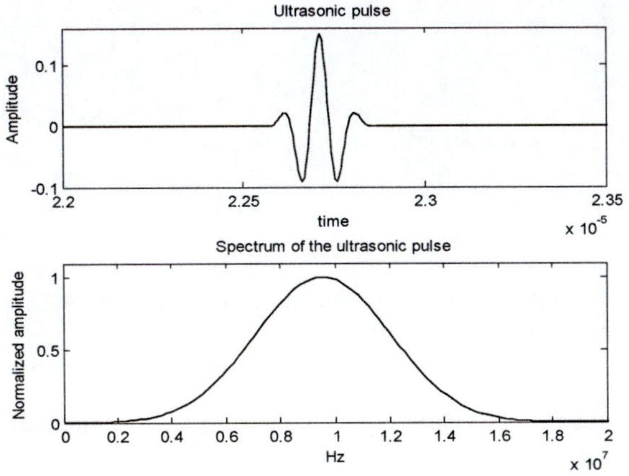

Fig. 5. Simulation of the characteristics in the time domain and the frequency of the transducer designed to be used in the TOFD technique. The resonance frequency, reported by the manufacturer, has a value of 9.55 MHz.

Then the mathematical expression that represents the shape of the ultrasonic pulse emitted by the transmitting transducer, is given by,

$$s(t) = \frac{1}{\sqrt{2\pi}\,\sigma} e^{-\frac{(t-t_0)^2}{2\sigma^2}} \cos(2\pi F_0 t) \tag{4}$$

The same expression was used to generate the pulses used in the simulation illustrated below.

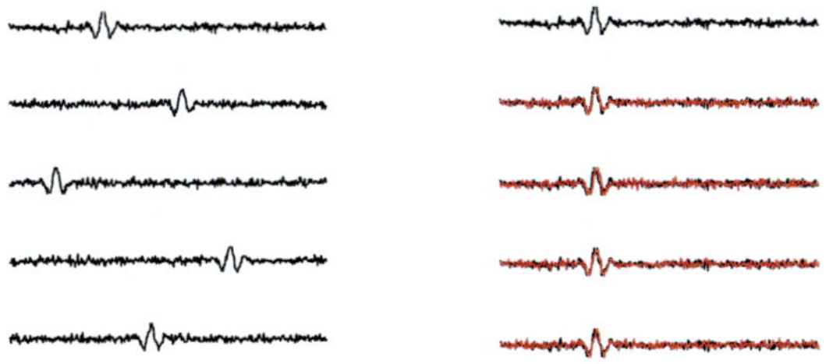

Fig. 6. The picture on the left illustrates examples of misaligned ultrasonic pulses, while the picture on the right shows the result of using the implemented cross correlation algorithm. The red color was chosen to highlight the pulses that have been aligned, but in order to visualize the contrast these have been superimposed with the black color pulse, which is the reference.

In reference to the additive noise included in the expressions 1.1, also in the simulations shown in Fig. 6, a Gaussian noise of zero mean was added with a variance of 0.333% of the amplitude of the sinusoid (4), aspect that to be considered when analyzing the images of the cited figure, permits concluding that although the pulses are immersed in random noise, the correlation algorithm corrects the problem of misalignment.

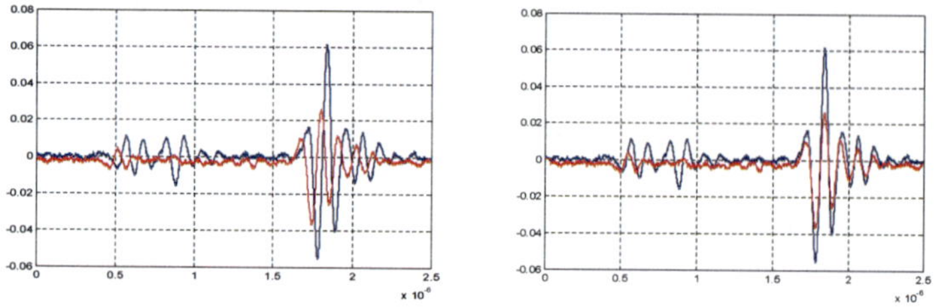

Fig. 7. The picture on the left illustrates the misalignment of a A-Scan signal, and its corresponding alignment in the picture on the right.

Below, there is the result of applying the correlation algorithm to align a real sweep signal type *A*, or A-Scan.

As in the cases presented in Fig. 6, in Fig. 7 shows the effect of the alignment algorithm applied to an amplitude sweep signal, A-Scan, immersed in noise. However, as mentioned before, the A-Scan signals constitute the raw material to construct the TOFD images, so it is reasonable to consider eliminating their current noise, procedure that is described below.

3 Proposed Approach

3.1 Wavelet Transform

A characteristic of the TOFD ultrasound signals is that they are not stationary [6], and consequently the frequency content of these changes with time, so that the analysis by wavelets adapts quite well because this technique is based on sliding windows of different duration. Therefore, just as the Fourier analysis allows to determine the similarity of a signal with sinusoidal prototype waves, the wavelet analysis allows to do something similar, but in terms of displaced and scaled (compressed or dilated) versions of a small prototype wave; that is to say, if $\psi(t)$ represents the small prototype wave, and $f(t)$ the signal with which this will be compared, then for an initial moment the similarity between them is determined by the value of their correlation, and the process is repeated for different temporal displacements. Subsequently, the prototype wave is scaled (compressed or dilated) and the process is repeated, so that finally a set of coefficients $C(s,\tau)$ will be obtained, which are a function of the scale, *s*, and from the temporary displacement, τ. For continuous time signals, these coefficients are obtained based on the continuous wavelet transform given by the analysis equation,

$$C(s,\tau) = \frac{1}{\sqrt{s}} \int_{-\infty}^{\infty} f(t)\Psi\left(\frac{t-\tau}{s}\right)dt \tag{5}$$

with *s* and τ real numbers, thus, the signal can be possibly reconstructed by performing the reverse process using the following synthesis equation [8],

$$f(t) = \int_{s}\int_{\tau} C(s,\tau)\Psi\left(\frac{t-\tau}{s}\right)dsd\tau \tag{6}$$

For the discrete wavelet transform (DWT) the parameters s and τ are discretized according to the ratios for the so-called dyadic scale,

$$s = 2^{m}, \text{ y } \tau = 2^{m}n, \tag{7}$$

where m and n are positive integers, such that if the values of m are large, that scale describes the low frequencies, and vice versa, small values of m, that scale s describes the high frequencies of the signal, therefore according to this, then the discrete wavelet transform will provide the analysis equation,

$$C(m,n) = \frac{1}{\sqrt{2^m}} \int_{-\infty}^{\infty} f(t)\Psi\left(\frac{t-2^m n}{2^m}\right)dt$$

$$C(m,n) = 2^{-\frac{m}{2}} \int_{-\infty}^{\infty} f(t)\Psi\left(2^{-m}t-n\right)dt \tag{8}$$

while the reconstruction will be by means of the synthesis equation,

$$f(t) = \sum_m \sum_n C(m,n)\Psi\left(2^{-m}-n\right) \tag{9}$$

which can also be expressed as,

$$f(t) = \sum_n c_m(n)\, 2^{-\frac{m}{2}} \Phi\left(2^{-m}t-n\right) + \sum_m \sum_n d_m(n) 2^{-\frac{m}{2}} \Psi\left(2^{-m}t-n\right)$$

$$f(t) = \sum_n c_m(n)\, \Phi_{m,n}(t) + \sum_m \sum_n d_m(n) \Psi_{m,n}(t) \tag{10}$$

where $\Phi_{m,n}(t) = 2^{-\frac{m}{2}}\Phi\left(2^{-m}t-n\right)$, $\Psi_{m,n}(t) = 2^{-\frac{m}{2}}\Psi\left(2^{-m}t-n\right)$, and the coefficients $c_m(n)$, $d_m(n)$, that contain the global and fine details of *f(t)*, respectively [8].

However; computational calculation, both for the continuous case (6), and for the discrete case (8), (9) and (10), must be calculated by means of the analysis and synthesis expressions for all time values and frequency or of the parameters of time and scale, which implies redundancy of data. So that for the discrete case there exists the method based on the multiresolution analysis that is recursive, and therefore suitable to be carried out computationally, since it obviates the problem of redundancy.

3.2 Multiresolution Analysis

The multiresolution analysis is possible based on linear filtering, in which a filtered signal *y(n)* involves the convolution of a signal *x(n)* with the impulse response $h_O(n)$ $y(k) = \sum_k h_O(k)x(n-k)$ [8], this fact allows to distinguish the rough (or global) details of the signal contained in coefficients $c_m(n)$, and the fine details contained in coefficients $d_m(n)$, because a low-pass filter with response to impulse *h(n)* allowing for the passage of the low frequency components of the signal (high scales), while a high-pass filter with response to impulse *g(n)* allowing the high frequency components of the signal (small scales) to pass. In this context a pair of filters, one lowpass *h(n)* and the other highpass *g(n)*, correspond to the analysis process, while another pair, one also lowpass and the other highpass make up the synthesis operation, but both groups are distinguished because the first pair subsamples the signal and the second pair oversamples, in order to preserve the number of samples, because the convolution alters their number.

Thus, taking into account the above, it is possible to represent the multiresolution filtering process as,

$$\Phi(t) = \sum_n h(n)\sqrt{2}\Phi(2t - n)$$
$$\Psi(t) = \sum_n g(n)\sqrt{2}\Phi(2t - n) \quad (11)$$

The following figure illustrates the process.

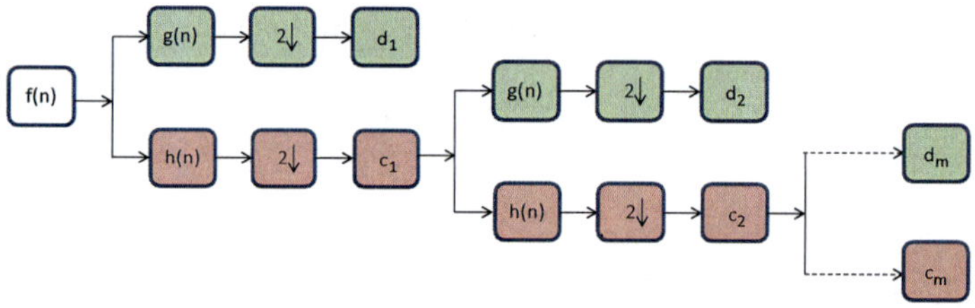

Fig. 8. Filtering process of a signal *f(n)*, when passing through *m* filtering stages. The orientation of the down arrow represents the subsampling operation.

Figure 8 illustrates the implementation of the process of analyzing a signal based on the discrete wavelet transform, where the concept of reiteration arises; due to this process is performed under the sequence of *m* stages.

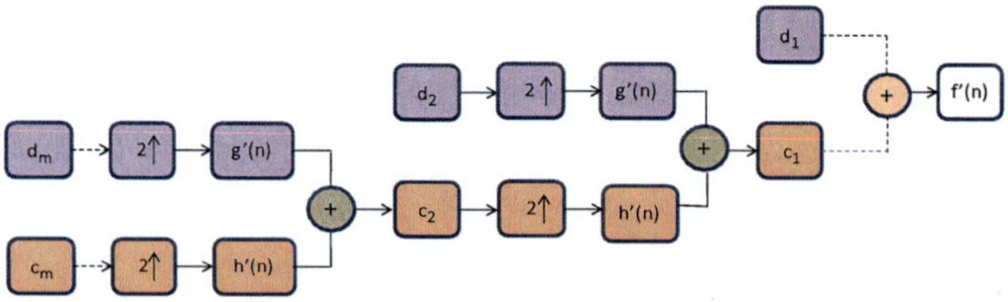

Fig. 9. Reconstruction process of a signal *f′(n)*, when passing through *m* filtering stages. The orientation of the arrow is vertical up, which means oversampling.

Figure 9 illustrates the synthesis process using the discrete wavelet transform. Note that the impulse responses of the filters highpass and lowpass are not the same as for the case under analysis [9, 10].

3.3 Multiresolution Filtering

Returning to the noise problem that contaminates the A-Scan signals (see Fig. 7), it was necessary to filter them, but with the purpose of preserving their useful properties to construct the TOFD images, the technique of the wavelet transform previously presented was used [9, 10], whose procedure is demonstrated forthwith.

As a prototype wavelet, the *bior3.7* pair was chosen because of its similarity to the ultrasound signals [11], (see Fig. 7), and because of its property of having a linear phase. What justifies that there are two is the fact that one of them is for the process of decomposition, and the other is for the reconstruction process.

The processing of the A-Scan signals was carried out using the aforementioned wavelet model, which after seven levels of decomposition, which was determined with the function of Matlab 'wmaxlev', supplies a value that depends on the number of samples of an A-Scan, and the wavelet model used [8].

The noise threshold was calculated considering the universal standard given by $\sqrt{2\log(N)}$ where N represents the number of samples of an A-Scan.

Taking into account the above, as well as the analysis and synthesis procedures by means of the discrete wavelet transform illustrated in Fig. 8 and Fig. 9, and their subsequent alignment by means of the cross correlation algorithm, in Fig. 10 two A-Scan signals are shown filtered, before and after being aligned.

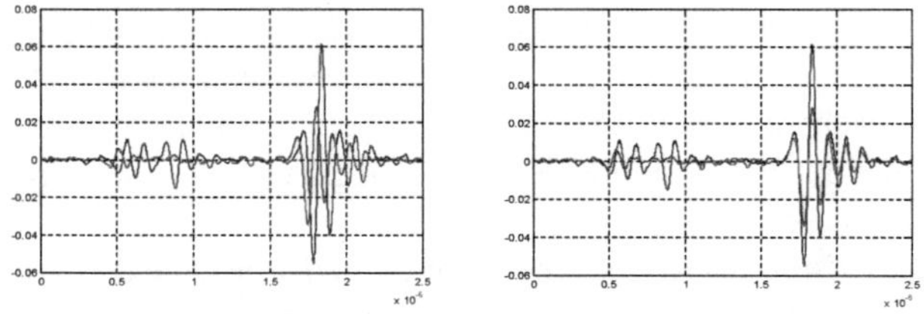

Fig. 10. In the image is illustrated the signal has been filtered before the alignment process.

Once each of the A-Scan scans are filtered and following their subsequent alignment, then the D-Scan image is constructed, following the procedure described in the section corresponding to the distortion of the TOFD images. The following figures show the results (Figs. 11 and 12).

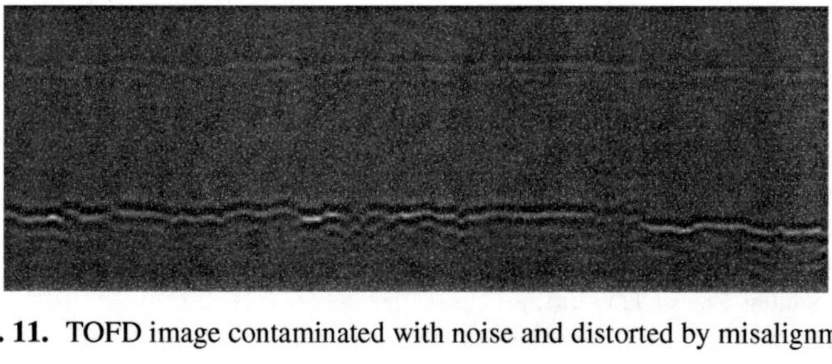

Fig. 11. TOFD image contaminated with noise and distorted by misalignment

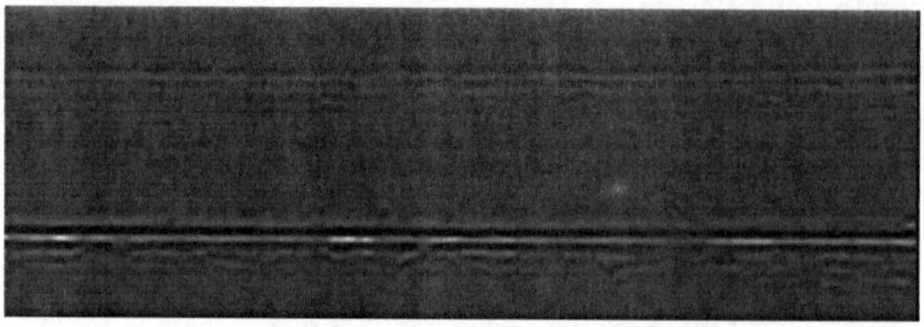

Fig. 12. TOFD image free of noise and distortion

4 Conclusions

The wavelet transform is a very useful technique for the purpose of eliminating noise in non-stationary signals.

The process of decomposing an ultrasonic signal in scales allows for the elimination of noise in this type of signals, a fact that guarantees the diagnoses issued by inspectors or personnel who perform an inspection task by means of ultrasound.

A correlation algorithm was implemented in order to eliminate the distortion in TOFD, D-Scan images, and its effectiveness was proven, both in simulated signals and in realistic signals.

The concepts of analysis and synthesis of signals were presented, based on the perspectives of the continuous wavelet, discrete wavelet and multiresolution filtering, showing the significant reduction of random noise, both in simulated and real signals, for which universal flexible thresholding was used, using the *bior 3.7*, as prototype wavelet with five levels of decomposition.

As a technique of non-invasive screening of welded joints, the flight time diffraction was detailed, and the manner in which a set of A-Scan signals are grouped to generate the TOFD D-Scan images but contrasting the distorted images with the distortion-free ones.

Acknowledgements. The author acknowledges the support received from the Universidad Santo Tomás by means of the project FODEIN 2019 code 1954505.

References

1. Vitola, J., Forero, E., Sandoval, S., Rodríguez, J.: Design and construction of a system for non-destructive inspection of failures and defects in metal using ultrasonic signals. EIA (2009)
2. Rodríguez, A., Segura, D.: TOFD imaging system in nondestructive test. EIA (2012)
3. Rodríguez, J., Forero, E.: Development of a system for detecting weld failures. IGI Global (2015)
4. Charlesworth, J.P.: Engineering Applications of Ultrasonic Time of Flight Diffraction. Research Studies Press LTD (2001)
5. Najarian, K., Splinter, R.: Biomedical Signal and Image Processing, 2 edn. (2012)
6. Angam, P., Vijayarekha, B.: De-noising of TOFD signals from austenitic stainless steel welds using stationary wavelet transform. Int. J. Comput. Appl. **66**(9) (2013)

7. Weeks, M.: Digital Signal Processing Using MATLAB and Wavelets. Infinity Science Press LLC (2006)
8. Sarkar, T., Adve, R., Salazar, M., Garcia, L., Castillo, R.: A Tutorial on wavelets from an electrical engineering perspective, part 1: discrete wavelet techniques. IEEE Antennas Propag. Mag. **40**(5), 49–68 (1998)
9. Dautov, P., Siraç, M.: Introduction to wavelets and their applications in signal denoising. J. Sci. Technol. **8**, 1–10 (2018)
10. Cohen, R.: Signal denoising using wavelets, department of electrical engineering technion. In: Israel Institute of Technology (2012)
11. Matz, V., Smid, R., Starman, S., Kreidl, M.: Signal-to-noise ratio enhancement based on wavelet filtering in ultrasonic testing, department of measurement. Faculty of Electrical Engineering, 166 27 Prague 6, Czech Republic (2009)

Proposal for the Implementation of a Business Intelligence Tool to Detect Cases of Student Desertion at the Francisco Jose de Caldas District University

Monica Lizeth Sánchez Arevalo(✉), Alexandra Abuchar Porras, Juan David Gutiérrez Herrera, and Roberto Ferro Escobar

Universidad Distrital Francisco José de Caldas, Grupo LIDER, Bogotá, Colombia
molsancheza@correo.udistrital.edu.co,
{aabucharp,rferro}@udistrital.edu.co, judgutierrezhe@unal.edu.co

Abstract. The methodology and tools of Business Intelligence (BI), present advantages to be taken into account in university academic processes, as a specific theme "university desertion". Case study: Faculty of Engineering of the Francisco José de Caldas District University (UDFJC). The information available to the University does not present an integration that allows detecting students with possible risk of desertion from the various subsystems. For which we have as a solution proposal the implementation of a BI tool, after analysis of the problem, identification of variables and correlations by means of a system dynamics model, application of the process of Extraction, Transformation and Loading (ETL) of data of the students in different moments of the university life, by means of a datawarehouse centralizing the university information systems, and allowing the construction of a dashboard with alerts of early detection.

Keywords: BI business intelligence · Systems dynamics · Predictive modeling · University dropout

1 Introduction

Social issues are called challenges in the economic and political development of a nation. Education is one of them, there is complexity in providing equal access to education and not only in aspects of quantity, but of quality, "there is an inverse relationship between poverty and academic level, among the lower level of schooling people are more likely that they access precarious jobs and live in poverty" [1]. In recent years, as a result of the increase in interest reflected in research studies on influential variables in the development of a nation, education is positioned as an element of transcendental importance in society, which is why the Pan American Health Organization determines education. and health as primordial axes in the development processes, impacting the social, economic and cultural circumstances of a country.

D. F. Cortes Tobar et al. (Eds.): AETA 2019, LNEE 685, pp. 482–489, 2021.
https://doi.org/10.1007/978-3-030-53021-1_49

The expansion of the levels of educational training is related to the improvement of population health conditions, the increase of productivity, the existence of social mobility, poverty reduction and citizen construction. By virtue of the above, the developing countries tend to respond to social problems, constituting a work area for the various complementary and alternative professions and disciplines [2], education and therefore the universities provide relevant conditions for modeling interactions present in society. For the exciting topic of education, a methodological proposal is presented which focuses on the proposal of a model of approach to the causes and components that generate the student's voluntary desertion in undergraduate programs of Engineering, and integration of the results through a tool of BI. To this end, the theme is theoretically based and contextualized through characterization, obtaining the variables of desertion, later an architectural design and a dashboard are proposed for analytical and decision making.

2 Problem Analysis

University desertion is one of the main focuses of education programs at the national level, The Organization for Development and Economic Cooperation (OECD) indicates that the desertion has implications for analysis from the economic and social perspective, and is reflected in the loss of public and private resources invested in an educational process that did not culminate successfully, reflecting losses in the life project of young people who bet on higher education and are the hope for the development of the country [3].

The entities in charge of regulating education in Colombia manifest the desertion as a problem, where the 52% of the students who start university careers, fail to complete, which indicates, of every two students enrolled in an undergraduate program, only one graduates [4]. In Colombia, about the year 2003, it begins to develop investigations of this problem, based on a first stage in a specialized review of literature, for the construction of the state of the art of student desertion, with the objective to identify the factors and variables to explain their causes empirically, this theoretical approach was achieved by the study conducted by the Ministry of National Education in 2005 and 2006, through the execution of the contract awarded to the "Universidad de los Andes", Along with the supervision by the University of Antioquia, it served to different institutions of higher education known In addition to this problem, in order to generalize the application of duration models, and thus achieve knowledge of the risk of dropping out in each student, along with institutional monitoring, consolidating the measurement tool SPADIES Defection Prevention System in Higher Education- allowing each institution to identify and classify students at risk of dropping out, by specifying the variables.

The Francisco José de Caldas District University UDFJC is no stranger to this problem, for 2015, present a dropout rate by cohort in the Professional level of 41% [4], for the foregoing, and for purposes of the case study, uses as a reference the programs of the Faculty of Engineering of the UDFJC, which has presented in recent years attrition by cohort of up to 49.8%, (Data extracted from reports generated in 2017 in Condor application - Academic University District System), alarming figures as they exceed the university average. The Faculty has five undergraduate Curricular Projects: Industrial Engineering, Systems Engineering, Cadastral Engineering and Geodesy, Electrical

Engineering and Electronic Engineering, and for the first period of 2015, it had 5,535 active students and an average dropout rate per cohort of the ten years (2005–2015) of 21.3%, established through the relation of the number of admitted persons and the number of defectors of the Faculty [5].

After a literature review of 71 studies conducted on this problematic situations, the factors that lead to this abandonment decision are shown, classified by the absolute frequency of each according to the experts consulted, choosing the variables that represent 80% (Pareto technique) in the cumulative relative frequency, according to the classification of the Ministry of National Education of Colombia and the mathematical validation, there are four subgroups [6] (Table 1):

Table 1. Factors that lead to this abandonment decision

1. Individual aspects:
a. Age
b. Gender
c. Familiar structure
d. University life adaptation
e. Demotivation
F. Personality
g. Place of origin
m. Migration

2. Academic aspects:
a. Academic performance
b. Vocation choice
c. Study habits
d. Type of school
e. College admission preparation
F. Satisfaction of the academic program
g. Academic program
h Difficulty subjects
i. Discomfort schedules
j. Time of permanence
k. Teacher satisfaction

3. Institutional aspects:
a. Teacher training and relationship
b. Institutional policies
c. University infrastructure
d. Abnormal public order

4. Social-economic aspects:
a. Labor insertion
b. Resources and economic income
c. Socioeconomic level
d. Educational level of parents
e. Family support
F. Dependents

The data for each of the variables were obtained from various sources, 1) Information provided by the Faculty of Engineering based on reports from the Systems Advisory Office (OAS), 2) Collection of minutes from the Faculty Council (2009–2017), and 3) Virtual Appreciation Tool applied to 343 deserters. The treatment of these variables allows the understanding of the problem, in terms of figures, validation and determination of correlations. The problem of desertion is considered a causal poly (77) which makes it necessary to review the variables one by one, indicating independence, percentage of occurrence or correlation equation according to the particularity, in addition, the correlations and the elimination of variables from the mathematical field are supported for the case study.

3 Methodology

Once the representative database of the students of the Faculty is constituted, we proceed to construct a model based on systems dynamics that allows to represent the behavior of the desertion. For each one of the subsystems previously defined, the correlations were defined mathematically and graphically based on the construction of a causal model, the mathematical model is defined around the determining variables of greater importance and weight in the problem of university dropout in the Faculty of Engineering, which generate an approximate representation of the real system, in order to experiment with modifications, identifying and evaluating policies for the representative reduction of dropout rates. In Fig. 1, there are 7 variables and 4 sub variables that affect the academic performance, in Fig. 2, 6 variables that interfere in the economic resources are presented,

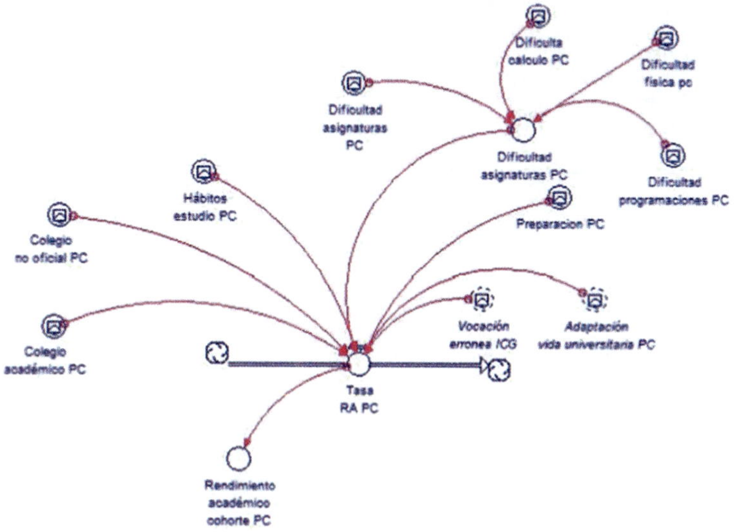

Fig. 1. Generic forrester diagram for the flow of academic performance RA

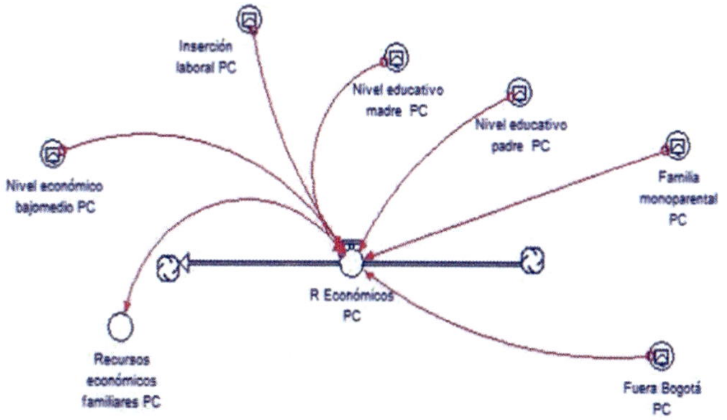

Fig. 2. Generic forrester diagram for the flow of economic resources.

in Fig. 3, 4 variables that define the perception are presented. about the university, and in Fig. 4, there are 6 variables that impact the student's motivation.

The academic variables based on academic performance taking as a measure of performance of the weighted average (Scale from 1 to 5), in addition to historical and behavioral data according to the perception survey applied to dropouts, the multiple correlation coefficient is 99%, which concludes a significant value of the linearity between the proposed variables see Fig. 1.

The proposed economic variable has a high coefficient of determination, between 84% and 91%, which explains a high coherence of variation based on the sub variables, which confirms the consistency of the relationship (Fig. 2).

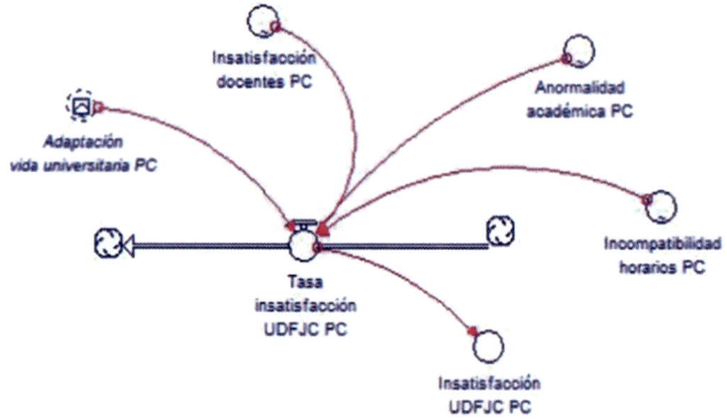

Fig. 3. Generic forrester diagram for the flow of institutional aspects.

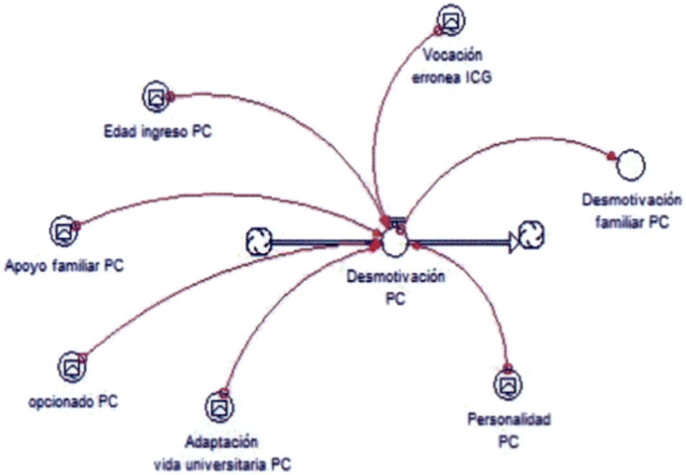

Fig. 4. Generic forrester diagram for the flow of personal aspects.

The linear relationship between independent and dependent variables presents a high degree of explanation of the variability of university dissatisfaction (Fig. 3) and demotivation (Fig. 4), in terms of the independent variables defined.

4 Experimental Results

In this section a Business Intelligent BI dashboard is proposed for the detection of possible defectors based on the previously identified and validated variables. To make use of this method it is necessary to define parameters such as the number of variables, students in the sample and the number of experiments.

To establish the number of variables, the truthful information was used as a reference, using data from 342 schools of the faculty, in order to validate the proximity of the prediction.

Within the tool proposal, a profiling is presented using the statistical technique of weighting, which for each of the variables has a relative importance or weight with respect to the comparison of the other elements [8], established by the number of incidents in the bibliographic review, said percentage values are input for the construction of the correlations.

The use of BI tools uses dashboards to display information related to studies or predictions conducted using business intelligence to interested people. A university can use dashboards as the main components of a BI solution. Eckerson defined a dashboard as "a multi-layered application built on a business intelligence and data integration infrastructure that enables organizations to more effectively measure, monitor and manage business performance" [9]. The results shown include metrics, graphic trend analysis, capacity indicators, geographic maps, shared percentage, semaphores and variance comparisons. These aspects are used in this article in order to show and interrelate the most important aspects found on the desertion of university students. Our project was developed in Microsoft's Power BI [10], which aims to connect student's information in a coherent and interactive manner, and to present the percentage of similarity to the profile of a high-risk deserter as a result. The feasibility of use in the University of this tool is high, since it allows the connection of various data sources regardless of the type

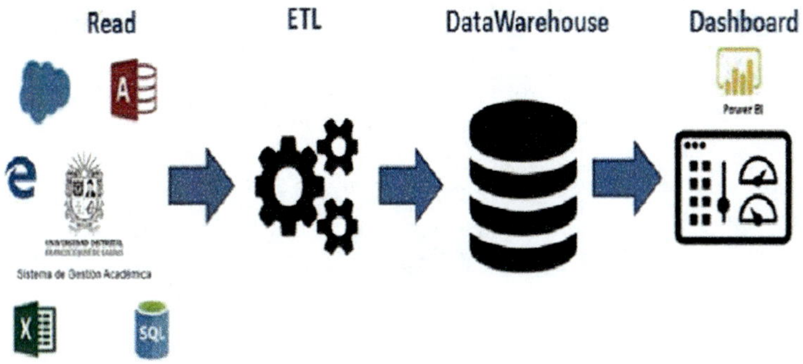

Fig. 5. Information integration power business intelligent BI dashboard.

of source, in addition to being stable and business style, being a business intelligence solution its analysis is can perform in real time, as presented in Fig. 5 (Fig. 6).

The second interface is based on the visualization of each student's own analysis, for which it indicates personal data such as name, code, curricular program and e-mail (confidential data). Subsequently, the total risk (per student) is indicated based on the sum of the characteristics by weight (Fig. 7).

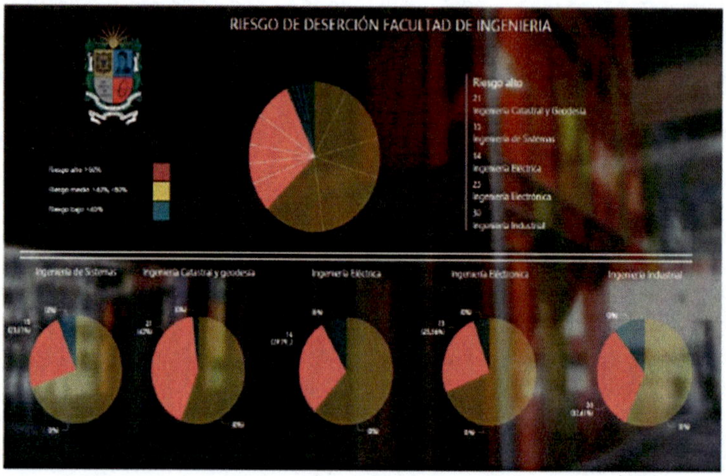

Fig. 6. Data interface faculty power BI - high risk.

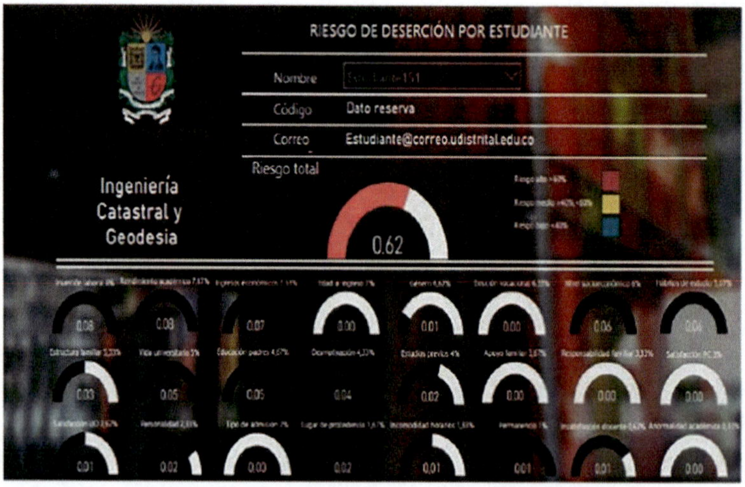

Fig. 7. Data interface per student power BI-high risk

References

1. Fideg: Campaña Social: La educación un arma contra la pobreza (2014). https://fideg.org/wpcontent/uploads/2017/02/Campaa_Social_2014_4.pdf
2. Méndez Giraldo, G.: Dinámica de sistemas y problemática social. Universidad Nacional de Colombia
3. MINEDUCACIÓN: mineducacion.gov.co, 13 Mayo 2015 (2015). https://www.mineducacion.gov.co/1759/articles-350451_recurso_6.pdf
4. Universidad Distrital Francisco José de Caldas. Camino a la Reacreditacion Institucional, Bogotá, Colombia: Pubicación del Comité de Autoevaluación y Acreditación (2017). www.udistrital.edu.co
5. Sánchez Arevalo, M.: Modelo representativo de deserción estudiantil voluntaria en carreras de pregrado de la Facultad de Ingeniería de la Universidad Distrital Francisco José de Caldas (2015)
6. MINEDUCACION: Metodología de seguimiento, diagnóstico y elementos para su prevención, Bogotá, Colombia (2009)
7. Universidad Pedagógica de Colombia: La deserción estudiantil: Reto investigativo y estratégico asumido de forma integral por la UPN (2005)
8. Mendoza, H., Bautista, G.: Probabilidad y Estadística (2002). http://www.virtual.unal.edu.co/cursos/ciencias/2001065/. Creative Commons BY-NC-ND
9. Turban, E., Aronson, J.E.: Decision Support Systems and Intelligent Systems, 8th edn. Prentice Hall, Upper Saddle River (2006)
10. Microsoft: Quées Power BI? Documentos de aprendizaje Microsoft (2018). https://docs.microsoft.com/es-es/power-bi/power-bi-overview

Resource Allocation Model for a Computer System

Joaquín F. Sánchez[1(✉)], Juan P. Ospina[2], Carlos Collazos[1], Henry Avendaño[1], Emiro De-la-Hoz-Franco[3], Zhoe Comas-Gonzalez[3], and N. Vanesa Landero[4]

[1] Universidad Manuela Beltran, Bogotá, Colombia
joaquin.sanchez@docentes.umb.edu.co,
{carlos.collazos,henry.avendano}@umb.edu.co
[2] Universidad Nacional, Bogotá, Colombia
jpospinalo@unal.edu.co
[3] Departamento Ciencias de Computación y Electrónica,
Universidad de la Costa-CUC, Barranquilla, Colombia
{edelahoz,zcomas1}@cuc.edu.co
[4] Universidad Politécnica de Apodaca, Nuevo León, Mexico
vlandero@upapnl.edu.mx

Abstract. This article describes a simulation for resource allocation. As a particular case, it describes how would be the assignment in an ad hoc network. We present the results of a simulation allocating resources from one system (communications network) to another system (agent collections). In the description of the model first shows the behavior of the allocation of resources towards the nodes depending on the usefulness of the network and the satisfaction of the agents. Based on the mathematical model of the tragedy of the commons, the behavior of the system is shown. The usefulness and satisfaction of the system are measured. Resource allocation is modeled as an interaction. The computational model Interaction Nets is used as the basis of the simulation.

Keywords: Resource allocation · Computational system · Mathematical model

1 Introduction

This article shows the design of a simulation for allocating resources in a computational system (decentralized network) to its component nodes. The components of the model are, on the one hand, the computational model of Interaction Nets, which is used to formalize the concept of interaction in the context of computation. The second component is the set of public goods as a vision of allocation of resources to the participants of a specific agglomerate.

The control of decentralized networks is an element that differentiates this type of networks because it is from each of the nodes that make up the network [4,12,14]. The application of decentralized networks with more significant

D. F. Cortes Tobar et al. (Eds.): AETA 2019, LNEE 685, pp. 490–500, 2021.
https://doi.org/10.1007/978-3-030-53021-1_50

potential for the design and implementation of solutions in different areas are ad hoc networks [1,3]. The operation of these networks is affected by a change in topology and a change in the number of participants as a function of time; since resources are limited, finite, techniques can be explored to optimize the use of these resources [8].

The analogies shown in the literature for assignment in traditional computational models are based on assignment in stable conditions, with centralized control schemes, or in current applications with distributed control schemes [9,10]. However, the change of paradigm of decentralized control in new generation networks makes it necessary to explore new techniques to perform the resources, such as memory, storage, and processing [2].

Decentralized networks have signs of complexity, and adaptability schemes to cope with these signs of complexity use various techniques to cope with this phenomenon. One of the possible ways to construct adaptation schemes is to understand the local interactions between network participants [17]. Thus, details of the functioning of decentralized networks can be covered.

The purpose of this work is to show how to generate a model of resource allocation for decentralized networks, the model is based on the game of public goods and is combined with Interaction Nets, looking for implementation in a computer tool (e.g., a programming language).

The distribution of the document is as follows: Sect. 2 presents the theoretical considerations of the model, first presents the concept of Interaction Nets, then describes the concept of the set of public goods and finally this the design of the model. Section 3 explains the simulation of the resource allocation model. Section 4 presents the conclusions of this work.

2 Considerations for the Model

The purpose of this section is to present the relevant concepts that have been worked on for the construction of the resource allocation model. First, the computational model is presented to represent the elements that are part of the experiment. Then the problem of resource allocation is described using the game of public goods; finally, there is the Interaction Nets model.

2.1 Interaction Nets

The Interaction Nets model is suitable for the construction of various systems, for example, the construction of ad hoc networks that fulfill telecommunications services such as data transfer, expansion of Internet coverage, information processing for sensors, and the processing of actuators, among others. The concept of opportunistic networks [19] requires a flexible infrastructure due to the dynamic environment in which it is going to be deployed. For example, the deployment of systems based on ad hoc networks such as the Internet of Things (IoT) [11], is an opportune scenario for the model serves as a design and analysis tool. The purpose is to have applications to solve different connectivity and coverage problems.

One of the conditions of highly interconnected environments is the number of nodes that exist in these systems. These nodes must interact with each other, passing signaling information or data traffic. The result of these interactions generates emerging behaviors in the systems, which are dominated by the resulting complexity [5,16]. One way to deal with this complexity is to provide the system with the ability to adapt.

The computational model based on interactions is referred to as Interaction Nets [6]. *A* node can have a number n of communication interfaces, depending on the application required. Each interface is used to communicate with other nodes. When a connection is made and information is passed from node to node, an interaction occurs. An example is the transmission of packets between a node A and a node B. Node A generates an interaction with node B when the packet transmission ends. The Interaction Nets model is based on the idea of "computing as interaction" [6,13].

The model is described as follows:

- A set Σ of symbols used to build the networks.
- A set R of rules called interaction rules.

Network: a network N of is a graph where the symbols of Σ are the nodes. A node is called an agent and the link between two agents is called a connector in such a way that networks are graphs that connect agents through connectors.

Active Pairs: an active pair is a pair of agents (A, B) that belong to a set of symbols. An interaction rule consists of an active pair on the left and a network on the right. The rules must have two conditions:

- The left side and the right side must have the same number of interfaces.
- In a set rule R, there is a maximum of one rule for each agent.

The Interaction Nets model is similar to other formal models for defining programming languages, as is the lambda calculus for functional programming languages and pi calculation for parallel processing. The model can have several notations:

- Graphical notation through graphs.
- $A \bowtie B \rightarrow N$: an active pair A and B produce a network interaction N.
- $foo(A, B)$: a function foo which receives parameters A and B creates an interaction.

2.2 Model of the Tragedy of the Commons

The model described below mathematically synthesizes the tragedy of the commons. The purpose of the simulation is to verify the behavior of the resource allocation of a collection of agents.

The problem of the voluntary provision of resources has been usually analyzed using the linear public good game [7]. Nevertheless, due to the nature of open self-organizing systems, this model presents some limitations that need to be considered before applying it in the context of ad hoc networks. For example, it assumes that the public payoff is equally distributed even when it is possible for the appropriation to exceed allocation; that there is a full disclosure of all information required for the process; that there is no cost related to monitoring; that the utility for all resources are the same no matter if they are needed or not. As a consequence, in order to get a more realistic model, we relax some of these conditions using a variation of this game [15]. In this case, n agents form a cluster in which each agent i owns a quantity of some divisible resource and freely decides if contribute or not to the public good. We assume that agents take their decisions under self-interest analysis, and the game is played in consecutive rounds. In each round, each agent i:

- Determines the resources it has available, $g_i \in [0, 1]$
- Determines its needs of resources, $q_i \in [0, 1]$
- Makes a demand for resources, $d_i \in [0, 1]$
- Makes a provision of resources, $p_i \in [0, 1]$ $(p_i \leq g_i)$
- Receives an allocation of resources, $r_i \in [0, 1]$
- Makes an appropriation of resources, $r'_i \in [0, 1]$

The total amount of resources owned by an agent at the end of the round is given by $R_i = r'_i + (g_i - p_i)$, in which R_i is the sum of resources appropriated by the agent and the ones that it keeps for itself. The contributions of all participants are summed and the payoff u_i for the agent i is given by:

$$U_i = \begin{cases} a(q_i) + b(R_i - q_i)\ if R_i \geq q_i \\ a(R_i) - c(q_i - R_i)\ otherwise \end{cases} \tag{1}$$

where a, b and c are coefficients in $\mathbb{R}$ that represent the relative utility of getting the resources that are needed, getting resources that are not needed, and not getting the resources that are needed.

Furthermore, independent of its utility and the cooperation pattern (the prisoner's dilemma or the linear public good game) each agent i makes a subjective assessment of its satisfaction S_i expressed as a value in $[0, 1]$ according to the relationship between its allocation and its demands. In this regard, we can define the satisfaction level of the agent i in the round $t + 1$ as follows:

$$S_i(t+1) = \begin{cases} S_i(t) + \alpha\,[1 - S_i(t)] & \text{if} \quad r_i \geq d_i \\ S_i(t) - \beta(q_i - R_i) & \text{otherwise}, \end{cases} \tag{2}$$

where α and β are coefficients in $\mathbb{R}$ which determine the rate of reinforcement of satisfaction and dissatisfaction of each agent. As a result, choosing different combinations of α and β allow us to model different behaviors in the agents. For example, high values of α and low values β enable us to model agents with a high level of tolerance to situations in which they do not get what they need. On the other hand, high values of β will make the agents be dissatisfied more quickly, and therefore, they would stop following the institutional rules. This scenario is modeled through a threshold value of τ and an interval value of m. If for m consecutive rounds the agent i evaluate $S_i < \tau$ as true, it will stop cooperate. In the case of the prisoner's dilemma, the agents appropriate several resources greater than the allocated (they turn into free-riders). In the linear public good game, the agent leaves the cluster.

2.3 Interaction Model for Resource Allocation

The purpose of this model is to have a notation that serves to generate functionalities in a programming language, a simple representation of the problem and a perspective to analyze a particular situation. Figure 1 presents the situation of resource allocation from a network to a group of agents. Following the convention described in the problem of public goods, in the interaction the set of Σ agents.

- $\Sigma = \{i, g, q, d, r, r'\}$
- Networks U, S

The rules of interaction are as follows:

- $i \bowtie g \rightarrow \lambda_1$ Determines available resources.
- $i \bowtie q \rightarrow \lambda_2$ Determines the need for resources.
- $i \bowtie d \rightarrow \lambda_3$ Makes a demand for resources to the network.
- $\lambda = \lambda_1 + \lambda_2 + \lambda_3$ The resources of the agent ask to the network.
- $i \bowtie r \rightarrow \mu_1$ Receive a resource allocation.
- $i \bowtie r' \rightarrow \mu_2$ Makes an appropriation of resources.
- $\mu = \mu_1 + \mu_2$ The resources that the network gives to the agent.

Taking into account the interactions between the agents and the network, U and S are created representing the contribution of the agents to the network and the satisfaction of each agent. So the network that represents the allocation of resources is given by:

- $(i \bowtie \lambda)^* \rightarrow U, S$
- $(i \bowtie \mu)^* \rightarrow U, S$

Figure 1 shows the concept of interaction for resource allocation. N represents the decentralized network, and A represents the collection of agents. Because a decentralized network has unique features (e.g., decentralized control), it is necessary to review how the model behaves in these cases.

Each interaction between the agents and the network is for a particular scenario. It is considering that there is a variation of nodes in each configuration that varies the time. In each interaction, the usefulness of the contribution of the agents to the network and the satisfaction of the agents varies. Figure 2 shows the variation in the value of the Utility and the satisfaction according to the possible configurations that a decentralized network can have. The calculation

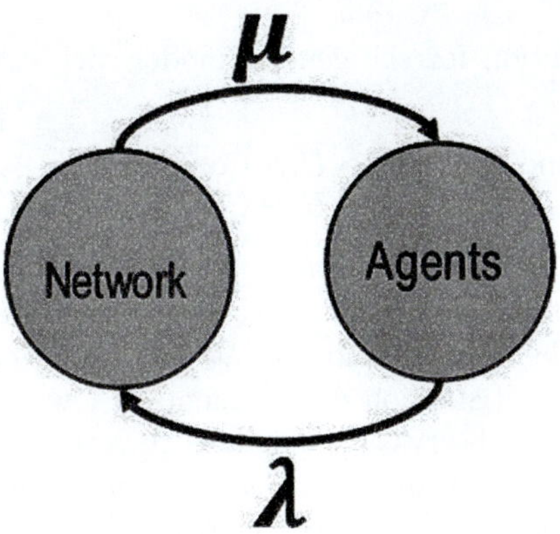

Fig. 1. μ: the network to accept agents, requests information from their resources. λ: the agents request to enter the network. Interaction occurs when network resources are assigned to agents.

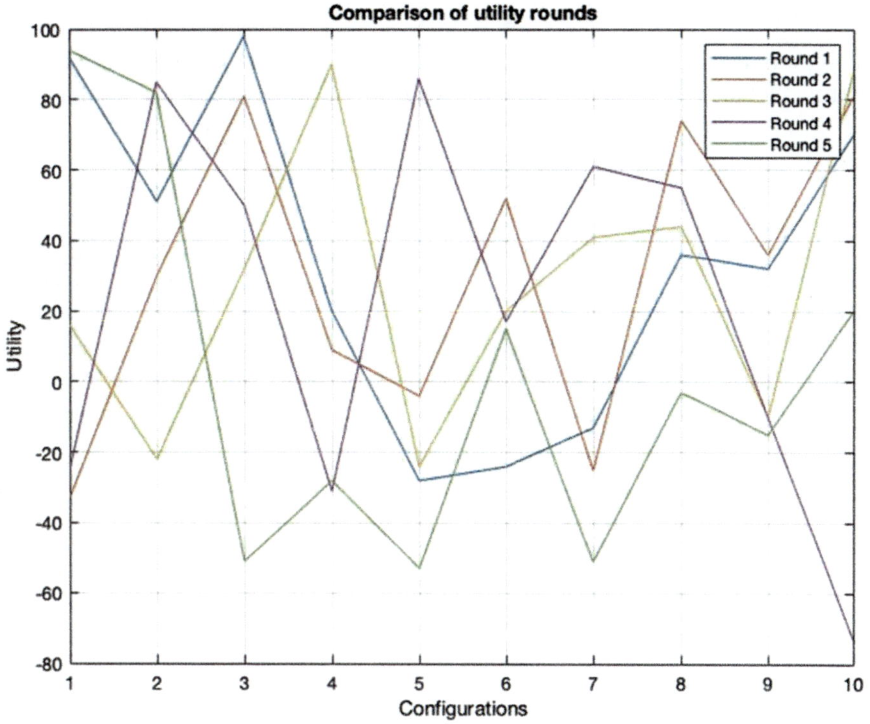

Fig. 2. Utility rounds comparation

of each contribution of the participants and the level of satisfaction is exposed. For this we use the model exposed Eqs. 1 and 2.

The rules of interaction for the general model with different configurations is:

- $(i \bowtie \lambda)^1 \rightarrow U_1, S_1$ and $(i \bowtie \mu)^1 \rightarrow U, S$ Configuration 1
- $(i \bowtie \lambda)^2 \rightarrow U_2, S_2$ and $(i \bowtie \mu)^2 \rightarrow U, S$ Configuration 2
- $(i \bowtie \lambda)^n \rightarrow U_n, S_n$ and $(i \bowtie \mu)^n \rightarrow U, S$ Configuration n

Figure 3 is present the comparison between the usefulness of the contribution of the participants of the network and the satisfaction for each agent. It is interesting to note that the satisfaction of the agents is low while the usefulness of the network is high. This observation indicates that the welfare of the majority superimposes the particular welfare. One conclusion is that the behavior shown is an emerging characteristic of network behavior as a function of the dynamics of interactions between participants.

3 Model Behavior

3.1 Simulation with Net Logo

Taking into account the description made in the previous section on the computation model and the resource allocation model, the purpose of this section is to

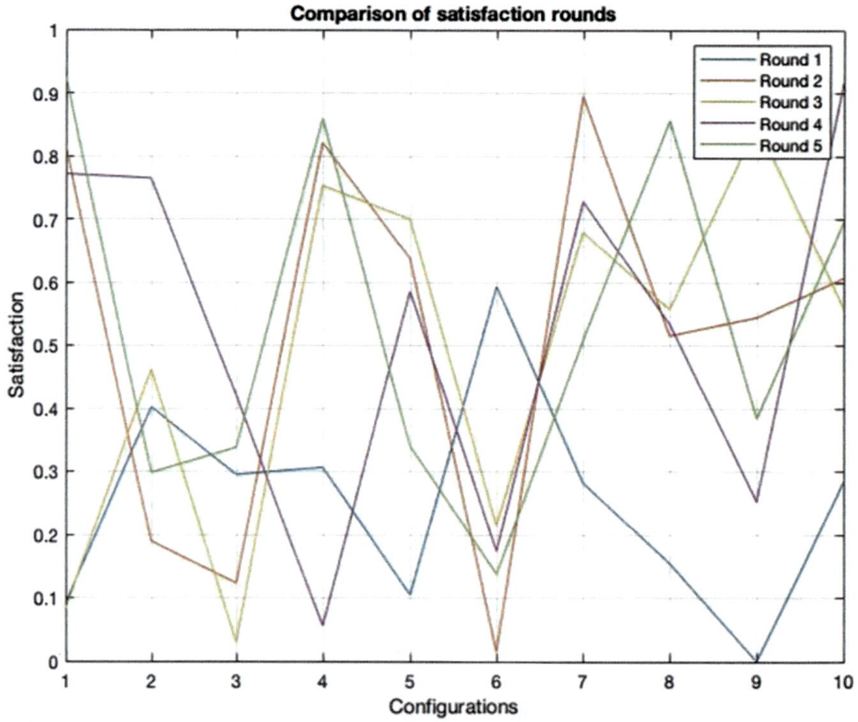

Fig. 3. Satisfaction rounds comparation

show how to allocate resources from an entity (a network of nodes) to different agents (nodes). NetLogo is used as a platform to simulate, and the variables used are the following: number of agents, degeneration rate, level of satisfaction, agents who do not cooperate. The idea is to make these agents interact between them and the resource, which is defined by a function that manages how much the network can offer and how much it receives from the agents. What is measured in this simulation is the evolution of the allocation of resources and the behavior of their proportion of the network.

Three scenarios are compared, depending on the number of agents interacting with the network. Figure 4 shows the results of the comparison.

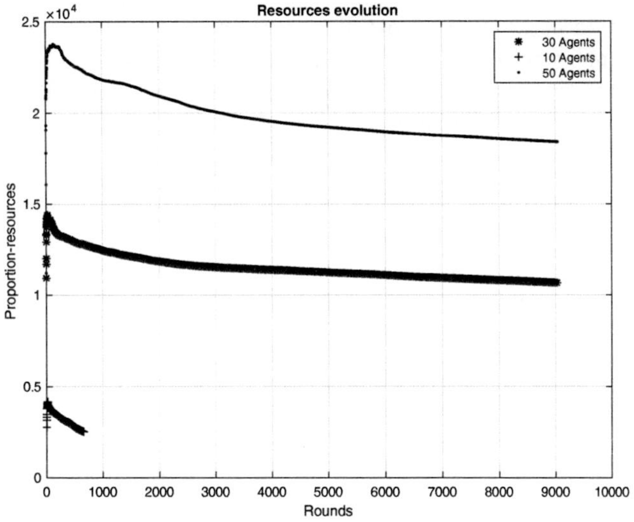

Fig. 4. Model of interactions

The comparison shows how the allocation of resources is stable from the network to the nodes. Below are the values of the simulation carried out:

- Number of agents: $[10, 30, 50]$
- Rate of degeneration of the resource: $0, 40$
- Scarcity: 1
- Satisfaction level: $S_{th} = 0.1$

The parameter values are used to adjust the simulation to a real situation, for example, an ad hoc network for the generation of a sensor network. In this application, the resource to share in the storage memory. The number of agents represents the number of nodes; the rate of degeneration is the rate at which the power of those nodes is decreasing (the battery of the devices). The scarcity implies the resource limit that the agent must maintain. Moreover, the parameter

S is the level of times that the node fulfills its purpose (to take samples with the sensor and to send the information).

The behavior of the simulation is shown in Figs. 2, 3. The idea is to review how the value of utility and satisfaction changes as time evolves. A 5 rounds exposition is made, where it is noted that both utility and satisfaction values are stable. In a network that changes with time with the number of nodes, this behavior of the model is desirable.

3.2 Implementation of System

The implementation of the interaction is done in the TL language, which is described in [17,18]. Sharing Internet from one node to the other nodes of the network. This interaction is the most powerful application in the ad hoc network. Several additional services can be offered in the Internet channel to convert the network into a more flexible system and meet the possible user requirements. The code for this interaction is:

```
import mas.__init__
function internet(identifier, description, times)
    x = ExecuteScript()
    x.start()
end
log("test Interaction")
a = internet(1, CycleCallBash, 1000000)
```

For this application, the ExecuteScript agent has been used. In this case the CycleCallBash agent is in charge of executing the native function, which modifies the configuration of the node to create a bridge interface and share the internet service. The graphical view can be seen in Fig. 5.

The interaction is described as follows:

- $internet(A, B) \rightarrow CycleCallBash(A)|ack| \sim (ack)$
- $internet(B, \otimes_n) \rightarrow internet(A, B)|ack| \sim (ack)$
- $CycleCallBash(A, device) \rightarrow ExecuteScript(A, device)$
- $ExecuteScript(A, device) \rightarrow data$

The resource that behaves in this application is the Internet (bandwidth). The parameters that intervene in the process are elements described in the previous sections, several nodes of the network, the utility of the network and the satisfaction of the nodes.

- Number of nodes N
- Contribution of the nodes $U(t)$
- Satisfaction of each agent $S(t)$
- Resource: bandwidth, Internet Bw
- WLAN interfaces w_0

The interaction of this application is:

- $(i \bowtie \lambda)^4 \rightarrow U_4, S_4 : Bw$
- $(i \bowtie \mu)^4 \rightarrow U_4, S_4 : w_0$

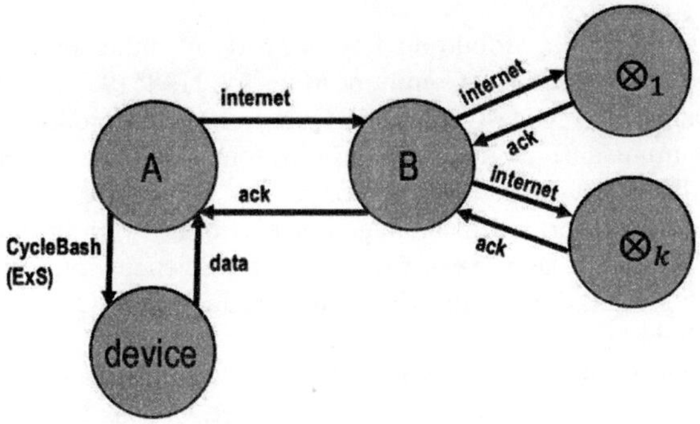

Fig. 5. Agent A shares the internet with agent B. The interaction of B between the n Nodes of the network is sharing the Internet that comes from the agent A. The function of A is to execute the script as many times as necessary on the device to share the internet.

4 Conclusions

This article has shown the process of designing a model for the allocation of resources for a decentralized network using two components, a computational Interaction Nets model and a set of public goods. The model is generalized to the situation where the configurations of the network participants change. The measurements made on the utility of the nodes (U(t)) and S(t) reflect a behavior where it is obtained that, the satisfaction of the nodes can be minimal, but the utility of the network is useful or high. However, if the utility is below zero, the satisfaction of the nodes is also low. To complement this analysis, in the description of the evolution of resource allocation, it is also evident that this allocation reaches a stable behavior when its allocation process evolves.

This behavior is the result of the dynamics of interactions between network participants and how the states of these participants are configured. There are signs of emerging behavior; this means that if the network can guarantee minimum operating resources, regardless of the level of personal satisfaction of the participants, the behavior of the system will be in good operating points.

On the other hand, the model can be coupled with a computational tool (programming language) and thus develop applications to solve problems in ad hoc networks, such as coverage extension or sensor networks. Some application scripts with the TL programming language are exposed.

References

1. Al-Sultan, S., Al-Doori, M.M., Al-Bayatti, A.H., Zedan, H.: A comprehensive survey on vehicular ad hoc network. J. Netw. Comput. Appl. **37**, 380–392 (2014)
2. Amadeo, M., Campolo, C., Iera, A., Molinaro, A.: Named data networking for IoT: an architectural perspective. In: 2014 European Conference on Networks and Communications (EuCNC), pp. 1–5. IEEE (2014)

3. Conti, M., Giordano, S.: Mobile ad hoc networking: milestones, challenges, and new research directions. IEEE Commun. Mag. **52**(1), 85–96 (2014)
4. De-La-Hoz-Franco, E., Ariza-Colpas, P., Quero, J.M., Espinilla, M.: Sensor-based datasets for human activity recognition-a systematic review of literature. IEEE Access **6**, 59192–59210 (2018)
5. Dressler, F., et al.: Self-organization in ad hoc networks: overview and classification. University of Erlangen, Department of Computer Science, vol. 7, pp. 1–12 (2006)
6. Fernández, M.: Models of Computation: An Introduction to Computability Theory. Springer, Heidelberg (2009)
7. Gächter, S.: Conditional cooperation: Behavioral regularities from the lab and the field and their policy implications. Technical report, CeDEx Discussion Paper, The University of Nottingham (2006)
8. Ju, H., Zhang, R.: Optimal resource allocation in full-duplex wireless-powered communication network. IEEE Trans. Commun. **62**(10), 3528–3540 (2014)
9. Li, M., Li, Z., Vasilakos, A.V.: A survey on topology control in wireless sensor networks: taxonomy, comparative study, and open issues. Proc. IEEE **101**(12), 2538–2557 (2013)
10. Lin, S., Miao, F., Zhang, J., Zhou, G., Gu, L., He, T., Stankovic, J.A., Son, S., Pappas, G.J.: ATPC: adaptive transmission power control for wireless sensor networks. ACM Trans. Sens. Netw. (TOSN) **12**(1), 6 (2016)
11. Liu, X., Li, Z., Yang, P., Dong, Y.: Information-centric mobile ad hoc networks and content routing: a survey. Ad Hoc Netw. **58**, 255–268 (2017)
12. Lo, C.H., Ansari, N.: Decentralized controls and communications for autonomous distribution networks in smart grid. IEEE Trans. Smart Grid **4**(1), 66–77 (2013)
13. Perrinel, M.: On context semantics and interaction nets. In: Proceedings of the Joint Meeting of the Twenty-Third EACSL Annual Conference on Computer Science Logic (CSL) and the Twenty-Ninth Annual ACM/IEEE Symposium on Logic in Computer Science (LICS), p. 73. ACM (2014)
14. Pinto, A.C., De la Hoz Franco, E., Pinto, D.C.: Las redes de sensores inalámbricos y el internet de las cosas. Inge Cuc **8**(1), 163–172 (2012)
15. Pitt, J., Schaumeier, J.: Provision and appropriation of common-pool resources without full disclosure. In: International Conference on Principles and Practice of Multi-Agent Systems, pp. 199–213. Springer (2012)
16. Prehofer, C., Bettstetter, C.: Self-organization in communication networks: principles and design paradigms. IEEE Commun. Mag. **43**(7), 78–85 (2005)
17. Sánchez, J.F., Quiñones, J., Corredor, J.M.: Interaction net as a representation model of a programming language. In: Current Trends in Semantic Web Technologies: Theory and Practice, pp. 57–83. Springer (2019)
18. Sánchez, J.F., Quiñones, J.A., Corredor, J.M.: A programming model for decentralised data networks. In: Workshop on Engineering Applications, pp. 468–479. Springer (2018)
19. Trifunovic, S., Kouyoumdjieva, S.T., Distl, B., Pajevic, L., Karlsson, G., Plattner, B.: A decade of research in opportunistic networks: challenges, relevance, and future directions. IEEE Commun. Mag. **55**(1), 168–173 (2017)

A Bumpless Transfer Control Scheme for Horizontal-Axis Wind Turbines Operating in Transition Region

Fernando Lozano, Edwin Villarreal-López, and Horacio Coral-Enriquez(✉)

Programa de Ingeniería Mecatrónica, Facultad de Ingeniería,
Universidad de San Buenaventura Bogotá, Carrera 8H No. 172-20, Bogotá, Colombia
falozano@academia.usbbog.edu.co, {evillareal,hcoral}@usbbog.edu.co

Abstract. In wind turbines is conventional to design controllers for each region of operation, i.e. wind energy maximization (region 2) or power regulation (region 3). Then, it is usual to switch between these controllers. In this paper, a bumpless transfer control approach is proposed to address the problem of switching between controllers for wind turbines operating in the transition region. The proposed scheme provides smooth transition between the torque controllers when the turbine is switching between regions 2 and 3.

Keywords: Wind turbines · Bumpless transfer · Standard torque control · Wind turbine control · Internal model principle

1 Introduction

Due to the adverse environmental effects of carbon emissions and other pollutants derived from the consumption of fossil fuels, developing of renewable energy technology is a mandatory trend in scientific and technological research. Future power resources must be considered in a sustainable perspective, where special attention will be held in environmental impact assessments.

One of the most promising technology in the field of clean production is wind energy. According to the World Wind Energy Association, in the beginning of the year 2018, the overall capacity of the wind turbines installed worldwide reached 539GW, and it has been growing at a rate around 10GW per year. Nowadays, around 5% of the global electricity demand is captured from the wind. However, according to the global wind energy council, by the year 2030 about 20% of world demand will be obtained from this renewable source. Much attention has been drawn to the design of these Wind Energy Conversion Systems (WECS) which include Horizontal Axis Wind Turbines (HAWT). These have been extensively used due to their high efficiency and low noise [20].

HAWTs operation can be associated primarily with wind speed. Prior studies [12,15] have established a partition into different regions of operation. Region 1 occurs when the wind speed is below the cut-in limit and the rotor cannot

D. F. Cortes Tobar et al. (Eds.): AETA 2019, LNEE 685, pp. 501–510, 2021.
https://doi.org/10.1007/978-3-030-53021-1_51

rotate at a useful speed. In region 2, system rotation occurs, and power can be extracted. The control objective in this region is to maximize energy capture, normally modifying the generator torque while pitch angles are fixed in their optimal value. Several approaches have been proposed to control wind turbines in this region. In [19] two different strategies are proposed using an aerodynamic torque observer to improve energy capture, one is based on the torque error and the other uses the estimated power loss. In [21] a two-step controller is proposed, one to track tip-speed ratio and individual blade pitch controllers to reduce cyclic blade bending moments. In [7] an adaptive sliding mode control is formulated to deal with system non-linearities, uncertainties and disturbances while reducing the structural loads. Also, [1,8] propose a sliding mode control for maximum power point tracking. In [6] a dual GPI observer-based control scheme is proposed to maximize energy capture under actuator faults.

Region 3 is reached when the HAWT rotates at rated speed while nominal power is extracted. Here the controller focuses on power regulation and load reduction adjusting the blades pitch angle. Typically, the same reference for each blade is given (collective pitch control) [16]. This approach exhibits limitations to control fatigue loads in structural components, leading to more complex strategies, considering individual pitch control. This strategy has been intensively explored in the last years to mitigate structural wear of blades and tower [11,18]. In [10] a three-loop controller is designed, the inner one controls the power train speed, the second loop considers individual bending moments of each blade for load reduction. Finally, the outer loop is intended to reduce tower bending moments. To avoid interference caused by individual tuning of each loop, a procedure to tune several controllers simultaneously is applied, balancing the three control goals, considering this task as a multi-objective optimization problem. In [4,5] several control schemes are developed based on the paradigm of active disturbance rejection, where periodic loads of the rotor and tower are attenuated through the design in the angular(azimuth) spatial-domain providing periodic disturbance rejection even under varying frequency conditions.

Nevertheless, most of the reported control strategies consider region 2 and 3 individually, leading to abrupt transitions of controllers caused by rapid wind speed variations, and unknown effects caused by the switching between controllers. This fact has been addressed in limited publications [2] including a transition region 2.5, which is the most used nowadays. The problem of switching between an on-line controller to another off-line might lead to undesired behaviors, fast degradation or even system instability, due to abrupt changes of the input signal [22]. This approach deals with the commutation of controllers between operation regions using the bumpless transfer method.

In this study, a bumpless transfer control scheme is proposed and developed in detail to provide a smooth transition between the torque controllers operating in regions 2 and 3. The proposed strategy smoothly switches between those controllers reducing rotor bending moments and damage equivalent loads.

This paper is organized as follows. Section 2 describes the wind turbine model. Section 3 details the control schemes used in region 2 and 3. Section 4 presents the control design for transitions between region 2 and 3 and vice versa. Section 5 is intended to present the experimental results of the controllers implemented in the FAST code environment. Finally, Sect. 6 shows the conclusions of the work.

2 Wind Turbine Model

2.1 Aerodynamic Model

HAWTs generate power from the interaction between wind flow and blades on the nacelle, generating kinetic energy captured by an electric generator. Power extracted from the rotor $P_r(t)$, rotation torque $T_r(t)$ and the thrust force on the nacelle transmitted to the tower $F_T(t)$, are given by the following equations:

$$P_r(t) = \frac{1}{2}\rho A V_w^3(t) C_P\left(\lambda(t), \beta(t)\right), \tag{1}$$

$$T_r(t) = \frac{1}{2}\rho A R V_w^2(t) C_Q\left(\lambda(t), \beta(t)\right), \tag{2}$$

$$F_T(t) = \frac{1}{2}\rho A V_w^2(t) C_T\left(\lambda(t), \beta(t)\right), \tag{3}$$

where ρ is the air density, A the rotor swept area, $V_w(t)$ the wind speed, R the rotor radius, $C_P\left(\lambda(t), \beta(t)\right)$, $C_Q\left(\lambda(t), \beta(t)\right)$ and $C_T\left(\lambda(t), \beta(t)\right)$ are the power, torque and thrust coefficients, as a function of the blades angle $\beta(t)$ and the tip speed ratio $\lambda(t)$, given by:

$$\lambda(t) = \frac{R\omega_r(t)}{V_w(t)} \tag{4}$$

where $\omega_r(t)$ is the rotor speed. C_P, C_Q and C_T values depend on the HAWT design features.

2.2 Drive-Train Model

The drive-train subsystem can be represented as two masses connected with a flexible shaft and a gear train, as described in the following equations:

$$J_r\dot{\omega}_r(t) = T_r(t) + \frac{B_{dt}}{N_g}\omega_g(t) - K_{dt}\theta_\Delta(t) - (B_{dt} + B_r)\omega_r(t) \tag{5}$$

$$J_g\dot{\omega}_g(t) = \frac{K_{dt}}{N_g}\theta_\Delta(t) + \frac{B_{dt}}{N_g}\omega_r(t) - \left(\frac{B_{dt}}{N_g^2} + B_g\right)\omega_g(t) - T_g(t) \tag{6}$$

$$\dot{\theta}_\Delta(t) = \omega_r(t) - \frac{1}{N_g}\omega_g(t) \tag{7}$$

where J_r, $\omega_r(t)$, J_g and $\omega_g(t)$ are the rotational inertia and angular speed of rotor and generator shafts, respectively, B_{dt}, B_r and B_g are the damping coefficients of the drive, rotor and generator, N_g is the gear ratio between the low and high speed shafts, K_{dt} is the drive-train stiffness, and $\theta_\Delta(t)$ is the torsion angle of the shaft [6].

2.3 Blade Pitch Model

An hydraulic actuator is connected to each blade for pitch movements. According to [3], this can be represented by a second order system, with damping ratio ζ, natural frequency ω_n and angular reference β_{ref}, as shown below:

$$\ddot{\beta}(t) = -2\zeta\omega_n\dot{\beta}(t) - \omega_n^2\beta(t) + \omega_n^2\beta_{ref}(t). \tag{8}$$

2.4 Generator Model

Generator torque T_g dynamics describes a first order behavior, given by:

$$\dot{T}_g(t) = -\frac{1}{\tau_g}T_g(t) + \frac{1}{\tau_g}T_{g,ref}(t) \tag{9}$$

with τ_g as the time constant, and $T_{g,ref}$ as the reference torque for the generator. Additionally, for the calculation of the generated power P_g, efficiency η_g is considered, as follows:

$$P_g(t) = \eta_g\omega_g(t)T_g(t). \tag{10}$$

3 Standard Control Schemes for HAWT

3.1 Standard Torque Control for Region 2

In this strategy the blade pitch angle is fixed to its optimum value β_{opt} and the generator torque control law is governed by

$$T_{g,ref} = k\omega_r^2, \tag{11}$$

$$k = \frac{1}{2}\rho\pi R^5\frac{C_{p_{\max}}}{\lambda_{opt}^3}, \tag{12}$$

where k is the optimum control gain [17].

3.2 Standard Collective Pitch Control for Region 3

A gain scheduling PI control is the industrial standard for collective pitch control of wind turbines operating in region 3. The baseline collective pitch Gain Scheduling (GS) PI controller is defined as [14]:

$$\Delta\beta_d(t) = K_P(\beta)N_g\omega_r(t) + K_I(\beta)\int_0^t N_g\omega_r(t)dt, \tag{13}$$

with,

$$K_P(\beta) = \frac{2\left(J_r + N_g^2J_g\right)\omega_N\zeta_{pi}\omega_{pi}}{N_g\left(-\frac{\partial P_r}{\partial\beta}\right)} \tag{14}$$

$$K_I(\beta) = \frac{\left(J_r + N_g^2J_g\right)\omega_N\omega_{pi}^2}{N_g\left(-\frac{\partial P_r}{\partial\beta}\right)} \tag{15}$$

where, ζ_{pi} is the desired damping ratio, ω_{pi} rad/s is the desired natural frequency and $\frac{\partial P_r}{\partial \beta}$ is the partial derivative of the rotor power P_r respect to the collective blade pitch angle β in watt/rad, which is scheduled by means of a look-up table. Notice that this control law schedules the control gains by means of β. On the other hand, the generator torque changes as function of the nominal power P_0 and the generator speed ω_g: $T_{g,ref} = P_0/\omega_g$.

4 Bumpless Transfer Control Design

In order to provide a smooth transition between the controllers of the wind turbine, a control scheme for bidirectional bumpless transfer is proposed. Figure 1 shows a detailed schema of the proposed control scheme.

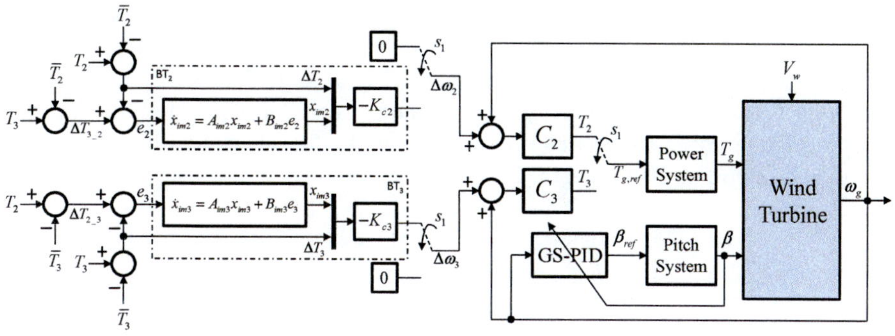

Fig. 1. Control scheme for bidirectional bumpless transfer in the wind turbine.

4.1 Controller Design for the Transition from Region 3 to Region 2

According to the bumpless transfer strategy shown in Fig. 1, a smooth transition from region 3 to region 2 can be obtained if the off-line controller C_2 is considered as a dynamic system designed to follow the reference signal T_3 provided by the on-line controller C_3, under multiple disturbances. To accomplish this task, the bumpless transfer block (BT_2) will be designed in the following.

For region 2, the Standard Torque Control (STC) is used for the Bumpless transfer design. Here, the blade pitch angles are fixed to their optimal value β_{opt}, while the generator torque is adjusted according to

$$T_2 = K_2\omega_2^2, \tag{16}$$

where ω_2 denotes the generator angular speed of the wind turbine and K_2 is the controller optimum gain. Then, a linear model of the controller is obtained using the Taylor series around the angular speed $\bar{\omega}_2$; which is the transition point between regions, leading to

$$T_2 = K_2\bar{\omega}_2^2 + (2K_2\bar{\omega}_2)(\omega_2 - \bar{\omega}_2). \tag{17}$$

Replacing $\bar{T_2} = K_2\bar{\omega}_2^2$ and rewriting in terms of torque and angular speed deviations, given by $\Delta T_2 = T_2 - \bar{T_2}$ and $\Delta\omega_2 = \omega_2 - \bar{\omega_2}$, it results in:

$$\frac{\Delta T_2}{\Delta\omega_2} = 2K_2\bar{\omega}_2. \tag{18}$$

In order to implement the bumpless transfer structure, a first order dynamics is added to (18), where τ_2 is chosen to place the new pole in a non-dominant location, obtaining:

$$C_2(s) = \frac{2K_2\bar{\omega}_2}{\tau_2 s + 1}. \tag{19}$$

From (19), a state-space representation is given by:

$$\dot{x}_{c_2} = A_{c_2}x_{c_2} + B_{c_2}\Delta\omega_2 \tag{20}$$

$$\Delta T_2 = C_{c_2}x_{c_2} + D_{c_2}\Delta\omega_2 \tag{21}$$

where $A_{c_2} = -1/\tau_2$, $B_{c_2} = 2K_2\bar{\omega}_2/\tau_2$, $C_{c_2} = 1$ and $D_{c_2} = 0$. In order to track/reject ramp type signals, a state-space controller with double integral action designed using the internal model principle [9]. Then, a state-space representation of the double integral is defined as follows:

$$\dot{x}_{im2} = A_{im2}x_{im2} + B_{im2}e_2 \tag{22}$$

$$y_{im2} = C_{im2}x_{im2} \tag{23}$$

where, $A_{im2} = \begin{bmatrix} 0\,1 \\ 0\,0 \end{bmatrix}$, $B_{im2} = \begin{bmatrix} 0 \\ 1 \end{bmatrix}$, $C_{im2} = \begin{bmatrix} 1\,0 \\ 0\,1 \end{bmatrix}$, and $e_2 = \Delta T_{3\_2} - \Delta T_2$. Now, an augmented system containing the state of the system (20) and the states of (22) is found:

$$\frac{d}{dt}\begin{bmatrix} x_{c_2} \\ x_{im2} \end{bmatrix} = A_{a2}\begin{bmatrix} x_{c_2} \\ x_{im2} \end{bmatrix} + B_{a2}\begin{bmatrix} \Delta\omega_2 \\ e_2 \end{bmatrix} \tag{24}$$

where $A_{a2} = \begin{bmatrix} A_{c_2} & 0 \\ -B_{im2}C_{c_2} & A_{im2} \end{bmatrix}$, and $B_{a2} = \begin{bmatrix} B_{c2}\; B_{im2} \end{bmatrix}$. Then, the following control law is proposed for the system (24):

$$\Delta\omega_2 = -K_{c_2}\begin{bmatrix} x_{c_2} \\ x_{im2} \end{bmatrix} \tag{25}$$

where K_{c_2} is selected such that the matrix $(A_{a2} - B_{a2}K_{c_2})$ is Hurwitz. In order to find appropriate values of K_{c_2}, the Linear Quadratic Regulator (LQR) approach is chosen, where Q_{c_2} and R_{c_2} are positive definite matrices with appropriate dimensions.

4.2 Controller Design for Transition from Region 2 to Region 3

For region 3, the base controller C_3 is defined by

$$T_3 = \frac{30}{\pi}\frac{P_0}{\omega_3}, \tag{26}$$

where ω_3 denotes the generator angular speed in the region 3 and P_0 is the nominal power of the wind turbine. To design the bumpless transfer from region 2 to region 3, a linear approximation of C_3 given by the first order Taylor expansion is obtained:

$$T_3 = \frac{30P_0}{\pi\omega_3} + \left.\frac{\partial T_3}{\partial \omega_3}\right|_{\omega_3=\bar{\omega}_3} (\omega_3 - \bar{\omega}_3), \tag{27}$$

$$T_3 = \bar{T}_3 - \frac{30P_0}{\pi\bar{\omega}_3} (\omega_3 - \bar{\omega}_3), \tag{28}$$

where $\bar{\omega}_3$ denotes the transition point between region 2 and region 3, and $\bar{T}_3 = \frac{30}{\pi}\frac{P_0}{\bar{\omega}_3}$. Replacing $\Delta T_3 = T_3 - \bar{T}_3$ and $\Delta\omega_3 = \omega_3 - \bar{\omega}_3$ leads to

$$\frac{\Delta T_3}{\Delta\omega_3} = -\left(\frac{30}{\pi}\frac{P_0}{\bar{\omega}_3}\right). \tag{29}$$

Then, a first order dynamics is added with time constant τ_3, thus:

$$C_3(s) = \frac{-\left(\frac{30}{\pi}\frac{P_0}{\bar{\omega}_3}\right)}{\tau_3 s + 1}, \tag{30}$$

which can be represented as:

$$\dot{x}_{c_3} = A_{c_3} x_{c_3} + B_{c_3}\Delta\omega_3 \tag{31}$$

$$\Delta T_3 = C_{c_3} x_{c_3} \tag{32}$$

where $A_{c_3} = -1/\tau_3$, $B_{c_3} = -\left(\frac{30}{\pi}\frac{P0}{\bar{\omega}_3}\right)/\tau_3$, and $C_{c_3} = 1$. Then, a state-space controller with double integral action is considered and defined as:

$$\dot{x}_{im3} = A_{im3} x_{im3} + B_{im3} e_3, \tag{33}$$

$$y_{im3} = C_{im3} x_{im3}, \tag{34}$$

where $A_{im3} = \begin{bmatrix} 0 & 1 \\ 0 & 0 \end{bmatrix}$, $B_{im3} = \begin{bmatrix} 0 \\ 1 \end{bmatrix}$, $C_{im3} = \begin{bmatrix} 1 & 0 \\ 0 & 1 \end{bmatrix}$, and $e_3 = \Delta T_{2\_3} - \Delta T_3$. Thus, an augmented system is then formulated:

$$\frac{d}{dt}\begin{bmatrix} x_{c3} \\ x_{im3} \end{bmatrix} = A_{a3}\begin{bmatrix} x_{c3} \\ x_{im3} \end{bmatrix} + B_{a3}\begin{bmatrix} \Delta\omega_3 \\ e_3 \end{bmatrix} \tag{35}$$

where $A_{a3} = \begin{bmatrix} A_{c_3} & 0 \\ -B_{im3}C_{c_3} & A_{im3} \end{bmatrix}$, $B_{a3} = \begin{bmatrix} B_{c3} \; B_{im3} \end{bmatrix}$. Then, a control law for the system (35) is proposed:

$$\Delta\omega_3 = -K_{c_3}\begin{bmatrix} x_{c3} \\ x_{im3} \end{bmatrix}; \tag{36}$$

where K_{c_3} is selected such that the matriz $(A_{a3} - B_{a3}K_{c_3})$ is Hurwitz, and the values of K_{c_3} can be found under the LQR tuning procedure.

5 Experimental Results

For validation purposes, the 5MW wind turbine simulator provided by FAST (Fatigue, Aerodynamics, Structure and Turbulence) Code [13] was used. The required design parameters for controller design and simulation are summarized in Table 1. To test the effectiveness of the proposed bumpless control scheme, several A-class Kaimal turbulent wind profiles generated with TurbSim were used. These wind profiles cause several fluctuations between region 2 and region 3. Simulation results reveal the smoother response of the proposed scheme vs the conventional torque control with region 2.5. This last exhibits stronger fluctuations in the rotor torque, that influence the structural damage due to fatigue loads. Figure 2 shows a summary of several simulations under different wind profiles, where Mlife package was used to calculate damage equivalent loads all over the rotor structure. The Figure shows that the proposed scheme provides reduced damage equivalent loads due to the smooth transition between regions.

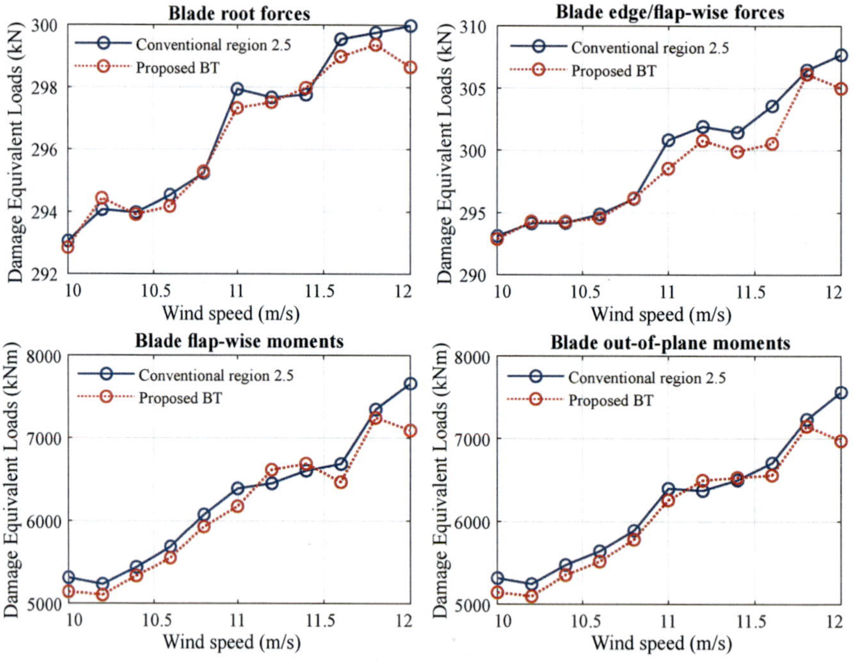

Fig. 2. Damage equivalent loads of rotor blades for different wind profiles.

Table 1. Design parameters.

Parameter	Value	Parameter	Value
$\bar{\omega}_2 = \bar{\omega}_3$	1161.9632 rpm	R_{c2}	9000
K_2	0.025576386 Nm/rpm^2	P_0	5.2966 MW
$\tau_2 = \tau_3$	1.25	Q_{c3}	diag([1, 1, 9000])
Q_{c2}	diag([0.01, 0.001, 80])	R_{c3}	900

6 Conclusions

In this work, a bumpless transfer control scheme was proposed to smoothly transfer between the most used torque controllers for operating region 2 and 3 in horizontal-axis wind turbines. The strategy showed important reductions of damage equivalent loads in the rotor structure, such as: blade root forces, edge-wise forces, flap-wise forces, out-of-plane moments and flap-wise moments.

References

1. Beltran, B., El Hachemi Benbouzid, M., Ahmed-Ali, T.: Second-order sliding mode control of a doubly fed induction generator driven wind turbine. IEEE Trans. Energy Convers. **27**(2), 261–269 (2012). https://doi.org/10.1109/TEC.2011.2181515
2. Chen, Q.: Improved wind turbine control strategies for maximizing power output and minimizing power flicker, Master Thesis (2014). http://citeseer.ist.psu.edu/viewdoc/citations;jsessionid=5DB9E7356A1DF09501D5BD6321088AC9?doi=10.1.1.682.7300
3. Sloth, C., Esbensen, T., Stoustrup, J.: Robust and fault-tolerant linear parameter-varying control of wind turbines. Mechatronics **21**(4), 645–659 (2011). https://doi.org/10.1016/j.mechatronics.2011.02.001
4. Coral-Enriquez, H., Cortés-Romero, J.: Spatial-domain active disturbance rejection control for load mitigation in horizontal-axis wind turbines. In: 2016 IEEE Conference on Control Applications (CCA), pp. 575–580 (2016). https://doi.org/10.1109/CCA.2016.7587891
5. Coral-Enriquez, H., Cortés-Romero, J., Dorado-Rojas, S.A.: Rejection of varying-frequency periodic load disturbances in wind-turbines through active disturbance rejection-based control. Renew. Energy **141**, 217–235 (2019). https://doi.org/10.1016/j.renene.2019.04.001. http://www.sciencedirect.com/science/article/pii/S0960148119304744
6. Coral-Enriquez, H., Cortés-Romero, J., Ramos, G.A.: Robust active disturbance rejection control approach to maximize energy capture in variable-speed wind turbines. Math. Probl. Eng. **2013**, 12 (2013). (Article ID 396740), https://doi.org/10.1155/2013/396740
7. Evangelista, C., Puleston, P., Valenciaga, F., Dávila, A.: Variable gains super-twisting control for wind energy conversion optimization. In: 2010 11th International Workshop on Variable Structure Systems (VSS), pp. 50–55 (2010). https://doi.org/10.1109/VSS.2010.5544713

8. Evangelista, C., Puleston, P., Valenciaga, F., Fridman, L.: Lyapunov designed super-twisting sliding mode control for wind energy conversion optimization. IEEE Trans. Ind. Electron. **PP**(99), 1 (2012). https://doi.org/10.1109/TIE.2012.2188256
9. Francis, B.A., Wonham, W.M.: The internal model principle of control theory. Automatica **12**(5), 457–465 (1976)
10. Gambier, A., Behera, A.: Integrated pitch control system design of a wind turbine by using multiobjective optimization. IFAC-PapersOnLine **51**(28), 239–244 (2018). https://doi.org/10.1016/j.ifacol.2018.11.708. http://www.sciencedirect.com/science/article/pii/S240589631833427X
11. Geyler, M., Caselitz, P.: Individual blade pitch control design for load reduction on large wind turbines. In: European Wind Energy Conference (EWEC 2007), Milano, Italy, May, pp. 7–10 (2007)
12. Johnson, L.P.K.: Control of wind turbines. Approaches, challenges, and recent developments. IEEE Control Syst. Mag. **58**(4), 44–62 (2011)
13. Jonkman, J.M., Buhl Jr., M.L., et al.: Fast user's guide. National Renewable Energy Laboratory, Golden, CO, Technical report No. NREL/EL-500-38230 (2005)
14. Jonkman, J.M., Butterfield, S., Musial, W., Scott, G.W.: Definition of a 5-MW reference wind turbine for offshore system development. Technical report NREL/TP-500-38060, National Renewable Energy Laboratory, 1617 Cole Boulevard, Golden, Colorado 80401-3393 (2009)
15. Laks, J.H., Pao, L.Y., Wright, A.D.: Control of wind turbines: past, present, and future. In: American Control Conference, 2009. ACC 2009, pp. 2096–2103. IEEE (2009)
16. Njiri, J.G., Söffker, D.: State-of-the-art in wind turbine control: trends and challenges. Renew. Sustain. Energy Rev. **60**, 377–393 (2016). https://doi.org/10.1016/j.rser.2016.01.110. http://www.sciencedirect.com/science/article/pii/S1364032116001404
17. Pao, L.Y., Johnson, K.E.: Control of wind turbines: approaches, challenges, and recent developments. IEEE Control Syst. Mag. **31**(2), 44–62 (2011). https://doi.org/10.1109/MCS.2010.939962
18. Petrović, V., Jelavić, M., Baotić, M.: Advanced control algorithms for reduction of wind turbine structural loads. Renew. Energy **76**, 418–431 (2015)
19. Pierce, K.: Control method for improved energy capture below rated power. Technical report NREL/CP-500-26322, National Renewable Energy Laboratory (1999)
20. Song, D., Fan, X., Yang, J., Liu, A., Chen, S., Joo, Y.H.: Power extraction efficiency optimization of horizontal-axis wind turbines through optimizing control parameters of yaw control systems using an intelligent method. Appl. Energy **224**, 267–279 (2018). https://doi.org/10.1016/j.apenergy.2018.04.114. http://www.sciencedirect.com/science/article/pii/S0306261918306755
21. Stol, K.A.: Disturbance tracking control and blade load mitigation for variable-speed wind turbines. J. Solar Energy Eng. **125**(4), 396–401 (2003)
22. Yame, J., Kinnaert, M.: Parameterization of linear controllers for bumpless switching in multicontroller schemes. In: AIAA Guidance, Navigation, and Control Conference and Exhibit, p. 5108 (2004)

Planar Cavity-Backed Antenna Prototype by Groove Waveguide Technique

H. F. Guarnizo Mendez[1](✉), M. A. Polochè Arango[2], T. A. Rubiano Suazo[1], S. H. Rojas Martínez[1], and F. J. Gutiérrez Bernal[2]

[1] Universidad El Bosque, Bogotá, Colombia
{hguarnizo,trubianos,shrojas}@unbosque.edu.co
[2] Universidad de San Buenaventura, Bogotá, Colombia
{mpoloche,fgutierrez}@usbbog.edu.co

Abstract. A new Cavity-Backed Antenna prototype compatible with planar technology is presented is this paper. The topology and circuit design are presented. The antenna is completely manufactured in hard-aluminum. The Cavity-Backed Antenna prototype is designed by using Groove waveguide technique. The new antenna prototype consists of a slotted cavity fed by a microstrip line. Two prototypes were designed and manufactured. The first prototype was designed with three metal walls and the second prototype has been manufactured with two metal walls. In this new antenna prototype the specific slot patterns on the cavity back face such as edge slots and meandered are combined. An example with a minimum 10 dB return loss bandwidth of 1.3%, 6.5 dBi gain, and 21.2 dB front to back ratio has been presented. The new Cavity-Backed Antenna prototype can be used in hyperthermia treatment.

Keywords: Cavity Backed · Low profile · Slot antenna · Groove waveguide technique

1 Introduction

Due to both the wireless communication development and the communication systems trend, robust and low profile antenna and with good radiation performance has a high demand [1, 2]. The printed patch antennas appear as a narrowband option initially [3]. Later, the slot antennas are used with the aim of improving the bandwidth without degrading the antenna performance [4]. Besides, slot antennas have been studied in many researches works due to low profile, easy integration with planar circuits, conformability with planar or curved surfaces and better isolation of the power supply network. A shallow cavity is placed behind the slot antenna to eliminate the backside radiation. In [5–7], methods for reducing the cavity dimensions of a slot antenna backed by a rectangular cavity are presented, methods for reducing the cavity dimensions of a slot antenna backed by a rectangular cavity are presented.

The advancement in the passive circuit's context has led to the new technology development such as the Groove waveguide technology. Groove waveguide technology

D. F. Cortes Tobar et al. (Eds.): AETA 2019, LNEE 685, pp. 511–520, 2021.
https://doi.org/10.1007/978-3-030-53021-1_52

emerged within research works about soft and hard surfaces [8]. Different contributions have been reported in [9–13]. For the slot antennas feed in Groove Waveguide Technology research works have been carried out in [14–16].

In this paper are presented two planar cavity-backed antenna prototypes by groove waveguide technique. The resonant frequency of two prototypes is 2.45 GHz. The two prototypes were designed and manufactured. In these prototypes the properties of the cavity TE_{101} mode are exploited to obtain the smallest possible cavity.

The prototypes developed can be use int hyperthermia treatment. In the hyperthermia treatment based on electromagnetic energy, waveguides or applicators (antennas) are used to couple the electromagnetic energy to the tissues. The applicators used are the array tapered microstrip patch antenna, array antipodal tapered slot, array of rectangular patch elements, antipodal Vivaldi antennas, miniaturized cavity-backed patch antenna, antenna with an embedded EBG structure, and applicators based on metamaterial technology [17–25].

2 Topology and Circuit Design

2.1 Rectangular Cavity Design

The rectangular cavity was designed by using the classic equations of waveguides and cavities [26]. The rectangular cavity was designed for a resonant frequency of 2.45 GHz. The TE_{101} mode is defined as the rectangular cavity resonant mode. A rectangular waveguide (see Fig. 1) short circuited at both ends can be seen as a cavity.

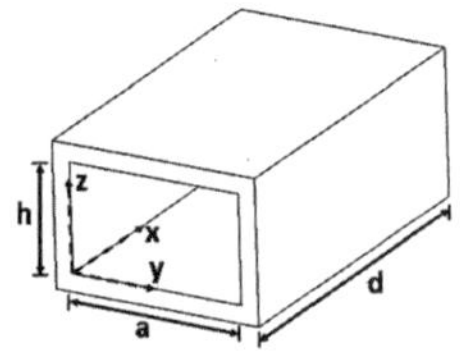

Fig. 1. A Cavity design procedure.

The width (a) of the cavity, the height of the substrate (h), propagation constant (β_{mn}), the guided wavelength (λ_g), the length (d) of the cavity was obtained according to method presented in [27].

The difference between higher order mode rectangular cavity resonant frequency and fundamental mode TE_{101} is presented in Fig. 2. In Fig. 2 the value of f_O remains fixed while the value of f_C varies.

In Fig. 2 can be seen that the TE_{102} and TE_{201} modes are higher modes closest to the TE_{101} mode. The rectangular cavity substrate is air ($\varepsilon_r = 1$, h = 1.6 mm).

2.2 External Coupling to the Rectangular Cavity Using CPW Access

In the cavity's upper face (see Fig. 3 and Fig. 5) is used a transition from microstrip to cavity, this transition is described in [28]. In the coplanar waveguide (CPW) rectangular

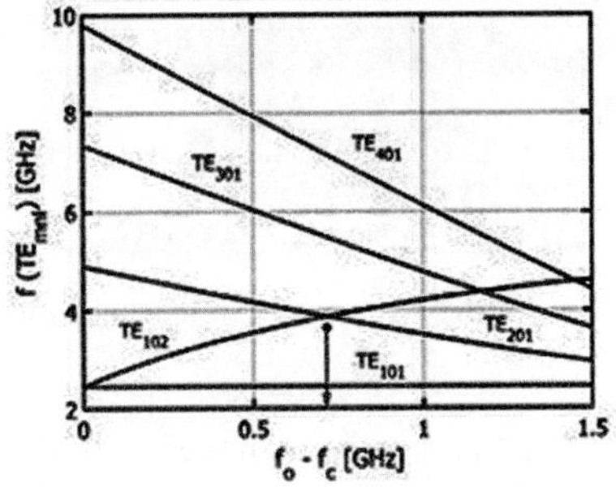

Fig. 2. Rectangular cavity resonant modes as a function of frequency difference.

cavity access a 50 Ω Microstrip line (length Sel) is coupled (see Fig. 3 and Fig. 5). To convert the quasi-TEM of the microstrip line into the TE_{101} of the Groove cavity is used the length Cl (see Fig. 3 and Fig. 5). The antenna is not shown in Fig. 3 because it is in patent process.

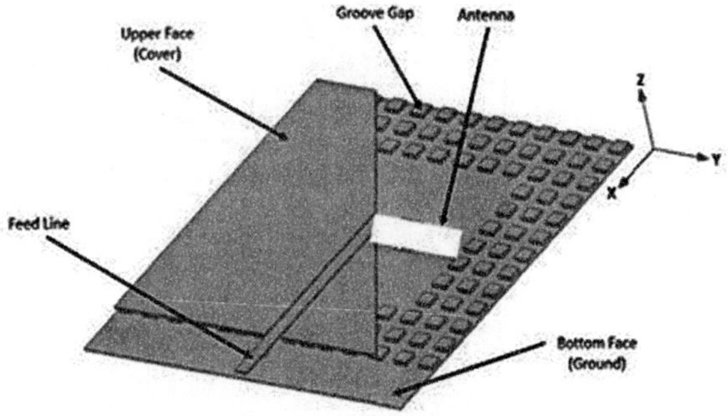

Fig. 3. 3D view of the implemented circuit.

In Fig. 3 and Fig. 5, it can be seen that the slot meandered is located in the cavity center. This location is for the purpose of exploiting the properties of the TE_{101} mode, that is, the highest of electric field concentration is in the cavity center.

2.3 Prototype Design Layout

A compact version of a slot antenna was inserted onto the cavity back metallic plane (bottom face).

2.4 Groove Waveguides Technology

The first step when designing a device in groove waveguide technology (see Fig. 4) is the selection of the periodic structure to be used as AMC (Artificial Magnetic Conductor). The analysis of the dispersion diagram of the unit cell of the structure gives the

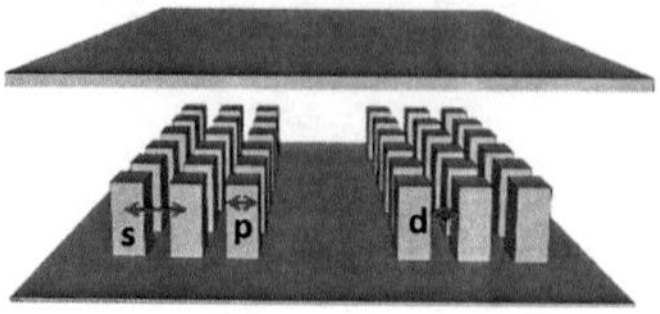

Fig. 4. 3D view of the groove waveguides technology [30].

information about the stop band that can be achieved with it. The target frequency (in this case 2.45 GHz) must be within that stop band [29].

The clearance (d, see Fig. 4 and Fig. 5) between two neighboring square posts and the s value are important factors in obtaining negligible leaks.

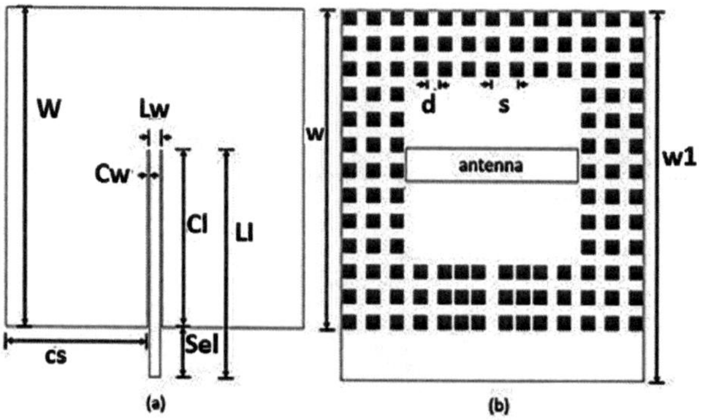

Fig. 5. Circuit topology. (a) Top view. (b) Bottom view.

3 Simulated and Experimental Results

Using the structures previously described, two prototypes were designed and manufactured.

The two prototypes were designed for a resonant frequency of 2.45 GHz. A square cavity of 86.58 mm was obtained from design procedure ($f_O - f_C = 717.5$ MHz) presented in [27]. Cavity lateral walls were implemented by using square metal (see Fig. 3).

The clearance (d) and width (p) were chosen according to [1, 31]. Electromagnetic simulations were carried out with the Ansoft High Frequency Structure Simulator (HFSS).

3.1 Prototype Design 1

The prototype design 1 was implemented with three cavity lateral walls. Figure 5 shows the prototype design 1 topology. The prototype design 1 was manufactured in hard-aluminum (see Fig. 6). The antenna is not shown in Fig. 5 and Fig. 6 because it is in patent process.

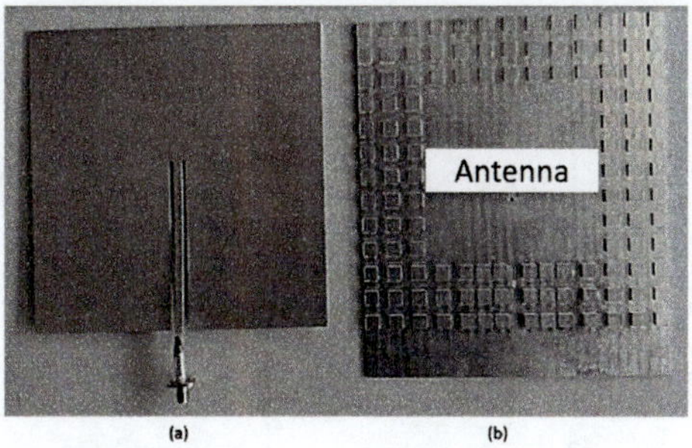

Fig. 6. Photography of the manufactured prototype 1. (a) Top view. (b) Bottom view.

The simulated S_{11} parameter of the prototype design 1 is presented in Fig. 9. From the simulation carried out of the prototype design 1, a resonance frequency of 2.45 GHz, a fractional bandwidth of 1.3%, a gain of 6.7 dBi, a VSW of 1.2 and a FTBR of 21.2 dB are obtained. The Table 2 summarizes simulation results. The prototype design 1 dimensions are listed in Table 1.

The photography of the manufactured prototype 1 is presented in Fig. 6.

Table 1. Prototype design 1: physical parameters [mm].

W	146.58	Cs	69.69
Lw	6.86	Sel	23.00
LI	104.00	d	6.00
Cw	1.00	s	11.58
Cl	81.00	W1	168.80

Table 2. Simulated electrical features of prototype design 1

	Frac. BW –10 dB	f_0 (GHz)	Gain (dBi)	FTBR (dB)	VSW (f_0)
Prototype Design 1	1.3%	2.450	6.7	21.2	1.2

3.2 Prototype Design 2

The prototype design 2 was implemented with two cavity lateral walls. The prototype design 2 was manufactured in hard-aluminum too (see Fig. 8). Figure 7 shows the prototype design 2 topology. The antenna is not shown in Fig. 7 and Fig. 8 because it is in patent process.

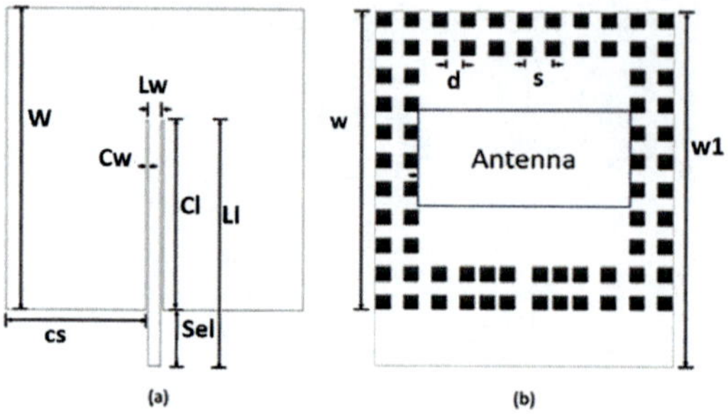

Fig. 7. Circuit topology: (a) Top view. (b) Bottom view.

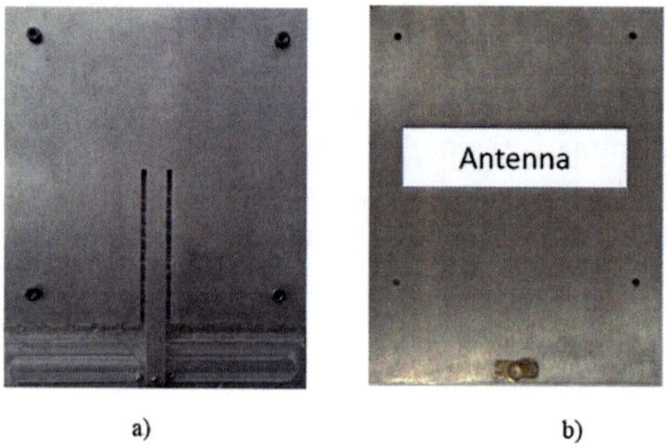

Fig. 8. Photography of the manufactured prototype 2. (a) Top view. (b) Bottom view.

In Table 3, the physical parameters of the prototype design 2 have changed with respect to the physical parameters obtained in the prototype design 1. This change is because when the two metal walls were implemented the electrical or magnetic energy in the cavity changed and therefore the resonance frequency also. To obtain the frequency resonance again to 2.45 GHz, a magnetic charge is introduced by the meandered slot when its the physical parameters was changed. Pozar presents the expressions for the change in the resonant frequency of an arbitrary cavity when its shape is slightly modified in terms of the stored energies [26]. The prototype design 2 dimensions are listed in Table 3.

Table 3. Prototype design 2: physical parameters [mm].

W	122.58	Cs	57.69
Lw	6.86	Sel	23.00
Ll	100.00	d	6.00
Cw	1.00	s	11.58
Cl	77.00	W1	145.22

The simulated S_{11} parameter of the prototype design 2 is presented in Fig. 9. From the simulation carried out of the prototype design 2, a resonance frequency of 2.45 GHz, a fractional bandwidth of 1.4%, a gain of 6.7 dBi, VSW of 1.2 and a FTBR of 25.4 dB are obtained. The Table 4 summarizes simulation results.

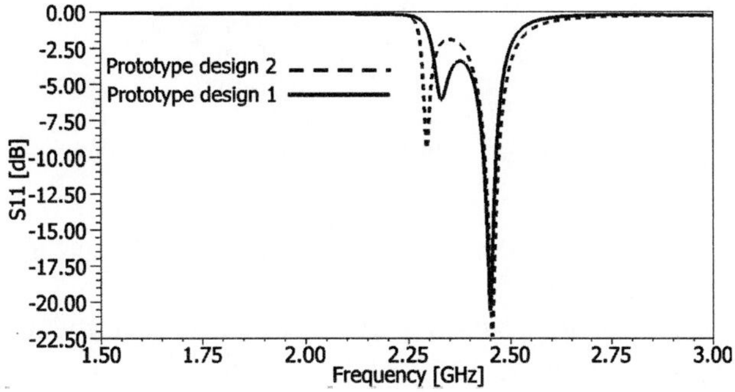

Fig. 9. Simulated return loss [S_{11}] for the prototype design 1 and prototype design 2.

Table 4. Electrical features of the prototype design 2.

	Frac. BW −10 dB	f_0 (GHz)	Gain (dBi)	FTBR (dB)	VSW (f_0)
Prototype Design 2	1.4%	2.454	6.7	25.4	1.12

In Fig. 9, it can be seen that a resonance frequency of 2.45 GHz is obtained with two metal walls or three metal walls. Figure 10 shows the simulated radiation pattern of prototype design 1 and 2 in the E plane.

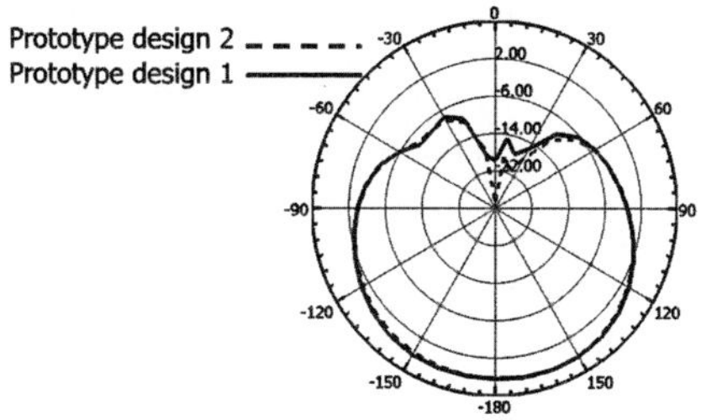

Fig. 10. Simulated radiation pattern (plane E) of prototype design 1 and prototype design 2

In the Fig. 10, it is observed that the radiation pattern obtained with two walls or three walls is the same. Figure 11 shows the simulated smith chart of prototype design 1 and 2.

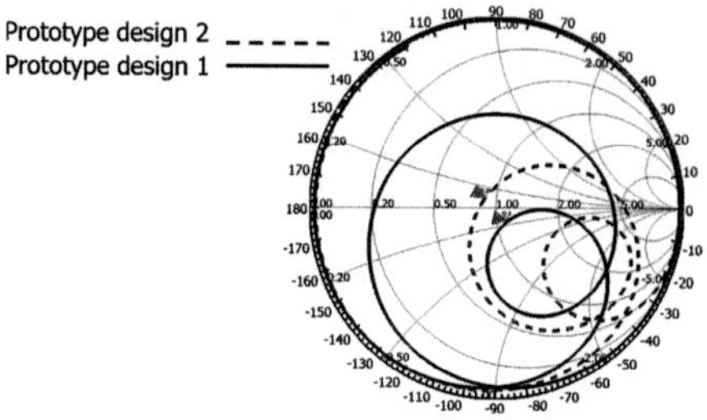

Fig. 11. Simulated smith chart of the prototype design 1 and prototype design 2

In the Fig. 11, it is observed that the decrease from 3 to 2 walls has an inductive effect on the input impedance. Moreover, there is also an increase in the coupling. In the case of prototype design 1, the simulated resonance frequency was 2.45 GHz and the measured resonance frequency was 2.4 GHz (see Fig. 12). The difference between simulated and measured resonance frequency is due to manufacturing errors, that is, there are differences between the length of the prototype design measured and the length of the prototype design 1 designed, the height of the square posts built is different from the height of the square posts designed.

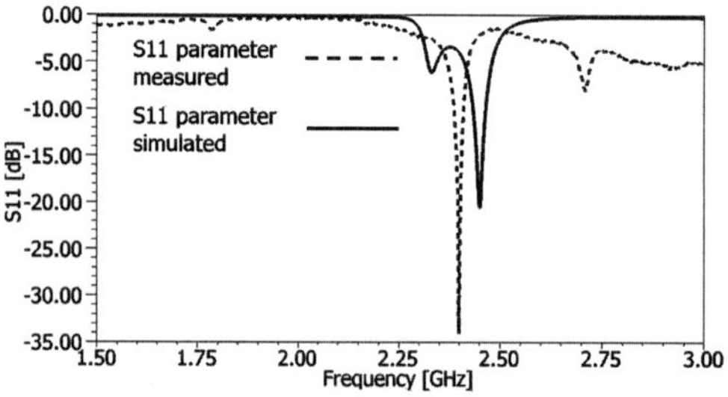

Fig. 12. S11 parameter simulated and measured of prototype design 1

The Table 5 presents the electrical features simulated and measured of the prototype design 1.

Table 5. Simulated and measured electrical features of prototype design 1

	Frac. BW −10 dB	f_0 (GHz)	VSW (f_0)
Simulated	1.3%	2.45	1.20
Measured	1.0%	2.40	1.05

4 Conclusions

Two cavity backed slot antenna prototype by using Groove waveguide technique at 2.45 GHz has been developed. The first prototype has been manufactured with three metal walls and the second prototype has been manufactured with two metal walls. The cavity backed slot antenna prototype has a simulated gain of 6.7 dBi, a fractional bandwidth measured of 1.0%, and a reflection coefficient measured lower than –20 dB. The electrical features measurements of the prototype design 2 are in process. Currently, the proposed prototype is being miniaturized with the aim of implementing it in a hyperthermia treatment.

References

1. Luo, G.Q., Hu, Z.F., Dong, L.X., Sun, L.L.: Planar slot antenna backed by substrate integrated waveguide cavity. IEEE Antennas Wirel. Propag. Lett. **7**, 236–239 (2008)
2. Yan, J., Xu, F., Cao, K., Qian, J.: Planar slot antenna based on triangle substrate integrated waveguide cavity. In: 2016 IEEE International Conference on Ubiquitous Wireless Broadband, pp. 1–3 (2016)
3. Pozar, D.M.: Microstrip antenna aperture-coupled to a microstripline. Electron. Lett. **21**, 49–50 (1985)
4. Azadegan, R., Member, S., Sarabandi, K.: A novel approach for miniaturization of slot antennas. IEEE Trans. Antennas Propag. **51**(3), 421–429 (2003)
5. Hong, W., Member, S., Behdad, N., Member, S., Sarabandi, K.: Size reduction of cavity-backed slot antennas. IEEE Trans. Antennas Propag. **54**(5), 1461–1466 (2006)
6. Andrés, C., et al.: Volume reduction of planar substrate integrated waveguide cavity-backed antennas. In: 2012 6th European Conference on Antennas Propagation, pp. 2919–2923 (2011)
7. Ghosh, B., Member, S., Haque, S.K.M., Yenduri, N.R.: Miniaturization of slot antennas using wire loading. IEEE Antennas Wirel. Propag. Lett. **12**, 488–491 (2013)
8. Kildal, P.S.: Artificially soft and hard surfaces in electromagnetics. Antennas Propag. IEEE Trans. **38**, 1537–1544 (1990)
9. Zaman, A.U., Kildal, P.: Wide-band slot antenna arrays with single-layer corporate-feed network in ridge gap waveguide technology. IEEE Trans. Antennas Propag. **62**(6), 2992–3001 (2014)
10. Al Sharkawy, M., Kishk, A.A.: Long slots array antenna based on ridge gap waveguide technology. IEEE Trans. Antennas Propag. **62**(10), 5399–5403 (2014)
11. Xi, J., Cao, B., Wang, H.: A novel 77 GHz circular polarization slot antenna using ridge gap waveguide technology. In: 2015 Asia-Pacific Microwave Conference, vol. 3, no. 61301024, pp. 1–3 (2015)
12. Sahu, A., Devabhaktuni, V., Aaen, P.H.: A slot antenna designed in ridge gap waveguide technology for v-band applications. In: 2015 IEEE MTT-S International Microwave RF Conference, pp. 385–387 (2015)
13. Zaman, A.U.: Slot antenna in ridge gap waveguide technology. In: 2012 6th European Conference on Antennas Propagation, pp. 3243–3244 (2011)
14. Zaman, A.U., Vukusic, T., Alexanderson, M., Kildal, P.: Design of a simple transition from microstrip to ridge gap waveguide suited for MMIC and antenna integration. IEEE Antennas Wirel. Propag. Lett. **12**, 1558–1561 (2013)
15. Zaman, A.U., Alfonso, E., Kildal, P.: Design of transition from coaxial line to ridge gap waveguide. In: 2009 IEEE Antennas and Propagation Society International Symposium, no. 1, pp. 1–4 (2009)

16. Giner, S.M., Valero-nogueira, A., Herruzo, J.I.H.: Excitation of untilted narrow-wall slot in groove gap waveguide by using a parasitic dipole. In: 2013 7th European Conference on Antennas Propagation, pp. 3082–3085 (2013)
17. Merunka, I., Fiser, O., Vojackova, L., Vrba, J., Vrba, D.: Utilization potential of balanced antipodal Vivaldi antenna for microwave hyperthermia treatment of breast cancer. In: 8th European Conference on Antennas Propagation, EuCAP 2014, vol. 6, no. EuCAP, pp. 706–710 (2014)
18. Korkmaz, E., Isık, O., Sagkol, H.: A directive antenna array applicator for focused electromagnetic hyperthermia treatment of breast cancer. In: 2015 9th European Conference on Antennas Propagation, vol. 1, pp. 1–4 (2015)
19. Chakaravarthi, G., Arunachalam, K.: Design and characterisation of miniaturised cavity-backed patch antenna for microwave hyperthermia. Int. J. Hyperth. **31**(7), 737–748 (2015)
20. Curto, S., Ruvio, G., Ammann, M.J., Prakash, P.: A wearable applicator for microwave hyperthermia of breast cancer: performance evaluation with patient-specific anatomic models. In: Proceedings of 2015 International Conference on Electromagnetics in Advanced Applications ICEAA 2015, pp. 1159–1162 (2015)
21. Merunka, I., Fiser, O., Vojackova, L., Vrba, J., Vrba, D.: Microwave hyperthermia treatment of neck cancer using eight UWB antennas. In: European Microwave Week 2014: Connecting the Future, EuMW 2014 - Conference Proceedings; EuMC 2014: 44th European Microwave Conference, pp. 790–793 (2014)
22. Li, J., Wang, X.: Comparison of two small circularly polarized antennas for focused microwave hyperthermia. In: 2019 13th European Conference on Antennas Propagation, no. EuCAP, pp. 5–8 (2019)
23. Stang, J., Haynes, M., Carson, P., Moghaddam, M.: A preclinical system prototype for focused microwave thermal therapy of the breast. IEEE Trans. Biomed. Eng. **59**(9), 2431–2438 (2012)
24. Vrba, D., Vrba, J.: Applicators for local microwave hyperthermia based on metamaterial technology. In: 8th European Conference on Antennas Propagation, EuCAP 2014, no. EuCAP, pp. 68–71 (2014)
25. Tao, Y., Wang, G.: Conformal hyperthermia of superficial tumor with left-handed metamaterial lens applicator. IEEE Trans. Biomed. Eng. **59**(12), 3525–3530 (2012)
26. Pozar, D.M.: Microwave Engineering, 3rd edn, pp. 298–303. Wiley, Hoboken (2005)
27. Bohórquez, J.C., et al.: Planar substrate integrated waveguide cavity-backed antenna. IEEE Antennas Wirel. Propag. Lett. **8**, 1139–1142 (2009)
28. Cheng Hao, Z., Hong, W., Ping Chen, X., Xin Chen, J., Wu, K., Jun Cui, T.: Multilayered substrate integrated waveguide (MSIW) elliptic filter. In: IEEE Microwave and Wireless Components Letters, vol. 15, no. 2, pp. 95–97 (2005)
29. Boria, V.E., Sánchez-Escuderos, D., Bernardo-Clemente, B., Berenguer, A., Baquero-Escudero, M.: Groove gap waveguide as an alternative to rectangular waveguide for H-plane components. Electron. Lett. **52**(11), 939–941 (2016)
30. Zaman, A.U., Kildal, P.: GAP Waveguides. In: Handbook of Antenna Technologies, Singapore, pp. 3273–3347 (2016)
31. Nawaz, M.I., Huiling, Z., Kashif, M.: Substrate integrated waveguide (SIW) to microstrip transition at x-band. In: Proceedings of 2014 International Conference on Circuits, Systems Control, pp. 61–63 (2014)

Design of Depth Control for Hybrid AUV

Ngoc-Huy Tran, Tan-Dat Huynh(✉), Thien-Phuong Ton, and Thai-Hoang Huynh

Ho Chi Minh City University of Technology, VNUHCM, Ho Chi Minh City, Vietnam
{tnhuy,1870464}@hcmut.edu.vn

Abstract. AUV and Glider are among the most well-known untethered autonomous underwater vehicles with their features and advantages. For the purpose of combining those individual outstanding properties, a new generation is released – Hybrid AUV. However, it is worth saying that hybrid AUV possesses its certain functionalities. This paper analyzes the motion characteristics in terms of a mathematical model and a design of depth control for hybrid AUV. The simulation results will provide evidence for the feasibility and effectiveness of the proposed method.

Keywords: Autonomous underwater vehicle · Depth control · Sliding mode control · Underwater glider · Hybrid

1 Introduction

Conquering the oceans and the other water zones has long been an appealing subject, attracting the interest of many organizations and countries as water occupies a very large area (71% of the surface of the earth is covered with water). Moreover, there are plenty of unexploited sources of energy and resources. In order to achieve that goal, the idea of developing underwater vehicles has been researched and deployed in the world early on. Initially, the equipment was built with a human pilot inside (Manned Underwater Submersible). However, the principle of safety for the operator as well as the requirements for optimizing the performance and time of operation led to the idea of specialized research for unmanned diving equipment (Unmanned Submersible).

These devices are classified into two types: wire control (tethered) and wireless control (untethered). For the first category, ROV is the most common and has recently played vital role in most of the underwater missions that were previously taken or even could not be done by human such as finding the wreck of the Titanic [1] or preventing the leakage of oil from the cracks [2]. However, the limitations of ROV such as factors from power supply and communication, cable length, dependence on operator, safety issues has resulted in the development of wireless diving equipment - typically AUV (Autonomous Underwater Vehicle). Due to its autonomous operating ability and wide range mobility, numerous applications of AUV have been realized in environmental measurements and sampling [3, 4]; regional survey and mapping [5]; surveillance [6, 7], etc. Nevertheless, the critical problem with AUV is the duration of operation since the amount of energy is limited and the batteries have to supply power continuously

D. F. Cortes Tobar et al. (Eds.): AETA 2019, LNEE 685, pp. 521–531, 2021.
https://doi.org/10.1007/978-3-030-53021-1_53

without being recharged. Therefore, a new type of underwater equipment called glider has been brought into service, which saves more energy to function in long periods of time using movable counterweights and ballast system. As a trade-off, without thruster, a glider must alternatively dive and climb to move forward. Hence, it is obvious that both AUV and glider have their own advantages and disadvantages. This paper will propose a new design called Hybrid AUV harnessing the advantages of both types. In addition, the paper also demonstrates how to mathematically model and design the depth controller for Hybrid AUV. Output performance and feasibility will be evaluated through simulation results.

2 Architecture of the Hybrid AUV

AUV2000 is a new platform into which the counterweight and ballast system (piston-cylinder) are integrated. This utilizes the form of both AUV and glider, hence we can treat AUV2000 as a Hybrid AUV (Fig. 2).

This type of AUV can control pitch angle by using a counterweight to adjust the center of gravity, much better than using stern planes as conventional types. On the other hand, performance and responsiveness are simultaneously improved.

Similarly, the ballast system composing of a piston and a cylinder allows AUV2000 to change the buoyancy, thus enhancing the diving performance. The traditional AUVs usually have to be designed with slightly greater buoyancy than their own weight, for example, the REMUS. Besides, this ballast system enables the Hybrid AUV to move vertically with relatively low velocity owing to the limited capacity of the cylinder. Detailed analysis of how the counterweight and ballast operate will be presented in the following sections.

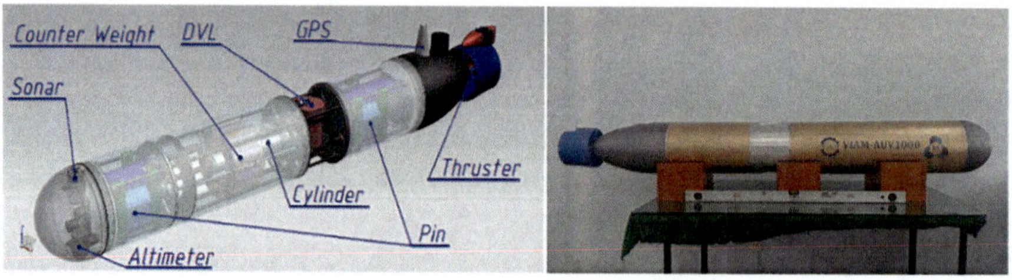

Fig. 1. Overview of AUV2000 system.

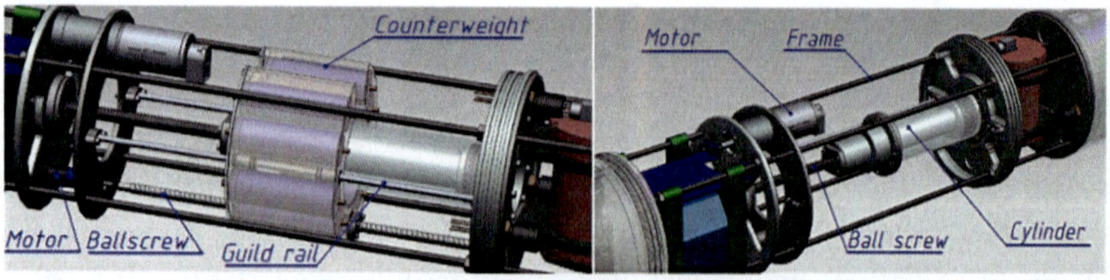

Fig. 2. Counter weight (left) and ballast (right) subsystems.

3 Dynamic Model of Hybrid AUV

A typical underwater robot with 6 degrees of freedom is described by non-linear differential equations using 12 state variables $\left[x,\ y,\ z,\ \phi,\ \theta,\ \psi,\ u,\ v,\ w,\ p,\ q,\ r\right]$ as shown in Fig. 1, according to Fossen [8] (Fig. 3).

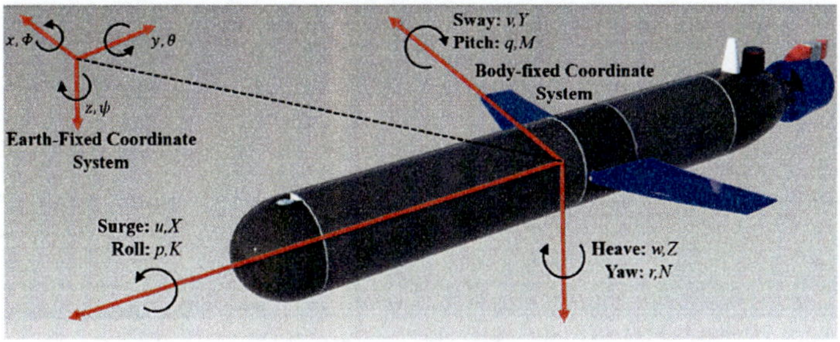

Fig. 3. The coordinate system for the designed vehicle

Due to the symmetric design of AUV2000, off-diagonal inertia moments can be considered as insignificant. The kinematic equation is simplified as:

$$\begin{array}{r} m\left[\dot{u}-vr+wq-x_G(q^2+r^2)+y_G(pq-\dot{r})+z_G(pr+\dot{q})\right]=\sum X_{ext} \\ m\left[\dot{v}-wp+ur-y_G\left(r^2+p^2\right)+z_G(qr-\dot{p})+x_G(qp+\dot{r})\right]=\sum Y_{ext} \\ m\left[\dot{w}-uq+vp-z_G\left(p^2+q^2\right)+x_G(rp-\dot{q})+y_G(rq+\dot{p})\right]=\sum Z_{ext} \\ I_{xx}\dot{p}+\left(I_{zz}-I_{yy}\right)qr+m\left[y_G(\dot{w}-uq+vp)-z_G(\dot{v}-wp+ur)\right]=\sum K_{ext} \\ I_{yy}\dot{q}+(I_{xx}-I_{zz})rp+m\left[z_G(\dot{u}-vr+wq)-x_G(\dot{w}-uq+vp)\right]=\sum M_{ext} \\ I_{zz}\dot{r}+\left(I_{yy}-I_{xx}\right)pq+m\left[x_G(\dot{v}-wp+ur)-y_G(\dot{u}-vr+wq)\right]=\sum N_{ext} \end{array} \quad (1)$$

In the preceding set of equations, the right-hand side values represent the external forces on the AUV [9], which are written as:

$$\begin{aligned} \sum X_{ext} =& X_{\dot{u}}\dot{u}+Z_{\dot{w}}wq+Z_{\dot{q}}q^2-Y_{\dot{v}}vr-Y_{\dot{r}}r^2 \\ &+X_{HS}+X_{u|u|}u|u|+X_{prop} \\ \sum Y_{ext} =& Y_{\dot{v}}\dot{v}+Y_{\dot{r}}\dot{r}+X_{\dot{u}}ur-Z_{\dot{w}}wp-Z_{\dot{q}}pq \\ &+Y_{HS}+Y_{v|v|}v|v|+Y_{r|r|}r|r|+Y_{uvl}uv \\ \sum Z_{ext} =& Z_{\dot{w}}\dot{w}+Z_{\dot{q}}\dot{q}-X_{\dot{u}}uq+Y_{\dot{v}}vp+Y_{\dot{r}}rp \\ &+Z_{HS}+Z_{w|w|}w|w|+Z_{q|q|}q|q|+\left(Z_{uwl}+Z_{uwf}\right)uw+Z_{uqf}uq \\ \sum K_{ext} =& K_{\dot{p}}\dot{p}+(Z_{\dot{w}}-Y_{\dot{v}})wv+\left(Z_{\dot{q}}+Y_{\dot{r}}\right)vq-\left(Z_{\dot{q}}+Y_{\dot{r}}\right)wr+\left(N_{\dot{r}}-M_{\dot{q}}\right)qr \\ &+K_{HS}+K_{p|p|}p|p|+K_{prop} \\ \sum M_{ext} =& Z_{\dot{q}}\dot{w}+M_{\dot{q}}\dot{q}-(Z_{\dot{w}}-X_{\dot{u}})uw-Y_{\dot{r}}vp+\left(K_{\dot{p}}-N_{\dot{r}}\right)rp-Z_{\dot{q}}uq \\ &+M_{HS}+M_{w|w|}w|w|+M_{q|q|}q|q|+\left(M_{uwl}+M_{uwf}\right)uw+M_{uqf}uq \\ \sum N_{ext} =& N_{\dot{v}}\dot{v}+N_{\dot{r}}\dot{r}-(X_{\dot{u}}-Y_{\dot{v}})uv+Z_{\dot{q}}wp-\left(K_{\dot{p}}-M_{\dot{q}}\right)pq+Y_{\dot{r}}ur \\ &+N_{HS}+N_{v|v|}v|v|+N_{r|r|}r|r|+N_{uvl}uv \end{aligned} \quad (2)$$

The hydrostatic terms of forces and moments are also important components:

$$\begin{aligned} X_{HS} &= -(W - B)\sin\theta \\ Y_{HS} &= (W - B)\cos\theta\sin\phi \\ Z_{HS} &= (W - B)\cos\theta cos\phi \\ K_{HS} &= (y_G W - y_B B)\cos\theta\cos\phi - (z_G W - z_B B)\cos\theta sin\phi \\ M_{HS} &= -(z_G W - z_B B)\sin\theta - (x_G W - x_B B)\cos\theta\cos\phi \\ N_{HS} &= (x_G W - x_B B)\cos\theta\sin\phi + (y_G W - y_B B)\sin\theta \end{aligned} \tag{3}$$

Notice that x_G and x_B will be controlled by a mechanism of counterweight and ballast piston - cylinder.

3.1 Piston - Cylinder Structure Analysis

In the hybrid AUV model, the center of gravity and center of buoyancy are not in a fixed position but can be changed thanks to the ballast system and the mechanism of counterweight shifting. Because those centers directly affect the moment that causes a great impact on the pitch angle, manipulating the center of gravity and center of buoyancy allows us to control pitch angle without using fins as conventional AUVs. The ballast system not only helps to change the center of buoyancy but also changing buoyancy B of AUV. The variation of buoyancy is the key to the control scheme of vertical movement.

Because the mechanical design is almost symmetrical about the x-axis and the moving direction of the piston is also co-axial, the center of gravity shifts back and forth almost exclusively along the x-axis when the ballast pump activates. In Fig. 4, O is the origin of the body-fixed frame, V_{fix} is the fixed water occupied volume (the volume when cylinder draws full of water), l_{fix} is the distance from the origin O to the center of buoyancy, conforming the following relationship.

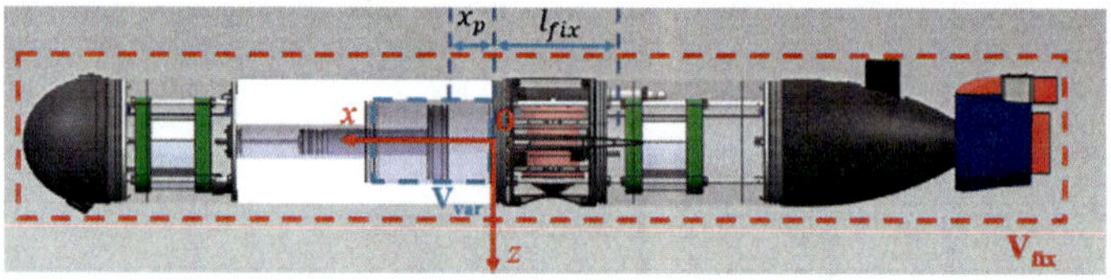

Fig. 4. Analysis of ballast subsystem

$$V_{\text{var}} = S_{cd} \times \left(x_c - x_p\right) \tag{4}$$

In formula 4, S_{cd} is the cross-sectional area of the cylinder, x_c is its length, and x_p is the distance from the piston to the origin O. After denoting all the elements, the formula to calculate the position vector of the center of buoyancy should be considered.

$$\overrightarrow{r_{cb}} = [x_B, 0, 0]^T = \left[\frac{V_{\text{var}}\frac{x_c+x_p}{2} + V_{fix}l_{fix}}{V_{\text{var}} + V_{fix}}, 0, 0\right]^T \tag{5}$$

The buoyancy B varies according to the displaced fluid as:

$$B = \rho g \left(V_{fix} + V_{\text{var}} \right) \tag{6}$$

The first multiplier ρ is the density of the occupied liquid, while g is the gravity acceleration. The volume of the piston is negligible compared to the total volume of AUV and the displacement of the piston is not large so the change of the center of gravity can be ignored when the piston changes its position.

3.2 Mechanism of Counterweight

Apart from the ballast system, the counterweight does not change the weight W or buoyancy B. It only contributes to locating a new location of the center of gravity. To achieve that goal, the counterweight mechanism depends on a heavy object that moves along the x-axis so that the position of the center of gravity can be changed as in Fig. 5.

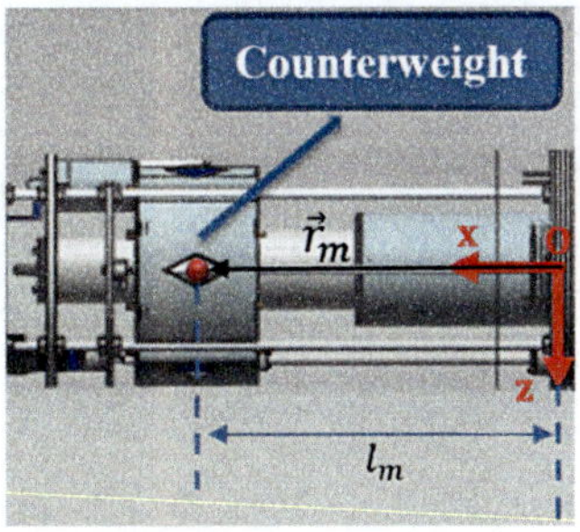

Fig. 5. The inner mechanism of the center of gravity shifting

We denote m_h the total mass of the hull, the position of its center is $\overrightarrow{r_h}$ and the mass of counterweight is m_m. The center position of the counterweight can be determined only by the relative displacement of the object l_m according to the formula:

$$\overrightarrow{r_m} = [l_m, 0, 0]^T \tag{7}$$

Then the center of gravity of AUV and its weight are now calculated according to the formula:

$$\begin{aligned} \overrightarrow{r_{cg}} &= \frac{m_h \overrightarrow{r_h} + m_m \overrightarrow{r_m}}{m_h + m_m} \\ W &= mg = (m_h + m_m) g \end{aligned} \tag{8}$$

4 Controller Design

Since this is a hybrid AUV type, the design of the controller will depend on its particular modes.

4.1 Glider Mode Control

In this mode, Hybrid AUV will make use of its inertia to alternatively dive and climb to be able to move forward. As a result, it is common to restrict glider activity in a zone with given depth and prioritize the possible distance so as to gather more data for other purposes. When Hybrid AUV dives, the ballast pump draws water from the surrounding environment until the pitch angle reaches 30° downward (Downward phase). Similarly, for the climbing movement, the ballast engine releases all the water and drives the pitch angle to 30° upward (Upward phase) (Fig. 6).

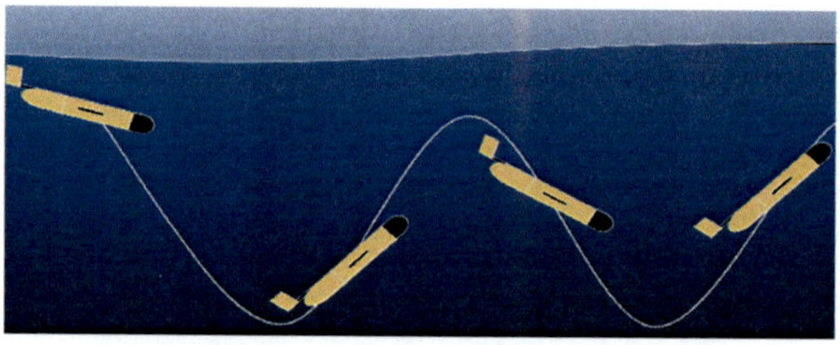

Fig. 6. A typical pattern of motion in glider mode

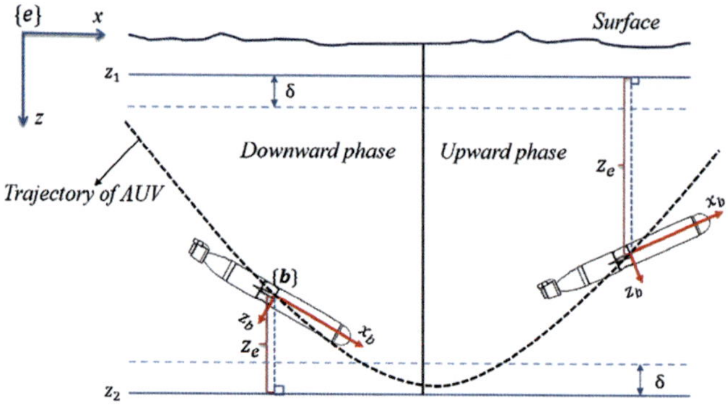

Fig. 7. Trajectory illustration in a cycle of movement

As shown in Fig. 7, we define z_e as the depth error and δ is the distance to compensate for the inertia when the vehicle dives or climbs. The final goal is to make AUV2000 travels as far as possible with a given energy capacity, hence we should not set the operation depth range too large. Doing so will greatly reduce the ability to move forward. On the other hand, a small range of operation depth will make the attitude control system work continuously, which leads to large consumptions of energy. In addition, the response of the piston-cylinder system that pumps water in and out is rather slow, thus affecting the vehicle when operating for a long duration.

4.2 AUV Mode Control

Hybrid AUV will use the thruster in this case, which reduces the influence of the ballast system on our vehicle during the diving process and the pitch angle will play a decisive role instead. Therefore, in order to save energy as well as avoid complexity in controlling, in this mode, we will barely control the ballast system to archive B = W and hold this state from the beginning to the very end of the moving process. A lot of benefits are brought into when B = W as w can be considered close to 0, hence if AUV has its pitch angle equal to 0, it will only move horizontally. Conventional AUVs without ballast system have to design B > W so that in case of emergency or uncontrolled problems, it is able to float on the surface itself. This design aims to have a negative pitch angle to compensate for factors that make AUV float in case they want to move horizontally at a specified depth, which is one of the disadvantages of conventional AUVs.

Suppose we want AUV to reach and remain at a certain depth as in Fig. 8, using the idea of Line of sight in the vertical dimension, we define a vector LOS to calculate the angle to drive our AUV to the desired depth.

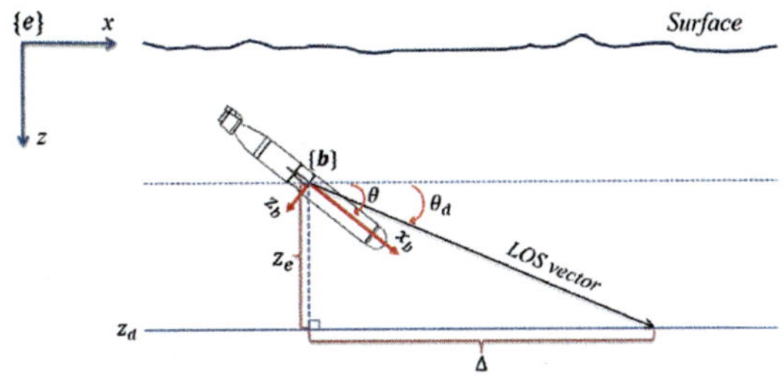

Fig. 8. Guidance scheme for depth tracking

From that figure, we have the formula: $\theta_d = \arctan\left(\frac{z_e}{\Delta}\right)$ in which $z_e = z - z_d$. We can see that in both modes, controlling pitch angle is very important and all control matters lead to controlling pitch angle to reach a preset value. We recall the pitch angle dynamic equation:

$$\begin{aligned}&\left(I_{yy} - M_{\dot{q}}\right)\dot{q} - \left(mx_G + Z_{\dot{q}}\right)\dot{w} + mz_G\dot{u}\\&= \left(X_{\dot{u}} - Z_{\dot{w}} + M_{uwl} + M_{uwf}\right)uw + (mx_G - Y_{\dot{r}})vp + \left(K_{\dot{p}} - N_{\dot{r}} - I_{xx} + I_{zz}\right)rp\\&\quad + \left(M_{uqf} - mx_G - Z_{\dot{q}}\right)uq - mz_G(wq - vr) + M_{HS} + M_{w|w|}w|w| + M_{q|q|}q|q| \qquad (9)\end{aligned}$$

As we are assuming pure depth-plane motion, we only focus on the body-relative surge velocity u, heave velocity w, pitch rate q, and the earth-relative vehicle forward position x, depth z, and pitch angle θ. We will set all other velocities (v, p, and r) to zero and drop the equations for any out-of-plane terms.

$$\begin{aligned}\dot{\theta} &= q \\ \left(I_{yy} - M_{\dot{q}}\right)\dot{q} &= \left(X_{\dot{u}} - Z_{\dot{w}} + M_{uwl} + M_{uwf}\right)uw + \left(M_{uqf} - Z_{\dot{q}}\right)uq - mz_G wq + M_{w|w|}w|w| \\ &\quad + M_{q|q|}q|q| - (z_G W - z_B B)\sin\theta + x_B B\cos\theta - x_G(muq + W\cos\theta)\end{aligned} \tag{10}$$

Define:

$$\begin{aligned}f_p &= \frac{\left(X_{\dot{u}} - Z_{\dot{w}} + M_{uwl} + M_{uwf}\right)uw + \left(M_{uqf} - Z_{\dot{q}}\right)uq - mz_G wq}{I_{yy} - M_{\dot{q}}} \\ &\quad + \frac{M_{w|w|}w|w| + M_{q|q|}q|q| - (z_G W - z_B B)\sin\theta + x_B B\cos\theta}{I_{yy} - M_{\dot{q}}} \\ g_p &= -\frac{muq + W\cos\theta}{I_{yy} - M_{\dot{q}}}\end{aligned} \tag{11}$$

Rewrite $\dot{q} = f_p + g_p x_G$ and define the error as $e = \theta - \theta_d$. The first and second derivative of that error can be obtained by:

$$\begin{aligned}\dot{e} &= \dot{\theta} - \dot{\theta}_d = q - \dot{\theta}_d \\ \ddot{e} &= \dot{q} - \ddot{\theta}_d = f_p + g_p x_G - \ddot{\theta}_d\end{aligned} \tag{12}$$

The sliding surface is designed according to the first-order system $s = \dot{e} + ke$. We construct a Lyapunov function in quadratic form $V = \frac{1}{2}s^2$ whose first-order derivative is:

$$\dot{V} = s\dot{s} = s(\ddot{e} + k\dot{e}) = s\left(f_p + g_p x_G - \ddot{\theta}_d + k\dot{e}\right) \tag{13}$$

In order to stabilize the system, the center of gravity should be chosen as:

$$x_G = \frac{1}{g_p}\left(\ddot{\theta}_d - f_p - k\dot{e} - \eta\,\mathrm{sat}(s)\right) \tag{14}$$

Substitute Eq. 14 into Eq. 13, the stability is evidently proved.

$$\dot{V} = -\eta\,\mathrm{sat}(s)s \le 0 \tag{15}$$

Once x_G is given, we can easily calculate the counterweight position.

5 Simulation Results

In glider mode, our concerned is whether AUV can operate properly at the specified depth and travel such a farthest distance with a given amount of energy. Thus, it indicates in Fig. 9 that the AUV works at depths between 10 and 40 m and the lateral displacement is around 200 m forward in 600 s. The velocity range is 0.4 m/s and pitch angle response is achieved as in the control strategy. When the vehicle changes its attitude, the forward velocity suddenly falls to 0.3 m/s and then gradually gets back on track. This result seems very satisfactory and consistent with the reality in comparison to other glider vehicles. It shows the effectiveness of the control strategy and the controller that has been designed (Fig. 10).

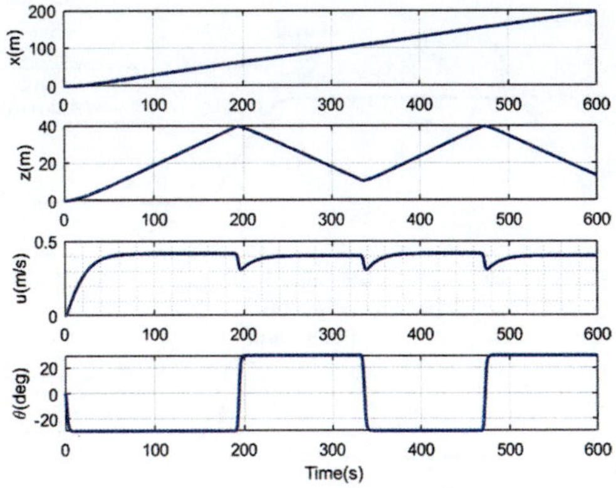

Fig. 9. State responses in glider mode

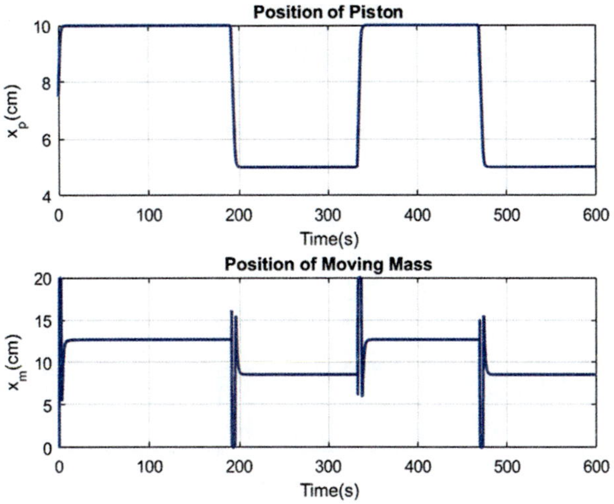

Fig. 10. Subsystem responses in glider mode

In AUV mode, we concern about the ability to move and remain at a specific depth. From the simulation results in Fig. 11, it is shown that the response of the depth control falls to about 25 s. Furthermore, the ability of diving and floating of AUV is different which means the responses when diving and floating are also inconsistent. Bear in mind that in order to keep the entire operation process out of uncontrollable circumstances, the set pitch angle is limited to 60°. Besides, the response of the controller adapts to the reference value very well without steady-state errors. This thereby shows the efficiency and quality of the controller (Fig. 12).

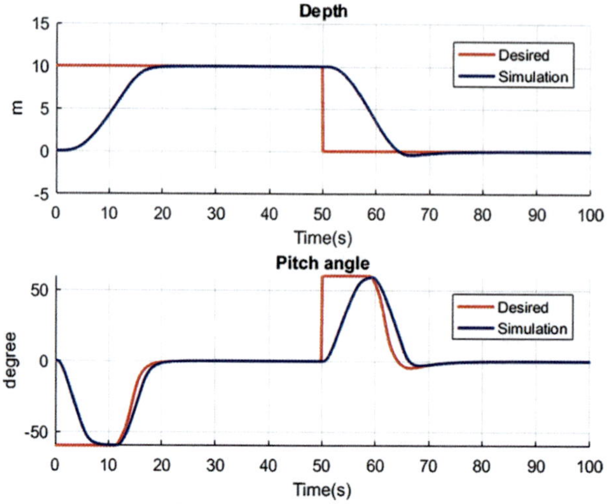

Fig. 11. Depth and pitch angle response in AUV mode

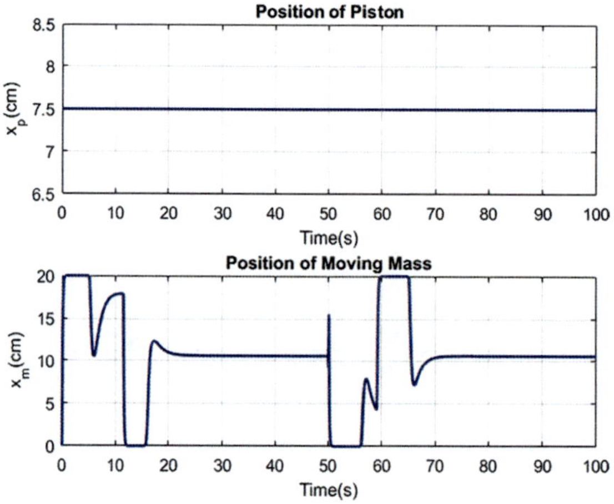

Fig. 12. Subsystem response in AUV mode

6 Conclusion

This paper refers to a new hybrid type of AUV as well as how it operates, analyzing the control characteristics based on its original subclasses. The dynamic model explained motions in six degrees of freedom mathematically and the way that subsystems affected the variables. Additionally, the paper demonstrated mode-switch control strategies to turn all the requirements into controlling the pitch angle and also design a pitch angle controller with the Sliding mode method.

The simulation results obtained were very optimistic, indicating that the proposed control strategy can help Hybrid AUV archive desired goals in each operating mode. In particular, it is possible to make the Hybrid AUV operate in a given depth range and travel a distance of approximately 200 m in 600 s with an energy conservative scheme, or drive it to reach the target destination and remain the movement with a thruster. All simulation results have proved the effectiveness and quality of the proposed method.

Acknowledgement. This research is supported by National Key Lab. of Digital Control and System Engineering (DCSELAB), HCMUT and Laboratory of Advance Design and Manufacturing Processes and funded by Vietnam National University Ho Chi Minh city (VNU-HCM) under grant number B2018-20b-01.

References

1. Wood Whole Oceanographic. https://www.whoi.edu
2. Scientific American. https://www.scientificamerican.com/article/deepwater-robot-sub
3. Fiorelli, E., Leonard, N.E., Bhatta, P., Paley, D.A., Bachmayer, R., Fratantoni, D.M.: Multi-AUV control and adaptive sampling in monterey bay. IEEE J. Oceanic Eng. **31**(4), 935–948 (2006)
4. Yu, X.: The application of autonomous underwater vehicles for interdisciplinary measurements in Massachusetts and Cape Cod Bays. Cont. Shelf Res. **22**(15), 2225–2245 (2002)
5. Kim, A., Eustice, R.M.: Toward AUV survey design for optimal coverage and localization using the Cramer Rao lower bound. In: OCEANS 2009, Biloxi, MS, pp. 1–7 (2009)
6. Ramos, P., Cruz, N., Matos, A., Neves, M.V., Pereira, F.L.: Monitoring an ocean outfall using an AUV. In: MTS/IEEE oceans 2001. An Ocean Odyssey. Conference Proceedings, Honolulu, HI, USA, vol. 3, pp. 2009– 2014 (2001)
7. Niu, H., Adams, S., Lee, K., Husain, T.: Applications of autonomous underwater vehicles in offshore petroleum industry environmental effects monitoring. J. Can. Petrol. Technol. **48**, 12–16 (2009)
8. Fossen, T.I.: Handbook of Marine Craft Hydrodynamics and Motion Control. Wiley, New York (2011)
9. Prestero, T.: Verification of a Six-Degree of Freedom Simulation Model for the REMUS Autonomous Underwater Vehicle. Thesis, September 2001
10. Joo, K.M., Qu, Z.: An autonomous underwater vehicle as an underwater glider and its depth control. Int. J. Control Autom. Syst. **13**, 1212–1220 (2015)
11. Tran, N.H., Choi, H.S., et al.: Design, control, and implementation of a new AUV platform with a mass shifter mechanism. Int. J. Precis. Eng. Manuf. **16**, 1599–1608 (2015)
12. Jeong, S., Choi, H.S.: Design and control of a high speed unmanned underwater glider. Int. J. Precis. Eng. Manuf. Green Technol. **3**, 273–279 (2016)
13. Yan, Y., Yu, S.: Sliding mode tracking control of autonomous underwater vehicles with the effect of quantization. Ocean Eng. **151**, 322–428 (2018)

Design and Implementation of an Indoor Guidance System for People with Visual Disabilities Consisting of an Intelligent Electronic Cane

Luis E. Pallares(✉), Arnaldo A. González, Roberto Ferro Escobar, and Helmer Muñoz Hernández

Corporación Unificada Nacional de Educación Superior CUN, Bogotá, Colombia
luis_pallares@gmail.com, {arnaldo_gonzalez,roberto_ferro, helmer_munoz}@cun.edu.co

Abstract. The article presents the most critical points of an electronic solution that addresses the problem of independently guiding a person with visual disability (partial or total) in a closed area, through the implementation of a guidance system in spaces closed for people with visual impairment [1], consisting of a series of guidelines and information points which can be perceived by an electronic cane used a sensor IR into the protective contact cover (see Fig. 2). The device can be connected wirelessly to a mobile device with a guided design with an Android application that provides audible information to the user (see Fig. 1). In Fig. 2, you can see the number of people who are benefited from the use of inclusive technologies and the reasons for use included in quantitative data. Finally, 65% of people with visual disabilities, use accessible technologies for their ease of use. Use when performing specific tasks or communes.

Keywords: Visual disability · Mobile application · Guide line · Electronic cane

1 Introduction

The difficulty to move with autonomy and independence is one of the most common obstacles for people suffering from a visual disability, a situation that in addition to affecting their mobility, can make them face emotional problems, as well as work and social.

This article provides information on an innovative, effective and economical solution to the problems described above; the present development refers to an indoor guidance system for blind people who use an electronic cane, a smartphone, guidance lines, and information points in the form of a bar code, allowing them to fulfill the objective of guiding the person always to his destiny; Inside the handle of the electronic stick is the electronics, mainly composed of a battery charging circuit, a rechargeable battery, a microcontroller, a wireless communication module and an infrared light sensor inside the protective contact tip [2, 3].

D. F. Cortes Tobar et al. (Eds.): AETA 2019, LNEE 685, pp. 532–540, 2021.
https://doi.org/10.1007/978-3-030-53021-1_54

2 Problem Description

According to the "Analysis of Situation of Visual Health in Colombia 2016 Agreement 519 of 2015", of the Ministry of Health, in Colombia, it is estimated that there are 7,000 blind Colombians per million (296,000 blind throughout the territory) the development of an electronic stick is related to some of the aids that people with visual impairments use to walk, replacing direct visual perception with another type of knowledge, in this case an audible persistence that can be heard through spoken messages using a mobile phone wirelessly linked to the electronic stick, classifying this technology within the category of electronics and human needs, understood as first level for this type of disabled population.

The blindness is a physical-sensory type disability; This limits the capacity of affected people to carry out activities of daily life and deteriorates their quality of life. It is usual for people with blindness to resort to alternative means of support to travel, such as the use of technical or auxiliary tools, among the most used are mobility poles and guide dogs, these tools provide solutions on the environment of the individual basically and in which it is intended to solve the problem of taking the individual from one place to another using tools such as GPS, where the precision of the devices that use this technology limits its use to outdoor, However, these are insufficient while it remains the problem of guiding the person who suffers from visual impairment, are not resolved with traditional aids, especially considering that the accuracy [4] of GPS technology for civil applications is approximately 3 m, so it is impossible to guide people with visual impairment indoors, the problem continues to exist because of the In the art of technology there is no solution to this problem [5].

The following graph shows the factors that influence as barriers in visual disability, measured on a scale from 0 to 5, where 5 is the highest level of importance. The use of this technology is based on the accessibility problems that a person with a visual disability presents when moving about an enclosed space, the ease with which they can manipulate the smart cane shown in Fig. 2, which is an inclusive technology of easy management

Fig. 1. Diagram of The Electronic Guide System in Closed Spaces 1, shows a diagram of the elements that make up the guidance system in closed spaces for people with visual disabilities, consisting of a series of guidelines and information points (bar codes) which they can be perceived by an intelligent electronic cane connected wirelessly to a mobile device with an app that provides audible information to guide the user.

and having and that corrects accessibility problems (see Fig. 3) this technology becomes a first level need for people with visual disabilities.

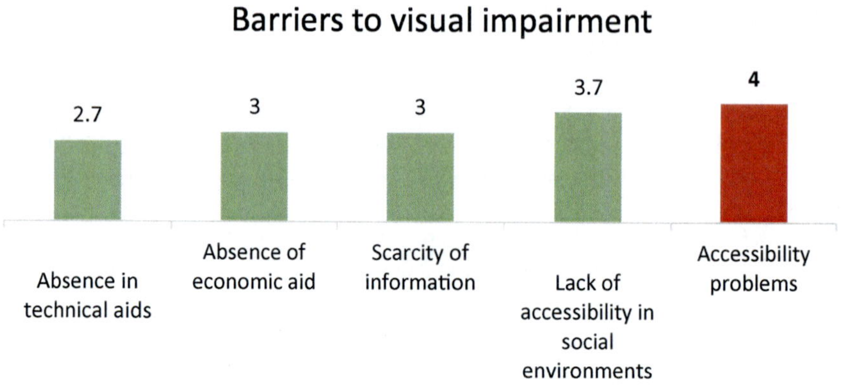

Fig. 3. Quantitative measurement of the degree of importance of the barriers that appear in a person with a visual disability weighted on a scale of 0 to 5 [1].

3 Design and Modeling

Next, a description of the proposed solution is made, according to Fig. 1, illustrating the main components that make up the guidance system in closed spaces for people with visual impairment, consisting of a series of guidelines (42) that create a network of roads to the different destinations within the enclosure; at the beginning, end, at the intersection points and intersections of this road network, there are information points (43) (bar codes) which can be perceived through an optical sensor (16) on the electronic cane (1) connected wirelessly to a mobile device (41) with an Android Guidance application that provides audible information to the user [5].

In Fig. 4, there is the electronic cane (1), composed of an aluminum cylinder (5), which is inserted at one of its ends, in the protective contact cover (2), inside the this is located the Infrared sensor (16) positioned at a maximum height of 5 mm, this distance has as justification the proper functioning of the sensor, plus a sphere (10) that facilitates the movement with the cane, this sphere (10) it is self-contained in the protective contact tip (2) that surrounds 66.6% of the volume thereof, allowing it to rotate freely in all directions, remaining contained in the protective contact tip (2); at the opposite end of the aluminum cylinder (5) the electronic handle (3) is connected, which is joined to the handle cover (4), and the connection is secured with the insertion of the coupling screws (22), forming inside the handle three cavities, in which there are (see Fig. 2.) the load control system (8) which communicates to the outside through a slot, which allows access to the micro connector -USB of the load control system (8), which connects with the rechargeable battery or FEM (7) just between the charging system (8) and the electronic control circuit (6) located in the compartment following the EMF (7), this arrangement allows to connect both loops to the FEM (7) easily using the least amount of wiring and allowing to contain the different electronic components inside the handle of the cane, which allows the size of the handle of the rod to be one size and for it is comfortable for users.

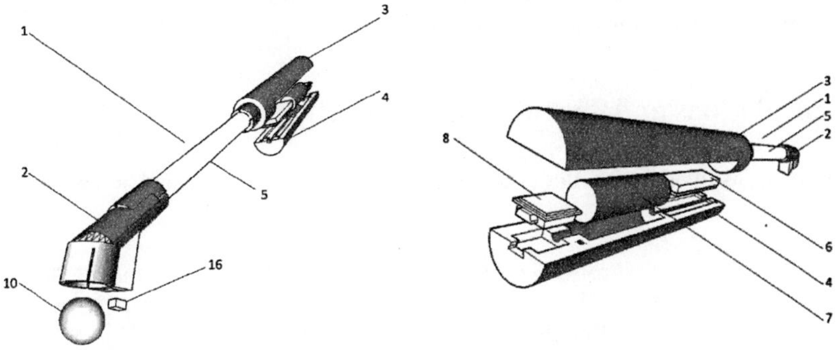

Fig. 4. Intelligent Station, on the left, shows a diagonal lower right side view of the device of the invention, showing its main components separated to provide information on the construction thereof. On the right, it shows a diagonal upper left side view of the device of the invention, showing its main components separated to provide information on the construction thereof.

In Fig. 5 we can see the electronics that are part of the device, where is the charging system (8) which is composed of an integrated load management module (reference TP4056), it is also written of a sensor infrared light (16) (reference QRD1114), which detects the guidance lines (42) and the information points (43), in the electronic control circuit (6) there is a microcontroller (reference Atmega 328P-MU) and the wireless connection module (reference HC-05) which allows communication between the microcontroller and the mobile device containing the guidance application.

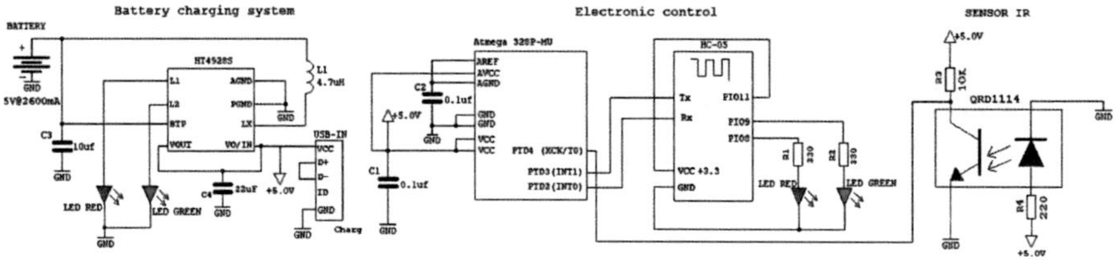

Fig. 5. Electronic Scheme Intelligent Station 1, a schematic diagram of the electronics used in the device of the invention where it is shown how the different elements of the electronics are interconnected, such as the charging system F.E.M. with micro USB, electronic control and IR sensor (infrared light sensor).

In order for the device to work, it is necessary to implement the method as a program inside the microcontroller and in the mobile application [5], Fig. 4 shows the block diagram showing to connect the electronic device embedded in the electric cane with a power supply that integrates radio-frequency wireless connection in the ISM band. The person with a visual disability is guided reception of the messages sent from the stick through the microcontroller, which in turn, processes the data acquired by the infrared light sensor located in the protective contact cover of the Smart Staff. You can use an APP for transcribing these messages that come from the microcontroller via wireless and are translated into audible words, which are easily interpreted by people with visual disabilities, this method is known as Text to Speech [6].

4 Calculations and Technical Considerations

The power consumed by the modules that make up the system are listed below (Table 1):

Table 1. Current consumption of the modules that make up the electronics of the device.

Device	Voltage	Current
TP4056	5 V	150 μA
QRD1114	5 V	20 mA
Atmega 328P-MU	5 V	105 mA
HC-05	5 V	50 mA

4.1 Calculation of Total Power Consumed by the System

The system generates a load for its operation that depends on the current or power required by each of the elements of the system, of which we have that the current required by the system is:

$$I_T = I_{TP4056} + I_{QRD1114} + I_{Atmega328P-MU} + I_{HC-05} \tag{1}$$

$$I_{TOTAL} = 150 \times 10^{-6}\text{A} + 20 \times 10^{-3}\text{A} + 105 \times 10^{-3}\text{A} + 50 \times 10^{-3}\text{A} \tag{2}$$

$$I_{TOTAL} = 175.15\,\text{mA} \tag{3}$$

4.2 Calculation of System Autonomy

Knowing the current consumed by the device we can estimate the independence, taking into account that to maximize the useful life of the battery, we should prevent it from discharging below 50% of its maximum capacity, we will then take that the battery can only supply half of its size, that is to say 1300 mAh, therefore carrying out the respective calculations:

$$Autonomy\ of\ the\ system\ (in\ hours) = \frac{50\%(\text{Capacidad de la FEM})}{I_{total}} \tag{4}$$

$$Autonomy\ of\ the\ system\ (in\ hours) = \frac{0.5 \times (2600\,\mathrm{mAh})}{175.15\,\mathrm{mA}} \tag{5}$$

$$Autonomy\ of\ the\ system\ (in\ hours) = \frac{1300\,\mathrm{mAh}}{175.15\,\mathrm{mA}} \tag{6}$$

$$Autonomy\ of\ the\ system\ (in\ hours) = 7.42\,\mathrm{h} \tag{7}$$

Then Fig. 6, shows the graphs where the frequency of cane or movement of the rod is analyzed from one side to another vs. the distance of different users, from the same we can infer that there is no pattern or correlation that allows us to generate some sampling frequency, because it is not possible to predict the speed with which the user performed the sweep with the cane since each user could perform the sweep at a different rate when using the rod and even the same user could change the Sweeping speed during the travel any attempt to calculate the average of said sweep and use that average to try to read the bar code would result in incorrect readings, therefore when reading the information points coded as barcode the number of flanks of rise and fall of the signal emitted by the optical sensor, then generates information points with different number of bars to identify the site in which the user is in addition to that they are built with a mirror effect which will give the same reading whether read from right to left or vice versa, (see Figs. 7) (Fig. 8).

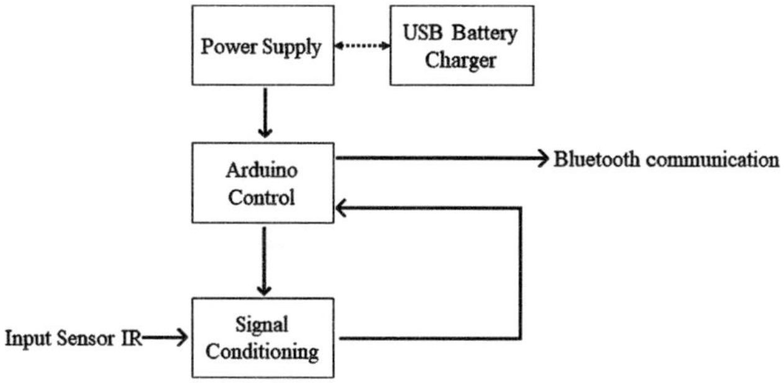

Fig. 6. Intelligent Cane Block Diagram, the block diagram of the smart pole is presented, the power and signal connections of the signal coming from the IR sensor are shown [6].

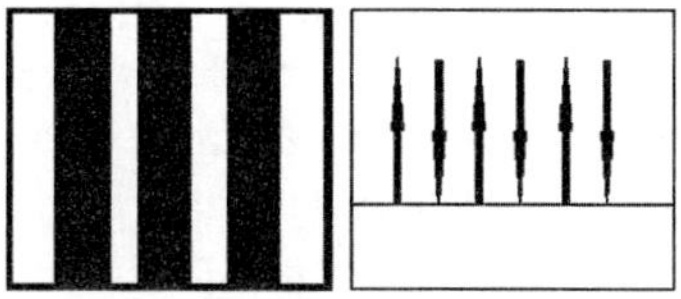

Fig. 7. Diagram of Barcode and Up and Down Flashes, the diagram of a barcode and the rising and falling flanks, notes that the reading of the sides would be the same if the sensor traversed the barcode from right to left or from left to right.

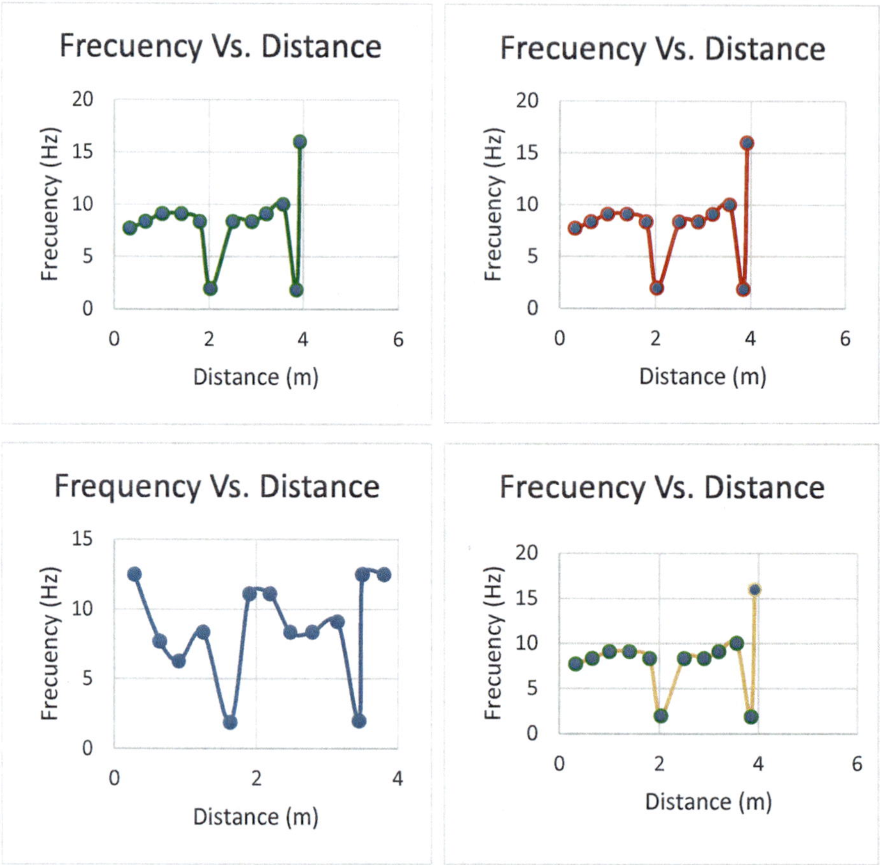

Fig. 8. Graph of Frequency Vs. Distance shows the figure where the frequency of cane movement or movement of the cane from one side to another vs. the range of different users is analyzed.

5 Tests and Data Collected

The tests are carried out when the person with visual impairment first uses the proposed system after training in the use of the same, testing the system during four attempts of displacement, the tracings were made in such a way that there are no obstacles in the road, then the blind man travels the road. With the help of a person to contrast, the times in the routes of the same, the summary of the acquired data is in Table 2.

As shown in Table 2 the user completes his journey, although the travel time is longer compared to humanitarian assistance, the user completes his way effectively, the present electronic solution helps guide in a closed enclosure autonomously a person with visual disability in a precise and constant way along the route. As can be seen in Table 2, human assistance is more effective. However, the use of the smart cane gives it a high degree of autonomy, and this last item is considered as more important (see Table 3).

Table 2. Summary of the average of the data acquired in the time measurements of the routes using the guidance system in closed spaces and without using the system, but with the assistance of a person

Proof#	Provenance	Arrival	Time	Method	Distance
3	Elevator	Bathroom	13.40 s	With cane	7.50 m
4	Bathroom	Elevator	12.22 s	With cane	7.50 m
2	Bathroom	Office	44.10 s	With cane	25.50 m
1	Office	Elevator	58.30 s	With cane	25.50 m
7	Elevator	Bathroom	6.11 s	Human assistance	7.50 m
8	Bathroom	Ascensor	6.13 s	Human assistance	7.50 m
6	Elevator	Oficina	26.51	Human assistance	25.50 m
5	Office	Elevator	27.30	Human assistance	25.50 m

Table 3. Levels of qualitative measurement in activities that require autonomy in people with visual impairment.

Variables involved in autonomy	Level of importance (Qualitative)
Physical health	High
Functional skills	Medium
State of mind	High
Motivations and interests	Medium
Daily activities	Very high

As the user becomes familiar with the use of the tool to support his movements within the enclosed space, he felt comfortable and confident that he could take it to his destination in a free and safe way.

The activities mentioned in Table 3 are linked to the level of satisfaction of independence or autonomy in daily routines and the health status of people with visual impairment when these activities are carried out independently [7]. Between the daily habits are the displacements of a place to another one and in closed enclosures using human assistance, guide dogs and the traditional white cane, that depend on the economy and the training in rehabilitation centers, the smart rod provides the levels of autonomy necessary for the displacement in enclosures of efficient way without considering the time used, since the degree of independence in this case is much more important (see Table 3).

6 Conclusions and Future Work

Calculating the autonomy of the electronic cane shows that it has a capacity of 7.42 h of continuous use, so we can conclude that the independence of the device provides more than enough time to perform the routes required by the user to play in the enclosure where use From the analysis the graphs of frequency of cambering vs. distance, it is evident that the rate of cambering can change according to the user or the range, it can tend to an average value depending on the user, but this can change at any time, therefore, can conclude that, if readings are generated based on the frequency of banding, it will lead to errors. When analyzing Table 2 we can see that the travel times assisted by a person are less than using the cane, but it can also be seen that as the tests progressed and the user became familiar with the use of the system, their trips took each timeless time. The system provides a solution to the problem of guiding a person in a state of visual disability to a location in an enclosed space in a free and safe way, with a light, minimalist and low-cost design, with the minimum of differences concerning the traditional canes.

At the level of future improvements, the integration of an ultrasound module for the detection of obstacles is recommended, something that would contribute to the safety of the trips, and the user's confidence. To avoid the constant use of Talkback, the application on the mobile device could have the ability to implement the voice command service, to facilitate user use.

References

1. Microsoft Corporation. All Rights Reserved. Study Commissioned by Microsoft, Conducted by Forrester Research, Inc. 2003 (2004)
2. Jeffs, T.: Virtual reality and special needs. Themes Sci. Technol. Educ. 253–268 (2010)
3. Gi, L.M.: Guide Device for Blind People Using Electronic Stick and Smartphone, vol. 0081675, pp. 1–24 (2013)
4. Wing, M.G., Eklund, A., Kellogg, L.D.: Consumer-grade global positioning system (GPS) accuracy and reliability. J. For. **103**, 169–173 (2018). https://doi.org/10.1093/jof/103.4.169
5. Xue, R.Z., Shao, M., Li, A., Yu Lijuan, G.F.: Intelligent Blindness Guide Walking Stick Based on GPS/GPRS.pdf., vol. 9 (2013)
6. Brown, D.J., McHugh, D., Standen, P., et al.: Designing location-based learning experiences for people with intellectual disabilities and additional sensory impairments. Comput. Educ. **56**, 11–20 (2011). https://doi.org/10.1016/j.compedu.2010.04.014
7. Díaz Veiga, P.: Discapacidad visual y autonomía: las posibilidades de las personas mayores. Análisis 26–33 (2008)

IMU Calibration Methods and Orientation Estimation Using Extended Kalman Filters

Xuan-Dung Trinh, Manh-Cam Le(✉), and Ngoc-Huy Tran

Ho Chi Minh City University of Technology, VNUHCM, Ho Chi Minh City, Vietnam
{1870006,tnhuy}@hcmut.edu.vn

Abstract. IMU sensor has long been developed to solve the problems with angular rotation of objects moving in space. There are variety of available IMU sensors which have built-in orientation estimation algorithms with high accuracy (AHRS) such as ADIS16480, RTxQ, XSENS. However, understanding of IMU calibration methodology and orientation estimation algorithms is still essential for further intervention or integration. This paper will focus on the orientation estimation algorithm, calibration methods and the IMU model. First, a simulation model is created to develop and evaluate the algorithm theoretically. We also build a turntable for testing, displaying, storing data from the real IMU while the orientation estimation algorithm is embedded into a microcontroller with 10-ms sampling time. Experimental results have shown that the proposed algorithm achieves accurate estimation of orientation. The angular RMS errors are all less than 2° in normal conditions as well as less disturbed in the conditions involving external acceleration and magnetic disturbance.

Keywords: Extended Kalman filters · Inertial Measurement Unit · External acceleration · External magnetism

1 Introduction

IMU (Inertial Measurement Unit) 9 DOF is a combination of three sensors: accelerometer, gyroscope and magnetometer. Measured data from sensors helps us determine the rotation of the object to which the sensor attaches to. The applications of IMU have widely spread in many areas such as space, balance control, human body motion simulation, etc.

The Kalman filter [1] has proved its great impact on the study of the rotation angle estimation. The problem of constructing the kinetic equations in this filter has been identified and analyzed by several authors using different types of mathematical models of IMU such as: Euler-angle [2], quaternion [3] or direction cosine matrix (DCM) [4–6]. Recent research has focused on some improvements that make estimation algorithms work well under conditions in which external acceleration and magnetism disturbance exist [2, 7].

The IMU calibration methods are divided into two types. One of them uses the reference angles [8] with high accuracy, but it depends much on costly reference equipment.

D. F. Cortes Tobar et al. (Eds.): AETA 2019, LNEE 685, pp. 541–551, 2021.
https://doi.org/10.1007/978-3-030-53021-1_55

In contrast, the calibration type that does not relate to reference angles [9, 10] is rather simple and precise at an acceptable range, though the outputs are not as reliable as the above method.

2 The Orientation Estimation Algorithms

2.1 Direction Cosine Matrix – DCM

Rotation matrix ${}_B^N C$ transfers a (3 × 1) vector attached to the body coordinate system to a (3 × 1) vector in the NED coordinate system according to the following formula:

$$k^N = {}_B^N C.k^B \tag{1}$$

The so-called DCM ${}_B^N C$ is defined as:

$${}_B^N C = \begin{pmatrix} {}_B^N C_{1i} \\ {}_B^N C_{2i} \\ {}_B^N C_{3i} \end{pmatrix} = \begin{pmatrix} \theta_c\psi_c & \phi_s\theta_c\psi_c - \phi_c\psi_s & \phi_c\theta_s\psi_c + \phi_s\psi_s \\ \theta_c\psi_s & \phi_s\theta_c\psi_s - \phi_c\psi_c & \phi_c\theta_s\psi_s - \phi_s\psi_c \\ -\theta_s & \phi_s\theta_c & \phi_c\theta_c \end{pmatrix} \tag{2}$$

The formula for calculating Euler angles from DCM:

$$\begin{aligned} \theta &= -\sin^{-1}({}_B^N C_{31}) \\ \phi &= \tan^{-1}({}_B^N C_{32}/{}_B^N C_{33}) \\ \psi &= \tan^{-1}({}_B^N C_{21}/{}_B^N C_{11}) \end{aligned} \tag{3}$$

Instead of directly estimating the Euler angles, we will only need to estimate the elements of DCM and then deduce the Eulerian angles according to the above formulas as trigonometric functions.

The kinematic equation based on DCM is discretized as:

$${}_B^N C_k^T = \left[I_3 - T.S\left(\omega_{k-1}^B\right)\right]{}_B^N C_{k-1}^T \tag{4}$$

where I_3 is the identity matrix; T is the sampling time and the skew-symmetric matrix is:

$$S\left(\omega_{k-1}^B\right) = \begin{pmatrix} 0 & -\omega_z^B & \omega_y^B \\ \omega_z^B & 0 & -\omega_x^B \\ -\omega_y^B & \omega_x^B & 0 \end{pmatrix} \tag{5}$$

2.2 IMU Mathematic Mode

Accelerometer, gyroscope and magnetometer models comprise these following components:

$$a_{meas}^B = a_0^B + a_e^B + v_a \tag{6}$$

$$\omega_{meas}^{B} = \omega^{B} + \upsilon_{\omega} \tag{7}$$

$$m_{meas}^{B} = m_{0}^{B} + m_{e}^{B} + \upsilon_{m} \tag{8}$$

Where υ is the measurement noise from the sensor (considered as white noise); a_e^B, m_e^B represents external acceleration and external magnetism signal respectively; a_0^B, m_0^B represents the gravitational vector and the earth magnetic field in the body coordinate system (where I_c, I_s are the cosine and sine functions of magnetic dip I):

$$m_0^B = {}_B^N C.m_0^N = {}_B^N C_{1i}^T I_c + {}_B^N C_{3i}^T I_s \tag{9}$$

$$a_0^B = {}_B^N C.a_0^N = {}_B^N C_{3i}^T \tag{10}$$

The angular rate in the body coordinate system is expressed as:

$$\omega^B = \begin{pmatrix} \dot{\phi} \\ 0 \\ 0 \end{pmatrix} + {}_B^N C_x \begin{pmatrix} 0 \\ \dot{\theta} \\ 0 \end{pmatrix} + {}_B^N C_x {}_B^N C_y \begin{pmatrix} 0 \\ 0 \\ \dot{\psi} \end{pmatrix} \tag{11}$$

Where:

$${}_B^N C_x = \begin{pmatrix} 1 & 0 & 0 \\ 0 & \phi_c & \phi_s \\ 0 & -\phi_s & \phi_c \end{pmatrix}, {}_B^N C_y = \begin{pmatrix} \theta_c & 0 & -\theta_s \\ 0 & 1 & 0 \\ \theta_s & 0 & \theta_c \end{pmatrix} \tag{12}$$

Based on the previous definition and relationship, we build the IMU simulation model in terms of the block diagram in Fig. 1.

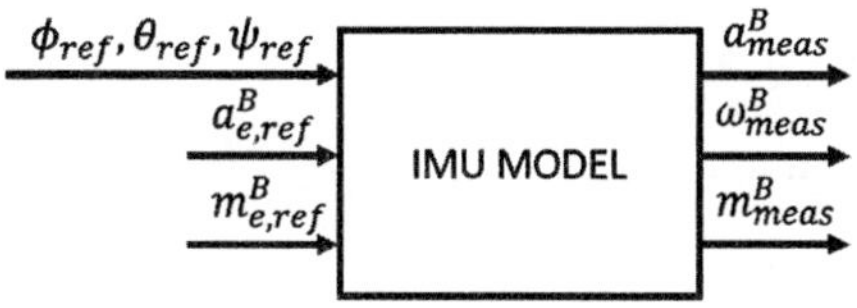

Fig. 1. IMU simulation block diagram

2.3 Orientation Estimation Algorithms

The orientation estimation algorithms shown below is based on the Extended Kalman filter. We firstly draw some following important remarks about the this filter:

1. Unlike the low-pass filter, through kinetic equations, Kalman filter can "understand" the system. The Kalman filter can eliminates noise and retains response of the system - while the low-pass filter can significantly slows down.

2. The measurement variance magnitude is inversely proportional to the "confidence" level of the Kalman filter for the measured values from the system. If this "trust" level is too great, it can lead to poor estimates. However, the "loss of faith" entirely on the measured value (the calculated Kalman filter is almost exclusively based on the kinetic equation) can also result in poor estimates. Therefore, the magnitude of the measurement variance must be corrected so that it is not too large, nor too small for the Kalman filter to combine two information flows (from actual measurement values, kinetic equations) and inferred prices. good estimation.
3. When implementing the Kalman filter, the most important problem is that we have to figure out the mathematical model of the system, which means that the Kalman filter "understands" the system.

The process of estimating orientation complies with the general structure as shown in Fig. 2.

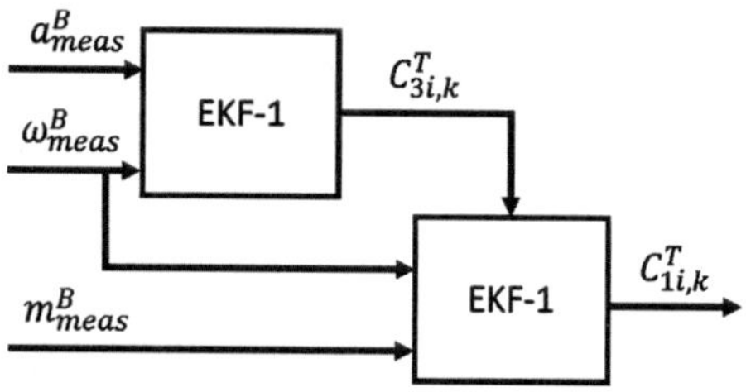

Fig. 2. The orientation estimation block diagram.

The parameters of the extended Kalman filter (Eq. 15) in standard mode (mode-3):

$$\begin{cases} x_k = f(x_{k-1}, w_{k-1}) \\ y_k = h(x_k, \upsilon_k) \end{cases} \tag{13}$$

$$\begin{aligned} &{}^1x_k = {}^N_B C^T_{3i,k} \\ &{}^1w_k = \left(w_{\omega x,k}\ w_{\omega y,k}\ w_{\omega z,k}\right)^T \\ &{}^1y_k = a^B_{meas,k} \\ &{}^1\upsilon_k = \left(\upsilon_{ax,k}\ \upsilon_{ay,k}\ \upsilon_{az,k}\right)^T \\ &{}^1f\left({}^1x_{k-1}, {}^1w_{k-1}\right) = \left\{I_3 - S\left(\omega^B_{k-1} + {}^1w_{k-1}\right)\right\}.{}^1x_{k-1} \\ &{}^1h\left({}^1x_k, {}^1\upsilon_k\right) = {}^1x_k + {}^1\upsilon_k \end{aligned} \tag{14}$$

$$\begin{aligned} &{}^2x_k = {}^N_B C^T_{1i,k} \\ &{}^2w_k = {}^1w_k \\ &{}^2y_k = m^B_{meas,k} - {}^N_B C^T_{3i} I_s \\ &{}^2\upsilon_k = \left(\upsilon_{mx,k}\ \upsilon_{my,k}\ \upsilon_{mz,k}\right)^T \\ &{}^2f\left({}^2x_{k-1}, {}^2w_{k-1}\right) = \left\{I_3 - S\left(\omega^B_{k-1} + {}^2w_{k-1}\right)\right\}.{}^2x_{k-1} \\ &{}^2h\left({}^2x_k, {}^2\upsilon_k\right) = {}^2x_k + {}^2\upsilon_k \end{aligned} \tag{15}$$

Next, we consider two filters (mode-1 and mode-2) commonly used to minimize the effect of external acceleration, external magnetic field on the estimator. For mode-2, we make improvements by changing the covariance matrix of the measurement from the accelerator/magnetometer based on the threshold level ε_a and ε_m. Note that the switching condition should be assured for a certain amount of time, i.e.: kΔt – where Δt is sample time

$$\begin{cases} {}^1R_k = diag\left(\sigma_{ax}, \sigma_{ay}, \sigma_{az}\right), if \left| \left\| a^B_{meas,\tau} \right\| - 1 \right| < \varepsilon_a \forall \tau [k - n_a, k] \\ {}^1R_k = diag(\infty, \infty, \infty), otherwise \end{cases} \tag{16}$$

$$\begin{cases} {}^2R_k = diag\left(\sigma_{mx}, \sigma_{my}, \sigma_{mz}\right), if \left| \left\| m^B_{meas,\tau} \right\| - 1 \right| < \varepsilon_m \forall \tau [k - n_m, k] \\ {}^2R_k = diag(\infty, \infty, \infty), otherwise \end{cases} \tag{17}$$

Where $\sigma_m, \sigma_m, \varepsilon_a, \varepsilon_m, n_a, n_m$ are constants. Mode-1 is implemented by changing the following parameters from mode-3:

$$\begin{cases} {}^1y_k = a^B_{meas,k} - c_a a^B_{e,k-1} \\ {}^1R_k = diag\left(\left(\sigma_{ax}\ \sigma_{ay}\ \sigma_{az}\right)^T + 3^{-1} c_a^2 \left| a^B_{e,k-1} \right|\right) \end{cases} \tag{18}$$

$$\begin{cases} {}^2y_k = m^B_{meas,k} - c_m m^B_{e,k-1} \\ {}^2R_k = diag\left(\left(\sigma_{mx}\ \sigma_{my}\ \sigma_{mz}\right)^T + 3^{-1} c_m^2 \left| m^B_{e,k-1} \right|\right) \end{cases} \tag{19}$$

The external acceleration and the external magnetic field has been modeled according to the following formula:

$$a^B_{e,k} = c_a a^B_{e,k-1} + w_{ae} \tag{20}$$

$$m^B_{e,k} = c_m m^B_{e,k-1} + w_{me} \tag{21}$$

Where c_a, c_m are constants and $c_a, c_m \in [0, 1]$.

3 IMU Calibration Methods

3.1 Using the Reference Angles

The calibrated output can be expressed as a function of measurements from the sensor:

$$c = Mr + o \tag{22}$$

Where variables defined by linear least mean square method are $o = \left(o_x\ o_y\ o_z\right)^T$ and $M = \begin{pmatrix} M_{xx}\ M_{xy}\ M_{xz} \\ M_{yx}\ M_{yy}\ M_{yz} \\ M_{zx}\ M_{zy}\ M_{zz} \end{pmatrix}$ which need to be calculated. The rest column vectors $r = \left(r_x\ r_y\ r_z\right)^T$ are measurements before and after calibration respectively. \$c\$ is defined by the transformation equation:

$$c = {}^N_B C\left(\phi_{ref,i}, \theta_{ref,i}, \psi_{ref,i}\right) x_0^N \tag{23}$$

Vector x_0^N can be either a_0^N or m_0^N, depending upon which one of accelerator and magnetometer is being calibrated. The DCM ${}^N_B C\left(\phi_{ref,i}, \theta_{ref,i}, \psi_{ref,i}\right)$ takes its arguments from the Euler angles measured by the turntable at the i-th measurement ($i \in [1, N], N \geq 12$).

3.2 Without Using the Reference Angles

Recall the standard form of the ellipsoid surface in the Cartesian coordinates Oxyz:

$$ax^2 + by^2 + cz^2 + 2gx + 2hy + 2iz = 1 \tag{24}$$

Where vector $\beta = \begin{pmatrix} a \; b \; c \; g \; h \; i \end{pmatrix}^T$ denotes coefficients derived from linear least mean square method. Then the measurement from the sensor after calibration is calculated by the equation:

$$c = \begin{pmatrix} 1/K_x & 0 & 0 \\ 0 & 1/K_y & 0 \\ 0 & 0 & 1/K_z \end{pmatrix} r + o \tag{25}$$

Where:

$$o = \begin{pmatrix} -a/g & -b/h & -c/i \end{pmatrix}^T \tag{26}$$

$$G = 1 + g^2/a + h^2/b + i^2/c \tag{27}$$

$$K = \begin{pmatrix} K_x \; K_y \; K_z \end{pmatrix}^T = \begin{pmatrix} \sqrt{a/G} \; \sqrt{b/G} \; \sqrt{c/G} \end{pmatrix}^T \tag{28}$$

4 Experimental Results

In this section, we use the IMU turntable system for the calibration purposes and filters evaluation. The layout of the components on the turntable system is shown in Fig. 3.

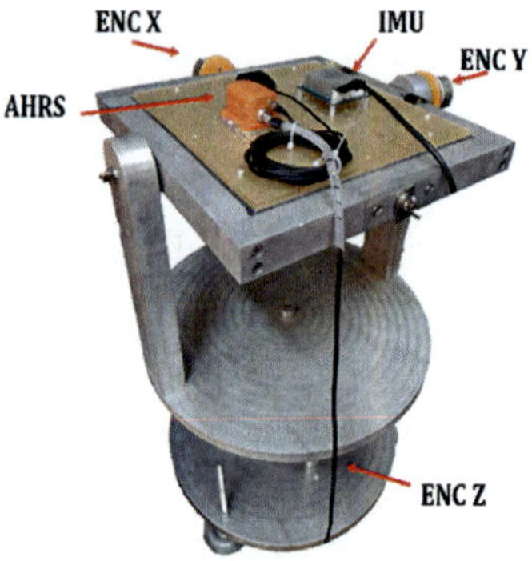

Fig. 3. IMU turntable system.

4.1 IMU Calibration Results

Data sets collected from the turntable experiments are thoroughly analyzed to calibrate accelerator and magnetometer according to methods in Sect. 3.1 and 3.2 respectively. The calibration results are presented as shown in Fig. 4. There are an extra parameter ζ used to detect static intervals. It acts as a logical switch with constraint.

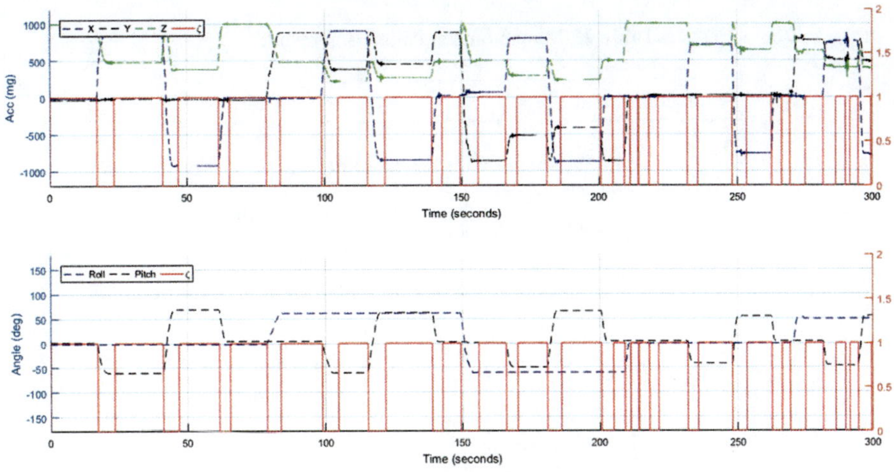

Fig. 4. Three-axis accelerator data for calibration.

$$\zeta_i = \begin{cases} 1, \; if\, \varepsilon(\tau) \leq \varepsilon_{threshold}, \forall\tau \in [i - n_d, i] \\ 0, \; otherwise \end{cases} \tag{29}$$

Where:

- $\varepsilon_{threshold}, n_d, N$ is constants.
- $\varepsilon(i) = \sqrt{\sum_{k=x,y,z} [\mathrm{var}_N(r_k(i))]^2}$
- $\mathrm{var}_N(r_k[i])$ is variance of the vector containing $r_k[i]$ and $(N-1)$ previous elements.

Each static point in calibration process is the average of the static interval. The processes of accelerometer and magnetometer result in Table 1 and Table 2 (Fig. 5).

Table 1. Accelerometer calibration result.

Parameter	Value	Unit
M	$\begin{pmatrix} 0.9991 & -0.0004 & -0.0375 \\ -0.0045 & 0.9985 & -0.0243 \\ 0.0444 & 0.0175 & 0.9997 \end{pmatrix}$	
o	$\begin{pmatrix} -0.0330 & 0.0031 & -0.0038 \end{pmatrix}^T$	g

Table 2. Magnetometer calibration result.

Parameter	Value	Unit
K	$\left(459.6600\ 459.0974\ 449.7460\right)^T$	
o	$\left(-24.6807\ 61.0005\ 11.8629\right)^T$	Milligauss

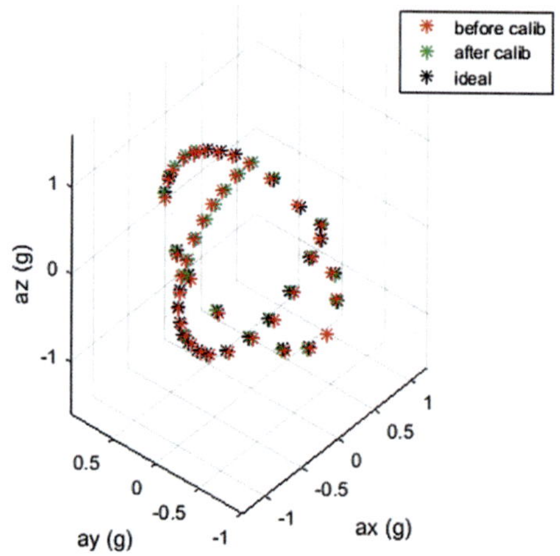

Fig. 5. Distribution of static acceleration points in space.

First, we compare the accelerometer before and after calibration with a simulation one, as obtained in the following RMS error table (Table 3).

Table 3. Accelerometer calibration method evaluation with RMS Errors.

Axis	Before	After
x	0.0261	0.0086
y	0.0270	0.0034
z	0.0176	0.0021

For the magnetometer calibration method evaluation, we rely on the criterion of dispersion based on their variance (Table 4), (Fig. 6).

Table 4. Magnetometer calibration method evaluation.

	Before	After
Value	19.4953×10^{-4}	3.7346×10^{-4}

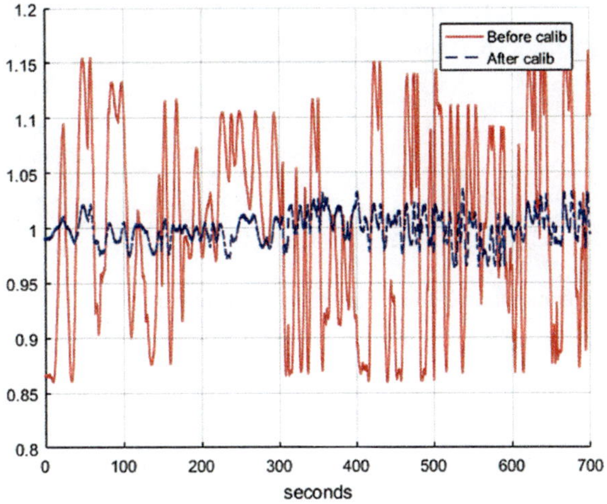

Fig. 6. The module of magnetic vectors before and after calibration.

4.2 Orientation Estimation Algorithms

The mode-1 filter built in Sect. 2 is realized by the microcontroller and afterwards, we evaluate the filter quality under different conditions by means of the turntable:

- Normal condition: STATIC, TURN_X, TURN_Y, TURN_Z, TURN_XYZ.
- External acceleration involved condition: MOVE_X.
- External acceleration magnetism: STATIC_MAG_EXT, TURN_Z_MAG_EXT (Table 5), (Fig. 7).

Table 5. RMS errors of the mode-1 filter.

Name	RMS error (degree)		
	ϕ	θ	ψ
STATIC	0.4055	0.0989	0.2977
TURN_X	0.2640	0.2892	0.3077
TURN_Y	0.4324	0.3495	0.3278

(*continued*)

Table 5. (*continued*)

Name	RMS error (degree)		
	ϕ	θ	ψ
TURN_Z	0.6066	0.6297	0.5540
TURN_XYZ	1.3928	1.0220	1.6283
STATIC_MAG_EXT	0.3729	0.3529	0.7769 (2.9158/2.8600)
TURN_Z_MAG_EXT	0.4903	0.5509	2.7880 (4.9982/4.2989)
MOVE_X	0.6613 (0.8310/0.2540)	0.8494 (1.4740/3.2868)	×

(X): Not considered.
(EKF2/EKF3): Data from mode-2 and mode-3 filter respectively

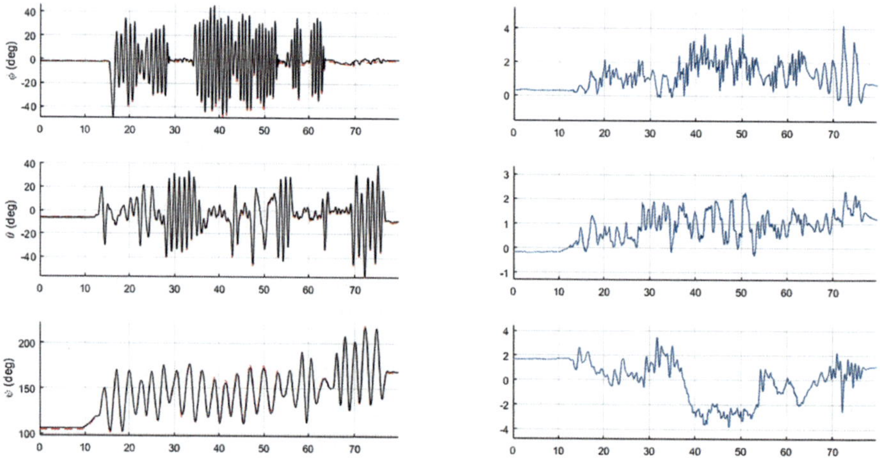

Fig. 7. Estimation result and error in TURN_XYZ.

The above data table and figure give us the final evaluation. It is noticed that the calibration methods help improve the quality of inertial angle information, which would be greatly effective in many related fields. For accelerometer, the RMS errors of all the three axes between the acceleration static points and the reference points decreases after calibration. Moreover, for magnetometer, the loci of the magnetic field is distributed around the unit sphere with a smaller dispersion after calibration. Mode-1 filter has performed better RMS errors, which are less than 2° in normal conditions, as well as less affected by external acceleration and magnetism than the mode-2, mode-3 filter.

5 Conclusion

The paper has developed IMU mathematical model with several modes of Kalman filter, followed by calibration methods and orientation estimation algorithms which yield optimistic outputs. The results show that our calibration methods contribute to reducing the error of the sensor, while the orientation estimation algorithm experiences the RMS error less than 2° in normal conditions and less influenced by the external acceleration/magnetism than the standard filter or the uncalibrated one.

Results from the paper can be applied to GPS/INS integrated positioning system, gimbal camera, virtual reality glasses…

Acknowledgement. This research is supported by National Key Lab. of Digital Control and System Engineering (DCSELAB) and Laboratory of Advance Design and Manufacturing Processes and funded by Vietnam National University Ho Chi Minh city (VNU-HCM) under grant number C2019-20b-02.

References

1. Bishop, G., Welch, G.: An introduction to the Kalman filter. Proc. of SIGGRAPH Course **8**, 41 (2001)
2. Suh, Y.-S., Park, S.-K., Kang, H.-J., Ro, Y.-S.: Attitude estimation adaptively compensating external acceleration. JSME Int. J. Ser. C Mech. Syst. Mach. Elements Manuf. **49**, 172–179 (2006)
3. Yun, X., Bachmann, E.R.: Design, implementation, and experimental results of a quaternion-based Kalman filter for human body motion tracking. IEEE Trans. Robot. **22**(6), 1216–1227 (2006). https://doi.org/10.1109/tro.2006.886270
4. Phuong, N.H.Q.: Study on orientation estimation with three different representations. In: Proceedings of the International Symposium on Electrical \& Electronics Engineering 2007, Vietnam, pp. 420–426 (2007)
5. Zihajehzadeh, S., Loh, D., Lee, M., Hoskinson, R., Park, E.: A cascaded two-step Kalman filter for estimation of human body segment orientation using MEMS-IMU. In: Engineering in Medicine and Biology Society (EMBC), 2014 36th Annual International Conference of the IEEE, pp. 6270–6273 (2014)
6. Tung, D.A., Hao, N.V.: DCM-based orientation estimation using cascade of two adaptive extended Kalman filters. In: 2013 International Conference on Control, Automation and Information Sciences (ICCAIS), pp. 152–157 (2013)
7. Lee, J.K., Park, E.J., Robinovitch, S.N.: Estimation of attitude and external acceleration using inertial sensor measurement during various dynamic conditions. IEEE Trans Instrum. Measure. **61**, 2262–2273 (2012)
8. Pedley, M.: High precision calibration of a three-axis accelerometer. Freescale Semiconductor Application Note, vol. 1 (2013)
9. Tedaldi, D., Pretto, A., Menegatti, E.: A robust and easy to implement method for IMU calibration without external equipments. In: 2014 IEEE International Conference on Robotics and Automation (ICRA), pp. 3042–3049 (2014)
10. Vitali, A.: Ellipsoid or sphere fitting for sensor calibration. STMicroelectronics (2016)

Motion Analysis and Fabrication of a Low-Cost Thruster Using Magnetic Coupling

Ngoc-Huy Tran(✉), Thanh-Hai Chau, and Thien-Phuong Ton

Ho Chi Minh City University of Technology, VNUHCM, Ho Chi Minh City, Vietnam
{tnhuy,1870441}@hcmut.edu.vn

Abstract. Underwater vehicle research is now well-developed around the world. Thrusters with O-ring seal and DC motor have a low waterproof rating and cannot work for a long interval of time. In this paper, we will introduce our designed thruster, using magnetic coupling and brushless DC motor. In addition, controllers used to drive BLDC motor are also mentioned such as the PID, Fuzzy and combined controllers. The experiments were also carried out and obtained consistent results with the simulation.

Keywords: Underwater thruster · Magnetic coupling · Fuzzy-PID controller

1 Introduction

Nowadays, the design and manufacture of underwater robots is a highly developed field of research. Unmanned Underwater Vehicles (UUVs) can support or replace people working in deep, polluted water for a long interval of time [1]. One of the most important devices in these vehicles is the thruster. In Vietnam, marine thrusters are sold at very high prices and must be imported. For example, a thruster of Tecnadyne costs $4000, and the thruster of Shenzhen KDWS can reach up to $4200. Such high costs are a huge obstacle for researches in Vietnam. Therefore, the research, design, and manufacture of underwater thruster are highly essential.

The disadvantages of the former thrusters come from O-ring seals and DC motors. The seals can be damaged at high-speed movements, which leads to the destruction of inner electrical components. DC motors with commutator segment and brush often need periodic maintenance, so it is challenging to work underwater over long durations. Our developed thruster takes advantage of magnetic coupling and brushless DC motor (BLDC) to improve its stability. Magnetic coupling uses contactless actuators to provide better protection. Not only assuring a longer operating life, but BLDC also has some other advantages than the conventional DC motor such as better torque characteristics, higher efficiency, noiseless operation, etc.

There are several methods to drive brushless DC motor. Prasad et al. in [2] present the method to build models and equations of BLDC, especially motors with trapezoidal back electromotive force (back EMF) using MATLAB/Simulink. Paper [3] introduces a strategy for the control of torque using the current observer and state feedback algorithm

D. F. Cortes Tobar et al. (Eds.): AETA 2019, LNEE 685, pp. 552–563, 2021.
https://doi.org/10.1007/978-3-030-53021-1_56

for a BLDC. Regarding the control technique, paper [4] compares the quality of PI and fuzzy controllers to find out which is more optimal. In this paper, we will improve the performance of the PID controller by combining it with fuzzy controllers. There are 3 types of controllers in this paper: PID, Fuzzy-PID, and Advanced Fuzzy-PID.

This paper highlights the design and experiment results of our thruster, including mechanical and electrical design. The mechanical study focuses on simulating the stress field and thrust of the propeller, calculating the thickness of the housing to allow safe operation at a depth of 100 m underwater and the magnetic coupling design. In the electrical part, we propose some BLDC control algorithms. The experiments were also taken and obtained good results.

2 Mechanism Design

With the advantages and disadvantages compared to DC motors and three-phase motor, brushless motors have been explored and studied. The thruster is designed and manufactured to follow these requirements:

- The thrust should be in range of 5 kgf to 6 kgf;
- The rated power of the thruster is 600 W;
- The thruster has a marginal depth rating of 100 m;
- The thruster must be able to work in temperature varying from 3 °C to 28 °C.

To meet the requirements of waterproof protection, power of the thruster, types of motors and propeller failures, the design solutions were selected as:

- The magnetic coupling is chosen because of frictionless property, highly waterproof, long-term operation and low maintenance requirement.
- A BLDC motor TBM (S)-7646-A that has its input voltage of 48 V, the capacity of 335 W, speed of 1885 rpm at its normal operating condition, manufactured by Kollmorgen [5], is selected to be the main thrust motor.

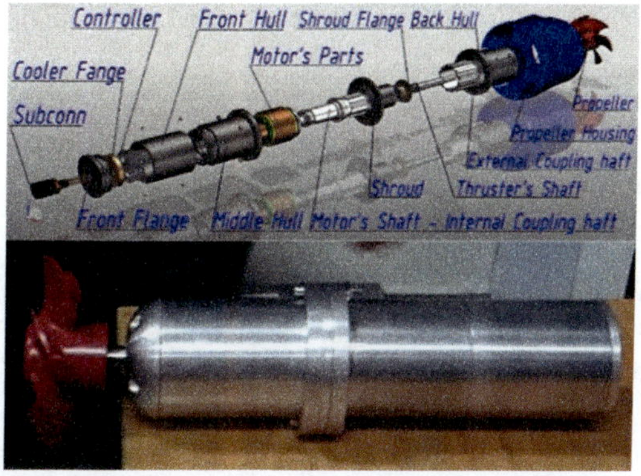

Fig. 1. The detailed structure of the designed thruster (above), BLDC Thruster (below).

- The propeller for the thruster can withstand the maximum thrust of 70 N.
- The nozzle reduces the effect of the fluid drag force and promotes the thrust via an optimal geometry.

Figure 1 describes the detailed structure of the designed thruster. The design is divided into several modules, including the thruster housing, magnetic coupling, propeller, nozzle, and the BLDC motor.

2.1 Thruster Housing

Aluminum alloy material of 6061-T6 can endure pressure up to 80 MPa, melting temperature of 585 °C, good corrosion resistance in seawater and atmospheric conditions. The thruster can operate at the depth of 100 m. It is necessary to ensure that the thruster housing can withstand a pressure of 1.11 MPa (depth of 100 m). Thus, we selected aluminum alloy material of 6061-T6 to manufacture the thruster housing. In the thruster 3D model, the most important parts of the housing are the front flange, front hull, and middle hull, which cover the motor's parts and controller board. The thruster hull is analyzed by NX Nastran with a meshing size of 2 mm 3D Tetra element. Finite element method helps predict the behavior of the thruster hull at 100 meters depth. In this case, 3 and 5 mm thickness hull is checked. The results are shown in Fig. 2 where maximum stress on 3 mm thickness hull is 55.7 MPa, on 5 mm thickness hull is 27.1 MPa

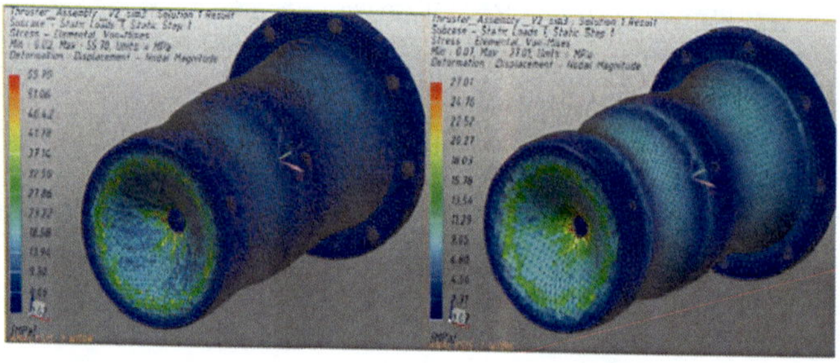

Fig. 2. Von Mises Stress on Thruster hull with a thickness of 3 mm (a) and 5 mm (b).

2.2 Magnetic Coupling

The magnetic coupling plays an important role in improving the waterproof capability of the motor, driver, stabilizing operation under high pressure, preventing overload using the conventional fixed magnet of torque through the containment shroud. With the operating principle based on the gravity of 2-pole magnets to transmit torque from the engine up to the shafts and are separated by a containment shroud made of metal, ceramic, etc. Figure 3 shows the detailed structure of the magnetic coupling.

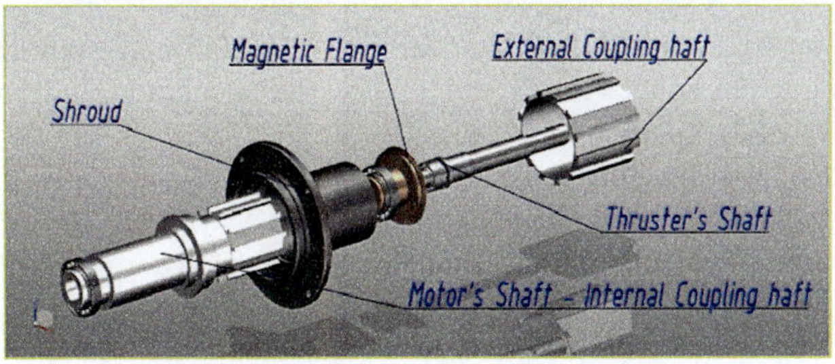

Fig. 3. Magnetic coupling model.

From [6, 7] we calculated the torque versus shifting angle θ between couplings:

As we can see from Fig. 4, the torque is 0 when the angle $\theta = 0°$. When the angle changes, the torque increases until it reaches the peak value Tmax = 7 Nm at $\theta \approx 10°$, then decreases to 0. If the torque on the coupling is greater than Tmax, the slip in magnetic coupling occurs. The driving coupling still rotates but the driven coupling will be restrained and stop working after all.

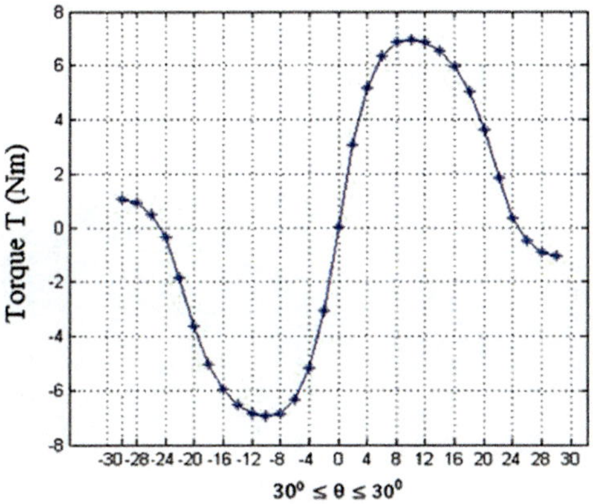

Fig. 4. Magnetic torque versus angular shifting.

From [5], the continuous stall torque of the motor is Tmotor = 2.18 Nm less than the maximum torque value; which proves that the design with magnetic coupling can make the thruster work well at the depth of 100 m below the sea level without slipping.

2.3 Propeller

Finite element methods are the most popular numerical method and used widely for solving problems of engineering and mathematical physics and in analyses of electric motors. From the existing propeller of Vetus [8], CFD analysis is based on the proposed sample wing profile. The boundary condition problem is hypothesized using the model K-ε realizable, solver: MRFSimpleFoam (inner rotation of the same propeller and fixed external domain), constant flow, incompressible. Specifically:

- The geometry of the propeller blade model is obtained through a screwdriver scan.
- Use CAD tools to reproduce geometry; this makes the calculation model more accurate (Fig. 5).
- Use ANSYS/CFX software to calculate and simulate velocity, pressure and thrust field (see Fig. 6 to 7).
- The sample size in calculation and simulation is taken according to the empirical model.
- The basic dimensions of the test tank in the calculation and simulation are taken according to Table 1.
- The results of the thrust simulation using v/p rotational speed were compared with the experiment in Table 2.

Table 1. Dimensions of the propeller and test tank.

Dimensions	Value
Diameter of the propeller	150 mm
Width of the test tank	750 mm
Height from the propeller to surface	195 mm
Height from bottom of the test tank to surface	450 mm

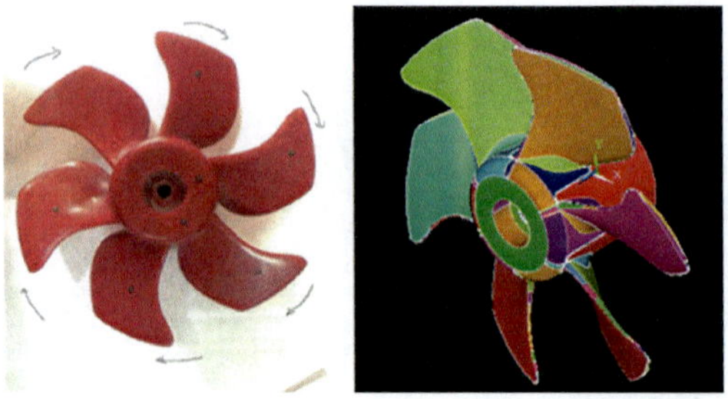

Fig. 5. Result of the 3D scan of the propeller.

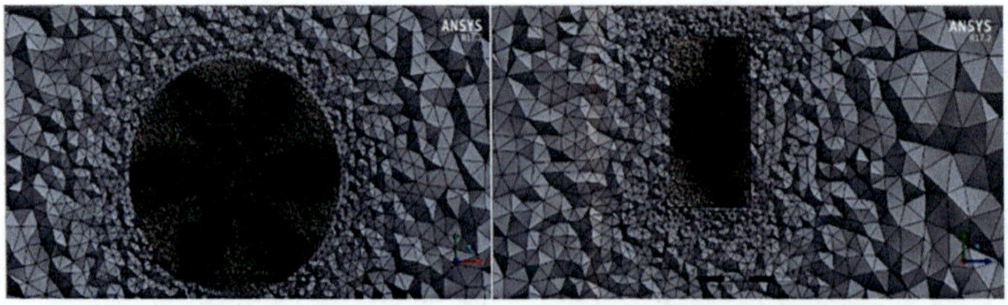

Fig. 6. Element blocks, with 4,646,912 elements and 1,220,709 nodes.

Table 2. Dimensions of the propeller and test tank.

Test	RPM	Thrust [N] (Simulation)	Thrust [N] (Experiment)
1	300	5.10148	5.8625
2	605	21.8028	23.4511
3	808	39.4012	42.2443
4	1008	61.8128	65.3155

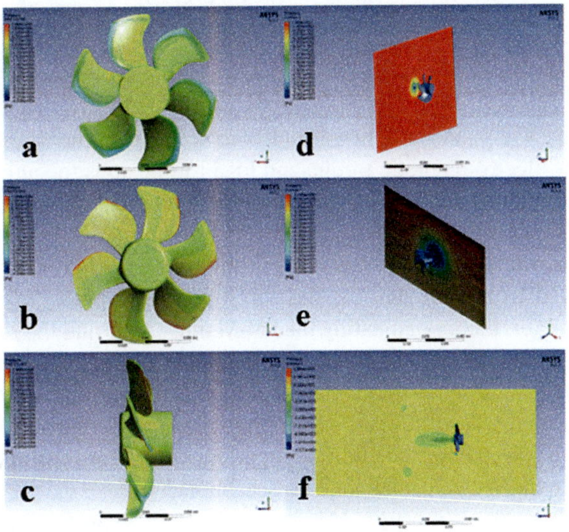

Fig. 7. Simulation results of pressure field effects on the front (a), back (b) and side (c), rear wing (d), front wing (e) and parallel wing (f) of the propeller.

3 Electrical Design

3.1 Simulation

The motor used in simulation and experiment is a three-phase trapezoidal BLDC motor. We can model it into a 3-phase circuit diagram as shown in Fig. 8. Each phase corresponds to a coil with resistance R and inductance (L-M), whereas L is self-inductance and M is the mutual inductance. When the motor is operating, on each phase will appear the back EMF, indicated in the figure by e_a, e_b, e_c, respectively. From [9], we obtain the electrical and mechanical equations of the BLDC:

$$\frac{d}{dt}\begin{bmatrix} i_a \\ i_b \\ i_c \end{bmatrix} = -\frac{R}{L-M}\begin{bmatrix} 1\,0\,0 \\ 0\,1\,0 \\ 0\,0\,1 \end{bmatrix}\begin{bmatrix} i_a \\ i_b \\ i_c \end{bmatrix} - \frac{1}{3(L-M)}\begin{bmatrix} 2V_a - V_b - V_c - 2e_a + e_b + e_c \\ 2V_b - V_a - V_c - 2e_b + e_a + e_c \\ 2V_c - V_a - V_b - 2e_c + e_a + e_b \end{bmatrix} \tag{1}$$

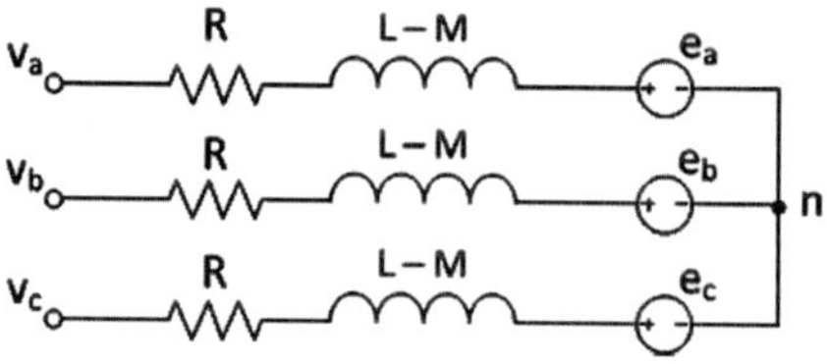

Fig. 8. Three-phase BLDC model.

$$T_e = \frac{1}{\omega}(e_a \cdot i_a + e_b \cdot i_b + e_c \cdot i_c) \tag{2}$$

$$T_e = T_l + J\frac{d\omega}{dt} + B\omega \tag{3}$$

Where i_a, i_b, i_c, V_a, V_b, V_c are the currents and voltages of the three phases, T_e and T_l are respectively electromagnetic torque and load torque, ω is the angular velocity of the rotor. The coefficient J and B are the rotor inertia and the damping coefficient.

According to paper [10], there are two BLDC control models: one-loop and two-loop control. In the first model (see Fig. 9a), there is only one speed-control loop. Based on the signals obtained from Hall sensors, the controller determines the current position and speed of the motor, thereby adjusting the PWM duty cycle and the rules for opening and closing semiconductor switches. The two-loop control model (see Fig. 9b), on the contrary, consists of two control loops: the inner loop is the current control based on the hysteresis current method, while the outer loop is the speed control. Compared to the one-loop model, this model has an additional current controller for three-phase current feedback, which reduces electromagnetic torque ripple.

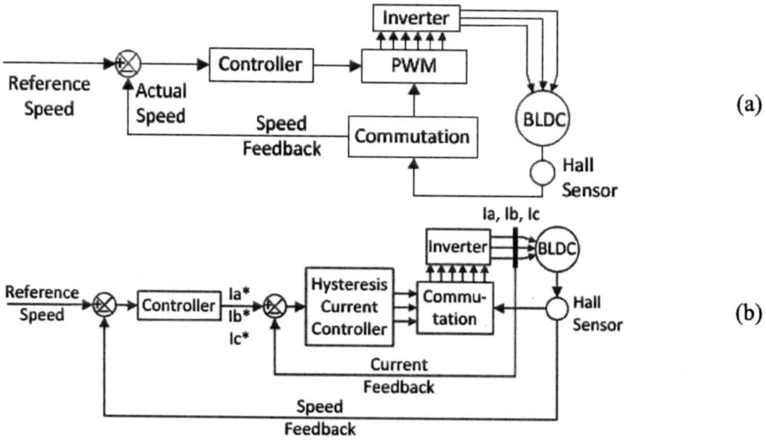

Fig. 9. One-loop control diagram (a), Two-loop control diagram (b).

In the simulation, we use three different controllers: PID, Fuzzy-PID and Advanced Fuzzy-PID. The proportional-integral-differential (PID) controller has been widely used thanks to its simple structure, but it only provides good response to linear models.

Practical systems often comprise nonlinear components. As a result, PID controllers usually perform well at some operating point. At the other different working points, the quality is worse. To overcome these weaknesses, we combined original PID with a fuzzy controller (Fuzzy-PID controller in Fig. 10). The fuzzy controller calculates K_p, K_i, K_d coefficients based on control error to improve the quality.

The Advanced Fuzzy-PID controller (Fig. 11) is studied to further boost the quality of the Fuzzy-PID controller. We use another fuzzy controller to adjust the PID coefficients based on the influences on the quality of response (rising time, settling time, steady-state error, overshoot, etc.). Among the controllers, Advanced Fuzzy-PID controller is the most complex and most time-consuming, but it trades off the best response quality (Fig. 12).

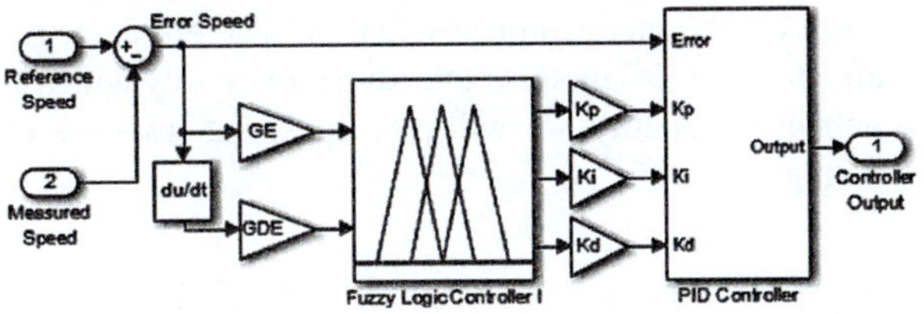

Fig. 10. Structure of the Fuzzy-PID controller.

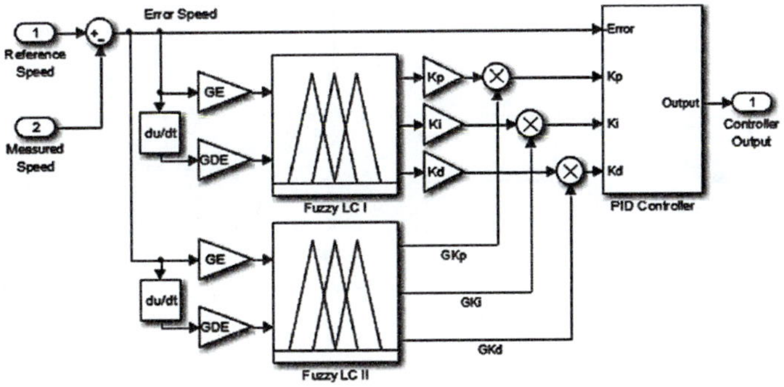

Fig. 11. Structure of the Adv. Fuzzy-PID controller.

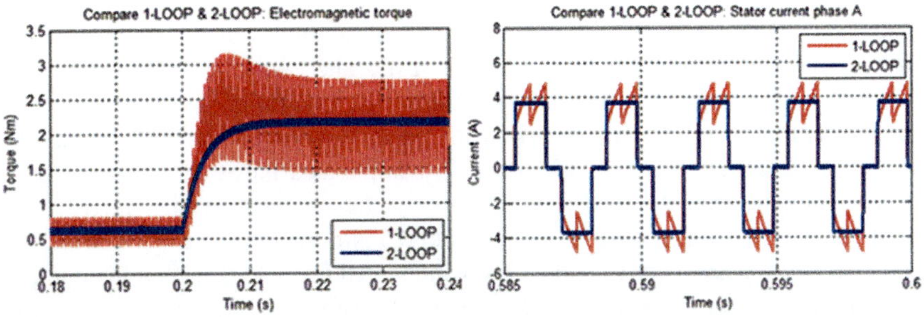

Fig. 12. Comparison between two models regarding electromagnetic torque and stator current.

MATLAB/Simulink is used to simulate the BLDC motor and analyze the results. In the simulation, we focus on comparing the quality of the one-loop and two-loop model and comparing the speed responses of the controllers. The model refers to the KOLLMORGEN TMB(S)-7646-A brushless DC motor [5] for simulation. and experiment as well. First, we compare the electromagnetic torque and phase current responses of two models:

As seen from the above, the two-loop model offers a better response with a smaller amplitude of oscillation (about 0.4 Nm, compared to 1.3 Nm of the one-loop model when the torque is big). As a result, the two-loop model can control the motor more stably in practical environment. Furthermore, the stator current shape in the two-loop model is rectangular, the same as the shape of theoretical back EMF. In the meantime, there are ripples in the current response of the one-loop control model, which leads to the greater magnitude of electromagnetic torque oscillation. Therefore, the two-loop control model will practically drive BLDC motors better and more stably with a longer lifespan. To compare the controllers' quality, we will plot the speed response of 3 controllers simultaneously in a single figure:

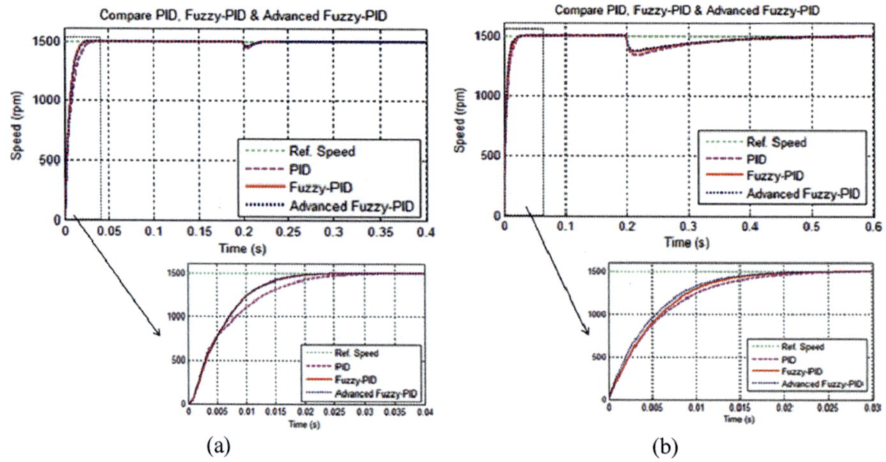

Fig. 13. (a) Speed response of the controllers using one-loop model, (b) two-loop mode

From the results, we can prove that the control quality is increasingly improved from PID to Fuzzy-PID and Advanced Fuzzy-PID. The two-loop control model has some further advantages in comparison with the one-loop counterpart such as smaller rising and settling time without oscillation at the steady state. On the other hand, weak points are here to stay. Not only the model is complicated but it also needs a longer recovery time whenever the load torque suddenly changes. As seen in Fig. 13a, 13b, when the load torque has a sudden change from 0.5 Nm to 2 Nm at t = 0.2 s, the two-loop model needs a recovery time of 0.2 s, which is 10 times longer than the one-loop model with the same rotor speed. At the next stage, we will observe the quality of the controllers with the existence of sensor noise based on the sum of squared error (SSE) (Table 3):

Table 3. SSE (rpm^2) value comparison with and without noise.

Model	Controller	Without noise	With noise	Difference
1-loop	PID	8685	8715	30
	Fuzzy-PID	8155	8179	24
	AF-PID	7893	7923	30
2-loop	PID	8356	8375	19
	Fuzzy-PID	7624	7637	13
	AF-PID	6760	6785	25

As it indicates in the table, the Fuzzy-PID controller has the best noise suppression ability, for both models. However, the Advanced Fuzzy-PID controller without noise can give better performance (regarding the rising time, settling time, overshoot and SSE) than the conventional and fuzzy-combined PID controller. Last but not least, the SSE in the two-loop control model is always smaller than that in the one-loop control model.

3.2 Experiments

The drive of the brushless DC motor is designed with a two-stage structure: the control stage and the power stage. This strategy avoids the effect of thermal noise from the power stage that probably affects the control results. In the control stage, the STM32F103C8T6 microcontroller with high processing speed allows for PID algorithm, Fuzzy algorithm, and advanced control algorithms. In addition, the control stage also has a current sensing circuit to break the power stage when the thruster overloads while temperature sensing components prevent MOSFETs from overheating. The power stage uses MOSFETS which have high temperature endurance and high power.

As shown in Fig. 14, the experiments were carried out and obtained consistent results with the simulation. Besides, experiment results show that the thruster is stable and less noise. At 1000 rpm, the thrust of the engine was able to reach more than 6 kgf corresponding to 55% of the motor's power output.

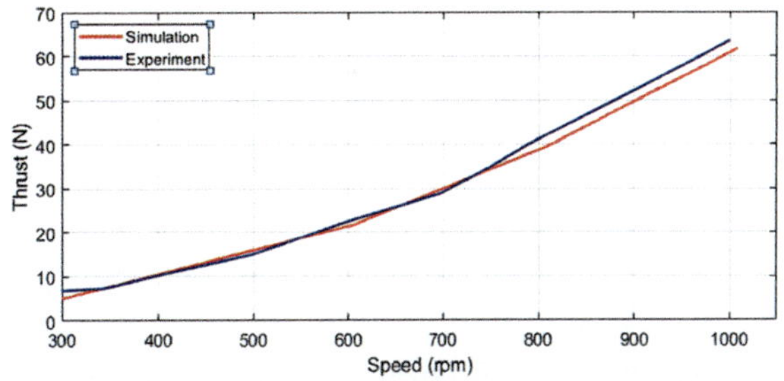

Fig. 14. Comparison between Thrust/Speed simulation (red) and experiment (blue).

Figure 15 depicts the relationship between speed and current of the thruster. Considering the motor speed of 1000 rpm, the circuit still withstands 10A currents because of its good heat dissipation.

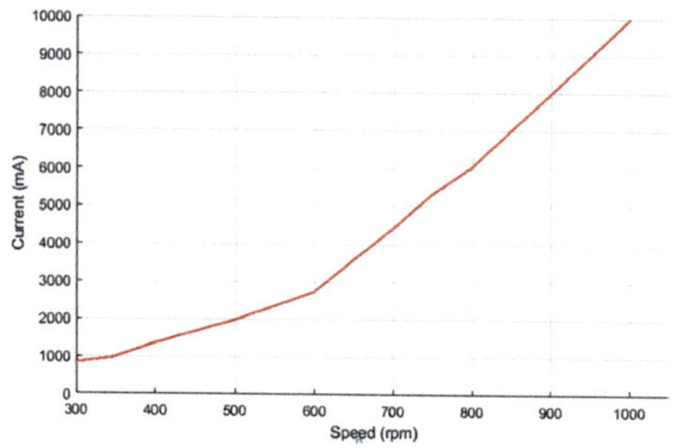

Fig. 15. BLDC Motor Current/Speed Curves.

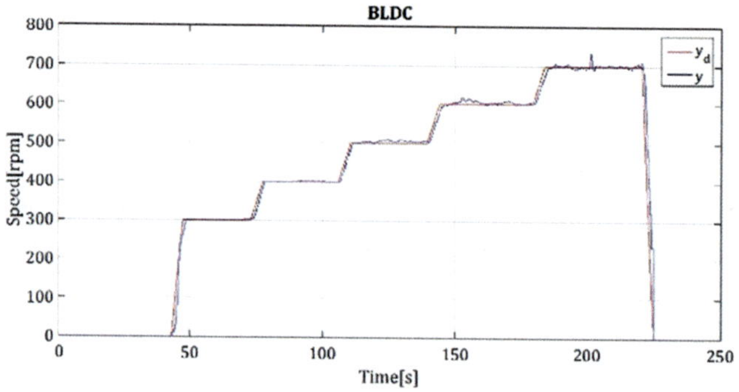

Fig. 16. PID speed response with different load.

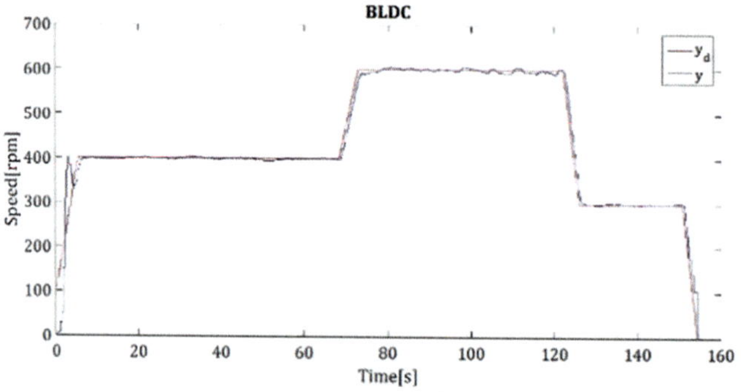

Fig. 17. Fuzzy speed response with different load.

From Fig. 16 and 17, the PID and Fuzzy algorithms generally give experimental results of the speed of the motor mounted at different set speeds loads. The PID algorithm outputs fast response time with many ripples while the Fuzzy algorithm has slower response time but smoother than PID. At the transient stage, the noise affects the actual speed response due to the thruster's non-linear property.

4 Conclusions

The association of magnetic coupling and brushless DC motor is an outstanding point in the design and manufacture of our thruster. Magnetic coupling uses neodymium magnets to transmit up to 7 Nm from a shaft to the other. It can also protect the motor from overload. Simulation and experimental results show that the chosen propeller is capable of delivering thrust up to 65 N at speed of 1000 rpm. Based on the finite element method, the 3 mm thick housing helps the thruster to work at 100 m water depth (pressure of 145psi). The BLDC increases the thruster's performance and operating time. Based on the simulation results, we can conclude that the two-loop model makes the motor's operation more stable with higher efficiency than the one-loop model. Comparing among the aforementioned controllers, we see that the controller combines a PID controller with a fuzzy logic controller is better than the conventional PID. Advanced Fuzzy-PID controller has the best overall response, but its noise suppression ability still needs to be improved. Finally, the practical experiment achieves good performance with small errors.

Acknowledgement. This research is supported by National Key Lab. of Digital Control and System Engineering (DCSELAB), HCMUT and Laboratory of Advance Design and Manufacturing Processes and funded by Vietnam National University Ho Chi Minh city (VNU-HCM) under grant number B2018-20b-01.

References

1. Ngoc-Huy, T., et al.: Design, control, and implementation of a new AUV platform with a mass shifter mechanism. Int. J. Precis. Eng. Manuf. **16**(7), 1599–1608 (2015)
2. Prasad, G., et al.: Modeling and simulation analysis of the brushless DC motor by using MATLAB. IJITEE **1**, 27–31 (2012)
3. Zhao, L., Zhang, X., Ji, J.: A torque control strategy of brushless direct current motor with current observer. In: 2015 IEEE ICMA, Beijing, pp. 303–307 (2015)
4. Sreekala, P., Sivasubramanian, A.: Speed control of brushless DC motor with PI and fuzzy logic controller using resonant pole inverter. In: IEEE PES Innovative Smart Grid Technologies, Kerala, pp. 334–339 (2011)
5. KOLLMORGEN TMB(S) Motor Selection Guide
6. Akoun, G., Yonnet, J.P.: 3-D analytical calculation of the forces exerted between two cuboidal magnets. IEEE Trans. Magn. **20**, 1962–1964 (1984)
7. Eliès, P., Lemarquand, G.: Analytical optimization of the torque of a permanent-magnet coaxial synchronous coupling. IEEE Trans. Magn. **34**, 2267–2273 (1998)
8. Vetus Shop Hompage. http://www.vetus-shop.com/
9. Ngoc-Huy, T., et al.: Study on design, analysis and control an underwater thruster for unmanned underwater vehicle (UUV). In: AETA 2017 - Recent Advances in Electrical Engineering and Related Sciences: Theory and Application, pp. 753–764 (2017)
10. Ngoc-Huy, T., et al.: A study on advanced fuzzy-PID speed control for underwater thruster. In: International Symposium on Electrical and Electronic Engineering, Ho Chi Minh city, Vietnam (2017)

Implementation of Path-Following Algorithm for an Unmanned Surface Vehicle Using Viam-Navi GPS/INS Module

Tu-Cuong Nguyen, Ngoc-Huy Tran(✉), and Xuan-Dung Trinh

Ho Chi Minh University of Technology, VNUHCM, Ho Chi Minh City, Vietnam
{1870012,tnhuy}@hcmut.edu.vn

Abstract. With the understanding and development of Unmanned Surface Vehicles (USV), will be beneficial to not only the environmental monitoring, military, but also in many commercial applications. The vehicle adopts a catamaran design making it easy to assemble, transport and integrated sensors for the desired applications. Sensor data is collected to estimate, improve the quality of the sensors and display real-time monitor environmental parameters. This paper describes the control system which provides path following, under external forces: wind, waves and currents. The guidance and control system are developed to guide the vehicle to follow waypoints using the Line of Sight (LOS) algorithm based on the input of the sensors. These waypoints are automatically received from ground station and stored on an on-board computer. The navigation system is one of the most important USV subsystems. To improve the ability of navigation for vehicles, Viam-Navi GPS/INS Module: integration of Inertial Navigation System (INS) and Global Positioning System (GPS) is developed with low-cost, highly accurate and stable navigation system. Beside, modelling and simulation are presented for developing an appropriate guidance and control systems, the results of which is applied to the designed platform. Finally, experiment of these navigation, controllers and guidance laws show that a low deviation between the vehicle and path.

Keywords: Unmanned Surface Vehicles (USV) · Line of Sight · Path following

1 Introduction

Today, a new generation of ships - Unmanned Surface Vehicle (USV) is the attention of experts in the world, because the diversity of functions, providing a safe, low investment costs, specially can operate on the surface without human intervention. With flexibility, USV's extensive activities to allow them to perform many different tasks: commercial, environmental monitoring, bathymetric mapping, scientific and military applications. Besides that vehicle support the work of the auto rescue, rescue is a breakthrough idea is more countries in the world adopt and increasingly popular in the world [1]. Then a large number of new USV have been developed in the world. However, how to control the USV efficiently is still challenging: it is hard to determine an accurate vehicle dynamic

D. F. Cortes Tobar et al. (Eds.): AETA 2019, LNEE 685, pp. 564–574, 2021.
https://doi.org/10.1007/978-3-030-53021-1_57

model, generally highly nonlinear, time-varying, and easy affected by environment such as ocean currents, waves and wind [2, 3]. In some countries, USV's technology has attracted scientists and engineers as a challenging field, because they have a system that is complex and involves many areas of technology. The researchers focused on aspects of navigation, controller, path following, avoid obstacle [4] to enhance the performance of USV. So, new USV platforms is built and developed for different purposes, especially in the field of the environment. The system including hardware and software architectures is developed in order to satisfy various needs of the experiment in real-life environment. Using the heading autopilot and Line Of Sight (LOS) guidance create tasks for the vehicle go to prescribed waypoints. The navigation system, which can provide the attitude, velocity and position for the vehicle, is one of the most important USV subsystems [5]. Therefore, Viam-Navi GPS/INS Module is designed use method of loosely coupled GPS/INS integrated navigation system and applied to this USV [6–9]. The Euler angles estimation and the velocity constraints are used to improve accuracy. Next, modelling and simulation [10–13] is presented for developing an appropriate guidance and control systems [11–14] under realistic simulation conditions. The controller associated with guidance system to create a desired angle, making the vehicle follows the trajectory and approaching the desired target position. Finally, numerical simulation and experiments for path-following of USVs were performed. This paper is organised as follows. Section 2: The design of an USV includes hardware and software, are introduced to provide common references for the relevant researchers. Beside, GPS/INS Module will be introduced with low-cost, highly accurate and stable navigation system. Section 3: The dynamic models of USV are presented to develop an appropriate control and control system, which will be applied to the designed platform. Section 4: Path-following using the Line of Sight (LOS) algorithm based on the input of the sensors. Section 5 presents experimental results, implemented in a real-life environment with path following algorithms. Finally, Sect. 6 concludes this paper.

2 System Design of USV

USV is designed to perform autonomous tasks: mainly in environmental monitoring, mapping and datastation. It is built with a propulsion system, a control system, data collection system and an image acquisition system.

The vehicle is a fully autonomous catamaran with high stability and easily integrated sensors for a range of environmental monitoring applications. Besides the size, hull of a ship and robustness of the vehicle are optimized to apply navigation, guidance and control algorithms efficiently. USV is about 1.2 m length, overall width about 0.85 m. It is equipped with 2 main DC thruster, 2 brushless motor, dedicated for underwater vehicle, which allows for a maximum speed about 2 knots. The control system is housed in a plastic, water-proof box and contains a single-board computer, radio frequency (RF) transceiver, module control thrusters.... The batteries used to provide the power for the propulsion and electronic systems are carried within the hulls. A sensor that plays the important role in perform environmental monitoring missions is the YSI - Multi-Parameter Water Quality Sonde. This will be installed in the center of the vehicle as illustrated in Fig. 1. This unit is capable of measuring several important parameters such as the oxygen, pH, conductivity and turbidity…, and quickly determine the water pollution level and hence detect of the polluting source. A website is designed to display, update and manager real-time monitor environmental parameters.

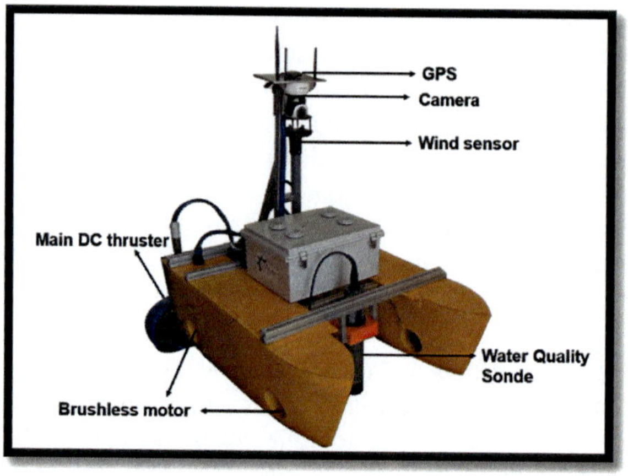

Fig. 1. Unmanned surface vehicle platform.

Hardware System Design
The generalized hardware architecture of the vehicle in Fig. 2, it consists of main control board and sub-boards. The main control board consists of a controller which communicates, processing and perform task from ground station. Sub-boards include: navigation, sensor and actuator board. A CAN bus network is developed to connect the sub-boards and allows connecting some systems with reduced cabling and reliability. By combining the USV hardware, the USV platform can work in two modes: manual mode by the RF or wireless modem and automatic mode by the navigation system. Besides, the vehicle has a video camera, using record the entire motion, images and video in different tasks. This section presents the hardware components which are essential to develop the USV platform and overview of the technical specifications of the main components.

Navigation Board
For USV platform to work in a stable and efficient manner, navigation is one of the most important issues to be aware of. So, an effective method is to integrate GPS with INS, in which the center is a nonlinear estimator (e.g. the Extended Kalman filter [15, 16]) to determine the navigation error, from which it can update the velocity and position the object more accurately. Low cost Viam-Navi GPS/INS Module is peculiar in their high stand-alone accuracy and run-to-run stability, which can result in low errors over long time intervals. Figure 3 show the performance of GPS/INS module. The experimental system is built on a low-cost IMU with tri-axis gyroscope, accelerometer and magnetometer and a GPS module to verify the model algorithm. The update rate of the integrated system is equal to the INS rate of 100 Hz and the rate of GPS is 10 Hz. The data acquisition and processing system is performed on an ARM Cortex-M4 microcontroller.

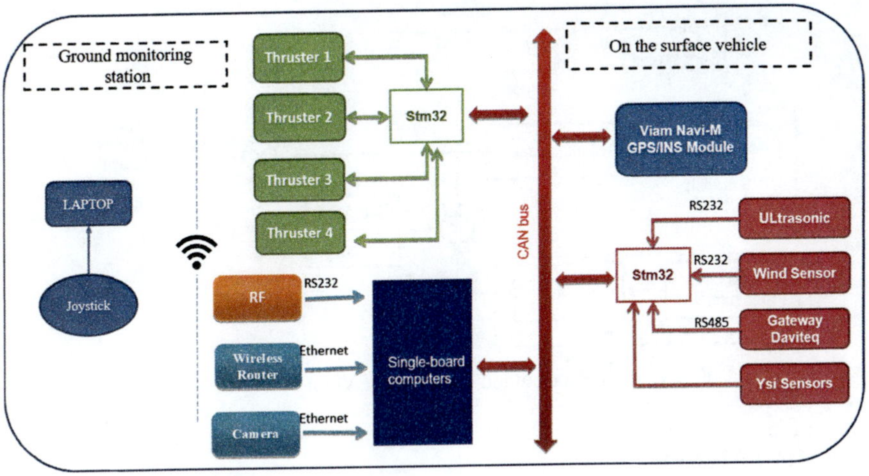

Fig. 2. Hardware functional block diagram.

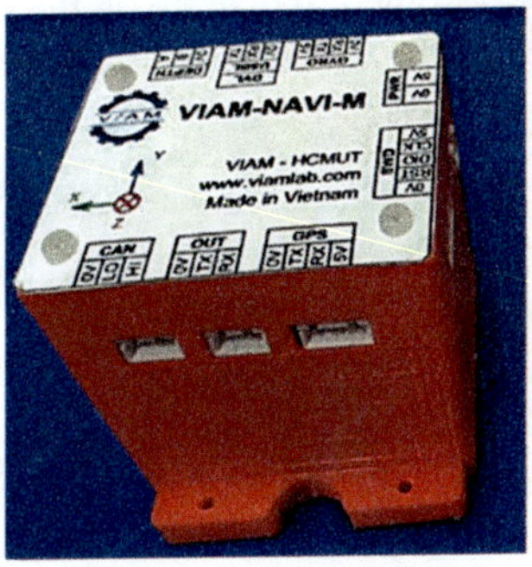

AHRS dynamic range	Roll	$\pm 180^0$
	Pitch	$\pm 90^0$
	Yaw	$\pm 180^0$
AHRS accuracy	Roll	1^0
	Pitch	1^0
	Yaw	2^0
GPS/INS position accuracy		2 m CEP
GPS/INS velocity accuracy		0.1 m/s
GPS/INS maximum update rate		100 Hz
AHRS maximum update rate		400 Hz
Interfaces		RS232, RS485, CANBUS

Fig. 3. Viam-Navi technical specification.

Viam-Navi GPS/INS Module is designed use method of loosely coupled GPS/INS integrated navigation system. The loosely coupled model also referred to as "decentralized" filtering, consists of two estimators. The first one is a nonlinear estimator. It combines the INS estimation results with the GPS results to estimate the position, velocity, attitude error and the IMU sensor's error. The second is the GPS filter. It uses the pseudorange and Doppler measurement values from GPS module to determine the position, velocity. Figure 4 shows the diagram of loosely coupled model. The Extended Kalman filter is used estimate states, which is suitable for nonlinear systems. Measurement values from IMU sensor (angular rate and acceleration) after being computed using the Euler angles estimation and INS mechanization, will be compared with the position and velocity of the GPS.

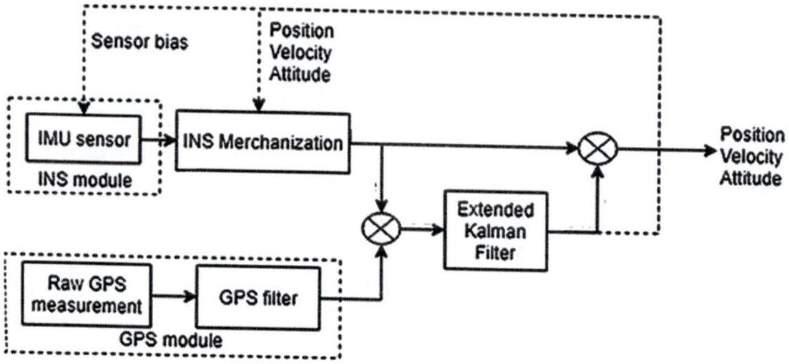

Fig. 4. Loosely Coupled GPS/INS of Viam-Navi Module.

3 System Model

The dynamic model of USV is built for use in developing path following, station-keeping, obstacle avoidance strategy.... Since these USV are designed to operate on the surface of the water, the vehicle motion can only move in horizontal plane: x and y direction and rotation about the z axis (Fig. 5).

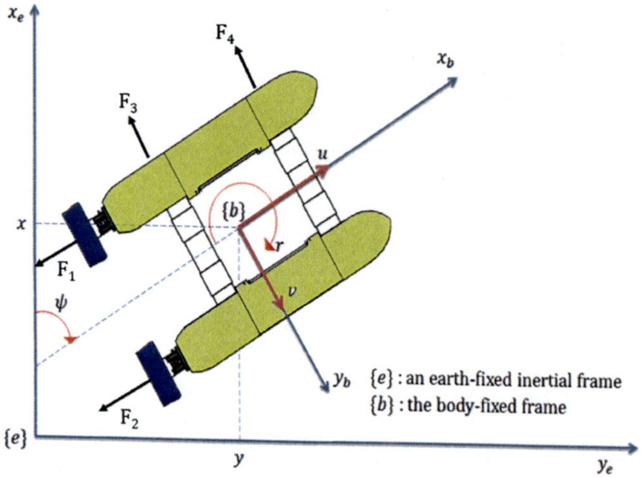

Fig. 5. Reference frame and the motor forces acting on the vehicle.

According to [11, 13] the dynamic model of USV is:

$$\begin{cases} M\dot{\upsilon} + C(\upsilon)\upsilon + D(\upsilon)\upsilon = \tau + \tau_{ext} \\ \dot{\eta} = R(\psi)\upsilon \end{cases} \tag{1}$$

$R(\psi)$ is the three DOF rotation matrix for converting from body-fixed to earth-fixed frames:

$$R(\psi) = \begin{pmatrix} cos(\psi) & -sin(\psi) & 0 \\ sin(\psi) & cos(\psi) & 0 \\ 0 & 0 & 1 \end{pmatrix}$$

Where M is the system inertia matrix, $C(v)$ is the Coriolis matrix, $D(v)$ is the damping matrix, and $\tau = [\tau_1, \tau_2, \tau_3]^T$ is the control input. The vector $\dot{\eta} = [\dot{x}, \dot{y}, \dot{\psi}]^T$ describes the vehicle's North ($\dot{x}$), East ($\dot{y}$) velocities and the angular velocity ($\dot{\psi}$) around the z axis of the ship in an earth-fixed inertial frame $\{e\}$, and $\upsilon = [u, v, r]^T$ contains the linear velocities: surge (u), sway (v) and yaw rate (r) in the body-fixed frame $\{b\}$. Environmental disturbances affect the vehicle through the disturbance vector τ_{ext}.

4 Guidance System

The vehicle need to follow desired trajectory such as zigzag, square… in different missions, so a guidance system is designed and developed. Guidance system is responsible for continuously creating and updating smooth and optimal trajectory commands to the control system according to the information provided by the navigation and sensor system. Then, USV can perform path following behavior by changing heading angle towards the desired paths. Popular method is Line of Sight (LOS) [17–19] method (Fig. 6).

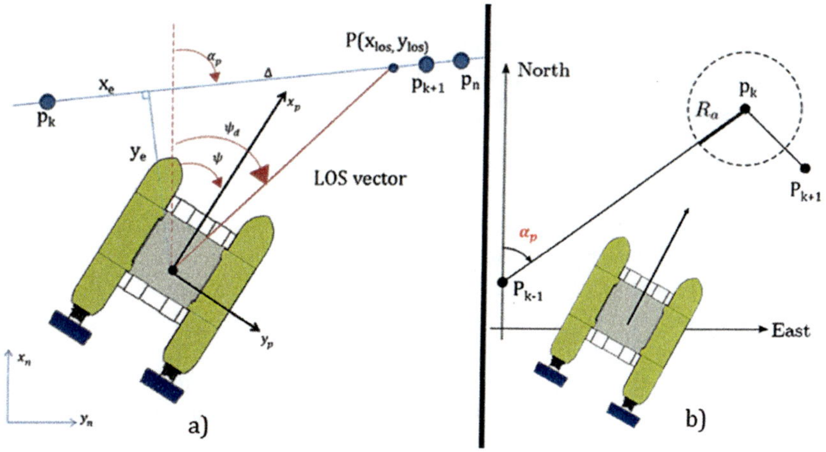

Fig. 6. Guidance algorithms. The LOS algorithm steers the vessel onto the line between the waypoints

In the case of decoupled horizontal plane path-following, the desired heading (or course) and surge velocity set-points are given as:

$$\psi_d = \alpha_p + \arctan(\frac{-y_e}{\Delta}),\ u_d = u_0 \tag{2}$$

Where,

$$\alpha_p = \text{atan2}(y_{k+1} - y_k, x_{k+1} - x_k) \tag{3}$$

is the path-tangential angle,

$$y_e = -(x - x_k)\sin\alpha_p + (y - y_k)\cos\alpha_p \tag{4}$$

is the cross-track error (the shortest distance to the path) and $\Delta > 0$ is the predefined lookahead distance, selected to fit in the experiment. $P_k(x_k, y_k)$ is the position of the k-th waypoint P_k expressed in the NED frame and u_0 is a predefined desired surge speed. Finally, the goal is making the cross track converge to zero:

$$\lim_{t \to \infty} y_e(t) = 0 \tag{5}$$

The LOS vector is considered a vector with tail at the origin of body-fixed frame and head is located at a point (x_{los}, y_{los}) on the tangent line connecting two way-points p_k and p_{k+1}.

When following a path described by a set of waypoints, one needs to know when to switch to the next waypoint. One such method is known as the "circle of acceptance" method (Fig. 6 b). Given the current waypoint $P_k = (x_k, y_k)$, the switching criterion is defined as follows:

$$(x_k - x)^2 + (y_k - y)^2 \le R_a^2 \tag{6}$$

5 Results

A mission is created to perform specific application. USV started out at the location indicated, and go to the first way-point, and is given several waypoints to follow. The experiments present results obtained in real environment conditions.

5.1 Navigation Results

The experimental system is performed to verify the algorithm of Viam-Navi Board. The hardware consists of Viam-Navi GPS/INS module and reference system is the GNSS/INS system from Xsens Technology. The MTi-G-700 [20] can give rotation angles estimate with a 1 degree accuracy, position error of 2 m and velocity error of 0.05 m/s.

For IMU sensors, the amplitude of its noise is large, not use the Euler angles estimator, the result is awful, the attitude, position, velocity errors are enormous. The estimated trajectory (red dots in Fig. 7) does not have the same shape with the reference one (black line). In contrast, when using the angles estimator, the errors are smaller, the accuracy is higher. The horizontal error of our GPS/INS system is 1.69 m, while the error of the individual GPS system is 1.93 m. For this reason, the GPS/INS algorithm can reduce over 10% of the error. And if the vehicle moves very fast, the GPS cannot describe the vehicle's trajectory accurately. Differently, the blue dots (GPS/INS) approximately form a continuous line. From the above results, it can be concluded that the angles estimator can improve the accuracy of the navigation system and the integrated GPS/INS system performs better than the single GPS system. Next, Viam-Navi GPS/INS module will be applied to the designed USV for path following application (Table 1).

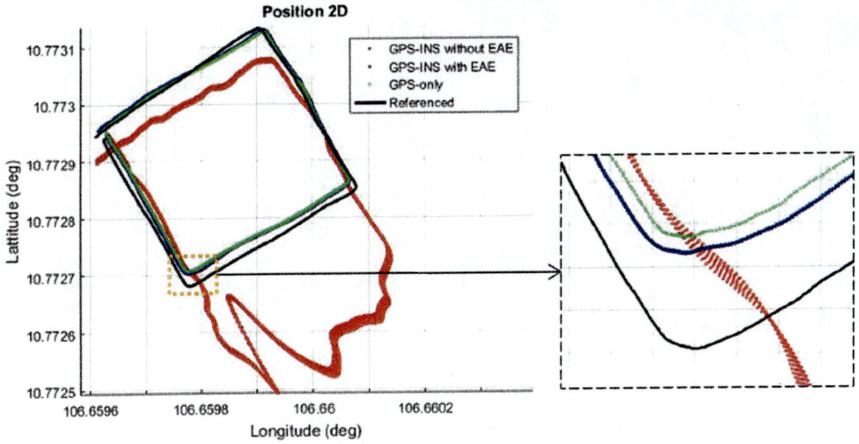

Fig. 7. Experimental results of Estimated 2D position

Table 1. Attitude, position and velocity errors in experiment

Error	GPS-INS without Euler angles estimation			GPS-INS with Euler angles estimation		
Attitude (r-p-y) (degree)	70	70	83	1.31	1.07	16.7
Position NED (m)	11	7.7	131	1.58	0.58	4.42
Velocity NED (m/s)	10	6.4	85	0.27	0.26	0.46

5.2 Path Following Results

In order to verify and check the effectiveness of the path following algorithm for USV, a series of simulation experiments have been performed on a dynamic models in Sect. 3.

The results of the two cases of simulation experiments (Fig. 8) show that the convergence of the vehicle to the desired path is quickly, effectively and reliably based on the above LOS guidance algorithm. In real-life environment, path following algorithms were tested. The mission is a zigzag and square-wave described in a NED coordinate system with points WP1, WP2, WP3…and implemented many times in order to test the credibility of the experiment. Figure 9 shows the results of the position USV during the experiments.

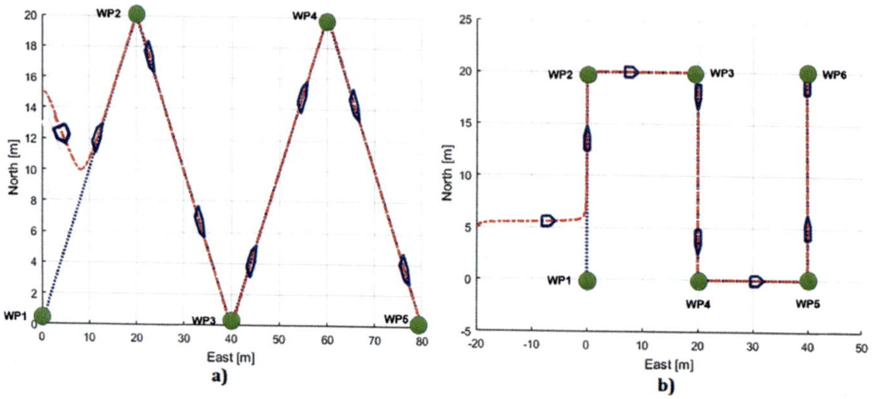

Fig. 8. Simulation of Path following mission a) Zig-Zag test b) Square-Wave test.

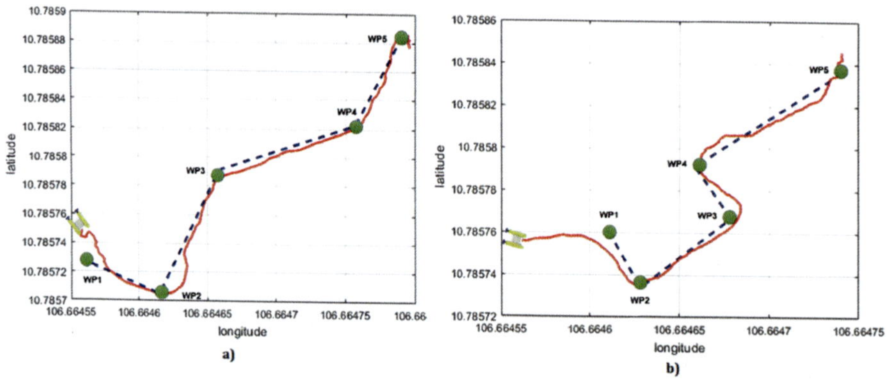

Fig. 9. The Path-following experimental results a) Zig-Zag test b) Square-Wave test.

When the vehicle starts moving, the heading controller is responsible to create force in order to compensate for the disturbances on the hulls due to the translatory motion. In Fig. 9 USV is far from the first waypoint (WP1), the LOS algorithm will create reference angle, which move vehicle follow path between two waypoint. Therefore, the first waypoint may not be passed. For every time step, the heading and distance to the target waypoint from the current position is calculated. During the period between waypoint transitions, the USV operate similarly to the constant heading and the vehicle follows this heading. After reaching the region of acceptance, USV will switch to the next waypoint. The results of this experiment is very good. Moreover, the experimental results, as in Fig. 10, show the Cross-track error ye is in the range of 0.7 m and heading is within $\pm 5°$ error when moving in the path.

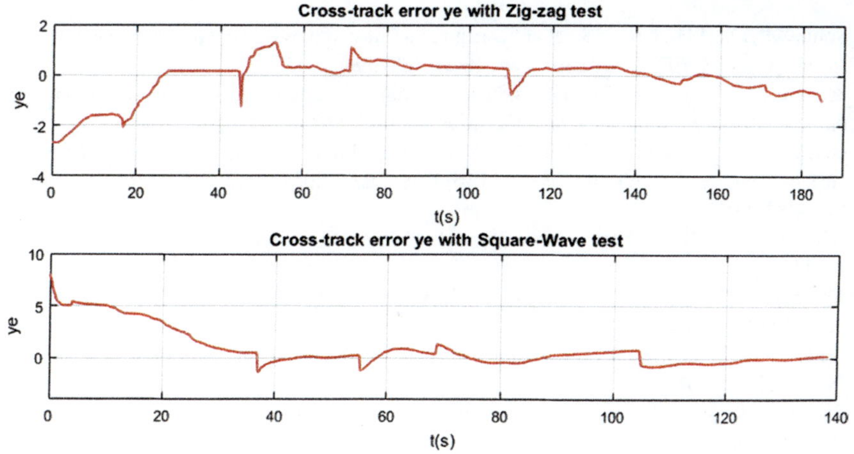

Fig. 10. Cross-track error ye with Zig-Zag test and Square-Wave test

6 Conclusion

This article introduces a model and controller designed for an innovative USV that is capable of environmental applications. The control system including hardware and software architectures are developed and implemented. The vehicle is very easy to deploy, launch it and a person can easily operate using the GUI interface. The autopilot and guidance system are developed that enables users to create mission scripts using local maps. Many real experiments have proven LOS guidance not only simple but also effective. This algorithms allow the vehicle at any initial position outside the desired path to converge on the path. Beside, Viam-Navi GPS/INS module is introduced with low-cost, highly accurate navigation system that enables the USV to operate correctly. The results of this experiment show good performance and the feasibility, reliability of the designed navigation, guidance and control system. Finally, a fully functional USV is developed, implemented in the lake and this platform makes an important step forward in developing the autonomous surface vehicle.

Acknowledgement. This research is supported by Laboratory of Advance Design and Manufacturing Processes and funded by Ho Chi Minh City University of Technology, VNU-HCM, under grant number BK-SDH-2019-1870012.

References

1. Manley, J.E.: Unmanned surface vehicles, 15 years of development. In: OCEANS 2008, pp. 1–4 (2008)
2. Aguiar, A.P., Hespanha, J.P.: Logic-based switching control for trajectory-tracking and path-following of underactuated autonomous vehicles with parametric modeling uncertainty. In: Proceedings of the 2004 American Control Conference, vol. 4, pp. 3004–3010, Boston, MA, USA (2004)

3. Sohn, S.I., Oh, J.H., Lee, Y.S., et al.: Design of a full-cell powered catamaran-type unmanned surface vehicle. IEEE J. Oceanic Eng. **40**(2), 388–396 (2015)
4. Zhuang, J.-y., Zhang, L., Zhao, S.-q., Cao, J., Wang, B.: Radar-based collision avoidance for unmanned surface vehicles. China Ocean Eng. **30**(6), 867–883 (2016)
5. Campbell, S., Naeem, W., Irwin, G.: A review on improving the autonomy of unmanned surface vehicles through intelligent collision avoidance manoeuvres. Ann. Rev. Control **36**, 267–283 (2012)
6. Kim, K.: GPS/INS with non-linear filters. In: ICCAS 2011, KINTEX, Gyeonggi-do, Korea (2011)
7. Schmidt, G.T., Phillips, R.E.: INS/GPS integration architectures. In: RTO Lecture Series RTO-EN-SET-116 (2010)
8. Angrisano, A.: GNSS/INS Integration Methods, Ph.D Thesis, The Parthenope University of Naples, Italy (2010)
9. Nguyen, H.-D., et.al.: Implementation of a GPS/INS tightly-coupled system in urban environment. In: The 3rd Vietnam Conference on Control and Automation, Ho Chi Minh city (2015)
10. Sun, X., Wang, G., Fan, Y., Mu, D., Qiu, B.: An automatic navigation system for unmanned surface vehicles in realistic sea environments. Appl. Sci. **8**, 193 (2018)
11. Fossen, T.I.: Guidance and Control of Ocean Vehicles. Wiley, New York (1994)
12. Tran, N.-H., Pham, N.-N.-T.: Design adaptive controller and guidance system of an unmanned surface vehicle for environmental monitoring applications. In: International Conference on Green Technology and Sustainable Development (GTSD), December 2018
13. Fossen, T.I.: Marine Control Systems, Trondheim. Tapir Trykkeri, Norway (2002)
14. Liu, T., Dong, Z.P., Du, H.W.: Path following control of the underactuated USV based on the improved line-of-sight guidance algorithm. Pol. Marit. Res. **24**(1), 3–11 (2017)
15. Ko, N.Y., Choi, H.T., Lee, C.M.: Navigation of unmanned surface vehicle and detection of GPS abnormality by fusing multiple sensor measurements. In: OCEANS 2016 MTS/IEEE Monterey, California USA, pp. 19–23 (2016)
16. dos Santos, D.S., Júnior, C.L.N., Cunha, W.C.: Autonomous navigation of a small boat using IMU/GPS/digital compass integration. In: SysCon 2013 (2013)
17. Fossen, T.I., Johansen, T.A., Perez, T.: A Survey of Control Allocation Methods for Underwater Vehicles (2009)
18. Tran, N.-H., Nguyen, T.-C.: The design of a VIAM-USVI000 unmanned surface vehicle for environmental monitoring applications. In: International Conference on Green Technology and Sustainable Development (GTSD), December 2018
19. Zheng, Z.W., Sun, L.: Path following control for marine surface vessel with uncertainties and input saturation. Neurocomputing **177**, 158–167 (2016)
20. Xsens Technologies, MTi User Manual: MTi 10-series and MTi 100-series 5th generation (2018)

Methodology for the Quantification of the Radio Spectrum Available for White Spaces in the Conditions of the Republic of Colombia

Gómez Carlos[1(✉)], Villarreal Martha[1], and Fonseca Valeria[2]

[1] Facultad de Ingeniería, Universitaria Agustiniana, Bogotá, Colombia
{carlos.gomez,martha.villareal}@uniagustiniana.edu.co
[2] Facultad de Ciencias, Universidad Nacional de Colombia, Bogotá, Colombia
vfonsecad@unal.edu.co

Abstract. Since the emergence of the possibility of using radio spectrum spaces not used in the TV band for the transmission of data channels, has begun to develop an industry of different products for the use of this technology, however, national regulations are needed to enable the use of this spectrum depending on the use given to the TV bands and the particular needs of each country. This document deals with the background of studies of methodologies for the quantification of spectrum available for TV White Spaces - TVWS, and then presents a methodological proposal for this quantification in the case of the Colombian territory. A statistical study is carried out on the propagation model with the best performance in Colombian TV stations, simulation of coverage of the network of digital TV stations, analysis of protection actions against interference and quantification of available spectrum in a TV channel, which can be replicated in more areas or countries with similar conditions.

Keywords: White Space · Quantification of radio spectrum · Interference protection

1 Introduction

In November 2008 the FCC (Federal Communications Commission of the United States) allowed the use of cognitive radios in white spaces within TV bands [1], which has created a market for white space devices -WSD- for transporting data on these frequencies. In addition, the blackout of analog TV in several countries has led to the reorganization of the radioelectric spectrum, where regulations have been created that allow the secondary use of TV frequencies for data transport applications, but always taking protective measures over the primary TV service. Most of these countries are located in Europe and North America.

In developing countries, regulations allowing the use of TV white space technology, TVWS, have not been massively promoted; however, among the cases to be highlighted is the first Latin American regulation made by Colombia to establish the conditions of use of white space devices [2].

D. F. Cortes Tobar et al. (Eds.): AETA 2019, LNEE 685, pp. 575–585, 2021.
https://doi.org/10.1007/978-3-030-53021-1_58

Because the a geographic, technical and regulatory conditions are particular in each country, it is necessary to create different spectrum quantification methodologies that allow for better local approaches to how much radio spectrum is available in each region to develop projects with TVWS devices.

2 Literature Review

In some countries, studies have been conducted to find out the amount of radio spectrum available that can be used to deploy white space technology networks.

Spectrum quantification for white spaces in the USA is discussed in [3]. The study was based on fixed devices authorized to use channels 2, 5-36 and 38-51, using ITU standard propagation models, and the census of the population by postal code specified in a polygon. This study was based on FCC recommendations on spectrum management for TVWS.

The FCC discusses 2 approaches to determining the distance at which secondary TVWS devices can be located from a primary TV station. The first approach assumes that the protected radio r_p and the communication radio r_n, are the distances that must exist for the secondary user to transmit, otherwise it will cause interference, to avoid this the formula $r_n - r_p$ is used.

In the second approach the FCC set as –114 dBm the ATSC signal detection threshold for cognitive radios, which do not have the ability to determine their location and consult the authorized geolocation information. This approach evaluated the amount of white space allowed by the –114 dBm rule in which the signal was at 50% of the time and the location also known as F (50.50). [3]

Mazinar N [4] presents a study in the United Kingdom, in which it talks about cognitive radio technology as a basis for avoiding interference from TVWS devices with primary TV service systems. The FCC stipulates that cognitive radio must have the ability to detect TV stations next to the devices they have inside, such as wireless microphones tolerated up to –114 dBm.

Mazir N [4] carried out a computer simulation taking into account OFCOM's database on transmission power, antenna height and transmission frequency for the 81 primary digital television transmissions in the UK. It is also taken into account as an important fact that the transmission power of digital TV ranges between 25 and 200 KW, for such reason the cognitive radios use a power (~100 mW) and must take into account $\frac{P_{cr}}{P_{Tv} \ll 1}$, in other terms the low consumption of power does that the upper limit of TVWS can be obtained simply with the maps of coverages or a few simple mathematical relations.

Cuevas-Ruiz J [5] presents a study on the quantification of spaces for TVWS in Mexico. For this study was taken into account only the digital TV service, due to the switch-off of analog TV in Mexico in 2015. The regulation in this country says that the DTT service should be offered in the main city taking into account the field strength F (50.90) with 48dBu reception power of channels 14 to 5. Samples were taken from 32 cities. [5] proposes some methods to obtain the distance between two points located on the earth's surface and that serve to identify if the frequency of a channel is available for TVWS in a specific place. This study identified six channels (14, 21, 40, 41, 50, 51) that can be identified as TVWS in each of the 32 cities used for the sample, for a total of 36 MHz of free spectrum.

Naik G [6] performs a quantitative and estimation assessment for the TV white space in India, using the frequency band between 470–590 MHz (UHF) in four areas of this country except for the northern area, this being the first step that is done in India to establish the data in a comprehensive way, The first method is done from the point of view of service protection and possible contamination, the second method is based on the regulations of the United States Federal Communications Commission (FCC).

In the results of [6] the author shows that unlike other more developed countries, the vast majority of spectrum in TV bands is not used in this country. Using the two approaches mentioned above, an available average of more than 100 MHz was found [6].

3 Method for Calculating Spectral Availability

3.1 Choice of the Propagation Model

There are a variety of propagation models designed to attempt to simulate the behavior of a radio frequency transmission in space. However, according to the geographical conditions of each space, it is necessary to identify the propagation model that best fits the reality.

In this work [7], a coverage study of DTT stations using the DVB-T2 standard in the territory of Romania is carried out. For this analysis, the authors use coverage simulations with 4 different propagation models: NTIA-ITS Longley Rice, ITU-R P.1546, Okumura-Hata-Davidson and ITU-R 525/526. Measurement results are compared with field measurements. Under the conditions studied they find that the Okumura-Hata-Davidson empirical model is the one that produces results closest to the values measured in the field.

In the Colombian case, a study of propagation models ITU-R P.526-11, Deygout, ITU-R P.1546, ITU-R P.1812, Okumura-Hata, and ITU-R O.526-1 has been carried out against digital television coverage measurements made by the National Television Authority of Colombia, ANTV [8].

In order to determine the simulation model that best predicts the real behavior of the transmission, an exploratory statistical evaluation has been performed based on the residuals of each model applied to three areas in Colombia: La Pita, Manjui and Monteria.. This evaluation consisted of calculating the square error and bias of each of the six models as well as visualizing the distribution of the residuals.

Let y_i be the obtained real measurements and $\hat{y}_i$ the corresponding predicted value with one particular model. The square error of the model and the bias are given by [9, 10]:

$$\text{Error} = \frac{1}{n}\sum\nolimits_{i=1}^{n} \cdot \varepsilon_i^2 = \frac{1}{n}\sum\nolimits_{i=1}^{n} \cdot \left(y_i - \hat{y}_i\right)^2 \tag{1}$$

$$\text{Bias} = \frac{1}{n}\sum\nolimits_{i=1}^{n} \cdot \varepsilon_i = \frac{1}{n}\sum\nolimits_{i=1}^{n} y_i - \hat{y}_i \tag{2}$$

Where ε represents the residuals and n the number of observations. The square error of the residuals represents a measure of dispersion while the bias represents a measure of the shift from zero. The best model is the one that produces an error and bias as close to zero as possible. However, a threshold of acceptance would allow us to determine all the models that can still be accepted. The thresholds were calculated as the error and bias combining all residuals per zone.

Figure 1 shows the performance results of each simulation model. From the figure, it can be seen that the performance according to the error is more similar for all the models in Monteria than the other two zones. Here, the differences in performance are detected in the bias. The models applied in Manjui showed the largest dissimilarities in performance across all models. This made the threshold of acceptance in Manjui be larger than the thresholds in La Pita and Monteria. Across the three regions, Okumura-Hata resulted in the smallest error in comparison to the other models. Particularly, in La Pita and Monteria, the deviations from the acceptance region are rather due to a high bias, meaning that the predictions tend to follow an overestimated or underestimated behavior. In these zones, the models U-526 and U-1546 presented the second-best performance in prediction.

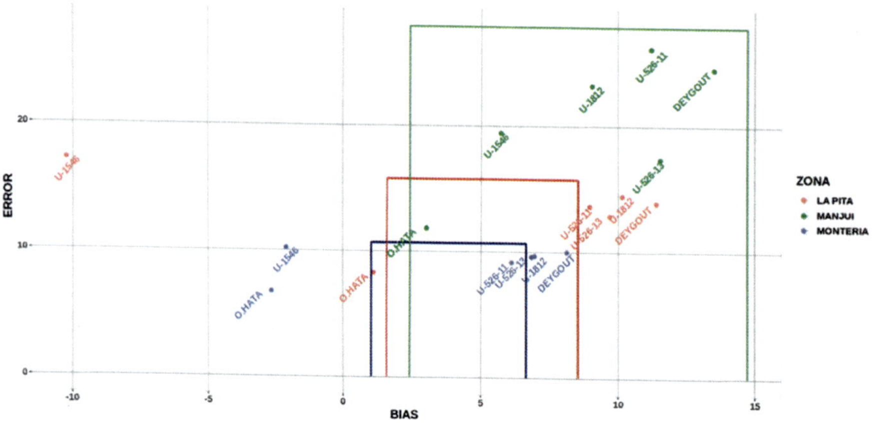

Fig. 1. Performance measures of each model with a threshold of acceptance per zone

In Manjui, Okumura-Hata and UIT-R P. 1546 were the two models that laid in the acceptance region. However, there is a clear distance between the performance measures of the two models. While the error and bias of Okumura-Hata are well inside the region, the error of UIT-R P. 1546 is almost borderline. Therefore in this zone still Okumura-Hata should be the only model accepted. In fact, by looking at the results of Table 1 for Manjui, there is no other model that could be comparable in error with Okumura-Hata as the other error values are much higher than this model.

To understand the differences in performance among the models, visualization of residuals was carried out based on density plots [11]. The dispersion of each plot is compared with the empirical three-sigma rule of statistical process control [12]. Figure 2 shows the density of residuals for each model applied in La Pita. It can be clearly seen how the distribution of residuals for Okumura-Hata are well centered around 0 and its dispersion is mostly contained inside the 3 deviations. UIT-R P. 1546 rendered highly dispersed residuals but also a clear shift towards the left. This means that this particular model in La Pita renders overestimated predictions. The rest of the models showed the opposite result, that is, the predictions are overall underestimated. Coverage study

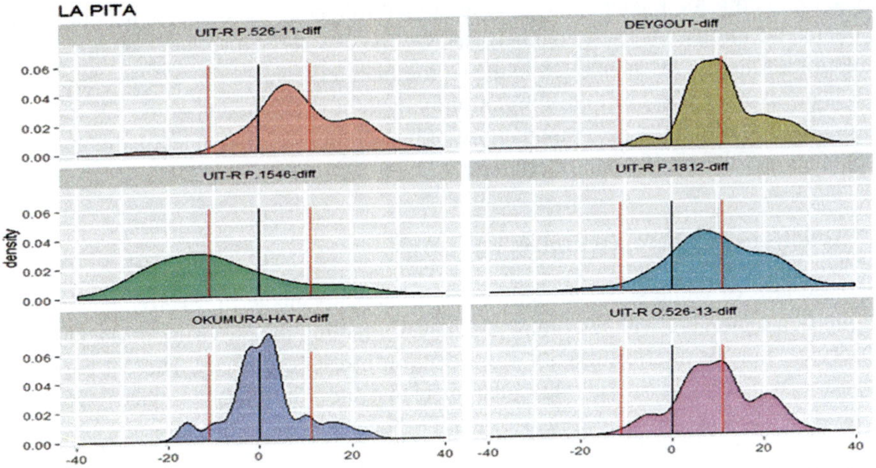

Fig. 2. Density plots of residuals per model for La Pita. The black vertical line is placed at 0 and red vertical lines are 3 deviations apart from 0. The deviation is the square root of the square error of La Pita.

Once the appropriate propagation model has been found, it is necessary to carry out a study on the coverage of each TV station. Computer tools are used that allow having as input the technical parameters of each TV station and as output, the graphical representation of the coverage of a TV station.

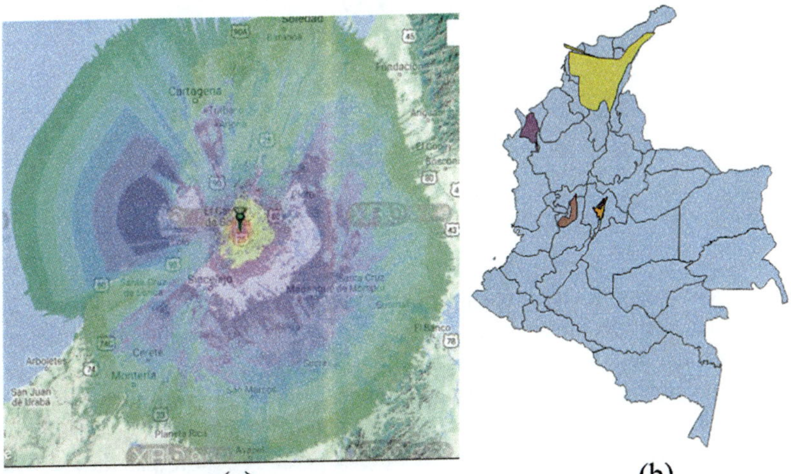

Fig. 3. (a) Simulated coverage of a TV station, using the Okumura-Hata propagation model. (b) Use of TV channel 28 in the Colombian territory, according to coverage simulation.

With the specific configuration data of each digital TV station, coverage simulations of each one of them are carried out. Figure 3a shows an example of coverage simulated with the Okumura-Hata propagation model.

By simulating all TV stations in an area or throughout the country, it is possible to understand the use of each channel used by the TV service. Figure 3b shows the simulation of the use of channel 28 in the Colombian territory.

It is necessary to perform the simulation of each of the channels of the TV service, in order to understand the geographical use of one of them, and then study the possible co-channel or adjacent channel interference.

3.2 Actions to Interference Protection

The basic idea of white space technology is to take advantage of TV frequencies not used in each region but with the premise of never affecting the provision of the primary TV service, i.e. not interfering with TV reception in users' devices. Thus, it becomes necessary to determine actions that guarantee that the white space devices WSD do not produce destructive interference in the TV receivers located within the coverage area of each TV station. Figure 4 shows the main protection action, which consists of determining a protection distance after which it is guaranteed that white station devices can be located, and will not be able to produce the interference.

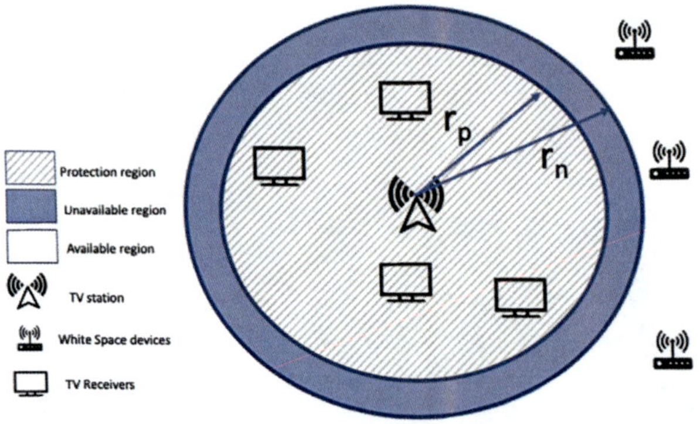

Fig. 4. Protection distances.

The protection region is the area of coverage of a TV station, and where TV receivers are present. In this region, the provision of the primary TV service cannot be affected and therefore there can be no interfering signals from white space devices WSD. In Fig. 4, the protection region is given by a r_p distance between the TV station and the boundary of the coverage area. To protect TV receivers, a minimum distance must be defined where white space devices can be located and given their operating conditions, they will not be able to achieve destructive interference in the protection region; this distance is called r_n in Fig. 4. The difference between $r_n - r_p$ is calculated so that a WDS device transmits at a distance of $r_n - r_p$ from the TV receiver that is located at r_p. Equation (3) allows the determination of the protection distance $r_p - r_n$ protection distance.

$$P_s - PL(r_n - r_p) = \Delta \qquad (3)[6]$$

Where Ps is the transmit power of the WDS device in dBm, $PL(r_n - r_p)$ represents the path loss in dB at a distance r from the transmitter and Δ is the TV reception threshold in dBm. [6]

The appropriate reception threshold for the correct reception of DVB-T2 signals has been studied by [13]. This study determines that the C/N ratio for the reception of DVB-T2 signals with 64 QAM modulation must be at least 12 dB.

To determine the reception threshold of TV receivers in Colombia, a field study was carried out to determine the C/N values found in the field at DTT stations. A Promax HDranger Ultralite receiver was used as a measuring instrument. Using experimental analysis it was possible to determine in a general way, in the conditions of transmission of the Colombian DTT stations, a C/N ratio >12 dB, is achieved when the reception power is greater than -61 dBm or field strengths greater than 47 dBuV

For the determination of the Ps and PL powers, the technical conditions defined in the national regulation and the technical conditions given by the equipment manufacturers are taken into account. Particularly in the Colombian case, the characteristics defined by the regulatory agency called Agencia Nacional del Espectro (National Spectrum Agency) are taken into account, some of which are:

- The power that a white space device delivers to your antenna may not exceed 12.6 dBm measured in any 100 kHz segment, corresponding to a transmitting power of 1 W per TV channel.
- The maximum gain of the antenna connected to a white space device shall not exceed 14 dB referred to a half-wave dipole (dBd).
- Unwanted emissions shall not exceed a power of –42.8 dBm measured in any 100 kHz segment.
- The height of the antenna above the ground level of White space devices shall not exceed 50 m.
- White space devices may only operate at geographical points whose height above ground average is less than 800 m. [2]

Given all the above information, the minimum protection distance has been calculated for the case of channel 28, centered on 555 MHz. In this case $r_n - r_p >= 15$ km, which is close to the separation distance recommendation suggested by the Federal Communications Commission FCC in [1]; this regulator says that the interference radius is 14.4 km for a co-channel and 0.74 km in an adjacent channel.

With respect to adjacent channel interference, given the tendency of DTT signals to produce it, the protection measures must extend towards the restriction of the use of frequencies adjacent to the central frequencies used in a TV station; this way the WSD will not be able to use adjacent frequencies within the protection regions discussed above.

3.3 Calculation of Available Spectrum by Regions

Once the protection distance has been calculated, it is possible to take the coverage study and apply the protection actions; this consists of taking the coverage footprints obtained in the simulation based on the Okumura-Hata propagation model, and adding distance to that coverage, corresponding to the protection distance. This new calculated area, corresponding to the area protected against interference, is the one that must be subtracted from the total area of the region to be analyzed (city, state, country, etc.).

The analysis exemplified in Fig. 5 should also take into account stations using adjacent channels; if one exists, coverage of those adjacent stations should also be included in the analysis of protection areas.

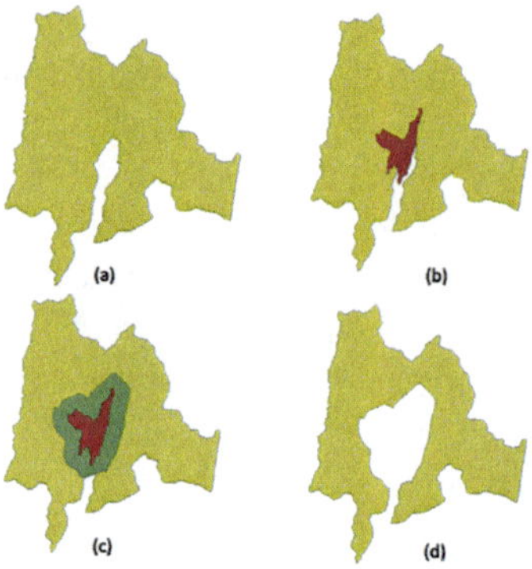

Fig. 5. (a) Coverage of a TV station in the area of the region., (b)Total area of the region of interest. (c) Application of the protection action based on the addition of a protection area of 15 km. (d) Difference of areas

Figure 5 shows the example of area analysis available for the use of white space devices, given the presence of a TV station located on channel 28.

Given the procedure shown in Fig. 5d, it is possible to obtain the area available to be used by WDS white space devices. The comparison of the total available area vs. the original area of interest allows concluding the total percentage of the area that can be used by the TVWS technology in each channel.

The analysis exemplified in Fig. 5 should also take into account stations using adjacent channels; if one exists, coverage of those adjacent stations should also be included in the analysis of protection areas. Table 1 presents the results of the available area analysis for the case of channel 28 in the Republic of Colombia.

Table 1. Available area for WSD in channel 28.

Region of interest	Percentage of available area (%)
Guajira	67,68
Cesar	8,56
Magdalena	28,27
Atlántico	8,62

(continued)

Table 1. (*continued*)

Region of interest	Percentage of available area (%)
Sucre	66,06
Bolivar	3,32
Antioquia	90,64
Chocó	89,52
Cundinamarca	73,91
Bogotá	46,43
Tolima	70,24
Caldas	80,36
Quindío	17,85
Valle del Cauca	98,34

4 Discussion

The state of the art analyzed for the beginning of this research shows the existence of some works that propose spectrum quantification methodologies, but always adapting to the particular conditions of each region, due to changes in the technical implementations of digital television networks, changes in national regulations about the use of blank space devices, or particular geographical conditions; In the same line of what was found in the state of the art, this work managed to develop a methodology of spectral quantification for the use of WDS devices, starting from the most significant contributions of the state of the art, but adaptations were made according to the conditions of implementation of digital television in Colombia, the national regulatory conditions and the local geographic conditions. The results of this work could be adapted to the technical and regulatory conditions of Colombia's neighboring countries, to obtain a rapid framework of spectral quantification adapted to each country.

Based on the above, the proposed methodology can be summarised in the following procedure:

- To carry out a study to identify the propagation model with the least error in the coverage productions of TV transmissions.

In the case of Colombia, account is taken of the propagation conditions of the TV stations established by the national authorities (National Television Authority and National Spectrum Agency) with regard to the general technical characteristics of the TV stations, and the particular implementation reports of each digital TV station (powers, losses and antenna arrangements).

- Based on the propagation model selected, carry out the coverage study of TV stations incidents in a selected area.

In the Colombian case, the Okumura-Hata model has been detected as the most successful propagation model.

- Set the conditions of the protection zone for each TV station.

In the case of Colombia, it has been recommended to add a protection zone of at least 15 km with respect to the limit of the coverage footprint of each TV station. The recommendation not to use channels adjacent to the TV channel used in the transmission of the TV station should be taken into account.

- Given the defined coverage and protection zones, identify the entire non-transmission area of WSD devices.
- Calculate the areas not covered by the protected area, as these are what will be referred to as "Available White Spaces".

In the Colombian case, spectral availability is calculated according to the limitations of the political divisions known as Departments, for greater adaptability to the regional coverage requirements imposed on Colombian TV operators.

5 Conclusions

It was possible to determine that the simulation model Okumura-Hata presents the closest predictions to the real measurements of TV coverage. While the dispersion of errors of this model is slightly higher in Monteria than the other two zones, this model renders predictions that are on average very close to the real value. No patterns of overall overestimation or underestimation were found in Okumura-Hata as in the other models. Yet, the precision of these predictions is more challenged in Monteria.

Using the proposed methodology, with data from public and private, national and regional TV networks, it is possible to precisely determine the radio spectrum available in each Colombian region.

The analysis of protection actions, taking into account experimental analyses and international recommendations, guarantees the determination of areas where white space devices can operate, without interfering with the primary TV service.

It is necessary to take into account the particular conditions of the national regulations, and the conditions of the lands of each country, for the determinations of the coverages, protection actions and available areas, since the application of the TV technology is not uniform at world-wide level.

References

1. FCC.: In the Matter of Unlicensed Operation in the TV Broadcast Bands: Second Report and Order and Memorandum Opinion and Order, Federal Communications Commision (2008)
2. ANE.: Por la cual se modifica la Resolución 711 de 2016 para establecer las condiciones de uso de los dispositivos de espacios en blanco. Colombia Patente Resolución No. 000461, 01 Agosto (2017)

3. Mubaraq, S.: How Much White Space has the FCC Opened Up?. Dept. of Electrical Engineering and Computer Sciences, U C Berkeley (2009)
4. Mazina N.: Quantifying the TV white spaces spectrum opportunity for cognitive radio access. In: Communications Infrastructure, Systems and Applications: First International ICST Conference, EuropeComm, London, UK, EuropeComm, pp. 46–57 (2009)
5. Cuevas-Ruiz, J.: A TVWS quantification approach for Mexico. In: Fourth International Conference on Advances in Computing, Electronics and Communication (2016)
6. Naik, G.: Quantitative assessment of TV white space in India. In: Twentieth National Conference on Communications (NCC) (2014)
7. Martian, A., Vladeanu, C.: DVB-T2 radio coverage analysis in Romania. In: 25th Telecommunication Forum (TELFOR), Belgrade, Serbia (2017)
8. ANTV Homepage. https://www.antv.gov.co/index.php/tdt/mediciones-de-cobertura-tdt/category/1093-cobertura-tdt-fases. Accessed 01 Mar 2018
9. Gomez, C., Fonseca, V., Valencia, G.: Use of non industrial environmental sensors and machine learning techniques in telemetry for indoor air pollution. ARPN J. Eng. Appl. Sci. **13**(8), 2702–2712 (2018)
10. Suykens J., et al.: Least Squares: Support Vector Machines, p. 10. World Scientific (2003). ISBN 9812381511
11. Wickham, H.: Elegant Graphics for Data Analysis ggplot2. Springer, New York (2016)
12. Goetsh, L., Davis, S.: Quality Management for Organizational Excellence: Introduction to Total Quality, 7th edn. Pearson, Harlow (2014)
13. Eizmendi, I., Berjon-Eriz, G., Velez, M., Prieto, G., Arrinda, A.: CNR requirements for DVB-T2 fixed reception based on field trial results. Electron. Lett. **47**(1), 57 (2011)

Morse Keyboard

Sergio Beltrán(✉), Sergio Mendoza, and Alfredo Espitia

Central University, Bogotá 110311, Colombia
{sbeltranp,smendozap,aespitiabl}@ucentral.edu.co

Abstract. This paper talks about the progress on developing a keyboard for people with severe motor paralysis, which will work with a single input on/off sensor. It exposes a brief history of Morse code and its importance as alternative and augmentative system of communication. The reasons why was redesigned the original structure of the Morse code will be explained.

Keywords: ICT · Keyboard · Paralysis · Morse code

1 Introduction

The human being wants to be independent by nature, from birth to death he has the need to perform different activities so that he can feel himself useful for society. People who suffer some disability and have not the same level of freedom can feel vulnerable. Allowing a person with severe motor paralysis communicating with his environment and be able to access the ICT is a way to improve their quality of life and minimize the impact generated because of the disability.

About disability in Latin America, the Economic Commission for Latin America and the Caribbean (ECLAC) offers the most updated information (Alméras, 2014). There are around 70 million people with some disability in Latin America and nearly 400 thousand in the Caribbean (because of the only source of data in the Caribbean are the population and housing censuses, the uncertainty in this quantity is high), which is equivalent to 12.5% of the total population, among them 1.7% have some type of severe disability.

2 A View on Morse Code

Human being evolution has gone hand in hand with communication, being able to express emotions and ideas has made him move towards development, this dates from near 40,000 years ago [4]. The great step for communications happened between 1794 and 1838, in this period it was passed from a data transmission using flags, which was limited by the visual field, capabilities of optical devices (spyglass) and weather conditions (it should be located in a visible place) [5]. In 1825 William Sturgeon, British physicist, invented the electromagnet, able to generate an electromagnetic field in a piece of iron with shape of horseshoe, itself wrapped by a coil wound. In 1830, Joseph Henry improved the Sturgeon's design, linking the electromagnet to a bell 1.6 km away, which was sounded

D. F. Cortes Tobar et al. (Eds.): AETA 2019, LNEE 685, pp. 586–592, 2021.
https://doi.org/10.1007/978-3-030-53021-1_59

each time an electric current was induced in the core of the coil. In 1835 Alfred Vail, while working with Samuel Morse, created a method able to ensure the transmission of each number and letter of the American alphabet in a consistent code of dots and dashes (telegraphic signals different from each other by duration of signal), so that they managed to transmit their first message in 1838 from a distance of 3 km [6].

Over the years, Morse code was losing importance as a telecommunication tool, however it has gained relevance as one of the Alternative and Augmentative Communication Systems (AACS) [1], which are methods developed to recover communication capabilities in people with disability. For this, different devices have been developed to adapt the environment and the needs of each person. For example, Morse code can be used from a sensor activation or pushing a button [2] by finger, toe, mouth [3], head, foot, eyelids [4] etc., depending the user's motor condition. Several methods have been used adapting the Morse code as an AACS [5].

3 Proposal

Several electronic devices used such as AACS that use Morse code to interact with the computer, for example DARCI [6], have retained the original Morse code letters and numbers, while adding functions and commands (e.g. Enter, escape, control, tab, etc.) with new combinations of dots and dashes, this creates a problem: The screen scan commands by keyboard must use very long and inefficient combinations (e.g. Tab: -.--.). This is the main reason why an alternative dictionary to the original Morse code is presented in this document, in which priority is given to screen scan by keyboard commands and is optimized for the Spanish language, based on the corpus of the Real Academia de la Lengua Española (RAE) [7] and the monitoring of frequency of use of the keyboard by a blind person, who does not use the mouse.

In addition to the above requirements, a device for this purpose must behave like a standard USB keyboard, which is automatically recognized by any operating system without the need to install addons; It must be compatible with text prediction applications that expedite the task of writing, and compliance requirements to take advantage of the web-accessibility features proposed by the WCAG (Web Content Accessibility Guidelines) [8].

4 Development and Logical Design

Based on the universal design methodology, which refers to a barrier-free design or easy-to-use product design for people no matter their physical conditions, it is decided to develop a USB keyboard with common computational keys. This keyboard has been designed in order to give higher priority to exploration than writing. This device will not only work on computers, but also it will be able to operate on mobile devices (Smartphones, Tablets, etc.). The use of the mouse is not necessary.

This keyboard will operate using a voluntary mobile part of the body, e.g. a hand, a foot, the head, etc. Particularly, it will be able to work by movements of the eyelid through the implementation of an infrared sensor which will have the maximum calculated radiant intensity of 4.46 mW/cm2 at 20 mm away from the eye. Taking as reference the declaration of the ICNIRP (2000) [9], the exposure to this radiant intensity does not represent any danger.

4.1 Logical Design

The algorithm is based on taking the times of activated pushbutton/sensor (*T_on*) and deactivated pushbutton/sensor (*T_off*), in order to compare times and deduce which type or command the user wants to execute. It will give preference to navigation.

Priority of codes is given by:
1. Higher priority: Lower number of signals
2. Higher priority: Greater number of points

The algorithm uses a rule to determine the type or command: All codes must have minimum one "dot" to work as base time. To be able determining this base time, an interruption is activated by falling edge, to have control over times while sensor/pushbutton is activated to be able to assign a memory space, where it will be saved. Later, these values can be compared to determine what type or command the user wants to send. To make this comparison is determined the base time or *T_min*. This base time is calculated in order not to have a predetermined *T_min*, since all people have different response times, this time is always recalculated in order to have a different *T_min* for each type or command.

Next, in order to use different types (e.g. A, B, C), special symbols (e.g. *, –, #) and commands (e.g. Ctrl + E, Win + D) were taken into account the three fundamental times for understanding the code: T_{min}, T_{on} and T_{off}.These times are calculated based on 3 equations, which are:

1. $If\ T_{on} \geq 2 * T_{min}\ and\ T_{on} \leq 2 * T_{min} \Rightarrow T_{on} = Dot$
2. $If\ T_{on} \geq 2 * T_{min} \Rightarrow T_{on} = Dash$
3. $If\ T_{off} > 4 * T_{min} \Rightarrow T_{on} = End\ of\ code$

The algorithm is able to determine it is time to send the character in two ways:

- The first way is indicated by the equation number (3). T_{off} is greater than 4 times the T_{min} or…
- The second way is the number of times the sensor/button was activated. If it is equal to the maximum number of signals allowed, i.e. 6.

The algorithm takes every times and store them in a vector, then multiplies them by mathematical power of 2, so that obtain an integral and unique value which would be equivalent to a code indicated in the designed dictionary Table 1 [10]. The algorithm through a "Switch Case" will look the value thrown by the vector and determine which type or command must be sent.

Table 1. Proposed dictionary

↑	– •
↓	• –
←	– • •
→	• • –
TAB	•
ENTER	• •
BACKSPACE	• • •
SPACEBAR	• • • •
DELETE	• • • • • •
ESCAPE	– – •
END	• – – • • •
HOME	• • • – – •
INSERT	• • • – – –
PAGE DOWN	– • • • – •
PAGE UP	– – – • • •
PAUSE/INTER	– • – • • •
PRINTSCREEN	• – • • • –
SCROLL LOCK	• • • – • –
CAPS LOCK	• • – • • •

A	– • • •
B	• – • • •
C	– – – •
D	– • • –
E	• • • –
F	• • – • –
G	• – – • •
H	• • – – •
I	– • – •
J	• – • – •
K	• – – – •
L	• – • •
M	– – • –
N	– – • •

Ñ	– • • – •
O	• • – •
P	– • – –
Q	• • – • •
R	• – – •
S	• • – –
T	• – – –
U	• – • –
V	• – • • –
W	• – – • –
X	• – • – –
Y	• • • – •
Z	– • – • •

0	– • • • –
1	• – – – –
2	• • – – –
3	• • • – –
4	• • • • –
5	• • • • •
6	– • • • •
7	– – • • •
8	– – – • •
9	– – – – •

5 First Validation

A strategy has been proposed to achieve a main objective, in which a person with severe motor paralysis can access Information and Communication Technologies (ICT). We tentatively pose the initial Morse taking into account the statistics obtained through the application WhatPulse (Data were taken from 3 people, including a person with visual impairment) which counts letters and commands most used by a person who does not use the mouse. These data were compared with the corpus of the Spanish language "Don Quijote De La Mancha" (Fig. 1) [7] obtaining some similarities, as for example the letter most used in the Spanish alphabet is the "E".

A dictionary is generated Table 1, which relates the different characters and commands used by a person with visual disability, giving priority to the most frequent insertion times based on the three rules mentioned in the logical design.

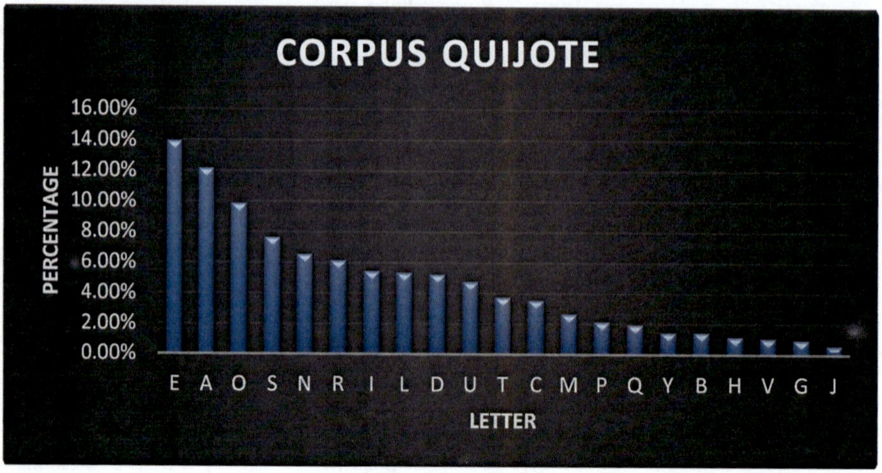

Fig. 1. Frequency corpus Don Quijote

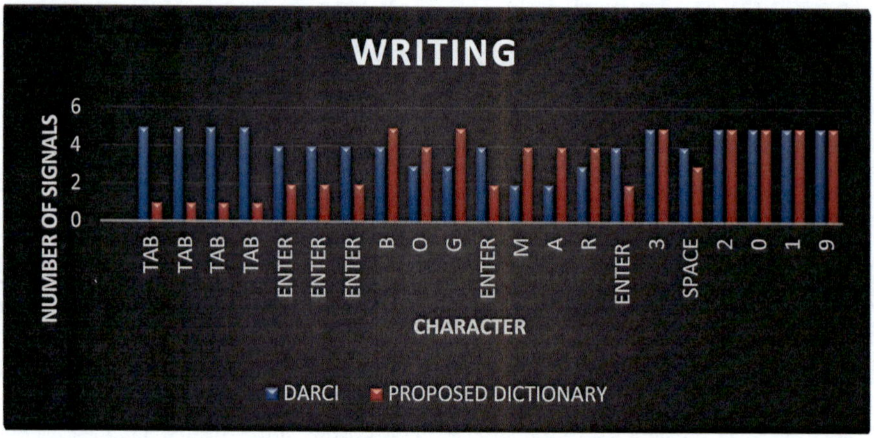

Fig. 2. Comparison Darci vs proposed in reading mode

Comparison tests are performed in access to the different current platforms (YouTube, Google, Facebook) in front of the Darci USB, obtaining a considerable advantage in the exploration approximately 66%, in topics focused on search and entertainment, taken on the YouTube platform, performing the search for the creep song of the band Radiohead (Fig. 3). In terms of writing, the advantage is not so noticeable, we reached a 9.8% advantage, writing the enunciation of a letter, which is headed by city, date and a paragraph of 20 words (Fig. 2).

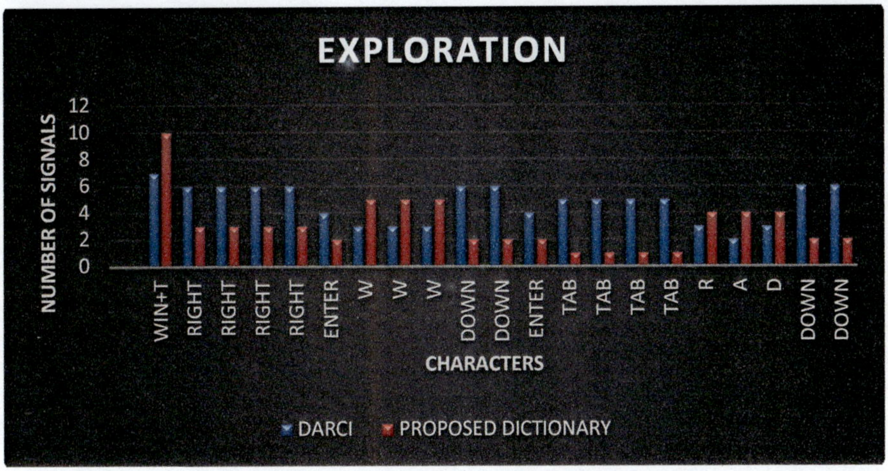

Fig. 3. Comparison Darci vs proposed in reading mode

6 Conclusion

The original Morse code fulfilled its purpose transmitting text messages in English very well. However, there is no reason to keep the original code when it will be used as an alternative communication system for people with disabilities. The original code did not meet the need for the special symbols and function commands of keyboards for electronic devices of today.

The new code presented in this document retains the characteristics of classical telegraphy that can be of great value in a tool designed to meet the needs of a person with disabilities: It only requires the voluntary movement of one part of body and can be read through different senses according to the way the writing is presented. Although the second feature was not treated in depth in this work and will be expanded in the future.

The tests showed that a keyboard that uses the original Morse key to interact with the computer will be efficient for writing in English, but will lose that advantage when the writing is in another language, such as Spanish, and when the user needs to explore the screen of his electronic device through the keyboard commands (excluding the use of the mouse, as it happens in several disabilities). The code that has been presented gives preference to screen scanning functions because, according to the usability features of modern electronic devices, in this way, the use of the system and access to computer tools is streamlined. Memory registers, search preferences and text predictors are some commonly used tools that allow the proposed code to improve its writing efficiency.

References

1. Castillo, S.V.T.: Lenguaje y parálisis cerebral: El uso de los SAAC como medio de comunicación. Montevideo, Uruguay (2016)

2. Baljko, M., Tam, A.: Indirect text entry using one or two keys. In: Proceedings of 8th International ACM SIGACCESS Conference on Computers and Accessibility - Assets 2006, p. 18 (2006)
3. Liang, C.H., Chen, S.C., Lok, W.K., Luo, C.H.: A ZigBee-based Electronic Aid for Daily Living for Quadriplegics (2010)
4. Alnfiai, M., Sampalli, S.: SingleTapBraille: developing a text entry method based on braille patterns using a single tap. Procedia Comput. Sci. **94**, 248–255 (2016)
5. Yang, C.H.: Adaptive Morse code communication system for severely disabled individuals. Med. Eng. Phys. **22**(1), 59–66 (2000)
6. Lynds, J.S.: DARCI TOO-a computer input device for people with disabilities, 1192. https://ieeexplore.ieee.org/abstract/document/217430
7. Española, R.A.: Corpus de Referencia del Español Actual (CREA) - Listado de frecuencias. http://corpus.rae.es/lfrecuencias.html
8. W3C. W3C Advisory Committee Elects Technical Architecture Group. https://www.w3.org/
9. ICNRP. UV RADIATION EXPOSURE - 2000. United king (2000)
10. Espitia, A.: Código morse alternativo para acceder al computador y a otros dispositivos electrónicos. Ingeciencia **2**(1), 5–12 (2017)

Analysis of Low-Cost Wireless Sensors Model for Weather Monitoring Based on IoT

Carlos Suarez[1], Jaime Parra[2], Paulo Gaona[2], and Sebastián Soto[3](✉)

[1] Fundación Universitaria Agraria de Colombia, Bogotá, Colombia
suarez.carlosre@uniagraria.edu.co
[2] Universidad Distrital Francisco José de Caldas, Bogotá, Colombia
{japarrap,pagaonag}@udistrital.edu.co
[3] Corporación Unificada Nacional de Educación Superior, Bogotá, Colombia
sebastian_soto@cun.edu.co

Abstract. This paper proposes a communication model based on implementation of weather stations with Internet of Things (IoT) transmission protocols. The research presents evaluation and the relevance current low-cost technology defined environmental conditions in Bogotá, Colombia and surroundings city area. Using wireless sensors (WSN), three meteorological stations distributed in strategic areas implemented where the performance packet error rate (PER) in the data transmission evaluated allowing to estimate the index of quality and efficiency link. Other aspects considered for research is description of latency and distance to confirm transmission standard compared to mathematical models. The results obtained reveal that is important consider the geographical conditions terrain, obstacles produced in buildings and vegetation. The appreciable yields arose in terms of latency, performance and distance losses evaluated in the three frequencies 900, 915 MHz and 2.4 GHz.

Keywords: Internet of things · Weather station · Transmission · Data

1 Introduction

At present, IoT devices used in new applications in areas with adverse environmental conditions because they acquire and transmit data continuously. It also stands out real time communication and its problems ending with a new system of IoT data flows, called information flow of things (IFoT) that processes and analyzes massive flows of IoT in real time using distributed processing between IoT devices. Another aspect that sought with new technologies IoT based is that they of fast configuration [1], presented an example of creation and implementation low cost hardware communication and transmission of data to websites offering free data hosting. The new trends reviewed by [2], also analyzes the IoT and communication systems based on 5G networks and presents the prospective of the green IoT in all the aspects that applied in the future. The purpose of this study establishes the behavior of low-cost technology in real and uncontrolled environments to verify the performance of parameters such as latency, distance and transmission of data because the available literature does not currently present it.

D. F. Cortes Tobar et al. (Eds.): AETA 2019, LNEE 685, pp. 593–605, 2021.
https://doi.org/10.1007/978-3-030-53021-1_60

One prototype that allow establishing several aspects of a convergent communication model capable of providing broadband transmission services that allow the acquestion of data from weather stations located in an urban environment.

1) analyze IoT low-cost technology available in Colombia for the construction of weather station and evaluate data transmission in real environment.
2) The geographical aspects present conditions that affect the behavior of data transmission and in this study, there urban environment with high density of buildings.
3) The communication pattern weather stations different for each, and they are 900, 915 MHz and 2.4 GHz, then the focus compare technologies, and in this way establish the real ranges of latency, distance and transmission of data.

Finally, the present research organized as follow. In Sect. 2, present literature review based on sensors para IoT. Section 3 presents materials and methods used. Section 4, protocols and mathematical models planned. Section 5, results. Section 6 discusses results whit other authors and Sect. 7, conclusions and identified future research.

2 Present Literature Review Based on Sensors IoT

Another aspect is the efficient use of energy in sensor networks and applications IoT and [3] a review art in terms of energy management in an efficient way in different fields of application and establishes a taxonomy proposal identifies the basic requirements of a sensor network (WSN) such as scalability, coverage, latency, QoS, security, mobility and robustness.

A practical example presented by [4] model based test and battery that allows to evaluate different IoT technologies and compares three, one open, one commercial and the third based IPV6 using different layers such as data acquisition, network and the result obtained comparative table open technology available but is less scalable and the opposite occurs with IPV6, commercial IoT proposal shows average availability but low mobility, result important to continue evaluating open technologies and configure IPV6.

Recent development using low cost technology and using IoT made [5], where they proposed the implementation and evaluation small scale meteorological station based STM32, the results obtained satisfactory due monitoring variables wind speed, wind direction, temperature, humidity and using wireless network radio packets server for analysis real time.

3 Methods and Materials

In the development project, the materials submitted to the tests and the standards of work in the environmental conditions in the city of Bogotá, Colombia described. Below is the method developed in the implementation process, the acquisition system and communication protocols, data analysis, among other aspects.

3.1 Devices

For research, various sensors integrated into data acquisition, transmission and communication systems. The sensors are standard and commercially acquired using references from meteorological stations factory with the objective of homologating them for environmental monitoring (Table 1).

Table 1. Description and ranges of measurement of the sensors

Sensor type and variable	Range	Units
Wind Speed	0–50	m/s
Direction of the wind	360°	Degree
External temperature	– 40 °C a 65 °C.	°C
Relative humidity	1%–99%	% RH
Atmospheric pressure	0–1013,25	hPa
Pluviometer	0–9,999	mm H_2O
Illuminance range	0–300,000	Lux/m^2

The hardware implementation and development based low cost, technologies such as Arduino, Raspberry Pi and LoRaWan are elaborate and their basic characteristics described below according standards related literature manufacturer and standards based protocols established date and regarding communications, the following architectures are available and listed Table 2 by [6].

3.2 Methodology

The methodology implemented based experimental process each type of network compared with the theoretical communication model, establishing differences between the experimental and theoretical data to know the behavior of each network in an urban environment with obstacles such as buildings and vegetation of the experimental area (Fig. 1).

To define the layers each part assembly and development communications, took [10] where established a model each layer and summary hardware case of application from the sensor layer data storage, being basis development and analysis model to propose.

For the type of network based analysis presented by [7] and [8], where he proposes various types of network development in wireless strategies, as well the sensor selection, evaluation and installation points.

3.3 Incorporation Hardware Components

The IoT elements of this project connect a set of measurement devices combined with a data acquisition and processing unit through a local network to the Internet and exchange

Table 2. Comparison of the existing communication technologies

Parameters	WIFi	WiMAX	LR-WPAN	Mobile communication	Bluetooth	LoRa
Standard	IEEE 802.11 a/c/b/d/g/n	IEEE 802.16	IEEE 802.15.4 (ZigBee)	2G-GSM, CDMA 3G-UMTS, CDMA2000 4G-LTE	IEEE 802.15.1	LoRaWAN R1.0
Frequency band	5–60 GHz	2–66 GHz	868/915 MHz, 2.4 GHz	865 MHz, 2.4 GHz	2.4 GHz	868/900 MHz
Data rate	1 Mb/s–6.75 Gb/s	1 Mb/s–1 Gb/s (Fixed) 50–100 Mb/s (mobile)	40–250 Kb/s	2G: 50–100 kb/s 3G: 200 kb/s 4G: 0.1–1 Gb/s	1–24 Mb/s	0.3–50 Kb/s
Transmission range	20–100 m	<50 km	10–20 m	Entire cellular area	8–10 m	<30 km
Energy consumption	High	Medium	Low	Medium	Bluetooth: Medium BLE: Very Low	Very low
Cost	High	High	Low	Medium	Low	High

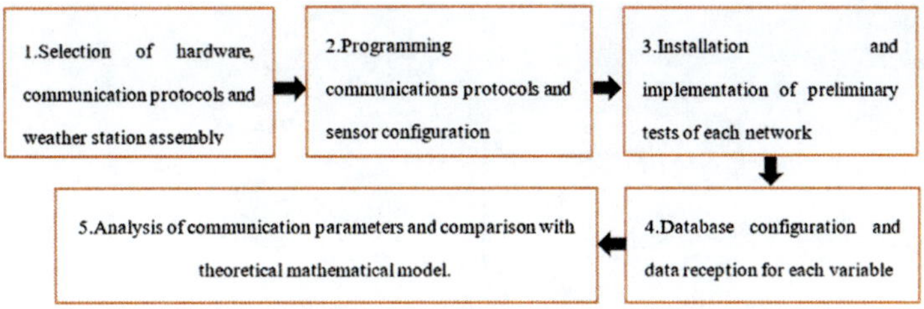

Fig. 1. Methodological proposal.

information about the three environment stations. The IoT gateway processes a large variety of data detected by various environmental sensors with the transmission device controlling the time of emission and then sends them to the server.

The hardware that makes up the automatic weather stations used in this project following systems (Fig. 2):

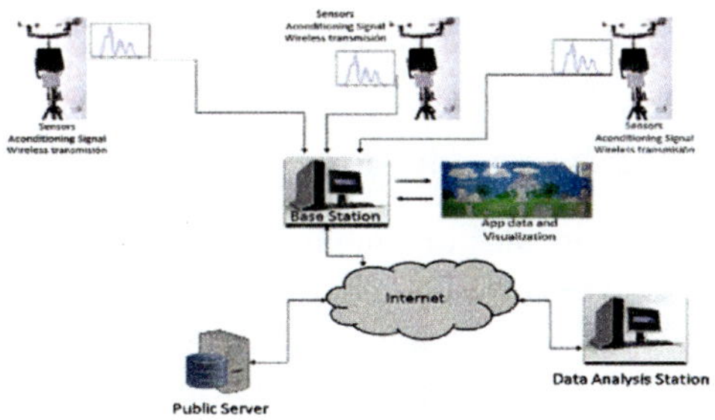

Fig. 2. Diagram of connection and adaptation of signals of each variable by means of hardware and software.

4 Testing and Proposal Model

The system was implemented in a place with buildings and obstacles of natural terrain in an urban area, where one seeks to know the performance of each of the LORAWAN and W-PAN models, which led to them being very real and these obstacles recreated a scenario high density to evaluate the two communication models (Fig. 3).

To determine the quality of a digital transmission system is to measure its Bit Error Ratio (BER). The BER is calculated by comparing the transmitted sequence of bits to the received bits and counting the number of errors. The ratio of how many bits received in error over the number of total bits received is the BER. For example, if the BER were suspected to be approximately 10–6, then the number of samples required for the above case would be estimated to be the number error relation by BER, [9] (Fig. 4).

Fig. 3. Topographical description of the communications model

Fig. 4. Photographs of the testing area

4.1 Communication Performance Measure Process

This section develops the communications model estimating the packet error rate (PER) in a wireless network. The types measures are executed using Fig. 5 illustrates the procedure of encoding and decoding in a communication model, where u is the information word emitted by sensors, v, is the codeword into frame, v' is the received word and u' is the decoded word. The encoder circuits of linear block codes and convolutional codes are implemented hardware.

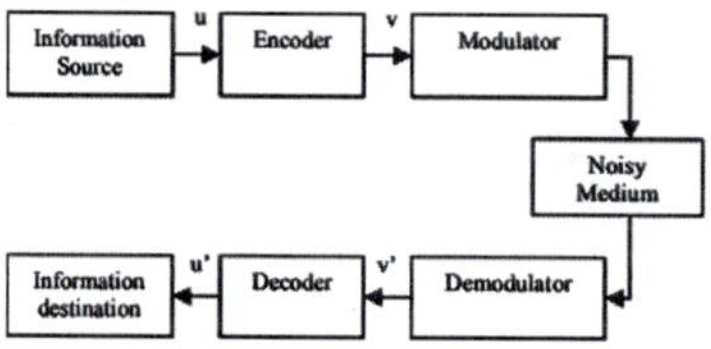

Fig. 5. Communication performance measure

This can be mixed with other models, for example, if there is an analytic expression for the Bit Error Ratio (BER) then a model can be used to find the symbol error rate, and the results can be used in conjunction with the other parts of this model to estimate the Bit Error Ratio (BER) and finally the performance metrics, like throughput and latency.

4.2 Calculation of Performance Metrics

Below briefly explain a mathematical model used in order to make a comparison with the experimental results, in this section, we will describe a process of estimating the impact on throughput and latency using our estimate of the packet error rate. The throughput is given by:

$$TP = (1 - PER)TPo \tag{1}$$

TPo is the base throughput without any interference.

The latency with interference is the latency without interference divided by the probability that a packet will be received correctly. Hence the latency with interference is given by:

$$\tau = \frac{\tau o}{(1 - PER)} \tag{2}$$

τo is the latency without any interference existing.

In most cases showing the impact of interference on the packet error rate, the throughput and the latency is often sufficient. In accordance with IEEE this SER will be used frequently we also call it p. It is a function of the SIR but often we will suppress the argument of γ to simplify the notation. If binary FEC is used it may be more practical to calculate the bit error rate (BER) after FEC. We define the BER after FEC as, [11].

$$Pb = Pb(\Upsilon) = BER = P(BE) \tag{3}$$

Relation Between BER and PER

The relationship between the BER and PER is valid for an ideal communication system that transmits data over binary symmetric channel with uncorrelated noise. The length of the packet in bits is n and the bit error probability for the channel is BP. Any packet that has one bit or more in error is discarded and the probability Pn of getting a packet in error is given by:

$$PER = 1 - (1 - BER)^n \tag{4}$$

A real communications system is considered equivalent to the ideal system if they have the same PER. The equivalency means that the real system has the same packet throughput as the reference ideal system having the known $BER = 10^{-6}$.

Measurement Packet Size and Error Count

To measure and evaluate BER by PER, n; packet size, should be defined first. The packet size is adopted as follows. For the BS receiver sensitivity, the packet length is chosen to be, after encoding, a single FEC block with a maximum data size allowed by the CTC sub-channel concatenation rule, i.e., 60/54 bytes depending on the MCS level. Each data packet consists of a 6-byte generic MAC header at the beginning and a 4-byte CRC- 32 at the end, leaving the remaining as payload bits. Random payload bits are used. The CRC is calculated based on MAC header and the payload.

4.3 Test

The test given here are chosen to illustrate the process. The second test is an extension of the first example, where the modulation is the Gaussian frequency shift coding (GFSK), the number of bits in the packet remains the same.

This demonstrates the effect of using modulations of different higher order with a signal to external interference ratio at different frequencies. We can now express the important ratio ES/N0 in terms of the signal power and the noise power after the receiver.

$$\frac{E_S}{N_0} = \frac{P_S T}{P_N T} = \frac{P_S}{P_N} \tag{5}$$

This project consists of three wireless stations with the distance variable between stations. In this investigation, each wireless station has one transmitter and one receiver, and the interfering wireless is the environment. Ahead in accordance whit summarizes the steps used in estimating the Bit Error Ratio (BER) and other performance metrics due to interference (Fig. 6).

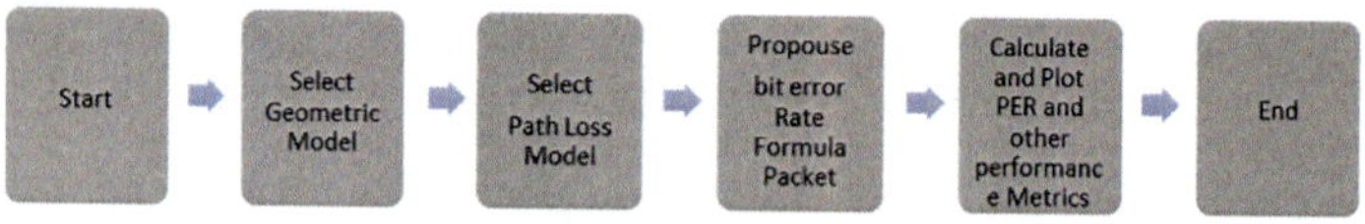

Fig. 6. Estimate BER/PER process

Select a Geometric Model: This consists of selecting the number of stations in both the affected wireless network and the interfering wireless network. It also involves selecting the locations of all stations. In this project we selected two stations per network, random interference from the environment and variable distance in the open field.

Select a Path Loss Model: There is a path loss model that has already been considered in the development of the proposed standard under development.

$$LOS = 32,4\,dB + 20\log D + 20\log F \tag{6}$$

$$D = Distance\ Km$$

$$F = Frecuencia\ Mhz$$

Propose Bit Error Rate Formula Packet Error Formula: In this instance we assume that we have co-channel interference with the same bandwidth (or smaller) than the bandwidth of the signal. So, we can set the SIR to be equal to the SNR in the formula for the symbol error rate of BPSK and GMFSK. The BER formula for both modulations obtained from any digital communication book are:

$$\text{QPSK}\ P(e) = \frac{1}{log_2 M}\text{erf}(z) \tag{7}$$

$$erf = error\, function$$

$$z = sen\left(\frac{\pi}{M}\right)\left(\sqrt{log_2 M}\right)(\sqrt{\frac{Eb}{No}})$$

$$GFSKP(e) = \frac{1}{2}erfc\sqrt{\frac{\lambda Eb}{No}}$$

$$B_{3dB}Tb = 0, 25\; then\; \lambda = 0, 25$$

Calculate and Plot PER and other Performance Metrics: Using the geometric model, we vary the distance parameter specifying the distance between stations in the affected wireless network and the interfering wireless network. Let us call this distance d.

For each value of d

1. Calculate the SIR at the station located at the origin
2. Using the SIR calculate the SER
3. Using the PER formula calculate the PER
4. Using the PER value calculate other performance metrics like throughput and latency.

Repeating the previous steps for a range of values of the graph and PER, below we can see the performance and latency as a function of the distance d.

5 Results

An overview and general comparison of performance, latency and BER is shown according to the measured data and the calculations proposed in this document, consider that this information provide a concise and precise description of the experimental results, their interpretation as well as the experimental conclusions. In Fig. 7 for performance nodes show the distance vs PER of three nodes, it shows the same PER but the distance transmission was the greatest rate until 16000 m of the LORAWAN station, while stations LR-WPAN the transmission distance fall until 50% of specs, in this point we can to say that although the mathematical method advises that for a certain distance the system should the highest PER, the system works in an acceptable way.

In Fig. 8 illustrate performance of all three nodes of the network vs distance, it shows the transmission rate was the greatest rate near to the stations, while we far of stations the transmission rate falls until 30% of specs of all manufacturers describe. Furthermore, with the three technologies the estimated throughput with the mathematical model must be 0%, but in the experimental process, it shows a high performance and shows that the system works.

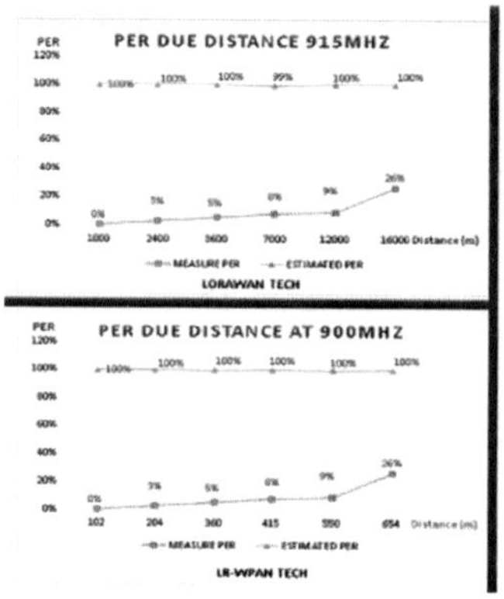

Fig. 7. General comparison of PER

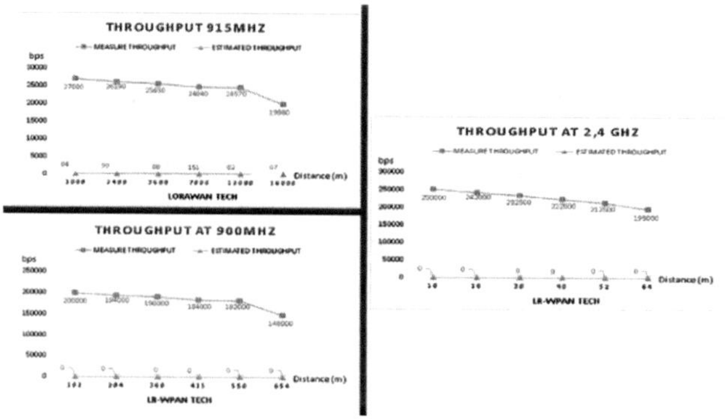

Fig. 8. General comparison of performance.

Electromagnetic interference is the disturbance that occurred caused by different sources of external electromagnetic radiation at the site of the tests. This disturbance could degrade or limit the performance.

In Fig. 9 prove the latency vs distance over three technologies used in the network, it shows the latency for LR-WPAN technologies was the lowest because were near the station, while we far as LORA-WAN station the latency up. By the other hand in all technologies, the measured latency is higher than the estimated latency with the mathematical model, this represents real process is uncertain because these parameters are experimental and determined or represent multiple processes.

On the site, fading occurred due to multipath propagation since the signals were received on roads with direct visibility and obstacles, which could cause distortion and attenuation of the signal.

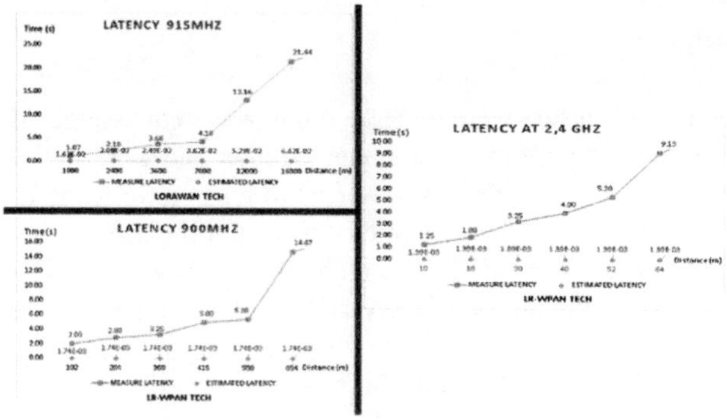

Fig. 9. General comparison of latency.

6 Discussion

Carry out the study of other authors with similar proposes to debate results of similar experimentations based on QoS results in literature found, while the manufacturer of xbeeproo at 900 MHz lists a direct visibility range of up to 9 miles (14 km) for the selected Xbee radios, the tests indicated a largest communication range of approximately 0.9 kms, other authors [12] have developed tests in similar conditions and reached 1 mile (1.6 km) in the transmission. Finally, the manufacturer of xbee at 2400 MHz mentions a range of 30 m at the interior building for the selected Xbee radios and the actual tests indicated a most communication range of approximately 15 m, another author [13] has similar results.

LoRaWAN technology offers long-range communication, LoRaWAN is the most adequate in coverage but not in the amount of information transmitted. We are according with [14] because they infer in this results the least interference range predicted by the capture threshold model is quite close to that under the additive interference model when the transmitter is close enough to the receiver. However, as the transmitter receiver separation increases, approximating the smallest interference range as a multiple of the transmitter-receiver separation because we have obtained similar result, with the observed data, the reach of the modules on the area is not a wide distance, although the received signal is in a moderate time since the sending of each frame is transmitted every two seconds, as it goes away, the delay of the signal increased a few seconds or, last, the total loss of the frames when it was 654 m away.

Although we consider that the model is good, it requires knowing and understanding the purposes of the model, we assume the mathematical model is unique and problems are dependent on data, both data used as input to the model as the experimental data, the other hand might be that model is not totally clear a user with the need to determine the exact characteristics of the version of the model being used before the results can be understood.

Finally, in future work, it is necessary to analyze the geographical aspects and the transmission difficulties, such as those presented by [15] who analyzes the difficulties caused by the obstacles and the quality of the signal.

7 Conclusions

Selected environmental sensors were integrated into a system assembled with different devices, technologies, and platforms that come together and work in a convergent way for the user, the system can consider it very different when developing it, but due to our work, it seems to have similar characteristics that scatter the differences for the reason that we achieve the integration of various technologies in a single system. Although developed this system to measure climate variables, can use it in an area that needs a WSN integrated with IoT.

The variety of technologies available for this type of project is wide and there are many options. In stage must identify the technologies that provide the best combinations of vendor specifications in the performance parameters of the technology. In stage 2, multiple attribute decisions must be made to select the technology from those identified in phase 1.

Design this system faced many limitations when choosing the right technology for this IoT application, but the hardest task was to integrate different types of hardware and software.

Within the work scenario and under the proposed conditions, we consider that the proposed model could be applied in environments under the following conditions: 1. Weather: Can be employed for any type of weather, 2. Topography: Can be used in open environments such as in the country areas, 3. Energy: It can remain without be connected to an electrical system.

In addition, although the average delay in data delivery can be greater than in the mathematical model, the end-to-end delay experienced by successfully received packets was high because the number of data packets that get transmitted in all technologies was higher than 70%.

Interoperability between network technologies is difficult due to work with different standards, in this case, layer physic adaptations do good setup end devices across the network, network protocols such as LoRa and Zigbee, was to be managed. As future work, there are several aspects of work: one of which is based on the treatment of data in order to predict the behavior of floods and have the following references, the one of [17] using methods based on fuzzy logic [16], which proposes the study under machine learning in the IoT, that raises the most frequent algorithms used in machine learning and some of these are virtual support machines, K-nearest neighbors, regression trees, analysis by principal components, analysis of canonical correlation among others. And another aspect to develop would be the one proposed by [18] that proposes the study and analysis for the optimization of data transmission in wireless networks, whit this prototype we can transmit either signal before analyze and conformation of the link layer and results of our analysis over experimental estimations in WSN, and extending the proposed framework utilizing different physical layer models.

References

1. Fisher, D.K.: Rapid deployment of internet-connected environmental monitoring devices. Adv. Internet Things **4**, 46–54 (2014)
2. Mukherjee, S., Biswas, G.P.: Networking for IoT and applications using existing communication technology. Egypt. Inform. J. **19**, 107–127 (2018)
3. Sodhro, A.H., Pirbhulal, S., Qaraqe, M., Lohano, S., Sodhro, G.H., Junejo, N.U.R., Luo, Z.: Power control algorithms for media transmission in remote healthcare systems. IEEE Access **6**, 42384–42393 (2018)
4. Goursaud, C., Gorce, J.M.: Dedicated networks for IoT: PHY/MAC state of the art and challenges. EAI Endorsed Trans. Internet Things **1**, 150597 (2015)
5. Perera, C., Liu, C.H., Jayawardena, S.: The emerging internet of things marketplace from an industrial perspective: a survey. IEEE Trans. Emerg. Top. Comput. **3**, 585–598 (2015)
6. Marques, G., Pitarma, R.: An indoor monitoring system for ambient assisted living based on internet of things architecture. Int. J. Environ. Res. Public Health **13**, 1152 (2016)
7. Saleem, Y., Crespi, N., Member, S., Husain Rehmani, M., Copeland, R.: Internet of Things-aided Smart Grid: Technologies, Architectures, Applications, Prototypes, and Future Research Directions, 1704.08977, pp. 1–30. IEEE (2017)
8. Li, H., Ochani, M.K., Zhang, H., Zhang, L.: Design of micro-automatic weather station for modern power grid based on STM32. J. Eng. **2017**, 1629–1634 (2017)
9. Bernas, M., Płaczek, B., Korski, W., Loska, P., Smyła, J., Szymała, P.: A survey and comparison of low-cost sensing technologies for road traffic monitoring. Sensors (Switzerland) **18**, 1–27 (2018)
10. Tzounis, A., Katsoulas, N., Bartzanas, T., Kittas, C.: Internet of Things in agriculture, recent advances and future challenges. Biosyst. Eng. **164**, 31–48 (2017)
11. Wang, F., Liu, J.: Networked wireless sensor data collection: issues, challenges, and approaches. IEEE Commun. Surv. Tutorials **13**, 673–687 (2011)
12. Parra, L., Karampelas, E., Sendra, S., Lloret, J., Rodrigues, J.J.P.C.: Design and deployment of a smart system for data gathering in estuaries using wireless sensor networks. In: IEEE CITS 2015 - 2015 International Conference Computer, Information and Telecommunication Systems, pp. 1–5 (2015)
13. Ojha, T., Misra, S., Raghuwanshi, N.S.: Wireless sensor networks for agriculture: the state-of-the-art in practice and future challenges. Comput. Electron. Agric. **118**, 66–84 (2015)
14. Aliev, K., Moazzam, M., Narejo, S., Pasero, E., Pulatov, A.: Internet of plants application for smart agriculture. Int. J. Adv. Comput. Sci. Appl. **9**, 421–429 (2018)
15. Shellhammer, S.J.: Estimation of packet error rate caused by interference using analytic techniques – a coexistence assurance methodology. In: IEEE P802.19 Wireless Coexistence (2005)
16. Hanatani, S., IMurata, C.: Packet error rate evaluation method. In: 5th_MEETING Working Group S ACP-WG S/5 (2014)
17. Jiao, W., Hagler, G., Williams, R., Sharpe, R., Brown, R., Garver, D., Judge, R., Caudill, M., Rickard, J., Davis, M., Weinstock, L., Zimmer-Dauphinee, S., Buckley, K.: Community air sensor network (CAIRSENSE) project: evaluation of low-cost sensor performance in a suburban environment in the southeastern United States. Atmos. Meas. **9**, 5281–5292 (2016)
18. Adelantado, F., Vilajosana, X., Tuset-Peiro, P., Martinez, B., Melià-Seguí, J., Watteyne, T.: Understanding the limits of LoRaWAN. IEEE Commun. Mag. **55**, 1–7 (2017)

A MLPG Formulation for Stress Analysis in Bi-dimensional Elastic Bodies

Luis Paternina(✉), Edgardo Arrieta, and Jairo Useche

Universidad Tecnológica de Bolívar, Cartagena, Colombia
lapc1917@hotmail.com, {earrieta,juseche}@utb.edu.co

Abstract. In this work, a meshfree method known as Meshless Local Petrov-Galerkin was implemented in a bidimensional linear elasticity problem. Bidimensional MLS shape functions were used in the polynomial approximation of the displacement and stress fields. The numerical integration was carried out by the Gauss-Legendre scheme and the quadrature points were located using the cartesian coordinate system. The local integration domain and the influence domain had circular shapes. A comparison with the analytical solutions of a plate with a circular hole subjected to traction was done. Numerical results showed a good agreement in the displacement and stress fields. Some schemes for improving the accuracy of the solutions were proposed.

Keywords: MLPG · Meshfree methods · Linear elasticity · Bidimensional

1 Introduction

The Finite Element Method (FEM) has been the standard method for computational simulation in the last decades because of its easy implementation, robustness, and accuracy. One of its main features is the discretization of the complex geometry of study into non-overlapping simple geometric elements through a mesh. This technique is not efficient when in fracture or breaking of the materials because a mesh is not the most reliable representation of the phenomenon. Besides, human intervention in the elaboration of meshes gives good results but subtract the efficiency of the method. This has motivated the accelerated development of meshless methods for the analysis of plates and shells [5–7, 14, 15]. Among of these, the Boundary-Element Method (BEM) has been extensively used in the analysis of isotropic and anisotropic plates and shells over the last decade [13–15].

Meshless methods have been proposed as an alternative to FEM in previous situations. Their complexity is greater than the standard method and their development was delayed because in the time they were proposed the computational power in the devices was not enough for achieving successful calculations.

Element Free Galerkin was introduced and became one of the most popular meshfree methods due to its fast convergence and its ability to solve fracture

D. F. Cortes Tobar et al. (Eds.): AETA 2019, LNEE 685, pp. 606–614, 2021.
https://doi.org/10.1007/978-3-030-53021-1_61

problems. Nevertheless, this method worked with a background mesh for the integration making this method not completely meshless.

Similar to the pure Galerkin scheme EFG, Meshless Local Petrov-Galerkin (MLPG) was proposed by Atluri [1] where the inner product is not done with the shape functions but the weight function used for the calculation of the shape functions. It was found that the method would not need a background mesh and that truly meshless methods finally conceived.

Some bidimensional and tridimensional elastostatic problems have been solved using MLPG. These works can be found in the following papers: [2–4,8,11].

2 Galerkin Methods for Elasticity

The global weak of the inner equilibrium equation is

$$\int_{\Omega} \psi \mathsf{D} \varepsilon d\Omega = - \int_{\Omega} \psi \mathbf{b}_j d\Omega - \int_{\Gamma_t} \psi \bar{\mathbf{t}} d\Gamma \tag{1}$$

where ψ is a test function used in the Galerkin method for weighted residuals and $\bar{\mathbf{t}}$ the external forces applied on the surface of the object.

Otherwise, the local weak form includes a boundary integral of the local domain used for the description of the geometry. The generalized local weak form equation has the form [2]

$$\begin{aligned} \int_{\partial\Omega_i} \psi \boldsymbol{\sigma}_k \cdot \widehat{\mathbf{n}}\, d\Gamma + \int_{\partial\Omega_u} \psi \boldsymbol{\sigma}_k \cdot \widehat{\mathbf{n}}\, d\Gamma \; - \int_{\Omega} \nabla\psi \mathsf{D} \frac{\partial \phi}{\partial x_k} u_i \, d\Omega \\ + \int_{\partial\Omega_i} \psi \boldsymbol{\sigma}_k \cdot \widehat{\mathbf{n}}\, d\Gamma + \int_{\partial\Omega_t} \psi \boldsymbol{\sigma}_k \cdot \widehat{\mathbf{n}}\, d\Gamma + \int_{\Omega} \psi \mathbf{b}\, d\Omega \; = 0. \end{aligned} \tag{2}$$

The integral applied to the $\partial\Omega_i$ is referred to as the boundary of the local domain. If a test function ϕ is selected such that its value becomes zero at its boundary, those integrals disappear. Usually, the test functions are selected with this feature to simplify the local form in a more easily form to work. The local weak form widely used in meshfree methods for elasticity is [2]

$$\int\limits_{\partial\Omega_u} \psi \boldsymbol{\sigma}_k \cdot \widehat{\mathbf{n}}\, d\Gamma \; - \int\limits_{\Omega} \nabla\psi \mathsf{D} \frac{\partial \phi}{\partial x_k} u_i \, d\Omega + \int\limits_{\partial\Omega_t} \psi \boldsymbol{\sigma}_k \cdot \widehat{\mathbf{n}}\, d\Gamma + \int\limits_{\Omega} \psi \mathbf{b}\, d\Omega \; = 0. \tag{3}$$

3 Numerical Implementation of Meshless Local Petrov Galerkin Method

3.1 Moving Least Squares Polynomial Approximation

Due to the absence of meshes in the description of the geometry in the meshless methods, different shape functions are needed to make the polynomial approximation. The shape functions proposed in the MLPG method were the Moving

Least Squares (MLS) [1]. These non-interpolant, rational functions ensure continuity and smoothness in the local support domains used for the approximation. A field variable u can be written as a matrix product like

$$\mathbf{u}(x) = \phi(x)\,\mathsf{U}_I \tag{4}$$

where U_I are the nodal parameters of the problem.

The MLS shape function has the form

$$\phi(x) = \mathbf{p}(x)\left(\mathsf{K}^{-1}\mathsf{B}\right) \tag{5}$$

where

$$\mathsf{K} = \mathbf{p}(x_I)\,\mathsf{W}_x\mathbf{p}^T(x_I) \tag{6}$$

$$\mathsf{B} = \mathbf{p}(x_I)\,\mathsf{W}_x. \tag{7}$$

$\mathbf{p}(x)$ or $\mathbf{p}$ is the monomial basis vector and its order defines the order of the MLS approximation. Linear polynomial basis have the form $\left[1\;x\;y\right]$ while quadratic polynomial basis can be written as $\left[1\;x\;y\;xy\;x^2\;y^2\right]$.

W_x is a diagonal matrix whose values are calculated with a weight function $w(x)$. The weight function can be defined as

$$w_i(\rho) = \begin{cases} 1 - 6\rho^3 + 12\rho^2 + 3\rho & \text{if } \rho \leq 1 \\ 0 & \text{if } \rho > 1 \end{cases} \tag{8}$$

where ρ is the normalized adimensional distance of a generic point to the center of the support domain

$$\rho = \frac{|\mathbf{x} - \mathbf{x}_i|}{R}. \tag{9}$$

The derivatives of the MLS function are

$$\partial_\alpha\phi = \partial\mathbf{p}^T\Gamma + \mathbf{p}^T\mathsf{K}^{-1}\left[\mathbf{p}(x_I)\partial_\alpha\mathsf{W}_x - \mathbf{p}(x_I)\partial_\alpha\mathsf{W}_x\mathbf{p}^T(x_I)\Gamma\right] \tag{10}$$

where Γ is calculated as

$$\Gamma = \mathsf{K}^{-1}\mathsf{B}. \tag{11}$$

3.2 Local Weak Formulation

The stress vector described in Eq. (3) can be written in terms of the displacements through

$$\sigma = \mathsf{DLU}_I \tag{12}$$

where L is a matrix of derivatives

$$\mathsf{L} = \begin{bmatrix} \frac{\partial\phi}{\partial x} & 0 \\ 0 & \frac{\partial\phi}{\partial y} \\ \frac{\partial\phi}{\partial y} & \frac{\partial\phi}{\partial x} \end{bmatrix}. \tag{13}$$

Combining Eqs. (3) with (12), the weak form equation for elasticity is

$$\left[\int_{\Gamma_u} \psi \mathsf{DL} \phi d\Gamma - \int_{\Omega} \nabla\psi \mathsf{DL} \phi d\Omega\right] u_I = -\int_{\Gamma_t} \psi \sigma_k \cdot \widehat{\mathbf{n}} d\Gamma - \int_{\Omega} \psi \, \mathbf{b}_k \, d\Omega \quad (14)$$

where Γ_u is the same as $\partial\Omega_u$, the boundary of the global domain.

When Eq. (14) is evaluated in a single node, it creates a $2x2N$ matrix that represents the contributions to the Galerkin stiffness matrix respectively. If this operation is done for all the N nodes of discretization, it is obtained a square matrix of $2N$. The algebraic form of Eq. (14) is

$$\mathsf{M}_{ik} \mathsf{U}_I = \mathsf{f}_k \quad (15)$$

where

$$\mathsf{M}_{ik} = \int_{\Gamma_u} \psi_i \mathsf{S}_k \mathsf{DL} \phi d\Gamma - \int_{\Omega} \nabla\psi_i \mathsf{S}_k \mathsf{DL} \phi d\Omega \quad (16)$$

$$\mathsf{f}_k = -\int_{\Gamma_t} \psi_i \sigma_{\mathbf{k}} \cdot \widehat{\mathbf{n}} d\Gamma - \int_{\Omega} \psi_i \mathbf{b}_k \, d\Omega. \quad (17)$$

3.3 Numerical Integration

The Gaussian quadrature is one of the most popular numerical integration method in FEM and meshfree methods for elasticity due its accuracy and robustness. In this study, the Gauss-Legendre scheme was used. For domain integrals of the weak form equation, the Gauss points were distributed over the x and y axis. The quadrature has the form

$$\int_a^b \int_c^d f(x,y) \, dy \, dx = \left(\frac{b-a}{2}\right)\left(\frac{d-c}{2}\right) \sum_{i=1}^{n} \omega_i \sum_{j=1}^{m} \omega_j f\left(\widetilde{\xi}_i, \widetilde{\xi}_j\right) \quad (18)$$

where $\widetilde{\xi}_i$ and $\widetilde{\xi}_i$ are the Gauss points adjusted to the generic interval $[a,b]$.

Figure 1 shows the spatial distribution of quadrature points inside a circular support domain in two situations. In those examples, 8 Gauss points and two subdivisions in each coordinate axis were used.

The distribution of the Gauss points can change because of the shape of the domain integration. Support domains can be located at any place and can intersect the global boundary giving as result complex domains. When there is no intersection, the integration domain can be determined easily.

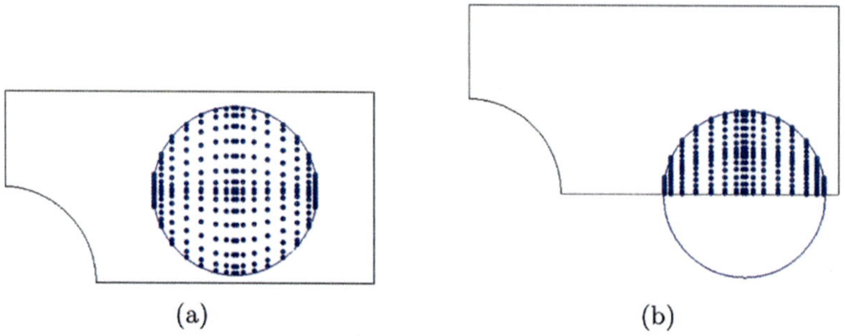

Fig. 1. Distribution of blue integration points in a support domain. **a** Inside the geometry. **b** Intersecting the boundary of the geometry

3.4 Enforcement of Boundary Conditions

Lagrange multipliers can be used to enforce displacement boundary conditions in meshless methods. Load conditions are implicitly applied in the boundary integral in the right-hand side of local weak formulation shown in Eq. (14). If there is a matrix of conditions C and a vector of described values d -with the same quantity of rows-, the assembling of the boundary equations is done as follows

$$\begin{bmatrix} \mathsf{M} & -\mathsf{C}^T \\ \mathsf{C} & 0 \end{bmatrix} \begin{bmatrix} \mathsf{U}_I \\ \lambda_{\mathsf{ML}} \end{bmatrix} = \begin{bmatrix} \mathsf{f} \\ \mathsf{d} \end{bmatrix} \tag{19}$$

where

$$C_u = \int_{\Gamma_u} N\phi d\Gamma_u. \tag{20}$$

4 Numerical Results

4.1 Finite Plate with Hole

A finite plate of 0.2 m of length and 0.1 m of height with a circular hole of radius 0.02 m is going to be solved. The elastic modulus of the material of the plate is $E = 210 \times 10^9$ Pa and the Poisson number is about 0.3. In this exercise, the weight of the plate will be neglected. The plate is subjected to a tension of 100 MPa in the horizontal direction.

Due to the load and geometrical symmetry, it is possible to take a quarter part of the plate, as it is shown in Fig. 2. Thus, the boundary conditions for this scenario are given by

- Displacement conditions
  - Inferior side. $u_y = 0$
  - Left side. $u_x = 0$

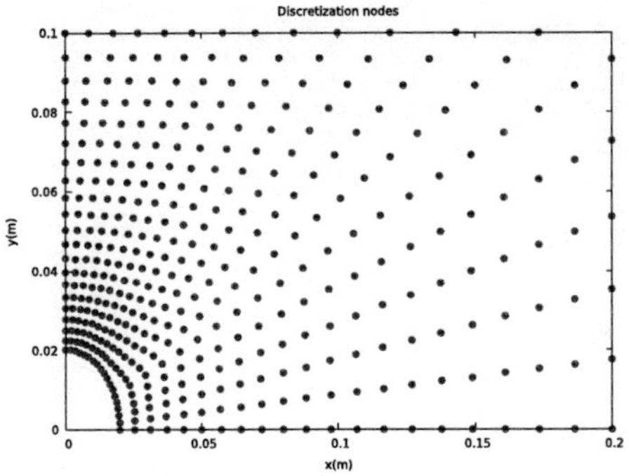

Fig. 2. Discretization of the general geometry with 399 nodes

- Stress conditions
  - Left side. $\tau_{xy} = 0$
  - Inferior side. $\tau_{xy} = 0$
  - Right side. $\sigma_x = 100$ MPa, $\tau_{xy} = 0$
  - Superior side. $\sigma_y = 0$, $\tau_{xy} = 0$
  - Hole side. $\sigma_x = 0$, $\sigma_y = 0$

For the discretization, 399 nodes (a transformed grid of 19 $\times$ 21 points) were used in the following configuration. The configuration was done in such a way that higher nodal density values would be found near the hole, which is desirable taking into account that zone presents stress concentration. The variation of the nodal distribution was achieved with a geometric progression of the distance between nodes.

Second-order MLS functions were selected. For the numerical integration, the domain integration was divided into 4 subdomains with 8 quadrature points each one. The sizes of the integration domain and the support domain will be considered equal. MLPG calculations were done in different routines using C++ codes made by the author of this work.

The analytical solutions for the stress fields for a plate with a circular hole subjected to traction can be found in [12]

$$\sigma_{\rho\rho} = \frac{\sigma_o}{2}\left(1 - \frac{a^2}{r^2}\right) + \frac{\sigma_o}{2}\left(1 + \frac{3a^4}{r^4} - \frac{4a^2}{r^2}\right)\cos 2\theta \tag{21}$$

$$\sigma_{\theta\theta} = \frac{\sigma_o}{2}\left(1 + \frac{a^2}{r^2}\right) - \frac{\sigma_o}{2}\left(1 + \frac{3a^4}{r^4}\right)\cos 2\theta \tag{22}$$

$$\tau_{\rho\theta} = -\frac{\sigma_o}{2}\left(1 - \frac{3a^4}{r^4} + \frac{2a^2}{r^2}\right)\sin 2\theta. \tag{23}$$

The analytical equations for displacements in cartesian coordinates are [3]

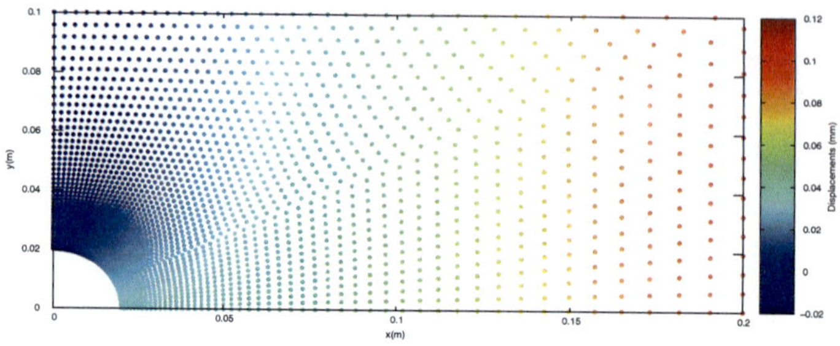

Fig. 3. Displacements in x axis given by MLPG method

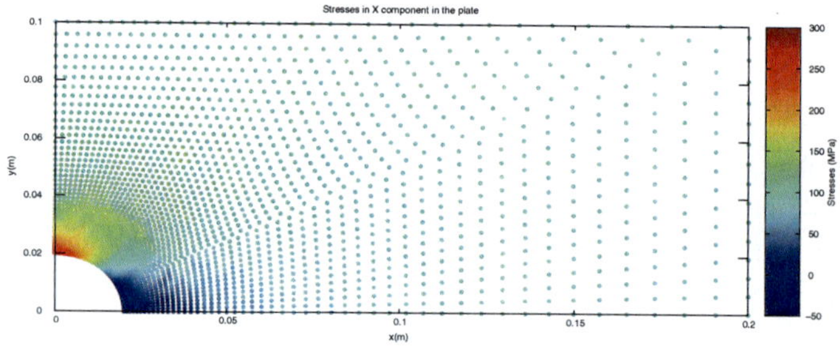

Fig. 4. Normal stresses in x axis given by MLPG method

$$u_x = \frac{\sigma_o\,(1+\nu)}{E}\left[\frac{r\cos\theta}{1+\nu} + \frac{2a^2\cos\theta}{r\,(1+\nu)} + \left(\frac{a^2}{2r} - \frac{a^4}{2r^2}\right)\cos 3\theta\right] \tag{24}$$

$$u_y = \frac{\sigma_o\,(1+\nu)}{E}\left[\frac{-\nu r\sin\theta}{1+\nu} + \frac{1-\nu}{1+\nu}\frac{a^2\sin\theta}{r} - \left(\frac{a^2}{2r} - \frac{a^4}{2r^2}\right)\sin 3\theta\right]. \tag{25}$$

Figures 3 and 4 have good agreement between the spatial distribution of the displacement and stress fields between the numerical and analytical solutions. One assumption that makes the analytical solution is that the plate is infinite respect to the size of the hole (this same situation can be understood as a finite plate with a tiny hole). Therefore the values of displacements near the zone of the applied force will be higher compared to the numerical values due to the no existence of explicit boundary in the analytical solution. In the case of the stress field, it is showed that the numerical method can reproduce accurately the behavior of the stresses near the hole.

Figure 5 shows a comparison between the displacement and stresses over a specific line (a horizontal line in the bottom of the geometry). As a confirmation of the aforementioned analysis in the previous paragraph, MLPG can reproduce the behavior of displacement and stresses in the geometry. In the case of the displacements, the numerical solution makes a good approximation because displacements are the main field approximated in the numerical method.

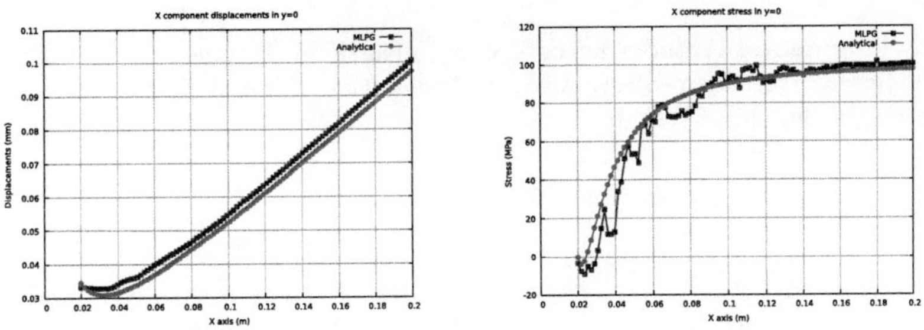

(a) (b)

Fig. 5. Fields along the line $y = 0$ in x component. **a** Displacements. **b** Normal stresses

For the stresses, even there are many irregularities, it is possible to find a tendency in the numerical values.

The identification of the possible sources of error in the stress field can be attributed to the calculation of the stress field, the implementation of the cartesian Gaussian quadrature andthe ill-conditioned nature of the algebraic equations. The stress field is calculated from the derivatives of the shape functions, which can be corrected using a different size for the support domain or with a mixed approach to the weak form of the differential equation. The integration scheme showed in Fig. 1 is not suitable for circular or irregular domains. Mazzia [9,10] explored other quadrature schemes for circular and composed geometries, proposing polar configurations and cubatures that achieve more accurate and stable results with fewer integration points. The solution of the algebraic equation in Eq. (19) is ill-conditioned. Thus, it is suggested for future work to implement iterative methods with preconditioning to alleviate the influence of the ill-conditioning in the solution of the numerical methods.

5 Conclusions

Meshless Local Petrov Galerkin (MLPG) was used to solve the bidimensional elastic problem of a rectangular plate with a central hole subjected to tensile strength. The displacement and stress fields were approximated with 2D MLS shape functions. Gauss-Legendre for numerical integration was implemented in the local integration domains. Both integration and support domains were circular and had the same size in the calculations.

Displacements obtained by MLPG showed excellent agreement with analytical equations because the meshless local weak form approximates directly the displacement field. Meanwhile, the stress field showed decent results but it is believed that a better scheme of integration and solution of the linear system of equations can give smooth fields. The collocation of integration points using the cartesian axis direction is not the most suitable quadrature for circular support domains. Other integration schemes were proposed.

Acknowledgments. Authors acknowledge Universidad Tecnológica de Bolívar for supporting this research project. Also the first author acknowledges the Mechanical Engineering Fellowship Program.

References

1. Atluri, S., Zhu, T.: A new meshless local Petrov-Galerkin (MLPG) approach in computational mechanics. Comput. Mech. **22**(2), 117–127 (1998). https://doi.org/10.1007/s004660050346
2. Atluri, S., Zhu, T.L.: The meshless local Petrov-Galerkin (MLPG) approach for solving problems in elasto-statics. Comput. Mech. **25**(2–3), 169–179 (2000). https://doi.org/10.1007/s004660050467
3. Atluri, S., Kim, H.G., Cho, J.: A critical assessment of the truly meshless local Petrov-Galerkin (MLPG), and local boundary integral equation (LBIE) methods. Comput. Mech. **24**(5), 348–372 (1999). https://doi.org/10.1007/s004660050457
4. Abdollahifar, A., Nami, M.R., Shafiei, A.R.: A new MLPG method for elastostatic problems. Eng. Anal. Bound. Elem. **36**, 451–457 (2012). https://doi.org/10.1016/j.enganabound.2011.08.008
5. Dinis, L.M.J.S., Jorge, R.M.N., Belinha, J.: A natural neighbour meshless method with a 3D shell-like approach in the dynamic analysis of thin 3D structures. Thin-Walled Struct. **49**(1), 185–196 (2011)
6. Fazzolari, F.A., Carrera, E.: Advances in the Ritz formulation for free vibration response of doubly-curved anisotropic laminated composite shallow and deep shells. Compos. Struct. **101**, 111128 (2013)
7. Ferreira, A.J.M., Roque, C.M.C., Jorge, R.M.N.: Static and free vibration analysis of composite shells by radial basis functions. Eng. Anal. Bound. Elem. **30**(9), 719–733 (2006)
8. Martinez, T.J.A., Arrieta, O.E.W.: Element Free Galerkin (EFG) sensitivity study in structural analysis. In: IOP Conference Series: Materials Science and Engineering (2019)
9. Mazzia, A., Ferronato, M., Pini, G., Gambolati, G.: A comparison of numerical integration rules for the meshless local Petrov-Galerkin method. Numer. Algorithm **45**(1–4), 61–74 (2007). https://doi.org/10.1007/s11075-007-9110-6
10. Mazzia, A., Pini, G.: Product Gauss quadrature rules vs. cubature rules in the meshless local Petrov-Galerkin method. J. Complex. **26**, 82–101 (2010). https://doi.org/10.1016/j.jco.2009.07.002
11. Paternina, L., Arrieta, E., Useche, J.: Analysis of plates through MLPG 3D solid elasticity. Sexto Simp. Nac. sobre Mec. de Mater. Estruct. Contin, pp. 179–187 (2018)
12. Sadd, M.H.: Elasticity: Theory, Applications, and Numerics. Academic Press, Cambridge (2009)
13. Useche, J.: Vibration analysis of shear deformable shallow shells using the boundary element method. Eng. Struct. **62**, 65–74 (2014)
14. Useche, J., Alvarez, H.: Elastodynamic analysis of thick multilayer composite plates by the boundary element method. CMES: Comput. Model. Eng. Sci. **107**(4), 277–296 (2015)
15. Useche, J., Medina, J.: Boundary element analysis of laminated composite shear deformable shallow shells. Compos. Struct. **199**, 24–37 (2018)

The Application of an Information System in the E-Government of Colombia to Improve Service to Citizens

G. Ortegon-Cortazar[1(✉)], J. Samper-Zapater[1], F. Garcia-Sanchez[1], O. Garcia-Bedoya[2], and C. O. Diaz[2(✉)]

[1] Universidad de Valencia, Valencia, Spain
luisgio@alumni.uv.es
[2] Universidad de Bogota Jorge Tadeo Lozano, Bogota, Colombia
cesaro.diazb@utadeo.edu.co

Abstract. One of the main objectives of e-government is the use of data to improve processes, identify opportunities and make better decisions, this article describes a success story in the government of Colombia in building an information system that makes use of data exploitation techniques in one of the most important state entities in the country that guides, inspects, monitors and controls registry offices and notaries. The article describes the context and reasons why the project was carried out along with the architecture that was used to design and develop the information system that is currently used to monitor and control in a timely, clear, effective and efficient service delivery to citizens in more than 1000 offices. After a year of producing the system, the article describes the administrative, economic and functional benefits that were reported by various areas and offices.

Keywords: Electronic government · Smart government · Innovation · Data exploitation · Business intelligence · Software design · Development

1 Introduction

The use of data is a key piece in the digital transformation and technological modernization of any country, which is why Colombia is the first country in Latin America with a public policy for the exploitation of data through the plan of the National Council for Economic and Social Policy (CONPES) 3920 of 2018 whose objective is to generate social and economic development along with the significant increase of timely, reliable and high quality data for decision making [1]. This plan supports the importance of managing timely, reliable and high-quality data through research conducted by the National Planning Department of Colombia (DNP) where it was found that only 4% of state entities have the required conditions to move consistently towards the use of data. According to the figures identified in this study, it is a problem for the government that

D. F. Cortes Tobar et al. (Eds.): AETA 2019, LNEE 685, pp. 615–625, 2021.
https://doi.org/10.1007/978-3-030-53021-1_62

decision making based on the use of data is so low, which is why information systems are required that effectively and reliably control the information that each entity is responsible for managing [2].

In accordance with DNP research, specifically recommendation number 4, paragraph A of CONPES 3920 of 2018, the purpose of it is "To design and implement a tool that facilitates the administration of data in public entities", the Superintendence of Notaries and Registry (SNR) formulated a technological application project, with the objective of improving the provision of service to citizens, since public administrations must determine, understand and integrate the demands and needs of citizens in the strategies for the design and provision of public services [3].

The purpose of the project is to improve the quality of service to citizens by optimizing administrative management and data analysis in a simple and effective way. In this sense the main objective is to improve the efficiency in the service to citizens that leads to the optimization of service procedures and the analysis of results through data visualization techniques. By easily and intuitively identifying and understanding results, public managers make better decisions for the benefit of citizens.

SNR applied a data management model that is articulated with institutional management and performance policies within the framework of the integrated planning and management model of public entities CONPES 3920, recommendation 5-a. This data-driven innovation corresponds to the exploitation of data through the application of analytical techniques to improve or create new goods, services or processes that contribute to the diversification and sophistication of the economy and the generation of social value, as a new source of growth [4]. Innovation in the entity is also important, but for innovation to happen effectively requires the interaction of various actors and institutions, which are geared to the development and dissemination of new technology and knowledge [5].

The data exploitation facilitates and accelerates innovation processes as it offers revelations that previously depended exclusively on human ingenuity [6]. In this sense, the project incorporates tools that improve the understanding of the provision of service to citizens such as histograms, heat maps, automatic notifications, among others. Thanks to the techniques of data consolidation and visualization, it increases the ease of making decisions.

The article begins by describing the context of the provision of services to citizens and how data exploitation techniques can improve the quality of service. The second part of the article describes the architecture of the system from the structure of the entity. Once the architecture is explained, the following section describes the results making use of user interfaces and then in the last section present the conclusions of the project.

2 Background

With the advent of information technologies, data have gone from scarce to superabundant [7], which is why data must be analyzed in terms of business needs and in terms of improving service delivery to citizens. In this way, it is possible to develop "intelligence of public value" (a social equivalent of "business intelligence"), which has the potential to be a strategic component for decision-making and the design, implementation, and evaluation of public policies within governments [8]. In this new fourth industrial revolution, data are an essential factor for production, such as fixed assets and human capital [9], which is why they should be seized in order to identify opportunities for improvement. It is there those information systems with advanced data analysis are needed that change the culture of treating all citizens in the same way, independent of their personal and group characteristics [7]. Understanding data as an asset that generates economic and social value involves defining [10], implementing, maintaining, and operating the data infrastructure in accordance with Conpes 3920 of 2018 [1]. The use of advanced analytics for decision making and the evaluation of public policies applying BI (business intelligence) allow increasing the level of satisfaction of the citizens when finding relevant information in data oceans that grow exponentially every day [11]. In the case of decision-making, the design, implementation, and evaluation of information produce evidence of what is relevant, of quality, and timely, to inform and guide decisions [12]. This means diagnosing problems that go unnoticed and for that reason are impossible to act upon [13]. This process is called "data-driven decision-making" [14]. Unprocessed bulk data has little value in itself, as it is only obtained after going through full processing [15]. This value is related to the possibility of building valuable knowledge, improving processes, and contributing to decision making by decreasing uncertainty, among others. It is there that they contribute greatly to improving the internal processes of government entities such as the provision of service to the public [16]. In terms of the purposes of smart governments, it assumes that it is expected to increase efficiency, effectiveness, and transparency in organizational management and service delivery, promote collaboration between city departments with other external organizations and citizens, thus facilitating the participation of non-governmental entities in decision-making and monitoring service delivery [13].

The digital transformation of society and the economy is radically changing service delivery practices. The private sector has introduced new ways of providing services and, as a result, citizens' expectations of public services have increased [3].

Mechanisms of innovation and intelligence in public management can be interpreted as intelligent governance, this allows a social system and its subjects to operate effectively in a dynamic and complex environment by rationally using its internal and external resources, making appropriate/pragmatic and advanced decisions related to the specific circumstances in order to create shared value [16].

Public data is a powerful resource for moving from a citizen-centered to a citizen-driven approach, enabling administrations to design public service delivery processes tailored to users' needs [17]. In this sense, public administrations must have a better understanding of citizens' needs and facilitate the digital transformation of the provision of public services [3].

3 Architecture of the Software of Data Management

Design According to the volume and magnitude of procedures and processes that the entity has, the information system was designed by components and modules, the resulting structure allows the separation of functionalities in order to separate responsibilities and improve the availability of the system [18]. Given the importance of the e-government process, the use of data and the optimization of service delivery to citizens, SNR designed and developed a robust, secure and reliable platform to manage procedures and processes. The main needs were focused on improving management efficiency, data exploitation, business intelligence for decision [11] making along with the integration of software quality attributes such as usability, reliability, availability, modularity, scalability, among others [19]. The design of the system was based on the functional and non-functional requirements of the Stakeholders, at the beginning of the project, the architecture drivers were established that was the guide to follow in the project. The main cross-cutting requirements identified for the system are:

- **No cost.** The technology used outside the database should be opensource as the system should not depend on suppliers or licenses. Technologies as Cloud Computing is good example of this [20]
- **High performance, Speed.** Techniques must be implemented that streamline the user experience in the system, to do so must use the browser cache, images, styles and compressed JavaScript functions, references to frameworks with Content Delivery Network (CDN).
- **Security.** The system must be integrated with the entity's Lightweight Directory Access Protocol (LDAP) since access to components must be based on user identification, role, and profile. This information is centralized in order to manage the access privileges to all the entity's systems from a single platform.
- **Usability.** The operating expertise of the system should be intuitive in order to increase the use of the system and decrease the creation and use of manuals for components and modules.

The volume of actors that manage the information system is structured by dependencies that directly belong to the entity and institutions that the entity supervises and controls. Table 1 describes this information

The actors of the system also interact with parameters such as the location segmented by regions, departments, municipalities and cities since they allow identifying geo-referential information with the types of procedures and processes at the national level.

The modules designed for the optimization of administrative management in the service to citizens were:

Generation of PQRSD: Allows to generate a PQRSD from several channels of attention like web, presential, mail, window, chat.

PQRSD Classification: Allows to classify the PQRSD in 184 options grouped in 4 hierarchical levels; Category, class, subject, reason.

Assignment of PQRSD to offices: Allows assigning the PQRS to areas of the main headquarters or to 195 registration offices.

Table 1. SNR actors

Institutions	Offices	Users	Sub. users
General Management, Central Level	25 units, 48 working groups	3900	3300 officials, 600 contractors
Registry offices for public instruments	195	195	195 Registrars
Notaries	330 first category, 181 second category, 392 third category	5903	903 Notaries, 5000 advisors
Curatorships	74	72	72 Curators

Assignment of PQRSD to officials: Allows assigning the process to a specific official.

PQRSD Requirements: Allows to require 903 notaries when the PQRSD is coming from that type of office.

Return of PQRSD: Allows to return the PQRSD to re-classify and re-assign this procedure.

The modules are related to each other according to the organizational structure of the entity, in this sense when the PQRSD is assigned to an area, it is automatically assigned to the head of the area or office and to the officers responsible for that issue.

Thanks to the high cohesion and low coupling between the components, an efficient and easy to maintain data structure is obtained whose main objective is to be able to apply descriptive analysis using business intelligence (BI) techniques through data visualization tools. Elements of electronic government focused on quantifying, analyzing and improving the delivery of service to citizens were requirements in the vision of architecture, monitoring tools, data visualization and a relational scheme of star based on the components and modules by offices and dependencies allowed the structure of the system to be efficient and effective when generating and presenting information. The granularity and relational structure of the components make it possible to obtain data models that can be grouped by the actors and parameters described above. This type of structure leads to the use of information through online analytical processes and data consolidation and visualization techniques. In the technology industry, there are many BI tools that are used for data visualization: Power BI, Oracle BI, Tableau, among others. Since the initial requirements stated that licensed software would not be used, these tools are not used.

Due to the high customization of the software, the control panels were designed in the interviews of the survey of requirements, in these work tables, it was identified that information is required to consolidate and visualize by means of histograms. At the technical level, open source data visualization tools were chosen, such as C3js, Chartist, Chartjs, among others.

The cases of use for the system depend on the role and the identification of the office, in this sense the users with chief role can perform actions such as return procedures to increase response times and other options that can not perform any user. According to the type of office, the system presents the work screens, if it is the type of office of the central headquarters, it has access to monitor all the country, if it is the type of office of a city, it can only see the information of that city.

Development
The information system was developed integrating the model architecture pattern, MVC controller view with the micro-services pattern. This type of architecture allows the separation of persistence, logic, and presentation of information through artifacts that have unique responsibilities and are agile to develop thanks to the efficiency of the system structure. Figure 1 describes the relationship between the layers of the system and the artifacts it uses.

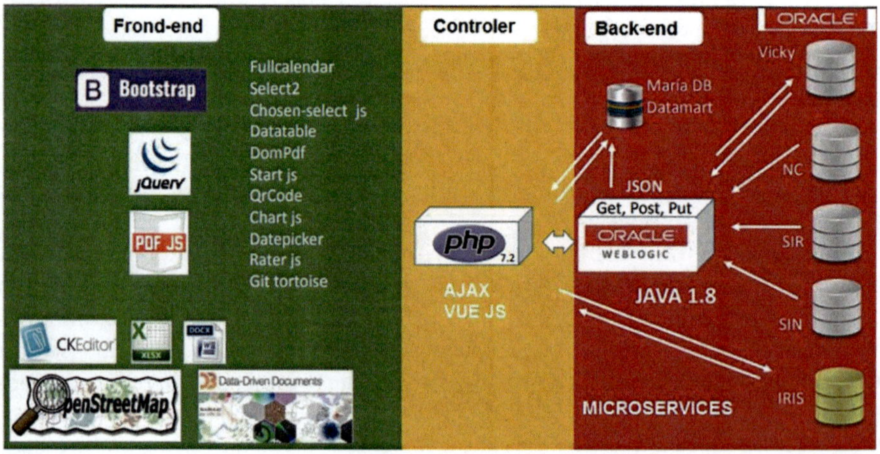

Fig. 1. System layers and artifacts

In order to improve speed and standardization in software development, the persistence layer is used dynamically. This means that through context instructions in a URL and using GET, POST and PUT methods, any database table is queried and used. The web services of the system were developed with REST technology, the exchange of data is done through JSON objects that allow operations such as insertion, updating, and consultation of the database by any component of the system. The security to use this module is done through access by IP address, this means that the module has a list of IP addresses that are the only ones to which the web service responds with JSON objects. The URL describes a complete query with pagination, conditions and order. Each of the properties when using GET method are separated by the symbol/where each parameter means:

1. DataSource Parameter "JSON": Name of the DataSource deployed on the Server.
2. Parameter Table "users": Name of the table to extract the data
3. Parameter Condition "id = 107": Consultation conditions
4. Parameter Order "asc id": Sorting condition of the results
5. Pagination Parameters "1/4": Indicate the number and segment of records to be presented.

The result of the web service uses the SoapUI testing tool, where the JSON objects are generated dynamically regardless of the characteristics and dimensions of the tables in the database, this is why it is a dynamic web service.

The layer that makes the system logic is developed with the integration of several technologies: PHP v7, Ajax and the framework JQUERY and VUE JS. These technologies are used in the system because of their high software build speed and agility to make changes to the system functionalities.

One of the key decisions in the agility of the development of the system is the standardization of the fields in the function of the tables of the database, this decision of architecture establishes rules in the development as:

1. Table names must be in the singular
2. The primary key field of each table must be called id-xx where xx is the name of the table.
3. There must be a field named name-xx where xx is the name of the table. This field will store descriptive information for each record.
4. There must be a field named status-xx where xx is the name of the table. This field must be boolean and represents the logical deletion of the records.

Thanks to the standardization of these rules throughout the system, access to and management of each of the tables can be performed dynamically since the table name behaves as a variable and the concatenation with the parameters id, name and status allows automatic functions to be generated such as:

- Dynamic logical deletions: update xx set status-xx = 0 where id-xx = y
- Registers: select count(id-xx) as z from xx where status-xx = 1
- Lists options: select name-xx from xx where status-xx = 1

The presentation layer uses 10 open source frameworks that allow to manage the system in a more efficient and intuitive way, the main functionalities are:

1. Generation of QR codes for document identification and tracking.
2. Generation of PDF files
3. Text editor to add attributes to the text.
4. Dynamic Calendar
5. Search for information in selection lists
6. Dynamic tables that allow searches and sorting by fields.
7. Automatic notifications
8. Asynchronous relations
9. Map
10. Statistical charts

The system's infrastructure is based on VMware virtual machines that enable dynamic vertical growth and horizontal growth by managing computational instances based on user demand. The advantage of using virtual servers and mounting points for storage lies in the high availability it generates and the ease of growth and maintenance.

4 Results

The information system has generated more than one million transactions according to the audit logs during one year of execution. These transactions represent all the management procedures that are part of the components, modules and functionalities of the system. The volume of transactions also includes the searches and accesses of the more than 7000 users of the system. The following table quantifies the number of records for each component (Table 2):

Table 2. Transactionality

Institutions	Offices	Quantity
Central level and ORIP	Citizens management	32.210
Central level and ORIP	Management of petitions, complaints, grievances, suggestions and complaints PQRSD	15.141
Central level and ORIP	Management of citizen perception	27.848
Central level and ORIP	Resolution management	19.002
Curatorships	Management of town planning licences	46.150
Curatorships	Management of costs of Curatorships	192
Notary	Management of possessions for Notaries	2.203
Notary	Management of forced retirements of Notaries	1.300
Notary	Management of permissions and licenses of Notaries	31.173
Notary	Management of requirements to Notaries	320

The panel allows the visualization of the figures associated with this process at the national level. It can display states, reasons, entry channels, classifications, offices, directions to offices, assignments to officials, returns from offices, citizens, among others.

The system's interfaces made it possible to identify statistically relevant information such as unfinished procedures, staff with pending tasks, number of procedures per unit, volume of procedures per month, categories of most recurrent procedures, among others. The panel allows the visualization of the figures associated with this process at the national level. It can display states, reasons, entry channels, classifications, offices, directions to offices, assignments to officials, returns from offices, citizens, among others.

Given the systematization of documents, the printing of procedures was reduced. This reduction was identified by comparing the volume of paper used before the project by dependency with that used today.

The evidence of the results obtained by the system can be found in an audit report made by the internal control office of the entity in which each one of the stages of the project was identified and the benefits generated for the entity. This audit report indicated that thanks to the construction of the system, the efficiency of service delivery to citizens improved given the control and traceability of each of the procedures by dependency and responsible official.

5 Discussion and Conclusions

Applying technological elements and attributes of electronic and intelligent government in the SNR improves the processes and control mechanisms that had been implemented some time ago, the simple obtaining of information by means of data visualization techniques and exact quantification of the same in real time improve the administrative efficiency of the offices and civil servants, to achieve the increase in the quality of the service provided to the citizen.

An audit report carried out by the internal control office of the entity noted the optimization of response time to procedures, recognized the positive impact of the system within the processes of the entity.

The use of data from management processes, control, and visualization of information that exists in the system allows the SNR to make better decisions to identify and quantify by periods, location, types of procedures, officials, among others, trends in existing data and observe the particular situations of each office. Thanks to the system's control panels, the directors of each unit can monitor the status of each procedure and the tasks of the employees.

The simple visualization of the data allows the public managers of each unit to make better decisions about the management of each office, thanks to this information can streamline procedures, change assignments of tasks, balance the performance of employees by assigning procedures according to their efficiency.

Applying data analysis techniques through star or snowflake type queries allowed to identify which cities and offices in Colombia are more prone to certain types of procedures under specific periods of time. The descriptive analysis that the system performs allows seeing the real state of the processes for each office, thus identifying which procedures are in management and which are overdue.

The level of ownership and appropriation of the system by officials was much higher than other systems implemented by the entity years ago or during the same period of construction of the system, the minutes of delivery of the components show this situation because they describe the number of officials who use the system satisfactorily. The technique used to identify this assertion is the satisfaction surveys applied to the entity's officials after training is conducted.

Thanks to the system's construction methodology and the periodic participation of expert officials in the entity's processes, it was possible to develop and implement an information system with a high degree of customization, innovation, and administrative efficiency. This system allowed to improve the efficiency of the service thanks to the control of each one of the procedures together with the time and deadlines to give answer.

The quality and reliability of the data are fundamental in order to ensure the reliability of the reports and control panels. Only in this way is it possible to guarantee the legitimacy, continuity, and relevance of the information that becomes strategically relevant, identifying quality attributes such as agility, efficiency, effectiveness, and transparency in public management.

The development from scratch was the best option to provide a solution to decision making from data since the entity is unique in its kind at the national level and implementation with the degree of detail and refinement required by the entity, would have been more costly to perform in time and money.

The project did not generate any kind of administrative effort since it is not part of any contract, it did not have a budget or costs associated with technology service providers.

In accordance with the first specific objective of Conpes 3920 "To massify the availability of data from public entities that are accessible, usable and of quality digital", the system created provides and brings this massification of data to citizens within the framework of the model of electronic government and intelligent government.

This work implements and responds to the eleventh action line of CONPES 3920 "Institutionalize the exploitation of data in public decision making" since better decisions can be made thanks to the control panels of the system.

The objective of the article was achieved because it was able to improve the administrative efficiency of the PQRSD procedures thanks to the controls developed and the visualization of data in a simple and clear way.

Finally, a smarter government seems to require a forward-looking approach to the use and integration of information, technology and innovation in government activities [5], which is why this article aims to set an example of innovation and what to make public for the benefit of citizens who are the reason for the state.

References

1. N. P. Department: National data exploitation policy (big data). CONPES, vol. 1, pp. 1–120 (2018)
2. Hilbert, M., López, P.: The world's technological capacity to store, communicate, and compute information. Science **332**(6025), 60–65 (2011)
3. OECD: Panorama of public administrations 2017, innovation in the public sector and digital government. OECD, vol. 1(2018)

4. Wolfe, R., Wright, P.M., Smart, D.L.: Radical HRM innovation and competitive advantage: the moneyball story. Hum. Resour. Manag.: Publ. Cooper. School Bus. Adm. Univ. Michigan Alliance Soc. Hum. Resour. Manag. **45**(1), 111–145 (2006)
5. Gil-Garcia, J.R., Helbig, N., Ojo, A.: Being smart: emerging technologies and innovation in the public sector. Govern. Inf. Q. **31**, I1–I8 (2014)
6. Analytics, M.: The age of analytics: competing in a data-driven world (2016)
7. Manyika, J., Chui, M., Brown, B., Bughin, J., Dobbs, R., Roxburgh, C., Byers, A.H.: Big data: the next frontier for innovation, competition, and productivity (2011)
8. Rodríguez, P., Palomino, N., Mondaca, J.: El uso de datos masivos y sus técnicas analíticas para el diseño e implementación de políticas públicas en latinoamérica y el caribe (2017)
9. Lopes, N.V.: Smart governance: a key factor for smart cities implementation. In: 2017 IEEE International Conference on Smart Grid and Smart Cities (ICSGSC), pp. 277–282. IEEE (2017)
10. Sanchez Riaño, V., Diaz, C., Ludeña, C., Suarez Baez, L.C., Sojo, J.: Towards automated advertising strategy definition based on analytics. In: Florez, H., Diaz, C., Chavarriaga, J. (eds.) Applied Informatics, pp. 45–59. Springer, Cham (2018)
11. Herden, O.: Data warehouse. Taschenbuch Datenbanken, pp. 430–460 (2015)
12. Chen, C.P., Zhang, C.-Y.: Data-intensive applications, challenges, techniques and technologies: a survey on big data. Inf. Sci. **275**, 314–347 (2014)
13. Mining, T.E.D.: Enhancing teaching and learning through educational data mining and learning analytics: an issue brief. In: Proceedings of Conference on Advanced Technology for Education (2012)
14. Provost, F., Fawcett, T.: Data science and its relationship to big data and data-driven decision making. Big data **1**(1), 51–59 (2013)
15. Gandomi, A., Haider, M.: Beyond the hype: big data concepts, methods, and analytics. Int. J. Inf. Manage. **35**(2), 137–144 (2015)
16. Gaulė, E., Šiugždinienė, J., Buškevičiūtė, J.: The need of smartness in public governance. In: Proceedings of the 8th Scientific Conference "Business and Managament" (2014)
17. Moreno, L.M.M., Páez, J.O.T., Parra, A., Campos, D.: The colombian government enterprise architecture framework. In: Proceedings of the 2014 Conference on Electronic Governance and Open Society: Challenges in Eurasia, pp. 38–41. ACM (2014)
18. McAbee, S.T., Landis, R.S., Burke, M.I.: Inductive reasoning: the promise of big data. Hum. Resour. Manag. Rev. **27**(2), 277–290 (2017)
19. Angel, A.: Arquitectura del software, vol. 1 (2017)
20. Díaz, C.O., Gómez, C.E., Castro, H.E., Barrios, C.J., Bolívar, H.D.: Federated campus cloud colombian initiative. In: 2016 16th IEEE/ACM International Symposium on Cluster, Cloud and Grid Computing (CCGrid), pp. 712–717, May 2016

System for the Measurement of sEMG and Angular Displacement of the Ankle-Foot Joint Complex for Muscle Co-activation Detection in the Diagnosis of Foot Drop Pathology

Santiago Noriega, Maria C. Rojas, and Cecilia Murrugarra(✉)

Electronic Engineering Program, Faculty of Engineering, Universidad El Bosque, Bogota, DC 110111, Colombia
{snoriega,mcrojass,cmurrugarra}@unbosque.edu.co
http://www.unbosque.edu.co

Abstract. The measurement of physiological variables for the assessment of pathologies is gaining a lot of strength and attention, even more so when it comes to wearable embedded systems able to offer comfort and precision in measurements. This work aimed to develop an electronic wearable sensing and wireless system that could measure the electrical activity of the Tibialis Anterior and Peroneus Longus muscles through bipolar surface electromyography; the system also identifies the muscle activations of antagonistic muscles, a phenomenon known as muscle co-activations. Sensing the angular displacement of the joint complex of the ankle in the sagittal and frontal planes through an Inertial Measurement Unit sensor system. A system with modular and smart architecture was designed and develop. It is based around five stages in charge of sensing, processing, and transmitting the data registers to a PC or a mobile device. The surface electromyography module is built around a two-channel amplifier sampling at 1 ksps/ch with a resolution of 10 bits; the angular displacement stage is based around an IMU sensor sampling at 1 ksps and 16 bits of resolution. Both data registers are transmitted wirelessly. A prototype with the architecture previously described was developed and tested. A statistical analysis of the data collected compared with commercial instruments was deployed, showing a Mean Square Error of $\leq \pm 5.5\%$ for the sEMG and an average error of $\leq \pm 1.5^{\circ}$ in the angular displacement measurements. The measurements made and the data verification protocol show that the equipment fully complies with all the technical and functional requirements of the project. Additionally, a record of muscle co-activation is presented, which can provide additional information, not only of the physiological state of the muscles but also the status of the pathology. By reducing the size of the device, improving the experimental setup and making small improvements to the hardware, the developed system opens a new panorama in the assessment and characterization of pathological conditions in real patients and therefore in the rehabilitation field.

D. F. Cortes Tobar et al. (Eds.): AETA 2019, LNEE 685, pp. 626–638, 2021.
https://doi.org/10.1007/978-3-030-53021-1_63

Keywords: Foot drop · Rehabilitation · Modular systems · sEMG amplifier · Smart sensing system

1 Introduction

Foot drop pathological condition can be generated through various causes. However, it has frequently attributed to accidents that compromise the dorsal extensor muscles of the foot and toes that are innervated by the deep peroneal nerve [1,2]. A person suffering this condition finds it complicated, depending on the degree of involvement, to move the foot through the joint complex of the ankle in the sagittal and frontal planes. Accurately perform the dorsiflexion/plantar flexion and eversion/inversion movements, which results in a deficiency of the Tibialis Anterior (TA) and Peroneus Longus (PL) muscles, muscular entities responsible for carrying out these movements [1,2]. The embedded and portable systems capable of establishing maximum comfort for the user and with enough accuracy in the data sensing for their respective study and analysis.

Nowadays, the study and development of surface electromyography (sEMG) are based on pairs of surface electrodes located over independent and individual muscles; this technique is known as *bipolar sEMG* [3]. Currently, there are many reasons to apply sEMG. However the most common are (i) analysis of muscular forces and characterization of muscular systems [4,5], (ii) in rehabilitation processes where biofeedback related to activation levels of the muscles is needed [6], (iii) in the control and activation of robotic prostheses or motion actuators [7], (iv) in sports medicine where it is sought to extend the knowledge of neuromuscular functions [8]. The characteristic feature of all the applications mentioned above is the fact that the success of the bipolar sEMG depends mainly on the characteristics and properties of the neuromuscular system of the person, both in its central and peripheral aspects [9]. Factors such as the anatomy of the muscle (size, class, type, structure), type of electrodes (intramuscular or surface) and electrode deployment techniques (the use of conductive electric transmission gel, shaving the area, the distance between electrodes) have significant repercussions on the effectiveness of bipolar sEMG [10–12]. A widely used tool to study muscle characteristics from sEMG, are muscle co-activations. By observing muscle co-activations, it is possible to determine, among other things, energy management or muscle stability [13].

Despite technological advances in the measurement of physical variables such as angular displacement, similar instruments like goniometers remain as the first option when medically evaluating a patient. However, the transition towards technological instruments and sensors is becoming faster and more global. Currently, there are several ways to sense and measure the displacement of a particular joint. One of these sensors is flexible sensors or *flex sensors*. Most of these kinds of sensors act as variable resistors depending on the bend grade over their surface [14]. Regardless of the joint to work with, flex sensors are commonly used to measure the angulation respect to a reference axis [15], in the control of prostheses or actuators according to a given angular movement [16], in rehabilitation applications [17]. Despite the benefits offered by flex sensors (including their low cost), facts like their low life cycle, imperative analog/digital conditioning and the critical measurement error, position them as an acceptable option but not the best [18]. Another alternative when sensing the angular displacement is the systems based on encoders [19]. Despite the excellent accuracy in the measurements, the encoder option involves additional mechanical systems around the joint to be evaluated, which negatively affects the user experience in terms of comfort and portability. Another option for measuring the angular displacement is the Inertial Measurement Unit sensors or commonly know IMU sensors. These types of sensors are embedded systems that included accelerometers, gyroscopes, and magnetometers. The IMU sensors perform a series of measurements (related to the built-in sensors) which can be used to determine additional magnitudes; this is known as indirect measurements [20]. One of the most significant benefits of IMU sensors is the fact that they can be placed anywhere on the body that undergoes the angular displacement, and not on the joint that is mechanically activated; this property makes them widely used in virtual immersion devices [21]. If we compare the three most effective methods to measure the angular displacement, (i) Brace&Encoder, (ii) Flex sensors and (iii) IMU System, the first one offers an average error of 3.25°, the second one 6.92° and the third one 0.08° [18].

We identify the following requirements for a new embedded sEMG and angular displacement measurement system: ***Modular architecture:*** Related to the comfort and portability to the operator; allows the integration of improved hardware modules that grant the opportunity to analyze more muscles or joint complexes. ***Wireless transmission:*** The movement dynamics that the person will carry out while using the prototype makes the implementation of wireless data transmission necessary. It is essential to carry out this without the loss of information, which could lead to misdiagnoses of the pathology. ***Mobile interface:*** Using mobile devices as data interface, such as smartphones and tablets, facilitates the experimental platform/setup when implementing the prototype. It also opens the doors to telemedicine and remote assessment of the patient without the need to be present in the evaluation laboratory or consulting room.

This research was to developed a wearable and wireless electronic system directly interfacing with PC's, smartphones or tablets, as a new tool in the diagnosis and assessment of the pathological condition of foot drop. Acquisition signals by sEMG of the TA and PL muscles and the measurement of the angular displacement in the joint complex of the ankle in the sagittal and frontal planes enable a non-invasive analysis of neuromuscular systems and characterization of medical conditions.

2 System Design

2.1 General System Architecture

The electronic prototype is developed around five modules, each one capable of carrying out specific and unique tasks. The main core or stage is called the

Central Processing Unit Module (CPU), which is responsible for acquisition and processing the data records from the two physiological variables to be measured; the electrical activity of the TA and PL muscles through sEMG and the angular displacement of the joint complex of the ankle in the sagittal and frontal planes, which provide decisive information in the treatment and rehabilitation of the pathological condition of foot drop. The sEMG data acquisition is carried out with a sampling frequency of 1 ksps per channel with a resolution of 10 bits, while the angular measurement is made with a sampling frequency of 1 ksps at a resolution of 16 bits. Once the data is processed, these are both stored and transmitted to either a smartphone or tablet for a real-time visualization. The stored data can be downloaded to personal computer for a much deeper deployment of information. Figure 1 shows the general architecture of the developed prototype and the experimental setup for data verification. Figure 1a shows the surface electrodes and the IMU sensor in the TA and PL muscles and in the user's instep, respectively; by the other hand, Fig. 1b shows the general system architecture; it consists of 5 modules responsible for particular tasks. The data records of the sensed physiological variables are also shown, that is the sEMG signals (raw, rectified and envelope) and the angular displacement of the ankle-foot complex. Table 1 shows the technical specifications of the prototype.

Table 1. Electronic system technical specifications

Description	Nomenclature	Value
sEMG channels	$sEMG_{ch}$	2
sEMG bandwidth	$sEMG_{bw}$	10 Hz–450 Hz
Max. gain	G	10000 V/V
Commom mode rejection ratio	CMRR	130 dB
sEMG ADC resolution	$sEMG_{ADC}$	10 bits
Sampling frequency	$sEMG_{fs}$	1 ksps/ch
IMU sensor resolution	IMU_{res}	16 bits
IMU sensor sampling frequency	IMU_{fs}	1 ksps
Range of motion arc	MOV_{ARC}	±90° in each plane
Communication protocol	–	Bluetooth
Reciever	–	Smartphone/tablet
Max. transmission distance	d_{TX}	9 m
Power supply	V_S	3 Ah 5-cell Lithium
Autonomy of the system	–	8 h
Dimensions	–	15.5 × 9 × 11 (cm)
Weight	–	600 g

sEMG Amplifier. The sEMG Conditioning Circuit refers to an electronic circuit constituted by stages of instrumentation, amplification and filtering necessary to obtain the EMG raw, rectified and envelope signals. The surface electrodes, the elements necessary to capture and conduce the muscular signals, needs gel to make a good impedance coupling; they also need to be installed in accordance with the SENIAM regulations or recommendations for both muscles [3].

The EMG system is built around a solid filtering architecture in charge of processing the signal collected by the electrodes. As main features, we used tunable gains, second order Chebyshev filters and a frequency spectrum of 10 Hz to 450 Hz (without Notch filter) to achieve the best possible treatment of the EMG signals.

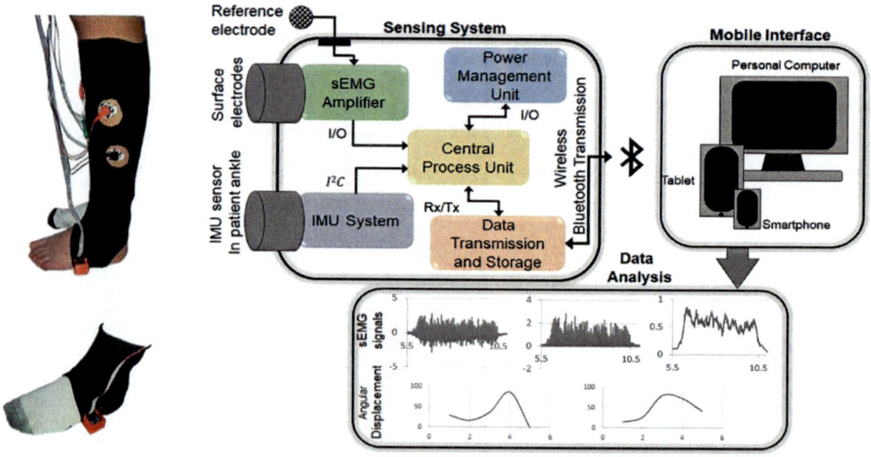

(a) Experimental platform.

(b) System's block diagram.

Fig. 1. Experimental platform and general system architecture.

IMU System. The IMU System Unit is elaborated around an IMU type sensor, which allows to obtain variables such as acceleration, speed or position. For this type of pathology we wanted to measure the angular position of the joint complex of the ankle in the sagittal and frontal planes. In this way, the angular positions are searched in the sensor registers to be read and processed by the CPU through an I^2C communication protocol. According to the pathological features of the foot drop, there is a pronounced limitation in these both planes, therefore an interest in measuring the angular displacement in them. The IMU sensor should be located on the outside of the foot, aligned with the horizontal; in this way when the foot is at rest and completely on the surface, the measurement will be 0° (See Fig. 1a).

Central Processing Unit (CPU). The Central Processing Unit is designed around a micro-controller which receives the information of both the electromyography of the two muscles and the angular variation of the foot in both planes. The EMG information are voltage signals, so it is necessary to use the ADC of the micro-controller (10 bits of resolution). We only process the envelope signal so 10 bits of resolution; in spite of this, the sampling frequency is still an important factor, because if the *Nyquist-Sannon theorem* [22] is not fulfilled, the processed signal would show different information than the real one; the sampling frequency must be at least twice the maximum analog frequency, in which case it is 450 Hz according to the frequency spectrum of the sEMG signals. Considering this, a sampling frequency of 1 ksps/ch was established to ensure a good digitalization of the signals.

On the other hand, the signals coming from the IMU sensor require an I^2C communication protocol in order to search and read the desired information in the sensor registers; in this case the desired information are the values of acceleration and the rotation angles from gyroscope over the 3 axis. As was mentioned in Sect. 1, an indirect measurement of the angular displacement is obtained through the values of acceleration (accelerometer), rotations (gyroscope) and earth's magnetic field (magnetometer). From these two values the angulation is determined with the following formulas.

$$\theta_{Acc} = \begin{bmatrix} tan^{-1}\frac{acc_x}{\sqrt{acc_y^2+acc_z^2}} \\ \\ tan^{-1}\frac{acc_x}{\sqrt{acc_y^2+acc_z^2}} \end{bmatrix} \tag{1}$$

$$\theta_{Gyro} = \begin{bmatrix} \theta_{xo} + \omega_x \Delta t \\ \\ \theta_{yo} + \omega_y \Delta t \end{bmatrix} \tag{2}$$

$$\theta_{Mag} = \begin{bmatrix} Mag_x * cos(\theta_{\mathrm{Acc[1]}}) + Mag_y * sin(\theta_{\mathrm{Acc[2]}}) * cos(\theta_{\mathrm{Acc[1]}}) + \\ Mag_z * cos(\theta_{\mathrm{Acc[2]}}) * sin(\theta_{\mathrm{Acc[1]}}) \\ \\ Mag_y * cos(\theta_{\mathrm{Acc[2]}}) - Mag_z * sin(\theta_{\mathrm{Acc[2]}}) \end{bmatrix} \tag{3}$$

$$\mathbf{Angle} = \begin{bmatrix} X(Angle[0] + Gyro_{ang_x} dt) + Y(Acc_{ang_x}) + Z(Mag_{ang_x}) \\ \\ X(Angle[1] + Gyro_{ang_y} dt) + Y(Acc_{ang_y}) + Z(Mag_{ang_y}) \end{bmatrix} \tag{4}$$

The presented formulas describe the expressions to determine the angle of inclination with the values of acceleration (Eq. 1), the angle of rotation with the values of the gyroscope (Eq. 2) and the magnetic field measurement (Eq. 3). The matrix presented in Eq. 4 describe the calculation of the angular displacement in terms of the previously determined values, known as a Complementary Filter (CF). This CF allows the implementation of information from the accelerometer, the gyroscope and the magnetometer, allowing to calculate an angle much closer to reality. This can be modeled as a low pass filter for the accelerometer and a high pass filter for the gyroscope and magnetometer data (X, Y and Z

elements in Eq. 4 are the filters coefficients); these filters allow to cushion the fast variations of acceleration and the detection of abrupt rotations.

The CPU is also responsible of establishing a Bluetooth communication protocol with the mobile device, so the information is displayed on the device screen. This module also carries out the algorithm necessary to store the information collected by the sensor network on a MicroSD card.

Data Transmission and Storage. The stage of data transmission and storage is responsible for receiving the information issued by the CPU, storing and transmitting it. The data storage feature is accomplished through a micro SD card allowing to store large amounts of information. The stored data are amplitudes in volts and angular displacement in degrees, which are stored as column vectors later to be processed with the desktop application (refer to the *mobile interface* section). Once the information is stored, the transmission of the data is carried out through the Bluetooth communication protocol. Due to the type of information and the real-time visualization aim, a baud rate of 125200 (the maximum baud rate of the Bluetooth module) was chosen.

2.2 Power Management Unit

It was decided to use a Lithium Lion battery. This battery is accompanied by a Battery Management System (BMS) which provides protection against overload, over discharge and short circuit scenarios. After the BMS, the battery voltage is supplied to a single to dual rail circuit thus achieving dual voltage, necessary for the *sEMG Conditioning Circuit* which handles positive and negative signals. In this case we used a 18.5 V battery to ensure a prolonged autonomy, however lower voltage batteries can be used with similar results. The battery is charged through a 3.5 mm DC jack with a 21 V power supply.

2.3 Mobile Interface

The Mobile Interface was developed on the Java programming language for smartphones or tablets; this tool can graph in real-time the values of the sensed physiological variables. The software is prepared to receive 4 different data segments at specific times; two segments from the bipolar sEMG and another 2 from the angle measurement. On the other hand, a desktop application has also been developed which works with the data stored by the electronic equipment through the micro SD card. The software extracts the information stored in the micro SD card, organizes and deploys it according to the presented interest needs. This application also performs a deeper statistical analysis, allowing to have more information about the variables registered.

3 Results

3.1 Angular Displacement Measurement

To obtain the results of angular displacement, a certified mechanical goniometer[1] was used. The methodology of the data verification was based on comparing measurement to measurement the value calculated by the IMU sensor against the angulation indicated by the goniometer. Five different arbitrary measurements were made for the sagittal plane; the chosen points were 27°, 17.5°, 36°, 90° and 0°. On the other hand, the measurements made in the frontal plane were 14°, 25°, 82°, 70° and 43°. The average error between the IMU sensor and the commercial goniometer for both planes did not exceed two degrees (1.5° and 1.38° for sagittal and frontal planes respectively). This can be due to a correct calibration of the sensor and a good installation of the card in the foot of the patient. Figure 2 shows a graphical comparison through a 2D line chart of the goniometer and the IMU sensor of our device in sagittal and frontal planes respectively, with the angle points previously mentioned. According to Nordin [23], the normal movement arc for movements in the sagittal plane (dorsiflexion and plantar flexion) is 75°, while for movements in the frontal plane (eversion and inversion) it is 40°. This suggests that the IMU sensor fully meets the functional requirements due its capability of measuring movements arcs of maximum 180° in each plane (−90° to 90°).

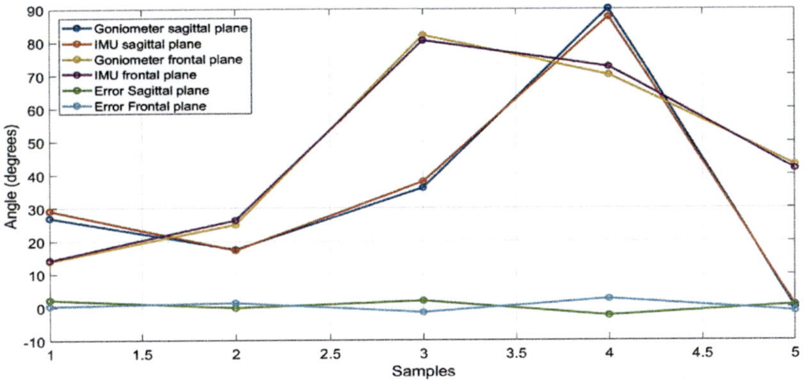

Fig. 2. Mechanical goniometer vs IMU sensor in sagittal and frontal planes. The angulation values taken in each sample, were chosen according to the arc of movement of a healthy person (without any pathology that limits movement), defined by Nordin [23].

[1] Ez Read Jamar Goniometer - Manual medical goniometer.

3.2 Bipolar sEMG

The envelope signal of the TA muscle was captured by both a professional electromyograph[2] and our device, with the intention of making a statistical comparison of both signals. To achieve this, it was necessary to parallelize the muscle signals to the inputs of both devices; the ground reference was the same for both instruments. We asked the patient to perform three muscle contractions (each one of around 1.5 to 2 s.) with a separation of around 2 s. As can be seen in Fig. 3, the muscle contractions were recorded by both devices. Despite the clear visual similarity presented by both registers, it is mandatory to perform a more in-depth statistical analysis to determine the clear data verification of our design, related to the professional device. The statistical analysis allows to determine parameters such as the Mean Square Error (MSE), average amplitudes of contraction, maximum amplitude of each contraction and contraction time intervals. As can be seen in Table 2, one of the most relevant indicator for statistically comparing the two registers is the MSE; this measure means that there is a difference between the estimator and what is estimated of *1.02%*, *7.7%* and *16.4%* for the first, second and third contraction respectively. But if we look at the whole record we would get a difference of *5.27%*. These results related to MSE can be better understood by overlapping the two registers, as shown in Fig. 3a. Figure 3b shows the contraction time recorded by both instruments; the biggest difference occurs in the third contraction. However, this difference is only 100 ms.

Table 2. Statistical analysis - sEMG

EMG data		
Measurement	Professional EMG	EMG module
MSE Contraction #1	1.02%	
MSE Contraction #2	7.71%	
MSE Contraction #3	16.40%	
Total MSE	5.27%	
Amplitude average	0.3191 mV	0.3306 mV
Contraction #1 max amplitude	0.8867 mV	0.9007 mV
Contraction #2 max amplitude	1.0050 mV	0.9007 mV
Contraction #3 max amplitude	0.9703 mV	1.0050 mV
Contraction #1 interval	1.65 s	1.644 s
Contraction #2 interval	1.649 s	1.696 s
Contraction #3 interval	1.756 s	1.908 s

[2] ADInstruments Powerlab recording unit, Teaching Series - 26T model.

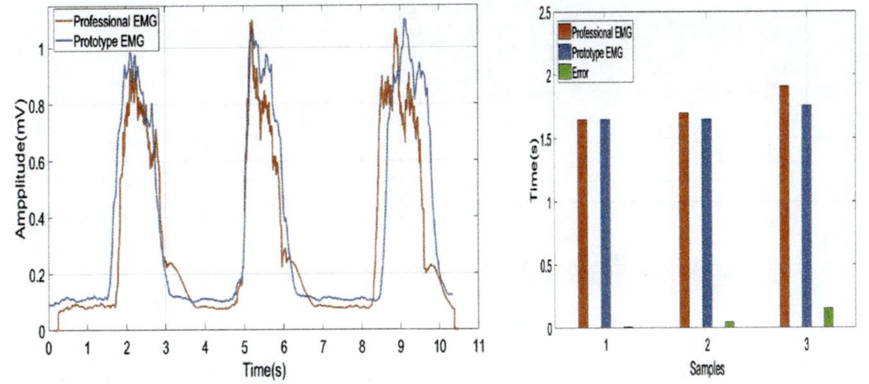

(a) sEMG envelope signal comparison - Professional vs sEMG amplifier.

(b) Time contraction comparison - Professional vs sEMG amplifier.

Fig. 3. sEMG data verification.

3.3 Muscular Co-activation

To study muscle co-activations, it is imperative to select co-activations indices; which are indicators that provide additional information related to the muscle [24]. For this case, two indices that directly relate the amplitude values of each muscle were selected; their mathematical expressions are presented below:

$$CI_1 = 2 * \frac{sEMG_{ANT}}{sEMG_{ANT} + sEMG_{AG}} * 100 \tag{5}$$

$$CI_2 = \frac{sEMG_{ANT}}{sEMG_{AG}} * 100 \tag{6}$$

The first index (CI_1) is defined as twice the ratio of the mean amplitude of the antagonist muscle ($sEMG_{ANT}$), over the sum of the averages of the antagonist and protagonist muscles ($sEMG_{AG}$). The second index (CI_2) is obtained by dividing the mean amplitude of the antagonist muscle and the mean amplitude of the protagonist muscle. As can be seen in Fig. 4a, the raw signals were recorded by the prototype. Muscle co-activations can be identified when the main muscle is stimulated. Figure 4b, shows the process to isolate the contraction of interest (first contraction of the PL muscle); due to the difference in amplitude presented during the contraction, between the main muscle and the antagonist muscle, the TA muscle (antagonist) amplitude was normalized (the y axis goes from 0 μV to 50 μV for the TA muscle). In this case it can be contemplated that, when the PL muscle is activated, the TA muscle undergoes a slight excitement. As can be seen in the figure, for this muscular complex (PL and TA), a too pronounced muscular co-activation is not manifested; on the contrary, it is quite slight. This behavior is also evidenced in the co-activation index taken for the research; the results of CI_1 and CI_2 were 20.53% and 11.44%, respectively. The results obtained are fully common, since the data was taken from a totally healthy

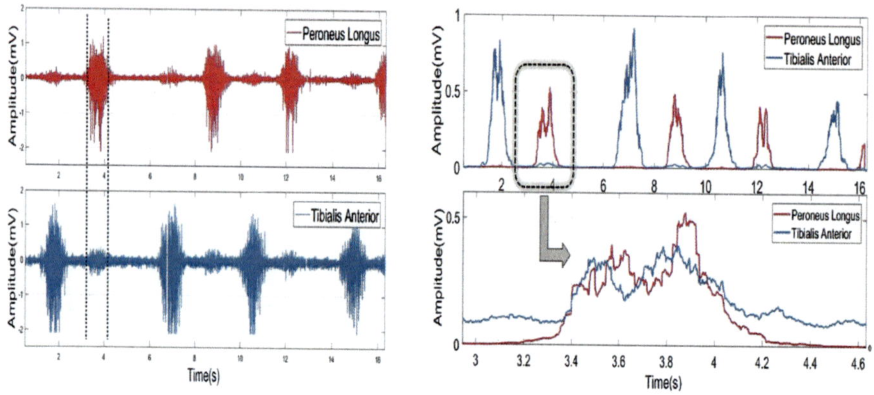

(a) RAW signals from Peroneus Longus (PL) and Tibialis Anterior (TA) muscles.

(b) Muscle co-activation identification.

Fig. 4. Muscle co-activation.

subject; in addition to this, the TA-PL muscle complex usually presents this behavior (unlike, for example, biceps and triceps muscle complex, where muscle co-activations are much more pronounced).

4 Conclusions

Currently, foot drop pathology is diagnosed and assessed by individually verifying the sEMG and the angular displacement of the joint complex of the ankle. In the same way, physiotherapists, orthopedists and specialists must autonomously integrate data records and achieve an accurate and precise verdict related to the pathological condition. This paper describes the design, development and testing of a modular, wearable and wireless electronic system capable of performing the previously mentioned measurements. The system captures and condition the muscle signals that are then transmitted to the CPU. These muscle signals, and the data of angular displacement, are processed, stored and transmitted in real-time to a mobile interface allowing observation of the behavior of them in an organized, fast and accurate way. The study of muscle co-activations allows to display additional information related to the physiological state of a patient. The tests carried out in different scenarios show that the equipment is portable, does not interfere with the movement of the person, is especially useful for dynamic and static activities and the accuracy of the measurements makes it a reliable instrument. In addition to the clear benefit of the portability of the device, the fact of transmitting the information to a smartphone/tablet through a mobile interface opens a new panorama related to remote activities such as telemedicine or home rehabilitation. The development of this device has the potential to establish new perspectives in the characterization and assessment of pathological conditions as well as serving as an instrument to build robust databases that provide support for further research.

Acknowledgment. This work was supported by the Faculty of Engineering and Electronic Engineering Program of Universidad El Bosque, with the research project PFI-2017-EL-011.

References

1. Aldemir, C., Duygun, F.: New and unusual causes of foot drop. Med. Sci. - Int. Med. J. **6**, 49–495 (2017)
2. Westhout, F.D., Paré, L.S., Linskey, M.E.: Central causes of foot drop: rare and underappreciated differential diagnoses. J. Spinal Cord Med. **30**(1), 62–66 (2007)
3. Stegeman, D., Hermens, H.: Standards for surface electromyography: the European project surface EMG for non-invasive assessment of muscles (SENIAM). Roessingh Res. Dev. 108–112 (2007)
4. Hof, A.L.: EMG and muscle force: an introduction. Hum. Mov. Sci. **3**(1–2), 119–153 (1984)
5. Merletti, R., Rainoldi, A., Farina, D.: Surface electromyography for noninvasive characterization of muscle. Electromyographie superficielle pour une caracterisation mesuree du muscle. Exercise Sport Sci. Rev. **29**(1), 20–25 (2001)
6. Neblett, R.: Surface electromyographic (SEMG) biofeedback for chronic low back pain. Healthcare **4**(2), 27 (2016)
7. Kutilek, P., Hybl, J., Kauler, J., Viteckova, S.: Prosthetic 6-DOF arm controlled by emg signals and multi-sensor system. In: Proceedings of 15th International Conference MECHATRONIKA (March 2017), pp. 1–5 (2012)
8. Massó, N., Rey, F., Romero, D., Gual, G., Costa, L., Germán, A.: Surface electromyography applications in the sport. Apunts Med. Esport **45**(165), 121–130 (2010)
9. Cerone, G.L., Botter, A., Gazzoni, M.: A modular, smart, and wearable system for high density sEMG detection. IEEE Trans. Biomed. Eng. **66**(12), 3371–3380 (2019). https://doi.org/10.1109/TBME.2019.2904398
10. Day, S.: Important factors in surface EMG measurement. Bortec Biomed. Ltd. 1–17 (2002)
11. Melaku, A., Kumar, D.K., Bradley, A.: Influence of inter-electrode distance on EMG, pp. 1082–1085, May 2005
12. Ghapanchizadeh, H., Ahmad, S.A., Ishak, A.J.: Effect of surface electromyography electrode position during wrist extension and flexion based on time and frequency domain analyses. Int. J. Control Theory Appl. **9**(5), 2643–2650 (2016)
13. Frey-Law, L.A., Avin, K.G.: Muscle coactivation: a generalized or localized motor control strategy? (2013)
14. Aneri, M., Rutvij, H.: A review on applications of ambient assisted living. Int. J. Comput. Appl. **176**(8), 1–7 (2017)
15. Beyaz, A.: Posture determination by using flex sensor and image analysis technique. Agric. Sci. Digest - Res. J. **37**(04), 257–262 (2017)
16. Vyawahare, V.M., Pardhi, D.: Design of a prosthetic arm using flex. Int. J. Electron. Commun. Eng. Technol. **8**(2), 1–6 (2017)

17. Masdar, A., Ibrahim, B.S.K.K., Hanafi, D., Jamil, M.M.A., Rahman, K.A.A.: Knee joint angle measurement system using gyroscope and flex-sensors for rehabilitation. In: BMEiCON 2013 - 6th Biomedical Engineering International Conference, pp. 5–9, October 2013
18. Khayani, S.B.: Development of wearable sensors for body joint angle measurement, p. 70, May 2011
19. Zhang, Z., Dong, Y., Ni, F., Jin, M., Liu, H.: A method for measurement of absolute angular position and application in a novel electromagnetic encoder system (2015)
20. Desa, H., Azfar, A.Z.: Study of inertial measurement unit sensor, May 2014
21. Kumar, K., Varghese, A., Reddy, P.K., Narendra, N., Swamy, P., Chandra, M.G., Balamuralidhar, P.: An improved tracking using IMU and vision fusion for mobile augmented reality applications. Int. J. Multimedia Appl. (IJMA) **6**(5), 13–29 (2014)
22. Adcock, B., Hansen, A., Roman, B., Teschke, G.: Generalized sampling: stable reconstructions, inverse problems and compressed sensing over the continuum. Adv. Imaging Electron Phys. **182**, 187–279 (2014). https://doi.org/10.1016/B978-0-12-800146-2.00004-7
23. Nordin, M.: Biomecanica Basica del Sistema Muscoesqueletico-Nordin.pdf (2004)
24. Ervilha, U.F., Graven-Nielsen, T., Duarte, M.: A simple test of muscle coactivation estimation using electromyography (2012)

Optimal Trajectory Planning with Dynamic Obstacles Avoidance for Mobile Robots Navigation

Jaime Arcos-Legarda(✉) and Morian Calderon-Diaz

Ingeniería Mecatrónica, Universidad de San Buenaventura, sede Bogotá, Bogota, Colombia
{warcos,mcalderon}@usbbog.edu.co

Abstract. This paper develops an optimal trajectory planning strategy for an omnidirectional mobile robot with dynamic obstacles avoidance. The proposed method uses the information of the obstacle's movement to design a collision-free trajectory. This collision-free trajectory generation approach uses an optimal path planning that minimizes the energy cost needed to move the robot from an initial point to a final position. The optimal trajectory planning considers the robot's model as a dynamic constraints in the optimization problem. Beyond it, constraints such as the physical saturation on torque and speed of the actuators are included as inequality constraints. The obstacles avoidance are included into the optimization problem to restrict the feasible path that the robot can take to achieve the target position. Additionally, a nonlinear feedback control strategy is used to guarantee the tracking trajectory. The trajectory generation technique and the control system are evaluated in a numerical simulation in a model of a Festo Robotino® platform. The numerical evaluation shown an effective trajectory planning with results into the feasible solutions.

Keywords: Self-optimizing systems · Self-organizing systems · Mobile robots navigation · Dynamic obstacles avoidance · Trajectory planning

1 Introduction

The optimal trajectory planning with dynamic obstacles avoidance is a fundamental problem in the navigation of mobile robotics. The navigation of mobile robotics is a task that can be classified into global and local navigation [7,12]. This paper studies the problem of local navigation with optimal trajectory planning. In order to find optimal trajectories, strategies to define paths with interpolating of smooth curves have been used by planners with Neural Network, Fuzzy Logic, and their combination, Neuro-Fuzzy [12]. Alternatively, bio-inspired trajectory planning algorithms have been developed based on the natural behavior of populations. Examples of such algorithms are the bug algorithm [11], the

D. F. Cortes Tobar et al. (Eds.): AETA 2019, LNEE 685, pp. 639–648, 2021.
https://doi.org/10.1007/978-3-030-53021-1_64

bat algorithm (BA) [7], the ant colony algorithm [5], and the swarm optimization algorithm [13]. These algorithms search for an optimal path to drive the mobile robots from a starting point to a target position with dynamic obstacles avoidance [14]. Additional methods based on mathematical approaches use potential field method (PFM) [9], vector field histogram (VFH) [4], and vision systems with machine learning algorithms [15–17]. This paper proposes a trajectory generation technique using a nonlinear constrained optimization algorithm that search for the coefficients of a set of Bézier polynomials, which guarantee bounded derivatives in the target trajectories. The Bézier polynomials are parametric curves that rely on control points to define their shape. The core of Bézier curves are the Bernstein polynomials and the advantage of this kind of curves is their low computational cost, because the curve behavior is defined by control points [3,6]. A classical multivariable feedback linearization control technique is used to guarantee an asymptotic tracking trajectory. This control strategy ensures that the robot approaches a target position with a target orientation [1,8]. The rest of this paper is organized as follows: Sect. 2 presents the omnidirectional mobile robot's dynamic model. Section 3 shows a nonlinear tracking control. Section 4 presents the optimal trajectory planning. In Sect. 5, a numerical evaluation of the trajectory generation strategy avoiding dynamic obstacles and the control system is performed. Finally, conclusions are draw in Sect. 6.

2 Dynamic Model

An omnidirectional mobile robot is considered in this paper to perform a trajectory tracking task. This robot has three wheels, each of them is coupled to the shaft of a DC motor. Each motor is controlled through an H-bridge driver that allows change the direction and the torque on each wheel. The robot has a set of sensor to measure the electrical current on each motor, the position, and the orientation of the robot's body. The mobile robot studied is shown in the Fig. 1.

Fig. 1. Omnidirectional mobile robot. Festo Robotino® platform.

In order to define a mathematical model of the mobile robot, a position vector of the center the mass with respect to a world frame, (x_w, y_w), is defined by $\overrightarrow{\mathbf{p}}$, this is shown in the Fig. 2. This figure also shows the traction forces produced by the electrical motors through the wheels. The relation between such forces and the generalized coordinates allows to define the robot's dynamics, which is mathematically described by the Euler-Lagrange equation,

$$D(\mathbf{q})\ddot{\mathbf{q}} + C(\mathbf{q}, \dot{\mathbf{q}})\dot{\mathbf{q}} = \Gamma(\delta)\tau, \tag{1}$$

where $\mathbf{q} := \begin{bmatrix} x_w \ y_w \ \delta \end{bmatrix}^{\mathrm{T}}$ is the vector of generalized coordinates that is defined with respect to the world frame, $D(\mathbf{q})$ has the inertial terms, $C(\mathbf{q}, \dot{\mathbf{q}})$ has the Coriolis and Centripetal effects, τ is the vector of torques produced by the motors, δ is the azimuth angle, which is measured with respect to a fixed frame in the robot's center of gravity, and $\Gamma(\delta)$ is a function that maps the motors torque, τ, from the relative coordinates to the absolute ones [10]. That is,

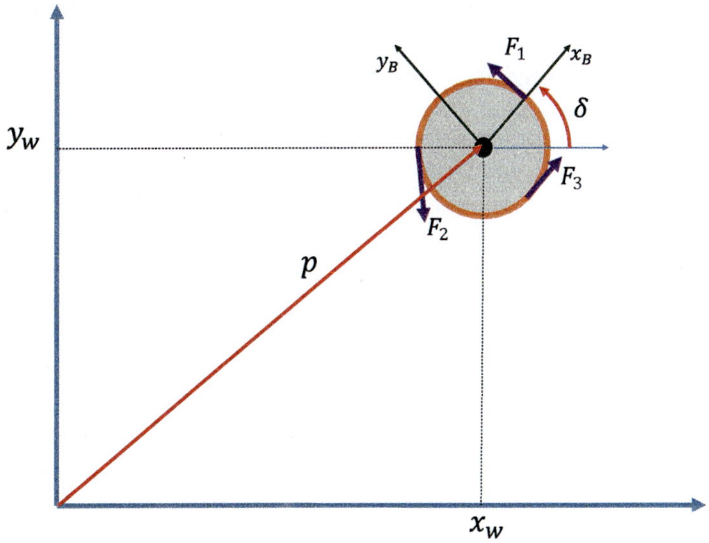

Fig. 2. Coordinate frames of the mobile robot.

$$\Gamma(\delta)\tau = \frac{1}{R}\begin{bmatrix} \sin\theta_1 & \sin\theta_2 & \sin\theta_3 \\ \cos\theta_1 & \cos\theta_2 & \cos\theta_3 \\ L & L & L \end{bmatrix}\begin{bmatrix} \tau_1 \\ \tau_2 \\ \tau_3 \end{bmatrix}, \tag{2}$$

where L is the distance between the center of the robot an the wheels, R is the wheel's radius, $\tau := \begin{bmatrix} \tau_1 \ \tau_2 \ \tau_3 \end{bmatrix}^{\mathrm{T}}$ are the motors' torque, and $(\theta_1, \theta_2, \theta_3)$ are the angles between the reference frame axis x_w and the tangential forces produced on each wheel. These angles are defined as,

$$\theta_1 = \delta + 90°, \tag{3}$$
$$\theta_2 = \delta + 210°, \tag{4}$$
$$\theta_3 = \delta + 330°. \tag{5}$$

The system represented in (1) can be simplified in a state space form as

$$\dot{\mathbf{x}} = \begin{bmatrix} \dot{\mathbf{q}} \\ -D^{-1}(\mathbf{q})C(\mathbf{q},\dot{\mathbf{q}})\dot{\mathbf{q}} \end{bmatrix} + \begin{bmatrix} 0 \\ D^{-1}(\mathbf{q})\Gamma(\delta)\tau \end{bmatrix}, \tag{6}$$

$$\dot{\mathbf{x}} = f(\mathbf{x}) + g(\mathbf{x})\mathbf{u}, \tag{7}$$

where $\mathbf{x} := \begin{bmatrix}\mathbf{q} & \dot{\mathbf{q}}\end{bmatrix}^{\mathrm{T}}$, and $\mathbf{u} := \tau$,

$$f(\mathbf{x}) = \begin{bmatrix} \dot{\mathbf{q}} \\ -D^{-1}(\mathbf{q})C(\mathbf{q},\dot{\mathbf{q}})\dot{\mathbf{q}} \end{bmatrix}, \quad \text{and} \quad g(\mathbf{x}) = \begin{bmatrix} 0 \\ D^{-1}(\mathbf{q})\Gamma(\delta) \end{bmatrix}. \tag{8}$$

3 Nonlinear Tracking Control

In order to define the tracking trajectory control strategy, let us define the output vector as the error produced by the difference between the target trajectory and the currently robot's position and orientation. Thus, the tracking error vector is defined as

$$\mathbf{h}(\mathbf{x}) := \mathbf{q} - \mathbf{q}_d = \begin{bmatrix} x_w \\ y_w \\ \delta \end{bmatrix} - \begin{bmatrix} x_d \\ y_d \\ \delta_d \end{bmatrix}, \tag{9}$$

where $\mathbf{q}_d := \begin{bmatrix} x_d \; y_d \; \delta_d \end{bmatrix}^{\mathrm{T}}$ is a vector with the target position and orientation.

Using the output vector, $\mathbf{h}(\mathbf{x})$, let us propose a valid diffeomorphism on an open set around $x_\circ$, which uses a local coordinate transformation to finally present the system into a normal form. This transformation is performed through successive time-derivatives of $\mathbf{h}(\mathbf{x})$, until the control input is explicit [8]. That is,

$$\begin{aligned} \frac{d}{dt}(\mathbf{h}(\mathbf{x})) &= \frac{\partial \mathbf{h}}{\partial \mathbf{x}}\dot{\mathbf{x}}, \\ &= \begin{bmatrix} \frac{\partial \mathbf{h}}{\partial \mathbf{q}} \; \frac{\partial \mathbf{h}}{\partial \dot{\mathbf{q}}} \end{bmatrix} \begin{bmatrix} \dot{\mathbf{q}} \\ -D^{-1}(\mathbf{q})C(\mathbf{q},\dot{\mathbf{q}})\dot{\mathbf{q}} \end{bmatrix} + \begin{bmatrix} \frac{\partial \mathbf{h}}{\partial \mathbf{q}} \; \frac{\partial \mathbf{h}}{\partial \dot{\mathbf{q}}} \end{bmatrix} \begin{bmatrix} 0 \\ D^{-1}(\mathbf{q})\Gamma(\delta), \tau \end{bmatrix}, \\ &= \nabla \mathbf{h} \cdot f(\mathbf{x}) + \nabla \mathbf{h} \cdot g(\mathbf{u}), \\ &= L_f \mathbf{h}(\mathbf{x}) + L_g \mathbf{h}(\mathbf{x})\mathbf{u}, \end{aligned} \tag{10}$$

where $L_f\mathbf{h}(\mathbf{x})$ and $L_g\mathbf{h}(\mathbf{x})$ are the Lie derivatives of $\mathbf{h}(\mathbf{x})$ along the vector fields $f(\mathbf{x})$ and $g(\mathbf{u})$, respectively. Since, the output $\mathbf{h}(\mathbf{x})$ is independent of $\dot{\mathbf{q}}$, then $\frac{\partial \mathbf{h}}{\partial \dot{\mathbf{q}}} = 0$; therefore, $L_g\mathbf{h}(\mathbf{x}) = 0$, thus the first time-derivative of the output $\mathbf{h}(\mathbf{x})$ is expressed as

$$\frac{d}{dt}(\mathbf{h}(\mathbf{x})) = L_f \mathbf{h}(\mathbf{x}). \tag{11}$$

The second time-derivative of the output can be expressed as,

$$\begin{aligned}
\frac{d^2}{dt^2}(\mathbf{h}(\mathbf{x})) &= \frac{d}{dt}\left(L_f\mathbf{h}(\mathbf{x})\right),\\
&= \left[\tfrac{\partial}{\partial \mathbf{q}}\left(\tfrac{\partial \mathbf{h}}{\partial \mathbf{q}}\dot{\mathbf{q}}\right) \; \tfrac{\partial \mathbf{h}}{\partial \mathbf{q}}\right]\left[\begin{bmatrix}\dot{\mathbf{q}}\\ -D^{-1}(\mathbf{q})C(\mathbf{q},\dot{\mathbf{q}})\dot{\mathbf{q}}\end{bmatrix} + \begin{bmatrix}0\\ D^{-1}(\mathbf{q})\Gamma(\delta)\tau\end{bmatrix}\right],\\
&= \underbrace{\left[\tfrac{\partial}{\partial \mathbf{q}}\left(\tfrac{\partial \mathbf{h}}{\partial \mathbf{q}}\dot{\mathbf{q}}\right) \; \tfrac{\partial \mathbf{h}}{\partial \mathbf{q}}\right]\begin{bmatrix}\dot{\mathbf{q}}\\ -D^{-1}(\mathbf{q})C(\mathbf{q},\dot{\mathbf{q}})\dot{\mathbf{q}}\end{bmatrix}}_{L_f^2\mathbf{h}(\mathbf{x})} + \underbrace{\frac{\partial \mathbf{h}}{\partial \mathbf{q}}D^{-1}(\mathbf{q})\Gamma(\delta)}_{L_gL_f\mathbf{h}(\mathbf{x})}\tau,\\
&= L_f^2\mathbf{h}(\mathbf{x}) + L_gL_f\mathbf{h}(\mathbf{x})\mathbf{u}.
\end{aligned} \tag{12}$$

Since $g(\mathbf{x})$ is involutive, the matrix $L_gL_f\mathbf{h}(\mathbf{x})$ is invertible around $x_\circ$. Thus, let us propose the following feedback control law:

$$\mathbf{u} = L_gL_f\mathbf{h}(\mathbf{x})^{-1}\left(-L_f^2\mathbf{h}(\mathbf{x}) + \mathbf{v}\right), \tag{13}$$

where $\mathbf{v}$ is an auxiliary control action that ensures the asymptotic convergence of the tracking errors to zero. Then, $\mathbf{v}$ is selected as

$$\mathbf{v} = -K_p\mathbf{h}(\mathbf{x}) - K_dL_f\mathbf{h}(\mathbf{x}). \tag{14}$$

In order to achieve an asymptotic convergence of the tracking errors to zero, the constants K_p and K_d are selected to be positive defined diagonal matrices. Hence, replacing the control law (13) into the transformed system (10), the error dynamics is described as

$$\frac{d^2}{dt^2}(\mathbf{h}(\mathbf{x})) + K_dL_f\mathbf{h}(\mathbf{x}) + K_p\mathbf{h}(\mathbf{x}) = 0, \tag{15}$$

which by definition is Hurwitz.

4 Optimal Trajectory Planning

In order to develop an optimal trajectory generation strategy, an optimization problem is defined such that the energy consumed by the motors is minimized and the task of moving from an initial to a final position is achieved. The solution of this problem is subject to several constraints that are imposed by the physical features of the robot, among them are: the closed-loop dynamics of the robot, motor's maximum and minimum speed, and motor's maximum torque.

An important set of constraints is imposed by the environment. These constraints are the robot's initial and target position and the obstacles in middle of the path. In order to define a strategy to generate trajectories, let us assume that the robot has to move in a highway from a point **A** to a point **B**. In this situation, it is imperative to perform smooth movements that follow a path. This path is determined by a smooth function that guarantees bounded changes in the speed and acceleration. In this case, such function is a Bézier polynomial, which guarantees bounded derivatives of the target trajectory [2]. In this way, the target trajectories are defined as,

$$\mathbf{q}_d^j(t) = \sum_{i=0}^{M} \beta_i^j \frac{M!}{i!(M-i)!} s^i (1-s)^{M-i}, \qquad j = 1, 2, 3, \tag{16}$$

where β_i^j are the coefficients of the Bézier polynomials, M is the degree of the polynomials, s is defined in the interval $[0, 1]$ and it is computed using the time period required to achieve the target position, T. This is,

$$s := \frac{t}{T}. \tag{17}$$

With the path defined by the Bézier polynomials, the following task is to find the coefficients β_i^j. In such a way, a cost function is defined with the integral of the square norm of the control signal vector as

$$J(\beta) = \int_{t_i}^{t_f} \left\| \mathbf{u}^{\mathrm{T}} \mathbf{u} \right\|^2 dt, \tag{18}$$

where t_i and t_f are the initial and final time of the test, respectively. Hence, an optimization problem can be stated as

$$\begin{array}{ll} \min\limits_{\beta} & J(\beta), \\ & \\ \text{subject to} & \text{Behavioral constraints,} \\ & \text{Nonlinear equality constraints,} \\ & \text{Nonlinear inequality constraints,} \\ & \underline{\beta} \leq \beta \leq \overline{\beta}, \end{array} \tag{19}$$

where $\underline{\beta}$ is the lower boundary vector and the $\overline{\beta}$ upper boundary vector. The behavioral constraints include the robot's dynamics (7) and its interaction with the control law (13), and the equality and inequality constraints are described below.

Nonlinear Equality Constraints

The nonlinear equality constraints consider the robot's initial and final positions to coincide with the initial and final points of the trajectory. This constraint is described as

$$\mathbf{q}(t_i) - \mathbf{q}_d(t_i) = 0, \tag{20}$$

$$\mathbf{q}(t_f) - \mathbf{q}_d(t_f) = 0. \tag{21}$$

Nonlinear Inequality Constraints

The nonlinear inequality constraints are used to delimit the feasible terrain in which the robot can move and the boundaries for the control action (e.g. motors' maximum torque and speed). In order to perform the obstacles avoidance, a time polynomial function is used to define the obstacle's position. Such position is defined as $x_{ob}(t)$ and $y_{ob}(t)$. These constraints are dynamic and the feasible region is function of the time. The method used to satisfy such constraint computes the euclidean distance between the center of the robot and the center of the obstacle. This difference must be bigger than a safe distance determine by the radius of the robot and the obstacle plus a security factor. In this case, the minimum distance is defined by ϵ, as it is shown below

$$\left\| \begin{bmatrix} x_{ob}(t) \\ y_{ob}(t) \end{bmatrix} - \begin{bmatrix} x_w(t) \\ y_w(t) \end{bmatrix} \right\|_2 \geq \epsilon, \quad t_i \leq t \leq t_f. \tag{22}$$

An additional constraint is imposed into the motors' torque; such that, the required control action is always under the maximum torque available in the motors. This constraint is set up as,

$$|\tau_j| \leq \tau_{max}, \qquad t_i \leq t \leq t_f, \qquad \forall\, j, \tag{23}$$

where τ_{max} represents the maximum permissible values for the torques. In the same way, the boundaries for the speed are described as

$$|v_j| \leq v_{max}, \qquad t_i \leq t \leq t_f, \qquad \forall\, j. \tag{24}$$

5 Numerical Evaluation

A numerical simulation is used to evaluate the performance of the control strategy and the trajectory generation approach. In this way, Simulink/Matlab is used to built a dynamic model of the mobile robot. The robot's parameters are shown in the Table 1. These parameters are used to built the robot's numerical model, which is used to perform a simulation that test the tracking controller. This evaluation allows to test the controller as well as to evaluate the dynamic constraint during the search of the optimal trajectory. In order to implement the controller evaluation, its gains are selected as,

$$K_p = 50 I_{3\times 3}, \qquad \text{and} \qquad K_d = 3 I_{3\times 3}, \tag{25}$$

where $I_{3\times 3}$ is an identity matrix. A trajectory tracking evidence is shown in the Fig. 3. This figure shows an effective tracking trajectory and the control signals (torques) required to complete the task.

In order to generate the optimal trajectory, the initial and target position are set as follows:

- Initial position: $\mathbf{q}(t_i) = [0 \quad 0 \quad 0]^{\mathrm{T}}$,
- Final position: $\mathbf{q}(t_f) = [2 \quad 8 \quad 1]^{\mathrm{T}}$,
- Safe distance between robot and obstacles: $\epsilon = 0.2$ m,
- Maximum torque: $\tau_{max} = 3$ N $\cdot$ m.

Table 1. Robot's physical parameters.

Parameters	Symbol	Units	Values
Mass	M	kg	15
Robot's inertia	I_z	kg·m^2	1
Radius of the robot's body	L	m	0.15
Wheel radius	R	m	0.04
Viscous friction	β	N·m·s/rad	0.3

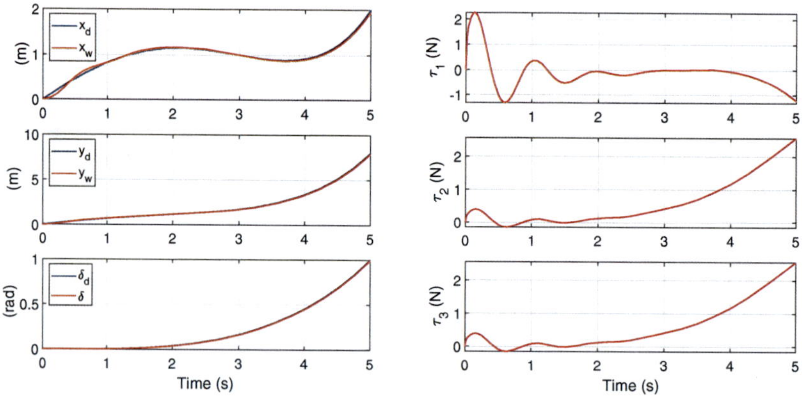

Fig. 3. Tracking trajectory and control signals

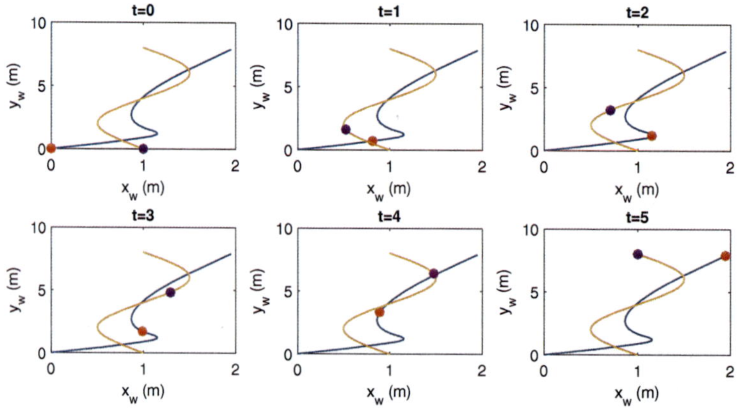

Fig. 4. Orange: Robot. Purple: Obstacle. Blue: Robot's trajectory. Yellow: Obstacle's trajectory

The evolution of the position of the robot is shown in a XY plot, which is presented in Fig. 4. This figure shows the evolution, through the time, of the position of the robot (Orange point) and the position of the obstacle (Purple point). The evolution of the position of the robot shows the computed trajectory, the initial position of the robot at $t = 0$ s, and its final position at $t = 5$ s. This simulation confirms that the robot approaches the target position avoiding the collision with a moving obstacle. The obstacle's path intercepts the required path to move from the initial position to the target one.

6 Conclusions

An optimal trajectory planning with dynamic obstacles avoidance has been developed to generate a collision-free path for an omnidirectional mobile robot. The mobile robot is driven through the target trajectory by a nonlinear control strategy that ensures an asymptotic convergence of the tracking error to zero. The trajectory generation method and the control system proposed in this paper minimizes the control effort of the actuators with the use of smooth movements. This feature guarantees the avoidance of sudden unwanted accelerations. Future work includes a physical implementation of the proposed trajectory generation strategy with on-line computations of the path.

References

1. Arcos-Legarda, J., Cortes-Romero, J., Beltran-Pulido, A., Tovar, A.: Hybrid disturbance rejection control of dynamic bipedal robots. Multibody Syst. Dyn. **46**(3), 281–306 (2019)
2. Arcos-Legarda, J., Cortes-Romero, J., Tovar, A.: Robust compound control of dynamic bipedal robots. Mechatronics **59**, 154–167 (2019)
3. Bézier, P.: Numerical Control: Mathematics and Applications. Wiley, Hoboken (1970)
4. Borenstein, J., Koren, Y.: The vector field histogram-fast obstacle avoidance for mobile robots. IEEE Trans. Robot. Autom. **7**(3), 278–288 (1991)
5. Garcia, M.P., Montiel, O., Castillo, O., Sepúlveda, R., Melin, P.: Path planning for autonomous mobile robot navigation with ant colony optimization and fuzzy cost function evaluation. Appl. Soft Comput. **9**(3), 1102–1110 (2009)
6. González, D., Pérez, J., Milanés, V., Nashashibi, F.: A review of motion planning techniques for automated vehicles. IEEE Trans. Intell. Transp. Syst. **17**(4), 1135–1145 (2015)
7. Ibraheem, I.K., Ajeil, F.H.: Path planning of an autonomous mobile robot in a dynamic environment using modified bat swarm optimization. arXiv preprint arXiv:1807.05352 (2018)
8. Isidori, A.: Nonlinear Control Systems. Springer, Heidelberg (2013)
9. Koren, Y., Borenstein, J.: Potential field methods and their inherent limitations for mobile robot navigation. In: Proceedings of 1991 IEEE International Conference on Robotics and Automation, pp. 1398–1404. IEEE (1991)
10. Liu, Y., Wu, X., Zhu, J.J., Lew, J.: Omni-directional mobile robot controller design by trajectory linearization. In: Proceedings of the 2003 American Control Conference, vol. 4, pp. 3423–3428. IEEE (2003)

11. Ng, J., Bräunl, T.: Performance comparison of bug navigation algorithms. J. Intell. Rob. Syst. **50**(1), 73–84 (2007)
12. Pandey, A., Pandey, S., Parhi, D.: Mobile robot navigation and obstacle avoidance techniques: a review. Int. Robot. Autom. J. **2**(3), 00022 (2017)
13. Qin, Y.Q., Sun, D.B., Li, N., Cen, Y.G.: Path planning for mobile robot using the particle swarm optimization with mutation operator. In: Proceedings of 2004 international conference on machine learning and cybernetics (IEEE Cat. No. 04EX826), vol. 4, pp. 2473–2478. IEEE (2004)
14. Raja, P., Pugazhenthi, S.: Optimal path planning of mobile robots: a review. Int. J. Phys. Sci. **7**(9), 1314–1320 (2012)
15. Sariff, N., Buniyamin, N.: An overview of autonomous mobile robot path planning algorithms. In: 2006 4th Student Conference on Research and Development, pp. 183–188. IEEE (2006)
16. Stone, P., Veloso, M.: Multiagent systems: a survey from a machine learning perspective. Autonom. Robot. **8**(3), 345–383 (2000)
17. Zohaib, M., Pasha, M., Riaz, R., Javaid, N., Ilahi, M., Khan, R.: Control strategies for mobile robot with obstacle avoidance. arXiv preprint arXiv:1306.1144 (2013)

Metaheuristics Applied to the Fleet Size and Mix Vehicle Routing Problems with Soft Time Windows and Stochastic Times

Helmer Muñoz Hernández[1](✉), Tobías A. Parodi Camaño[1], Diego A. Soto de la Vega[2], and Jorge M. López Pereira[2]

[1] Corporación Unificada Nacional de Educación Superior CUN, Bogotá, Colombia
{helmer_munoz,tobias_parodi}@cun.edu.co
[2] Universidad de Córdoba, Córdoba, Argentina
die-soto@hotmail.com, jotamlopez@hotmail.com

Abstract. This work explores a complex new vehicle routing problem in which transport times and services derive from probability distributions, and a time window constraint is associated with each client. Vehicles of different capacity are routed from the warehouse to a set of geographically dispersed customers, late delivery of the order is allowed by adding a penalty cost, routes should be determined in such a way that the courses are minimized total system costs, and all customers are satisfied. Such routing problems are NP-hard in deterministic environments, and, in our context, even more given the stochastic nature. The objective of the research is to compare the efficiency of a new metaheuristic (Ebola) adapted to the problem, with the best metaheuristics found in the specific topic literature (a genetic algorithm and an improved taboo search algorithm). For each algorithm, random and mapped encodings were developed in the generation of initial solutions. Several instances were randomly generated, the results show that the genetic algorithm gets the best results; on the contrary taboo and Ebola prove to be more efficient algorithms in computational times.

Keywords: The problem of vehicle routing · Stochastic times · Flexible time windows · Metaheuristics

1 Introduction

The Vehicle Routing Problem (VRP) is a complex problem of combinatorial optimization, already considered a paradigm in specialized literature (Hermosilla 2005), which emerged, according to Olivera (2004), since 1959. The problem is the scheduling of a set of routes on which one or a collection of vehicles, must leave one or more warehouses to supply with goods or services to a finite set of customers (Russell and Urban 2008). Practical examples for such problems include school routes, fuel or food transport trolleys, beverage trucks, courier cars, and more. Consideration stake in each of its main components must be considered in scheduling routes: the configuration of the transport network, customers, warehouses, and the type of vehicles (Scheffermann 2009). On the

D. F. Cortes Tobar et al. (Eds.): AETA 2019, LNEE 685, pp. 649–659, 2021.
https://doi.org/10.1007/978-3-030-53021-1_65

other hand, the costs associated with the distribution of goods in an organization usually represent between 10% and 20% of the total cost of goods (Toth and Vigo 2002) in a stricter case, for those organizations whose main activity is the transport of raw materials, finished products, among others, these costs are undoubtedly the most relevant.

Depending on the nature of the VRP, there are several models, from the simplest ones such as the trained VRP to the more complex ones that mix stochastic components (Gao Jie 2009). The trend in recent years tends to combine several types of VRP in the same system; this is how VRP problems arise with heterogeneous fleet and time windows (Malekly et al. 2009), stochastic demand with delivery and collection, among other combinations. As for the nature of the stochastic components of the VRP, in most of the contexts studied, we work with stochastic demands and customers (Olivera 2004), However, in practical situations, you cannot know the travel and service times due to the rise of uncertainness in weather or transit conditions. Also, as Xiang (2010) states in his research, service times can be affected by distribution technology, driver skills, parking conditions, and cargo area conditions. This article compares the behavior of a new metaheuristic (Ebola), an improved taboo search algorithm (TS), and a genetic algorithm (AG) by solving the STFSMVRPSTW. The behavioral study was conducted under the analysis of the target cost function of the routing system and computational time. The focus is on determining whether there are significant differences between response quality and computational efficiency between algorithms, as well as verifying their behavior when heuristics are adapted for the generation of initial solutions and when these solutions are randomly generated, the research result provides input for the design of a computational application that allows for more significant minimization of transport costs and is also executable in times reasonable. The rest of the article is organized as follows: Sect. 2 is intended to be presented by GA, TS, and Ebola. Section 3 includes the explanation of the problem to be solved. In Sect. 4, experimental results and discussions are displayed. Finally, Sect. 5 draws some conclusions, as well as some prospects for future research.

2 Metaheuristics Used

2.1 Genetic Algorithm (AG)

GA has received a lot of attention regarding its potential as an optimization technique for complex functions (Settles et al. 2003; Vesterstrom and Thomsen 2004). Genetic algorithms were formally introduced to the United States in the 1970s by Holland (1992) of the University of Michigan. They have been used mainly in some types of optimization; in particular, GA works very well in continuous and discrete combinatorial optimization issues. This metaheuristic is less likely to get caught up in local optimal, compared to local search methods. However, it has an expensive computational cost. For the execution of the genetic algorithm, the simple genetic algorithm proposed by Mitchell (1999) has been followed. The following is a description of the steps executed in the algorithm. **Step1:** generate initial population; the size of this Pob population is a parameter of the algorithm; the community can be generated randomly or using heuristic algorithms. The type of coding used in this algorithm was Random Key, which is characterized by presenting an integer part and a decimal, where the first represents the vehicle by which the client will be served and the decimal part in ascending order, will allow knowing the sequence in which they will be visited. This generates a population of random numbers, between one and the total number of vehicles plus 99 tenths, which are considered the chromosomes of the individual.

INDIVIDUAL	1.34	2.75	3.90	3.14	2.67

Each column represents a customer; in this sense, the customer 3 and 4, will be served by vehicle 3, the order in which they will be visited is 4-3 due to the corresponding tenths. **Step2:** evaluation of the objective function and selection of the best individual, compares the best global solution with the current resolution of the best individual, and updates the best overall individual. **Step3:** a selection of parents, for the collection of parents has performed the procedure of mourning in which two individuals are randomly selected from the population, and the one with the best fitness is selected as parent 1, the process is repeated for the selection of father 2. **Step4:** crossbreeding. Under a chance of crossbreeding the possibility of crossing the parents is evaluated, if this is fulfilled, individuals with characteristics similar to their parents are generated on the contrary the children will be identical copies of the parents (Fig. 1).

To cross the crossing, a random number is generated between 1 and the number of customers; the first child consists of the first part of father 1 with the second part of father 2; whereas for the latter it will be done oppositely.

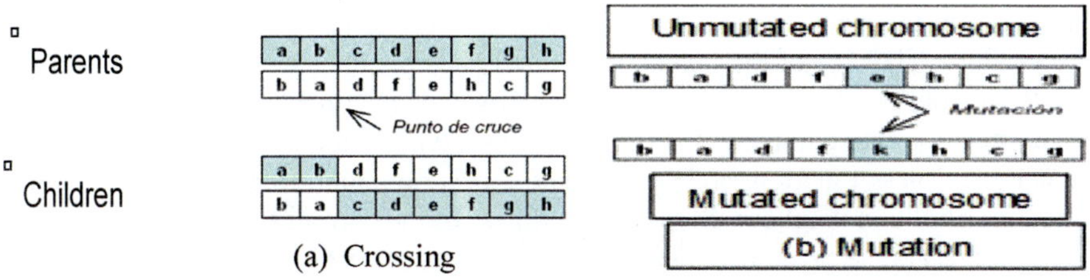

Fig. 1. Cross-linking and mutations operators

Step5: mutation, under a probability of variation, the possibility of randomly changing a chromosome for the generated children is evaluated (Fig. 1). Having the position then a random number is made between 1 and the total vehicles plus 99 tenths, which will replace the previous chromosome at the selected location. **Step6:** Repeat step3, 4 and five until the population of children is equal to the community of parents. Check if stop criterion is met (maximum number of iterations or fitness value reached), if not go to step 2.

New Taboo Search Algorithm (TS)

The taboo search algorithm was initially proposed by Glover (1986). The basic idea of this method is that it explores the solution space by moving in each iteration from the best solution to its neighboring solutions, and taboo marking the starting point to avoid cycles and stagnate in the best premises. A review of very literature reveals that several exact and approximate algorithms have been used to solve the problem. While exact algorithms can only solve relatively small problems in size (Malekly 2009). In 2008 a new Taboo Search algorithm was designed proposed by QI et al. applied for VRPTW (Beach 2008; Maury and Perez 2010). Below is the description of the algorithm as well as the adaptation to the STFSMVRPSTW.

Stepl: Generate initial solution (Solution generated based on heuristics or randomly), make an initial solution better solution from the taboo list. **Step2:** For the best individual in the list, cause n1 neighbors, create a random number to choose the neighborhood operator to use. The representation of a solution in the taboo search algorithm was done as follows (Fig. 2).

Individual

Route 1	1	0	0	6	3	0	8	0	0	0	4	0
Route 2	0	0	9	0	0	10	0	0	0	12	0	7
Route 3	0	5	0	0	0	0	0	11	2	0	0	0

Fig. 2. Individual coding for TS

Where each of the rows in the array represents a route and columns to clients to be served, zeroes imply that in that position, there are no customers to help. In this sense, course one would be the sequence: deposit -1-6 -3-8 -4 -deposit. In this algorithm, neighbors can be generated from one of the six neighborhood operators available in Beach (2008). The size of the neighborhood (n1) is a parameter of the algorithm, and the selection of the operator to be performed is based on its probability of being chosen, such expectations are parameters of the algorithm. **Step3:** For each neighbor generated, evaluate the fitness function, and determine the best neighbor, make a solution of the best current solution neighbor. **Step4:** If the fitness of the current solution is better than the value saved in the taboo list, then update taboo list replacing the better solution with the current solution. **Step5:** Check if the stop criterion is met (maximum number of iterations or fitness value reached), if not go to step 2.

2.2 New Ebola Algorithm

Ebola is a new method of optimization that makes use of the behavioral-inspired biological analogy of the Ebola virus in the body, this technique was tested in several nonlinear functions, some of them multimodal as are the Sphere functions, (separable unimodal), the Rastrigin purpose, (separable multimodal) and the Ackley function (non-separatable multimodal), the results obtained by the algorithm were compared to the classical genetic algorithm found in the literature Scientific. Ebola achieved significantly better results than the genetic algorithm, as stated by Lopez and Madrid (2010).

Stepl: generation of the initial population, the size of the initial population Pob is a parameter of metaheuristics and just like in the genetic algorithm, the number of individuals is maintained throughout all iterations. The community can be generated in a randomized way or by the use of heuristics. The encoding used for this algorithm is the same as that used for the genetic algorithm. **Step2:** evaluation of the objective

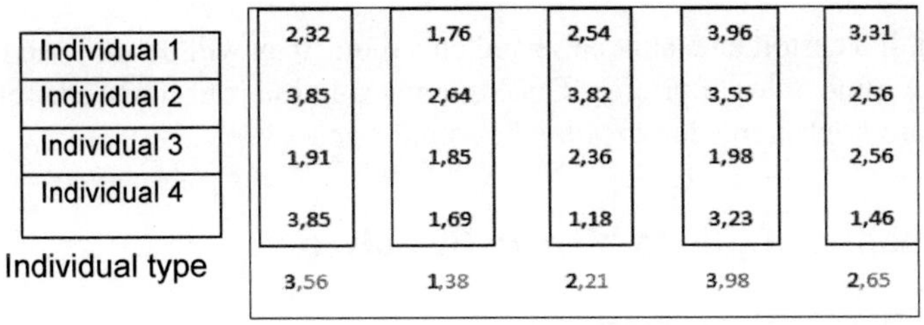

Fig. 3. Generation of the type individual from fashion as a recombination operator.

function and selection of individuals with better fitness. To identify the best ones, the average of the solutions is determined, the individuals with the best fitness than the average population are selected in the best matrix, the best individuals are saved as the best global solution. **Step3:** the creation of individual type, after the selection of the best individuals, has determined an individual whose characteristics resemble the general characteristics of the population of better individuals (recombination operator), this new individual is called a population type individual. The selection of the statistic is a decision that depends on the problem to be dealt with. For nonlinear functions, it was shown to obtain better results than those obtained with the genetic algorithm (Lopez and Madrid 2010) when the mean was used as a recombination operator. In the problem, STFSMVRPSTW was best achieved with fashion as a recombination operator (Fig. 3).

The fashion of the integer values of the chromosome column generates the whole

Cost of the respective chromosome for the individual type, the decimal values are obtained randomly and will allow organized upstream determine the sequence in which they will be served customers when they are part of the same route. **Step4:** crossing and mutation of initial solutions, just as the genetic algorithm are used crossbreeding and mutation operators, but unlike that probabilities are not parameters of the algorithm. Each of the initial individuals is selected, if their fitness is better than the average of the general population, then this is crossed with the type individual, the crossing procedure is similar to the genetic randomly generating a cut-off point and passing the Parts. **Step5:** determine new fitness for new solutions, upgrade better overall if you get a better solution. **Step6:** Check if the stop criterion is met (maximum number of iterations or fitness value reached), if not go to step 2.

2.3 Time Sales Priority Heuristics (TW)

This procedure is to assign service priorities for each of the clients, depending on these priorities will be assigned to each route, in this regard the customer whose priority is higher will be served as soon as possible. To appoint these prelates, the following algorithm was followed:

Clients are divided into subgroups, Customers who have nearest opening time windows will be served first in their respective subset, In the event of a tie, customers with most adjacent closing time windows will be served first, If the relationship still prevails, one whose demand is higher is served first on the condition that the vehicle is faster with fewer quantities transported, Once the sequences are set, capacity and time constraints are checked. If a customer cannot be served on a route, they will be assigned to the next course in the same order of priority, Once the first solution is established, optimization algorithms are implemented to explore the solution space.

3 Explanation of the Problem to Be Solved

3.1 STSFMVRPSTW Approach

The problem of study in the research is a new non-deterministic modification of the traditional VRP, called STFSMVRPTW, whose components correspond to stochastic transport and service times. Also, there is a heterogeneous fleet of vehicles and known demand. In this model, each client must be served within a set time interval. The above context closely resembles reality, yet no research on the specific subject was found in the literature review.

Few research is carried out on the subject, within which it is worth citing research for the VRP with windows of time and/or heterogeneous fleet, such as Qing et al. (2010), Zheng-Yu (2010) Maury and Pérez (2010), other exciting publications include VRP research with stochastic considerations, Chaug-Ing et al. (2007), Lecluyse et al. (2009), Kok et al. (2011).

3.2 STSFMVRPSTW Mathematical Modulation

3.2.1 Defining Variables

There are vehicles in a given distribution center; each car is limited in capacity. For the problem, the fleet is assumed heterogeneous, so the cars are different in both function and fixed costs (per lease, insurance, maintenance, among others). The number of cars can be considered as a constant or an exit variable.

Stochastic programming models can be addressed and modeled in various ways, depending on the purposes of it, some authors have proposed in the literature, several mathematical models. For the approach of our objective function, we are based on a limited probability model (CCP) and a resource stochastic programming model (SPR) (Xiang 2010).

$$\text{Minimizer } Z = (f)\sum\nolimits_{j}^{m}\sum\nolimits_{k}^{m} x_{ij}^{k} + \sum\nolimits_{k}^{m} C_k + \sum\nolimits_{k}^{m} P_k$$

4 Experimental Results and Discussions

To make the comparisons of efficiency (target function) and computational time between the different algorithms, 12 instances of the specific problem were designed, these problems were adaptations of Solomon instances for the VRPTW and reconfigured by the authors to deal with the STFSMVRPSTW. It was necessary to incorporate the proportional unit costs and the respective conversions of distances between vertices at transport times, the times provided were considered the average values of a normal distribution with deviation 10% of the average. As a standard value, the number of iterations for all

Table 1. Parameters used for algorithms

Genetic Algorithm (GA)	Taboo Search (TS)	Ebola algorithm
Population size 100	Neighbor's number 100	Population size 100
Probability of crossbreeding - 0.95 Prob. mutation - 0.3	Probability of use of operators [0.2/0.1/0.25/0.25/0.1 0.1] Respectively Alpha 0.01	Percentage of genes that mutate 0.7
Genetic Algorithm (GA) Population size 100 Probability of crossbreeding - 0.95		

algorithms was set to 500 so that they were as equivalent as possible. As for the parameters used were set based on experimental runs, in which the metaheuristics showed good behaviors in both solution efficiency and execution time (Table 1).

The algorithms were programmed in MatLab and run on computers with Intel Pentium 4 3.2 GHz processor and 1 GB of RAM; each algorithm was executed twice, one for the generation of initial solutions from time sales heuristics and another for the production of random initial solutions. To illustrate the results obtained, below is the best value based on objective and computational time that each algorithm presented in each of the proposed instances, the first columns correspond to the solutions when it does not adopt heuristics to algorithms and solutions are generated in randomly, the second columns correspond to the incorporation of the window priority heuristics (Table 2).

The value of the target function and the runtime for each of the algorithms are standardized, to perform the efficiency comparisons regardless of the instance treated:

$$Z_i = \frac{x_i - \bar{x}}{s}$$

Where it represents each of the cases, the average x bar represents the mean and standard deviation of each situation. A simple variance analysis (ANOVA) is performed to check for statistically significant differences between the solutions obtained by one or the other algorithm; this analysis also provides information on the significance of heuristics in the dependent variable value, target function, and computational time.

4.1 Result Analysis for the Cost Target Function as a Dependent Variable

The variance analysis was performed; the p-value indicates that there are statistically significant differences in the application of the algorithms on the target function of the cost (Table 3). The assumptions of normality, independence, and homoscedasticity of the waste were not rejected. To determine which means were significantly different from others, a multi-range test was performed.

Five homogeneous groups were identified based on the alignment of the X's in columns. There are no statistically significant differences between those levels that share the same column of X's (Table 4). The method used to discriminate between means is

Table 2. Better results in cost and time for each algorithm

INST	GENETIC ALGORITHM				TAB SEARCH				Ebola			
	RANDOM		HEURISTIC TW		RANDOM		HEURISTIC TW		RANDOM		HEURISTIC TW	
	F.O.	Time	F.O.	Time	F.O.	Time	F.O.	Time	F.O.	Time	F.O.	Time
R101	*242867*	*8750*	*263364*	*9839*	*243178*	*2824*	*231832*	*2766*	*325046*	*6154*	*327490*	*4799*
R102	*240157*	*7585*	*235876*	*8411*	*263140*	*2600*	*249571*	*2584*	*304987*	*5390*	*315739*	*15000*
R103	*273432*	*6894*	*329184*	*7846*	*354754*	*2562*	*350500*	*2532*	*427757*	*4864*	*435573*	*4143*
R104	*263808*	*6802*	*220433*	*7274*	*287693*	*2465*	*268686*	*1982*	*307962*	*4534*	*316067*	*3451*
R105	*507209*	*8773*	*512903*	*8496*	*526772*	*2724*	*495223*	*2385*	*684111*	*5425*	*688173*	*4837*
R106	*269367*	*6599*	*271299*	*6692*	*306919*	*2623*	*290543*	*2589*	*373525*	*3904*	*380983*	*4490*
C201	*474498*	*7085*	*477213*	*7625*	*574923*	*1816*	*554533*	*1743*	*902878*	*2815*	*982713*	*2770*
C202	*442150*	*6655*	*422076*	*5412*	*535307*	*1595*	*537260*	*1629*	*838597*	*2402*	*915478*	*2338*
C203	*421408*	*4506*	*396470*	*4273*	*504635*	*1490*	*470466*	*1540*	*755091*	*2112*	*784639*	*2084*
C204	*158166*	*3915*	*113237*	*3750*	*144105*	*1211*	*145100*	*1365*	*227589*	*1945*	*257408*	*1927*
R201	*355346*	*4837*	*330667*	*5107*	*382828*	*1671*	*390192*	*1819*	*522120*	*2553*	*557068*	*2500*
R202	*240317*	*3771*	*230692*	*4020*	*331546*	*1628*	*303025*	*1765*	*397247*	*2357*	*392560*	*2249*

Table 3. ANOVA table for FO COST by METAHEURISTICS

Source	Sum of squares	Gl	Middle square	F-reason	P-value
Algorithm	55,8271	5	11,1654	176,6	0
Error	4,1729	66	0,0632258		
Importe (Corr.)	60	71			

Fisher's Minimum Significant Difference Procedure (LSD). With this method, there is a 95% confidence level.

Table 4. Multiple Range tests for FO COST by METAHEURISTICS

Medium	Cases	Middle	Homogeneous groups
Genétic TW	12	−0,898006	X
Genétic RD	12	−0,787787	X
Taboo TW	12	−0,469576	X
Taboo RD	12	−0,258431	X
Ebola RD	12	1,0957	X
Ebola TW	12	1,3181	X

The results position the genetic algorithm as the most efficient algorithm as solutions, this result is contrary to previous research (see Sect. 2), it can also be observed that there is no statistical difference significant in the application of heuristics to obtain initial solutions with this algorithm, in this sense the adaptation of heuristic processes to this algorithm does not necessarily generate better solutions. On the other hand, there is a significant difference between GA and TS, then taboo turns out to get better results than Ebola significantly, and the implementation of heuristics to the algorithm affects the response variable. It can be noted that Taboo turns out to be better at including heuristics in the generation of initial solutions in comparisons with random solutions. Ebola turns out to be the least efficient algorithm compared to the algorithms treated; the multi-range analysis reveals that the algorithm is more efficient when initial solutions are generated randomly, on the contrary when they are incorporated heuristics such a procedure tends to converge stagnation rapidly in optimal premises. Figure 4 and Fig. 5 show the graphical behavior of the algorithms throughout the iterations, showing that GA and Ebola converge quickly to a solution, while taboo always tries to escape the answers.

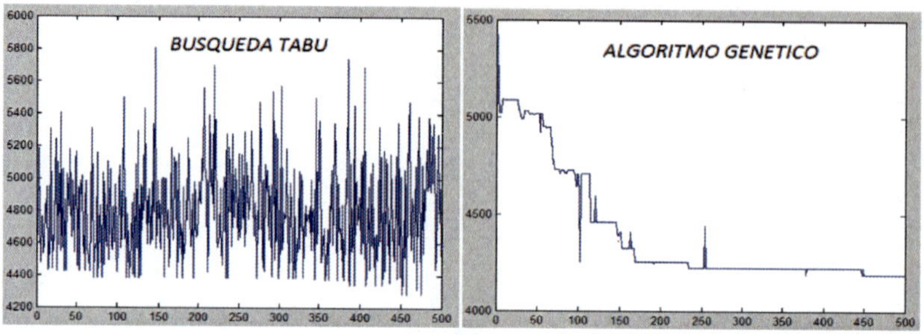

Fig. 4. Graphical (F.O. vs. iterations) taboo and genetic behavior

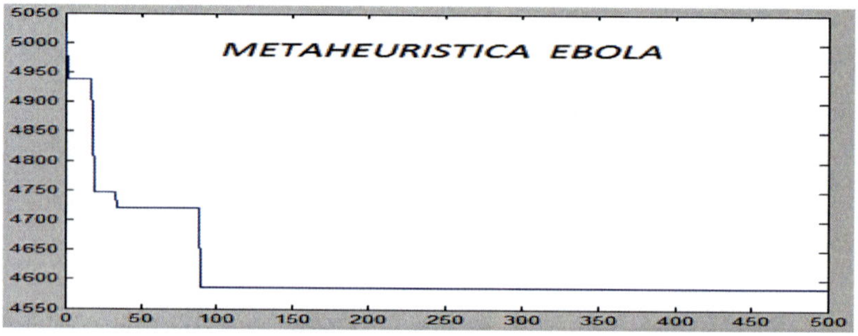

Fig. 5. Comportamiento grafico (F.O. vs iteraciones) Ebola.

4.2 Result Analysis for Cost Computational Time as a Dependent Variable

The variance analysis was performed. However, the assumptions of normalcy were rejected. In the case where the assumption of normality is not justified, the Kruskal-Wallis test is a nonparametric alternative that is used to check the hypothesis that there is no significant difference between the different factor levels (Table 5).

Table 5. Kruskal-Wallis time test by METAHEURISTICS

Kruskal-Wallis for time test by METAHEURISTICS	
Statistical = 60,8311	Value-P = 8,18379E−12

Since the P-value is less than 0.05, there is a statistically significant difference

5 Conclusions

This work was studied a non-deterministic modification of the traditional between medians with a level of 95.0% confidence. VRP, called STFSMVRPTW, whose components correspond to stochastic transport and service times, with a heterogeneous fleet of vehicles and known demand. In this model, each client must be served within a set time interval; this is an NP-hard combinatorial optimization problem quite close to reality.

The behaviors of three metaheuristics (Ebola, GA, and TS) were compared for the solution of the problem above by analyzing the target function variables (cost of the routing system) and execution time of the algorithm. To achieve this, 12 theoretical instances were adapted to the STSFMVRPSTW, a window priority heuristic was also designed, and the impact of this procedure on the output variables was analyzed. IT has been shown that GA is the most efficient algorithm is the solution of this stochastic problem, making a vast difference over TS and the new Ebola algorithm. As far as the time of the Taboo and Ebola solution prove to be the best, on the contrary AG gets considerably high time. On the other hand, window priority heuristics do not significantly affect algorithm execution times, in terms of response quality improving results for TS significantly, while for the AG it does not generate better solutions necessarily, and Ebola proves to be better when initial individuals are randomly generated. As future work, it is hoped to incorporate new stochastic and dynamic variables to the problem, to bring it as close as possible to the realities of the environment.

References

Beach, L.: A new tabu search heuristic algorithm for the vehicle routing problem with time windows. In: Management Science, pp. 1648–1653 (2008)

Cordeau, J., Gendreau, M., Laporte, G., Potvin, J., Semet, F.: A guide to vehicle routing heuristics. J. Appl. Res. Soc. **53**, 512–522 (2002)

Holland, J.: Genetic algorithms. Sci. Am. **267**(1), 66–67 (1992)

Malekly, H., Haddadi, B., Tavakkoli, R.: A fuzzy random vehicle routing problem: the case of Iran, pp. 1070–1075. IEEE (2009)

Mitchel, M.: An Introduction to Genetic Algorithm, London (1999)

Scheffermann, R., Bender, M., Cardeneo, A.: Robust solutions for vehicle routing problems via evolutionary multiobjective optimization. IEEE (2009)

Settles, M., Rodebaugh, B., Soule, T.: Comparison of genetic algorithm and particle swarm optimizer when evolving a recurrent neural network. In: Proceedings of the Genetic and Evolutionary Computation Conference 2003 (GECCO 2003), Chicago, IL, USA, pp. 151–152 (2003)

Toth, P., Vigo, D.: The Vehicle Routing Problem. SIAM: Monographs on Discrete Mathematics and Applications (2002)

Methods of Data Encryption for Use in Safe Space Information

Yenny Espinosa Gómez(✉), Javier Felipe Moncada Sánchez(✉), and Roberto Ferro Escobar(✉)

Universidad Distrital Francisco José de Caldas, Bogota, Colombia
{yespinosa,jfmoncada,rferro}@udistrital.edu.co

Abstract. The security in the transfer of Raster-type image data using different open computer networks has been extended in recent years, in order to avoid alterations of the information it is necessary to use different encryption methods of the Pixelated data. In order to carry out this purpose, it is required at least to fulfill two objectives in the encryption of Images; the first is to eliminate the correlation between neighboring pixels that have a high similarity and the second is to encrypt the result obtained. The encryption methods of conventional text formats cannot be applied to Raster data, that is why different methods of encryption have been developed, where transformation methods and Chaos theory are used which, due to their high mathematical complexity, allow us to obtain high security encryption. According to the above, the spatial databases require security mechanisms, access control and encryption to guarantee the integrity of the information, therefore there are several types of procedures and techniques oriented to encryption and the security of the databases. Data; some of them will be presented in this article in order to open a new specific field of work in this important area of knowledge.

Keywords: Chaotic maps · Ciphers · Databases · Encryption · Image · Security

1 Introduction

In today's digital world, security in the transmission of digital images and databases becomes increasingly important, as communications of digital products through networks occur more and more frequently. On the other hand, the special and reliable security in the storage and transmission of digital images and in general of information in a database is necessary in many applications.

In the Geomatics part, the Raster images are of great importance for the input and analysis of the information. The transfer of Raster Images through the different open networks becomes of great importance for the security and reliability of the data.

In the present article, we describe the different methods of image encryption and we focus on transformation methods and chaos schemes, which are the ones that offer greater reliability in the face of third party attacks. Different techniques, processes and ciphers for databases are also presented in order to guarantee the security, access and quality of the data.

D. F. Cortes Tobar et al. (Eds.): AETA 2019, LNEE 685, pp. 660–669, 2021.
https://doi.org/10.1007/978-3-030-53021-1_66

2 Encryption of Images

Digital images represent 70% of the information transmitted over the Internet. However, the information in the images are different from text messages, have a larger scale, greater redundancy and higher correlations between the pixels [1]. This valuable multimedia content, however, is vulnerable to unauthorized access during storage and transmission over a network. The transmission of digital images also requires a large bandwidth for transmission. An effective transmission of images through the Internet, should therefore consider the security and bandwidth problems [2].

During the last decade, source coding and channel coding have been extensively studied to overcome bandwidth problems and transmission errors. Source coding is used to compress information as much as possible by eliminating redundancy, while channel coding strives to offer the transmission more reliable within a given bandwidth. According to Shannon, the combination of a source encoder and an optimal channel encoder can create an optimal system, therefore, the joint design of source and channel encoder is necessary to optimize communication systems. But in addition to the study of source coding and channel coding techniques, coding techniques for multimedia applications have been studied and developed [3].

Advances in space science, data analysis and communication technologies present new opportunities for users to increase productivity, reduce costs, facilitate innovation and create collaborative virtual environments to face new challenges. GIS and remote sensing technologies, together with technologies related to geospatial data analysis, provide powerful tools for the preservation and protection of a nation's resources. In such systems, a space platform collects data on surface features and transmits them to a ground station, where a series of imaging products are processed and made available to researchers or commercial organizations. To comply with the privacy needs in various applications, the encryption of images is very important to minimize malicious attacks by unauthorized third parties [3].

There are three basic problems for the transmission of images through the Internet. The bandwidth, number of errors, and security problems. There are many mechanisms to protect digital satellite images, such as compression, digital watermark, stenography and cryptography. Here are some relevant aspects of these techniques:

Compression
The compression eliminate the redundancy of the data and converts them to another form. Encryption converts the data to some illegible form and reduces the size of the data. In general, compression techniques, such as: cosine transformations, Wavelets, and fractals present some limitations such as loss of resolution and distortion while the decompression process is being done. The result is that the image loses clarity, definition, and the appearance of artifacts.

Digital Watermark
Digital watermarks are widely used to protect digital images. It simply incorporates some known data bit patterns into digital images as a digital signature. The digital signature is used as an identifier to protect digital images and is incorporated into images such as noise or random information in the file, making the detection and extraction without authoriza-

tion of watermarks difficult. The purpose of this technique is the protection of copyright, fingerprints, dissemination control and covert communication (stenography) [4].

Stenography
Stenography is an act of secret writing and uses the same idea of the digital watermark. Stenography is a hidden message writing process in the real message so that the sender and receiver can understand it. The hidden message incorporation is typically parameterized by a secure key, without knowing this key it is not possible to detect or eliminate the actual embedded material.

Cryptography
Cryptography is the process of converting or encoding the message through encryption to produce an illegible or incomprehensible encrypted text. This process is controlled by the algorithm that uses secure keys for encryption and decryption. Cryptography is widely used in many applications and has different standards with respect to the parameters used in this process, such as security, performance, key size, and block size. Some of the most used algorithms are: Data Encryption Standard (DES), Triple DES, International Data Encryption Algorithm (IDEA) and Rivest's Code 5 (RC5).

2.1 Schemes Proposed for the Encryption of Images

For the treatment, compression and transmission of image data different alternatives have been proposed, which mostly apply different compression processes, to eliminate data redundancy, and encryption, for the secure transmission of data.

Discrete Fourier Transform and DES
Consists of the compression of the image to achieve a low bitrate in the digital representation of an image with a minimum perceived loss in image quality [2]. In order to achieve the high compression ratio with less distortion, wavelet compression with zero tree coding scheme is applied. The combination of compression and encryption techniques improves security for image transmission and also improves transmission speed.

The proposed algorithm includes the discrete wavelet transform (DWT), an Embedded Zero-tree Wavelet (EZW) coding scheme, an asynchronous transmission method (ATM) and the DES algorithm and transposition operations.

This algorithm presents advantages from the point of view of security, cryptanalysis is usually based on the redundancy of plain text to break the code. In order to reduce the calculation time, the encryption must be processed after compression. The proposed approach improves security for data transmission and also speeds up the calculation process for encryption. Another characteristic of the proposed scheme is that using wavelet techniques for image compression provides high compression ratio and high quality reconstructed images.

Theory of Chaos and DES.
Encryption techniques based on chaos theory use sequences generated by random chaotic fields to encrypt the image. The encryption system uses a logistic chaos sequencer to make the pseudo-random sequence carry the RGB data and then apply a double DES encryption [1]. The theoretical analysis and the simulation indicate that this method

enjoys high security and offers a high coding speed. The algorithm can be used in the encryption of images in real time.

Encryption of Images on the WEB and Mobile Devices

The simplest way to encrypt an image is to consider the 2-D images as 1-D data and then encrypt any available encryption system [5]. Encryption is used to transmit data securely from different open networks.

Encryption schemes consider the special characteristics of digital images, such as bulky size and great redundancy in uncompressed images. Each type of data has its own characteristics; therefore different techniques must be used to protect confidential data in unauthorized access images.

Transformation Method

In most of the images the values of the neighboring pixels are strongly correlated, that is, the value of any pixel can be predicted from the values of its neighbors, then the objective of the image encoders in the internet is to dissipate the high correlation between the pixels and increase the value of the entropy. The image encryption algorithms must contain two types of secret keys: the first whose purpose is to increase entropy and the second to encrypt the image.

The most common methods for encrypting images via WEB are block-based transformation, which consists in combining the transformation of the image with the encryption of this transformation [6]. See Fig. 1.

The first part of this method, which is called transformation, is carried out to obtain a lower correlation between the neighboring pixels and, therefore, greater entropy, for this the original image is divided into blocks, which are subsequently reorganized and encrypted.

The more number of blocks the image is divided, that is, the smaller blocks are obtained, the lower the correlation between the pixels and the higher the entropy.

An encryption method of this type is the one developed by Younes and Jantan [6], which proposes a transformation-based encryption method in blocks with an encryption called Blowfish.

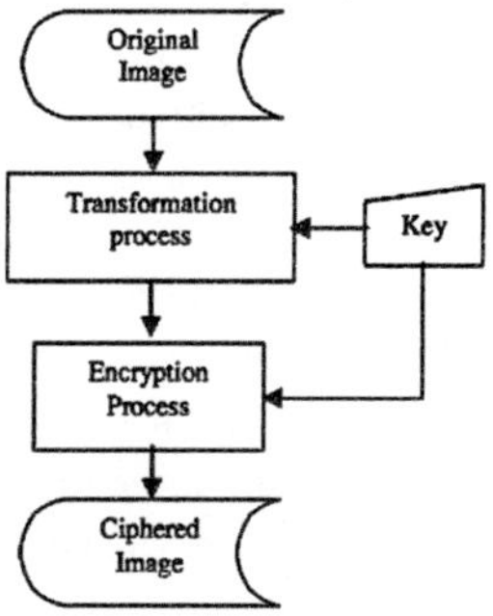

Fig. 1. Diagram of the algorithm of the transformation method [6].

Chaotic Systems

Chaotic systems have many important properties, such as dependence on initial conditions and system parameters, most of the properties are related to some essential requirements of cryptography, that's why chaotic cryptosystems have more useful and practical applications. Any method used in chaotic systems is based on analytical floating-point calculations. The security in the transfer of image information can be increased by combining conventional methods and chaotic systems.

The chaotic system consists of a sequence where the input image is divided into Blocks, then by Diffusion Methods (XOR) is transformed by a sequence of Spatial Permutation and Substitution by a chain of points or chaotic information is encrypted data [7]. The block diagram is show in Fig. 2.

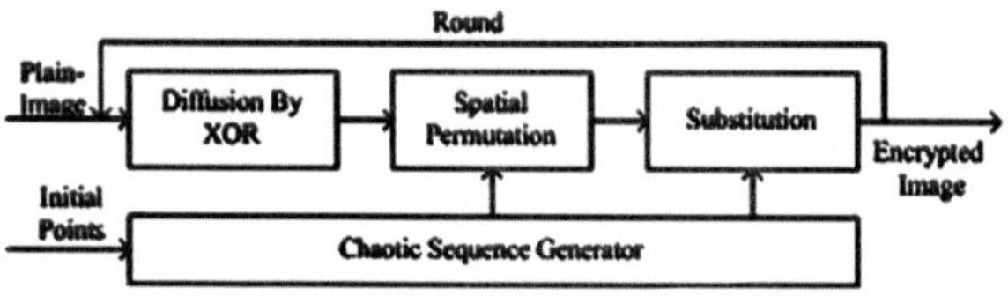

Fig. 2. Block diagram of the encryption of data [9].

2.2 Access Control, Security and Encryption of Databases

Users and organizations have increased the use of computer systems for the development of their work, for the management of their information they have increased the use of database systems, as a key system for operations and decision making of the day a day.

Access control and confidentiality in database management systems has been based on two different kinds of models; those based on discretional access control policies and those based on mandatory access control policies, These models, and especially discretionary models, introduced some important principles that distinguish access control models for database systems from access control models adopted by other operating systems or file systems.

Universally, an enormous amount of effort, time and resources have been invested in trying to make database systems comply with security requirements. These security requirements usually include:

- Prevention of unauthorized disclosure of information
- Prevention of unauthorized modification of information
- Avoid denial of service
- Prevent unauthorized penetration of the person system
- Prevent abuse of special privileges

Encrypting the data stored in a database can prevent its disclosure to attackers. A cryptographic technique can guarantee excellent security for databases, reducing the entire security process to the protection of encryption keys only a few. However, the cost time involved in encrypting and decrypting data can degrade the performance of a database system.

Characteristics of a Database Encryption
A security system for a database must be flexible and must not force a change in its structure. It is usual for a database to impose a series of restrictions on an encryption mechanism, such as:

1. The encryption system should be secure or require extremely high work at the time to break it.
2. Encryption and decryption must be fast enough to not unacceptably degrade system performance.
3. The encrypted data must not have a significantly larger volume than the decrypted data.

4. The encryption must be oriented to the registry. If a database is relational, networked, hierarchical or oriented to files, the basic unit of information for addressing is the registry. It must be possible to perform encryption and decryption in individual records without taking into account their physical or logical position in the database.
5. The encryption system must support the logical approach of the subschema of the Database [13]. Users with different logical views of the data must be presented with corresponding views of the database by the SMBD, so the SMBD will present a user only with the fields he requests and for which he is entitled.
6. The encryption scheme should be as flexible as possible with respect to the combinations of the read and write privileges that can be granted and the conditions under which the reading and writing should take place.
7. Ideally, the SMBD should be able to detect and reject the records that are created or updated by virtue of a false encryption key without being able to determine what the data is.
8. The SMBD should not be forced by the encryption system to keep duplicate copies of the data elements in order to allow the subschema to be represented.

System System of Encryption of Databases with Subkeys
Essentially, what is needed is a register-oriented encryption scheme in which an encrypted record is a unique function of all its field values, and in which each separate field is encrypted/decrypted by means of a separate key. This is called a scheme of an encryption system using subkeys. The keys are subkeys in the sense that, while the encrypted register is a function of all the encryption keys and field values, each decryption key will decrypt only a subset (a field) of the original record [8].

This cipher system for Databases is the one that has subkeys that allow the encryption and decryption of fields within a registry and is based on the Chinese Residual Theorem (CRT) [9]. The Chinese Residual Theorem (CRT) is also used as a basis for various works that propose high capacity embedding or embedding schemes and high visual quality of hidden information.

Encryption in Databases as a Service (Database as a Service - DaaS)
A database as a service operates the databases in the cloud reducing operating costs through the automation of: provisioning, support, monitoring, cloning and restoration of the database.

In Daas it is necessary to establish a model based on policies that allow the generation of resource consumption, security and data protection. But the cloud model can make some aspects more difficult to handle, such is the case of encryption that constitutes the method to preserve the integrity, confidentiality and security of the data, but computationally it is excessively expensive for the service provider. In the case of a relational database the encryption has had several advances, the first approaches use extensions for the Query Language that simply apply encryption before writing a value and executing a decryption function before reading a value [10], these extensions also work together with advanced homologadores. This type of homologadores and ciphers are based on having an absolute trust of the Database, but in the case of the Cloud this does not happen because the passage of all the information on the Internet and for various applications means that it cannot be trust in the security of the Database and that it is necessary to propose alternatives such as binning technology, encryption of privacy homomorphism, encryption of indexes.

Encrypted and Lightened of a Database Using the TSFS Algorithm
The TSFS algorithm (Transposition, Substitution, Folding, Shifting - Substitution, Substitution, Bending, Displacement) is a symmetric key block cipher, that is, the same key is used to encrypt and decrypt and its security depends on the length of the the key

TSFS can be used for the encryption of an efficient lightweight database, in which only the sensitive data in the database is encrypted using this algorithm, which will allow a fast and efficient execution of the queries to give an Timely response to users. In the proposal of Manivannan and Sujarani three keys are used for encryption and decryption and these three (3) keys are expanded in twelve (12) sub-keys to use the Key Expansion Technique in order to provide more security and effectiveness for the database [11]. This algorithm works for the following types of data: alphanumeric, numeric, and numeric with decimal point; it does not work with any symbol, only with the previous data.

The TSFS algorithm uses four types of transformations in the encryption and decryption to guarantee the security of the data: Transposition, Substitution, Bending and Displacement.

DES Encryption for Databases.
One of the encryption techniques that has been considered and used in databases is the Data Encryption Standard (DES) [12], but this scheme is also applicable to cryptographic and standard techniques. To reduce the time spent on encryption and decryption, the database schema is divided into sensitive and non-sensitive data. Its decryption process is very fast, since only one key is needed to decrypt a full column of classified data encryption. In addition, although encrypted private access data has to be deciphered separately using its unique keys, requests for private data very rarely, takes place only once in an instant.

Encryption Technique: Order of Preservation
Another specialized encryption technique that has been proposed is the preservation order of Agrawal [13]. This raises the encryption of data belonging to a subset of the integers (pmin, pmax), although in recure to the possibility of handling the floating-point numbers as if they were integers, because the positive ones maintain the order as it is, and for the negative ones, which would have the inverted order, simply subtract the resulting integer from the largest possible negative. In this way, data that follows a certain distribution is transformed into cryptograms that maintain order and also follow a different distribution chosen by the user. To generate the encryption function, they make use of all the data to be encrypted in the database, which can be a problem when the database is empty or is very new, in addition the user must give a list of samples of which will be the distribution to simulate. From these samples, the base of the auxiliary information to be used will be created that will allow to encrypt/decrypt the data.

Anti-Tamper: Query Processing on Encrypted Databases
Another of the encryption schemes used in databases is the Anti-Tamper [14] which is done by adding a set of pseudo-random numbers. One of its biggest drawbacks is that it is given by the encryption time that increases exponentially if the flat text value is large.

This encryption method allows an integer x to be encrypted by adding the first p values of a secure pseudo-random sequence. The biggest problem with this system is that the encryption obtained is too predictable.

OPES and COPE Ciphers

Another of the ciphers is the Order-Preserve Encryption (OPES) system, this system of data encryption identifies the order of a plain text. This feature can cause some problems. In the first place, if the basic text is already organized on an ordinal scale (ascending or descending), for example the identity number of the persons, the data encryption is very vulnerable to the attack of inference. Secondly, if the text is numeric, the encryption can cause an error of precision.

Parallel to the OPES, an encryption scheme known as Chaotic Order Preserving Encryption (COPE) [15] has been proposed. In this cryptosystem, the order of the buckets depends on the key, so, to see it more closely, it is not a purely ordered cryptosystem. All cryptosystems of this type are necessarily symmetric, since knowing the encryption function allows us to approximate, as much as we want, the decryption function.

Bucketization Technique

Bucketization is another of the cryptosystems recently used in databases [16]. In this the tuples are not encrypted using conventional encryption, but an additional identifier bucket is created for each of the values of an attribute. In this way each identifier that is made represents the partition to which the non-encrypted value belongs, these can be indexed. The constants that appear in a query are replaced by their identifiers of the corresponding bucket. In this way it is almost obvious that the result of executing a query will contain false entries that must be deleted in a step after the processing to decrypt the tuples returned by the query.

Index B+ - Trees.

The Trees B+ indexes [Balancing Confidentiality and Efficiency in Untrusted Relational DBMSs]. It allows for range queries, as well as indexing of information, but it requires many SQL execution commands, even a non-complex query. Even so, this system is the most widespread in the ordering and indexing of databases, in DBMS such as Oracle or PostgreSQL have it as the default algorithm. Unfortunately, it can become too recursive. For this reason, this method causes a system overload and an expense of additional resources for the system. In addition, it is necessary to decipher the complete data and reconstruct index B+ trees if users perform data insertion operations repeatedly and/or constantly, or tend to delete or update the views [17].

3 Conclusions

In this article, we present several methods that have been developed for the security in the transfer of data and images used in spatial databases, the combination of transformation of the images based on blocks and the theory of Chaos. The encryption seeks that the correlation between pixels is reduced and thus improve the process of transmission of information in open networks.

When encryption algorithms are compared it will result that the best algorithm must have the lowest correlation and the highest entropy. The Methods Based on the transformation and encryption of Images by Chaos Theory due to their high complexity allow obtaining very good results in the security in the transfer of Images in the Open Networks. Although the first DBMS, among them the relational ones, did not have authorizations with respect to the information of the scheme, today several products have such characteristics. In such a context, access control policies may also need to be protected, since they can reveal sensitive information, such as data encryption and access control.

These models often need to have metadata information in order to rely on access control content for multimedia data and rely on access control credentials to deal with external users. Recent efforts in this direction include the development of comprehensive access control models for XML.

References

1. Yun-peng, Z., Wei, L., Shui-ping, C., Zheng-jun, Z., Xuan, N., Wei-di, D.: Digital image encryption algorithm based on chaos and improved DES. In: 2009 IEEE International Conference on Systems, Man and Cybernetics, pp. 474–479 (2009)
2. Dang, P.P., Chau, P.M.: Image encryption for secure Internet multimedia applications. In: 2000 Digest of Technical Papers. International Conference on Consumer Electronics. Nineteenth in the Series (Cat. No.00CH37102), pp. 6–7 (2000)
3. Usama, M., Khan, M.K.: Classical and chaotic encryption techniques for the security of satellite images. In: 2008 International Symposium on Biometrics and Security Technologies, pp. 1–6 (2008)
4. Bo, W., Cui, X.M., Zhang, C.: Realization of digital image watermarking encryption algorithm using fractional fourier transform, pp. 822–825 (2011)
5. Hernandez Encinas, L., Rey, Á., Encinas, A.: Encryption of Images with 2-dimensional cellular automata (2019)
6. Bani Younes, M.A., Jantan, A.: Image encryption using block-based transformation algorithm. IAENG Int. J. Comput. Sci. **35** (2008)
7. Zhang, L., Liao, X., Wang, X.: An image encryption approach based on chaotic maps. Chaos, Solit. Fractals **24**, 759–765 (2005)
8. Davida, G.I., Wells, D.L., Kam, J.B.: A database encryption system with subkeys. ACM Trans. Database Syst. **6**, 312–328 (1981)
9. Zheng, X., Huang, C.-T., Matthews, M.: Chinese remainder theorem based group key management, pp. 266–271 (2007)
10. Weis, J., Alves-Foss, J.: Securing database as a service: issues and compromises. IEEE Secur. Priv. **9**, 49–55 (2011)
11. Manivannan, D., Rajendran, R.: TechnologiesLight weight and secure database encryption using TSFS algorithm (2010)
12. Madany, Y.M., El Kamchouchi, H.M., Halim, B.I.: Investigation and design of rectangular waveguide using composite right/left handed transmission line (CRLH-TL) structure for microwave and communication consumer applications. In: 2016 13th IEEE Annual Consumer Communications Networking Conference (CCNC), pp. 292–294 (2016)
13. Bertino, E., Sandhu, R.: Database security - concepts, approaches, and challenges. IEEE Trans. Dependable Secur. Comput. **2**(1), 2–19 (2005)
14. Chung, S.S., Ozsoyoglu, G.: Anti-Tamper Databases: Processing Aggregate Queries over Encrypted Databases. In: 22nd International Conference on Data Engineering Workshops (ICDEW 2006), p. 98 (2006)

15. Lee, S., Park, T.-J., Lee, D., Nam, T., Kim, S.: Chaotic order preserving encryption for efficient and secure queries on databases. IEICE Trans. **92-D**, 2207–2217 (2009)
16. Hacigümüs, H., Iyer, B.R., Li, C., Mehrotra, S.: Executing SQL over encrypted data in the database-service-provider model. In: Proceedings of the ACM SIGMOD International Conference on Management of Data, pp. 216–227 (2002)
17. Damiani, E., di Vimercati, S., Jajodia, S., Paraboschi, S., Samarati, P.: Balancing confidentiality and efficiency in untrusted relational DBMSs. In: Proceedings of the ACM Conference on Computer and Communications Security, pp. 93–102 (2003)

BRT Station Door Fare Evasion Control Through Image Recognition Using IoT Approaches

Edward Steven Burgos-Prada[1], Laura Alejandra Rosero-Sanchez[1], Nicolas Velásquez Martínez[1], Diego Bermudez[1], Luis Andres Marentes[2], Juan David Rojas[1], and Luis Felipe Herrera-Quintero[1(✉)]

[1] Universidad Piloto de Colombia, Bogotá, Colombia
{diego-bermudez,luis-herrera}@unipiloto.edu.co
[2] Universidad de los Andes, Bogotá, Colombia

Abstract. Recently some cities have been adopting bus rapid transit systems to decrease public travel times at a relative low cost. The system operation is based on an off-board fare subsystem that collects financial resources from transit passenger. Yet, BRT systems face a fare evasion problem, where users gain access by evading fare controls. Fare evasion through doors not only is a risky practice for the passenger, but also affects bus time scheduling for arrivals to be programmed in an effective way. This paper proposes a control system using image recognition with an IoT and Cloud computing approach for detecting possible bus fares evasions in real time. The application provides a new alternative to detect and report when and where the infraction is committed. The information gathered from the proposed application can be used in the future for analytic analysis allowing to predict possible attempts of bus fares evasions.

Keywords: Fare evasion · Video monitoring · Image recognition · IoT · ITS

1 Introduction

One of the main problems that deals most people and cities around the world is urban transportation. Transport plays a vital role in country development terms because it provides access for people to education, markets, recreation and health care [1]. Governments have developed diverse ways to cope with mobility problems, a successful alternative is the implementation of bus rapid transit (BRT) systems [2], which use exclusive lanes for public transport buses and an offboard changing scheme. These features result in a higher average speed when compared to that in private vehicles.

BRT systems have been implemented across the globe and, for 2011, authors in [3] reported at least 41 systems implemented in an equal number of cities. In Bogota, Colombia, it is implemented a worldwide know BRT system called

D. F. Cortes Tobar et al. (Eds.): AETA 2019, LNEE 685, pp. 670–680, 2021.
https://doi.org/10.1007/978-3-030-53021-1_67

Transmilenio, which is the main transportation mode for the city, it supports 69% of the city's population [4] and by the time of its introduction, it showed outstanding results; such as, "reduced travel times, considerable passenger satisfaction, high capacity, accident and emission reduction, and operation without subsidies" [4].

However, nowadays, like many other BRT systems across the globe, Transmilenio is suffering the fare evasion problem. This misconduct occurs when a user accesses the system in inconsistent ways to evade the fare payment. Authors in [5] report evasion estimates around the world. In Melbourne, for example, the evasion problem represents as much as 12% of total rides. In Santiago, Chile, that number reaches 28%. Therefore, their studies suggest that evasion has important consequences in system's sustainability. In addition, authors studied the determinants of fare evasion, which are related to socio-demographical aspects (i.e., age and gender) as well as controlling measures; such as, those promoting actions to increase the probability of being catch.

For transmilenio, fare evasion has been reported as a major problem and it is closely related with passenger unsatisfaction. In fact, Transmilenio argues that because fares evasion, the entity cannot have a real knowledge of the number of passengers within a bus station, which has a direct consequence on wrong bus scheduling.

In BRT systems, there are many different alternatives for evasion, but one of the most dangerous is to get access through station doors. In fact, local reports indicate that in the first three months of 2019, three people died and 41 others were injured trying to avoid payment at the bus station [6].

This paper addresses the problem of evasion through door station by implementing a real-time surveillance system. The system is designed to help the BRT security group by identifying, registering and informing evasion events. Thus, people in charge can take immediate actions and, by this way, reduce the fare evasion. In particular, the following is the list of contributions: a) a proposal for an algorithm to evade detection, b) an embedded system design for image processing and communications, c) a system for checking evasion events for security agents within the bus station and the control center and, d) an analysis of requirements in bandwidth for the real time operation.

The rest of the paper is organized as follows. In Sect. 2 we review previous alternatives for evasion detection. In Sect. 3, we provide a deeper review for the problem that we want to cope. Section 4 shows our proposal and its requirements as architecture. Section 5 presents a prototype and its results. Finally, Sects. 6 and 7 presents some conclusions and future work.

2 Literature Review

There are many aspects that determine fare evasion including income level, service quality, and the level of success in the punish strategies implemented, to name a few. Even though, transportation authorities can only create public policies for enforcement and punishment. For this reason, previous proposals [7] for fare evasion are focused mainly in mechanisms to improve the inspection process and alternatives to detect evaders. In what follows, we present the most relevant previous contributions on both categories and explain how this paper is different from them.

Most of the alternatives for improving the inspection process create a mathematical model for optimizing the cost/benefit ratio when implementing inspections. In particular, authors in [8] introduced a new methodology to determine the optimal strategy for ticket inspection to be performed in bus stations and bus routes. In this case, instead of controlling people in access sites, inspectors control at any time within the system to check whether or not a passenger paid for the trip. Likewise, in [9] authors create a model that consider the probability of being catch as well as inspection costs to decide the number of inspectors to hire for a particular day.

Algorithms to detect evaders are complementary measures to inspection efforts. The main idea is to use a technology, mostly image recognition, to classify passengers accessing the system in a fraudulent form. In [10] it is presented a non-supervised machine learning algorithm for detecting evaders in buses operating in regular routes with on-board charging. For subways, in [11] it is presented the integrated approach used for controlling evasion on New York City. The system includes the use of cameras for passenger identification by using a classification algorithm. Then, the system registers those entries classified as illegal.

This paper follows a similar approach as those contributions in the second category, but it is different from the previous literature in the following aspects: (1) it considers a evasion mechanism not addressed before, the access though door stations, (2) it considers and a real-time operation to complement the efforts of inspection teams, (3) the system has been designed to be low cost, so it can be implemented in all stations of the BRT system.

3 Bus Fare Evasion in Transmilenio

Transmilenio is a transport massive system that was created to respond to increasing demands for trips in Bogota. Right now, Transmilenio system has 112.9 lane kilometers, eleven principal trunks, one hundred thirty-four bus stations and nine portals in full operation.

When Transmilenio started, there were almost five million people in the city. Now, this number has grown to more than seven million people. That population growth has implied buses and stations more crowded, consequently this overflowed service is causing annoyance, which in turn is promoting bus fares evasions. According with Transmilenio administration and different analysts, the feeling of a bad service that has been growing is due to the bus fare evaders. The authority argues that "when there are bus fare evaders, we cannot schedule bus arrival times in an effective way, as a consequence of fewer people in the scheduling system than they really are".

There are two possible ways of bus fares evasions, the first one is by jumping the turnstiles and the second one is by crossing the avenue and entering on the wagon side doors. In September 2017, five deaths and thirty-one wounded fare evaders were announced by Transmilenio [6], also in the same year the aggression cases against to Transmilenio workers by the evaders were at one thousand ninety-two.

As we have seen the bus fare evasion is a real problem in Transmilenio that is causing annoyance between its users, this practice does not let Transmilenio know how many people are at bus stations in a specific time driving a feeling of a bad service.

4 Proposal

4.1 General Description

Our proposal is to develop a system that alerts not only police and Transmilenio workers at the bus station, but also the Transmilenio operation center for resource management. The information that police and Transmilenio workers receive from the alert is in the form of bus fare evaders photos, the specific location where the evader got into and the date/hour when the evasion was committed.

4.2 System Design

The system design (see Fig. 1) is focused in providing a video surveillance service at different stations of Transmilenio (1) where evaders can be reported to the police (2), to the Transmilenio workers within the bus station and (3) Transmilenio operation center (4). Additionally, we use different cloud computing services (5) to strengthen the proposed system.

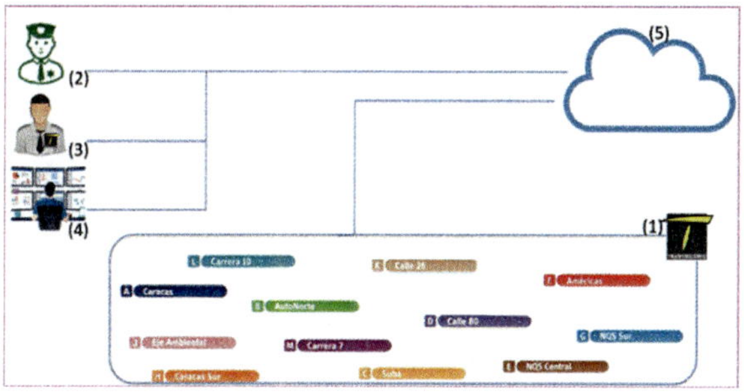

Fig. 1. General system design.

In this proposal, the video capture and image pre-processing for the possible bus fare evaders is done by an embedded computing system. Taking advantage of the exclusivity Transmilenio lane, the system will differentiate possible bus fares evaders from other buses or vehicles by calculating their specific area.

Once a possible bus fare evader has been detected, the embedded system will also populate a NoSql relational database running on the cloud putting valuable information as pictures of the bus fare evaders, the specific location where the evader got into and the date/hour where the evasion was committed, in this way, the different control entities will be notified, through different platforms, either Web or Mobile.

The distribution of the embedded devices, in each bus station, is planned so that each device handles three cameras, each of them associated with different access side doors, see Fig. 2. The connectivity of embedded devices is made through Wi-Fi allowing to connect to the access point of the bus station and routing the traffic to the cloud.

A mobile application, that will be handled by the police and/or Transmilenio workers, is able to notify in real time about the bus fare evasion and the door through which the infraction was committed. Notifications are limited to the station in which they are working. In addition, for reporting purposes, it is possible to confirm whether the bus fare evader was captured or not.

Finally, the operation center accesses the system through the Web interface, which is able to review evasion alarms for the entire system.

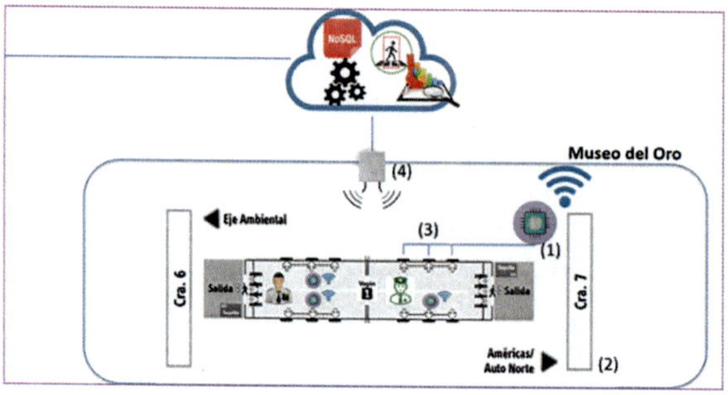

Fig. 2. Embedded systems distribution and connection to the cloud.

5 Prototype and Results

This section presents the prototype solution for the bus fares evasions problem that Transmilenio faces, describing the different technologies used and its results. For a better understanding, the section was divided in four main components; Mobile Application, Web Application, Embedded system and the NoSql database.

5.1 Mobile Application

The mobile application runs in Android devices, but it can work on different vendors, offering an easy-to-use interface that makes it easy to report catches of evaders and get alarms each time a possible bus fare evader is detected on the bus station in which the employee is working. After every detection of a evader, the application presents the information of each event showing a picture of the evader, a station map pointing to the door and wagon and, the date/hour when the infraction was committed. See Fig. 3.

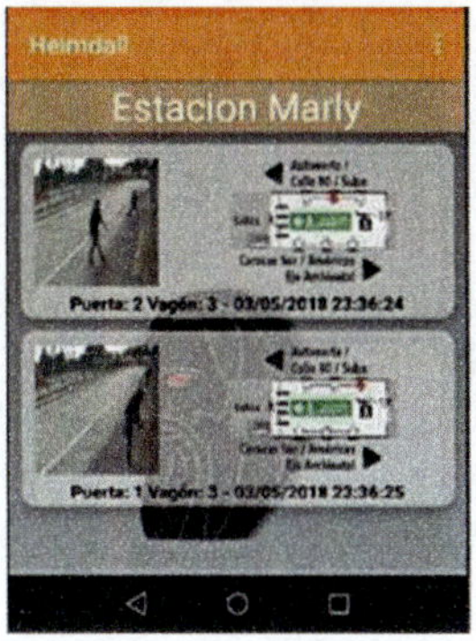

Fig. 3. Main screen of the app.

5.2 Web Application

The web application was designed for monitoring all the Transmilenio buses stations or just specific ones and for reporting generation. The information gathered from the mobile application can help Transmilenio to manage in a better way its staff in charge of the bus station security. It is worth mentioning that this module was only designed for the operation center, which has access to all evaders notifications of different bus stations (Fig. 4).

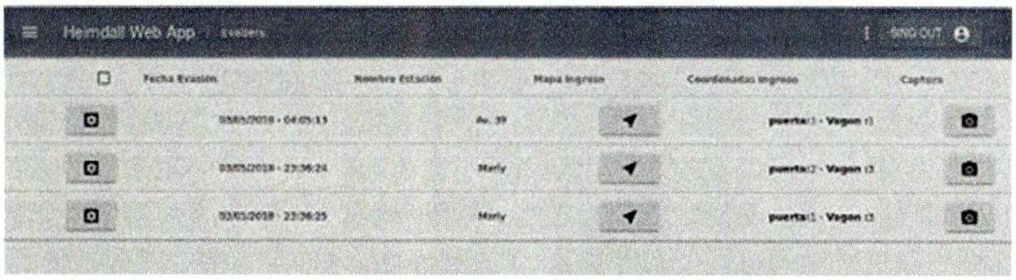

Fig. 4. Evaders web page.

For analysis purposes, the web page can generate useful information to the operation center like the amount of bus fares evaders at specific date, hour, station, wagon and/or location. Also with a large amount of data and a third-party analysis software, it will give the ability to predict where and when an evasion attempt will be.

5.3 Embedded System

The embedded device used for the prototype implementation was the Raspberry Pi 3 due to its hardware characteristics, its low cost and the OpenCV support [12–14]. The Raspberry pi is in charge of oversee the video capture, execute the evader recognition algorithm and data upload to the cloud. As aforementioned, each Raspberry device has three cameras connected to it. Once the evader recognition program has a video source, it extracts each frame to be analyzed by the algorithm, in this case we are using the k-nearest neighbors algorithm (k-NN) [15]. The k-NN algorithm is a background substractor. This algorithm was complemented with different threshold and masking techniques to improve the reliability detection level. Finally, when a possible bus fare evader is detected, the application will upload the image of the evader and all related information as door, wagon, station and date where the infraction was committed to the system. Figure 6 shows the original image and the result after applying the k-nearest neighbors algorithm. This algorithm was used not only by its low-medium hardware impact on the Raspberry but also its machine learning capabilities. As we can note the algorithm learns about the image environment and will extract the background highlighting moving objects (Fig. 5).

Fig. 5. Reporting web page.

Fig. 6. Pre and post-processed frame using the k-NN algorithm.

Fig. 7. Learning process of the k-NN algorithm.

Fig. 8. Shadows detection.

For detecting moving objects the algorithm works quite well (more than 85% successful detection). It also detects and removes the object shadows that represents future problems for evaders detection, since depending on the angle of the light this shadow can be larger by increasing the object size.

In addition, we have applied a transformation to each video-frame (Figs. 7, 8, 9 and 10) in order to extract the object contour and remove non-useful information. Once we have processed the image we proceed to take advantage of the Transmilenio lane for the evaders recognition, making use of a defined detection-zone (border line) in which the evader will be reported when the zone is crossed into the bus-station direction.

Fig. 9. Shadow removing.

Fig. 10. Masking and area detection.

5.4 NoSql Database

Due to the needs of real time notifications, the mobile application was integrated with firebase which is a cloud-hosted NoSQL database that enables the building of serverless apps [16]. The main firebase functions for this project are the trigger events, cloud storage, authentication methods, real time database and hosting.

5.5 Results

The main characteristics of the application is that it is intended to work in real time and control three video sources per embedded device.

To understand network requirements different test measuring the performance of network latency were performed by using different ISP and network-connections; such as, a conventional home connection, a university Internet connection and a 4G mobile connection. For the sake of shortness, the test concluded that for each camera it is necessary an internet connection of 2.8 Mbps for uploads and 4.5 Mpbs for downloads.

0 instances	GUI enabled	GUI disabled
CPU use	0.5%	0.3%
RAM use	120M/927M	44.7M/927M
SWAP use	6.50M/100M	0M/100M

(a) Detecting no evaders

1 instances	GUI enabled	GUI disabled
CPU use	28.2%	25.9%
RAM use	167M/927M	83.1M/927M
SWAP use	6.50M/100M	0M/100M

(b) Detecting one evaders

2 instances	GUI enabled	GUI disabled
CPU use	57.1%	53.1%
RAM use	208M/927M	120M/927M
SWAP use	6.50M/100M	0M/100M

(c) Detecting two evaders

3 instances	GUI enabled	GUI disabled
CPU use	81.9%	77.4%
RAM use	249M/927M	158M/927M
SWAP use	6.50M/100M	0M/100M

(d) Detecting three evaders

Fig. 11. Raspberry hardware load with different evader detection program instances running on it and both enabled and disabled GUI.

Another important aspect evaluated in the system is the hardware performance when running different evader recognition instances in the Raspberry Pi 3 device. For these tests, we have measure the amount of resources used for each number of instances and the impact that has the Graphical User Interface when it is enabled on the device (See Fig. 11). As we can note, three simultaneously instances of the evaders detection program are supported per device without overloading the CPU. It is important to note that this proposal is using a prototyping and, in a real implementation, we will need different devices to bear the environments conditions that also are more robust on hardware characteristics.

6 Conclusions

This work presents an innovative solution to solve the bus fare evasion control that is facing Transmilenio.

The video monitoring and alerting system proposed allows improving the Transmilenio bus-schedulling system detecting and notifying the bus fare evasion making use of a low-cost solution that can be deployed in any site of the Transmilenio bus-station.

One of the main advantages of this solution, among others, is that it is not tied to a specific hardware to get the input video, that is, it allows the integration of any camera-system providing a input source. In the same way, we can include any sensor to the system to extend and improve the presented solution.

This proposal takes full advantage of the available hardware resources of the Raspberry Pi 3 to provide a good performance in the video-image processing for the fare evasion. It is important to know that many of the problems that society is facing can be solved, or faced, using IoT approach and the cloud computing services.

Finally, although this proposal focuses on the bus fares evasions that Transmilenio faces right now, this implementation is not limited to that scenario, the system can fit to any customer requirements where a video monitoring and alerting system is needed.

7 Future Work

We are currently working on a new implementation of this prototype that uses the graphics processing unit (GPU) included in the embedded device to speed-up the analysis and increase the number of instances of the software handled per device.

References

1. Shanghai Manual: A Guide for Sustainable Urban Development in the 21st Century. China International Publishing Group (2011)
2. Levinson, H.S., Zimmerman, S., Clinger, J., Rutherford, S., Smith, R.L., Cracknell, J., Soberman, R.: Bus rapid transit. Transportation Research Board (2003)
3. Deng, T., Nelson, J.D.: Recent developments in bus rapid transit: a review of the literature. Transp. Rev. **31**(1), 69–96 (2011)
4. Infraestructura: Transmilenio.gov.co (2018). http://www.transmilenio.gov.co/Publicaciones
5. Cools, M., Fabbro, Y., Bellemans, T.: Identification of the determinants of fare evasion. Case Stud. Transp. Policy **6**(3), 348–352 (2018). ISSN 2213-624X
6. Local report: Injured persons for fare evasion at bus-station control. https://www.elespectador.com/noticias/bogota/en-2019-ya-van-tres-muertos-y-20-heridos-por-intentar-colarse-en-transmilenio-articulo-848633
7. de Smedt, F.: Pedestrian detection for real-life applications, 1st edn., p. 200. KU Leuven – Faculty of Engineering Technology, Belgium (2015)
8. Guarda, P., Galilea, P., Handy, S., Muñoz, J.C., de Dios Ortúzar, J.: Decreasing fare evasion without fines? A microeconomic analysis. Res. Transp. Econ. **59**, 151–158 (2016). ISSN 0739-8859
9. Barabino, B., Salis, S., Useli, B.: A modified model to curb fare evasion and enforce compliance: empirical evidence and implications. Transp. Res. Part A: Policy Pract. **58**, 29–39 (2013). ISSN 0965-8564
10. Delfau, J., Pertsekos, D., Chouiten, M.: Optimization of control agents shifts in public transportation: tackling fare evasion with machine-learning. In: 2018 IEEE 30th International Conference on Tools with Artificial Intelligence (ICTAI), Volos, pp. 409–413 (2018). https://doi.org/10.1109/ICTAI.2018.00070
11. Reddy, A.V., Kuhls, J., Lu, A.: Measuring and controlling subway fare evasion: improving safety and security at New York City transit authority. Transp. Res. Rec. **2216**(1), 85–99 (2011). https://doi.org/10.3141/2216-10
12. Minichino, J.: Learning OpenCV 3 Computer Vision with Python, 2nd edn. Packt Publishing, Birmingham (2015). p. 148, 179
13. OpenCV library: Opencv.org (2018). https://opencv.org
14. Coombs, J., Prabhu, R.: OpenCV on TI's DSP+ARM platforms: mitigating the challenges of porting OpenCV to embedded platforms. White paper, p. 2. Texas Instruments, Dallas, Texas (2018)

15. Okfalisa, Gazalba, I., Mustakim, Reza, N.G.I.: Comparative analysis of k-nearest neighbor and modified k-nearest neighbor algorithm for data classification. In: 2nd International conferences on Information Technology, Information Systems and Electrical Engineering (ICITISEE), Yogyakarta, pp. 294–298 (2017). https://doi.org/10.1109/ICITISEE.2017.8285514
16. Li, W., Yen, C., Lin, Y., Tung, S., Huang, S.: JustIoT internet of things based on the firebase real-time database. In: 2018 IEEE International Conference on Smart Manufacturing, Industrial and Logistics Engineering (SMILE), Hsinchu, pp. 43–47 (2018). https://doi.org/10.1109/SMILE.2018.8353979

An Approach on a Balance for a Small Humanoid Robot by Using Movable Mass

Xuan Tien Nguyen[1,3], Huy Hung Nguyen[1], Trong Hai Nguyen[3], Tan Tien Nguyen[2(✉)], and Thanh Phuong Nguyen[3]

[1] Faculty of Electric and Electronic Communication Engineering, Sai Gon University, Ho Chi Minh City, Vietnam
tien.nx@sgu.edu.vn, nghhung74@gmail.com

[2] Key Lab of Digital Control and System Engineering, University of Technology, VNU-HCM, Ho Chi Minh City, Vietnam
nttien@hcmut.edu.vn

[3] Faculty of Mechanical and Electrical Electronic, HUTECH, Ho Chi Minh City, Vietnam
haintmitu@yahoo.com, nt.phuong@hutech.edu.vn

Abstract. In this paper, an approach on a balance for a small humanoid robot by using movable mass. A standard humanoid robot can stand stably when its center of mass (CoM) is in a zero moment point (ZMP) zone. However, CoM's robot can go out the ZMP zone when the robot moves. To overcome this issues, a movable mass is added in the robot at the abdomen to balance. By controlling position of the movable mass, CoM's robot can be kept in the ZMP zone when it moves. Simulation results show effectiveness of the proposed method.

Keywords: Humanoid robot · Movable mass · Balance

1 Introduction

A significant problem in walking control of a biped humanoid robot is balance. To retain the balance, a robot's Center of Mass (CoM) is demanded and this demand is usually difficult to obtain. There existed several researches in this field. In [1], Kwon and Oh proposed a method to estimate the Center of Mass of a humanoid robot by utilizing a closed-loop observer to extract reliable estimates of humanoid CoM. Then a discrete Kalman filter was adopted as the estimation framework to minimize estimation error. Kim et al. in [2] calculated CoM's the robot from the force/torque sensors integrated with the foot in reality. The CoM location had a deviation from the ideal CoM location and the ZMP controller was proposed to adjust the CoM's robot. Kenji H. and Atsuo T. focused on the determination of an algorithm finding the real position of the Center of Mass (CoM) of a biped humanoid robotic platform using only 2 force/torque sensors located on the feet of the robot in [3].

The humanoid robot can stand or move stably as its CoM is always in the ZMP zone. There are several strategies to track CoM's robot to ZMP zone such as a control strategy using COG Jacobian [4], a torso position compliance control strategy [5] and so on.

D. F. Cortes Tobar et al. (Eds.): AETA 2019, LNEE 685, pp. 681–691, 2021.
https://doi.org/10.1007/978-3-030-53021-1_68

Another approach was shown in [6] by controlling generalized inertial forces derived from the linear and angular momentum of the whole the humanoid robot. In [7, 8], Suleiman et al. described a dynamic algorithm allowing the real-time calculation of the gradient function of joint torques with respect of joint position, velocity and acceleration to smooth the shape of ZMP and avoid oscillations from robot.

This paper proposes a mechanical hardware adopted in the humanoid robot by a movable mass at the abdomen to balance as walking. The proposed method focuses on controlling a mass location inside the robot in order to CoM's robot tracks to ZMP zone when the robot moves.

2 Dynamics of Robot with Movable Mass

2.1 Describe Small Humanoid Robot

In this paper, a small humanoid robot includes 16° of freedom at the arms and legs joints, and 1° of freedom at the abdomen to balance. It is 31.7 cm tall and 2.1 kg weight. The robot utilizes RC servo motors in the form of module to control joints as shown in Fig. 1.

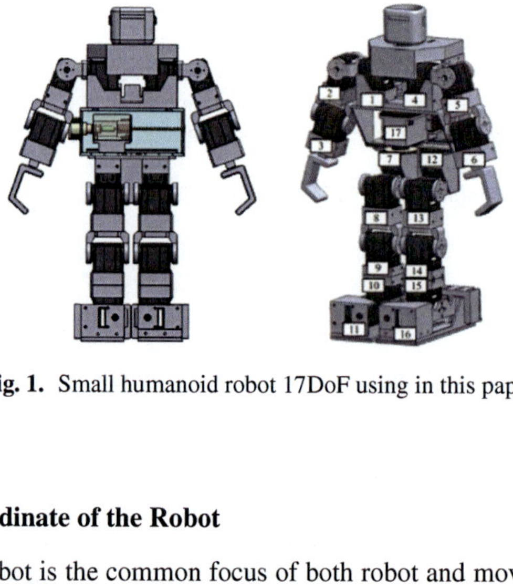

Fig. 1. Small humanoid robot 17DoF using in this paper

2.2 Central Coordinate of the Robot

The focus of the robot is the common focus of both robot and movable mass, denoted by CoM (Center of Mass).

According to the document [8] the focus coordinates are calculated

$$x_{com} = \frac{\sum_i m_i x_i}{\sum_i m_i} \tag{1}$$

$$y_{com} = \frac{\sum_i m_i y_i}{\sum_i m_i} \tag{2}$$

$$z_{com} = \frac{\sum_i m_i z_i}{\sum_i m_i} \tag{3}$$

In order for the stable robot to not fall, the ZMP point must be placed in the base and center of gravity (COP) at the foot to coincide with ZMP [9]. ZMP (Zero moment point) is the point on the ground where the total torque of the forces acting on the robot is zero. However, the robot model designed above has not been met. Therefore, movable mass is added to the robot model and the focus of the robot is the common focus of both robot and movable mass.

In the method presented in this paper, the movable mass work to balance weight when the robot moves, so for simplicity, we convert the system into two main parts, robot (m_R) and movable mass (m_D). Then, the CoM focus of the robot is calculated:

$$x_{com} = \frac{m_R x_R + m_D x_D}{m_R + m_D} \tag{4}$$

$$y_{com} = \frac{m_R y_R + m_D y_D}{m_R + m_D} \tag{5}$$

$$z_{com} = \frac{m_R z_R + m_D z_D}{m_R + m_D} \tag{6}$$

2.3 Dynamics of Robot with Movable Mass

When the robot is supported by one leg, the dynamics of the robot can be approximated by a simple 3D inverted pendulum whose base is the foot of the robot and its head is CoM of the robot [10]. Figure 2 represent a 3D model of an inverted pendulum based on 3 parameters: the length of the pendulum r can be changed, rotation angle θ_r around x axis and rotation angle θ_p around y axis in the Cartesian coordinate system (Oxyz). Therefore, the position of the inverted pendulum $C(x_c, y_c, z_c)$ in the Cartesian coordinates is determined by vector $\mathbf{q} = [\theta_r, \theta_p, r]^T$.

$$x_c = r\sin\theta_p = rS_p \tag{7}$$

$$y_c = r\sin\theta_r = rS_r \tag{8}$$

$$z_c = r\cos\theta_r\cos\theta_p \approx r\sqrt{1 - \sin^2\theta_r - \sin^2\theta_p} = rD \tag{9}$$

where, $S_r = \sin\theta_r$, $S_p = \sin\theta_p$, $C_r = \cos\theta_r$, $D = \sqrt{1 - \sin^2\theta_r - \sin^2\theta_p}$, $C_p = \cos\theta_p$

Assume $[\tau_r, \tau_p, f]^T$ are defined as the actuator torques and force associated with the variables $[\theta_r, \theta_p, r]^T$ in 3D inverted pendulum, as show in Fig. 2.

The Lagrange of the 3D inverted pendulum is:

$$L = \frac{1}{2}m\left(\dot{x}_c^2 + \dot{y}_c^2 + \dot{z}_c^2\right) - mgz_c \tag{10}$$

where m is the total mass of the robot, g is the gravitational acceleration.

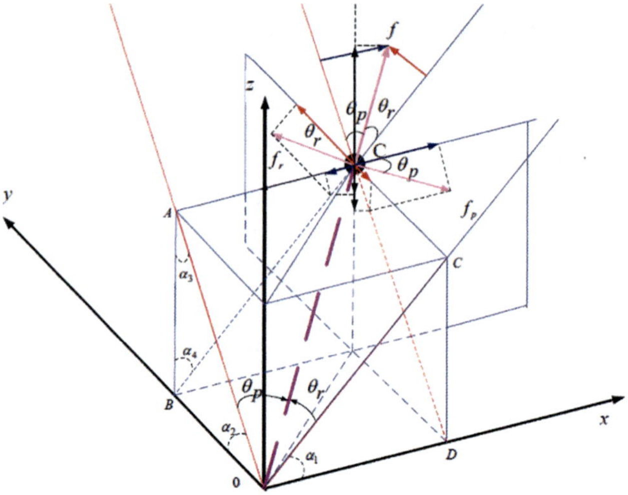

Fig. 2. The three dimension (3D) inverted pendulum

Based on the Lagrange's equation, the dynamics of the 3D inverted pendulum can be obtained in the Cartesian coordinate as follows:

$$m\begin{bmatrix} \ddot{x}_c \\ \ddot{y}_c \\ \ddot{z}_c \end{bmatrix} = \begin{bmatrix} -\frac{S_r S_p}{rC_r} & \frac{C_p}{r} & S_p \\ -\frac{C_r}{r} & \frac{S_r S_p}{rC_p} & S_r \\ -\frac{S_r D}{rC_r} & \frac{S_p D}{rC_p} & D \end{bmatrix} \begin{bmatrix} \tau_r \\ \tau_p \\ f \end{bmatrix} + \begin{bmatrix} 0 \\ 0 \\ -mg \end{bmatrix} \tag{11}$$

Then, the dynamics equation of the inverted pendulum along the x axis and along the y axis:

$$m(z_c\ddot{x}_c - x_c\ddot{z}_c) = \tau_y + mgx_c \tag{12}$$

$$m(-z_c\ddot{y}_c + y_c\ddot{z}_c) = \tau_x - mgy_c \tag{13}$$

where, $\tau_x = \tau_r \frac{D}{C_r}$ and $\tau_y = \tau_p \frac{D}{C_p}$ is the torque around x axis and y axis.

There are many classes of moving pattern of the inverted pendulum. For selecting one of them, a constraint is applied to limit the motion of the inverted pendulum. That is, the motions of the COM of the inverted pendulum are constrained on the plane whose normal vector is $\mathbf{v_{cp}} = [k_x, k_y, -1]^T$, and z intersection is z_{cd} as shown in Fig. 3.

It is assumed that the constraint plane intersects the z axis at $Q(0, 0, z_{cd})$, the coordinates of CoM is $C(x_c, y_c, z_c)$. We have $\mathbf{v_{cp}}$ is perpendicular to vector $\overrightarrow{\mathbf{QC}}$:

$$z_c = x_c k_x + y_c k_y + z_{cd} \tag{14}$$

where, k_x, k_y, z_{cd} are constants.

When biped robot walks on rugged terrain, the normal vector $\mathbf{v_{cp}}$ of the constraint plane should be perpendicular to the slope of the ground, and z intersection z_{cd} in the

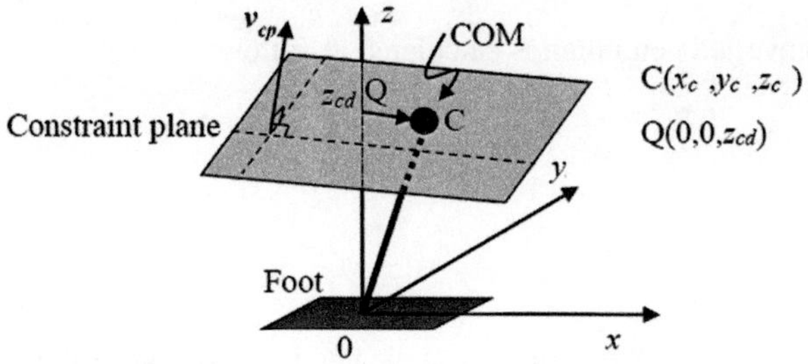

Fig. 3. Motion of inverted pendulum on constraint plane [11]

coordinate system is set as distance between the CoM and the ground:

$$\ddot{z}_c = \ddot{x}_c k_x + \ddot{y}_c k_y \quad (15)$$

Substituting into the dynamics equation above, we get the equation to represent the motion of the 3D inverted pendulum in the constraint plane as follows:

$$\ddot{y}_c = \frac{g}{z_{cd}} y_c - \frac{1}{m z_{cd}} \tau_x - \frac{k_x}{z_{cd}} (x_c \ddot{y}_c - \ddot{x}_c y_c) \quad (16)$$

$$\ddot{x}_c = \frac{g}{z_{cd}} x_c + \frac{1}{m z_{cd}} \tau_y + \frac{k_y}{z_{cd}} (x_c \ddot{y}_c - \ddot{x}_c y_c) \quad (17)$$

It is assumed that the biped robot walks on the flat floor with the horizontal plane. In this case: $k_x = 0$, $k_y = 0$, CoM of the pendulum moves on the horizontal plane which has height $z_c = z_{cd}$.Then:

$$\ddot{x}_c = \frac{g}{z_{cd}} x_c + \frac{1}{m z_{cd}} \tau_y \quad (18)$$

$$\ddot{y}_c = \frac{g}{z_{cd}} y_c - \frac{1}{m z_{cd}} \tau_x \quad (19)$$

(x_{zmp}, y_{zmp}) is defined as the location of zero moment point (ZMP) on the floor as shown in Fig. 4:

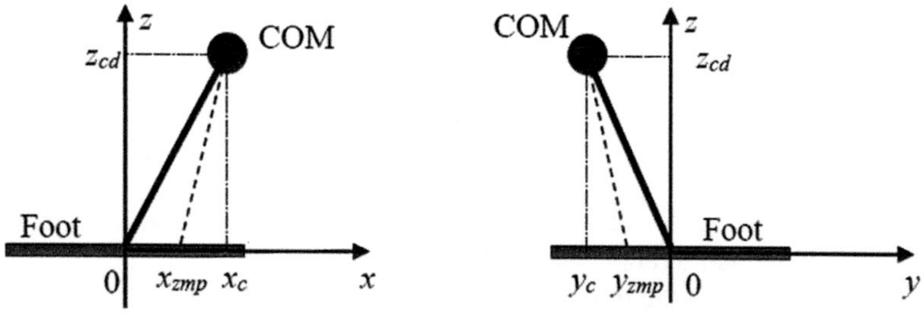

Fig. 4. ZMP of inverted pendulum

ZMP of 3D inverted pendulum is calculated as follows:

$$x_{zmp} = -\frac{1}{mg}\tau_y \tag{20}$$

$$y_{zmp} = \frac{1}{mg}\tau_x \tag{21}$$

In which x_{zmp}, y_{zmp} is the value that indicates the position of the foot is changed when considering the ground. Replace the value of x_{zmp}, y_{zmp} in Eqs. (18) and (19) we get the relationship between COM and ZMP position set.

$$\ddot{x}_c = \frac{g}{z_{cd}}x_c + \frac{1}{mz_{cd}}\tau_y = \frac{g}{z_{cd}}(x_c - x_{zmp}) \tag{22}$$

$$\ddot{y}_c = \frac{g}{z_{cd}}y_c - \frac{1}{mz_{cd}}\tau_x = \frac{g}{z_{cd}}(y_c - y_{zmp}) \tag{23}$$

Thus, it yields

$$x_{zmp} = x_c - \frac{z_{cd}}{g}\ddot{x}_c \tag{24}$$

$$y_{zmp} = y_c - \frac{z_{cd}}{g}\ddot{y}_c \tag{25}$$

When the robot moves with slow speed, it is considered accelerations in both x and y directions go to zero, $\ddot{x}_c \approx 0$, $\ddot{y}_c \approx 0$. From Eqs. (24) and (25), it obtains

$$x_{zmp} = x_c \tag{26}$$

$$y_{zmp} = y_c \tag{27}$$

From Eqs. (22), (23), (26) and (27), it yields

$$\ddot{y}_{zmp} = \ddot{y}_c = \frac{g}{z_{cd}}y_c - \frac{1}{mz_{cd}}\tau_x \tag{28}$$

$$\ddot{x}_{zmp} = \ddot{x}_c = \frac{g}{z_{cd}}x_c + \frac{1}{mz_{cd}}\tau_y \tag{29}$$

3 Simulation Result

Simulated with MATLAB to find an effect of movable mass balance in a stable step control for small humanoid robots like human walking.

First, we will generate the trajectory of the robot leg for x, y and z directions. But because we will control the movable mass to balance this movement, we only need to orbit in the x and z directions. Trajectory moves the robot in the x and z directions over time [7]:

$$\tilde{x}_A(t) = (b+f)\left[\frac{t-t_1}{t_2-t_1} - \frac{1}{2\pi}\sin\left(2\pi\frac{t-t_1}{t_2-t_1}\right)\right] - b \quad (30)$$

In order to walk flexibly, the robot's legs will fall to a 18 mm section, in other words, the hips are about 140 mm away from the ground.

The duration of a step will be 1.15 s, a cycle consisting of two steps will be 2.30 s. In this case, we plan 6 steps. The first swing leg from the stationary position is the right leg. However, the movement in the X and Z directions of the legs and hips will be delayed compared to the movement in the Y direction of the hip by about 0.2 s before each step to ensure the time of the Stance phase and Swing Phase of one leg is approximately 40% and 60% compared to the walking cycle.

The basic parameter for generating walking pattern will be presented in the following Table 1:

Table 1. Basic parameter

Parameter	Step 1	Step 2
b	0 mm	30 mm
f	30 mm	30 mm
S_y	15 mm	15 mm
V_y	0 mm/s	10 mm/s

where,

t_1: time when the orbit begins when the step-foot places behind the base-foot
t_2: the end of the orbit when the step-foot places in front of the base-foot
Δt: the duration of one step, $\Delta t = t_2 - t_1 = 1.15$ s
b: the absolute distance between two ankles at time t_1
f: the absolute distance between two ankles at time t_2
S_y: the movement distance of the hip in the Y direction
V_y: the velocity in the Y direction

Size of the movable mass: 50 mm × 50 mm × 30 mm
Mass of the movable mass: 506 g = 0.506 kg
When mounted on the robot frame
The translational distance of the movable mass: ≈60 mm
Lead screw: Picth 2 mm

3.1 Simulation Results Without Using Balance System

In case of without a movable mass, for the robot to walk stably according to ZMP standard, the robot's hips must move in a Y direction around 25 mm to 35 mm to ensure balance when walking (Fig. 5):

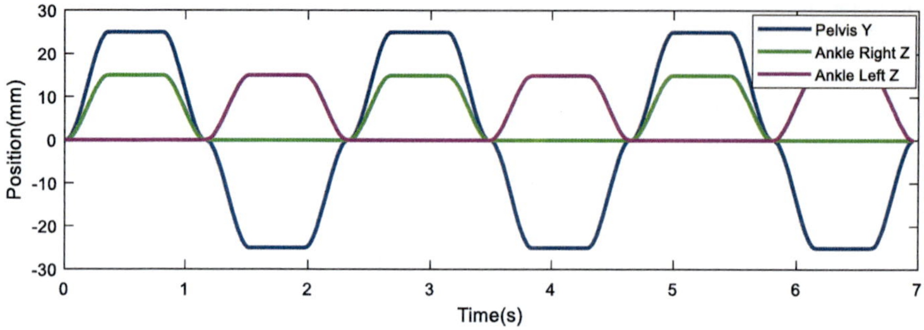

Fig. 5. Graph of coordinates of robot pelvis in the Y direction and ankles in the Z direction

Based on the graph, we see that, in each step, the robot leg must be tilted before lifting its foot.

3.2 Simulation Results When Using Balance System

In case of having a movable mass, assume the robot's hip do not move in the Y direction, only the movement of the movable mass in the Y direction about 30 mm as shown in the graph below (Fig. 6).

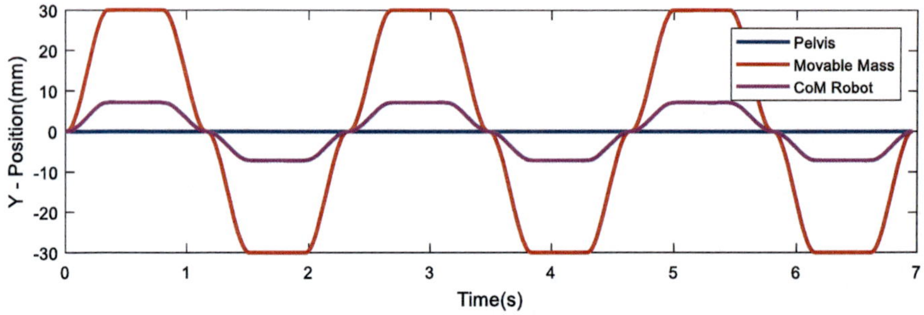

Fig. 6. Graph of coordinates of robot hip, movable mass and CoM in the Y direction

The coordinate of the center of mass CoM of the system consists of robot and movable mass is the pink graph, built on the function:

$$y_{com} = \frac{m_R y_R + m_D y_D}{m_R + m_D} \tag{31}$$

where,

m_R: mass of robot, $m_R = 1,6\,\text{kg}$
m_D: mass of movable mass, $m_D = 0,5\,\text{kg}$
y_R: coordinates of robot hip in the Y direction
y_D: coordinates of movable mass in the Y direction

We have coordinates of CoM reach maximum in the Y direction:

$$y_{com\_max} = \frac{m_R y_R + m_D y_D}{m_R + m_D} = \frac{1,6.0 + 0,5.30}{1,6 + 0,5} = 7,15\,\text{mm}$$

$$\text{CoM}_{\max} < \frac{\text{ft}}{2} = 8\,\text{mm}$$

where ft is the distance between the two feet of the robot, therefore not responding balance according to ZMP standard, ft = 16 mm.

Several studies had shown that ZMP of robot coincides with the projection of the mass of the center of gravity on the base at standing status. When a human moves, the projection of the center of mass may be outside the base at several times but the ZMP is always in the base [12]. Thus, if the ZMP is in the base, the human robot can go without falling. In Fig. 7, ZMP indicated by blue line is outside the base of the human robot indicated by red line. It means that the human robot is easy to fall when it walks.

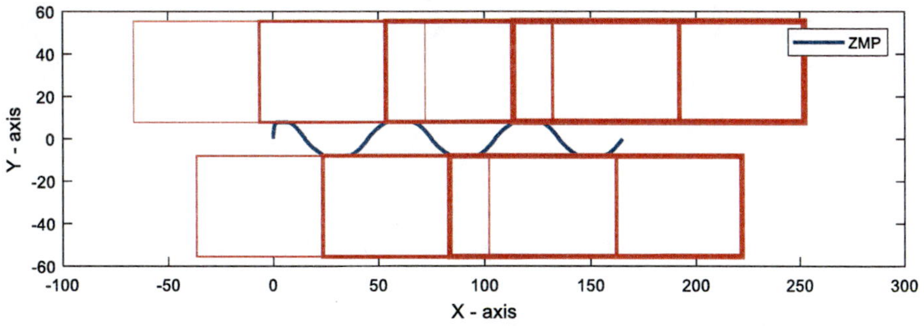

Fig. 7. Graph of checking ZMP standard

To responding the balance condition according to the ZMP standard, we can either increase the distance moved in the Y direction of the movable mass, or increase the weight of the movable mass, or allow the robot's hips to move in the Y direction about a distance instead of the hip do not moving as assume.

Since space at the robot's abdomen is limited, it is not possible to increase the distance moved of the movable mass, as well as to increase the weight of the movable mass. Therefore, we allow the robot's hip to move about 15 mm in the Y direction so that the robot achieves the balance conditions according to ZMP standard. So that (Figs. 8, 9):

$$y_{com\_max} = \frac{m_R y_R + m_D y_D}{m_R + m_D} = \frac{1,6.15 + 0,5.30}{1,6 + 0,5} = 18,57\,\text{mm}$$

$$\text{CoM}_{\max} > \frac{\text{ft}}{2} = 8\,\text{mm}$$

In Fig. 10, ZMP indicated by blue line is inside the base of the human robot indicated by red line. It means that the human robot can walk without falling.

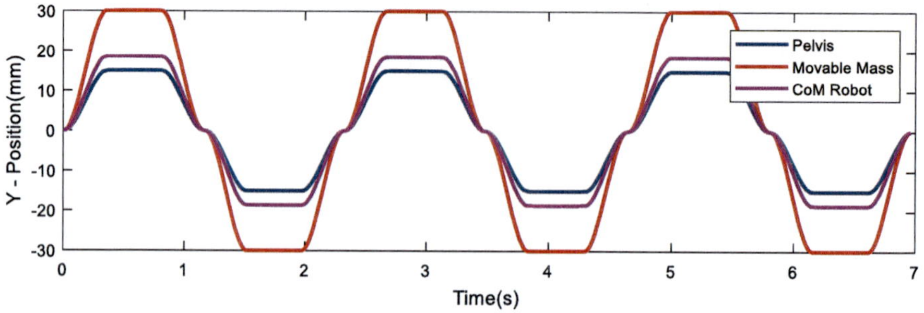

Fig. 8. Graph of coordinates of robot hip, movable mass and CoM in the Y direction

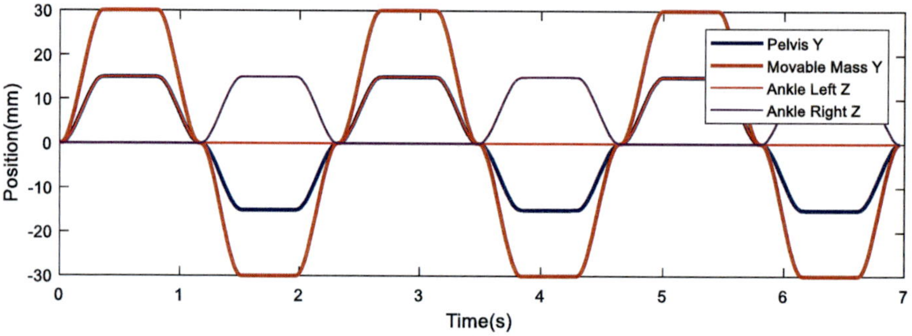

Fig. 9. Graph of coordinates of robot hips in the Y direction and ankles in the Z direction

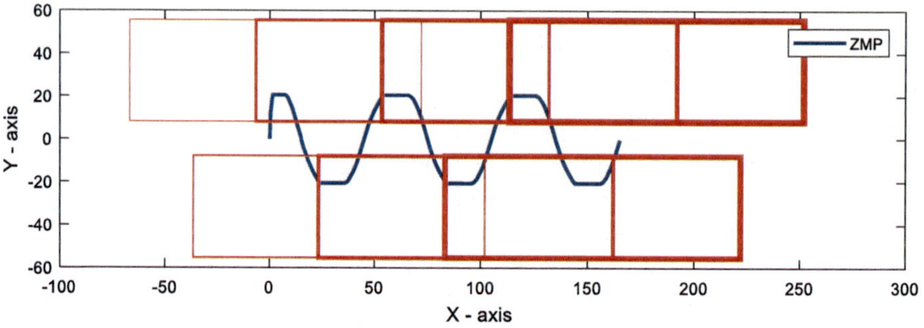

Fig. 10. Graph of checking ZMP standard

Based on the graph, we see that, in each step, the robot leg is also must to tilt before lifting its foot in the same way as when there is no movable mass. However, there is a balanced action of movable mass so the robot does not need to tilt the hip and ankles as much as there is no movable mass.

Thus, through two options that do not use movable mass and use movable mass, we found that when using movable mass will help the robot to tilt the hip and tilt the ankle less when walking, so the walking pattern of the robot will be closer to the human gait.

4 Conclusion

This paper offered a study of balancing small humanoid robot. Simulation results showed that movable mass had the ability to balance and keep stably for robot as walking. These results were the basis for conducting the robot control and the other studies on balancing when walking for the small humanoid robot in the future.

Acknowledgements. This work was fully supported by Key Laboratory of Digital Control and System Engineering (DCSELAB), HCMUT, VNU-HCM under grant number TX2019-20B-01.

References

1. Kwon, S., Oh, Y.: Estimation of the center of mass of humanoid robot. In: Proceedings of the International Conference on Control, Automation and Systems (2007). (in Korea)
2. Kim, J.H., Kim, J.Y., Oh, J.H.: Adjustment of home posture of biped humanoid robot using an inertial sensor and force torque sensors. Proceedings of the 2007 IEEE/RSJ International Conference on Intelligent Robots and Systems, 29 October-2 November 2007, San Diego, CA, USA (2007)
3. Hashimoto, K., Takanishi, A.: A method for the calculation of the effective center of mass of humanoid robots. In: International Conference on Humanoid Robots Bled (2011)
4. Sugihara, T., Nakamura, Y.: Whole-body cooperative balancing of humanoid robot using COG Jacobian. In: Proceedings International Conference on Intelligent Robots and Systems (2002)
5. Nagasaka, K., Inaba, M, Inoue, H.: Stabilization of dynamic walk on a humanoid using torso position compliance control. In: Proceedings 17th Annual Conference on Robotics Society of Japan (1999)
6. Nunez, V., Gauthier, N.N., Yokoi, K., Blazevic, P., Stasse, O.: Inertial forces posture control for humanoid robots locomotion. In: Hackel, M., (ed.) Humanoid Robots: Human-like Machines, p. 642. Itech, Vienna (2007)
7. Suleiman, W., Yoshida, E., Laumond, J.P., Monin, A.: On humanoid motion optimization. In: 7th IEEE-RAS International Conference on Humanoid Robots (2007)
8. Suleiman, W., Yoshida, E., Laumond, J.P., Monin, A.: Optimizing humanoid motions using recursive dynamics and lie groups. In: 3rd International Conference on Information and Communication Technologies: From Theory to Applications (2008)
9. Han, J.: Bipedal walking for a full-sized humanoid robots utilizing sinusoidal feet trajectories and its energy consumption, PhD thesis, Virginia Polytechnic Institute and State University, USA (2012)
10. Kajita, S., Hirukawa, H., Harada, K., Yokoi, K.: Introduction to Humanoid Robotsics. Springer, Heidelberg (2014)
11. Kajita, S., Kanehiro, F.o., Kaneko, K., Fujiwara, K., Harada, K, Yokoi, K., Hirukawa, H.: Biped walking pattern generation by using preview control of zero-moment point. In: International Conference on Robotics and Automation (2003)
12. Nguyen, T.P.: Stable walking control of biped robot, Thesis for Degree of Doctor of Philosophy, Pukyong National University, Korea (2008)

Design of Palletizing Robot Using Series Elastic Actuator

Tan Tien Nguyen[1(✉)], Quang Dung Le[1], Thien Phuc Tran[1], and Sang Bong Kim[2]

[1] Key Lab of Digital Control and System Engineering, Ho Chi Minh City University of Technology, VNU-HCM, Ho Chi Minh City, Vietnam
{nttien,ttphuc.rectie}@hcmut.edu.vn, lqdung95ntp@gmail.com
[2] Pukyong National University, Busan, Korea
kimsb@pknu.ac.kr

Abstract. For along time, palletizing robot use stiffness motor to perform its movement. Its motor will be damaged easily when robot stops suddenly. This paper is about new approachment, applying series elastic actuator to control the robot because of its pros: the vibration will be controlled, controlling force without force sensor which make the robot low cost and easy to control.

Keywords: Palletizing robot · Series Elastic Actuator · FSEA

1 Overview

For along time, palletizing robot uses stiffness motors which placed on the machine base leading to the significant reduction of mass of the robot arm, smaller inertia moment. Therefore, this allows the robot to operate at high speed, increase the load capacity. By operating at high speed, the robot will be vibrated when stop suddenly which makes its motors damage easily.

Series Elastic Actuator (SEA) is an actuator combined with elastic elements using for force sensing and was developed by MIT University (1995). With many traits such as anti-shock ability when having impacts applying and transforming from controlling force to become controlling position result in the simplification of the control system. SEA has many structures which are Force-sensing Series Elastic Actuator (FSEA) and Reaction Force-sensing Elastic Actuator (RFSEA). The SEA, however, was not studied to be applied to control robot arm even though it is mostly applied in controlling humanoid.

Replacing motors by SEAs, the vibration that made when robot stops suddenly will be controlled by displacement of springs of the SEAs, therefore, the robot will operate safer.

2 Problem Setting

The designed robot has joint 1 is prismatic which increases the operational range of the robot and revolute joint 2 and 3 are controlled by SEA through the corresponding moment arms. In this paper, joint 1 will be ignored (Figs. 1 and 2).

With the properties of each series, SEA was designed to be FSEA combined with the advantages of RFSEA (Fig. 3).

D. F. Cortes Tobar et al. (Eds.): AETA 2019, LNEE 685, pp. 692–702, 2021.
https://doi.org/10.1007/978-3-030-53021-1_69

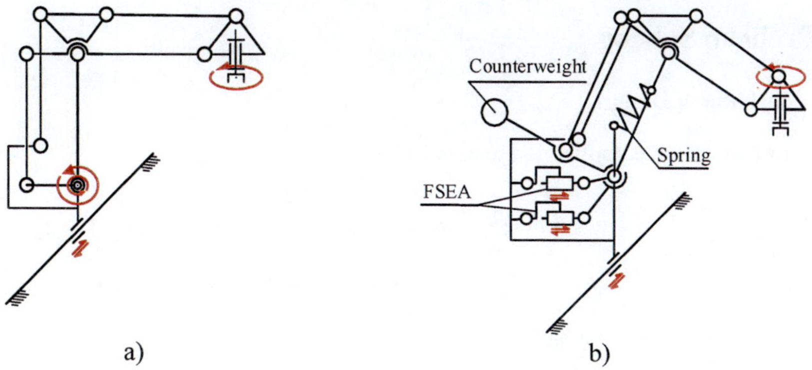

Fig. 1. a) The schematic of the normal palletizing robot, b) The schematic of palletizing robot using balancing mechanisms and SEA.

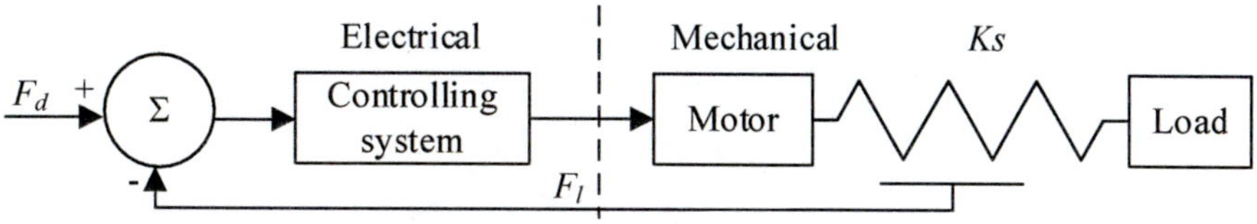

Fig. 2. The schematic of the SEA.

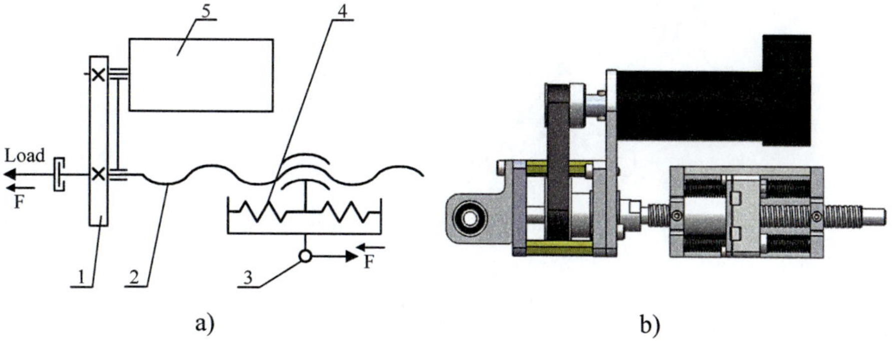

Fig. 3. a) The schematic of SEA: 1) Timing Belt, 2) Ball Screw, 3) Base 4) Compressed springs, 5) Motor, b) 3D model of SEA

To reduce the load for SEA, the balancing mechanism for the robot will be study and designed to passive balance for the robot including a spring to balance link 2 and a counterweight to balance link 3.

3 Problem Solving

3.1 Mechanical Design

Sum of moment force applied to the axis 1 (Fig. 4):

$$\sum M_1 = \sin\theta_{31}\left(m_2 g\frac{l_2}{2} + m_3 g l_2 + m_{21} g\frac{l_2}{2} - K(x - x_0)\frac{ab}{x}\right) + \sin\theta_{32}(m_3 g l_{32} - m_{21} g l_{31} - m_C g l_C) \quad (1)$$

For the robot being passive balance:

$$\sum M_1 = 0 \Leftrightarrow \begin{cases} m_2 g\frac{l_2}{2} + m_3 g l_2 + m_{21} g\frac{l_2}{2} - K(x - x_0)\frac{ab}{x} = 0 \\ m_3 g l_{32} - m_{21} g l_{31} - m_C g l_C = 0 \end{cases} \quad (2)$$

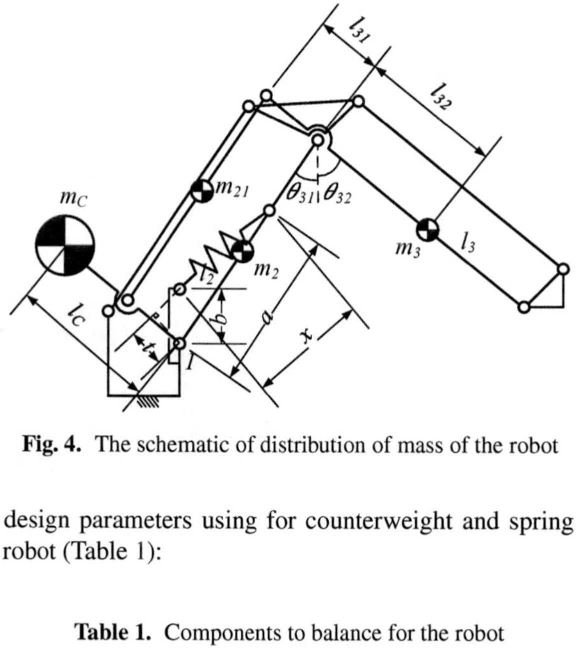

Fig. 4. The schematic of distribution of mass of the robot

Hence, the design parameters using for counterweight and spring components to balance for the robot (Table 1):

Table 1. Components to balance for the robot

Components	Parameter
Spring	Stiffness: $K = \left(\frac{1}{2}m_2 + m_3 + \frac{1}{2}m_{21}\right)\frac{g l_2}{ab}$ (2)
Counterweight	Mass: $m_C = \frac{m_3 l_{32} - m_{21} l_{31}}{l_C}$ (3)

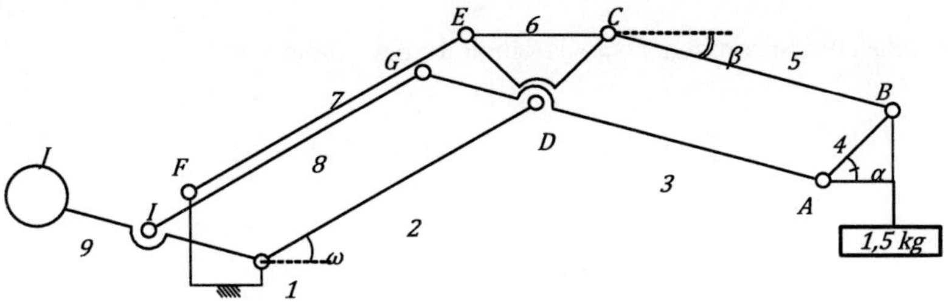

(a)

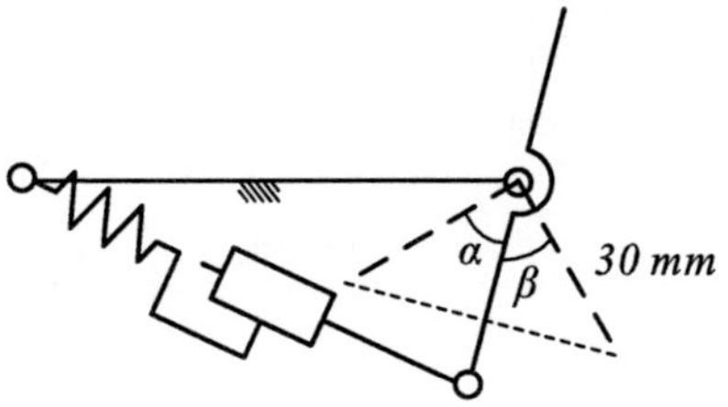

(b)

Fig. 5. a) The robot configuration reaching the farest while lifting the maximum load, b) The schematic of controlling joint 2 and joint 3.

Given Parameters

Robot requirements:

- Range: 500 mm
- Maximum load: 1.5 kg
- End-effector accuracy: ±2.5 mm
- Degrees of freedom: 3.

The dimensions of link 2 and 3 of the robot arm were scaled from the IRB 460 of ABB Company. Therefore, the dimensions of the robot arm will be: link 1: $l_1 = 500$ mm, link 2: $l_2 = 240$ mm, link 3: $l_3 = 260$ mm, the length of the moment arm: $d = 80$ mm, the length of the beam having counterweight: $l_{dt} = 180$ mm, the moment arm connect to SEA: $r = 30$ mm.

Designing Link 2 and Link 3

To design FSEA, the following input parameters need calculating: power, operating path, … The power of FSEA must satisfied the requirement of holding the robot when it reach the farest with the maximum load (1.5 kg) .

With the maximum rotational speed of 90°/s for each joint, the parameters will be calculated as follow (Table 2):

Table 2. The parameters of joint 2 and joint 3

	Maximum moment	Minimum thrust of FSEA	Minimum power of FSEA
Joint 2	$M_2 = 3,06$ Nm	$F_2 = 144,25$ N	$P_2 = 7,4$ W
Joint 3	$M_3 = 3,7$ Nm	$F_3 = 174,42$ N	$P_3 = 4,8$ W

Thus, the chosen components to assemble FSEA (Table 3):

Table 3. Components of FSEA

Components	Parameter	Quantity
Motor RE25	Power: 20 W Norminal speed: 9660 rpm	1
Timing Belt	Belt type: $S2M - 140$ Ratio: 0, 5	1
Ball screw	Pitch: 2 mm Thread length: 60 mm	1
Spring	Stiffness: 5000 N/m	8

3.2 Modeling FSEA

To control FSEA, modeling FSEA will be constructed as (Fig. 6):

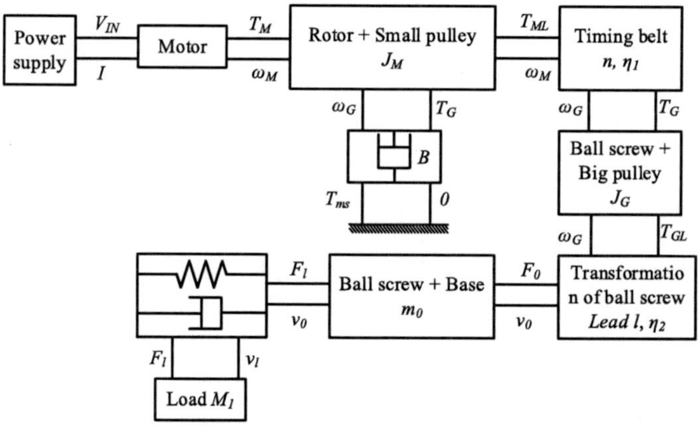

Fig. 6. The scheme of the links between components that make FSEA.

The scheme of the links between components that make FSEA (Fig. 7)

$$\frac{v_0(s)}{U_{virt}(s)} = \frac{K_T \frac{l}{n2\pi}}{J_{td}Ls^2 + (RJ_{td} + LB)s + (RB + K_{EMF}K_T)} \tag{5}$$

With 2 results gained from mathematical model (the parameter taken from the datasheet of the motor) and system identifying by toolbox Identification of Matlab (input the voltage and measure the output velocity):

Figure 5 shows that the estimated model and identification model are nearly equivalent (Fig. 8).

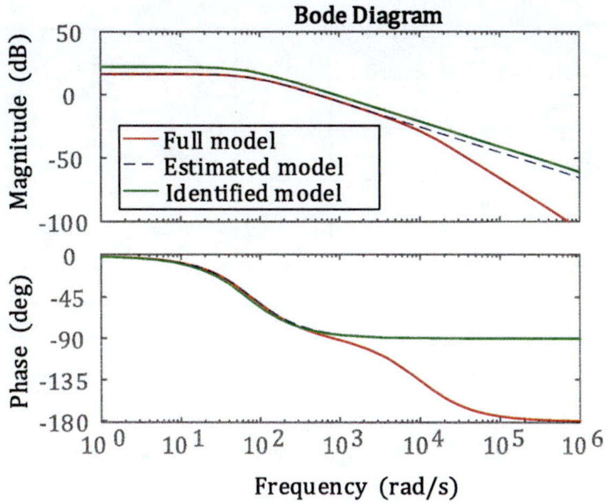

Fig. 7. Bode diagram of full model (second order), estimated model (first order) and identification model (first order)

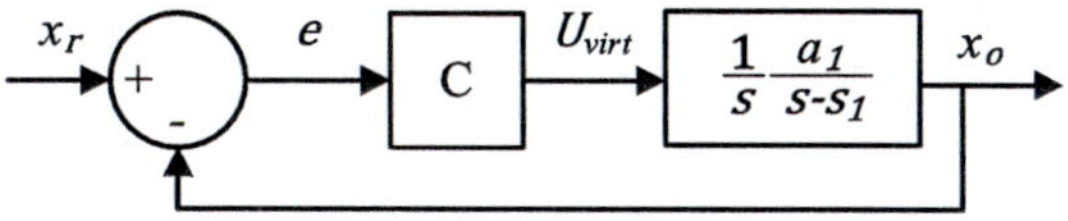

Fig. 8. Schematic of position control

PD controller was applied on position control of the motor:

$$C = K_P + K_D s \tag{4}$$

Requirement of the controller:

- Settling – time: $t_{qd} < 0,05$ s
- Percent of Overshoot: $POT \leq 10\%$

After calculating, $K_P = 195.4$ and $K_D = 0.9$ (Fig. 9).

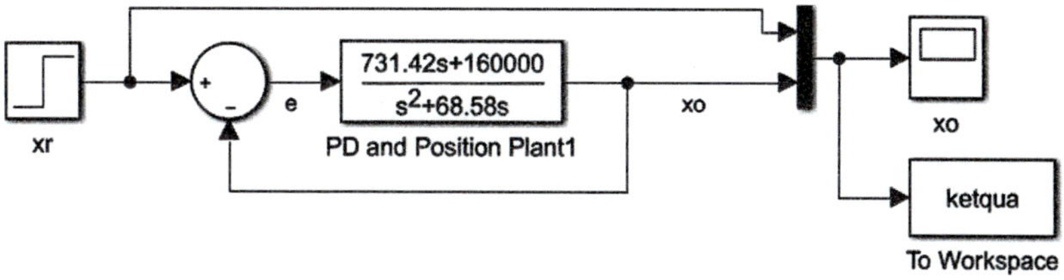

Fig. 9. Schematic of controller and plan in Simulink Matlab

Steady state response simulation in Simulink Toolbox of Matlab (Fig. 10.a):

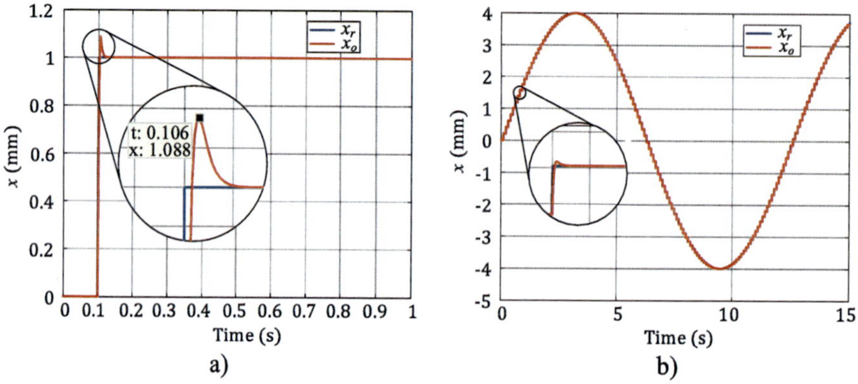

Fig. 10. a) Steady state respond satify the requirement of the controller, b) Decrete sine input and respond of the PD controller

Applying motor's position control on end – effector's position control of FSEA with decrete sine signal input (Fig. 10.b)

3.3 Modeling Robot

To control the robot, its modeling is calculated (Fig. 11):

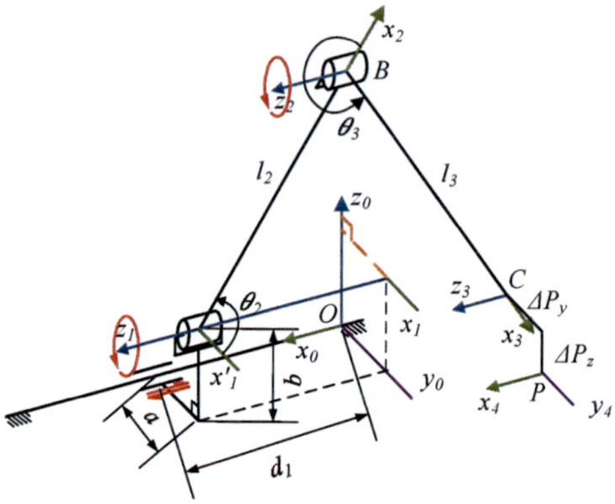

Fig. 11. Coordinates of robot arm type PRR

A D – H parameter table of robot arm is established (Table 4):

Table 4. D – H parameter

	$a_i(mm)$	α_i	$d_i(mm)$	θ_i
1	a	90°	b	90°
2	l_2	0°	d_1	θ_2
3	l_3	0°	0	$180° + \theta_3$

Transformation matrix

$$
{}_3^0T = \begin{bmatrix} 0 & 0 & 1 & d_1 \\ -c_{23} & s_{23} & 0 & a + l_2c_2 - l_3c_{23} \\ -s_{23} & -c_{23} & 0 & b + l_2s_2 - l_3s_{23} \\ 0 & 0 & 0 & 1 \end{bmatrix} \quad (5)
$$

where,

$c_i = \cos(\theta_i)$
$s_i = \sin(\theta_i)$
$c_{23} = \cos(\theta_2 + \theta_3)$
$s_{23} = \sin(\theta_2 + \theta_3)$
$0 \leq d_1 < 500\text{mm}$
$30^\circ \leq \theta_2 \leq 120^\circ$
$45^\circ \leq \theta_3 \leq 145^\circ$

Kinematic of the robot arm

$$
P = \begin{bmatrix} x_P \\ y_P \\ z_P \end{bmatrix} = \begin{bmatrix} d_1 \\ a + l_2c_2 - l_3c_{23} + \Delta P_y \\ b + l_2s_2 - l_3s_{23} - \Delta P_z \end{bmatrix} (mm) \quad (6)
$$

Inverse kinematic of the robot:

$$
\begin{cases} d_1 = x_p \\ \theta_2 = atan2\left(\pm\sqrt{1 - \left(\frac{A(l_2 - l_3c_3) - Bl_3s_3}{A^2+B^2}\right)^2}, \frac{A(l_2 - l_3c_3) - Bl_3s_3}{A^2+B^2} \right) \\ \theta_3 = atan2\left(\pm\sqrt{1 - \left(\frac{(l_2^2+l_3^2) - (A^2+B^2)}{2l_2l_3}\right)^2}, \frac{(l_2^2+l_3^2) - (A^2+B^2)}{2l_2l_3} \right) \end{cases} \quad (7)
$$

where,

$A = y_p - a - \Delta P_y$
$B = z_p - b + \Delta P_z$
$c_3 = \cos(\theta_3)$
$s_3 = \sin(\theta_3)$

4 Simulation and Experiments

4.1 Simulation

Applying the FSEA component to control the position of the robot arm (Fig. 12):

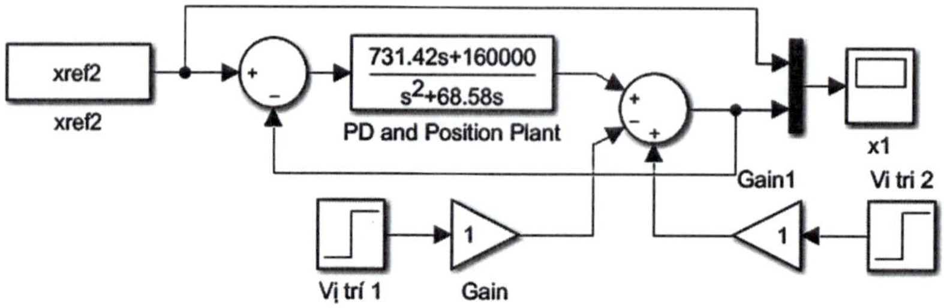

Fig. 12. Simulation for position response of FSEA

The practical length of FSEA depends on the length of the end-effector and the displacements of the spring. When the robot pick or place load, due to the rapid change of the deformation of the spring the practical length of FSEA also changes rapidly but the control system will control the end-effector to obtain the desired length of FSEA.

Hence, FSEA enable robot to move when raising or dropping the load or being impacted from surrounding environment without having the hard component damaging like other conventional types. Therefore, the results showed as Fig. 13.a and Fig. 13.b:

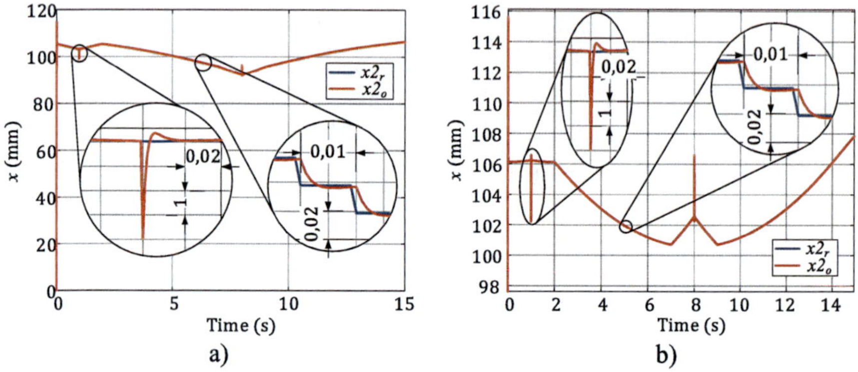

Fig. 13. a) Response of FSEA controlling link 2, b) Response of FSEA controlling link 3

4.2 Experiments

The FSEA and Palletizing robot were built for experiment (Fig. 14.a and Fig. 14.c)

From the Fig. 14.b, the differential of SEA's end – effector and its reference is about ±0, 25 mm which mean it can be perform satisfy the accuracy of the robot.

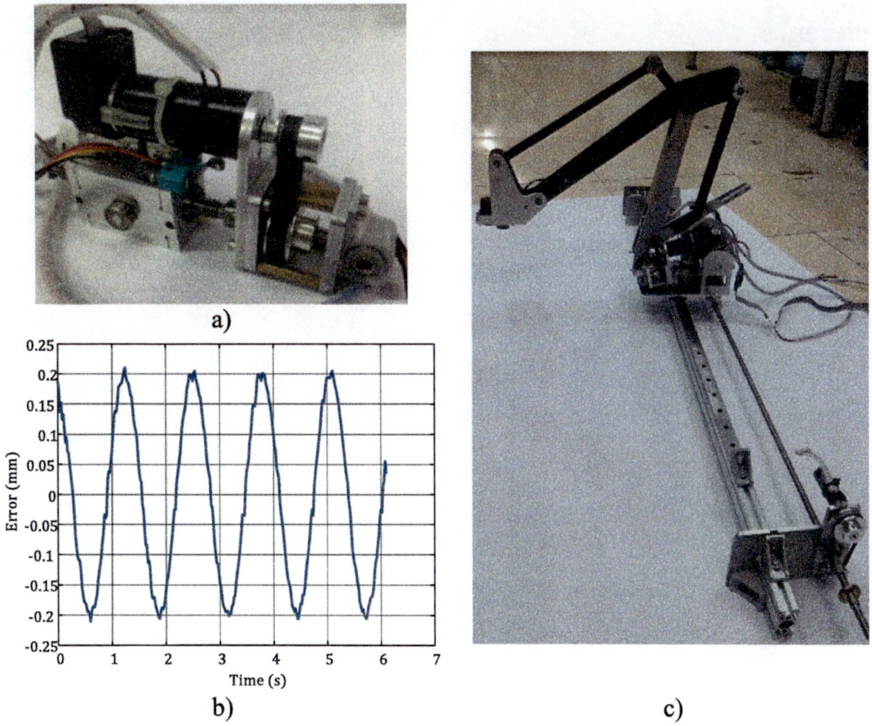

Fig. 14. a) FSEA, b) Error between reference and feedback position of FSEA with $f = 0, 5\,\text{Hz}$, c) Palletizing robot

5 Conclusion

The paper has solved the following issues:

- The design of Series Elastic Actuator applying to palletizing robot and passive balancing mechanism for the robot.
- Experiment results confirmed the balancing mechanism and the algorithm for controlling FSEA and robot.

Development of the project:

- Designing and conduct experiment of algorithm for controlling force on FSEA.
- Applying force controlling to simulate and control robot in case of having load.

Acknowledgments. This work was fully supported by HCMUT, VNU-HCM under grant number B2019-20-09.

References

1. Wikipedia. https://en.wikipedia.org/wiki/Palletizer. Accessed 27 Mar 2018
2. Pratt, G.A., Williamson, M.M.: Series Elastic Acutator (1995)
3. Paine, N.: Book title Design and Control Considerations for High – Performance Series Elastic Actuator, vol. 19, No. 3, June 2014
4. Arakelian, V., Briot, S.: Balancing of Linkages and Robot Manipulators. Mechanism and Machine Science, vol. 27. Springer, Cham (2015)
5. Jazar, R.N.: Theory of Applied Robotics Kinematics, Dynamics and Control, 2nd edn. Springer, New York (2010)
6. Spong, M.W., Hutchinson, S., Vidyasagar, M.: Robot Dynamics and Control, 2nd edn. Wiley, New York (2008)
7. Lee, C., Kwak, S., Kwak, J., Oh, S.: Generalization of Series Elastic Actuator Configurations and Dynamic Behavior Comparison, Department of Robotics Engineer, DGIST, Korea, 22 Aug 2017

A Novel Approach for Determining a Hit Point Based on Estimating Target Movement and Ballistic Table

Anh Son Nguyen[1], Van Dong Nguyen[2], Huy Hung Nguyen[3], and Tan Tien Nguyen[2](✉)

[1] Le Quy Don Technical University, Ha Noi, Vietnam
[2] DCSE Lab, Ho Chi Minh City University of Technology, Ho Chi Minh City, Vietnam
ladoga001@gmail.com, nttien@hcmut.edu.vn
[3] Sai Gon University, Ho Chi Minh City, Vietnam
nghhung@sgu.edu.vn

Abstract. This paper proposes a novel approach for determining a hit point based on estimating target movement and ballistic table. Computing the hit point is a process of calculating and determining the hit point between a shell and a target. To do this task, the following steps are done. Firstly, estimate the movement target position based on previous target movement data collected by a radar on a ship. Secondly, compute the movement time of the shell based on the ballistic table. Thirdly, estimate the hit point between a shell and a target by combine two above steps. Finally simulation results verify effectiveness of the proposed approach.

Keywords: Bullets · Fire control system · Movements of target · Meeting point · Ballistic

1 Introduction

Artillery shells are not controlled by any artillery control system after they are launched and reach a target. For shooting at a mobile target, a target position has to be estimated in order to a shell launched from an artillery hits the mobile target. If both an artillery and its target move along all directions and their velocities also change randomly, the target position also varies with the artillery position. So both the artillery and the target velocities are also estimated together the target position. In fact, it is assumed that artillery and target velocities do not change for computing simply in an artillery control system.

To solve a problem computing a hit point between a shell launched from an artillery and a target, there exist two methods including estimation method based on ballistic table and balance calculation method based on similar transformation [1–4].

With the above assumption, computing methods for estimating the hit point between the shell and the target have a big error, long computing time and unreachable computing convergence [5, 6]. Another method was proposed by changing parameters as shown in [7] to reduce the error due to estimation process.

D. F. Cortes Tobar et al. (Eds.): AETA 2019, LNEE 685, pp. 703–713, 2021.
https://doi.org/10.1007/978-3-030-53021-1_70

To overcome the aforementioned problems, this article proposes an algorithm estimating target movement and a method determining the hit point between a shell and a target based on the estimated target position and a ballistic table.

Estimated target positions and a real hit point are shown in Fig. 1

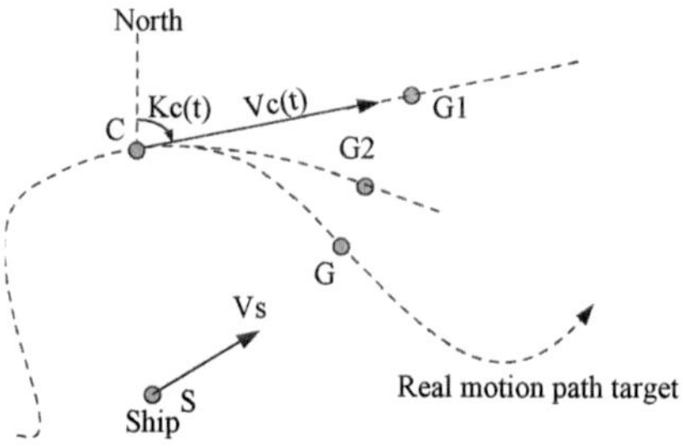

C: Target position
G_1: Estimated target position in case only target movement
G_2: Estimated target position in case both artillery and target movement.
G: Real colliding position
K_C: Target direction
V_C: Target velocity
V_S: Artillery velocity

Fig. 1. The model takes into account the change in movement speed and the speed of the target over time

In Fig. 1, G_1 is an estimated target position in case the target moves with constant velocity and straight direction. G_2 is an estimated target position in case both artillery and target move. G_2 is closer to G being a real hit point than G_1. So the target can be destroyed with high probability.

2 Algorithms

2.1 Position of a Mobile Target in Cartesian and Local Coordinate System

This session builds a moving model of a target based on data on a previous target. The proposed model has simple calculations for unaffected calculation speed.

Movement of a target is described generally by Cartesian equations with Y axis toward the North direction, X axis toward the East direction and Z axis toward up as follows

$$\begin{cases} q(t) = \sum_{i=0}^{r} \left(q_i \times T_c^i\right) \\ D(t) = \sum_{i=0}^{r} \left(d_i \times T_c^i\right) \\ H(t) = \sum_{i=0}^{r} \left(h_i \times T_c^i\right) \end{cases} \quad (1)$$

where $D(t)$: inclination distance from ship to hit point
$H(t)$: target height at the hit point
$q(t)$: the target azimuth at the hit point

T_c^i: time for shell movement to hit the target with the ith model
r: total target data models and r is chosen 2 to reduce calculations

Due to $r = 2$, parameters q_0, q_1, q_2, d_0, d_1, d_2, h_0, h_1, h_2 are determined by the previous target motion data collected from a radar. The movement model of the target is built and the target position can be predicted.

When the target is detected by the radar, another coordinate system named the local coordinate system is given and the target position is considered as the origin of the local coordinate system. The target movement is described by the following equation system:

$$\begin{cases} V_c(t) = v_0 + v_1 t \\ \alpha(t) = \alpha_0 + \alpha_1 t \\ \beta(t) = \beta_0 + \beta_1 t \end{cases} \tag{2}$$

where $V_c(t)$ is the target velocity, v_0 is the initial target velocity, v_1 is acceleration, $\alpha(t)$ is the target angle compared to the North direction, α_0 is the initial target angle, α_1 is the angle acceleration, $\beta(t)$ is the dive angle of the target, β_0 is the initial dive angle of the target and β_1 is the dive angle acceleration.

The coefficients v_0, v_1, α_0, α_1, β_0, β_1 are computed by analyzing target movement at the previous time and shown in Sect. 2.2.

To determine the target position at the time t_0+dt in the Cartesian coordinate system, the target position at the time t_0 and the time interval dt are given as shown in Fig. 2.

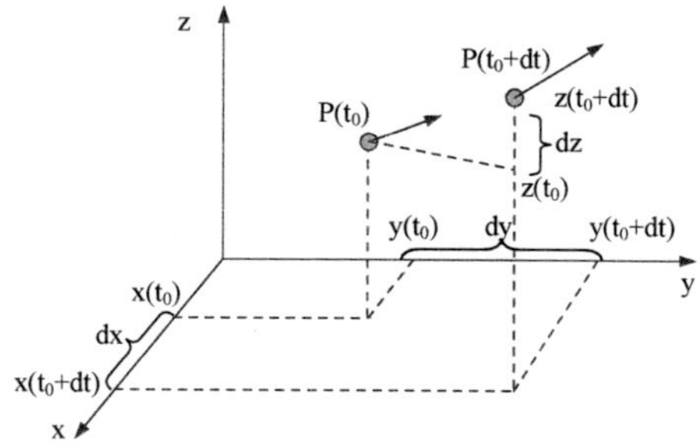

Fig. 2. Target position before and after *dt* period

By projecting the target position on the coordinate axes, the following equations are given

$$x(t_0 + dt) = x(t_0) + (v_0 + v_1 t)\cos(\alpha_0 + \alpha_1 t)\cos(\beta_0 + \beta_1 t)dt \tag{3}$$

$$y(t_0 + dt) = y(t_0) + (v_0 + v_1 t)\sin(\alpha_0 + \alpha_1 t)\cos(\beta_0 + \beta_1 t)dt \tag{4}$$

$$z(t_0 + dt) = z(t_0) + (v_0 + v_1 t)\cos(\beta_0 + \beta_1 t)dt \tag{5}$$

To determine the target position at the estimated time $t_0 + dt$, Eqs. (3), (4) and (5) are integrated by time as follows

$$x(t_0 + t) = x(t_0) + \int_{t_0}^{t_0+t} (v_0 + v_1 t)\cos(\alpha_0 + \alpha_1 t)\cos(\beta_0 + \beta_1 t)dt \tag{6}$$

$$y(t_0+t) = y(t_0) + \int_{t_0}^{t_0+t} (v_0 + v_1 t)\sin(\alpha_0 + \alpha_1 t)\cos(\beta_0 + \beta_1 t)dt \tag{7}$$

$$z(t_0+t) = z(t_0) + \int_{t_0}^{t_0+t} (v_0 + v_1 t)\cos(\beta_0 + \beta_1 t)dt \tag{8}$$

From Eqs. (6) and (7), it yields

$$\begin{aligned} x(t_0+t) = & x(t_0) + \frac{1}{2}\int_{t_0}^{t_0+t} (v_0 + v_1 t)\cos[\alpha_0 + \beta_0 + (\alpha_1 + \beta_1)t]dt \\ & + \frac{1}{2}\int_{t_0}^{t_0+t} (v_1 + v_0 t)\cos[\alpha_0 - \beta_0 + (\alpha_1 - \beta_1)t]dt \end{aligned} \tag{9}$$

$$\begin{aligned} y(t_0+t) = & y(t_0) + \frac{1}{2}\int_{t_0}^{t_0+t} (v_0 + v_1 t)\sin[\alpha_0 + \beta_0 + (\alpha_1 + \beta_1)t]dt \\ & + \frac{1}{2}\int_{t_0}^{t_0+t} (v_0 + v_1 t)\sin[\alpha_0 - \beta_0 + (\alpha_1 - \beta_1)t]dt \end{aligned} \tag{10}$$

From Eqs. (8), (9) and (10), the target position in Cartesian coordinate system is given at the estimated time $t + t_0$

$$\begin{aligned} x(t_0+t) = & x(t_0) + \frac{1}{2}[(v_0 + v_1(t_0+t)) \\ & \times \left(\frac{\sin A}{\alpha_1+\beta_1} + \frac{\sin B}{\alpha_1-\beta_1}\right) + v_1\left(\frac{\cos A}{(\alpha_1+\beta_1)^2} + \frac{\cos B}{(\alpha_1-\beta_1)^2}\right) \\ & - \frac{1}{2}\left[(v_0 + v_1 t_0)\left(\frac{\sin C}{\alpha_1+\beta_1} + \frac{\sin D}{\alpha_1-\beta_1}\right)\right] + v_1\left[\frac{\cos C}{(\alpha_1+\beta_1)^2} + \frac{\cos D}{(\alpha_1-\beta_1)^2}\right] \end{aligned} \tag{11}$$

$$\begin{aligned} y(t_0+t) = & y(t_0) + \frac{1}{2}[(v_0 + v_1(t_0+t)) \\ & \times \left(\frac{\sin A}{\alpha_1+\beta_1} + \frac{\sin B}{\alpha_1-\beta_1}\right) + v_1\left(\frac{\cos A}{(\alpha_1+\beta_1)^2} + \frac{\cos B}{(\alpha_1-\beta_1)^2}\right)\Bigg] \\ & - \frac{1}{2}\left[-(v_0 + v_1 t_0)\left(\frac{\cos C}{\alpha_1+\beta_1} + \frac{\cos D}{\alpha_1-\beta_1}\right)\right] + v_1\left[\frac{\sin C}{(\alpha_1+\beta_1)^2} + \frac{\sin D}{(\alpha_1-\beta_1)^2}\right] \end{aligned} \tag{12}$$

$$\begin{aligned} z(t_0+t) = z(t_0) & - \left[v_0\frac{\cos E}{\beta_1} + v_1\left(\frac{\sin E}{\beta_1^2} - t\frac{\cos E}{\beta_1}\right)\right] \\ & + \left[v_0\frac{\cos F}{\beta_1} + v_1\left(\frac{\sin F}{\beta_1^2} - t\frac{\cos F}{\beta_1}\right)\right] \end{aligned} \tag{13}$$

where

$A = \alpha_0 + \beta_0 + (\alpha_1 + \beta_1)(t_0 + t)$ $B = \alpha_0 - \beta_0 + (\alpha_1 - \beta_1)(t_0 + t)$
$C = \alpha_0 + \beta_0 + (\alpha_1 + \beta_1)t_0$ $D = \alpha_0 - \beta_0 + (\alpha_1 - \beta_1)t_0$
$E = \beta_0 + \beta_1(t_0 + t)$ $F = \beta_0 + \beta_1 t_0$

The advantages of a model estimating the target movement are the fast computing speed and computing result at all target positions. However, the process of computing target movement factors depends on determining the previous target position. The target position is shown in the local coordinate system. It means that the target position must be changed to the Cartesian coordinate system.

2.2 Algorithm for Estimating Parameters of a Mobile Target

To determine parameters of the mobile target, A previous input data set is given during a period T_w. A function $Y(t)$ is built from the input data set. It assumes that the current time t_0 is zero. An estimation function $F(t)$ is built at time $t > t_0$ by extrapolating from function $Y(t)$. The algorithm idea is shown in Fig. 3

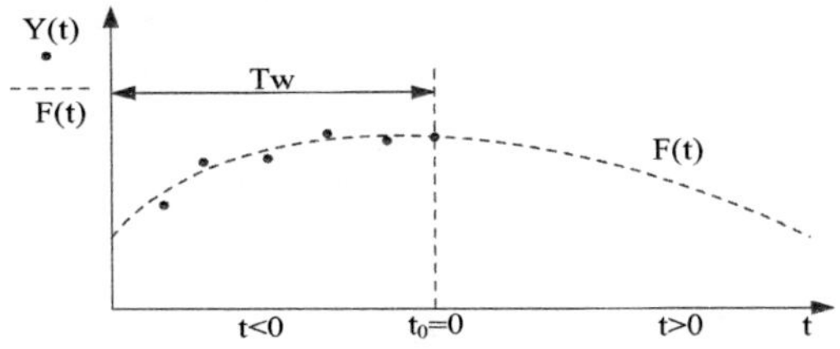

Fig. 3. Figure shows the algorithm for finding motion parameters

Assuming that $F(t)$ is a combination of linear functions $\varphi_i(t)$

$$F(t) = \sum_{i=0}^{r}(w_i \times \varphi_i(t)) \tag{14}$$

where r: polynomial degree
w_i: the ith weighting

By using the least square method to select the best matching curve for the data range. The content of this method is to find values w_i so that the following expression is minimal.

$$\sqrt{\sum_{i=1}^{n}(F(T_i) - Y_i)^2} = min \tag{15}$$

where n is number of measurement points during the T_w

2.3 Algorithm for Estimating a Ballistic Table and Computing a Hit Point

To computing the hit point between a shell and a target, estimating the target position is necessary. In additional, an approximate ballistic function is also necessary to determine movement time of the shell from the current position to the hit point.

A Ballistic table is given as follows

$$Tp_{ij} = F\left(D_i, H_j\right)i = 1\ldots n, j = 1\ldots m, i, j \in N \tag{16}$$

where Tp_{ij}: movement time of shell to a measured point

F: the function of the ballistic table

n, m: number of measured points

D_i: inclined distance of the ith measured point

H_j: height of the jth measured point

Ballistic table for AK630 cannon on ship is shown in Fig. 4.

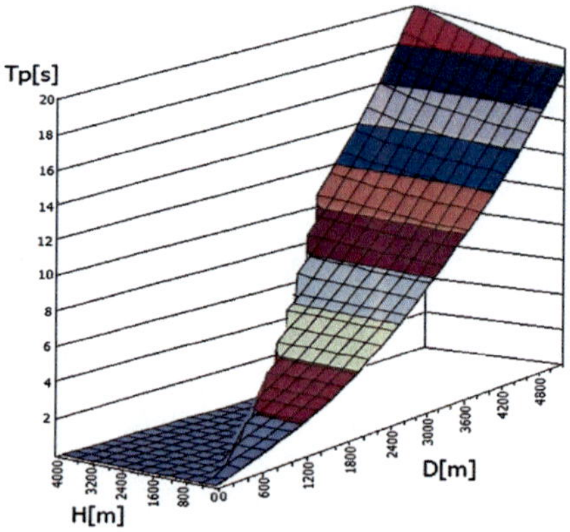

Fig. 4. Performing ballistic table AK630 cannon

To convert the ballistic table to a continuous function, a function G is given as follows

$$T_p = G(D, H)T_p, D, H \in R \tag{17}$$

where T_p is movement time of a shell, D is inclined distance of the target, H is height of the target and the following condition satisfies

$$\sqrt{\sum_{i=1}^{n}\sum_{j=1}^{m}\left(F\left(D_i, H_j\right) - F\left(D_i, H_j\right)\right)^2} \le C_{min} \tag{18}$$

It yields

$$G(D, H) = \sum_{i=0}^{k}\sum_{j=0}^{l}\left(a_{ij} \times D^i \times H^j\right), i, j, k, l \in N \tag{19}$$

where a_{ij} is estimate factors

To reduce amount of calculations, k and l are chosen 2 and 1, respectively to obtain the function G with the selected suitable value C_{min}.

Now, an error function (t) is defined as follows

$$(t) = t - G(D(t), H(t)) \quad (20)$$

where $D(t) = \sqrt{x^2(t) + y^2(t) + z^2(t)}$, $H(t) = z(t)$, and $x(t)$, $y(t)$, $z(t)$ are the target coordinate values in the Cartesian coordinate system

The error function (t) obtains zero when t equals T_a. It means that

$$(T_a) = 0 \quad (21)$$

where T_a is movement time of the shell to the hit point

The coordinate of the hit point is given as follows

$$\begin{cases} x_a = x(T_a) \\ y_a = y(T_a) \\ z_a = z(T_a) \end{cases} \quad (22)$$

In Fig. 5, it shows calculation steps for shooting the artillery.

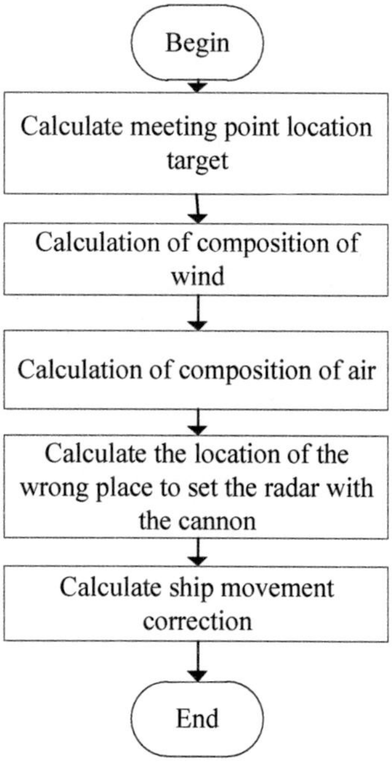

Fig. 5. The order of calculation of the algorithm for calculating the entire angle of view of the cannon

3 Simulation Results

To evaluate the algorithm for determining the target position, several movement target models are proposed in simulation with straight, curved orbit. The hit point position is computed by algorithms at Sect. 2 and compared with simulated position as shown in Fig. 6.

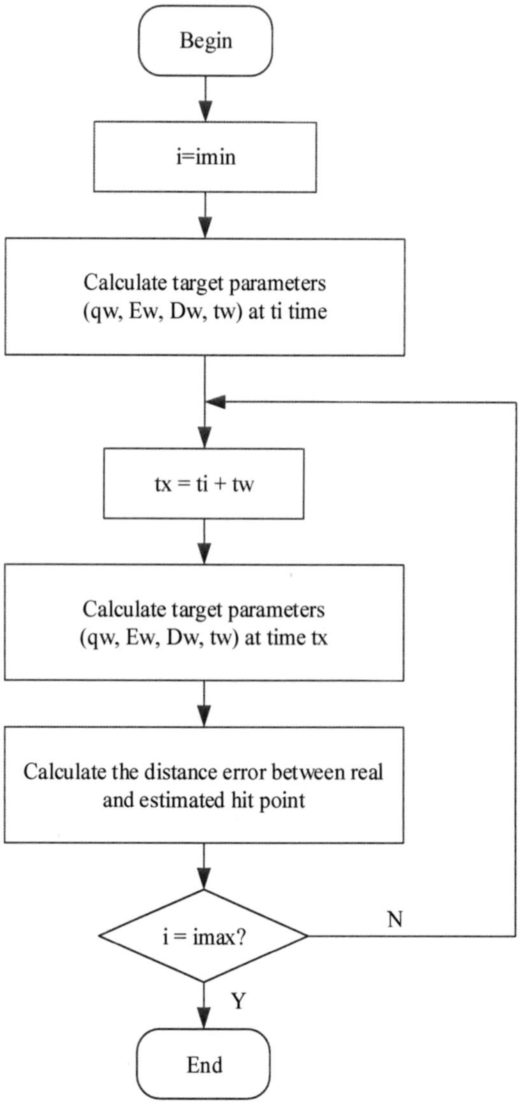

Fig. 6. Flowchart for determining the distance error between real and estimated hit point

Parameters of movement targets estimated in several cases with different target velocities and orbits are shown in Figs. 7, 8, 9 and 10.

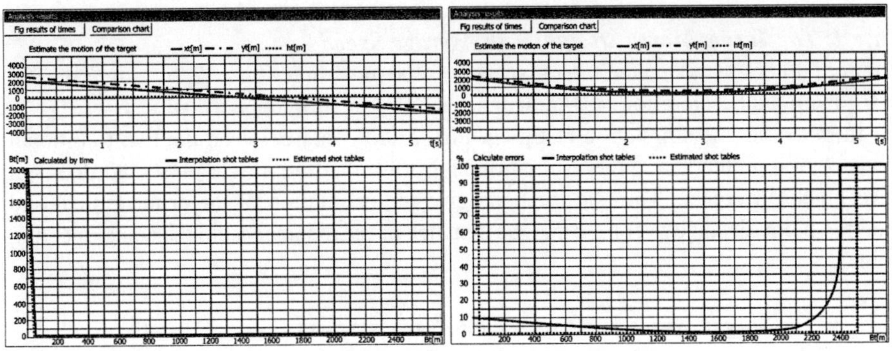

Fig. 7. The window estimates the target movement and meeting point over time and wrong analysis of distance points meets the actual meeting point

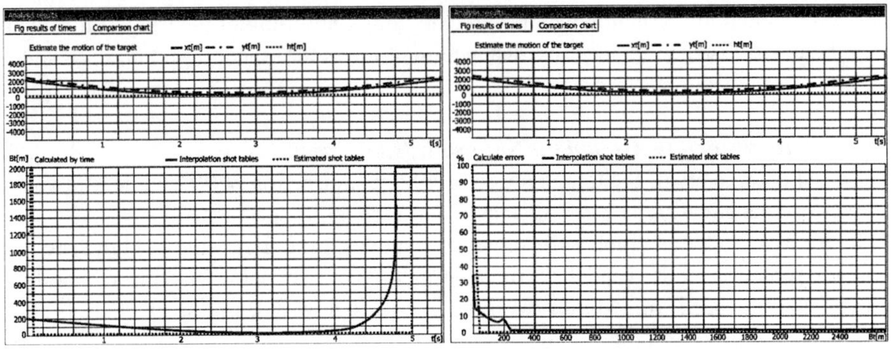

Fig. 8. The window estimates the target movement and meeting point over time and wrong analysis of distance points meets the actual meeting point

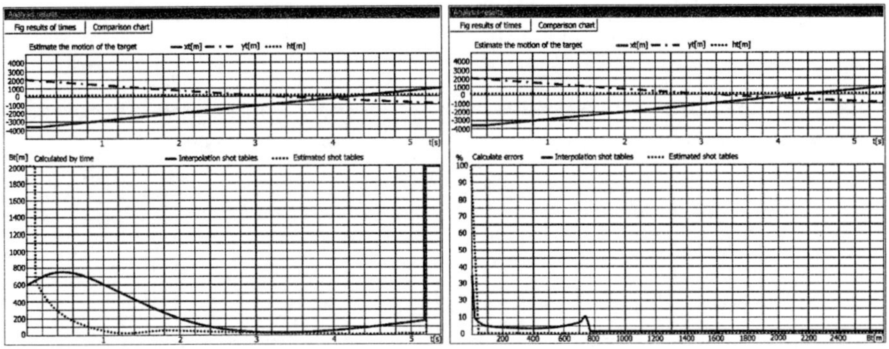

Fig. 9. The window estimates the target movement and meeting point over time and wrong analysis of distance points meets the actual meeting point

Case 1: The Target Moves Straight with Average Speed

The initial parameters of the target: speed = 200 m/s, direction = 225°, angle of dive = 0°, overload acceleration = 0 G, distance = 2.7 km, altitude = 100 m.

In this case, the simulation results in Fig. 7 show that between the two traditional interpolation methods and the estimation algorithm that the article presented does not have much difference. The two methods give the estimation of target movement and meeting point distance over time and the distance of the distance from the actual meeting point.

Case 2: The Target Moves in a Curve with Average Speed

The initial parameters of the target: speed =200 m/s, direction = 270°, angle of dive = 0°, overload acceleration = 2 G, distance = 3 km, altitude = 100 m

Analysis results are shown in Fig. 8.

In this case, the use of new algorithms has advantages in estimating target motions with better actual interpolation results and extending the timing of the trigger. It is also much better in distance of meeting distance than the actual meeting point at close to 300 m.

Case 3: Target Mobility Around the Ship with High Speed

The initial parameters of the target: speed = 400 m/s, direction = 90°, angle of dive = 0°, overload acceleration = 8.5G, distance = 4.3 km, altitude = 100 m

In the case of actual combat the target ship moves around the ship at high speed if using interpolation method and the estimation method clearly knows the time of stability and calculation speed. In addition, with the goal of below 800 m, the estimation plan gives results with superior accuracy.

Case 4: Small-Range Arc Flying Target with High Acceleration

The initial parameters of the target: speed 200 m/s, direction = 22°, angle of dive = –1°, Overload acceleration = 17 G, distance = 3.8 km, altitude = 500 m

Analysis results are shown in Fig. 10.

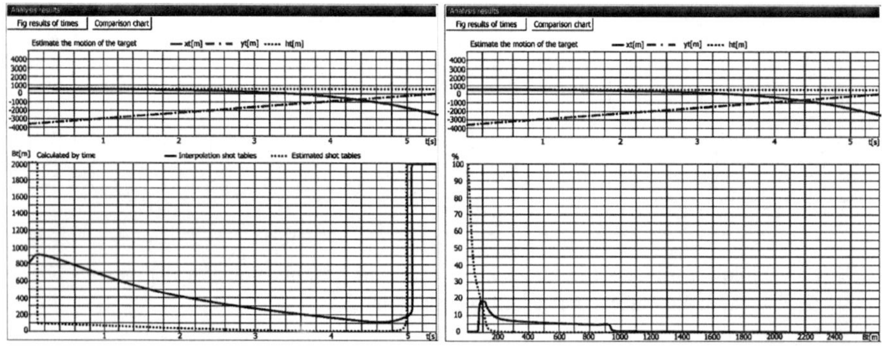

Fig. 10. The window estimates the target movement and meeting point over time and wrong analysis of distance points meets the actual meeting point

The case of small-range arc-flying targets with high acceleration is the case of real-time artillery fire. For this case, the new algorithm shows outstanding advantages of the new algorithm compared to the interpolation to estimate the meeting point distance over time. Also under 1000 m errors encountered point range compared to the actual meeting place is also significantly improved.

4 Conclusion

This paper proposed a novel approach for determining a hit point based on estimating target movement and a ballistic table. The movement target position was computed by previous target movement data and using two coordinate systems. Utilizing a ballistic table for determining movement time of the shell, the hit point was estimated. Estimate error between the computing hit point and the real hit point was tiny approximately 2 m. Simulation results verified the effectiveness of the proposed approach

Acknowledgements. This work was fully supported by Key Laboratory of Digital Control and System Engineering (DCSELAB), HCMUT, VNU-HCM under grant number TX2019-20B-01.

References

1. Vietnam Navy: Radar Control Gun ВЫПЕЛ-АМЭ - Part 5: Calculating Device, Technical Document (2005)
2. Scott, L.R.: Numerical Analysis, 2nd edn. Princeton University Press, Princeton (2016)
3. Pejsa, A.J.: Modern Practical Ballistics. Kenwood publishing, Oklahoma (1991)
4. Bliss, G.A.: Mathematics for Exterior Ballistics. Wiley, New York (1944)
5. McCarty Jr, R.J., Willingham, M.R.: Continuous Alignment System for Fire Control, US7870816B1 (2011)
6. Lee, Y.W.: Neural solution to the target intercept problems in a gun fire control system. Neurocomputing **70**, 689–696 (2007)
7. Liu, H., Mei, W., Shan, G.: Shooting control algorithm based on emendation of model error of assumed relative target motion. WSEAS Trans. Syst. Control **9**, 492–499 (2014)

Study on Velocity Control of Gymnotiform Undulating Fin Module

Van Hien Nguyen[1], Canh An Tien Pham[2], Van Dong Nguyen[2], and Tan Tien Nguyen[2](✉)

[1] PetroVietnam Camau Fertilizer Joint Stock Company, Camau City, Vietnam
hiennv@pvcfc.com.vn
[2] Key Lab of Digital Control and System Engineering, Hochiminh City University of Technology, VNU-HCM, Hochiminh City, Vietnam
antienpham@gmail.com, nvdong@dcselab.edu.vn, nttien@hcmut.edu.vn

Abstract. Inspired by the fish swimming motion of gymnotiform type, an undulating fin module has been developed in the effort to replace the traditional propulsion system of autonomous underwater vehicles. Base on the modelling of force generated in the undulating process of gymnotiform undulating fin module, the force has the non-linear relationship with frequency and velocity of the fin. This paper proposes an estimated function of force generated and design a back stepping sliding mode model for velocity control of gymnotiform undulating fin module.

Keywords: Gymnotiform undulating fin · Velocity control · Back stepping sliding mode

1 Introduction

Recently, Autonomous Underwater Vehicles (AUV) has developed rapidly. In order to enhance the flexibility and the efficiency of AUV, a variety of types of propulsion system have been studied [3–14]. Biomimetic propulsion system is a potential approach in this field with the advantages of natural propulsion of species. Gymnotiform is a type of fish swimming which only use a long anal fin as the main propeller in the swimming process (Fig. 1).

According to previous study [15], an undulating fin module which only uses 1 DC motor for all fin-rays in module have been created. Besides, the force generated when its undulating process is considered. The relation between force generated and others parameters is extremely non-linear. As a result, control issue of this system is rather complex. In this paper, the modelling of force generated in the swimming process of fin will be considered and develop the feasible approach to deal with velocity control issue.

2 Modeling

2.1 Fin Modeling

The fin motion is defined via the motion of all fin-rays. In this paper, the fin-rays swing with the angle defined by (Fig. 2)

D. F. Cortes Tobar et al. (Eds.): AETA 2019, LNEE 685, pp. 714–722, 2021.
https://doi.org/10.1007/978-3-030-53021-1_71

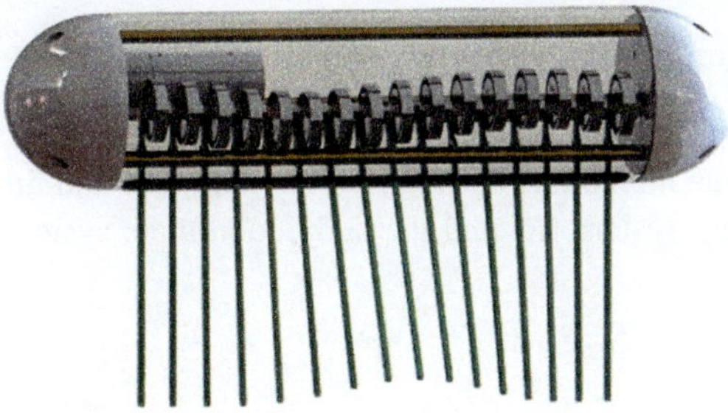

Fig. 1. Gymnotiform undulating fin module

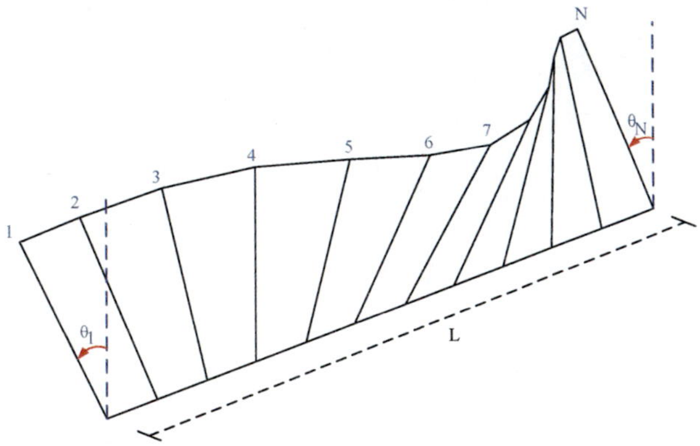

Fig. 2. Modelling of fin's membrane

$$\theta(n, t) = \theta_{max} \sin\left(2\pi ft + \phi_0 - \frac{2\pi n}{N-1}\frac{L}{\lambda}\right)$$

where,

$\theta(n, t)$: the angle of n-th ray at moment t
ϕ_0: initial phase, in this paper, $\phi_0 = 0$
θ_{max}: amplitude of $n - th$ ray
λ: wavelength, L is the length of fin
f: frequency of the fin
N: the number of fin-rays

The module has one full sine wave, therefore, $\lambda/L = 1$. In addition, the most reasonable number of fin-rays in one wave length is 16. Equation (1) is accordingly

$$\theta(n, t) = \theta_{max} \sin\left(2\pi ft - \frac{2\pi n}{15}\right) \tag{1}$$

2.2 Force Exerted on the Fin

The modelling is adjusted from the model which is presented by Sfakiotakis [1, 2]. Consider point q on the membrane between fin-ray number i and fin-ray number $i + 1$. The inertial coordinate system $\{P\}$ and the body coordinate system of the fin $\{O\}$ are shown in Fig. 3.

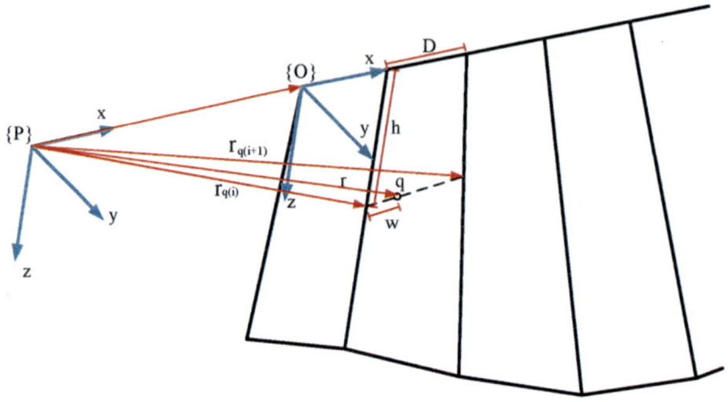

Fig. 3. Coordinate systems of the fin

where,

$\vec{r}_{qi}, \vec{r}_{q(i+1)}$: the trajectory vector of point q on the inertial coordinate system
h: the distance between the rotated axis of fin-ray and point q
w: the width of membrane at q
D: the distance between two adjacent fin-rays

Coordinate of point q in the inertial coordinate system $\{P\}$ is obtained as

$$\vec{r}_q = \vec{r}_{qi} + \frac{\vec{r}_{q(i+1)} - \vec{r}_{qi}}{\left\|\vec{r}_{q(i+1)} - \vec{r}_{qi}\right\|} w \tag{2}$$

The modelling of interactions of undulating fin with surrounding fluid bases on the simplified quasi-steady approach which could be applied in the large Reynolds Number flow regimes. In this model, the tangential force of membrane is ignored. The force that fluid exerts on point q

$$\vec{f} = \overrightarrow{f_{nq}} = \frac{1}{2}\rho C_n \left\|\overrightarrow{V_n}\right\| \overrightarrow{V_n} = \frac{1}{2}\rho C_n \left\|\overrightarrow{V_n}\right\|^2 \vec{n}_0 sgn\left(\overrightarrow{V_n}\right) \tag{3}$$

where, $\overrightarrow{V_n} = \overrightarrow{\dot{r}_q}$

$$\overrightarrow{n_0} = \frac{\frac{\partial \vec{r}_q}{\partial h} \frac{\partial \vec{r}_q}{\partial w}}{\left\|\frac{\partial \vec{r}_q}{\partial h} \frac{\partial \vec{r}_q}{\partial w}\right\|} \tag{4}$$

Integrating $\vec{f}$ on the whole membrane, the force exerted on the membrane is

$$\vec{F}_{\text{fluid}} = \overrightarrow{F_n} = \sum_{i=0}^{N-1} \int_{h_{min}}^{h_{max}} \int_{0}^{w_{max}} \vec{f}\, dwdh \quad (5)$$

The force generated by the fin in the undulating process is

$$\vec{F}_{\text{fin}} = -\vec{F}_{\text{fluid}} \quad (6)$$

At this stage, the force is calculated on MATLAB, the constants need to be determined are shown in Table 1. Simplify the results, the force is a function of frequency f, velocity of the module v and time t

$$\vec{F}_{\text{fin}} = \begin{bmatrix} F_x(f, v, t) \\ F_y(f, v, t) \\ F_z(f, v, t) \end{bmatrix} \quad (7)$$

Table 1. Necessary constants for force calculation in MATLAB

Parameter	Name	Value	Unit
θ_{max}	Amplitude of the fin motion	$\pi/12$	rad
$[h_{min}, h_{max}]$	Positions of the membrane's boundary on the fin-ray	[0, 0.1]	m
L	Length of the membrane	0.195	m
D	Distance between 2 fin-ray	0.013	m
ρ	Density of fluid	1000	kg/m^3
C_d	Drag co-efficiency of membrane and fluid	2.8	
m	The mass of module	0.7	kg

3 Velocity Control

3.1 Modeling

In order to control the velocity of fin module, the state equation have to be established. First of all, the direction of movement needs to be considered. In this paper, only the direction along the fin (x-direction) are free for the movement of the fin. Other directions are fixed, therefore, the module is placed on the Slider – Guide (LM guide) (see in Fig. 4). All of the module are immersed in water. The forces exerted on the module are shown in Fig. 4.

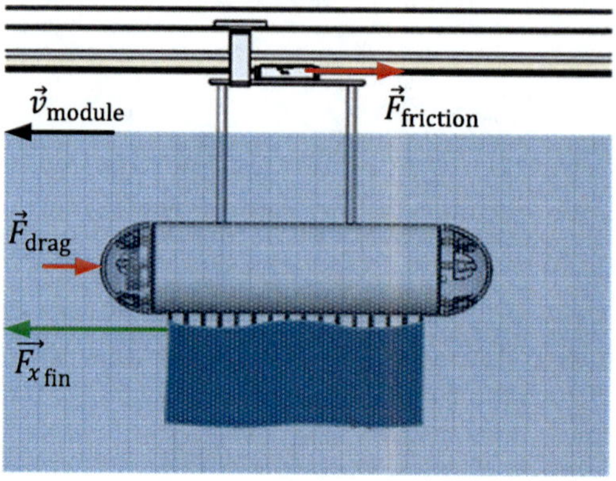

Fig. 4. Modelling of forces exerted on the module

From the Newton's Second Law

$$m\dot{v} = F_{x\,fin} - F_{drag} - F_{friction} \tag{8}$$

Denote $x_1 = x, x_2 = \dot{x}_1, x$ is the position of module. The state equation of system is

$$\begin{cases} x_2 = \dot{x}_1 \\ \dot{x}_2 = \frac{F_x(f, x_2, t)}{m} - \frac{1}{2m}\rho A x_2^2 - \frac{F_{friction}}{m} \end{cases} \tag{9}$$

The problem in this state equation is that $F_x(f, x_2, t)$ is a non-linear function and its parameters include control parameter f and output parameter x_2. In this paper, an estimated function is established for $F_x(f, x_2, t)$ in the range of f [0; 10], x_2[0; 1].

$$F_x(f, x_2, t) = G(f) + \theta_f(t) \tag{10}$$

where $G(f) = 0.006754f^2 + 0.00823f, \theta_f(t)$ is a random noise with the maximum value is 0, 01.

In the modelling of force generated, the movement of water is ignored. This could lead to an error of results. Therefore, the compensation of this error is presented by $\theta_k(t)$ with the same form of $\theta_f(t)$.

Derived from state equation

$$\begin{cases} x_2 = \dot{x}_1 \\ \dot{x}_2 = \frac{1}{m}G(f) - \frac{1}{2m}\rho A C_d x_2^2 + d_x(t) \end{cases} \tag{11}$$

where,

$G(f)$: the control parameter of system
$d_x(t)$: the sum of unmatched uncertainties and disturbances, $|d_x(t)| \leq 0.02/m$

$$d_x(t) = \frac{1}{m}\left(-F_{friction} + \theta_f(t) + \theta_k(t)\right)$$

3.2 Controller Design

The following procedure is based on back-stepping sliding mode method introduced by Jinkun Liu et al. [16].

Let $e_1 = x_1 - r$

where r is the desired trajectory. The Lyapunov function is selected by

$$V = \frac{1}{2}e_1^2 + \frac{1}{2}s^2$$

where s is the sliding variable, $s = x_2 + c_1 e_1 - \dot{r}$

Derivate the Lyapunov function

$$\dot{V} = e_1\dot{e}_1 + s\dot{s}$$

Because

$$\dot{s} = \dot{x}_2 + c_1\dot{e}_1 - \ddot{r}$$

$$= \frac{1}{m}G(f) - \frac{1}{2m}\rho A C_d x_2^2 + d_x(t) + c_1\dot{e}_1 - \ddot{r}$$

Hence,

$$\dot{V} = e_1 s - c_1 e_1^2 + s\left(\frac{1}{m}G(f) - \frac{1}{2m}\rho A C_d x_2^2 + d_x(t) + c_1\dot{e}_1 - \ddot{r}\right)$$

In order to realize $\dot{V} \le 0$, a controller is designed as

$$G(f) = m\Big[(-\frac{1}{2m}\rho A C_d x_2^2 + d_x(t) - c_2 s - e_1$$

$$-c_1\dot{e}_1 + \ddot{r} - \eta sign(s)\Big] \quad (12)$$

where $c_2 > 0$, $\eta \ge D$. Therefore,

$$\dot{V} = -c_1 e_1^2 - c_2 s^2 + s d_x(t) - \eta|s| \le 0$$

3.3 Modeling

3.3.1 Simulation

The simulation program is shown in Fig. 5. In this simulation, the disturbances are added in the plan system.

However, $G(f)$ is not the control parameter of the real system. Therefore, after calculate $G(f)$ value by back-stepping sliding mode controller, frequency of the fin f has to be calculated by $G(f) = 0.006754f^2 + 0.00823f$. Because the estimated function is established in the range of $f \in [0, 10]$, the saturation block needs to be added to limit the value of f (Fig. 6).

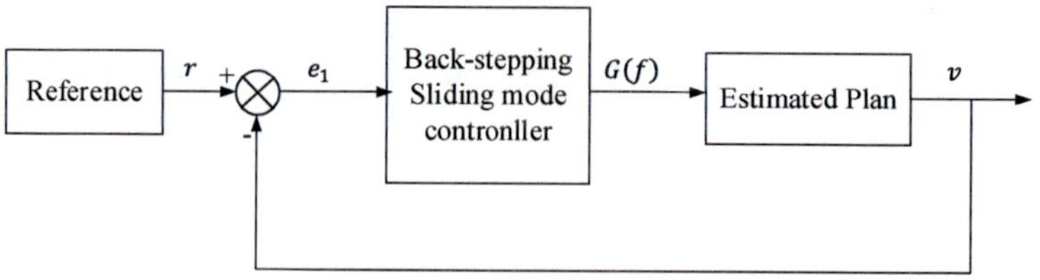

Fig. 5. Simulation diagram of back-stepping sliding mode control

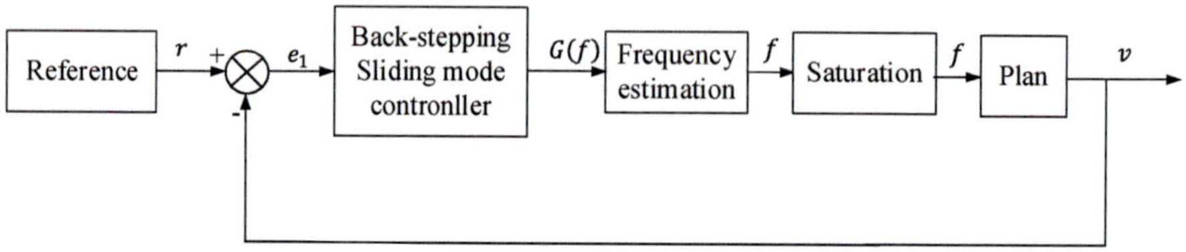

Fig. 6. Simulation diagram of back-stepping sliding mode control of real system

3.3.2 Results

The results of simulation are shown in Fig. 7. The simulations of estimated system and the real system. This result is an evidence for the performance of estimated function established in modelling (Fig. 8).

The results illustrate that the system has been controlled and the responses of estimated system and real system are matched with each other

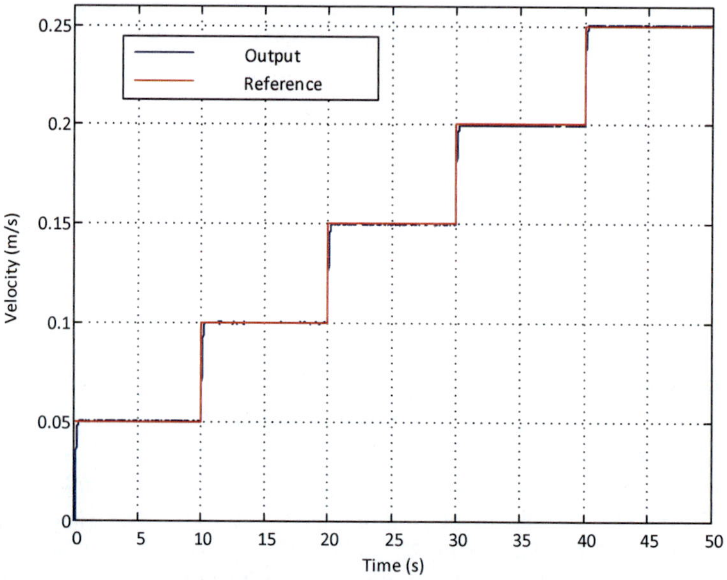

Fig. 7. Simulation results of estimated system

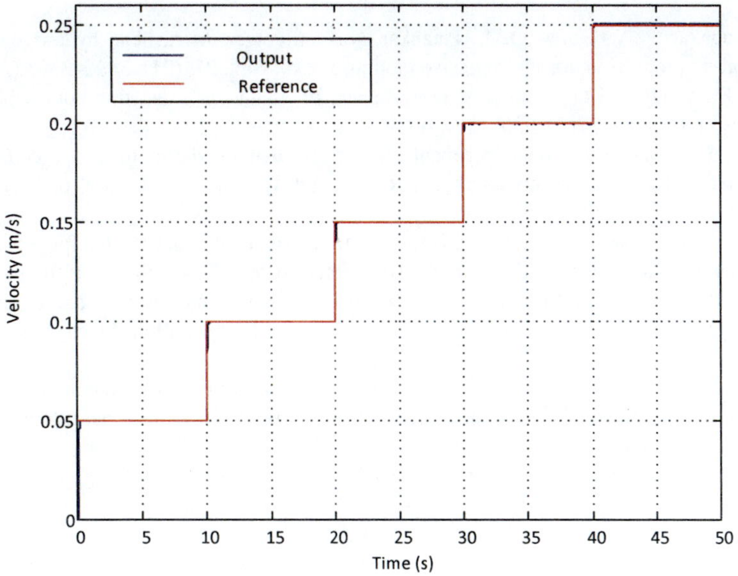

Fig. 8. Simulation result of real system

4 Conclusion

This paper has modelled the force generated in undulating process of fin module. Besides, a back-stepping sliding mode controller has designed to control the velocity of module in one direction. This paper has established an estimation function of forced generated to simplify the system for control issue.

Acknowledgments. This work was fully supported by Key Laboratory of Digital Control and System Engineering (DCSELAB), HCMUT, VNU-HCM under grant number TX2019-20B-01.

References

1. Sfakiotakis, M., Fasoulas, J., Development and experimental validation of a model for the membrane restoring torques in undulatory fin mechanisms. In: 2014 22nd Mediterranean Conference of Control and Automation, MED, 2014, pp. 1540–1546 (2014)
2. Sfakiotakis, M., Fasoulas, J., Gliva, R.: Dynamic modeling and experimental analysis of a two-ray undulatory fin robot. In: 2015 IEEE/RSJ International Conference on Intelligent Robots and Systems, IROS, Hamburg, Germany, pp. 339–346 (2015)
3. Wang, S., Dong, X., Shang, L.-J.: Thrust analysis of the undulating ribbon-fin for biomimetic underwater robots. In: 2011 2nd International Conference on Intelligent Control and Information Processing, ICICIP, vol. 1, pp. 335–340 (2011)
4. Lamas, M., Rodríguez, J., Rodríguez, C., González, P.: Three-dimensional CFD analysis to study the thrust and efficiency of a biologically-inspired marine propulsor. Pol. Marit. Res. **18**(1), 10–16 (2011)

5. Shirgaonkar, A.A., Curet, O.M., Patankar, N.A., MacIver, M.A.: The hydrodynamics of ribbon-fin propulsion during impulsive motion. J. Exp. Biol. **211**(21), 3490–3503 (2008)
6. Liu, H., Curet, O.M.: Propulsive performance of an under-actuated robotic ribbon fin. Bioinspir. Biomim. **12**(3), 036015 (2017)
7. Willy, A., Low, K. H., Initial experimental investigation of undulating fin. In: 2005 IEEE/RSJ International Conference on Intelligent Robots and Systems, IROS 2005, pp. 1600–1605 (2005)
8. Liu, F., Lee, K.-M., Yang, C.-J.: Hydrodynamics of an undulating fin for a wave-like locomotion system design. IEEEASME Trans. Mechatron. **17**(3), 554–562 (2012)
9. Lighthill, J., Blake, R.: Biofluiddynamics of balistiform and gymnotiform locomotion. Part 1. biological background, and analysis by elongated-body theory. J. Fluid Mech. **212**, 183–207 (1990)
10. Hu, T., Shen, L., Lin, L., Xu, H.: Biological inspirations, kinematics modeling, mechanism design and experiments on an undulating robotic fin inspired by Gymnarchus niloticus. Mech. Mach. Theory **44**(3), 633–645 (2009)
11. Peter, B., Ratnaweera, R., Fischer, W., Pradalier, C., Siegwart, R.Y.: Design and evaluation of a fin-based underwater propulsion system. In: 2010 IEEE International Conference on Robotics and Automation, ICRA, pp. 3751–3756 (2010)
12. Xie, H., Shen, L., Dynamic analysis on the bionic propulsor imitating undulating fin of aquatic animals. In: IEEE International Conference on Robotics and Biomimetics, ROBIO 2007, pp. 268–273 (2007)
13. Epstein, M., Colgate, J.E., MacIver, M.A.: Generating thrust with a biologically-inspired robotic ribbon fin. In: 2006 IEEE/RSJ International Conference on Intelligent Robots and Systems, pp. 2412–2417 (2006)
14. Curet, O.M., Patankar, N.A., Lauder, G.V., MacIver, M.A.: Mechanical properties of a bio-inspired robotic knifefish with an undulatory propulsor. Bioinspir. Biomim. **6**(2), 026004 (2011)
15. Nguyen V.D., Phan D.K., Pham C.A.T., Kim D.H., Dinh V.T., Nguyen T.T.: Study on determining the number of fin-rays of a gymnotiform undulating fin robot. In: Recent Advances in Electrical Engineering and Related Sciences: Theory and Application. AETA 2017. Lecture Notes in Electrical Engineering, vol 465. Springer, Cham (2018)
16. Liu, J., Wang, X.: Advanced Sliding Mode Control for Mechanical Systems. Springer, Heidelberg (2011)

Study on Hybrid Method for Grasping Objects in 3D Environment Using Stereo 3D Camera

Trong Hai Nguyen[1] and Le Nhat Binh[2](✉)

[1] Hutech Institute of Engineering,
Ho Chi Minh University of Technology, Ho Chi Minh City, Vietnam
nt.hai75@hutech.edu.vn

[2] Faculty of Aviation Electronics and Telecommunication, Vietnam Aviation Academy,
Ho Chi Minh City, Vietnam
binhln@vaa.edu.vn

Abstract. This paper proposes a hybrid method for grasping objects in a picking robot system in 3D environment using stereo 3D camera. To do this task, the followings are done. Firstly, an image processing system including stereo 3D camera sensor is described. Secondly, recognizing of the object is obtained by using stereo 3D camera sensor. Thirdly, for grasping object, a hybrid method to obtain the orientation of the handle is proposed. The main idea is to search the point cloud for neighborhoods that satisfy handle-like grasp affordances and can be grasped by the end-effector of the manipulator. Finally, the effectiveness and the applicability of the proposed algorithms is verified by using experiment. The experimental results show that the proposed algorithm successfully detects an object and finds its grasping points with an acceptable small error.

Keywords: Mobile robot · Object recognition · Object localization · Grasping object

1 Introduction

A robotic manipulation of objects typically involves object detection/recognition and grasping control. The most dominant methods are based on single camera to receive color image and laser sensor for depth information. In real-world grasping, the full 3D shape of the object is hard to perceive [1]. Other methods [2] focus on grasping 2D planar objects using edges and contours to determine form and force closures for simple objects such as squares, triangles, etc. For RGBD image, several learning algorithm [3, 4] have shown promise in handling incomplete and noisy data and variations in the environment as well as grasping novel objects. D. Fischinger et al. [3] presented a learning approach for grasping unknown objects in a basket. In [4], a new rectangle representation algorithm based on the Kinect camera was proposed to localize the grasping point of the object. For 3D point cloud image, C. Papazov et al. [5] utilized a Kinect stereo camera sensor to acquire depth images of the scene. F. Bley et al. [6] proposed another approach of grasp selection by fitting learned generic object models to point cloud data. C. Choi et al. [7]

D. F. Cortes Tobar et al. (Eds.): AETA 2019, LNEE 685, pp. 723–731, 2021.
https://doi.org/10.1007/978-3-030-53021-1_72

proposed a Hough voting-based approach that extended point-pair features, which was based on oriented surface points, by boundary points with directions and boundary line segments. In most of these mobile manipulation demonstrations, the handled objects are well-separated. However, they are complex and expensive.

To solve this problem, this paper proposes a hybrid method for grasping objects in a picking robot system in 3D environment using stereo 3D camera. Recognizing of the object is obtained by using stereo 3D camera sensor. For grasping object, a hybrid method to obtain the orientation of the handle is proposed. The main idea is to search the point cloud for neighborhoods that satisfy handle-like grasp affordances and can be grasped by the end-effector of the manipulator. Finally, the effectiveness of the proposed algorithms are verified by experiment. The experimental results show that the mobile picking robot successfully reaches the goal point with an acceptable small error.

2 System Description

Figure 1 shows the workspace of a picking robot system consisting of manipulator platform, stereo camera and a horizontal coffee shop table with an object at the manipulator workspace. The stereo camera sensor is placed on the table with 60 cm height.

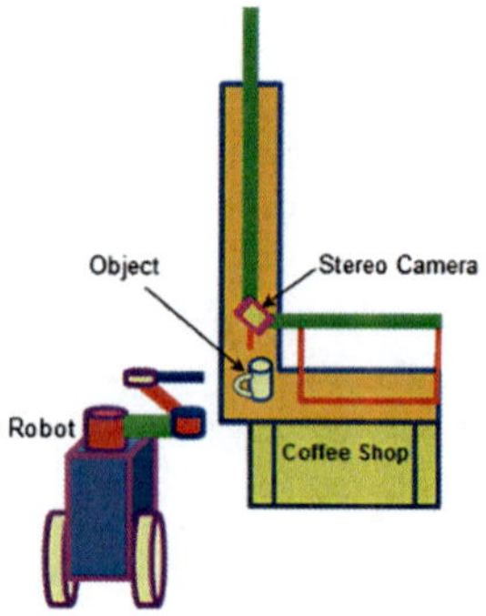

Fig. 1. Shows the workspace of a picking robot system.

3 Proposed Algorithm

3.1 Database

Local Binary Patterns (LBP) is the non-parametric operator that describes the local spatial structure of images and is invariant to illumination change with fast calculation. The value of the center pixel x is used as a threshold and compared with spatial neighborhoods to obtain a binary code for texture feature description. After defining a neighborhood radius r, N pixels of radius r around x are processed to construct the texture feature. Given the intensity P_x of the center pixel x and P_n ($n = 1, 2, \ldots, N$) of spatial neighborhoods, we obtain a binary pattern by comparing P_n with P_x clockwise or counter-clockwise. Each digit of the binary pattern is expressed as

$$s(n) = \begin{cases} 1 \; P_n \geq P_x \\ 0 \; P_n < P_x \end{cases} \tag{1}$$

LBP is achieved by converting the binary pattern to a decimal number:

$$LBP(x) = \sum_{n=0}^{N} s(n).2^n \tag{2}$$

The LBP texture feature for x is shown as follows

$$f_{LBP}(x) = \left[LBP(x_1), LBP(x_2), \ldots, LBP(x_{h\times w})\right]^T \tag{3}$$

where $xj, j \in [1, h \times w]$ is the spatial neighborhood of x in the feature window.
Figure 2 shows the basic of LBP operator

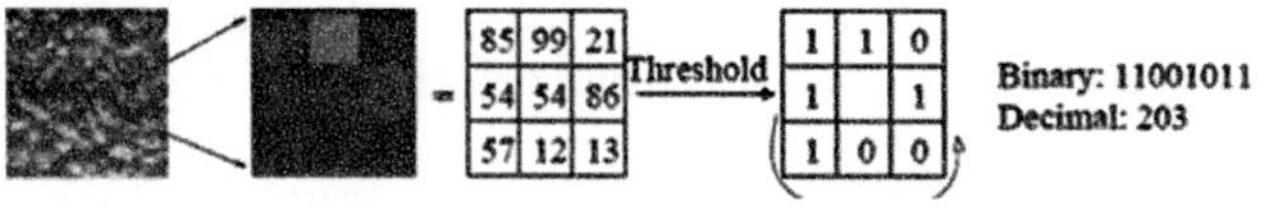

Fig. 2. The basic of LBP operator.

The AdaBoost algorithm performs effectively in various fields. This algorithm combines weak classifiers with a strong classifier using a weighted voting mechanism, defined by:

$$C(x) = \begin{cases} 1; \sum_{t=1}^{T} \alpha_t h_t(x) \geq \frac{1}{2} \sum_{t=1}^{T} \alpha_t \\ 0; \quad \text{otherwise} \end{cases} \tag{4}$$

where h and C are weak and strong classifiers, respectively, and α is the weight coefficient for each h. The AdaBoost algorithm can be described as Table 1:

The training process [8] consists on stages of an ensemble of weak learners, trained using boosting. On every stage, a sliding window goes over every input image and classifies this region as positive or negative. If the region is labeled as positive, this region goes to the next stage, rejecting the negative ones. The process finishes when a certain number of stages, established by the users, are reached, or when there are no more negative samples for the next stage. The trained detector will be returned as an XML file.

One advantage of this mechanism is that it extracts automatically the LBP features from the images when it trains and tests the detector, making the process faster.

Table 1. The AdaBoost algorithm for classifier learning.

- Given example images $(x_1, y_1),\ldots, (x_n, y_n)$ where y_i=0, 1 for negative and positive examples respectively.
- Initialize weights $w_{1,i} = \frac{1}{2m}, \frac{1}{2l}$ for y_i=0, 1 respectively, where m and l are the number of negative and positive respectively.
- For t=1,...,T:
  1. Normalize the weights
  $$w_{t,i} \leftarrow \frac{w_{t,i}}{\sum_{j=1}^{n} w_{t,j}}$$
  So that w_t is a probability distribution.
  2. For each feature, j, train a classifier h_j which is restricted to using a single feature. The error is evaluated with respect to w_t, $\varepsilon_j = \sum_i w_i |h_j(x_i) - y_i|$.
  3. Choose the classifier, h_t, with the lowest error ε_t.
  4. Update the weights: $w_{t+1,i} = w_{t,i}\beta_t^{1-e_i}$ where e_i=0 if example x_i is classified correctly, e_i=1 otherwise, and $\beta_t = \frac{\varepsilon_t}{1-\varepsilon_t}$.
- The final strong classifier is:
$$C(x) = \begin{cases} 1; & \sum_{t=1}^{T} \alpha_t h_t(x) \geq \frac{1}{2}\sum_{t=1}^{T} \alpha_t \\ 0; & otherwise \end{cases}$$
where $\alpha_t = \log\frac{1}{\beta_t}$

3.2 Cascade Object Detector

The Cascade Object Detector uses the Viola-Jones algorithm [9] to detect objects. Many platforms provide trained Cascade Object Detectors of nose, frontal face, upper-body, etc. For detecting the handle of coffee cup, paper proposes two sets of negative and positive samples and training the Cascade Object Detector.

The trained detectors are then used to detect objects by sliding a window over the image. Cascade detectors are a concatenation of many weak object classifiers that are placed sequentially, such that the output of a given classifier is sent as additional information to the next classifier. Each weak linear classifier forms a stage, and the final object detector combines these stages along with a sliding window detector. Figure 3 shows the schematic depiction of a the detection cascade.

By using this characteristic, several scanning windows are defined (Fig. 4) with the following scale factor:

$$\text{ScaleFactor} = \frac{w_1 - w_0}{y_1 - y_0} \tag{5}$$

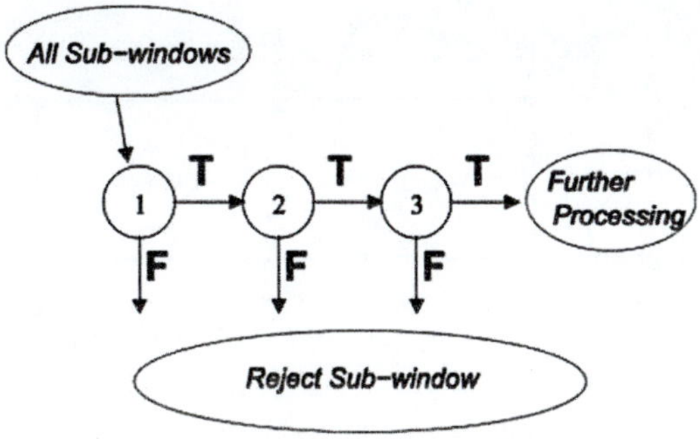

Fig. 3. Schematic depiction of a the detection cascade.

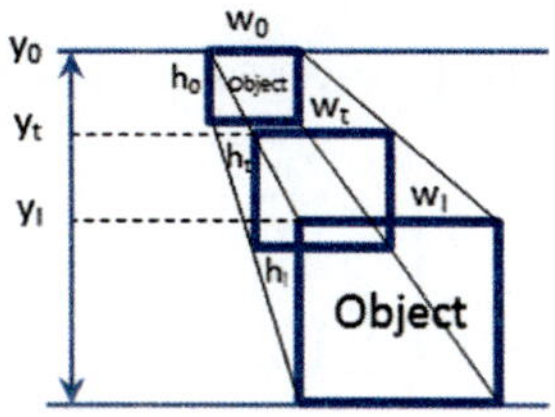

Fig. 4. Size of window detector at different position of image coordinate.

where w_0 and w_1 denote the smallest and largest object size, respectively. y_0 and y_1 are their corresponding positions in image coordinate. The sliding window size at t-position is defined in the following equation:

$$w_t = w_0 + (y_t - y_0) \times \text{Scale Factor} \tag{6}$$

By using w_t, the unreasonable size of object detector, whose width is excessively large or small for a particular position, can be filtered out by the tolerance range for acceptable width and final filtering rule as follows:

$$w_t - \varepsilon < \mathit{AdaBoostobjectdetectorresult} < w_t + \varepsilon \tag{7}$$

where ε is the empirical threshold for efficient falsification.

3.3 Depth Calculation

Figure 5 shows disparity and depth information. Given the disparity map, the baseline and the Focal length (calibration): triangulation computes the position of the correspondence in the 3D.

From Fig. 5 the position of the correspondence in the 3D is obtained as follows:

$$x = \frac{x_l.z}{f};\ y = \frac{y_l.z}{f};\ z = \frac{f.b}{x_l - x_r} = \frac{f.b}{d} \tag{8}$$

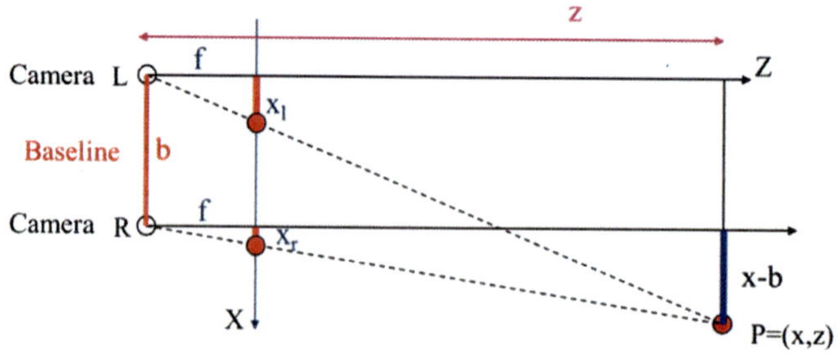

Fig. 5. Disparity and depth

3.4 Determines Angle

The standard deviational ellipse [10] is given as:

$$C = \begin{pmatrix} \mathrm{var}(x) & \mathrm{cov}(x,y) \\ \mathrm{cov}(y,x) & \mathrm{var}(y) \end{pmatrix} = \frac{1}{n} \begin{pmatrix} \sum_{i=1}^{n} (x_i - \bar{x})^2 & \sum_{i=1}^{n} (x_i - \bar{x})(y_i - \bar{y}) \\ \sum_{i=1}^{n} (x_i - \bar{x})(y_i - \bar{y}) & \sum_{i=1}^{n} (y_i - \bar{y})^2 \end{pmatrix} \tag{9}$$

where x and y are the coordinates for feature i, $\{\bar{x}, \bar{y}\}$ represent the Mean Center for the features and n is equal to the total number of features.

The sample covariance matrix is factored into a standard form which results in the matrix being represented by its eigenvalues and eigenvectors.

To obtain the orientation of the ellipse, we simply calculate the angle of the largest eigenvector towards the x-axis:

$$\varphi = \arctan \frac{V_1(y)}{V_1(x)} \tag{10}$$

where V_1 is the eigenvector of the covariance matrix that corresponds to the largest eigenvalue.

3.5 The Algorithm Determines the Picking Point

Block diagram for a proposed method for object grasp detection using stereo camera sensor is shown in Fig. 6.

4 Experimental Results

Figures 7, 8 show the result of coffee cup detection.

Figure 9, 10 show some grasping the position of the correspondence in the 3D and the orientation of the handle results of the proposed method. Table 2 show the results of object detection and angle representation of the proposed method. The experimental results show that the proposed algorithm successfully detects an object with accuracy 93% when the background distance is 60 cm.